Complex Potential Theory

NATO ASI Series

Advanced Science Institutes Series

A Series presenting the results of activities sponsored by the NATO Science Committee, which aims at the dissemination of advanced scientific and technological knowledge, with a view to strengthening links between scientific communities.

The Series is published by an international board of publishers in conjunction with the NATO Scientific Affairs Division

A Life Sciences **B Physics**	Plenum Publishing Corporation London and New York
C Mathematical and Physical Sciences **D Behavioural and Social Sciences** **E Applied Sciences**	Kluwer Academic Publishers Dordrecht, Boston and London
F Computer and Systems Sciences **G Ecological Sciences** **H Cell Biology** **I Global Environmental Change**	Springer-Verlag Berlin, Heidelberg, New York, London, Paris and Tokyo

NATO-PCO-DATA BASE

The electronic index to the NATO ASI Series provides full bibliographical references (with keywords and/or abstracts) to more than 30000 contributions from international scientists published in all sections of the NATO ASI Series.
Access to the NATO-PCO-DATA BASE is possible in two ways:

– via online FILE 128 (NATO-PCO-DATA BASE) hosted by ESRIN, Via Galileo Galilei, I-00044 Frascati, Italy.

– via CD-ROM "NATO-PCO-DATA BASE" with user-friendly retrieval software in English, French and German (© WTV GmbH and DATAWARE Technologies Inc. 1989).

The CD-ROM can be ordered through any member of the Board of Publishers or through NATO-PCO, Overijse, Belgium.

Series C: Mathematical and Physical Sciences - Vol. 439

Complex Potential Theory

edited by

Paul M. Gauthier
Département de Mathématiques et de Statistique,
and Centre de Recherches Mathématiques,
Université de Montréal,
Montréal, Québec, Canada

and Technical Editor

Gert Sabidussi
Département de Mathématiques et de Statistique,
Faculté des Arts et des Sciences,
Université de Montréal,
Montréal, Québec, Canada

Springer-Science+Business Media, B.V.

Proceedings of the NATO Advanced Study Institute and
Séminaire de mathématiques superiéures on
Complex Potential Theory
Montréal, Canada
July 26–August 6, 1993

A C.I.P. Catalogue record for this book is available from the Library of Congress.

ISBN 978-0-7923-3005-9 ISBN 978-94-011-0934-5 (eBook)
DOI 10.1007/978-94-011-0934-5

Printed on acid-free paper

Table of Contents

Preface

The objective of this ASI was to bring together specialists in several complex variables (many of whom have contributed to complex potential theory) and specialists in potential theory (all of whom have contributed to several complex variables) together with young researchers and graduate students for an interchange of ideas and techniques. Not only was the status of current research presented, but also the relevant background, much of which is not yet available in books. The following topics and interconnections among them were discussed:

1. **Real and Complex Potential Theory.** Capacity and approximation, basic properties of plurisubharmonic functions and methods to manipulate their singularities and study their growth, Green functions, Chebyshev-type quadratures, electrostatic fields and potentials, propagation of smallness.

2. **Complex Dynamics.** Review of complex dynamics in one variable, Julia sets, Fatou sets, background in several variables, Hénon maps, ergodicity, use of potential theory and multifunctions.

3. **Banach Algebras and Infinite Dimensional Holomorphy.** Analytic multifunctions, spectral theory, analytic functions on a Banach space, semigroups of holomorphic isometries, Pick interpolation on uniform algebras and Von Neumann inequalities for operators on a Hilbert space.

The basic notion of complex potential theory is that of a plurisubharmonic function. In his lectures, C.O. Kiselman begins by comparing convex, subharmonic, and plurisubharmonic functions. He goes on to show that certain sets associated to plurisubharmonic functions are analytic varieties. One of the important attributes of an entire function is its rate of growth. Kiselman studies, more generally, the growth of plurisubharmonic functions and generalizes the notions of order and type of an entire function of finite order to functions of arbitrarily fast growth.

A major theme of several of the lecturers was approximation. N.N. Tarkhanov considers the general problem of approximation of a function defined on a compact set by solutions of a partial differential equation $Pu = 0$, where P is a linear elliptic partial differential operator with analytic coefficients. J. Verdera considers finer problems by restricting his attention to the case where the operator P is homogeneous with constant coefficients. He devotes particular attention to the case of the Cauchy-Riemann operator – that is, holomorphic approximation in one complex variable. In this setting, P.M. Gauthier considers the approximation problem when the set on which the approximation occurs is no longer necessarily compact but is rather allowed to be a (possibly unbounded) closed set. This same problem is investigated by T. Bagby and Gauthier, but in the context of harmonic approximation. The problem of approximation on unbounded sets by solutions of more general elliptic equations has been considered elsewhere and is mentioned in the lectures of Tarkhanov. Both Verdera and Tarkhanov treat the relation between approximation and removability of singularities for solutions of PDEs. Tarkhanov's lectures are greatly motivated by boundary value problems in PDEs.

The subject of complex dynamics, that is, iteration of holomorphic mappings, has attracted a lot of attention in recent years from a wide public, in part (but not only) because of its beautiful pictures and connections with chaos. The dynamics for a function of a single complex variable have been the subject of a large number of studies. Recently, however, new methods from pluripotential theory have produced many new interesting results in the higher dimensional case. J.E. Fornæss and N. Sibony present an overview of this timely topic.

B. Aupetit surveys the subject of analytic multifunctions. This new theory which has its origins in both several complex variables and spectral theory grew out of such problems as the following. How do the eigenvalues of a family of matrices behave if the coefficients of these matrices depend analytically on a parameter? Aupetit presents a remarkable array of applications of this theory: to spectral theory, to the joint spectrum, to uniform algebras in connection with approximation, to spectral interpolation, to local spectrum, to non-associative Jordan algebras, and to complex dynamics.

The lectures of E. Vesentini on semigroups of holomorphic isometries and hyperbolic domains begin with a review of finite-dimensional hyperbolic complex analysis, but mainly, treat infinite-dimensional complex analysis. In fact, infinite-dimensional complex analysis arises naturally in finite-dimensional complex analysis, since, for example, spaces of holomorphic functions (of even a single variable) are infinite-dimensional. The Kobayashi pseudodistance is a very natural pseudodistance on a domain of $\mathbb{C}^n$ which is invariant for automorphisms. If it is a distance, the domain is said to be hyperbolic. In the theory of a single complex variable, there are two domains which are of outstanding importance: the plane itself, $\mathbb{C}$, and the unit disc. Hyperbolic domains are a higher dimensional analog of the unit disc. Vesentini discusses holomorphic mappings on infinite-dimensional hyperbolic domains in complex Banach spaces, devoting particular attention to automorphisms of a domain. For these the basic algebraic operation is composition.

T.W. Gamelin lectures on analytic functions on a Banach space. Here, the target space is usually one-dimensional, the complex plane $\mathbb{C}$. However, he also occasionally discusses analytic functions with values in a normed space. The basic algebraic operations on functions to $\mathbb{C}$ are addition and multiplication. These functions form an algebra. The spectrum of a uniform algebra, which consists of the non-zero complex-valued homomorphisms of the algebra, has played an important role in various problems in analysis. Gamelin studies the spectra of various algebras of holomorphic functions. An interesting aspect of the subject is that natural problems of approximation that are trivial in the plane become difficult in the infinite-dimensional setting.

The paper by B. Cole and J. Wermer on Pick interpolation, Von Neumann inequalities and hyperconvex sets, was presented by Wermer. The authors investigate a class of convex bodies in $\mathbf{C}^n$ which they call hyperconvex. These arise naturally in many interpolation problems – for example, in the problem of interpolating by bounded holomorphic functions in the unit disc. They also arise in problems in operator theory on Hilbert space. Von Neumann proved the following inequality: if T is a contraction on a Hilbert space and if P is a polynomial, then

$$||P(T)|| \leq \sup_{|z|\leq 1} ||P(z)||.$$

D. Sarason has shown that these two beautiful topics (complex interpolation and Von Neumann inequalities) are in fact related. Cole and Wermer embellish this relationship for us.

J. Korevaar's lectures were on Chebyshev-type quadratures: use of complex analysis and potential theory. A Chebyshev-type quadrature formula with nodes $\zeta_1, \ldots, \zeta_N$ for a set E and a measure σ on E is an approximation formula

$$\int_E f(x)d\sigma(x) \approx (1/N)\sum_{j=1}^{N} f(\zeta_j)$$

for integrals over E. In his lectures, Korevaar surveys and extends fundamental quadrature formulas. Of course, the choice of nodes is crucial and Korevaar shows how this is related to electrostatics (distribution of point charges), potential theory, and complex analysis (one and several variables). As an offshoot of his investigation on the "social habits of electrons" Korevaar rediscovered the phenomenon of "propagation of smallness" of harmonic functions. This phenomenon had been observed by Armitage, Bagby, and Gauthier but in a purely qualitative way. Now, Korevaar presents a very elegant quantitative formulation. Indeed, he shows that if Ω is a domain in $\mathbf{R}^n$, Ω_0 a non-empty open subset of Ω, and E a compact subset of Ω, then there is a constant α in $(0,1]$ such that for any harmonic function u on Ω,

$$\sup_E |u| \leq (\sup_{\Omega_0} |u|)^\alpha (\sup_{\Omega} |u|)^{1-\alpha}.$$

Notice the striking resemblance to the Nevanlinna two-constants theorem. Since it is not assumed that the domain is bounded, the phenomenon of propagation of smallness has an impact on the possibility of approximation on unbounded sets, the theme of the lectures of Bagby and Gauthier.

Open problems were also a major component of the conference. All speakers formulated such problems and the very last event of the conference was a problem session at which all participants were invited to submit and discuss their favorite problems.

I wish to express my sincere thanks to all the lecturers and participants for having helped to make this ASI a success. Special thanks are due to Aubert Daigneault, director of the ASI, and to Ghislaine David, secretary of the SMS, both of whom contributed immeasurably to the preparation, mise-en-scène, and "aftermath". Also, my thanks go to Gert Sabidussi and Guogang Gao for their excellent work in editing the present volume.

Last not least, I wish to express on behalf of the Organizing Committee our gratitude to NATO whose financial support has made this ASI possible, and especially to Dr. L. Veiga da Cunha, the Director of the ASI programme, for his help, advice, and understanding.

Paul M. Gauthier

Scientific Director of the ASI

Montreal, April 21st, 1994.

Participants

Kuzman ADZIEVSKI
Department of Mathematics
University of South Carolina
Columbia, SC 29208
USA

John T. ANDERSON
Department of Mathematics
College of the Holy Cross
Worcester, MA 01610–2395
USA

Federica ANDREANO
Department of Mathematics
Brown University
Box 1917
Providence, RI 02912
USA

Ayse Z. AROGUZ
Department of Chemistry
Faculty of Engineering
Istanbul University
34459 Avcilar – Istanbul
Turkey

Jonas AVELIN
Matematiska Institutionen
Uppsala Universitet
Box 480
S-751 06 Uppsala
Sweden

Ruben AVETISYAN
402 Ocean Parkway, Apt. 309
Brooklyn, NY 11218
USA

Sahbi AYARI
Département de mathématiques
et de statistique
Université de Montréal
C.P. 6128-A, Montréal, Qué., H3C 3J7
Canada

Aydin AYTUNA
Department of Mathematics
Middle East Technical University
06531 Ankara
Turkey

Ulf BACKLUND
Department of Mathematics
University of Michigan
Ann Arbor, MI 48109–1003
USA

Esther BARRABÉS VERA
Dept. de Matemàtica Aplicada i Anàlisi
Universitat de Barcelona
Gran Via 585
E-08071 Barcelona
Spain

Riadh BEN GHANEM
Département de mathématiques
et de statistique
Université de Montréal
C.P. 6128-A, Montréal, Qué., H3C 3J7
Canada

Charaf BENSOUDA
Département de mathématiques
et de statistique
Université de Montréal
C.P. 6128-A, Montréal, Qué., H3C 3J7
Canada

Anders BJÖRN
Department of Mathematics
Linköping University
S-581 83 Linköping
Sweden

Pierre BLANCHET
19 rue Ste-Catherine
Lauzon (Comté Lévis)
Québec, Qué., G6V 2W4
Canada

Zbigniew BLOCKI
Institute of Mathematics
Jagiellonian University
ul. Reymonta 4
PL-30059 Kraków
Poland

André BOIVIN
Department of Mathematics
University of Western Ontario
London, Ont., N6A 5B7
Canada

James BRENNAN
Department of Mathematics
715 Patterson Office Tower
University of Kentucky
Lexington, KY 40506–0027
USA

Gregory T. BUZZARD
Department of Mathematics
University of Michigan
Ann Arbor, MI 48109–1003
USA

Jean-Paul CALVI
Département de Mathématiques
U.F.R. - M.I.G.
Université Paul Sabatier
118, route de Narbonne
F-31062 Toulouse Cédex
France

Seddik CHACRONE
Département de mathématiques
et de statistique
Université de Montréal
C.P. 6128-A, Montréal, Qué., H3C 3J7
Canada

Ramón COVA
Dept. of Mathematical Sciences
Science Laboratory
University of Durham
South Road
Durham, DH1 3LE
UK

Chiara DE FABRITIIS
SISSA-ISAS
Via Beirut 2/4
I-34014 Trieste
Italy

Driss DRISSI
Département de mathématiques
et de statistique
Université Laval
Cité Universitaire
Québec, Qué., G1K 7P4
Canada

El Kettani M. ECH-CHÉRIF
Département de mathématiques
et de statistique
Université Laval
Cité Universitaire
Québec, Qué., G1K 7P4
Canada

Abdelkrim EZZIRANI
Lab. de mathématiques appliquées
Université de Pau
Ave de l'Université
F-64000 Pau
France

Juan Carlos FARINA GIL
Dpto. de Análisis Matemático
Universidad de La Laguna
E-38271 La Laguna-Tenerife
Spain

Manuel FLORES MEDEROS
Dpto. de Análisis Matemático
Universidad de La Laguna
E-38271 La Laguna-Tenerife
Spain

Jacques FORTIN
Département de mathématiques
et de statistique
Université Laval
Cité Universitaire
Québec, Qué., G1K 7P4
Canada

El Mostapha FRIH
Département de Mathématiques
Faculté des Sciences
Université Mohammed V
B.P. 1014
Rabat
Morocco

Estela GAVOSTO
Department of Mathematics
University of Michigan
Ann Arbor, MI 48109–1003
USA

Louis-Philippe GIROUX
Département de mathématiques
et de statistique
Université de Montréal
C.P. 6128-A, Montréal, Qué., H3C 3J7
Canada

Ian GRAHAM
Department of Mathematics
University of Toronto
Toronto, Ont., M5S 1A1
Canada

Sandrine GRELLIER
Mathématiques-Bâtiment 425
Université de Paris-Sud
F-91405 Orsay Cédex
France

Allal GUESSAB
Lab. de mathématiques appliquées
Université de Pau
Ave de l'Université
F-64000 Pau
France

Stefan HALVARSSON
Matematiska Institutionen
Uppsala Universitet
Box 480
S-751 06 Uppsala
Sweden

Osvaldo HOSSIAN
Département de mathématiques
et de statistique
Université Laval
Cité Universitaire
Québec, Qué., G1K 7P4
Canada

Alexander IZZO
Department of Mathematics
Brown University
Box 1917
Providence, RI 02912
USA

Hakki T. KAPTANOGLU
Department of Mathematics
Middle East Technical University
06531 Ankara
Turkey

Oleg KAREPOV
Institute of Physics
Siberian Section
Russian Academy of Sciences
Akademgorodok
660036 Krasnoyarsk
Russia

Ognyan KOUNCHEV
FB 11 Mathematik
Universität Duisburg
D-4100 Duisburg 1
Germany

Arno KUIJLAARS
Faculteit der Wiskunde en Informatica
Universiteit van Amsterdam
Pl. Muidergracht 24
NL-1018TV Amsterdam
The Netherlands

Per E. MANNE
Department of Mathematics
University of Oslo
P.O. Box 1053 Blindern
N-0316 Oslo 3
Norway

Abdelaziz MAOUCHE
Département de mathématiques et de statistique
Université Laval
Cité Universitaire
Québec, Qué., G1K 7P4
Canada

Joan MATEU
Dept. de Matemàtica Aplicada
ETSEJB
Univ. Politècnica de Catalunya
Diagonal 647
E-08028 Barcelona
Spain

Thanh Van NGUYEN
Département de Mathématiques
U.F.R. - M.I.G.
Université Paul Sabatier
118, route de Narbonne
F-31062 Toulouse Cédex
France

Marco PELOSO
Dpto. di Matematica
Politecnico di Torino
Corso Duca degli Abruzzi 24
I-10129 Torino
Italy

Karen PINNEY
Department of Mathematics
715 Patterson Office Tower
University of Kentucky
Lexington, KY 40506-0027
USA

Wiesław PLESNIAK
Institute of Mathematics
Jagiellonian University
ul. Reymonta 4
PL-30059 Kraków
Poland

Eugeny POLETSKY
Department of Mathematics
Syracuse University
Syracuse, NY 13244-1150
USA

Analogyj PRYKARPATSKYJ
Department of Nonlinear Mathematical Analysis
Ukrainian Academy of Sciences
290052 Lviv
Ukraine

Alexander RASHKOVSKII
Mathematics Division
Institute for Low Temperature Physics and Engineering
47 Lenin Ave.
310164 Kharkov
Ukraine

Alexander RUSSAKOVSKII
Mathematics Division
Institute for Low Temperature Physics and Engineering
47 Lenin Ave.
310164 Kharkov
Ukraine

Leszek RZEPECKI
Department of Mathematics
University of South Carolina
Columbia, SC 29208
USA

Selim SEKER
Department of Electrical and Electronic Engineering
Bogaziçi University
80815 Bebek - Istanbul
Turkey

Nikolay SHCHERBINA
Departament de Matemàtiques
Universitat Autònoma de Barcelona
E-08193 Bellaterra (Barcelona)
Spain

Rafat N. SIDDIQI
Department of Mathematics
Kuwait University
P.O. Box 5969
13060 Safat
Kuwait

Ragnar SIGURDSSON
Science Institute
University of Iceland
Dunhaga 3
107 Reykjavik
Iceland

Sankhata SINGH
Department of Mathematics
and Statistics
Memorial University
St. John's, Newfoundland, A1C 5S7
Canada

Dan SIRBU
Str. Lirei, Mr15–Bl1–Sc A–Et 4
Bucureşti Sect II
Roumania

Mikhail M. SMIRNOV
Department of Mathematics
Princeton University
Princeton, NJ 08544
USA

Manfred STOLL
Department of Mathematics
University of South Carolina
Columbia, SC 29208
USA

Synne STORLIEN
Department of Mathematics
University of Oslo
P.O. Box 1053 Blindern
N-0316 Oslo 3
Norway

Jerzy SZCZEPANSKI
Institute of Mathematics
Jagiellonian University
ul. Reymonta 4
PL-30059 Kraków
Poland

Jan SZYNAL
Institute of Mathematics
M. Curie-Sklodowska University
PL-20031 Lublin
Poland

Roberto TAURASO
Classe di Scienze
Scuola Normale Superiore
Piazza dei Cavalieri
I-56100 Pisa
Italy

B. Alan TAYLOR
Mathematics Department
University of Michigan
Ann Arbor, MI 48109-1003
USA

Adnan TAYMAZ
Department of Nuclear Physics
Faculty of Science
University of Istanbul, Vezniceler Campus
34459 Istanbul
Turkey

Gülsen TOKAT
Department of Mathematics
Faculty of Sciences
Istanbul Technical University
80626 Maslak – Istanbul
Turkey

Toma TONEV
Dept. of Mathematical Sciences
University of Montana
Missoula, MT 59812–1032
USA

Daniel TURCOTTE
Département de mathématiques
et de statistique
Université Laval
Cité Universitaire
Québec, Qué., G1K 7P4
Canada

M. van FRANKENHUYSEN
Mathematisch Inst.
Katholieke Universiteit
Toernooiveld
NL-6525 ED Nijmwegen
The Netherlands

Dror VAROLIN
Department of Mathematics
University of Toronto
Toronto, Ont., M5S 1A1
Canada

Bert G. WACHSMUTH
Department of Mathematics
and Computer Science
Seton Hall University
South Orange, NJ 07079–2696
USA

James Li-Ming WANG
Department of Mathematics
University of Alabama
Box 870350
Tuscaloosa, AL 35487
USA

Abdoul O. WATT
École Polytechnique
B.P. 10
Thiès
Sénégal

Georges WEILL
Department of Mathematics
Polytechnic University
6 Metrotech Center
Brooklyn, NY 11201
USA

Tim WILKINS
Department of Pure Mathematics
and Mathematical Statistics
University of Cambridge
16 Mill Lane
Cambridge CB2 1SB
UK

Vyaceslav ZAHARIUTA
Department of Mathematics
Middle East Technical University
06531 Ankara
Turkey

Contributors

Bernard AUPETIT
Département de mathématiques
et de statistique
Université Laval
Cité Universitaire
Québec, Qué., G1K 7P4
Canada

S. Thomas BAGBY
Department of Mathematics
Rawles Hall
Indiana University
Bloomington, IN 47405
USA

Brian J. COLE
Department of Mathematics
Brown University
Box 1917
Providence, RI 02912
USA

John Erik FORNÆSS
Department of Mathematics
University of Michigan
Ann Arbor, MI 48109–1003
USA

Theodore W. GAMELIN
Department of Mathematics
405 Hilgard Ave.
University of California, Los Angeles
Los Angeles, CA 90024–1555
USA

Paul M. GAUTHIER
Département de mathématiques
et de statistique
Université de Montréal
C.P. 6128-A, Montréal, Qué., H3C 3J7
Canada

Christer O. KISELMAN
Matematiska Institutionen
Uppsala Universitet
Box 480
S-751 06 Uppsala
Sweden

Jacob KOREVAAR
Faculteit der Wiskunde en Informatica
Universiteit van Amsterdam
Pl. Muidergracht 24
NL-1018TV Amsterdam
The Netherlands

Nessim SIBONY
Mathématiques-Bâtiment 425
Université de Paris-Sud
F-91405 Orsay Cédex
France

Nikolay N. TARKHANOV
Max-Planck-Gesellschaft
Arbeitsgruppe Analysis
Universität Potsdam
Pf. 60 15 53
D-14415 Potsdam
Germany

Joan VERDERA
Departament de Matemàtiques
Universitat Autònoma de Barcelona
E-08193 Bellaterra (Barcelona)
Spain

Edoardo VESENTINI
Istituto Matematico
Scuola Normale Superiore
Piazza dei Cavalieri
I-56100 Pisa
Italy

John WERMER
Department of Mathematics
Brown University
Box 1917
Providence, RI 02912
USA

Analytic multifunctions and their applications

Bernard AUPETIT

Département de mathématiques et de statistique
Université Laval
Québec, G1K 7P4
Canada

Abstract

This long survey article is a quick introduction to the theory of analytic multifunctions and their numerous applications. This new theory originates both from spectral theory and several complex variables. In the introduction we describe its historical evolution from its very beginnings to the present day. Then are given the main results of the general theory of analytic multifunctions like Liouville's theorem, the localization principle, the holomorphic variation of isolated points, the scarcity theorems for finite or countable analytic multifunctions, Słodkowski's theorem, the Oka–Nishino theorem and finally the open mapping theorem and Picard's theorem on distribution of values. In the rest of the text are first given very striking applications to spectral theory, then to joint spectrum, to uniform algebras in connection with problems of analytic structure and polynomial and entire approximation in $\mathbb{C}^n$, to spectral interpolation and to local spectrum. Recently, important applications were given to non-associative Jordan-Banach algebras and to complex dynamics, that is, the study of the variation of Julia sets depending on a parameter, which are described in the last two chapters.

Chapter 1
Historical introduction

The theory of analytic multifunctions originates from two different fields, spectral theory and the theory of functions of several complex variables. Even if analytic multifunctions were implicitly contained in some old work of K. Oka, as we shall see below, it took a long time to realize that they had any relevance to spectral theory.

1.1 Origins in spectral theory

In spectral theory it is a natural question to ask how behave the eigenvalues of a family of matrices whose coefficients depend analytically on a parameter. This was studied for the first time by A. Cauchy (see [Haw]). In the 1830s, A. Cauchy, C. Sturm and J. Liouville studied the behaviour of the solutions of a differential equation whose coefficients depend analytically on a parameter z. For instance, considering a Sturm–Liouville equation of the type

$$y'' - q_z(x)y + \lambda z = 0$$

P. M. Gauthier (ed.) and G. Sabidussi (techn. ed.), Complex Potential Theory, 1–74.

with boundary conditions

$$\alpha_1 y(a) + \beta_1 y'(a) = \alpha_2 y(b) + \beta_2 y'(b) = 0,$$

where q_z depends analytically on z, there are non-trivial solutions for some eigenvalues $\lambda_1(z) < \lambda_2(z) \cdots < \lambda_n(z) < \cdots$ tending to infinity and the $\lambda_n(z)$ depend analytically on z in some sense, except that they have branching points (see [Lü] for instance). In his fundamental memoir [Pu], V. Puiseux studies the multifunction $x \mapsto K(x)$ associated to an algebraic equation $f(x,z) = 0$ (in fact this problem is equivalent to studying the variation of the eigenvalues of a family of matrices whose coefficients are polynomials). He determines the branching points and the local series representing the solutions

$$\sum_{k=m}^{\infty} a_n (x - x_0)^{k/q}$$

around a branching point x_0 of multiplicity q. These ideas go back in fact to I. Newton who studied the branches of a real algebraic curve $F(x,y) = 0$ around a singular point and invented the *Newton polygons* to determine the exponents in the previous series (see [D1], pp. 106–112). The ideas of Puiseux influenced very much B. Riemann in his creation of *Rieman surfaces* and the parametrization of the different sheets (see [Ne] for more details). In 1896, G.F. Frobenius introduced some analytic methods for the study of matrices, methods which were improved in more general situations, first by I. Fredholm in 1903 in the case of integral equations

$$f(s) - \lambda \int_a^b k(s,t) f(t) dt = g(s),$$

by D. Hilbert in 1904, who converted problems with boundary conditions to integral equation problems and linear systems with an infinite number of unknowns, that is, the study of infinite matrices on ℓ^2, and finally by F. Riesz in 1913 who gave a very abstract setting to these problems. He introduced the resolvent of an operator and Cauchy's formula for some operators. This was later extended to general operators on Banach spaces by N. Wiener in 1923 and M.H. Stone in the 1930s. The great years for spectral theory are the years 1932–1943 where the most important contributions were obtained by M. Nagumo (1936), S. Mazur and A.E. Taylor (1938), E. Hille (1939), E.R. Lorch (1943) and mainly I.M. Gelfand in the 1940s. Later, important contributions on spectral perturbation were obtained by F. Rellich (see [Re]), K.O. Friedrichs, T. Kato (see [Ka]), B. Sz.–Nagy, N. Dunford, etc. A new important tool was introduced into spectral theory in 1955 by G.E. Shilov, when he used several complex variables and generalized Cauchy formulas proving for instance what is now called the *Shilov idempotent theorem*. The reader interested to know more about the historical evolution of spectral theory until the beginning of the 1960s should read [D3], the papers of J.-L. Verley (Chapter 5) and J. Dieudonné (Chapter 8) in [D2], [Mo] and [T].

1.2 Origins in several complex variables

Subharmonic functions were implicitly introduced in their work by H. Poincaré in his theory of balayage and F. Hartogs in his studies of holomorphic functions of several complex variables. The modern concept was introduced by F. Riesz in the 1920s (see his *Oeuvres complètes*). F. Riesz, as we saw before, is also the creator of modern spectral theory, but at that time he did not realize that the theory of subharmonic functions and spectral theory are equivalent in the following sense: given a subharmonic function φ defined on a domain D and a relatively compact subdomain D' of D, there exists a family $T(\lambda)$ of operators in $\mathcal{B}(\ell^2)$ depending analytically on $\lambda \in D'$ such that $\varphi(\lambda) = \operatorname{Log} r(T(\lambda))$, where r denotes the spectral radius; and conversely, for any analytic family $T(\lambda) \in \mathcal{B}(X)$, where X is a Banach space, then $\operatorname{Log} r(T(\lambda))$ is subharmonic (see Theorem 2.1.16, Theorem 2.5.1 and Corollary 3.1.3).

The fundamental work of F. Hartogs on analytic continuation, separate analyticity, singularities of holomorphic functions, was done at the beginning of the 20th century (see [Har] for instance). It is very well described in Chapters 2, 3, 4 of [Na]. The first important result of Hartogs in relation with analytic multifunctions is the following: given a continuous function defined on a domain of $\mathbb{C}$ with values in $\mathbb{C}$, it is holomorphic if and only if its graph in $\mathbb{C}^2$ is a set of singularities, or in other words if its complement is an open pseudoconvex set (see Theorem 2.1.4 and Theorem 2.1.15). He also determined the radius of a polycylindral pseudoconvex set (see Corollary 2.1.17).

The ideas of F. Hartogs were extended by K. Oka in 1934 in a rather obscure paper [O], where he vaguely introduces the concept of analytic multivalued function. He refers to E.E. Levi (1910) and G. Julia (1926) as precursors of his ideas. Outside of the algebroid case he gives no example, and on the last page he gives a list of theorems that could be proved (log-subharmonicity of the diameter of fibres, corresponding to Theorem 2.1.3 in this paper, scarcity of finite fibres, corresponding to Theorem 2.1.12, and scarcity of countable fibres, corresponding to Theorem 2.2.8) without any proofs. These results were proved later by T. Nishino [Ni] and H. Yamaguchi [Ya], but their arguments are very far from being satisfactory. As we shall see below, these papers of Oka, Nishino and Yamaguchi were forgotten for some time and rediscovered at the beginning of the 1980s when the theory of analytic multifunctions came to life.

It is also possible that the topological theory of set-valued functions, created in Poland in the 1930s (see [Ku1,2] for instance), had some influence on the theory. But this is not clear. The only noticeable one is the use of transfinite induction in the first class of ordinals to prove the Oka–Nishino theorem.

1.3 The birth of analytic multifunctions

In 1966, A. Brown and R.G. Douglas considered the problem of formulating a maximum principle for the multifunction $\lambda \mapsto Sp\, f(\lambda)$, where f is an analytic function from a domain D of $\mathbb{C}$ into a Banach algebra. In 1968–1970, E. Vesentini solved this question by proving that $\lambda \mapsto \operatorname{Log} r(f(\lambda))$ is subharmonic, where r denotes the spectral radius. This result was also obtained by V. Istrăţescu in 1969 and B. Schmidt in 1970, but E. Vesentini was the first to give interesting applications (see [Ve1,2]) like Theorems 2.1.6, 2.1.8. Very curiously this important result, which contained implicitly many spectral consequences and the theory of analytic multifunctions, was not even mentioned in the classical book of F.F. Bonsall and J. Duncan [BD]. Many applications were given in [Au1–6] and recent ones in [Au7], Chapter 3 and 5 and [Au8].

Inspired by I. Kaplansky's theorem characterizing the finite-dimensional Banach algebras (Theorem 3.2.2), the scarcity theorem for finite spectrum (Theorem 2.1.12) is really the turning point in the creation of analytic multifunctions. Originally it was proved in 1975 using only subharmonic functions in a very technical way and submitted to J. Wermer for publication in the Journal of Functional Analysis where it appeared [Au1]. These arguments gave J. Wermer the idea that they could be used for fibres in uniform algebras in order to improve Bishop's and Basener's theorems on analytic structure. These generalizations where published in [AW] (see Chapter 5 for more details).

Another problem also provoked the birth of the new child. In [PS] A. Pełcyńzski and Z. Semadeni settled the following problem : let A be a C^*-algebra and suppose the spectrum of every self-adjoint element finite or countable, is it true that the spectrum of every element of A is finite or countable and that the algebra has a precise algebraic structure? (they are called *scattered C^*-algebras*). In his lectures at the Stefan Banach Center in Warsaw (September–December 1977), the author explained how it would be possible to solve this conjecture in the general setting of Banach algebras with involution, knowing a scarcity theorem for countable spectrum (now known, it is Theorem 2.2.8), and he proved it when the spectrum is finite or a sequence going to zero (this is the first step in the ordinal argument, see [Au2], pp. 84–86). The original conjecture was solved for C^*-algebras by T. Huruya [Hu] and, in 1979, E. Kirchberg claimed to have solved the general conjecture [Ki] but his paper contains several obscure points (it never appeared because of the long time spent by the author in the DDR jails).

During July–August 1978 the author and J. Wermer exchanged several letters in which they tried to introduce the good concept of *analytic multifunction* using subharmonic functions. It is during his stay in Canada (1978–1979), that Z. Słodkowski introduced the convenient definition using the more geometric notion of pseudoconvex set (see Theorem 2.1.15). Later on it was discovered that this definition is in fact due to K. Oka, as mentioned previously, but of course K. Oka never saw any relation and applications to spectral theory. So the very beginnings of the theory are the papers [Sł1] and [Au3]. To be fair it must be said that this convenient definition is also contained in germ in the big survey of M.G. Zaidenburg, S.G. Krein, P.A. Kuchment and A.A. Pankov [ZKKP], where they prove that in a Banach algebra the open set of invertible elements is pseudoconvex (this is equivalent to Theorem 3.1.2).

1.4 Conclusion

Since the beginning of the 1980s the baby has very much grown. Every year sees new results and new applications to theories which at first sight seem to be very different. Probably a lot has still to be done, for instance in local spectrum theory, in cluster set theory (some partial applications are given in section 5.4), in the theory of differential inclusions, etc. But an important point must be kept in mind, mainly for future researchers in the field : analytic multifunctions use several complex variables and are useful in many fields, but their main origin and their main motivation are in spectral theory.

The reader needs a few prerequisites which are briefly summarized in the Appendix of [Au7]. Concerning several complex variables he can read [Hö], [Vl] and all the standard textbooks on the field. Concerning subharmonic functions there are [HK], [Ts] and the forthcoming [Ra]. For a quick introduction to modern spectral theory via subharmonic functions and analytic multifunctions see [Au7].

Chapter 2
General properties of analytic multifunctions

2.1 Definitions and general properties

As we explained in the historical introduction, the convenient definition of analytic multifunctions was discovered by Z. Słodkowski [Sł1]. In his paper, Z. Słodkowski gave two definitions, the first one being valid only for analytic multifunctions from $\mathbb{C}$ into $\mathbb{C}$, the second one for analytic multifunctions from $\mathbb{C}^n$ into $\mathbb{C}^m$, and he proved their equivalence for $m = n = 1$. We shall use the second definition as it leads to interesting results very quickly.

The majority of results and ideas mentioned in this section are due to B. Aupetit, T.J. Ransford, Z. Słodkowski, H. Yamaguchi and A. Zraïbi.

Definition Let K be a mapping from an open subset D of $\mathbb{C}^n$ into the set of non-empty compact subsets of $\mathbb{C}^m$. We shall say that K is an *analytic multifunction* on D if it satisfies the following properties:

(i) K is upper semicontinuous on D,

(ii) for every relatively compact open subset V of D and every function ϕ plurisubharmonic on a neighbourhood of the restriction of the graph of K to V, the function

$$\psi(\lambda) = \mathrm{Max}\{\phi(z) : z \in K(\lambda)\}$$

is plurisubharmonic on V.

A function is plurisubharmonic if and only if it is locally plurisubharmonic, and consequently a multifunction is analytic if and only if it is locally analytic. In (ii) we may suppose that ϕ is a C^∞-strictly plurisubharmonic function since ϕ can be approximated by such functions.

Examples

1. If h is holomorphic on $D \subset \mathbb{C}^m$, with values in $\mathbb{C}^n$, then $\lambda \mapsto \{h(\lambda)\}$ is an analytic multifunction on D.

2. Let K_0, K_1 be two compact subsets of $\mathbb{C}^n$. For $\lambda \in D \subset \mathbb{C}$, the multifunction $K(\lambda) = \lambda K_0 + K_1$ is analytic on D.

3. Let f be an analytic function from $D \subset \mathbb{C}$ into $M_n(\mathbb{C})$. Then $\lambda \mapsto Sp\, f(\lambda)$ is an analytic multifunction on D.

Much more important results will be given later. An interesting class of multifunctions is the following.

Definition An upper semicontinuous multifunction K from $D \subset \mathbb{C}^n$ into $\mathbb{C}^m$ is said to have *holomorphic selections* if, for each $\lambda_0 \in D$ and each $z_0 \in \partial K(\lambda_0)$, there exists h holomorphic on a neighbourhood U of λ_0, with values in $\mathbb{C}^m$, such that $z_0 = h(\lambda_0)$ and $h(\lambda) \in K(\lambda)$, for $\lambda \in U$.

Such multifunctions having holomorphic selections are continuous analytic multifunctions. Examples 1 and 2 are in this class, but not Example 3 globally on D because of the branching points. The following theorem has a very technical proof which will not be given here (we refer the reader to the papers of B. Aupetit, B. Aupetit & A. Zraïbi, Z. Słodkowski, and T.J. Ransford).

Theorem 2.1.1 *The following properties hold.*

(i) *If (K_p) is a sequence of analytic multifunctions defined on $D \subset \mathbb{C}^n$ with values in $\mathbb{C}^m$, such that $K_{p+1}(\lambda) \subset K_p(\lambda)$ for each $\lambda \in D$, then $K = \bigcap_{p=1}^{\infty} K_p$ is an analytic multifunction from D into $\mathbb{C}^m$.*

(ii) *If $K_1, \cdots, K_p$ are analytic multifunctions from $D \subset \mathbb{C}^n$ into $\mathbb{C}^m$ then $K = K_1 \cup \cdots \cup K_p$ is an analytic multifunction from D into $\mathbb{C}^m$.*

(iii) *If K is an analytic multifunction from $D \subset \mathbb{C}^n$ into $\mathbb{C}^m$ and if L is an analytic multifunction from $G \subset \mathbb{C}^m$ into $\mathbb{C}^k$, where G is an open set containing all the $K(\lambda)$ for $\lambda \in D$, then $L \circ K$ defined by*

$$(L \circ K)(\lambda) = \{L(z) : z \in K(\lambda)\}$$

is an analytic multifunction from D into $\mathbb{C}^k$.

(iv) *If $K_1, \cdots, K_p$ are analytic multifunctions from $D \subset \mathbb{C}^n$ into $\mathbb{C}^m$, then $K = K_1 \times \cdots \times K_p$ is an analytic multifunction from D into $\mathbb{C}^{mp}$.*

(v) *Let K be an upper semicontinuous multifunction from $D \subset \mathbb{C}^n$ into $\mathbb{C}^m$. Then K is an analytic multifunction on D if and only if $t \mapsto K(at+b)$ is an analytic multifunction on $\{t : t \in \mathbb{C},\ at+b \in D\}$, for every $a, b \in \mathbb{C}^n$.*

Let $E \subset \mathbb{C}^m$ we recall that the *polynomially convex hull of E*, denoted by $\hat{E}$, is the intersection of all the sets $\{z \in \mathbb{C}^m : |p(z)| \leq \max_{w \in E} |p(w)|\}$, for all the polynomials of m variables. If $m = 1$, $\hat{E}$ is the union of E and its holes but if $m > 1$ it may be bigger.

Theorem 2.1.2 *Let K be an analytic multifunction from $D \subset \mathbb{C}^n$ into $\mathbb{C}^m$ and suppose that L is an upper semicontinuous multifunction from D into $\mathbb{C}^m$ such that $\partial L(\lambda) \subset K(\lambda) \subset L(\lambda)$, for each $\lambda \in D$. Then L is an analytic multifunction. In particular, $\hat{K}$ is an analytic multifunction.*

Given a compact set E we denote its *radius* by

$$r(E) = \operatorname*{Max}_{\lambda \in E} |\lambda|,$$

its *n-th diameter* by

$$\delta_n(E) = \operatorname*{Max}_{\lambda_1, \cdots, \lambda_{n+1} \in E} \left(\prod_{1 \le i \le j \le n+1} |\lambda_i - \lambda_j| \right)^{2/n(n+1)}$$

and its *capacity* by $c(E)$, the logarithmic capacity of E.

Theorem 2.1.3 *Let K be an analytic multifunction from $D \subset \mathbb{C}$ into $\mathbb{C}$. Then $\lambda \mapsto \operatorname{Log} r(K(\lambda))$, $\lambda \mapsto \operatorname{Log} \delta_n(K(\lambda))$ for $n \ge 1$, and $\lambda \mapsto \operatorname{Log} c(K(\lambda))$ are subharmonic on D.*

For the first function, the proof is immediate from the definition using the plurisubharmonic function $\phi(\lambda, z) = \operatorname{Log}|z|$. For the second one, let

$$\phi(\lambda, z_1, \cdots, z_{n+1}) = \frac{2}{n(n+1)} \sum_{1 \le i < j \le n+1} \operatorname{Log}|z_i - z_j|,$$

which is plurisubharmonic on $\mathbb{C}^{n+2}$. By Theorem 2.1.1 (iv), we know that

$$\lambda \mapsto K(\lambda) \times \cdots \times K(\lambda), \ n+1 \text{ times},$$

is an analytic multifunction. So, by definition,

$$\operatorname{Log} \delta_n(K(\lambda)) = \operatorname{Max}\{\phi(\lambda, z_1, \cdots, z_{n+1}) : z_1 \in K(\lambda), \cdots, z_{n+1} \in K(\lambda)\}$$

is plurisubharmonic. It is known that $c(K(\lambda))$ is the decreasing limit of $\delta_n(K(\lambda))$, when n goes to infinity. So, by the analogue of Harnack's theorem for subharmonic functions, we get the last result.

Let D be an open subset of $\mathbb{C}^n$, h be holomorphic on D and $\Omega = \{(\lambda, z) : \lambda \in D, z \neq h(\lambda)\}$ be the complement of the graph of h. Considering the function $g(\lambda, z) = 1/(z - h(\lambda))$, which is holomorphic on Ω, it is obvious that Ω is an open set of holomorphy and so it is pseudoconvex. F. Hartogs proved the converse, namely that if h is a locally bounded function from D into $\mathbb{C}$ such that Ω is an open set of holomorphy, then h is holomorphic on D. The original proof uses subharmonic arguments (see R. Narasimhan, [Na], p. 56). With Theorem 2.1.15, Hartogs's result can be reduced to the following generalization.

Theorem 2.1.4 *Let h be a function from $D \subset \mathbb{C}^n$ into $\mathbb{C}^m$. Then h is holomorphic on D if and only if $\lambda \mapsto \{h(\lambda)\}$ is an analytic multifunction on D.*

From the previous results we can obtain the following theorems whose proofs are given in Chapter 3 of [Au7] in the situation of the spectrum. The general results have identical proofs.

Theorem 2.1.5 (Liouville's theorem for analytic multifunctions) *Let K be an analytic multifunction from $\mathbb{C}^n$ into $\mathbb{C}^m$. Suppose there exists a bounded set E such that $K(\lambda) \subset E$ for all $\lambda \in \mathbb{C}^n$. Then $\hat{K}(\lambda)$ is constant.*

Unfortunately $K(\lambda)$ is not constant in general. For instance on $\ell^2(\mathbf{Z})$ with the orthonormal basis $(e_n)_{n\in\mathbf{Z}}$ the two weighted shifts

$$ae_n = \begin{cases} 0, & \text{if } n = -1 \\ e_{n+1}, & \text{if } n \neq -1 \end{cases} \qquad be_n = \begin{cases} e_0, & \text{if } n = -1 \\ 0, & \text{if } n \neq -1 \end{cases}$$

define an analytic multifunction $K(\lambda) = Sp\,(a + \lambda b)$ such that $K(0)$ is the unit disk and $K(\lambda)$ is the unit circle for $\lambda \neq 0$.

The next theorems where proved by E. Vesentini [Ve2] in the case of the spectrum. The first one has a proof which contains a nice geometrical argument which is also used in the proof of Lemma 7.2.2. We recall that the *peripherical part* of E is by definition the set of λ such that $|\lambda| = r(E)$.

Theorem 2.1.6 *Let K be an analytic multifunction from a domain D of $\mathbb{C}^n$ into $\mathbb{C}^m$. Suppose there exists $\lambda_0 \in D$ such that $r(K(\lambda)) \leq r(K(\lambda_0))$ for all $\lambda \in D$. Then the peripherical part of $K(\lambda)$ is constant on D.*

Corollary 2.1.7 *If $K(\lambda) \subset \mathbf{R}^m$ for all $\lambda \in D$ then $K(\lambda)$ is constant on D.*

Theorem 2.1.8 (Maximum principle) *Let K be an analytic multifunction from a domain D of $\mathbb{C}^n$ into $\mathbb{C}$. Suppose that there exists $\lambda_0 \in D$ such that $K(\lambda) \subset K(\lambda_0)$, for all $\lambda \in D$. Then $\partial K(\lambda_0) \subset \partial K(\lambda)$ and $\hat{K}(\lambda_0) = \hat{K}(\lambda)$, for all $\lambda \in D$. Consequently if K is an analytic multifunction from a domain $D \subset \mathbb{C}^n$ into $\mathbb{C}^m$ and if there exists $\lambda_0 \in D$ such that $K(\lambda) \subset K(\lambda_0)$ then $\hat{K}(\lambda_0) = \hat{K}(\lambda)$ on D.*

The first part of the proofs comes from the maximum principle for subharmonic functions. The second part derives from the first using the multifunctions $\lambda \mapsto p(K(\lambda))$ where p is a polynomial of n variables.

The next theorem is a very important one. In some sense it improves the use of holomorphic functional calculus in spectral theory.

Theorem 2.1.9 (Localization principle) *Let D be a domain of $\mathbb{C}^n$ and let K be an analytic multifunction from D into $\mathbb{C}^m$. Suppose that there exist two disjoint open sets U, V in $\mathbb{C}^m$ such that $K(\lambda) \subset U \cup V$, for every $\lambda \in D$. Then either $K(\lambda) \cap U = \emptyset$, for every $\lambda \in D$, or $K(\lambda) \cap U \neq \emptyset$, for every $\lambda \in D$. In the latter case $L(\lambda) = K(\lambda) \cap U$ defines an analytic multifunction on D.*

The J.D. Newburgh's theorem on continuity of the spectrum can be extended in the following way.

Corollary 2.1.10 *Let K be an analytic multifunction from a domain D of $\mathbb{C}^n$ into $\mathbb{C}^m$. Suppose that $K(\lambda_0)$ is totally disconnected for some $\lambda_0 \in D$. Then K is continuous at λ_0.*

This implies in particular that K is continuous at all λ where $K(\lambda)$ is finite or countable or more generally of zero capacity.

Let f be an analytic function form $D \subset \mathbb{C}$ into the algebra $\mathcal{K}(X)$ of compact operators on the Banach space X, and let $\lambda_0 \in D$, $\alpha_0 \in Sp\, f(\lambda_0)$ with $\alpha_0 \neq 0$. To simplify suppose that α_0 is an eigenvalue with multiplicity one, or equivalently that the projection associated to $\mathcal{N}(f(\lambda_0) - \alpha_0 I)$ has rank one. Then there exist $r, \delta > 0$ such that $|\lambda - \lambda_0| < \delta$ implies that $Sp\, f(\lambda) \cap B(\alpha_0, r)$ contains only one eigenvalue $\alpha(\lambda)$. What can be said about this function α? In this particular case it is known that α is holomorphic on $B(\lambda_0, \delta)$. The classical proof depends strongly on the fact that $f(\lambda) \in \mathcal{K}(X)$ (see for instance the book by I.C. Gohberg and M.G. Krejn [GoK], Chapter 2).

In the next theorem we shall see that this result is true in general.

Theorem 2.1.11 (Holomorphic variation of isolated points) *Let D be a domain of $\mathbb{C}^n$ and K be an analytic multifunction from D into $\mathbb{C}^m$. Suppose that U is an open subset of $\mathbb{C}^m$ such that $\#(K(\lambda) \cap U) = 1$ and $K(\lambda) \cap \partial U = \emptyset$, for $\lambda \in D$. Then there exists h holomorphic on D, with values in $\mathbb{C}^m$, such that $K(\lambda) \cap U = \{h(\lambda)\}$.*

The next theorem, which was proved in the case of the spectrum for the first time in [Au1], has many important applications. We shall meet several of these applications in the next chapters.

Theorem 2.1.12 (Scarcity of elements with finite values) *Let D be a domain of $\mathbb{C}^n$ and K be an analytic multifunction from D into $\mathbb{C}^m$. Then either $\{\lambda\colon \lambda \in D, \#K(\lambda) < \infty\}$ is pluripolar in D or there exist an integer N and a closed analytic subvariety F of D such that $\#K(\lambda) = N$ on $D \setminus F$ and $\#K(\lambda) < N$ on F. Moreover, in this last situation, for each $\lambda_0 \in D \setminus F$ there exist N functions $h_1, \cdots, h_N$ with values in $\mathbb{C}^m$, holomorphic on a neighbourhood of λ_0, such that $K(\lambda) = \{h_1(\lambda), \cdots, h_N(\lambda)\}$ on this neighbourhood.*

In order to prove Z. Słodkowski's result we need two theorems, the second one resulting from the classical theorem on the exhaustion of pseudoconvex open sets by regular strictly pseudoconvex open sets. The following definition extends the notion of holomorphic selections. For instance, Example 3 has no holomorphic selections at the branching points but has good selections everywhere, a good selection being defined as follows.

Definition An upper semicontinuous multifunction K from $D \subset \mathbb{C}^n$ into $\mathbb{C}$ is said to have *good selections* if, for each $\lambda_0 \in D$ and each $z_0 \in \partial K(\lambda_0)$, either there exist $r > 0$ and h holomorphic for $|\lambda - \lambda_0| < r$, with complex values, such that $h(\lambda_0) = z_0$ and $h(\lambda) \in K(\lambda)$ for $|\lambda - \lambda_0| < r$, or there exist $s > 0$ and k holomorphic for $|z - z_0| < s$, with values in D, such that $k(z_0) = \lambda_0$ and $z \in K(k(z))$ for $|z - z_0| < s$.

Theorem 2.1.13 *Let K be an upper semicontinuous multifunction from $D \subset \mathbf{C}$ into $\mathbf{C}$, having good selections. Then K is a continuous analytic multifunction on D.*

Theorem 2.1.14 *Let K be an upper semicontinuous multifunction from $D \subset \mathbf{C}$ into $\mathbf{C}$ such that the complement of its graph is pseudoconvex in $\mathbf{C}^2$. Then there exist an increasing sequence of relatively compact open subsets D_n of D, exhausting D, and a sequence (K_n) of analytic multifunctions, respectively defined on D_n, having good selections, and such that $K_{n+1}(\lambda) \subset K_n(\lambda)$ for each $\lambda \in D_n$ and $K(\lambda) = \lim_{n\to\infty} K_n(\lambda)$ for each $\lambda \in D$. In particular, K is an analytic multifunction on D.*

We now arrive at the fundamental theorem, which was proved in [Sł1].

Theorem 2.1.15 (Z. Słodkowski) *Let K be an upper semicontinuous multifunction from $D \subset \mathbf{C}$ into $\mathbf{C}$. Then the following are equivalent:*

(i) *K is an analytic multifunction;*

(ii) $-\mathrm{Log}\,\mathrm{dist}(z, K(\lambda))$ *is plurisubharmonic on Ω, the complement of its graph;*

(iii) *Ω is pseudoconvex in $\mathbf{C}^2$.*

Z. Słodkowski later extended this theorem to an analytic family of closed operators replacing the spectrum by the extended spectrum which is a subset of the Riemann sphere $\mathbf{C}_\infty$ (see [Sł3]).

The following result is in fact equivalent to a classical result due to F. Hartogs. It gives interesting examples of discontinuous analytic multifunctions.

Theorem 2.1.16 *Let D be open in $\mathbf{C}$ and ϕ be defined on D with values in $\mathbf{R} \cup \{-\infty\}$. Then the multifunction K, defined by $K(\lambda) = \{z : |z| \le e^{\phi(\lambda)}\}$ on D, is an analytic multifunction on D if and only if ϕ is subharmonic on D.*

Corollary 2.1.17 (F. Hartogs) *Let D be open in $\mathbf{C}$ and ψ a positive function on D such that $\{(\lambda, z) : \lambda \in D, |z| < \psi(\lambda)\}$ is pseudoconvex in $\mathbf{C}^2$. Then $-\mathrm{Log}\,\psi(\lambda)$ is subharmonic on D.*

Taking

$$\phi(\lambda) = \sum_{n=1}^{\infty} \frac{\mathrm{Log}\,|\lambda - 1/n| + \mathrm{Log}\, n}{n^3},$$

which is well-defined and subharmonic on $\mathbf{C}$, we have $\phi(1/n) = -\infty$ for $n = 1, 2, \cdots$, and $\phi(0) = 0$. The corresponding analytic multivalued function K satisfies $K(1/n) = \{0\}$ and $K(0) = \{z : |z| \le 1\}$, so it is discontinuous at 0. By condensation of the singularities it is even possible to construct an analytic multifunction K discontinuous on a dense set.

Using concrete examples in $\mathcal{B}(H)$, where H is a Hilbert space, or pseudoconvex subsets of $\mathbf{C}^2$, it is easy to build many examples of very discontinuous analytic multifunctions.

Despite this discontinuity there is nevertheless some boundary continuity for the fine topology (that is, the topology for which all subharmonic functions are continuous) and this fact can be used in some cases like for instance in Chapter 9. This result was proved for the first time in [Au1], see also [Au7], p. 60.

For the definition of a *non-thin set* see the standard books on potential theory like [Br] or [He]. A Jordan curve is non-thin at each of its points, a domain is non-thin at any of its boundary points.

Theorem 2.1.18 (Weak lower semicontinuity of the boundary) *Let K be an analytic multifunction from a domain D of $\mathbb{C}$ into $\mathbb{C}$. Suppose that $E \subset D \setminus \{\lambda_0\}$ is non-thin at $\lambda_0 \in D \cap \overline{E}$. Then there exists a sequence (η_k) converging to λ_0, such that $\eta_k \in E$ and*

$$\partial K(\lambda_0) \subset \partial K(\eta_k) + B\left(0, \frac{1}{k}\right), \textit{for } k \geq 1.$$

Corollary 2.1.19 *With the previous hypotheses, suppose that there exists a closed subset F on $\mathbb{C}$ such that $\partial K(\lambda) \subset F$, for all λ in E. Then $\partial K(\lambda_0) \subset F$. If moreover F has no interior points and does not separate the plane then $K(\lambda_0) \subset F$.*

This corollary can be applied to $F = \mathbb{R}$ or any Jordan arc.

2.2 The Oka–Nishino theorem and its applications

We now intend to give the proof of the very important Oka–Nishino theorem. The original proof due to T. Nishino [Ni] contains many obscure points. The following presentation is largely inspired by the ideas of B. Aupetit, T. Nishino, T.J. Ransford, Z. Słodkowski and J. Zemánek. For more details see [Au7] or [Ra1].

Definition Let K be an analytic multifunction from $D \subset \mathbb{C}$ into $\mathbb{C}$. We say that $z_0 \in K(\lambda_0)$ is a *good isolated point* of $K(\lambda_0)$ if there exist $r, s > 0$ such that $\overline{B}(z_0, s) \cap K(\lambda_0) = \{z_0\}$, and $B(z_0, s) \cap K(\lambda)$ is finite for $|\lambda - \lambda_0| < r$.

This definition implies that z_0 is isolated in $K(\lambda_0)$. But conversely, an isolated point of $K(\lambda_0)$ is not necessarily a good isolated point. To see this, consider the following example of the analytic multifunction defined for $\lambda \in \mathbb{C}$ by

$$K(\lambda) = \{0\} \cup \left\{z: \frac{|\lambda|}{2} \leq |z| \leq |\lambda|\right\}.$$

Of course 0 is isolated in $K(0)$, but it is not a good isolated point. In fact this definition of a good isolated point is purely geometrical. It means that in a neighbourhood of (λ_0, z_0), the graph of K is an analytic variety. This notion is closely related to that of extension points for pseudoconvex sets which we now define.

Definition Let Ω be a pseudoconvex open subset of $\mathbb{C}^2$. We say that a point $a \in \mathbb{C}^2 \setminus \Omega$ is an *extension point* for Ω if there exist an open neighbourhood U of a and a non-zero holomorphic function f on U, with values in $\mathbb{C}$, such that

$$U \setminus \Omega = \{z : z \in U, f(z) = 0\}.$$

We define Ω' to be the union of Ω and its extension points. Clearly Ω' is open and contains Ω.

Theorem 2.2.1 *Let K be an analytic multifunction from $D \subset \mathbb{C}$ into $\mathbb{C}$. If z_0 is a good isolated point of $K(\lambda_0)$ then (λ_0, z_0) is an extension point for $\Omega = \{(\lambda, z) : \lambda \in D, z \notin K(\lambda)\}$. Conversely, if $z_0 \in \partial K(\lambda_0)$ and (λ_0, z_0) is an extension point for Ω, then z_0 is a good isolated point of $K(\lambda_0)$.*

The proof of Oka–Nishino theorem is based on the following technical lemma.

Lemma 2.2.2 *Let M,N be open disks in $\mathbb{C}$ and $U = N \times M$. Let f be holomorphic on U, with values in $\mathbb{C}$, not identically zero on U. Let Z be the set of $(z, \eta) \in \mathbb{C}^2$ such that*

(i) $D(z, \eta) = \{w : w \in M, z + \eta w \in N\} \neq \emptyset$;

(ii) $g(w) = f(z + \eta w, w)$ *is identically zero on* $D(z, \eta)$.

Then Z is at most countable.

Theorem 2.2.3 (K.Oka–T.Nishino) *Let Ω be a pseudoconvex open subset of $\mathbb{C}^2$. Then Ω' is also pseudoconvex.*

For more details on this proof, see [Au7], Chapter 7 or [Ra1,2].

By definition, we denote by $DK(\lambda)$ the set of points of $K(\lambda)$ which are not good isolated points. It is easy to prove that $DK(\lambda)$ is compact and satisfies $K(\lambda)' \subset DK(\lambda) \subset K(\lambda)$, where $K(\lambda)'$ denotes the set of limit points of $K(\lambda)$. By transfinite induction we can define $D^\alpha K(\lambda)$ for all ordinal numbers α by:

$$D^\alpha K(\lambda) = \begin{cases} D(D^{\alpha-1}K(\lambda)), & \text{if } \alpha \text{ is not a limit ordinal} \\ \bigcap_{\beta<\alpha} D^\beta K(\lambda), & \text{if } \alpha \text{ is a limit ordinal} \end{cases}$$

with the convention that $D^0 K(\lambda) = K(\lambda)$.

G.Cantor introduced the notion of an *α-derived set* of a closed set C defined by:

$$C^{(\alpha)} = \begin{cases} (C^{(\alpha-1)})', & \text{if } \alpha \text{ is not a limit ordinal} \\ \bigcap_{\beta<\alpha} C^{(\beta)}, & \text{if } \alpha \text{ is a limit ordinal} \end{cases}$$

with the convention that $C^{(0)} = C$.

By transfinite induction it is easy to prove that $K(\lambda)^{(\alpha)} \subset D^\alpha K(\lambda)$, for all ordinal numbers α.

Theorem 2.2.4 (K.Oka–T.Nishino) *Let K be an analytic multifunction from a domain $D \subset \mathbb{C}$ into $\mathbb{C}$ and let α be an ordinal number. Then either $D^\alpha K(\lambda) \neq \emptyset$ for all $\lambda \in D$ and $D^\alpha K\colon \lambda \mapsto D^\alpha K(\lambda)$ is an analytic multifunction on D, or $D^\alpha K(\lambda) = \emptyset$ for all $\lambda \in D$. In the latter case let γ be the smallest ordinal such that $D^\gamma K(\lambda) = \emptyset$, for all $\lambda \in D$. Then γ is not a limit ordinal and there exist an integer n and a closed discrete subset F of D such that $\#D^{\gamma-1}K(\lambda) = n$, for $\lambda \in D \setminus F$, and $\#D^{\gamma-1}K(\lambda) \leq n-1$, for $\lambda \in F$.*

In fact we shall see in the next theorem that it is not necessary to consider $D^\alpha K(\lambda)$ for all $\alpha \geq \omega_1$, where ω_1 denotes the first uncountable ordinal number, because $D^\alpha K(\lambda)$ stabilizes after some $\gamma < \omega_1$. This was proved by B. Aupetit and J. Zemánek [AZ] for $K(\lambda)$ finite or countable, but the proof is the same in the general case.

Using a result of K. Kuratowski on the stabilization of a family of closed sets indexed by the ordinal numbers we can obtain the following.

Theorem 2.2.5 *Let K be an analytic multifunction from $D \subset \mathbb{C}$ into $\mathbb{C}$. Then there exists $\gamma < \omega_1$ such that $D^\gamma K(\lambda) = D^\alpha K(\lambda)$ for all α such that $\gamma \leq \alpha < \omega_1$ and for all $\lambda \in D$.*

The classical Cantor–Bendixson theorem says that every closed subset of $\mathbb{C}$ is the disjoint union of a perfect set and a finite or countable set. It can be generalized in the following form.

Corollary 2.2.6 *Let K be an analytic multifunction from a domain $D \subset \mathbb{C}$ into $\mathbb{C}$. Then for each $\lambda \in D$, $K(\lambda)$ is the disjoint union of two sets $L(\lambda)$, $M(\lambda)$ such that:*

(i) *either $L(\lambda) = \emptyset$, for all $\lambda \in D$ or L is an analytic multifunction from D into $\mathbb{C}$ such that $DL(\lambda) = L(\lambda)$, for all $\lambda \in D$;*

(ii) *$M(\lambda)$ is finite or countable for all $\lambda \in D$.*

Using condensation of singularities and Theorem 2.1.12 we can get the following.

Theorem 2.2.7 (B. Aupetit–J. Zemánek) *Let K be an analytic multifunction from $D \subset \mathbb{C}$ into $\mathbb{C}$ and let F be a closed subset of D having non-zero capacity. Suppose that $\lambda \in F$ implies $K(\lambda)$ finite or countable. Then there exists $\lambda_0 \in F$ such that $DK(\lambda_0) \neq K(\lambda_0)$.*

Corollary 2.2.6 and Theorem 2.2.7 imply immediately the following scarcity theorem.

Theorem 2.2.8 (Scarcity of elements with countable values) *Let K be an analytic multifunction from a domain $D \subset \mathbb{C}$ into $\mathbb{C}$. Then either the set of λ, for which $K(\lambda)$ is finite or countable, has zero capacity, or $K(\lambda)$ is finite or countable for all $\lambda \in D$. In the latter situation there exists $\gamma < \omega_1$ such that $D^\gamma K(\lambda) = \emptyset$, for all $\lambda \in D$.*

Remark This result is best possible. Let F be a compact set having zero capacity. By Evans's theorem for compact sets having zero capacity there exists u subharmonic on $\mathbb{C}$, such that $F = \{\lambda : \lambda \in \mathbf{C}, u(\lambda) = -\infty\}$. We define the multifunction K by

$$K(\lambda) = \{z : z \in \mathbf{C}, |z| \leq e^{u(\lambda)}\}.$$

It is an analytic multifunction defined on $\mathbf{C}$ which satisfies $K(\lambda) = \{0\}$ on F and which is uncountable on $\mathbf{C} \setminus F$.

We now give an application of the Oka–Nishino theorem to the identity principle.

Using Theorem 2.1.12 and the argument of the proof of Theorem 3.4.26 in [Au7], it is easy to prove the following result which has interesting consequences (see Chapter 3).

Theorem 2.2.9 *Let K be an analytic multifunction from a domain D of $\mathbf{C}$ into $\mathbf{C}$. Suppose that for all $\lambda \in D$ the set $K(\lambda)$ has at most 0 as a limit point. Let $z \neq 0$ be given. Then either $Z = \{\lambda : \lambda \in D, z \in K(\lambda)\}$ is a closed discrete subset of D or it is all of D.*

The same argument even proves the following.

Corollary 2.2.10 *Let K be an analytic multifunction from D into $\mathbf{C}$ and let $z \in \mathbf{C}$ be fixed. Then every point of the set*

$$Z = \{\lambda : \lambda \in D, z \in K(\lambda) \setminus DK(\lambda)\}$$

is either isolated or interior.

If K is a countable analytic multifunction, the analogue of Theorem 2.2.9 cannot be true. For instance, let $K_0 = \{1/n : n = 1, 2, \cdots\} \cup \{0\}$ and let $K(\lambda) = \lambda + K_0$, which is an analytic multifunction on $\mathbf{C}$. Then $Z = \{\lambda : 1 \in K(\lambda)\}$ is neither discrete nor $\mathbf{C}$. Nevertheless we have the following result which was proved in [AZ] for the first time.

Theorem 2.2.11 (B. Aupetit–J. Zemánek) *Let K be a finite or countable analytic multifunction from a domain D of $\mathbf{C}$ into $\mathbf{C}$ and let $z \in \mathbf{C}$ be fixed. Then the set*

$$Z = \{\lambda : \lambda \in D, z \in K(\lambda)\}$$

is either finite or countable or it is all of D.

Recently in [Sa], A. Sadullaev proved the following important result.

Lemma 2.2.12 *Let Ω be a pseudoconvex open subset of $\mathbf{C}^2$. Suppose that for every λ in U, the first projection of Ω on the complex plane, the intersection of $\{\lambda\} \times \mathbf{C}$ with $\mathbf{C}^2 \setminus \Omega$ has zero capacity. Then $\mathbf{C}^2 \setminus \Omega$ is completely polar, that is, there exists φ plurisubharmonic on $U \times \mathbf{C}$ such that $(\lambda, z) \notin \Omega$ is equivalent to $\varphi(\lambda, z) = -\infty$.*

With this result we can get the following theorem.

Theorem 2.2.13 *Let K be an analytic multifunction from a domain D of $\mathbb{C}$ into $\mathbb{C}$. Suppose that $K(\lambda)$ has zero capacity on a non-zero capacity subset of D. Given a fixed $z \in \mathbb{C}$, then the set*

$$Z = \{\lambda : \lambda \in D, z \in K(\lambda)\}$$

is either of zero capacity or it is all of D.

Obviously Theorems 2.2.11 and 2.2.13 are not true if K is a general analytic multifunction. For example, if we take

$$K(\lambda) = \{z : |z| \leq 1\} \cup \{z : |z| \leq |\lambda|\},$$

then we have $2 \in K(\lambda)$ if and only if $|\lambda| \geq 2$.

2.3 Distribution of values of analytic multifunctions

The open mapping theorem for holomorphic functions defined on a domain of the complex plane says that their images are either open sets or singletons. What is happening for analytic multifunctions?

It is easy to see that the former property cannot be true, taking for instance the analytic multifunction $K(\lambda) = \{1, \lambda\}$, defined on the unit disk Δ, whose image is $\Delta \cup \{1\}$. If the image of K is very flat, it is constant, as seen in Corollary 2.1.7. The first attempts to obtain a satisfactory open mapping theorem for general analytic multifunctions were given in [Ra2]. But they are very far from being satisfactory. A lot of work has to be done in this direction because any progress would have important consequences in the theory of local spectrum (see Chapter 7).

Given an analytic multifunction K defined on a domain D of $\mathbb{C}$ we denote by U the union of the $K(\lambda)$ for $\lambda \in D$. If the geometry of $K(\lambda)$ is nice, several results are known. For instance we have $U \cap \partial U$ included in $\bigcap_{\lambda \in D} K(\lambda)$, if $K(\lambda)$ is finite or countable. More generally, using Lemma 2.2.12 it is possible to prove the following.

Theorem 2.3.1 *Let K be an analytic multifunction from a domain F of $\mathbb{C}$ into $\mathbb{C}$ whose values $K(\lambda)$ have zero capacity. Then $U \cap \partial U \subset \bigcap_{\lambda \in D} K(\lambda)$.*

Using Theorem 2.4.5 (to which we shall come later) and Rouché's theorem for analytic multifunctions [Ra3], L. Baribeau and S. Harbottle [BaH] have been able to prove the following.

Theorem 2.3.2 *Let K be an analytic multifunction from a domain D of $\mathbb{C}$ into $\mathbb{C}$ whose values $K(\lambda)$ are convex sets. Then $U \cap \partial U \subset \bigcap_{\lambda \in D} K(\lambda)$.*

This result is not very useful for applications. Nevertheless it is easier to understand this last result than the former one, because convexity of $K(\lambda)$ implies that these sets are rather big, consequently the interior of U must be big.

The next theorems were proved by T.J. Ransford [Ra2]. We denote by I the set $\bigcap_{\lambda\in D} \partial K(\lambda)$.

Theorem 2.3.3 *Let K be an analytic multifunction from a domain D of $\mathbb{C}$ into $\mathbb{C}$. Then $(U \cap \partial U) \setminus I$ is a F_σ-set with empty interior. Moreover $\mathbb{C} \setminus U$ is thin at every point of $(U \cap \partial U) \setminus I$.*

Corollary 2.3.4 *If $I = \emptyset$, then U is finely open.*

Corollary 2.3.5 *Let F be a subset of $\mathbb{C}$ and suppose that $K(\lambda) \subset F$, for all $\lambda \in D$.*

(i) *If $\mathbb{C} \setminus F$ is thin at z_0, then $z_0 \in K(\lambda)$ either for no $\lambda \in D$ or for all $\lambda \in D$;*

(ii) *If $\mathbb{C} \setminus F$ is non-thin at every point of F, then K is constant on D.*

He was also able to prove that in general we do not have $U \cap \partial U \subset I$ by proving the following result.

Theorem 2.3.6 *Let F be any compact subset of $\mathbb{C}$ with empty interior. Then there exists an analytic multifunction K from $\{\lambda : 0 < |\lambda| < 1\}$ into $\mathbb{C}$ such that $I = \emptyset$ and $U \cap \partial U = F$.*

Taking E a dense subset of the unit disk and φ subharmonic on this disk such that $E = \{\lambda : \varphi(\lambda) = -\infty\}$, $0 < \varphi(0) < \operatorname{Log} 2$, then taking

$$K(\lambda) = \{z : |z| = 1\} \cup \{z : |z| < 1,\ u(z) \geq \operatorname{Log}|\lambda|\},$$

he showed that K is analytic, that $K(C)$, where C denotes the corona $\{\lambda : 1 \leq |\lambda| \leq 2\}$, is a compact set with empty interior and nevertheless K is non-constant for $1 \leq |\lambda| \leq 2$.

All these problems suggest the following conjectures.

Conjecture 1 Let K be an analytic multifunction from a domain D of $\mathbb{C}$ into $\mathbb{C}$ whose values are polynomially convex. Is it true that $U \cap \partial U \subset \bigcap_{\lambda\in D} K(\lambda)$?

Conjecture 2 Let F be a closed subset of $\mathbb{C}$ with empty interior and let K be an analytic multifunction from a domain D of $\mathbb{C}$ into $\mathbb{C}$ such that $K(\lambda) \subset F$ for all λ in D. Is it true that $\bigcap_{\lambda\in D} K(\lambda) \neq \emptyset$?

Even if these conjectures are not true, any improvement of Theorem 2.3.3 and Corollary 2.3.5 would have a important consequences in section 7.2.

Now what is happening for analytic multifunctions defined on the whole complex plane ?

The famous theorem of Picard asserts that a non-constant entire function takes all the values of the complex plane except perhaps one point. But what happens for the union of all the spectral values of $f(\lambda)$ if f is an analytic function from $\mathbb{C}$ into $M_n(\mathbb{C})$? This problem was partly studied by E. Borel, G. Valiron and G. Rémoundos, but their arguments are not always very convincing (even H. Cartan gave some insights on the general situation, but with a false conclusion on the number of exceptional points). First we shall describe the work of A. Zraïbi on the solution of this problem with the help of Nevanlinna theory.

Later we show the intimate connection between such analytic multifunctions and pseudoconvex open subsets of $\mathbb{C}^2$. This connection reduces many problems on analytic multifunctions — and hence many spectral problems — to purely geometrical problems on pseudoconvex sets. This geometrical idea provides a very simple proof of the generalization of Picard's theorem to arbitrary analytic multifunctions.

Using Nevanlinna's theory it is possible to prove the following lemma which is a weak form of a theorem due to E. Borel.

Lemma 2.3.6 *Let $\phi_1, \cdots, \phi_n$ be n linearly independent entire functions such that $\phi_1 + \phi_2 + \cdots + \phi_n = 1$. Then at least one of the ϕ_i has a zero.*

Now we intend to generalize Picard's theorem to finite analytic multifunctions by exploiting an idea of G. Rémoundos. First we introduce the notion of spectral multiplicity.

Lemma 2.3.7 *Let K be an analytic multifunction defined on an open subset D of $\mathbb{C}$ such that $K(\lambda)$ is finite for all λ in D. Let $K(\lambda_0) = \{\alpha_1, \cdots, \alpha_p\}$ and $\epsilon > 0$ be such that $B(\alpha_i, \epsilon) \cap B(\alpha_j, \epsilon) = \emptyset$ for $i \neq j$. Then there exist $\alpha > 0$ and integers $n_1, \cdots, n_p$ such that $\#(K(\lambda) \cap B(\alpha_i, \epsilon)) = n_i$ for $0 < |\lambda - \lambda_0| < \alpha$ and $i = 1, \cdots, p$.*

The integer n_i is called the *spectral multiplicity* of α_i.

Lemma 2.3.8 *Let $F(\lambda, u) = u^n + A_1(\lambda)u^{n-1} + \cdots + A_n(\lambda)$ be defined on $\mathbb{C}^2$, where the $A_i(\lambda)$ are non-constant entire functions. Then F has at most $2n - 1$ exceptional values in the sense of Picard, that is, for every $u \in \mathbb{C}$ there exists $\lambda \in \mathbb{C}$ such that $F(\lambda, u) = 0$, except perhaps for at most $2n - 1$ values of u.*

The proof is based on some arguments involving Vandermonde determinants. See [Au7], Chapter 7, §3, for more details, or [Zr].

Theorem 2.3.9 (Picard's theorem for finite-valued analytic multifunctions) *Let K be a non-constant analytic multifunction on $\mathbb{C}$. Suppose that $K(\lambda)$ is finite on a set E having non-zero capacity. Then there exists a smallest integer n such that $\#K(\lambda) \leq n$ for all $\lambda \in \mathbb{C}$ and $\mathbb{C} \setminus \bigcup_{\lambda \in \mathbb{C}} K(\lambda)$ has at most $2n - 1$ points.*

The first part of the proof comes from Theorem 2.1.2. So outside F we have $K(\lambda) = \{\alpha_1(\lambda), \cdots, \alpha_n(\lambda)\}$, where the α_i are locally holomorphic. Let

$$F(\lambda, u) = \prod_{i=1}^{n} (\alpha_i(\lambda) - u) = u^n + A_1(\lambda)u^{n-1} + \cdots + A_n(\lambda) \text{ for } \lambda \notin F.$$

This function can be extended analytically to all of $\mathbb{C}^2$ by Lemma 2.3.7, counting each

$\alpha_i(\lambda)$ with its multiplicity if $\lambda \in F$. The $A_i(\lambda)$ are well-defined in all of $\mathbb{C}$, and they are entire because they can be expressed as symmetric functions of the α_i (in fact we use Radó's extension theorem at that level). Moreover, they are not all constant since K is not constant. So u is not in $\bigcup_{\lambda\in\mathbb{C}} K(\lambda)$ if and only if u is exceptional for F. We then apply Lemma 2.3.8.

This result is best possible because, given $2n-1$ arbitrary distinct points, it is possible to construct a finite analytic multifunction on $\mathbb{C}$ avoiding these points.

Let $a_1, \cdots, a_{2n-1}$ be given distinct points, and consider the following analytic function from $\mathbb{C}$ into $M_n(\mathbb{C})$ defined by

$$f(\lambda) = \begin{pmatrix} C_1e^\lambda + a_1 & -C_2e^\lambda & C_3e^\lambda & \cdots & (-1)^n C_{n-1}e^\lambda & (-1)^{n+1}C_ne^\lambda \\ 1 & a_2 & 0 & \cdots & 0 & 0 \\ 0 & 1 & a_3 & \cdots & 0 & 0 \\ \vdots & \vdots & \vdots & \ddots & \vdots & \vdots \\ 0 & 0 & 0 & \cdots & 1 & a_n \end{pmatrix}$$

We have

$$\det(f(\lambda) - z) = (a_1 - z)\cdots(a_n - z) + e^\lambda \left[\sum_{i=1}^{n-1} C_i(a_{i+1} - z)\cdots(a_n - z) + C_n\right].$$

Let

$$P(z) = \sum_{i=1}^{n-1} C_i(a_{i+1} - z)\cdots(a_n - z) + C_n.$$

By induction it is possible to choose the constants $C_1, \cdots, C_n$ in such a way that we have $P(z) = (a_{n+1} - z)\cdots(a_{2n-1} - z)$, and consequently

$$\det(f(\lambda) - z) = (a_1 - z)\cdots(a_n - z) + e^\lambda(a_{n+1} - z)\cdots(a_{2n-1} - z).$$

Then the analytic multifunction $\lambda \mapsto Sp\, f(\lambda) = \{z: \det(f(\lambda) - z) = 0\}$ avoids exactly the $2n-1$ points $a_1, \cdots, a_{2n-1}$.

We shall now be interested in improving Picard's theorem when the analytic multifunction assumes an infinite number of values.

A. Zraïbi and the author (see [AZ]) obtained the following generalization of Picard's theorem to analytic multifunctions: if K is an analytic multifunction on $\mathbb{C}$, then either $\hat{K}(\lambda)$ is constant or the complement of the union of the sets $\hat{K}(\lambda)$ is a G_δ-set having zero capacity. The original proof uses Frostman's theorem and is rather complicated. We now intend to give an easy and more geometric method.

The following lemma will show that it is always possible to associate plenty of analytic multifunctions to a pseudoconvex open subset of $\mathbb{C}^2$.

Lemma 2.3.10 *Let Ω be a non-empty pseudoconvex open subset of $\mathbb{C}^2$ and let $(\lambda_0, a) \in \Omega$. Denote by D the open set of $\lambda \in \mathbb{C}$ such that $(\lambda, a) \in \Omega$. Then the multifunction K defined on D by*

$$K(\lambda) = \left\{ \frac{1}{z-a} + a : (\lambda, z) \notin \Omega \right\} \cup \{a\}$$

is analytic.

Theorem 2.3.11 *Let Ω be a pseudoconvex open subset of $\mathbb{C}^2$ and let U be a domain of $\mathbb{C}$ such that $U \times \{0\} \subset \Omega$. Then we have the following properties:*

(i) *either the set of $\lambda \in U$ such that $\{\lambda\} \times \mathbb{C} \subset \Omega$ is a G_δ-set of capacity zero, or $U \times \mathbb{C} \subset \Omega$;*

(ii) *either the set of $\lambda \in U$ such that $\{\lambda\} \times \mathbb{C} \subset \Omega$, except for a finite number of points, is a G_δ-set of zero capacity, or $(U \times \mathbb{C}) \cap \Omega$ is the complement of an analytic variety.*

We are now able to give a generalization of Picard's theorem to analytic multifunctions.

Theorem 2.3.12 (Picard's theorem for analytic multifunctions) *Let K be an analytic multifunction on $\mathbb{C}$. If U is a component of $\mathbb{C} \setminus K(\lambda_0)$, for some $\lambda_0 \in \mathbb{C}$, then either U is a component of $\mathbb{C} \setminus K(\lambda)$, for all $\lambda \in \mathbb{C}$, or $U \setminus \bigcup_{\lambda \in \mathbb{C}} K(\lambda)$ is a G_δ-set of zero capacity. In particular, if we consider the analytic multifunction $\hat{K}$, then either $\hat{K}(\lambda)$ is constant or $\mathbb{C} \setminus \bigcup_{\lambda \in \mathbb{C}} \hat{K}(\lambda)$ is a G_δ-set of zero capacity. Moreover, if $\hat{K}$ is not constant and is not algebroid, then the set F of z for which $\{\lambda : z \in \hat{K}(\lambda)\}$ is finite, is a G_δ-set of zero capacity.*

Is this result the best one? Given a compact set C of capacity zero, is it possible to construct an analytic multifunction K on $\mathbb{C}$ such that $\mathbb{C} \setminus \bigcup_{\lambda \in \mathbb{C}} \hat{K}(\lambda) = C$? Is it even possible to do this if $K(\lambda)$ is finite or a sequence converging to zero for every λ. We have the following particular cases:

- If C is a subset of $\mathbb{C}$ not containing 0 and having at most 0 as a limit point, then there exists an analytic multifunction K such that $\mathbb{C} \setminus \bigcup_{\lambda \in \mathbb{C}} \hat{K}(\lambda) = C$.

- If C is a compact subset of $\mathbb{C}$ of capacity zero, then there exists an analytic multifunction K such that $\mathbb{C} \setminus \bigcup_{\lambda \in \mathbb{C}} K(\lambda) = C$ (but the problem is that $K(\lambda)$ has holes and the sets $\hat{K}(\lambda)$ cover all the plane!).

It is interesting to note that Theorem 2.3.11 gives a new proof of Tsuji's theorem concerning the distribution of values of entire functions of two complex variables, see [Au7], Theorem 7.3.11, or [Au5], Theorem 2.15 for a proof using companion matrices on $\mathcal{B}(H)$. The original proof given in [Ts], pp. 329-331, is complicated and uses conformal mapping.

Theorem 2.3.13 (M. Tsuji) *Let $G(\lambda, \mu)$ be an entire function on $\mathbb{C}^2$ which is not of the form $G(\lambda, \mu) = e^{H(\lambda,\mu)}$, with H entire on $\mathbb{C}^2$. Then there exists a G_δ-set E having zero capacity such that for $\mu \notin E$ there exists λ in $\mathbb{C}$ satisfying $G(\lambda, \mu) = 0$. Moreover, if G is not algebroid — that is there are no entire functions $a_1, \cdots, a_n$ such that*

$G(\lambda,\mu) = a_n(\mu)\lambda^n + \cdots + a_1(\mu)\lambda + a_0(\mu)$ — then there exists a G_δ-set F having zero capacity such that for $\mu \notin F$ there exist an infinite number of λ satisfying $G(\lambda,\mu) = 0$.

Tsuji's theorem and Theorem 2.3.11 are the essential ingredients in the proof of Lemma 8.5.8 which has important consequences in Jordan-Banach algebra theory.

Given K an analytic multifunction on $\mathbb{C}$ and $0 \leq \alpha \leq 1$, it is easy to verify that $\lambda \mapsto \alpha K(\lambda) + (1-\alpha)K(\lambda)$ is analytic. This implies that $\lambda \mapsto \text{co } K(\lambda) = \bigcup_{0\leq\alpha\leq 1}[\alpha K(\lambda) + (1-\alpha)K(\lambda)]$ is also an analytic multifunction on $\mathbb{C}$. For convex analytic multifunctions and even for connected analytic multifunctions it is possible to improve Picard's theorem.

Using covering spaces and lifts of multifunctions T.J. Ransford [Ra3] proved the following result.

Theorem 2.3.14 *Let K be an analytic multifunction on $\mathbb{C}$ and suppose that $K(\lambda)$ is connected for all $\lambda \in \mathbb{C}$. Then either $\hat{K}(\lambda)$ is constant or the union of all $\hat{K}(\lambda)$ covers all the plane except perhaps one point.*

As a corollary we immediately obtain an earlier result of J.P. Williams on numerical range.

Corollary 2.3.15 *Let a,b be two non-commuting elements of a Banach algebra A. We define $W(x) = \{f(x): f \in A', \| f \| = f(1) = 1\}$ to be the numerical range of x. Then $\bigcup_{\lambda\in\mathbb{C}} W(e^{\lambda b}ae^{-\lambda b}) = \mathbb{C}$.*

This result implies in particular that for a convex analytic multifunction defined on the complex plane, only one of the following possibilities occurs:

(i) if $\bigcup_{\lambda\in\mathbb{C}} K(\lambda)$ avoids two points of $\mathbb{C}$, then $K(\lambda)$ is constant on $\mathbb{C}$;

(ii) if $\bigcup_{\lambda\in\mathbb{C}} K(\lambda)$ avoids one point $\alpha \in \mathbb{C}$ then $K(\lambda)$ has the form $K(\lambda) = \alpha + e^{h(\lambda)}K_0$, where h is an entire function and K_0 is a fixed compact convex set;

(iii) $\bigcup_{\lambda\in\mathbb{C}} K(\lambda) = \mathbb{C}$.

2.4 The selection problem for analytic multifunctions

Let K be an analytic multifunction defined on a domain D of $\mathbb{C}$. For spectral reasons it is important to know when such an analytic multifunction has a local holomorphic selection at $\lambda_0 \in D$ going through $z_0 \in \partial K(\lambda_0)$. Unfortunately this question is very far from being solved. For instance we know, by Theorem 2.1.11, that finite and countable analytic multifunctions have a lot of local holomorphic selections for points which are not branching points. For analytic multifunctions with zero capacity values the situation is very bad. In section 5.3, we shall give an example of this kind having no local continuous selections at all.

All of this suggests very general questions:

1. If K is continuous and if the geometry of $K(\lambda)$ is nice, does K have local holomorphic selections?
2. If K behaves extremely well (for instance it is locally lipschitzian) and has a slow growth at infinity, does K has local holomorphic selections?
3. If K is continuous with convex values is it possible to find holomorphic functions which are not too far from $K(\lambda)$?

The first elementary result is due to L. Baribeau [Ba1].

Theorem 2.4.1 *Suppose that K is a continuous analytic multifunction on a domain D and suppose moreover that $K(\lambda)$ is always a segment. Then the two vertices of the segment vary holomorphically on a neighbourhood of every $\lambda_0 \in D$ for which $K(\lambda_0)$ is not a single point.*

By Radó's extension theorem this implies in particular that the subset of D where $K(\lambda_0)$ is reduced to a point is discrete, and that $\frac{1}{2}(\alpha(\lambda)+\beta(\lambda))$ is always in $K(\lambda)$, where $\alpha(\lambda)$, $\beta(\lambda)$ denote the two vertices of the segment. The best example illustrating this theorem is given by $K(\lambda) = \left[-\sqrt{\lambda}, \sqrt{\lambda}\right]$.

This result was extended in [Ba2] to the more general situation where $K(\lambda)$ is a polygon having n vertices. For $n \geq 3$ the situation becomes more difficult as is shown by the following example.

Example 1 Let $D = \{z : \operatorname{Im} z > 0\}$ and let the multifunction K be defined by

$$K(\lambda) = \begin{cases} [0,1], & \text{if } |z| \leq 1 \\ co\left\{0, 1, -\frac{i}{\pi}\operatorname{Log} z\right\}, & \text{if } |z| > 1, \end{cases}$$

where co denotes the convex hull and $\operatorname{Log} z = \operatorname{Log}|z| + i\operatorname{Arg} z$ with $0 < \operatorname{Arg} z < \pi$. The multifunction K has plenty of holomorphic selections so it is continuous and analytic. For $|z| > 1$, $K(\lambda)$ is a triangle but for $|z| \leq 1$ its degenerates to a segment, and outside of this big set the vertex $-\frac{i}{\pi}\operatorname{Log} z$ is holomorphic. In general for polygons the situation is similar.

Theorem 2.4.2 *Suppose that K is a continuous analytic multifunction on a domain D and suppose moreover that $K(\lambda)$ is a polygon having at most n vertices. Denote by E the set of $\lambda \in D$ for which $K(\lambda)$ has less than n vertices. Then $D \setminus E$ is open and the n vertices of $K(\lambda)$ vary locally holomorphically on $D \setminus E$.*

If K is a finite analytic multifunction it is known, from Theorem 2.1.12, that the number of points of $K(\lambda)$ is uniformly bounded. Is there a similar theorem concerning the number of vertices of a polygonal analytic multifunction? Unfortunately the answer is no as shown by the next example.

Example 2 Let Δ be the unit disk and $K(\lambda) = co\{0, \lambda, \lambda^2, \cdots, \lambda^4, \cdots\}$. This multifunction is continuous and analytic. It is not difficult to see that $K(\lambda)$ is a polygon. Nevertheless the number of vertices is not uniformly bounded as is easily seen by taking $\lambda = re^{2\pi i/n}$ with $r < 1$ and $n \geq 1$ arbitrary, in which case $K(\lambda)$ has n vertices.

Concerning question 1, what is happening if the $K(\lambda)$ are disks? Are the centres of these disks moving holomorphically? Alas! the answer is no.

In [BR], B. Berndtsson and T.J. Ransford used analytic multifunctions to give a very nice proof using a selection theorem of Carleson's corona theorem (which says that evaluations are dense in the set of characters of $H^\infty(U)$, where U is simply connected in the complex plane). Z. Słodkowski obtained similar results in [Sl6,9] and extended the result to U having a finite number of holes (a result which was previously known using different arguments). This beautiful geometric argument raised the hope that it could solve the same problem for an arbitrary open set U. But unfortunately it failed. We extract from [BR] a few interesting general results.

Theorem 2.4.3 *Denote by Δ the unit disk. Let $c : \overline{\Delta} \to \mathbb{C}$ and $r : \overline{\Delta} \to [0, +\infty[$ be C^2-functions. Then the disk-valued multifunction K defined by*

$$K(\lambda) = \{z : |z - c(\lambda)| \leq r(\lambda)\}$$

is analytic if and only if

$$\frac{\partial^2}{\partial\lambda\partial\overline{\lambda}}(\operatorname{Log} r(\lambda)) \geq \frac{1}{r(\lambda)}\left|\frac{\partial^2 c(\lambda)}{\partial\lambda\partial\overline{\lambda}} - 2\frac{\partial c(\lambda)}{\partial\overline{\lambda}}\frac{\partial(\operatorname{Log} r(\lambda))}{\partial\lambda}\right| + \frac{1}{r(\lambda)^2}\left|\frac{\partial c(\lambda)}{\partial\overline{\lambda}}\right|^2.$$

In the C^2-case, this is of course an improvement of Theorem 2.1.16, when $c(\lambda) = 0$, because in this case it says that $\operatorname{Log} r$ is subharmonic. This theorem implies in particular the following result.

Theorem 2.4.4 *Let E be a compact polar subset of Δ and let f be any function analytic on a neighbourhood of E. Then there exists a disk-valued multifunction K on $\overline{\Delta}$, which is analytic on Δ and such that $K(\lambda) = \{f(\lambda)\}$, for $\lambda \in E$.*

From this last theorem we get a very surprising example.

Example 3 Take

$$E_0 = \left\{\frac{1}{2}\right\} \cup \left\{\frac{n}{2n+1}, n \geq 1\right\}, \quad E_1 = \left\{-\frac{1}{2}\right\} \cup \left\{-\frac{n}{2n+1}, n \geq 1\right\}$$

which are two disjoint countable compact subsets of Δ, so $E_0 \cup E_1$ is polar. Taking two disjoint neighbourhoods N_0, N_1 respectively of E_0, E_1 and taking $f = 0$ on N_0 and $f = 1$ on N_1, then by the previous theorem there exists a disk-valued analytic multifunction K

such that

$$K(\lambda) = \begin{cases} \{0\} \text{ on } E_0 \\ \{1\} \text{ on } E_1 . \end{cases}$$

Certainly K has no holomorphic selection h on Δ because otherwise we would have $h(\lambda) = 0$ on E_0 and $h(\lambda) = 1$ on E_1, and this would violate the identity principle.

In [Be], E. Behrends gave a very concrete example of a disk–valued analytic multifunction defined on the whole complex plane for which c is bounded but not constant (so it is not entire!).

In [Sh1], N.V. Shcherbina proved that it is possible to fibre into analytic curves a C^1-smooth hypersurface in $\mathbb{C}^2$ on both sides of which lie domains of holomorphy (see also [Sh2]). Is it possible to extend this result to some analytic multifunctions replacing the C^1-condition by a Lipschitz condition on the Hausdorff distance between the $K(\lambda)$ and a slow growth at infinity for the $K(\lambda)$?

In [AlW2], H. Alexander and J. Wermer improved a former result of [AlW1] about approximation of singularity sets. The result they obtained was also independently proved by Z. Słodkowski [Sł6] (see [AlW3] for further generalizations). This result precisely describes the polynomially convex hull of a set over the unit circle whose sections are convex.

Theorem 2.4.5 (H. Alexander–Z. Słodkowski–J. Wermer) *Denote by Δ the unit disk. Let Γ be a compact subset of $\mathbb{C}^2$ which is above $\partial\Delta$, with convex fibres over $\partial\Delta$ (that is to say $\Gamma(\lambda) = \{z : (\lambda, z) \in \Gamma\}$ in non-empty for $|\lambda| \leq 1$ and convex for $|\lambda| = 1$). Then the polynomially convex hull of Γ is the union af all the graphs of the holomorphic functions $h \in H^\infty(\Delta)$ such that $h(\lambda)$ is almost everywhere in $\Gamma(\lambda)$ for $|\lambda| = 1$.*

This result was used in [BR] and [Sł6,9] to prove the corona theorem.

If $\Gamma(\lambda)$ is not assumed convex, but it is a Jordan domain which varies smoothly with λ, for $\lambda \in \partial\Delta$, the problem of determining $\hat{\Gamma}$ is much harder. It was solved by F. Forstnerič [Fo] with the restriction that 0 is interior to $\Gamma(\lambda)$, for all $\lambda \in \partial\Delta$, but this condition was removed independently by Z. Słodkowski [Sł10] and by J.W. Helton and D.E. Marshall [HM].

Theorem 2.4.6 (J.W Helton–D.E. Marshall–Z. Słodkowski) *If Γ is a compact subset of $\mathbb{C}^2$ which is above $\partial\Delta$, such that the fibres $\Gamma(\lambda)$ are simply connected and connected, then $\hat{\Gamma} \backslash \Gamma$ is a union of bounded analytic graphs over Δ.*

For more details see [We4].

Let K be an analytic multifunction on a domain D containing $\overline{\Delta}$ having simply connected and connected values. Denote by Γ the set $\{(\lambda, z) : z \in K(\lambda), |\lambda| = 1\}$. As we shall see in Chapter 3, the polynomially convex hull $\hat{\Gamma}$ defines on $\overline{\Delta}$ a multifunction $L(\lambda) = \{z : (\lambda, z) \in \hat{\Gamma}\}$ which is analytic on Δ and which is maximal in the sense that $K(\lambda) \subset L(\lambda)$ for $\lambda \in \overline{\Delta}$ and $L(\lambda) = K(\lambda)$ for $|\lambda| = 1$ (it plays the rôle of the least harmonic majorant of a subharmonic function). This new multifunction has plenty of selections by Theorem 2.4.6; in particular, it is analytic. So if we take $\lambda_0 \in \Delta$ and $z_0 \in \partial K(\lambda_0)$ there exists $h \in H^\infty(\Delta)$ such that $z_0 = h(\lambda_0)$ and $h(\lambda) \in L(\lambda)$ for $\lambda \in \Delta$. Unfortunately $h(\lambda)$ goes out of $K(\lambda)$ at some points. Nevertheless we have dist $(h(\lambda), K(\lambda)) \leq \operatorname{Max}_{|\lambda|=1} \delta(K(\lambda))$. Is it possible to have a similar estimation globally on all of D?

Theorem 2.4.5 was extended in [AlW4] to prove that a continuous n-sheeted cover of Δ can be approximated by an n-sheeted analytic cover. Theorem 2.4.5 and its generalizations have given interesting applications to several complex variables but unfortunately none to spectral theory.

2.5 Representation theory for analytic multifunctions

We finish this chapter with two interesting results due to Z. Słodkowski and M.C. White. Unfortunately, until now these results have given no interesting applications.

In Chapters 3 and 5 we shall give two important examples of analytic multifunctions. The first one is $Sp\, f(\lambda)$, where f is an analytic family of bounded operators on a Banach space, defined on a domain D. The second is given by the fibres $g(f^{-1}(\lambda))$ associated to a uniform algebra and two of its elements f, g. Very strangely the general situation can be reduced to these situations, this was proved by Z. Słodkowski in [Sł1].

Theorem 2.5.1 *Let K be an analytic multifunction from a domain D of $\mathbb{C}$ into $\mathbb{C}$. Given an arbitrary relatively compact subdomain Δ of D, then*

(i) *there exists a separable uniform algebra $\mathcal{A}$ and $f, g \in \mathcal{A}$, such that $f(\mathcal{S}) \subset \partial\Delta$, where $f(\mathcal{S})$ denotes the Shilov boundary of $\mathcal{A}$, and such that*

$$K(\lambda) = g\left(f^{-1}(\lambda)\right), \text{ for } \lambda \in \Delta;$$

(ii) *there exists an analytic family $T(\lambda)$ of operators on ℓ^2, defined on Δ such that*

$$K(\lambda) = Sp\,T(\lambda), \text{ for } \lambda \in \Delta.$$

What is happening in some particular useful cases, for instance $K(\lambda)' \subset \{0\}$? The author suggested this problem to M.C. White who solved it completely [Wh], using the earlier ideas of companion matrices defined in [Au5].

Theorem 2.5.2 *Let K be an analytic multifunction from a domain D of $\mathbb{C}$ into $\mathbb{C}$. Suppose moreover that $K(\lambda)' \subset \{0\}$, for every λ in $\mathbb{C}$. Then there exists an analytic family $T(\lambda)$ of compact operators on ℓ^2, defined on D, such that*

$$K(\lambda) = Sp\,T(\lambda), \text{ for } \lambda \in D.$$

In this case the family T can be globally defined on all of D, and the graph of K is the set of zeros of a holomorphic function defined on $D \times (\mathbb{C}_\infty \setminus \{0\})$.

If $K(\lambda)$ is supposed to be countable for all $\lambda \in D$, then using transfinite induction on ordinals, as we did in section 2.2, it is possible to prove that the graph of K is a countable union of algebraic varieties. This was done previously by H. Yamaguchi [Ya].

Using this result and deep theorems on cluster sets it is probable that the following result can be proved: given a G_δ-subset E of zero capacity in $\mathbb{C}$, there exists an analytic family $T(\lambda)$ of compact operators on ℓ^2, defined on the complex plane, such that $\bigcup_{\lambda\in\mathbb{C}} Sp\,T(\lambda) = \mathbb{C} \setminus E$.

Chapter 3
Applications to Banach algebras and spectral theory

3.1 General results

For the standard definition and properties of spectrum, holomorphic functional calculus, etc., in Banach algebras please consult [BD], [Ri] and [Au7], Chapter 3. If A is a Banach algebra, for instance $\mathcal{B}(X)$ where X is some Banach space, it is well-known that the multifunction $x \mapsto Sp\,x$ is upper semicontinuous, but it is extremely discontinuous in general. The first interesting topological result is due to K. Kuratowski [Ku1].

Theorem 3.1.1 *Let A be a Banach algebra. Then the set of points of continuity of $x \mapsto Sp\,x$ is a dense G_δ-subset of A.*

Let D be an open subset of the complex plane and f an analytic function from D into a Banach algebra. Even if $\lambda \mapsto Sp\,f(\lambda)$ is very discontinuous it must have in some sense some analytic properties (it is an algebroid function in the sense of Puiseux, for instance, if $A = M_n(\mathbb{C})$). As we explained in the historical introduction, the first result in this direction is Corollary 3.1.3 which was obtained in the period 1968–1970. But the main result was proved in 1980 and published in [Au3] and [Sł1].

Theorem 3.1.2 (B. Aupetit–Z. Słodkowski) *Let f be an analytic function from an open subset D of the complex plane into a Banach algebra. Then $\lambda \mapsto Sp\,f(\lambda)$ is an analytic multifunction on D.*

The proof is not very difficult. It uses the fact that $\varphi(\lambda, z) = \|(z1 - f(\lambda))^{-1}\| - \operatorname{Log}\operatorname{dist}(\lambda, \partial D)$ is plurisubharmonic on the complement of the graph and that it goes to $+\infty$ when (λ, z) goes to its boundary.

Corollary 3.1.3 (V. Istrăţescu–B. Schmidt–E. Vesentini) *With the same hypotheses, $\lambda \mapsto \operatorname{Log} r(f(\lambda))$ and $\lambda \mapsto r(f(\lambda))$ are subharmonic on D, where r denotes the spectral radius.*

Corollary 3.1.4 (B. Aupetit–Z. Słodkowski) *With the same hypotheses, $\lambda \mapsto \delta_n(f(\lambda))$, $\lambda \mapsto \operatorname{Log}\delta_n(f(\lambda))$, $\lambda \mapsto c(f(\lambda))$ and $\lambda \mapsto \operatorname{Log} c(f(\lambda))$ are subharmonic on D.*

These two corollaries are deduced immediately from Theorems 3.1.2 and 2.1.3. For an elementary proof of Corollary 3.1.3 see [Au7], pp. 52–53. Originally, Corollary 3.1.4 was proved by the author for $n = 1$ and in general by Z. Słodkowski [Sł] using tensor products of operators, see [Au7], p. 62.

These two corollaries have very important consequences. The reader will find some of these consequences for instance in [Au7], Chapters 3 and 5. We only select two of these applications. The first one because it is new [Au8], and the second because it is very striking.

The next theorem is in fact an improvement of Zemánek's characterization of the radical which was proved previously by the use of representation theory (see Theorem 5.2.1 of [Au7]).

Theorem 3.1.5 *Let a be an element of a Banach algebra A. Then a is in the Jacobson radical of A if and only if* $\sup\{r(x+ta): t \in \mathbb{C}\} < +\infty$ *for every x in A.*

If a is in the Jacobson radical, it is well-known that $r(x+ta) = r(x)$. Conversely if the functions $t \mapsto r(x+ta)$ are bounded for every x in A, then by Corollary 3.1.3 these functions are subharmonic and bounded, so by Liouville's theorem they are constant. Consequently $r(x+ta) = r(x)$ for every x in A and t in $\mathbb{C}$. Let $|\mu| > r(x)$. Then $\mu - (x+ta)$ is invertible for every t in $\mathbb{C}$. The relation

$$\mu - (x+ta) = (\mu - x)(1 - t(\mu - x)^{-1}a)$$

implies that $1/t$ is not in the spectrum of $(\mu - x)^{-1}a$, for every t in $\mathbb{C}$, so that $r((\mu-x)^{-1}a) = 0$. Now let y be arbitrary in A and $|\lambda| > 2r(y)$. Then $y - \lambda = (\mu - x)^{-1}$, where

$$x = \frac{y}{\lambda^2}\left(1 + \frac{y}{\lambda} + \frac{y^2}{\lambda^2} + \cdots\right), \mu = -\frac{1}{\lambda}.$$

Consequently $r(x) < 1/|\lambda| = |\mu|$, and so, by the first part, $r((y-\lambda)a) = 0$. But $r((y-\lambda)a) = r(ya - \lambda a) = r(ya)$, so that $r(ya) = 0$ for every y in A. Consequently $1 - ya$ is invertible for every y in A, hence a is in the Jacobson radical of A.

It is easy to see that on a commutative semisimple Banach algebra, all the Banach algebra norms are equivalent. In the 1950s, I. Kaplansky conjectured that the same result is true for non-commutative semisimple Banach algebras. This problem was solved only in 1967 by B.E. Johnson. His proof, which is not so easy, uses mainly representation theory (see [BD], pp. 128–131 or [Au2], pp. 161–163). Using subharmonic functions we now give a very simple proof of an extension of this result.

Let A and B be two Banach algebras and let T be a linear mapping from A into B. We define the *separating space* of T by

$$\mathcal{S}(T) = \{a : a \in B, \exists (x_n) \text{ in } A, \lim_{n\to\infty} x_n = 0 \text{ and } \lim_{n\to\infty} Tx_n = a\}.$$

It is a closed linear subspace of B and, by the closed graph theorem, T is continuous if and only if $\mathcal{S}(T) = \{0\}$.

Theorem 3.1.6 (B. Aupetit [Au4]) *Let A and B be two Banach algebras. Suppose that T is a linear mapping from A into B such that $r(Tx) \le r(x)$ for every $x \in A$. Then $a \in \mathcal{S}(T)$ implies $r(Tx) \le r(a + Tx)$, for all $x \in A$. In particular, $\mathcal{S}(T) \cap T(A)$ is included in the set of quasi-nilpotent elements of B.*

Let $a \in \mathcal{S}(T)$ and (x_n) be such that $\lim_{n\to\infty} x_n = 0$ and $\lim_{n\to\infty} Tx_n = a$. Let $a \in A$ and $\lambda \in \mathbb{C}$ be arbitrary. Then $\lim_{n\to\infty}(\lambda x_n + x) = x$ and $r(T(\lambda x_n + x)) = r(\lambda Tx_n + Tx) \le r(\lambda x_n + x)$ by hypothesis. So

$$\overline{\lim_{n\to\infty}}\, r(\lambda Tx_n + T_n) \le \overline{\lim_{n\to\infty}}\, r(\lambda x_n + x) \le r(x),$$

by upper semicontinuity of r on A. We set $\phi_n(\lambda) = r(\lambda Tx_n + Tx)$, which is subharmonic. Consequently,

$$\phi(\lambda) = \overline{\lim_{n\to\infty}}\, \phi_n(\lambda) \le r(x)$$

satisfies the mean inequality on $\mathbb{C}$, but in general is not upper semicontinuous. We set

$$\psi(\lambda) = \overline{\lim_{\mu\to\lambda}}\, \phi(\mu)$$

to be its upper regularization, which is subharmonic on $\mathbb{C}$. We have $\phi(\lambda) \le \psi(\lambda) \le r(x)$, for all $\lambda \in \mathbb{C}$. So by Liouville's theorem for subharmonic functions, ψ is constant. So $r(Tx) = \phi(0) \le \psi(0) = \psi(\lambda)$ for all $\lambda \in \mathbb{C}$. By upper semicontinuity of r on B we have

$$\phi(\lambda) \le r(\lambda a + Tx)$$

and consequently

$$\psi(\lambda) \le \overline{\lim_{\mu\to\lambda}}\, r(\mu a + Tx) \le r(\lambda a + Tx).$$

So we conclude that $r(Tx) \le r(\lambda a + Tx)$ for all $\lambda \in \mathbb{C}$, and in particular for $\lambda = 1$. If $a \in \mathcal{S}(T) \cap T(A)$, then $a = Tu$ for some $u \in A$. Taking $x = -u$, we get $r(a) = 0$, hence the result.

Corollary 3.1.7 *Suppose that we have the hypotheses of Theorem* 3.1.6 *with B semisimple, and moreover that T is onto. Then T is continuous.*

Corollary 3.1.8 (B.E. Johnson) *Let A and B be two Banach algebras, with B semisimple. Suppose that T is a morphism from A onto B. Then T is continuous.*

If T is a morphism we obviously have $Sp\,Tx \subset Sp\,x$, so $r(Tx) \le r(x)$ for all $x \in A$. We then apply Corollary 3.1.7.

We finish with an application to spectral theory.

In the period 1952–1955, F.V. Atkinson, B.Sz.-Nagy and Ju.L. Šmul'jan proved independently the following result: let $\lambda \mapsto f(\lambda)$ be an analytic function from a domain $D \subset \mathbb{C}$

into the algebra of compact operators on a Banach space and let $z \neq 0$. Then the set of $\lambda \in D$ such that $z \in Sp\, f(\lambda)$ is a closed and discrete subset of D. Their argument was essentially based on the fact that the projections associated with isolated eigenvalues of compact operators have finite rank. If $D = \mathbb{C}$ and $f(\lambda) = \lambda K$ for some fixed compact operator K, then this result says nothing more than the fact that the spectrum of K is a sequence converging to zero. For a general f, B.Sz.-Nagy believed that the result was deeper than Riesz's theorem. Actually it does not depend on the fact that $f(\lambda)$ is compact but only on the geometry of the graph of the multifunction $\lambda \mapsto Sp\, f(\lambda)$, namely that $Sp\, f(\lambda)$ has at most 0 as a limit point for all $\lambda \in D$. So it can be used with an analytic family of Riesz operators. This result is an immediate consequence of Theorem 2.2.9.

3.2 Spectrally finite Banach algebras

Let A be a Banach algebra such that $A/\mathrm{Rad}\, A$ is finite-dimensional. For all $x \in A$ the coset $\dot{x}$ is algebraic in $A/\mathrm{Rad}\, A$ and consequently $Sp\, x$ is finite. Surprisingly, the converse is true even supposing that the spectrum is finite on a very small part of the algebra. This result was used by K. Kaplansky in 1954 to prove the following result: if ϕ is a ring morphism from a semisimple Banach algebra A onto a Banach algebra B there exist three two-sided ideals A_1, A_2, A_3 in A such that $A = A_1 \oplus A_2 \oplus A_3$, A_1 is finite-dimensional, ϕ is linear on A_2 and antilinear on A_3.

The following lemma is a generalization of the Wedderburn–Artin theorem; it is used at the end of the proof of Theorem 3.2.2.

Lemma 3.2.1 *Let A be a semisimple Banach algebra. Suppose there exists an integer $n \geq 1$ such that for all $x \in A$, x is algebraic of degree $\leq n$. Then A is the direct sum of at most n algebras isomorphic to some $M_k(\mathbb{C})$, with $k \leq n$.*

In a real vector space X we say that a set U is *absorbing* if there exists $a \in U$ such that for all $x \in X$, there exists $r > 0$ such that $a + \lambda x \in U$ for $-r \leq \lambda \leq r$. For instance, an open set is absorbing but the converse is not true in general.

Theorem 3.2.2 *Let A be a Banach algebra containing an absorbing set U such that $Sp\, x$ is finite for all $x \in U$. Then $A/\mathrm{Rad}\, A$ is finite-dimensional.*

Replacing A by $A/\mathrm{Rad}\, A$ and U by its image under the canonical mapping from A onto $A/\mathrm{Rad}\, A$, we may suppose without loss of generality that A is semisimple. Let $a \in U$ be such that for all $x \in A$ there exists $r > 0$ such that $a + \lambda x \in U$ for $-r \leq \lambda \leq r$. Considering the analytic function $\lambda \mapsto a + \lambda(x - a) = f(\lambda)$, we have $Sp\, f(\lambda)$ finite for λ in some real interval which has a non-zero capacity. So, by Theorems 3.1.2 and 2.1.12, $\#Sp(a + \lambda(x - a)) < +\infty$ for all $\lambda \in \mathbb{C}$. In particular, $\#Sp\, x < +\infty$ for all $x \in A$. Let $A_k = \{x : x \in A,\, \#Sp\, x \leq k\}$ which is closed. So by Baire's theorem there exists a smallest integer n such that $\#Sp\, x \leq n$ for x in a ball $B(b, s)$. Applying again the argument at the beginning of this proof, with the absorbing set $B(b, s)$, we conclude that $\#Sp\, x \leq n$ for all $x \in A$. The rest of the argument is purely algebraic, it uses Lemma 3.2.1 and the

Cayley–Hamilton theorem.

Corollary 3.2.3 *Let A be a Banach algebra with involution. Suppose that the real vector subspace H of self-adjoint elements contains an absorbing subset U such that $Sp\, h$ is finite for all $h \in U$. Then $A/\mathrm{Rad}\, A$ is finite-dimensional.*

As U is an absorbing set, there exists $h_0 \in U$ which satisfies the following: for $h \in H$ given, there exists $r > 0$ such that $h_0 + \lambda(h - h_0) \in U$ for $0 \leq \lambda \leq r$. By Theorems 3.1.2 and 2.1.12 we conclude that $\#Sp\, h < +\infty$, for all $h \in H$. Now let $x = h + ik \in A$ be arbitrary, with h, $k \in H$. Considering, as before, the analytic function $\lambda \mapsto h + \lambda k$ we have $\#Sp(h + \lambda k) < +\infty$, for $\lambda \in \mathbf{R}$. So $\#Sp(h + \lambda k) < +\infty$ for all $\lambda \in \mathbf{C}$, and in particular for $\lambda = i$. Then by Theorem 3.2.2, $A/\mathrm{Rad}\, A$ is finite-dimensional.

This last result was recently used by V. Runde [Ru] to obtain interesting results concerning the group algebras associated with some topological groups.

3.3 Elements with finite spectrum, the socle and inessential elements

Using holomorphic functional calculus and Theorem 2.1.12 it is possible to prove the following.

Theorem 3.3.1 *Let f be an analytic function from a domain $D \subset \mathbf{C}$ into a Banach algebra. Suppose that for every $\lambda \in D$ the element $f(\lambda)$ is algebraic. Then there exist an integer $n \geq 1$ and n holomorphic functions on D, denoted by $\alpha_1, \cdots, \alpha_n$, such that*

$$f(\lambda)^n + \alpha_1(\lambda) f(\lambda)^{n-1} + \cdots + \alpha_n(\lambda) 1 = 0,$$

for all $\lambda \in D$.

Corollary 3.3.2 *Let X be a Banach space and let f be an analytic function from a domain $D \subset \mathbf{C}$ into $\mathcal{B}(X)$. Suppose that for every $\lambda \in D$ the element $f(\lambda)$ is polynomially compact. Then there exist n holomorphic functions on D, denoted by $\alpha_1, \cdots, \alpha_n$, such that*

$$f(\lambda)^n + \alpha_1(\lambda) f(\lambda)^{n-1} + \cdots + \alpha_n(\lambda) 1 \in \mathcal{K}(X),$$

for all $\lambda \in D$.

For a given Banach algebra A denote by $\mathcal{F}$ the set of elements of A which have finite spectrum. This set may be extremely complicated. It contains in particular the set of quasi-nilpotent elements and the set of projections. We investigate some properties of $\mathcal{F}$, using Theorem 2.1.12 and representation theory in their proofs.

Theorem 3.3.3 *Suppose that $a + \lambda b \in \mathcal{F}$ for all $\lambda \in \mathbf{C}$. Then we have $(a - \alpha 1)(b - \beta 1)^{-1} \in \mathcal{F}$ for all $\alpha \in \mathbf{C}$ and $\beta \in \mathbf{C} \backslash Sp\, b$.*

Corollary 3.3.4 *If for some $a \in A$ we have $a + \mathcal{F} \subset \mathcal{F}$, then $a\mathcal{F} \subset \mathcal{F}$.*

Theorem 3.3.5 *If for some $a \in A$ we have $a + \mathcal{F} \subset \mathcal{F}$, then a is algebraic modulo the radical of A.*

Applying the previous theorem to the Calkin algebra $\mathcal{B}(H)/\mathcal{K}(H)$ which is semisimple we obtain:

Corollary 3.3.6 *Let T be a bounded linear operator of the Hilbert space H which is not polynomially compact. Then there exists $U \in \mathcal{B}(H)$ such that $Sp_e U$ is finite and $Sp_e(T+U)$ is infinite, where Sp_e denotes the essential spectrum.*

If $(e_n)_{n\geq 0}$ is the standard basis of $\ell^2(\mathbf{N})$, then considering the two operators a, b defined by

$$ae_n = \begin{cases} e_{n+1}, & \text{if } n \text{ is odd} \\ 0, & \text{if } n \text{ is even} \end{cases} \qquad be_n = \begin{cases} 0, & \text{if } n \text{ is odd} \\ e_{n+1}, & \text{if } n \text{ is even} \end{cases}$$

we have $a^2 = b^2 = 0$ and $(a+b)e_n = e_{n+1}$ for $n \geq 0$. So $a+b$ is the unilateral shift whose spectrum is the unit circle (see [H], Problem 85). For a general Banach space X, is it possible to build two quasi-nilpotent operators whose sum has infinite spectrum? This problem is difficult because in general X has no topological basis so it is impossible to give an explicit construction. Nevertheless, we can solve the problem using a circuitous method which is based on a lemma of S. Grabiner and the use of the scarcity theorem for elements with finite spectrum.

Lemma 3.3.7 (S. Grabiner) *Let A be a Banach algebra such that its set of nilpotent elements contains a linear subspace on which the degree of nilpotency is unbounded. Then A contains a non-nilpotent quasi-nilpotent element which is a limit of nilpotent ones.*

Theorem 3.3.8 *Let A be a semisimple Banach algebra. Suppose that $q_0 \in A$ is a non-nilpotent quasi-nilpotent element. Then there exists another quasi-nilpotent element $q_1 \in A$ such that $Sp(q_0 + q_1)$ is infinite.*

If X is a Banach space of infinite dimension it is easy to prove that $A = \mathcal{K}(X) + \mathbb{C}1$ is semisimple and satisfies the hypotheses of Lemma 3.3.7, so by Theorem 3.3.8 we get:

Theorem 3.3.9 *Let X be a Banach space of infinite dimension. Then there exist two quasi-nilpotent and compact operators T_1, T_2 on X such that $Sp(T_1 + T_2)$ is infinite.*

If a Banach algebra A has minimal left ideals (resp. minimal right ideals), then by definition its *socle*, denoted by $\mathrm{soc}(A)$, is the sum of the minimal left ideals (it is also equal to the sum of minimal right ideals, so it is a two-sided ideal). The reader will find more information on the socle in [Au2], pp. 78-87. Every element of the socle is algebraic, consequently of finite spectrum.

If $\dim A < +\infty$, then the socle of A is non-zero because $A = \mathrm{soc}(A)$. Conversely, if A

is semisimple and $A = \mathrm{soc}(A)$, then by Theorem 3.2.2, A is finite-dimensional. If X is a Banach space and $A = \mathcal{B}(X)$, then the socle is non-zero because it contains all finite-rank operators. It would be interesting to have more examples of Banach algebras with non-zero socle.

The next result was proved by B.A. Barnes [Bar1]. He first obtained the commutative case using a deep result called the Shilov idempotent theorem (see [We2], Chapter 8). In [Au7], Theorem 5.7.8, we gave a proof based only on subharmonic functions.

Theorem 3.3.10 (B.A. Barnes) *Let A be a semisimple Banach algebra such that the spectrum of every element of A is finite or countable. Then* $\mathrm{soc}(A) \neq \{0\}$.

Banach algebras for which the spectrum of every element is finite or countable are called *scattered Banach algebras.* In the next section, using Theorem 3.3.10, we shall give the precise algebraic structure of these scattered Banach algebras.

In Theorem 3.1.5 we gave a purely spectral characterization of the radical. It is possible to give a similar one for the socle. The proof depends on Theorem 2.1.12 and was published in [AM].

Theorem 3.3.11 (Multiplicative caracterization of $\mathrm{soc}\, A$ and $kh(\mathrm{soc}\, A)$) *Let A be a semisimple Banach algebra and let $a \in A$. Then we have:*

(i) *$a \in \mathrm{soc}\, A$ if and only if $Sp(xa)$ is finite for all $x \in A$,*

(ii) *$a \in kh(\mathrm{soc}\, A)$ if and only if $Sp(xa)$ has at most 0 as a limit point for every $x \in A$.*

Theorem 3.3.12 (Additive characterization of $\mathrm{soc}\, A$ and $kh(\mathrm{soc}\, A)$) *Let A be a semisimple Banach algebra and let $a \in A$. Then we have :*

(i) *$a \in \mathrm{soc}\, A$ if and only if there exists an integer $n \geq 1$ such that*

$$\bigcap_{t \in F} \sigma(x + ta) \subset \sigma(x)$$

for every $x \in A$ and every $(n+1)$-element subset $F \subset \mathbb{C}\backslash\{0\}$,

(ii) *$a \in kh(\mathrm{soc}\, A)$ if and only if for every $x \in A$ and for every subset $F \subset \mathbb{C}\backslash\{0\}$ having only a nonzero limit point we have*

$$\bigcap_{t \in F} \sigma(x + ta) \subset \sigma(x),$$

where σ denotes the full spectrum, that is, the polynomially convex hull of the spectrum.

This interesting additive characterization of the socle and the kh-socle implies in particular that they are invariant by linear mappings preserving the spectrum or the full spectrum. This will have interesting consequences in section 3.5.

We shall denote by $\mathcal{F}_n(A)$, the set of rank n elements of A, which is the set of elements satisfying condition (i) in the previous theorem. Again, using Theorem 2.1.12, this is equivalent to saying that $\#(Sp(xa)\backslash\{0\}) \le n$ for all x in A.

There are many results in spectral theory concerning the relation between the spectrum of an operator and its essential spectrum, that is, the spectrum of the coset of this operator in the quotient algebra obtained from the closed two-sided ideal of compact operators. These include the theorems of B.A. Barnes, I.C. Gohberg, D.C. Kleinecke and A.F. Ruston which are given below.

We now show that the hypothesis that the elements of the closed two-sided ideal are compact is irrelevant. The essential assertion is that these elements have a spectrum which is either finite or a sequence converging to zero. With this point of view many results in spectral theory can be extended and greatly simplified. The main ingredient in these arguments is Theorem 2.2.9.

Let I be a two-sided ideal (not necessarily closed) of a Banach algebra A. We say that I is *inessential* if, for every $x \in I$, the spectrum of x has at most 0 as a limit point. For instance in $\mathcal{B}(X)$ the finite-rank operators and the set $\mathcal{K}(X)$ of compact operators are two-sided inessential ideals. Given a two-sided ideal I of A we denote by $kh(I)$ the intersection of all kernels of continuous irreducible representations π of A such that $I \subset \ker \pi$. It is easy to see that $I \subset \bar{I} \subset kh(I)$, and that $kh(I)$ is the inverse image of the radical of $A/\bar{I}$.

Let x be in A and α be isolated in the spectrum of x. We define the *projection associated to x and α* by

$$p = \frac{1}{2\pi i}\int_\Gamma (\lambda 1 - x)^{-1} d\lambda,$$

where Γ is a curve surrounding α and separating α from the remaining spectrum of x. In fact, p does not depend on the contour Γ, as long as Γ separates α from the rest of the spectrum. Thus we can suppose that Γ is a small circle with centre at α.

Lemma 3.3.13 *Let I be a two-sided ideal of A and let $x \in kh(I)$. Suppose that $a \neq 0$ is isolated in the spectrum of x. Then the projection associated to x and α is in I.*

Let Γ be a circle centered at α, separating α from 0 and from the rest of the spectrum. For $\alpha \in \Gamma$ we have

$$(\lambda 1 - x)^{-1} = \frac{1}{\lambda} + \frac{1}{\lambda} x(\lambda 1 - x)^{-1}.$$

So we have

$$p = \frac{1}{2\pi i}\int_\Gamma \frac{d\lambda}{\lambda} + \frac{x}{2\pi i}\int_\Gamma \frac{1}{\lambda}(\lambda 1 - x)^{-1} d\lambda.$$

The first term is zero and the second term is in $kh(I)$, so $p \in kh(I)$. Let $\dot{p}$ denote the coset of p in $A/\bar{I}$. Then $\dot{p} \in \mathrm{Rad}(A/\bar{I})$ and so $r(\dot{p}) = 0$, where p denotes the spectral radius. But $\dot{p}$ is also a projection, consequently $\dot{p} = 0$, and hence $p \in \bar{I}$. Moreover $p\bar{I}p$ is a closed subalgebra of A, hence a Banach algebra with identity p, in which pIp is a dense two-sided ideal, and so $pIp = p\bar{I}p$. Then $p = p^3 \in p\bar{I}p = pIp \subset I$.

The argument shows that I and $kh(I)$ have the same set of projections, and from that remark we can obtain the following improvement of a classical result of D.C. Kleinecke.

Theorem 3.3.14 *Let I and J be two-sided inessential ideals of A having the same set of projections. Denoting by $x+I$ (resp. $x+J$) the coset of x in A/I (resp. A/J), then $x+I$ is invertible in A/I if and only if $x+J$ is invertible in A/J. If moreover I and J are closed, then $Sp(x+I) = Sp(x+J)$, for all $x \in A$.*

Corollary 3.3.15 (D.C. Kleinecke) *Let X be a Banach space and let T be a bounded linear operator on X. Then we have $Sp(T+\bar{\mathcal{F}}) = Sp(T+\mathcal{K}(X)) = Sp_e(T)$, where $\bar{\mathcal{F}}$ denotes the closure of the ideal of finite-rank operators.*

We shall see below that if I is a two-sided inessential ideal then $\bar{I}$ and $kh(I)$ are also inessential. Thus Theorem 3.3.14 can be used in that case.

Let I be a fixed inessential two-sided ideal of A. For x in A, we define $D(x)$ in the following way:

$$\lambda \notin D(x) \Longleftrightarrow \begin{cases} \lambda \notin Sp\,x \\ \text{or} \\ \lambda \text{ is an isolated spectral value of } x \text{ with} \\ \text{the corresponding projection in } I. \end{cases}$$

It is easy to verify that $D(x)$ is compact and that $Sp\,x \backslash D(x)$ is discrete, and hence finite or countable. It is also obvious that $D(x-\lambda 1) = D(x) - \lambda$ for every $\lambda \in \mathbb{C}$.

The next result is a strong improvement of a theorem obtained previously by I.C. Gohberg for $A = \mathcal{B}(X)$ and $I = \mathcal{K}(X)$ (see for instance [GoK], Chapter 1, Theorem 5.1 and Lemma 5.2). Its proof is essentially based on Theorem 2.2.9.

Theorem 3.3.16 (Perturbation by inessential elements) *Let I be a two-sided inessential ideal of a Banach algebra A. For $x \in A$ and $y \in I$ we have the following properties:*

(i) *if G is connected component of $\mathbb{C}\backslash D(x)$ intersecting $\mathbb{C}\backslash Sp(x+y)$, then it is a component of $\mathbb{C}\backslash D(x+y)$;*

(ii) *the unbounded connected components of $\mathbb{C}\backslash D(x)$ and $\mathbb{C}\backslash D(x+y)$ coincide, in particular, $D(x)$ and $D(x+y)$ have the same external boundaries;*

(iii) *if $\dot{x}$ denotes the coset of x in $A/\bar{I}$ then we have $Sp\,\dot{x} \subset D(x)$ and $D(x)\hat{} = (Sp\,\dot{x})\hat{}$, where $\hat{}$ denotes the polynomially convex hull of the set. Moreover we have*

$$D(x) = \bigcap_{\substack{y \in I \\ xy = yx}} Sp(x+y)\,.$$

Let H be a Hilbert space. Taking $x \in \mathcal{B}(H)$ and $y \in \mathcal{K}(H)$, it is false in general that $D(x) = D(x+y)$. By inessential perturbations, some holes may appear. For instance on

$H = l^2(\mathbf{Z})$, taking the two weighted shifts

$$ae_n = \begin{cases} 0, & \text{if } n = 1 \\ e_{n+1}, & \text{if } n \neq -1 \end{cases} \qquad be_n = \begin{cases} e_0, & \text{if } n = -1 \\ 0, & \text{if } n \neq -1 \end{cases}$$

we have b of rank one, and so in $\mathcal{K}(H)$, and we have $D(a) = Sp\,a = \{z: |x| \leq 1\}$, $D(a+b) = Sp(a+b) = \{z: |z| = 1\}$.

In 1954, using a rather complicated argument, A.F. Ruston [Ru] proved that if $T \in \mathcal{B}(X)$ has an essential spectral radius equal to zero, then the spectrum of T is either finite or a sequence converging to zero, and the projections associated with the non-zero spectral values have finite rank. By Corollary 3.3.15, the condition that $r(\dot{T}) = 0$ in the Calkin algebra $\mathcal{B}(X)/\mathcal{K}(X)$ is equivalent to saying that for every $\epsilon > 0$ there exists an integer N such that for every $n \geq N$ there exists T_n of finite rank with $\|T^n - T_n\| < \epsilon^n$. A.F. Ruston called such an operator *asymptotically quasi-compact.* This result derives immediately from the following:

Corollary 3.3.17 *Let I be a two-sided inessential ideal of a Banach algebra A. Let $x \in A$ and suppose that $r(\dot{x}) = 0$, where $\dot{x}$ denotes the coset of x in $A/\bar{I}$. Then the spectrum of x has at most 0 as a limit point and, for every non-zero spectral value of x, the associated projection is in I.*

Corollary 3.3.18 *Let I be a two-sided inessential ideal of a Banach algebra A. Then $kh(I)$ is inessential, so in particular $\bar{I}$ is inessential.*

Using a rather complicated method, B.A. Barnes [Ba1] proved that every element of $kh(\mathrm{soc}(A))$ has at most 0 as a limit point in its spectrum. This proof was simplified by J.C. Alexander and M.R. Smyth. In fact, this result derives from Corollary 3.3.18.

Corollary 3.3.19 *In a Banach algebra with minimal left (or right) ideals, $kh(\mathrm{soc}(A))$ is an inessential ideal.*

3.4 Scattered Banach algebras

As we said before, a Banach algebra A is a *scattered* Banach algebra if for every x in A the spectrum of x is finite or countable.

As we explained in the introduction, the general Pełczyński conjecture is the main motivation for the proof of the following result.

Theorem 3.4.1 *Let A be a Banach algebra containing an absorbing set U such that $Sp\,x$ is finite or countable for all $x \in U$. Then $A/\mathrm{Rad}\,A$ is a scattered Banach algebra.*

Corollary 3.4.2 *Let A be a Banach algebra with involution. Suppose that the real vector subspace H of self-adjoint elements contains an absorbing subset U such that $Sp\,h$ is finite* *or countable for all $h \in U$. Then $A/\mathrm{Rad}\, A$ is a scattered Banach algebra.*

The proofs of these two results are very similar to the proofs of Theorem 3.2.2 and Corollary 3.2.3, except that Theorem 2.1.12 is replaced by Theorem 2.2.8.

Is it possible to give the precise algebraic structure of scattered Banach algebras, at least in the separable case? The answer is yes by Theorem 3.4.3.

In the case where every element of A has a spectrum with at most 0 as a limit point, B.A. Barnes [Bar2] has proved that they are modular annihilators (the converse is also true) in the following sense which was introduced by B. Yood in 1964.

A left ideal (resp. right ideal) I of A is said to be *modular* if there exists $e \in A$ such that $ex - x \in I$ for all x in A (resp. if there exists $f \in A$ such that $xf - x \in I$ for all x in A). If E is a subset of A the left annihilator, denoted by $L(E)$ (resp. the right annihilator of E, denoted by $R(E)$) is the set of $x \in A$ such that $xE = 0$ (resp. $Ex = 0$).

A Banach algebra A is called a *modular annihilator algebra* if $L(A) = R(A) = 0$ and $R(I) \neq 0$, $L(J) \neq 0$ for all left modular ideals I and right modular ideals J.

There are many examples of modular annihilator algebras which are not compact algebras. But in the situation of C^*-algebras modular annihilator algebras coincide with all $\mathcal{K}(H)$, where H is a Hilbert space.

In the following, $\alpha \in \Omega$ will mean that α is an ordinal of the first or second class (see [Si]).

Let A be an arbitrary Banach algebra. We take $A_0 = A/\mathrm{Rad}\, A$ and inductively we define $A_n = A_{n-1}/kh(\mathrm{soc}(A_{n-1}))$. The corresponding morphisms of A onto $A_0, A_1, \cdots, A_n, \cdots$ are denoted by $\phi_0, \phi_1, \cdots, \phi_n \cdots$, and their kernels by

$$I_0 = \mathrm{Rad}\, A, \quad I_1 = kh(\mathrm{soc}(A)), \quad \cdots, \quad I_n = \ker \phi_n, \quad \cdots.$$

We then define A_ω, with $I_\omega = kh\left(\bigcup_{n\geq 1} I_n\right)$, and ϕ_ω, where ω is the first infinite ordinal. For every $\alpha \in \Omega$ it is possible to define A_α and ϕ_α, by transfinite induction, in the following way:

- if α is not a limit ordinal, $A_\alpha = A_{\alpha-1}/kh(\mathrm{soc}(A_{\alpha-1}))$ and $\phi_\alpha = \pi_{\alpha-1} \circ \phi_\alpha$, where $\pi_{\alpha-1}$ is the canonical morphism from $A_{\alpha-1}$ onto A_α;
- if α is a limit ordinal we take $I_\alpha = kh\left(\bigcup_{\beta<\alpha} I_\beta\right)$, $A_\alpha = A/I_\alpha$ and ϕ_α the corresponding canonical morphism.

By definition we shall say that A_α is the *α-Calkin algebra* associated to A. It is easy to verify that it is semisimple.

Using transfinite induction it is possible to prove the following result (for the proof see [Au7], Theorem 5.7.9).

Theorem 3.4.3 *Let A be a separable Banach algebra such that the spectrum of every element of A is finite or countable. Then there exist $\alpha_0 \in \Omega$ and an ordinal composition sequence $(I_\alpha)_{\alpha \leq \alpha_0}$ of closed two-sided ideals of A such that $I_0 = \mathrm{Rad}\, A$, $1 \leq \mathrm{codim}\, I_{\alpha_0} < +\infty$, $I_{\alpha_0+1} = A$ and $I_{\alpha+1}/I_\alpha$ is a modular annihilator algebra for $\alpha \leq \alpha_0$.*

3.5 Spectrum-preserving mappings

The theory of spectrum preserving linear mappings originates from Hua's theorem on fields which has very interesting geometrical applications. This theorem says that an additive mapping $\sigma: K_1 \longrightarrow K_2$, where K_1, K_2 are two fields, such that $\sigma(1) = 1$, and $\sigma(x^{-1}) = \sigma(x)^{-1}$ for $x \neq 0$, is an isomorphism or an anti-isomorphism. If ϕ is a linear mapping from a Banach algebra A_1 into another one A_2 such that $\phi(1) = 1$ and $\phi(x)^{-1} = \phi(x^{-1})$, for x invertible, using exponentials it is easy to prove that ϕ is a Jordan morphism, that is $\phi(x^2) = \phi(x)^2$, for every x in A. In the situation of Banach algebras the problem was enlarged by I. Kaplansky to the following one: if ϕ is linear, satisfies $\phi(1) = 1$ and ϕ maps invertible elements into invertible elements, is it true that ϕ is a Jordan morphism? Due to Lemma 4, p.30 of [Au2], this question is equivalent to studying linear mappings which preserve the spectrum.

Almost at the same time, in 1967-1968, A. Gleason, J.-P. Kahane and W. Żelazko proved that if A and B are Banach algebras, with B commutative and semisimple, and if $\phi: A \to B$ is a linear mapping that satisfies $\phi(1) = 1$ and x invertible in A implies $\phi(x)$ invertible in B, then ϕ is a homomorphism (See [Au7], p. 69–70, for the simple and elegant proof given by M. Roitman and Y. Sternfeld).

In the case of matrices the general problem is justified by a result of M. Marcus and R. Purves which says that if $\phi : M_n(\mathbb{C}) \to M_n(\mathbb{C})$ is a linear mapping which preserves eigenvalues and their multiplicity, then ϕ is either of the form $\phi(T) = ATA^{-1}$ or $\phi(T) = AT^tA^{-1}$ (incidentally, we mention that the same conclusion is true if ϕ preserves only the greatest eigenvalue, see [Au8]).

Unfortunately Kaplansky's problem is too general to be true, as the following example (taken from [Au2]) shows. Let A be the subalgebra of $M_4(\mathbb{C})$ built up with matrices of the form

$$\begin{pmatrix} a & , & b \\ 0 & , & c \end{pmatrix}$$

with $a, b, c \in M_2(\mathbb{C})$, and define a linear mapping ϕ from A onto A by

$$\phi\left(\begin{pmatrix} a & , & b \\ 0 & , & c \end{pmatrix}\right) = \begin{pmatrix} a & , & b \\ 0 & , & c^t \end{pmatrix}.$$

Then ϕ is bijective, $\phi(1) = 1$, and ϕ maps invertible elements onto invertible elements. However,

$$\phi\left(\begin{pmatrix} a & , & b \\ 0 & , & c \end{pmatrix}^2\right) - \phi\left(\begin{pmatrix} a & , & b \\ 0 & , & c \end{pmatrix}\right)^2$$

is in general not zero but just in the radical of A. So the natural question is the following: if A and B are two semisimple Banach algebras and if $T: A \to B$ is a surjective spectrum-preserving linear mapping, is T Jordan?

In this direction, A.A. Jaffarian and A.R. Sourour [JS] generalized the Marcus-Purves theorem proving the following result. If $\phi: \mathcal{B}(X) \to \mathcal{B}(Y)$ is a surjective spectrum-preserving linear mapping, then either

1. there exists a bounded invertible linear operator A from X onto Y such that $\phi(T) = ATA^{-1}$ for every $T \in \mathcal{B}(X)$; or

2. there exists a bounded invertible operator B form the dual X^* onto Y such that $\phi(T) = BT^*B^{-1}$ for every $T \in \mathcal{B}(X)$.

Denoting by σ the full spectrum, that is, the polynomially convex hull of the spectrum, in this section we shall study a slightly more general problem: if A and B are two semisimple Banach algebras and if T is a surjective linear mapping with the property that $\sigma(Ta) = \sigma(a)$ for every a in A, is T Jordan?

We solve this problem for two extremal classes of Banach algebras, first the primitive algebras with minimal ideals (this class contains $\mathcal{B}(X)$, consequently we get the Jafarian–Sourour result as a corollary), second the scattered algebras for which every element has a finite or countable spectrum. Unfortunately the general problem is still unsolved until today even for the class of C^*-algebras. The following results appeared in [AM].

We assume throughout this section that A is semisimple. In the previous section we have defined the rank one elements of A as the set $\mathcal{F}_1(A) = \{a \in A: Sp(xa)$ contains at most one nonzero point for every $x \in A\}$.

Clearly the set $\mathcal{F}_1(A)$ is closed under multiplication by elements of A, and by Theorem 3.3.12 we have $\mathcal{F}_1(A) \subset \operatorname{soc} A$. Examples of rank one elements are the minimal idempotents of A. Furthermore, every minimal left ideal of A is of the form Ap, where p is a minimal idempotent, and hence soc A is equal to the set of all finite sums of rank one elements of A.

If A is not isomorphic to $\mathbb{C}$ and $a \in \mathcal{F}_1(A)$, then $Sp(a)$ consists of 0 and possibly one other point. We define a map $t: \mathcal{F}_1(A) \to \mathbb{C}$ by $Sp(a) = \{0, t(a)\}$.

Lemma 3.5.1 *Let $a, b \in \mathcal{F}_1(A)$ such that $a + \lambda b \in \mathcal{F}_1(A)$ for all $\lambda \in \mathbb{C}$. Then $t(a+b) = t(a) + t(b)$.*

By Theorem 2.1.11, the map $h: \mathbb{C} \to \mathbb{C}$, $h(\lambda) = t(a + \lambda b)$ is entire and

$$\lim_{|\lambda| \to \infty} \frac{h(\lambda)}{|\lambda|} = \lim_{|\lambda| \to \infty} t\left(\frac{a}{\lambda} + b\right) = t(b),$$

by Newburgh's theorem (see [Au7], Corollary 3.4.5). Hence, by Liouville's theorem we have $t(a + \lambda b) = t(a) + \lambda t(b)$ and the result follows.

The condition $a+\lambda b \in \mathcal{F}_1(A)$ will be automatically satisfied if a and b are left multiples of the same element in the socle. This fact is used in the proof of the next theorems.

Assume now that A and B are semisimple Banach algebras and that $T: A \to B$ is a surjective linear mapping with the property that $\sigma(Ta) = \sigma(a)$ for every $a \in A$. Using properties proved in sections 3.1 and 3.3 it is easy to prove the following.

Theorem 3.5.2 *With the previous hypotheses we have:*

(i) T *is injective;*

(ii) $T1 = 1$;

(iii) $T(\mathcal{F}_1(A)) = \mathcal{F}_1(B)$;

(iv) $T(\operatorname{soc} A) = \operatorname{soc} B$;

(v) $T(kh(\operatorname{soc} A)) = kh(\operatorname{soc} B)$.

Using Lemma 3.5.1 and some intricate calculations we get the following.

Theorem 3.5.3 *With the previous hypotheses, for every* $x \in \operatorname{soc} B$ *and* $a \in A$ *we have* $(Ta^2 - (Ta)^2)x = 0$.

Corollary 3.5.4 *If* B *has the property that* $b \operatorname{soc} B = \{0\}$ *implies* $b = 0$, *then* T *is Jordan.*

A Banach algebra A is said to be *prime* if $aAb = \{0\}$ implies $a = 0$ or $b = 0$. By Jacobson's density theorem it can easily be seen that every primitive Banach algebra is prime. In ([He], p. 47–51) it is shown that if A and B are prime rings, then every Jordan morphism $T: A \to B$ is either a morphism or an antimorphism. Furthermore, if A is a primitive Banach algebra with minimal ideals, then A has the property that $a \operatorname{soc} A = \{0\}$ implies $a = 0$ (see [Ri], p. 73). Hence we have the following result.

Corollary 3.5.5 *If* B *is primitive Banach algebra with minimal ideals, then* T *is either a morphism or an antimorphism.*

The following is an improvement on the result of Jaffarian and Sourour.

Corollary 3.5.6 *Let* $\phi: B(X) \to B(Y)$ *be a surjective linear mapping such that* $\sigma(\phi(T)) = \sigma(T)$ *for every* $T \in B(X)$, *then either*

(i) *there is a bounded invertible operator* $A: X \to Y$ *such that* $\phi(T) = ATA^{-1}$, *or*

(ii) *there is a bounded invertible operator* $B: X^* \to Y$ *such that* $\phi(T) = BT^*B^{-1}$.

By Corollary 3.5.5, ϕ is either a homomorphism or an antimorphism. If ϕ is a homomorphism, then (i) follows from the fundamental isomorphism theorem ([Ri], Theorem 2.5.19,

p. 76). If ϕ is an antimorphism, then (ii) follows from the fundamental isomorphism theorem and the fact that the Banach algebra $C = \{T^*: T \in \mathcal{B}(X)\}$ is a strictly dense subalgebra of $\mathcal{B}(X')$ which is anti-isomorphic to $\mathcal{B}(X)$ under the map $T \mapsto T^*$.

Lemma 3.5.7 *If $a \in A$ and $u = Ta^2 - (Ta)^2 \in kh(\operatorname{soc} B)$, then $u = 0$.*

If B is a modular annihilator algebra, then $B = kh(\operatorname{soc} B)$ so, by the previous lemma, T is Jordan. In this situation the spectrum of every element has at most zero as a limit point. This result can be extended in the more general situation of scattered Banach algebras.

By using the structure theorem on scattered algebras (Theorem 3.4.3) and transfinite induction on Lemma 3.5.7 we can get the following.

Theorem 3.5.8 *If B is a separable scattered Banach algebra, then T is Jordan.*

Chapter 4
Applications to joint spectrum theory

4.1 Introduction

Given a Banach algebra A with identity and an m-tuple $a = (a_1, \cdots, a_m)$ of elements of A, we can define the *left joint spectrum* of a, denoted by $\sigma_L(a)$, as the set of $(\lambda_1, \cdots, \lambda_m) \in \mathbb{C}^m$ such that the left ideal generated by the elements $\lambda_1 - a_1, \cdots, \lambda_m - a_m$ is different from A. Analogously the *right joint spectrum* of a, denoted by $\sigma_R(a)$, is the set of $(\lambda_1, \cdots, \lambda_m) \in \mathbb{C}^m$ such that the right ideal generated by the elements $\lambda_1 - a_1, \cdots, \lambda_m - a_m$ is different from A. By definition the *joint spectrum* of a, denoted by $\sigma(a)$, is the union of the two previous sets. It is a compact subset of $\mathbb{C}^m$ but, contrary to the situation for $m = 1$, it may be empty even in the case of $A = M_2(\mathbb{C})$. For instance, in $M_2(\mathbb{C})$ the two matrices

$$a_1 = \begin{pmatrix} 0 \,, & 1 \\ 0 \,, & 0 \end{pmatrix}, \qquad a_2 = \begin{pmatrix} 0 \,, & 0 \\ 1 \,, & 0 \end{pmatrix}$$

have an empty joint spectrum. In fact, Banach algebras for which the joint spectrum of an arbitrary m-tuple is always non-empty are Banach algebras with characters, by a result of C.-K. Fong and A. Sołtysiak [FS], and many examples of algebras have no characters, for instance $M_n(\mathbb{C})$, $\mathcal{B}(H)$ for a Hilbert space H. Denoting by B_m (resp. B'_m) the subset of A^m on which $\sigma_L(a)$ (resp. $\sigma_R(a)$) is non-empty, the function $a \mapsto \sigma_L(a)$ (resp. $a \mapsto \sigma_R(a)$) is upper semicontinuous on B_m (resp. B'_m). Moreover B_m and B'_m contain all the commuting m-tuples and we have in this case $\sigma_L(p(a)) = p(\sigma_L(a))$, $\sigma_R(p(a)) = p(\sigma_R(a))$ for all polynomials in m variables (see [Ha] for more details). These are the reasons why almost all the papers concerned with the joint spectrum consider only commuting m-tuples of elements of a Banach algebra.

4.2 The results

If A is a Banach algebra and if f is an analytic family of non-commuting elements of A, then $\lambda \mapsto Sp\, f(\lambda)$ is an analytic multifunction but the corresponding multifunctions for the left and the right spectra are not analytic. What is happening for the left and right joint spectra of analytic families of commuting m-tuples? The following results have been proved by M. Klimek [Kl].

Theorem 4.2.1 *Let A be a Banach algebra with identity and let $f_1, \cdots, f_m$ be commuting analytic functions defined on a domain D of $\mathbb{C}$ with values in A. Then $\lambda \mapsto \hat{\sigma}_L(f(\lambda))$, $\lambda \mapsto \hat{\sigma}_R(f(\lambda))$, $\lambda \mapsto \hat{\sigma}(f(\lambda))$ are analytic multifunctions, where* $\hat{}$ *denotes the polynomially convex hull and $f(\lambda) = (f_1(\lambda), \cdots, f_m(\lambda))$. Moreover if A is commutative, then $\lambda \mapsto \sigma(f(\lambda))$ is analytic.*

Applying Theorem 2.1.3 to $\hat{\sigma}_L$, $\hat{\sigma}_R$ and $\hat{\sigma}$ we obtain immediately:

Corollary 4.2.2 *With the previous hypotheses the functions $\lambda \mapsto \operatorname{Log} d^*(\sigma_L(f(\lambda)))$, $\lambda \mapsto \operatorname{Log} d^*(\sigma_R(f(\lambda)))$ and $\lambda \mapsto \operatorname{Log} d^*(\sigma(f(\lambda)))$ are subharmonic on D, where d^* denotes the upper regularization of the transfinite diameter.*

This comes from the fact that $d(\hat{K}) = d(K)$, see [Zah].

Corollary 4.2.3 *With the previous hypotheses, denoting by K one of the three multifunctions σ_L, σ_R, σ, we have $\{\lambda \in D : \operatorname{ord} K(\lambda) < \infty\}$ of capacity zero or there exists an integer p and a discrete subset E of D such that $\operatorname{ord} K(\lambda) = p$ for $\lambda \in D \setminus E$ and $\operatorname{ord} K(\lambda) < p$ for $\lambda \in E$.*

This also comes from the fact that $\operatorname{ord} \hat{\sigma} = \operatorname{ord} \sigma$. E. Vesentini [Ve3] proved that for a commutative Banach algebra the polynomially convex hull of the joint spectrum satisfies the maximum principle, but this follows immediately from Theorem 4.2.1 and Theorem 2.1.8.

In 1970 Joseph L. Taylor introduced a new joint spectrum for commuting operators acting on a Banach space which is more difficult to define but which is more natural in this situation. Suppose that a_1, a_2 are commuting elements of $\mathcal{B}(X)$, consider the chain complex

$$0 \longrightarrow X \xrightarrow{\delta_1} X \oplus X \xrightarrow{\delta_0} X \longrightarrow 0,$$

where $\delta_1(x) = (-a_2 x, a_1 x)$, $\delta_0(x_1, x_2) = a_1 x_1 + a_2 x_2$ satisfy $\delta_0 \circ \delta_1 = 0$. If $a_1 b_1 + a_2 b_2 = 1$ this sequence is exact, so it is natural to say that (a_1, a_2) is nonsingular if the previous sequence is exact and to define a new joint spectrum to be all (λ_1, λ_2) for which $(\lambda_1 - a_1, \lambda_2 - a_2)$ is singular. In that way using exactness of the Koszul complex it is possible to define the Taylor joint spectrum for an m-tuple $(a_1, \cdots, a_m)$, which is denoted by $\sigma_T(a_1, \cdots, a_n)$ (see [Ta] for more details).

Z. Słodkowski [Sl5] proved that the Taylor joint spectrum behaves well analytically:

Theorem 4.2.4 *Let X be a Banach space and let $T_1, \cdots, T_m$ be commuting analytic families of operators acting on X, defined on a domain D of $\mathbb{C}$. Then $\lambda \mapsto \sigma_T(T_1(\lambda), \cdots, T_m(\lambda))$ is an analytic multifunction.*

These results of M. Klimek and Z. Słodkowski seem to be interesting, but until now they gave no convincing application. It is easy to understand why, the hypothesis that the $f_1, \cdots, f_m$ commute for all λ in D being too strong.

In the case of $M_n(\mathbb{C})$ the sets B_m and B'_m are algebraic subvarieties of $\mathbb{C}^m$ $(m < n)$. So it would be good to know what happens for these sets in the case of non-commutative Banach algebras. Do they have some analytic structure? Are they, at least in some cases, analytic sets? If the answer would be positive it would be of great interest to know if the functions σ_L, σ_R and σ behave analytically when they are restricted to these sets.

Chapter 5
Applications to uniform algebras

5.1 General properties

A *uniform algebra* is a complex commutative Banach algebra for which the norm satisfies $\|f^2\| = \|f\|^2$, for every element f of the algebra. A uniform algebra can be identified, by the Gelfand transform, with a closed subalgebra of $C(\mathcal{M})$ separating the points of $\mathcal{M}$, where $\mathcal{M}$ denotes the set of characters of the algebra provided with the Gelfand topology ($\mathcal{M}$ is compact if the algebra has an identity, otherwise it is locally compact). We recall that the Gelfand transform is defined by

$$\hat{f}(\chi) = \chi(f), \quad \text{for } \chi \in \mathcal{M}.$$

By definition, the *Shilov boundary* of a uniform algebra A, denoted by $\mathcal{S}$, is the smallest maximizing subset of $\mathcal{M}$ for the algebra. In the unitary case, $\mathcal{S}$ is a compact subset of $\mathcal{M}$ and it is easy to see that A is isometrically isomorphic to a closed separating subalgebra of $C(\mathcal{S})$.

For more details on these notions and the rest of the chapter, please consult [We2].

Given an arbitrary uniform algebra, from a classical example due to G. Stolzenberg, it is well-known that $\mathcal{M}$ may not contain any analytic structure. Nevertheless the two following theorems prove that in general there is in $\mathcal{M}$ some coarse analytic structure.

Theorem 5.1.1 (H. Rossi local maximum principle) *Let A be a uniform algebra. Denote by $\mathcal{M}, \mathcal{S}$ its set of characters and its Shilov boundary. Suppose that $\chi_0 \in \mathcal{M} \backslash \mathcal{S}$ and that U is an open neighbourhood of χ_0, for the Gelfand topology, which is disjoint from $\mathcal{S}$. Then for every $f \in A$ we have*

$$|\chi_0(f)| \leq \operatorname*{Max}_{\chi \in \partial U} |\chi(f)|.$$

The proof of this theorem is difficult. It is based on Cousin's problem, that is, $\overline{\partial}$-theory. For instance it is given in [We2], Chapter 9, pp. 52–55.

Using Rossi's theorem, J. Wermer [We1] proved in 1976 the following result whose proof can be found in [We2], Chapter 20.

Theorem 5.1.2 (J. Wermer) *Let A be a uniform algebra. Denote by $\mathcal{M}$ the set of its characters, S its Shilov boundary, f, g two elements of A. Then*

$$\lambda \mapsto \operatorname{Log} \operatorname*{Max}_{\chi \in f^{-1}(\lambda)} |\chi(g)|$$

is subharmonic on $\hat{f}(\mathcal{M}) \backslash \hat{f}(S)$, where $f^{-1}(\lambda)$ denotes the set of $\chi \in \mathcal{M}$ for which $\chi(f) = \lambda$.

The set $f^{-1}(\lambda)$ is called the *fibre* over λ. This result was also independently obtained by V.N. Senichkin [Se]. Z. Słodkowski gave a more elementary proof avoiding Rossi's local maximum principle (see [Sł4] or [Au5]).

Let $K_g(\lambda)$ denote the set of $\chi(g)$ for $\chi \in f^{-1}(\lambda)$. In [AW] we proved that $\lambda \mapsto \operatorname{Log} \delta(K_g(\lambda))$ is subharmonic on $\hat{f}(\mathcal{M}) \backslash \hat{f}(S)$, where δ is the diameter. We also conjectured that $\lambda \mapsto \operatorname{Log} \delta_n(K_g(\lambda))$ and $\lambda \mapsto \operatorname{Log} c(K_g(\lambda))$ are subharmonic. D. Kumagai gave a partial answer to this question when A satisfies the condition

$$\partial^1(A \hat{\otimes} A) = (\partial^0 A \times \partial^1 A) \cup (\partial^1 A \times \partial^0 A),$$

where $A \hat{\otimes} A$ is the projective tensor product, $\partial^0 A$ the ordinary Shilov boundary and $\partial^1 A$ the generalized Shilov boundary of order 1 (see below for a definition).

In fact, these questions are consequences of the very general theorem which was proved by Z. Słodkowski in 1981 [Sł1].

Theorem 5.1.3 (Z. Słodkowski) *Let A be a uniform algebra. Denote by $\mathcal{M}$ the set of its characters, S its Shilov boundary, f, g two elements of A. Then $\lambda \mapsto K_g(\lambda)$ is an analytic multifunction on $\hat{f}(\mathcal{M}) \backslash \hat{f}(S)$.*

Having this fundamental result and applying Theorem 2.1.3 we immediately get:

Corollary 5.1.4 *With the hypotheses of Theorem 5.1.3 we conclude that $\lambda \mapsto \operatorname{Log} \delta_n(K_g(\lambda))$ and $\lambda \mapsto \operatorname{Log} c(K_g(\lambda))$ are subharmonic on $\hat{f}(\mathcal{M}) \backslash \hat{f}(S)$.*

Even these two results can be improved in a much more general situation. For the details see [Sł1] and [Au5].

First let us give some notations. Let A be a uniform algebra, $\mathcal{M}$ be its set of characters and we denote now by $\partial^0 A$ its classical Shilov boundary. Let $F = (f_1, \cdots, f_n) \in A^n$ and $V(F) = \{\chi \in \mathcal{M} | \chi(f_1) = \cdots = \chi(f_n) = 0\}$. It is well known that $V(F)$ is A-convex or equivalently that

$$A(V(F)) = \{g \in C(V(F)) | \exists f \in A \text{ such that } \chi(f) = \chi(g), \forall \chi \in V(F)\}$$

satisfies $\mathcal{M}(A(V(F))) = V(F)$.

By definition the *n-generalized Shilov boundary* is

$$\partial^n A = \overline{\bigcup_F \partial^0(A(V(F)))}\,, \text{ for all } F \in A^n.$$

It can be characterized by some maximum principle.

Theorem 5.1.5 *Suppose we have $g \in A$, $F = (f_1, \cdots, f_n) \in A^n$. Denote by $F^{-1}(\lambda) = \{\chi \in \mathcal{M}: \chi(f_1) = \lambda_1, \cdots, \chi(f_n) = \lambda_n\}$ and by $K_g(\lambda) = \{\chi(g): \chi \in F^{-1}(\lambda)\}$, where $\lambda = (\lambda_1, \cdots, \lambda_n)$. Then $\lambda \mapsto K_g(\lambda)$ is analytic multivalued on $F(\mathcal{M})\backslash F(\partial^{n-1}A)$.*

5.2 Analytic structure

If Γ is a bounded Jordan arc of the complex plane, J.L. Walsh, in 1926, proved that every continuous function on Γ can be uniformly approximated on Γ by polynomials. But this result is no longer true if $\Gamma \subset \mathbb{C}^n$ with $n \geq 2$, except if the arc is rather smooth.

This was proved by J. Wermer in 1958 if the arc is analytic, by E. Bishop (unpublished) and G. Stolzenberg in 1966 in the case of a piecewise C^1 Jordan arc, and finally by H. Alexander and J.-E. Björk for rectifiable arcs in 1971.

The main difficulty in the argument is to prove that the arc is polynomially convex. The basic idea goes back to the pioneering work of J. Wermer on analytic arcs. It is to prove that $\hat{\Gamma}\backslash\Gamma$ has some analytic structure and then by some argument principle to conclude that it must be empty, so $\Gamma = \hat{\Gamma}$. All these proofs are extremely difficult; for a quick introduction consult [We2], Chapters 12–13, [St], Chapter 6 and [Ga]. Some ideas contained in [Bj1,2] simplify very much the arguments.

An important step in the proof that C^1 or rectifiable arcs are polynomially convex is the following theorem which in its original formulation was proved by E. Bishop in 1963 [Bi]. Using subharmonic functions, Bishop's theorem was extended in [AW]. But now, using Theorem 5.1.3 its proof is very simple.

Theorem 5.2.1 (E. Bishop–B. Aupetit–J. Wermer analytic structure theorem) *Let A be a uniform algebra. Denote by $\mathcal{M}$ the set of its characters, $\mathcal{S}$ its Shilov boundary, f an element of A. Suppose that $\hat{f}(\mathcal{M})\backslash\hat{f}(\mathcal{S})$ is non-void and let W be a component of this set. Suppose now that W contains a set G such that:*

(i) *the outer capacity of G is positive;*

(ii) *the fibres $f^{-1}(\lambda)$ are finite on G.*

Then there exists and integer n such that $\#f^{-1}(\lambda) \leq n$ for every $\lambda \in W$ and $f^{-1}(W)$ has the structure of a complex analytic manifold of dimension 1 on which the elements of A are analytic.

The classical theorem of E. Bishop contains the stronger hypothesis that G has positive planar measure. Its classical proof is arduous; see for instance [We2], Chapter 11, where it is already simplified.

As in [AW] we can obtain the following generalization of a result of R. Basener.

Theorem 5.2.2 *Let $A, \mathcal{M}, \mathcal{S}, f, g, W$ be as in Theorem 5.2.1. Suppose that W contains a set G such that:*

(i) *G has positive outer capacity;*

(ii) *the fibres $f^{-1}(\lambda)$ are countable on G.*

Then $\mathcal{M}$ contains a non-void open set with the analytic structure of a complex analytic manifold of dimension 1 (it is even a polydisk) on which all the elements of A are analytic.

R. Basener supposes that $G = W$ and he concludes that $f^{-1}(W)$ contains a dense open set with an analytic structure. If we suppose only that G has positive planar measure, the same proof, as remarked by B. Cole, shows the existence of an analytic polydisk in $f^{-1}(W)$, but to obtain Theorem 5.2.2 it is necessary to use a subharmonic argument or Theorem 5.1.3 with Theorem 2.2.8. For $G = W$ it is also possible to give a purely topological proof. It is even possible to globalize Theorem 5.2.2 when A is separable, in the following form.

Theorem 5.2.3 *Let $A, \mathcal{M}, \mathcal{S}, f, g, W, G$ be as in the previous theorem, with A separable. Then $f^{-1}(W)$ contains a dense open set with the analytic structure of a complex analytic manifold of dimension 1 on which the elements of A are analytic.*

P. Jakóbczak obtained extensions of Bishop's and Basener's analytic structure theorems but they are weaker than the results of [AW]. His paper is interesting because it clarifies Alexander's result concerning polynomial approximation on rectifiable curves.

Now we give a generalization of an n-dimensional analytic structure theorem obtained by R. Basener and used by N. Sibony [Si] to get several applications. Its proof comes immediately from Theorem 5.1.5.

Theorem 5.2.4 (Generalization of several dimension analytic structure theorem of R. Basener) *Suppose $F \in A^n$ and let W be a component of $F(\mathcal{M})\backslash F(\partial^{n-1}A)$. Suppose that W contains a subset G such that:*

(i) *G is not pluripolar;*

(ii) *the fibres $F^{-1}(\lambda)$ are finite on G.*

Then there exists an integer n such that

(a) *$W = \bigcup_{k=1}^{n} W_k$;*

(b) *$\bigcup_{k=1}^{n-1} W$ is a proper analytic subvariety of W;*

(c) $\mathcal{L} = (F^{-1}(W), F, W)$ *is an analytic cover, then* $F^{-1}(W)$ *has the structure of an analytic complex manifold of dimension* n *on which all the elements of* A *are analytic.*

5.3 A pathological analytic multifunction

Denote by Δ the unit disk and let Γ be a compact subset of $\mathbb{C}^2$ which is above $\partial\Delta$. Let A be the uniform algebra $P(\Gamma)$, that is, the closure in $C(\Gamma)$ of the algebra of polynomials on Γ. Let $f, g \in A$ defined by $f(\lambda, z) = \lambda$ and $g(\lambda, z) = z$. The map $\chi \mapsto (\chi(f), \chi(g))$ is a homomorphism from $\mathcal{M}$ onto $\hat{\Gamma}$. Then $f(\mathcal{M}) \subset \overline{\Delta}$ and $f(X) \subset \partial\Delta$, where $\mathcal{S}$ is the Shilov boundary. So $f(\mathcal{M})\backslash f(\mathcal{S})$ is open, hence it is empty (in which case $\hat{\Gamma}$ is above $\partial\Delta$) or $f(\mathcal{M}) = \overline{\Delta}$ and $f(\mathcal{S}) = \partial\Delta$. In the latter case it is easy to see that $K_g(\lambda) = g(f^{-1}(\lambda)) = \{z \in \mathbb{C}: (\lambda, z) \in \hat{\Gamma}\}$. So, by Theorem 5.1.3, we immediately get the next result.

Theorem 5.3.1 *Let* Γ *be a compact subset of* $\mathbb{C}^2$ *which is above* $\partial\Delta$, *where* Δ *denotes the unit disk in the complex plane. For* $\lambda \in \Delta$ *denote by*

$$L(\lambda) = \{z \in \mathbb{C}: (\lambda, z) \in \hat{\Gamma}\}.$$

Then either $L(\lambda)$ *is identically empty or* L *is an analytic multifunction on* Δ.

Using the previous theorem and maximum principle it is easy to see that we can obtain the next corollary.

Corollary 5.3.2 *Let* K *be an analytic multifunction defined on a domain containing* $\overline{\Delta}$. *Denote by* Γ *the part of the graph of* K *which is above* $\partial\Delta$ *and by* L *the corresponding multifunction as defined before. Then* $K(\lambda) \subset L(\lambda)$ *for* $\lambda \in \overline{\Delta}$.

In [We3] J. Wermer gave an example, based upon one due to B. Cole, proving that there exists a compact subset Γ of $\mathbb{C}^2$ above $\partial\Delta$ such that:

(i) for $\lambda \in \Delta$ the set $L(\lambda) = \{z \in \mathbb{C}: (\lambda, z) \in \hat{\Gamma}\}$ is non-empty and totally disconnected;

(ii) there exists no continuous function f defined on an open and non-empty subset of Δ such that $(\lambda, f(\lambda)) \in \hat{\Gamma}$.

Modifying slightly the argument of J. Wermer, with some ε_n decreasing to zero very quickly it is possible to build Γ in such a way that the $L(\lambda)$ have zero capacity. So this modified example gives the following pathological result.

Theorem 5.3.3 *There exists a continuous analytic multifunction defined on the unit disk having zero capacity values and no local continuous selections.*

Because these $L(\lambda)$ have zero capacity they are totally disconnected. By Corollary 2.1.10 this multifunction is continuous. By the remark following Theorem 2.5.2 these $L(\lambda)$ are never countable, except possibly on a polar subset of Δ.

5.4 Cluster sets

To finish this chapter let us now show how some subharmonic or analytic multivalued methods can be used to get theorems about cluster sets. Probably by such methods it is possible to get more, for example, results concerning the Iversen–Gross theorem, the Weiss theorem, etc.

Let Δ be the unit disk. By an *inner function* $z \mapsto f(z)$ on Δ we mean a bounded analytic function on Δ such that $\lim_{r\to 1} |f(re^{i\theta})| = 1$ a.e. Seidel and Frostman have proved the following result.

Theorem 5.4.1 *If f is inner, then either f is a finite Blaschke product or every value in Δ, except perhaps for a closed set of zero capacity, is taken by f infinitely often in Δ.*

In the case of $H^\infty(\Delta)$, Δ can be identified topologically with an open subset of the set $\mathcal{M}$ of characters of $H^\infty(\Delta)$. Then the previous theorem is strangely similar to Theorem 5.2.1, with $f^{-1}(\lambda)$ replaced by $f^{-1}(\lambda) \cap \Delta$. In [Au5] we gave a purely subharmonic form of this theorem (pp. 40–41), the proof of which can now be simplified using Theorems 5.1.3 and 2.1.12.

In the case of $H^\infty(\Delta^n)$, with $n > 1$, this method shows that we have almost the same result, namely: if f is inner on Δ^n, then there exists E of zero capacity in Δ such that every a not in E is taken by f. If f is a good inner function, this follows from a standard result; if f is only inner there exists $a \in \Delta$ such that $(g-a)/(1-\bar{a}g)$ is good and then we consider the conformal mapping $x \mapsto (z+a)/(1+a\bar{z})$ on Δ onto Δ, which transforms a set of zero capacity to a set of zero capacity .

A similar subharmonic proof can be used to obtain Tsuji's theorem mentioned before as Theorem 2.3.13 (see [Au5], Theorem 7.3.11). It is interesting to note that it is also possible to give a purely spectral proof of this result using companion matrices (see [Au5], p. 42).

Chapter 6
Applications to spectral interpolation

6.1 Introduction

As Yitzhak Katznelson says at the beginning of Chapter 4 of his book [Kat] 'Interpolation of norms and of linear operators is really a topic in functional analysis rather than harmonic analysis proper; but, though less so than ten years ago, it still seems esoteric among authors in functional analysis...'. This difficult theory originates from many parts of mathematics: harmonic analysis, matrices, summation methods, multipliers, certain singular integrals, and so on.

To get precise information on this theory, see [Ca], [Ra4], [Sn], [Za] and their extensive lists of references.

Let us start with a concrete example. Let G be a locally compact topological group equipped with a fixed left Haar measure (for instance $\mathbb{R}^n$ with the Lebesgue measure). For every real p satisfying $1 \leq p \leq +\infty$ we can define on the Banach space $L^p(G)$ the operator

$$(T_f)_p(g) = f * g, \quad g \in L^p(G).$$

In this example the Banach spaces $L^p(G)$ interpolate from $L^1(G)$ to $L^\infty(G)$, and the convolution operator is interpolated by the $(T_f)_p$. The first natural question is: how does $Sp(T_f)_p$ vary with p? Is it true in some cases that it is constant and equal to $Sp(T_f)_1 = Sp_{L^1(G)} f$? Due to a theorem of N. Wiener this is true if G is abelian. This last result was extended by B. Barnes [Bar3] to groups which are symmetric and amenable. We shall comment on this result at the end of section 6.2, but we explain now why it is a generalization of the abelian case. If G is abelian, then the group algebra $L^1(G)$ is symmetric, see for instance [GRC], page 132 and the group is amenable, see [BD], page 241.

We now give the abstract setting of interpolation theory. Let B_0, B_1 be two Banach spaces, respectively for the norms $\| \ \|_0$ and $\| \ \|_1$, which are continuously embedded in some Handsdorff topological vector space. Then $B_0 \cap B_1$ and $B_0 + B_1$ become Banach spaces for the norms $\mathrm{Max}(\|x\|_0, \|x\|_1)$ and $\mathrm{Inf}\, \{\|y\|_0 + \|z\|_1 : y \in B_0,\ z \in B_1,\ y + z = x\}$. We shall always assume that $B_0 \cap B_1$ is dense in B_0 and B_1. Let D be the strip $\{\lambda \in \mathbb{C} : 0 < \mathrm{Re}\,\lambda < 1\}$. We now create the interpolation family (B_λ) in the following way. Define $\mathcal{F}$ to be the family of all continuous functions f from $\overline{D}$ into $B_0 + B_1$, which are analytic on D and satisfy the following properties:

(i) $f(it) \in B_0$ is continuous for $\| \ \|_0$ and goes to zero at infinity;

(ii) $f(1 + it) \in B_1$ is continuous for $\| \ \|_1$ and goes to zero at infinity.

Then $\mathcal{F}$ is a Banach space for the norm $\underset{k=0,1}{\mathrm{Max}}\, \underset{t \in \mathbb{R}}{\mathrm{Sup}}\, \|f(k + it)\|$. By definition, for $\lambda \in \overline{D}$, we have

$$B_\lambda = \mathcal{F}/\{f \in \mathcal{F} : f(\lambda) = 0\}.$$

To understand this awkward definition it suffices to see what happens for instance for $\mathrm{Re}\,\lambda = 0$. In that case it corresponds to the quotient of the algebra of continuous functions going to zero at infinity with values in B_0 by some maximal ideal of functions annihilating λ, so it can be identified with B_0. This B_λ is a Banach space for the quotient norm and depends only on $\mathrm{Re}\,\lambda$.

If T is a linear operator on $B_0 + B_1$ whose restrictions to B_0 and B_1 are bounded operators, then T extended to B_λ is also a bounded operator, and by an abstract version of the Riesz–Thorin theorem we have a good estimate on its norm given by

$$\|T\|_\lambda \leq \|T\|_0^t \, \|T\|_1^{1-t}$$

where $\|T\|_\lambda$ denotes the norm of the extension of T to B_λ, in $\mathcal{B}(B_\lambda)$, and $t = \mathrm{Re}\,\lambda$.

6.2 The results

The results of this section are essentially due to T.J. Ransford [Ra4]. He proved Theorem 6.2.1 supposing that the operator T satisfies an extra condition called 'local uniqueness-of-resolvent condition', but this condition has been shown redundant by K. Saxe (private communication from J. Zemánek). This can also be found implicitly in a paper of Z. Słodkowski [Sł7]. We denote by $Sp_\lambda T$ the spectrum of the operator T extended to the interpolation space B_λ.

Theorem 6.2.1 *With the notations of section* 6.1 *the multifunction* $\lambda \mapsto Sp_\lambda T$ *is analytic on the strip* D.

The proof of this theorem is completely analytic. Even upper-semicontinuity depends on an analytic argument, see [Sn].

The next result is a spectral analogue of the Riesz–Thorin theorem. It is a generalisation of a result of J.D. Stafney.

Corollary 6.2.2 *Let* φ *be a subharmonic function defined on an open set containing all the* $Sp_\lambda T$. *For* $-\infty \leq \alpha < +\infty$ *we set* $E_\alpha = \{z : \varphi(z) \leq \alpha\}$. *If* $Sp_0 T \subset E_\alpha$ *and* $Sp_1 T \subset E_\beta$, *then* $Sp_\lambda T \subset E_{(1-s)\alpha + s\beta}$, *where* $s = \operatorname{Re}\lambda$.

This implies in particular that $Sp_\lambda T$ is constant if $Sp_{\lambda_0} T$ has zero capacity for some λ_0, a result which extends an earlier one of R. Sarnak in the case of a convolution operator on $L^p(G)$, where G is a locally compact abelian group.

Suppose now that G is a symmetric and amenable group. The first condition says that $Sp_{L^1(G)}(f) \subset \mathbb{R}$ for $f = f^*$. The second condition implies that $r(f^* * f)^{\frac{1}{2}} = \|(Tf)_2\|$ which rather easily implies that $Sp_{L^1(G)}(f) = Sp(T_f)_1 = Sp(T_f)_2$. Taking $f = f^*$ we have $Sp(T_f)_1 = Sp(T_f)_2 \subset \mathbb{R}$, so applying Corollary 6.2.2 to $\varphi(z) = |\operatorname{Im} z|$ with the interpolation couple $[L^1(G), L^2(G)]$ and applying Theorem 6.2.1 and Corollary 2.1.7 we conclude that $Sp_{L^1(G)}(f) = Sp(T_f)_p$ for all $p \geq 1$. The same is true for all $f \in L^1(G)$ by Hulanicki's theorem, see [Bar3].

Chapter 7
Applications to local spectrum theory

7.1 General results

Given an analytic family $T(\lambda)$ of operators on a Banach space we saw in Chapter 3 many applications of the fact that $SpT(\lambda)$ is an analytic multifunction. Is it also possible to use analytic multifunctions in the theory of local spectrum introduced in the 1950s by N. Dunford (see [Du] and his other papers)?

Let $T \in \mathcal{B}(X)$ and $x \in X$. We define Ω_x to be the set of $\alpha \in \mathbb{C}$ for which there exists a neighbourhood V_α and u analytic on V_α with values in X such that $(\lambda - T)u(\lambda) = x$ on

V_α. This set is open and contains the complement of the spectrum of T. The function u is called a *local resolvent* of T on V_α. By definition, the *local spectrum* of T at x, denoted by $Sp_x(T)$, is the complement of Ω_x, so it is a compact subset of $Sp(T)$.

In general this set may be empty even for $x \neq 0$ (see Example 1 below). But for $x \neq 0$, the local spectrum of T is non-empty if T satisfies the *uniqueness property for the local resolvent*, that is, $(\lambda - T)v(\lambda) = 0$ implies $v = 0$ for any analytic function v defined on any domain D of $\mathbb{C}$ with values in X. N. Dunford [Du] noticed easily that T has this property if the spectrum of T has no interior points. So this happens if T has a finite, countable or real spectrum. For the operators satisfying the uniqueness property for the local resolvent there is a unique local resolvent which is the analytic extension of $(\lambda - T)^{-1}x$. Using this property and Liouville's theorem it is easy to conclude that $Sp_x(T) \neq \emptyset$ for $x \neq 0$. Also in this case the *local spectral radius*

$$r_x(T) = \max\{|\mu| : \mu \in Sp_x(T)\}$$

is equal to $\overline{\lim\limits_{k\to\infty}} \|T^k x\|^{\frac{1}{k}}$. In general this last property is false, as is shown by the following examples.

Example 1 On the Hilbert space l^2 we consider the left shift operator

$$T: (\xi_1, \xi_2, \cdots) \mapsto (\xi_2, \xi_3, \cdots).$$

It is well known that the spectrum of T is the closed unit disk (see [H], problem 82). Denoting by (e_n) the canonical orthonormal basis we have $T^k e_k = 0$, for $k \geq 1$, so for $\lambda \neq 0$ the relation

$$e_k = (\lambda^k - T^k)\frac{e_k}{\lambda^k} = (\lambda - T)(\lambda^{k-1} + \cdots + T^{k-1})\frac{e_k}{\lambda^k}$$

implies that $Sp_{e_k}(T) \subset \{0\}$. If we consider the right shift operator defined by $S: (\xi_1, \xi_2, \cdots) \mapsto (0, \xi_1, \xi_2, \cdots)$ which is an isometry and satisfies $TS = I$, it is easy to verify that we have for $x \in l^2, x \neq 0, |\lambda| < 1$,

$$(\lambda - T)\left(-\sum_{n=0}^{\infty} \lambda^n S^{n+1} x\right) = x.$$

The series defines an analytic function in λ, so we have $Sp_x(T)$ included in the unit circle. Consequently, $Sp_{e_k}(T)$ is empty for $k \geq 1$, which implies that $Sp_x(T) = \emptyset$ if x is in the linear subspace F generated by $e_1, e_2, \cdots$. There are some x not in F for which $Sp_x(T)$ is non-empty, in which case $Sp_x(T)$ is included in the unit circle. For instance we may choose $x = \sum_{n=1}^{\infty} \frac{1}{n} e_{2^n}$ for which $\rho_x(T) = 1$, consequently $r_x(T) = 1$ with the notations we are now giving.

By convention, for $x \neq 0$, we set

$$r_x(T) = \begin{cases} \max\{|\alpha| : \alpha \in Sp_x(T)\}, & \text{if } Sp_x(T) \neq \emptyset \\ 0, & \text{if } Sp_x(T) = \emptyset \end{cases}$$

and $\rho_x(T) = \overline{\lim}_{k\to\infty} \|T^k x\|^{\frac{1}{k}} \leq r(T)$. It is well known that in general $\|T^k x\|^{\frac{1}{k}}$ has no limit. From the formal identity $(\lambda - T)\frac{1}{\lambda}\sum_{k=0}^{\infty}\left(\frac{T}{\lambda}\right)^k x = x$, we conclude that

$$r_x(T) \leq \rho_x(T) \leq r(T).$$

The next example shows that in general $r_x(T)$ is less than $\rho_x(T)$ even if $Sp_x(T)$ is non-empty.

Example 2 Let $X = l^2 \oplus \mathbb{C}^2$, with the norm $\|(x_1, x_2)\| = \text{Max}(\|x_1\|, \|x_2\|)$, on which we consider the bounded linear operator

$$A(x_1, x_2) = ((I+T)x_1, Mx_2),$$

where T is the left shift operator on l^2 and M a 2×2 matrix having eigenvalues $\frac{1}{2}$ and $\frac{3}{2}$. We have $Sp(A) = Sp(I+T) \cup \left\{\frac{1}{2}, \frac{3}{2}\right\} = \overline{D}(1,1)$, the closed unit disk centered at 1. So $r(A) = 2$. Let u be an eigenvector of M corresponding to $\frac{1}{2}$. Then we have $\rho_{(e_1,u)}(A) = \text{Max}(\rho_{e_1}(I+T), \rho_u(M)) = 1$. Moreover,

$$Sp_{(e_1,u)}(A) = Sp_{e_1}(I+T) \cup Sp_u(M) = Sp_u(M) = \left\{\tfrac{1}{2}\right\}$$

because $Sp_{e_1}(I+T)$ is empty. So $r_{(e_1,u)}(A) = \frac{1}{2}$.

Nevertheless the local spectrum has a good holomorphic functional calculus.

Theorem 7.1.1 (Holomorphic functional calculus for local spectrum) *Let $T \in \mathcal{B}(X), x \neq 0$, and f holomorphic on a neighbourhood D of $Sp(T)$. Then $f(Sp_x(T))$ is included in $Sp_x(f(T))$. If f is injective on D, then $f(Sp_x(T)) = Sp_x(f(T))$. Moreover if T has the uniqueness property for the local resolvent, then we have the same equality for any f holomorphic.*

The proof of this theorem is essentially given in [EL], Theorem 1.6, p.6. The injective case is obtained by applying the first case to f^{-1} and the operator $f(T)$.

Corollary 7.1.2 *Let $T \in \mathcal{B}(X)$, T invertible. Then*

$$\text{dist}(0, Sp_x(T)) = \frac{1}{r_x(T^{-1})}.$$

Supposing that T has the uniqueness property for the local resolvent, N. Dunford proved that $Sp(T) = \bigcup_{x\in X} Sp_x(T)$. The argument is very simple. Let $\alpha \in Sp(T) \setminus \bigcup_{x\in X} Sp_x(T)$. From the definition of Ω_x this implies that $\alpha - T$ is onto, so it is not one-to-one, consequently there exists $u \neq 0$ such that $(\alpha - T)u = 0$, in which case we have $(\lambda - T)\frac{u}{\lambda-\alpha} = u$, for $\lambda \neq \alpha$, so $Sp_u(T) \subset \{\alpha\}$, but $Sp_u(T)$ is non-empty and this implies that $Sp_u(T) = \{\alpha\}$ which gives a contradiction. If $Sp_x(T)$ is empty for some $x \neq 0$, then Example 1 shows that this result cannot be true in general. This contradicts earlier false results of J.D. Gray which are unfortunately reproduced in [I].

P. Vrbová [Vr] improved Dunford's result when T has the uniqueness property for the local resolvent. She proved that there exists $x \in X$ such that $Sp(T) = Sp_x(T)$, a result which is obviously false in the general situation. She also improved a result of J. Daneš [Da] proving that the set of $x \in X$ for which $Sp_x(T)$ contains the boundary of $Sp(T)$ is nonmeagre (i.e. it is not of the first category) in X.

It is possible to improve Vrbová's theorem using elementary arguments, so in [AD] we obtained the following.

Theorem 7.1.3 *Let $T \in \mathcal{B}(X)$. There exists a subset E of X, which is a countable union of linear subspaces of X invariant by any operator commuting with T, such that $X \setminus E$ is a dense G_δ-subset of X and such that $x \notin E$ implies $\partial Sp(T) \subset Sp_x(T)$.*

Remark If instead of one operator $T \in \mathcal{B}(X)$ we have a sequence (T_n) of operators, then for each n there exists some subset E_n as defined by Theorem 7.1.3, consequently taking E to be the union of all these E_n, it satisfies the properties of Theorem 7.1.3, for any T_n. Consequently, Theorem 7.1.3 generalizes the main theorem of J. Daneš [Da] on $\rho_x(T)$.

We say that $T \in \mathcal{B}(X)$ is *locally algebraic* if for every $x \in X$ there exists a non-zero polynomial p such that $p(T)x = 0$. It is a well-known result due to I. Kaplansky [K] that a locally algebraic bounded linear operator on a Banach space is in fact algebraic. Several different proofs of this important result are known (see for instance [Au7], p. 86 and [RR]). Using Theorem 7.1.3 it is easy to obtain a new one (see [AD]).

7.2 Analytic results

Usually generalized spectral operators on a Banach space X are defined in terms of spectral distributions (see [CF]). But this notion is equivalent to the following one: we say that $N \in \mathcal{B}(X)$ is a *generalized spectral operator* if $N = H + iK$, where $H, K \in \mathcal{B}(X)$, $HK = KH$ and H, K are in the class $\mathcal{S}$ of *real generalized spectral* operators defined by the growth condition

$$\|e^{itT}\| \leq C(1+|t|)^n, \qquad t \in \mathbb{R},$$

for constant C and some integer $n \geq 0$. Such operators of class $\mathcal{S}$ have a real spectrum, consequently they satisfy the uniqueness property for the local resolvent. In 1960, C. Foiaş [Fo] proved that these generalized spectral operators also have the uniqueness property for the local resolvent.

In [AD] we introduced new analytic arguments on the local spectrum improving the result of C. Foiaş. Supposing that T satisfies the uniqueness property for the local resolvent and given an analytic function u from D into X, we proved that $Sp_{u(\lambda)}(T)$ behaves rather well analytically. It would be very nice to prove that $Sp^*_{u(\lambda)}(T)$ is an analytic multifunction (for the definition of this multifunction see below), but unfortunately we now only know the following weaker results.

Theorem 7.2.1 *Suppose $T \in \mathcal{B}(X)$ satisfies the uniqueness property for the local resolvent, and let u be an analytic function from an open subset D of $\mathbb{C}$ into X. Then*

$g: \lambda \mapsto \operatorname{Log} r_{u(\lambda)}(T)$ satisfies the mean inequality property, consequently g^*, its upper regularization, is subharmonic on D. Moreover, the set of λ such that $g^*(\lambda) > g(\lambda)$ is a polar subset of D.

The multifunction $\lambda \mapsto Sp_{u(\lambda)}(T)$ is not upper semicontinuous, even in the case of matrices. This is the reason why we introduce the *upper semicontinuous regularization* of the local spectrum by

$$Sp^*_{u(\lambda)}(T) = \bigcap_{r>0} \overline{\bigcup_{\mu \in D, |\mu-\lambda|<r} Sp_{u(\mu)}(T)}.$$

It is easy to see that the graph of $\lambda \mapsto Sp^*_{u(\lambda)}(T)$ is the closure of the graph of $\lambda \mapsto Sp_{u(\lambda)}(T)$, consequently the radius of $Sp^*_{u(\lambda)}(T)$ is the regularization of $r_{u(\lambda)}(T)$. Moreover, $\lambda \mapsto Sp^*_{u(\lambda)}(T)$ is upper semicontinuous. Let K be a compact subset of the complex plane. We saw in section 2.1 that the peripherical part of K is the intersection of K with the smallest closed disk centered at 0 containing K. This peripherical part contains at least one point and all its points have the same modulus.

Lemma 7.2.2 *Let $T \in \mathcal{B}(X)$, satisfying the uniqueness property for the local resolvent, and let u be an analytic function from a domain D of $\mathbb{C}$ into X. Suppose there exists $\lambda_0 \in D$ such that $r_{u(\lambda)}(T) \le r_{u(\lambda_0)}(T)$, for all $\lambda \in D$. Then there exists $c > 0$ such that $r^*_{u(\lambda)}(T) = c$ on D and the peripherical part of $Sp^*_{u(\lambda)}(T)$ is constant on D.*

The geometrical proof of this lemma is almost identical with the proof of Theorem 3.4.11 in [Au7], which is originally due to E. Vesentini [Ve2] (see also Theorem 2.1.6).

We shall say that a compact subset K of the complex plane is *nice* if for every $\alpha \in \partial K$ there exists $a \notin K$ such that $K \subset \{z: |z-a| \ge |\alpha - a|\}$. This condition is automatically satisfied if K is contained in a smooth curve.

Theorem 7.2.3 *Let $T \in \mathcal{B}(X)$ and let u be an a analytic function from a domain D of $\mathbb{C}$ into X. Suppose moreover that $Sp(T)$ is nice and without interior points. Then $Sp^*_{u(\lambda)}(T)$ is constant on D.*

If K is finite it is nice, it has no interior points, so the next result improves the finite-dimensional situation.

Theorem 7.2.4 *Let $T \in \mathcal{B}(X)$ be such that $Sp(T)$ is nice, without interior points, and let u be an analytic function from a domain D of $\mathbb{C}$ into X. Then there exists a compact set K and a polar subset E of D such that $Sp_{u(\lambda)}(T) = K$ for $\lambda \in D \setminus E$ and $Sp_{u(\lambda)}(T) \subset K$, $Sp_{u(\lambda)}(T) \neq K$, for $\lambda \in E$.*

The proof depends on Theorem 7.2.1, 7.2.3 and Runge's theorem on rational approximation.

The two next results are generalizations of Foiaş's result [Fo].

Theorem 7.2.5 *Let $T, S \in \mathcal{B}(X)$. Suppose that $TS = ST$, that the spectrum of T is nice and without interior points, and that the spectrum of S has no interior points. Then $T + S$ has the uniqueness property for the local resolvent.*

Corollary 7.2.6 *Let $T, S \in \mathcal{B}(X)$. Suppose $TS = ST$ and $Sp(T), Sp(S)$ real, then $T + iS$ has the uniqueness property for the local resolvent.*

Let $T, S \in \mathcal{B}(X)$. The *generalized derivation* $C(T,S)$ defined by T, S is by definition the operator defined on $\mathcal{B}(X)$ by $C(T,S)A = TA - AS$. If $T = S$ it is an inner derivation. In [FoV], Proposition 2.2, C. Foiaş and F.-H. Vasilescu proved that $C(T,S)$ has the uniqueness property for the local resolvent if T and S are decomposable operators. They applied this fact to obtain two very interesting inclusions for the local spectra (see Corollaries 7.2.9, 7.2.10 below). At the end of their paper they noticed that the uniqueness property for $C(T,S)$ could be obtained supposing only that S is decomposable and that T has the uniqueness property and the property that $\{x \in X : Sp_x(T) \subset F\}$ is closed for every closed subset F of $\mathbb{C}$ (Dunford's property).

It is easy to see that in general generalized derivations, even inner derivations, do not have the uniqueness property for the local resolvent. Taking the left shift operator T on ℓ^2, then by Example 1 of section 7.1 there exists for $0 < |\lambda| < 1$ an analytic family $x(\lambda)$ of vectors such that $(\lambda - T)x(\lambda) = 0$, where

$$x(\lambda) = \frac{e_1}{\lambda} + e_2 + \lambda e_3 + \cdots + \lambda^{n-1} e_n + \cdots.$$

So denoting by $A(\lambda) \in \mathcal{B}(\ell^2)$ the operator defined by $A(\lambda)y = (y|x(\lambda))x(\lambda)$ we obtain an analytic family of operators such that on $\ell^2 \oplus \ell^2$ we have

$$\begin{pmatrix} T , & 0 \\ 0 , & 0 \end{pmatrix}\begin{pmatrix} 0 , & A(\lambda) \\ 0 , & 0 \end{pmatrix} - \begin{pmatrix} 0 , & A(\lambda) \\ 0 , & 0 \end{pmatrix}\begin{pmatrix} T , & 0 \\ 0 , & 0 \end{pmatrix} = \lambda \begin{pmatrix} 0 , & A(\lambda) \\ 0 , & 0 \end{pmatrix}$$

which implies that the inner derivation defined by $\begin{pmatrix} T , & 0 \\ 0 , & 0 \end{pmatrix}$ on $\ell^2 \oplus \ell^2$ does not have the uniqueness property for the local resolvent.

We now give some results deriving from Theorem 7.2.5 and Corollary 7.2.6, saying that $C(T,S)$ has the uniqueness property for the local resolvent, if T and S have spectra with good topological properties.

Corollary 7.2.7 *Let $T, S \in \mathcal{B}(X)$. Suppose that the spectrum of one of these operators is nice and without interior points and that the spectrum of the other has no interior points. Then $C(T,S)$ has the uniqueness property for the local resolvent, consequently $Sp_A C(T,S)$*

is non-empty for every non-zero $A \in \mathcal{B}(X)$, and there exists a local resolvent at A.

Corollary 7.2.8 *Let $T = T_1 + iT_2$, $S = S_1 + iS_2 \in \mathcal{B}(X)$. Suppose that $T_1T_2 = T_2T_1$, $S_1S_2 = S_2S_1$ and that T_1, T_2, S_1, S_2 have real spectra. Then $C(T,S)$ has the uniqueness property for the local resolvent.*

This last result can be applied in particular to generalized spectral operators.

Using the arguments of C. Foiaş and F.-H. Vasilescu ([FoV], Theorems 2.3 and 2.4) it is possible to obtain the following.

Corollary 7.2.9 *Suppose $T, S \in \mathcal{B}(X)$ satisfy the hypotheses of Corollary 7.2.7 or of Corollary 7.2.8, let $A \in \mathcal{B}(X)$ and let $x \in X$. Then*

$$Sp_{Ax}(T) \subset Sp_A C(T,S) + Sp_x(S).$$

Corollary 7.2.10 *Suppose $T, S, R \in \mathcal{B}(X)$ satisfy the hypotheses of Corollary 7.2.7 or of Corollary 7.2.8 and let $A, B \in \mathcal{B}(X)$. Then*

$$Sp_{BA}C(R,T) \subset Sp_A C(R,S) + Sp_B \subset C(S,T).$$

If it were possible to prove that $\lambda \mapsto Sp^*_{u(\lambda)}(T)$ is an analytic multifunction, using Corollary 2.3.5 it would be possible to extend Theorems 7.2.3, 7.2.4 in the case where the spectrum of T has no interior points and no points where $\mathbb{C}\backslash Sp(T)$ is thin. Consequently Theorem 7.2.5 could be extended to a bigger class of operators. Nevertheless the problem would remain to extend Theorem 7.2.5 only supposing that T, S commute, and that their spectra have empty interior. To do this it would be necessary to improve the open mapping theorem in the form mentioned in section 2.3.

Chapter 8
Applications to Jordan-Banach algebras

8.1 Introduction

A *Jordan algebra* is a non-associative algebra in which the product satisfies the two conditions $xy = yx$ and $(xy)x^2 = x(yx^2)$, for all x, y in the algebra. Such algebras were introduced, in the period 1932-1934, by P. Jordan, J. von Neumann and E. Wigner in order to improve the quantum mechanics formalism (see [JNW]). In the subsequent years they were studied by A.A. Albert [Al,2] who gave a complete classification of finite-dimensional Jordan algebras. Since that time, Jordan algebras have been intensively investigated by algebraists and the bibliography on this subject is very huge (see for instance [J1-3], [Zh], and [Mc3]).

Given an associative algebra A we can build a Jordan algebra, denoted by A^+, defining the Jordan product $x \cdot y = \frac{1}{2}(xy + yx)$. Any Jordan subalgebra of A^+ is consequently a Jordan algebra. Such algebras are called *special Jordan algebras.*

Another interesting example is constructed in the following way. Let V be a linear vector space and let φ be a symmetric bilinear form on V. We define a product on $J = \mathbb{C} \times V$ by

$$(\alpha, x) \cdot (\beta, y) = (\alpha\beta + \varphi(x, y), \beta x + \alpha y).$$

Then J is a Jordan algebra for this product, and every element $u = (\alpha, x)$ is algebraic of degree two because it satisfies the equation $u^2 - 2\alpha u - \varphi(x, x) = 0$. This algebra is in fact a special algebra (see [Zh], Exercise 1, p. 57). Unfortunately there are *exceptional Jordan algebras*, that is to say, non-special algebras, and this is the main cause of the difficulty of the theory (see for instance Albert's theorem on $H(C_3)$ in [Zh], pp. 55–57).

On a Jordan algebra A we define the important *quadratic operator* U_a defined by $U_a(x) = 2a(ax) - a^2x$. It satisfies the following property

$$U_x U_y U_x = U_{U_x(y)}, \text{ for } x, y \in A.$$

N. Jacobson introduced the notion of invertibility in Jordan algebras, a notion which of course generalizes the standard notion of invertibility in associative algebras. Given x in A we say that x is *invertible* if there exists y in A such that $xy = 1$ and $x^2y = x$. This element y is unique and denoted by x^{-1}. This notion is intimately related to the quadratic operator.

Theorem 8.1.1 *If $x \in A$ then x is invertible if an only if U_x is invertible in $\mathcal{B}(A)$, the algebra of linear operators on A, in which case $U_{x^{-1}} = (U_x)^{-1}$. If $x, y \in A$ they are both invertible if and only if $U_x(y)$ is invertible in A. In particular, x is invertible if and only if x^n is invertible for every integer $n \geq 1$.*

This theorem implies that the set of invertible elements in A is invariant by powers, but unfortunately it is not stable for products.

8.2 Analytic properties of the spectrum

A *Jordan-Banach algebra* A is a Jordan algebra with a complete norm satisfying $|xy| \leq |x|\,|y|$, for x, y in A. Adjoining an identity to A if necessary, we may suppose without loss of generality that A has an identity, denoted by 1, and that $|1| = 1$. Because we shall be concerned mainly with analytical tools we shall suppose, throughout this paper, that A is a complex algebra with identity. Very often the study of real Jordan-Banach algebras can be reduced to the study of complex ones using complexification and the identity principle for analytic functions, but in some cases this study may be very tricky.

As seen before, x is invertible if and only if U_x is invertible and $U_{x^{-1}} = (U_x)^{-1}$. This implies the following.

Theorem 8.2.1 *The set U of invertible elements of A is open and is invariant under the differentiable homeomorphism $x \mapsto x^{-1}$.*

For x in A we can define the *spectrum of* x, denoted by $Sp(x)$, to be the set of $\lambda \in \mathbf{C}$ such that $\lambda 1 - x$ is not invertible in A. The spectral radius $r(x)$ is by definition the maximum of $|\lambda|$ for all λ in the spectrum of x. As for Banach algebras it is possible to obtain the following.

Theorem 8.2.2 *Let x be an element of a Jordan-Banach algebra. Then*

(i) *$Sp(x)$ is compact and non-empty;*

(ii) *$\lambda \mapsto (\lambda 1 - x)^{-1}$ is analytic on the complement of $Sp(x)$;*

(iii) $r(x) = \lim_{n \to \infty} |x^n|^{1/n}$.

The standard holomorphic functional calculus in Banach algebras can be extended to the situation of Jordan-Banach algebras. This comes from the fact that the closed subalgebra generated by 1 and x is associative, consequently a Banach algebra.

Theorem 8.2.3 (Holomorphic functional calculus for Jordan-Banach algebras) *Let A be a Jordan-Banach algebra, x in A and U a neighbourhood of the spectrum of x. If h is holomorphic on U, then we can define*

$$h(x) = \frac{1}{2\pi i} \int_\Gamma h(\lambda)(\lambda 1 - x)^{-1} d\lambda,$$

where Γ is a positively oriented curve included in U and surrounding $Sp(x)$. Then we have the following properties:

(i) *$h(x)$ is independent of the choice of Γ in U and surrounding $Sp(x)$;*

(ii) *$\varphi: h \mapsto h(x)$ is an algebraic morphism from $H(U)$ into the smallest closed strongly associative subalgebra containing 1 and x; moreover, if $h(\lambda) = 1$ then $\varphi(h) = 1$, and if $h(\lambda) = \lambda$ then $\varphi(h) = x$;*

(iii) *φ is continuous on $H(U)$ for the uniform convergence on every compact subset of U;*

(iv) $Sp\, h(x) = h(Sp(x))$.

For a detailed exposition of this theorem see [M]. The three previous theorems were until the beginning of the 1980s the only analytic spectral tools known in the theory of Jordan-Banach algebras.

The next result is the fundamental tool which give us the opportunity to use systematically the deep theory of analytic multifunctions.

Theorem 8.2.4 (B. Aupetit–A. Zraïbi [AZ]) *Let $\lambda \mapsto f(\lambda)$ be an analytic function defined on an open subset D of the complex plane with values in a Jordan-Banach algebra. Then the multifunction $\lambda \mapsto Sp\, f(\lambda)$ is analytic.*

By Theorem 2.1.3, this result implies in particular that the functions

$$\lambda \mapsto \operatorname{Log} r(\lambda), \quad \lambda \mapsto \operatorname{Log} \delta_n(f(\lambda)), \quad \lambda \mapsto \operatorname{Log} c(f(\lambda))$$

are subharmonic, where δ_n and c denote respectively the n-th diameter and the capacity of the spectrum of $f(\lambda)$. This theorem has a large number of consequences, some of them elementary, like the analogues of Newburgh's theorems on spectral continuity and some of them very deep.

In 1976 I. Kaplansky introduced an interesting class of complex Jordan-Banach algebras with involution, which is a natural extension of the class of B^*-algebras. This is the class of JB^*-algebras. By definition we say that A is a *JB^*-algebra* if for every x in A we have $\|U_x(x^*)\| = \|x\|^3$.

It is not difficult to prove that $\|x^*\| = \|x\|$, that every closed associative $*$-subalgebra of A is a C^*-algebra, consequently it is isomorphic to $C(K)$ for some compact set K if it is commutative. In particular we have $\|\exp(ih)\| = 1$ for $h = h^*$.

Consider the set H of the self-adjoint elements of a JB^*-algebra. It has the following properties:

(i) H is a real Jordan-Banach algebra;

(ii) $\|h^2\| = \|h\|^2$, for all h in H;

(iii) $\|h^2\| \le \|h^2 + k^2\|$, for all h, k in H.

Real Jordan-Banach algebras satisfying these conditions were called JB-algebras by Alfson, Shultz and Størmer [ASS]. They are natural generalizations of formally real finite-dimensional Jordan algebras. In fact, J.D.M. Wright [W] proved that every JB-algebras is the self-adjoint part of a unique JB^*-algebra.

Consequently the representation theory of JB^*-algebras is quite satisfactory, due to the following result.

Theorem 8.2.5 (E.M. Alfsen–F.W. Shultz–E. Størmer [ASS]) *Let A be a JB^*-algebra. Then A contains a closed Jordan ideal J such that A/J has a faithful $*$-representation as a special subalgebra of $\mathcal{B}(H)$, where H is some Hilbert space, while every irreducible representation of A not annihilating J is onto $H(C_3)$, the 27-dimensional exceptional algebra over the octonions.*

If A is a Banach algebra with involution, the Shirali–Ford theorem states that every self-adjoint element of A has real spectum if and only if x^*x has positive spectrum for every x in A (see [BD], Theorem 4.1.5 or [Au1], Theorem 4.2.3). This theorem implies that such algebras are spectrally similar to C^*-algebras and have many Hilbert space representations. The corresponding problem for Jordan-Banach algebras is important because in this case we can use representation theory for JB^*-algebras given by Theorem 8.2.5.

Theorem 8.2.6 was proved for the first time by H. Behncke [Be], using a nice but long argument. In [AY] we gave a new proof of this theorem and also of Corollary 8.2.7 using subharmonic functions. Using the localization principle (Theorem 2.1.9) it is even possible to give a very easy proof.

Theorem 8.2.6 (H. Behncke [Be], B. Aupetit–M.A. Youngson [AY]) *Let A be a Jordan-Banach algebra with involution $*$. The following properties are equivalent:*

(i) *if $h = h^*$ then $Sp\, h \subset \mathbb{R}$;*

(ii) *if $h = h^*$ and $k = k^*$ have positive spectra then so has $h + k$;*

(iii) *x^*x has positive spectrum for x in A.*

Corollary 8.2.7 [AY] *A is a JB^*-algebra for an equivalent norm if and only if there exists a constant $C \geq 1$ such that $|\exp(ih)| \leq C$ for all h satisfying $h = h^*$.*

JB^*-algebras are an important tool which has been systematically studied (see [HS] for more details).

Theorem 8.2.4 is also important for philosophical reasons. By Theorem 2.5.1, it is known that every analytic multifunction can be locally represented as $Sp\, T(\lambda)$, where T is an analytic function with values in the algebra of bounded operators on the Hilbert space ℓ^2. So, in some sense, it reduces a non-associative problem to an associative one. This is the reason why we still believe in the following creed which, until now, has never been contradicted : any Banach algebra result which is proved by a purely spectral argument must be true and provable in a similar way, modulo some technical modifications, in the situation of Jordan-Banach algebras.

The first striking applications will be given in the next section. We proved the first one in 1982 in order to extend B. Johnson's theorem on the equivalence of the complete norms.

8.3 The radical

The notion of Jacobson radical for associative algebras has been generalized by K. McCrimmon to Jordan algebras (see [Mc1,2]). In a Jordan algebra we say that an ideal I is *quasi-invertible* if for every $x \in I$ we have $1 - x$ invertible. McCrimmon proved that in any Jordan algebra there exists a unique maximum quasi-invertible ideal. By definition, this ideal is the *Jacobson radical* of A, and it is denoted by Rad A. If A is a special Jordan algebra he proved in that case that the radical coincides with the associative Jacobson radical. From this definition it is easy to see that the radical of $A/\mathrm{Rad}\, A$ is zero, in other words, we say that this quotient algebra is *semisimple* (some authors say semiprimitive).

We mentioned in section 3.1, that in 1967, using representation theory, B.E.Johnson proved that for a semisimple Banach algebra all the complete algebra norms are equivalent. Johnson's argument depends heavily on representation theory and on the fact that for Banach algebras the Jacobson radical is the intersection of the kernels of irreducible representations, that is, primitive ideals. Unfortunately in the case of Jordan-Banach algebras the analogous result is not true (as noticed by J.M. Osborn [O]) so Johnson's argument cannot be adapted (except in some rather easy situations like H^*-algebras).

At first sight the extension of Johnson's result to Jordan-Banach algebras seemed to be impossible to get at the end of the 1970s, as mentioned by J. Martínez Moreno in the introduction to his thesis [M].

Finally in [Au4] we proved a very beautiful extension of Johnson's result for Banach algebras (see Theorem 3.1.6) and in the case of Jordan-Banach algebras the following.

Theorem 8.3.1 *A semisimple Jordan-Banach algebra has a unique complete norm topology and every Jordan morphism from a Jordan-Banach algebra onto a semisimple Jordan-Banach algebra is continuous.*

Corollary 8.3.2 *Every involution on a semisimple Jordan-Banach algebra is continuous.*

These two results simplify considerably the results and proofs of [PY]. The argument of the proof is identical to the one given after Theorem 3.1.6.

At that time we were not able to give Johnson's extension in full generality for non-associative algebras, because we had no equivalent of Zemánek's characterization of the radical (see Theorem 3.1.5) which was, until 1992, based on representation theory. But now there is a spectral characterization of the radical for Jordan-Banach algebras (see Theorem 8.3.4 below). Consequently we have:

Theorem 8.3.3 *Let A and B be two Jordan-Banach algebras. Suppose that T is a linear mapping from A into B which is spectrally contractive, that is, $r(Tx) \leq r(x)$ for every x in A. Denoting by $\mathcal{S}(T)$ the separating space of T, then $a \in \mathcal{S}(T)$ implies $r(Tx) \leq r(a+x)$, for all x in A. In particular, if T is onto we have $\mathcal{S}(T) \subset \operatorname{Rad} B$, so T is continuous if B is semisimple.*

In [Au8] we gave a purely spectral proof of Zemánek's characterization of the radical in the situation of Banach algebras. Using Theorem 8.2.4 and sophisticated arguments on analytic multifunctions we succeeded in obtaining the following.

Theorem 8.3.4 (Spectral characterization of the radical) *Let a be an element of a Jordan-Banach algebra A. Then a is in the radical of A if and only if $\sup\{r(x+ta) : t \in \mathbb{C}\}$ is finite for every x in A.*

Corollary 8.3.5 *Let a be an element of a Jordan-Banach algebra A. Then a is in the radical of A if and only if $r(U_x(a)) = 0$ for every x invertible in A.*

Corollary 8.3.6 *Let a be an element of a Jordan-Banach algebra A. Then a is in the radical of A if and only if there exists $C \geq 0$ such that $r(x) \leq C|x-a|$ on a neighbourhood of a.*

If A is a semisimple Jordan algebra N. Jacobson proved that $U_a = 0$ for some $a \in A$ implies $a = 0$. In the case of semisimple Jordan-Banach algebras this result comes immediately from Corollary 8.3.5 because $(U_x(a))^2 = U_{U_x(a)}(1) = U_x U_a U_x(1) = U_x U_a(x^2) = 0$, so $r(U_x(a)) = 0$ for every x in A, consequently a is in the radical of A, hence it is zero.

From Theorem 8.3.4 it is possible to deduce the following nice result wich improves an earlier one given in [AZ].

Theorem 8.3.7 *Let A be a Jordan-Banach algebra. The following conditions are equivalent:*

(i) $A/\mathrm{Rad}\,A$ *is associative, consequently* $A/\mathrm{Rad}\,A$ *is a commutative Banach algebra;*

(ii) *the spectral radius is subadditive, that is, there exists* $C > 0$ *such the* $r(x+y) \leq C(r(x) + r(y))$, *for every* x, y *in* A;

(iii) *the spectral radius is uniformly continuous that is there exists* $M > 0$ *such that* $|r(x) - r(y)| \leq M\|x - y\|$, *for every* x, y *in* A.

Looking carefully at the subharmonic arguments contained in [AZ] we see that the main difficulty is to prove in both cases that the radical coincides with the set of quasi-nilpotent elements. Suppose we are in situation (ii) (the third situation is treated similarly). Let a be a quasi-nilpotent element. Then we have $r(x + ta) \leq Cr(x) < +\infty$ for every t in $\mathbb{C}$ and x in A, so by Theorem 8.3.4, a is in the radical. The converse is obvious.

For a semisimple Banach algebra it is known that every derivation is bounded. This beautiful result, due to A. Sinclair and B.E. Johnson, is proved using representation theory. It would be very interesting to give a purely spectral proof of that result, probably using analytic arguments as we did in the proof of Theorem 3.1.6, because it would give some hope to solve the same problem for semisimple Jordan-Banach algebras.

8.4 Spectrally finite Jordan-Banach algebras

In section 3.2, we mentioned a result of I. Kaplansky which suggested the proof of the local spectral characterization of finite-dimensional Banach algebras. Consequently, is it possible to prove the same result in the situation of Jordan-Banach algebras? The same result cannot be true in general because if we take an infinite-dimensional Banach space V with Jordan product defined on $\mathbb{C} \times V$ using a symmetric bilinear form, as explained in the introduction, then $A = \mathbb{C} \times V$ is a Jordan-Banach algebra for which every element has a spectrum with at most two points. The first attempt in this direction is the following result.

Theorem 8.4.1 [AZ] *Let U be a non-empty open subset of a Jordan-Banach algebra A. Suppose that for every x in U the spectrum of x has one point. Then* $A/\mathrm{Rad}\,A = \mathbb{C}$.

We suggested to A. Kaïdi the study of this question and finally it was solved by his student M. Benslimane. The final result is the following.

Theorem 8.4.2 (M. Benslimane–A. Kaïdi [BeFK, BeK]) *Let U be a non-empty open subset of a semisimple Jordan-Banach algebra A. Suppose that for every x in U the spectrum of x is finite. Then we have a finite decomposition $A = I_1 \oplus \cdots \oplus I_n$, where the I_k are simple closed ideals of A having one of the following two properties:*

(i) *I_k is finite-dimensional;*

(ii) *I_k is infinite-dimensional and the Jordan product on $\mathbb{C}1 \oplus I_k$ is defined by a symmetric bilinear form.*

Obviously the proof depends heavily on Theorem 8.2.4 and Theorem 2.1.12. With this last result is it possible to extend Kaplansky's theorem for ring morphisms of Jordan-Banach algebras?

Of course the analogue of Corollary 3.2.3 can be obtained in a similar way.

8.5 The socle

As we saw in section 3.3, if A is a complex Banach algebra, the socle is by definition the sum of the minimal left (resp. right) ideals of A. For Jordan-Banach algebras the definition of the socle is slightly more difficult. To simplify we shall suppose that the Jordan-Banach algebra A is semisimple.

We say that an idempotent p is *minimal* if $U_p(A) = \mathbb{C}p$, in which case $U_p(A)$ is a minimal quadratic ideal. But there is another class of minimal quadratic ideals Q, those which have the properties $Q = U_q(A)$ for every $q \in Q$, and $Q^2 = 0$ (this derives from a more general result of N. Jacobson showing that in any Jordan algebra there are only three classes of minimal quadratic ideals).

By definition the *socle* of A, denoted by Soc A, is the sum of all minimal quadratic ideals of A.

Theorem 8.5.1 (J.M. Osborne–M.L. Racine [OR]) *The socle of A is an ideal and it is the sum of all ideals generated by the minimal projections.*

It is not clear at all for which algebras the socle is non-zero. Extending the arguments involved in the proof of Theorem 3.3.10, we obtained the following result which improves Barnes's theorem.

Theorem 8.5.2 (B. Aupetit–L. Baribeau [AB]) *Let A be a semisimple Jordan-Banach algebra. Suppose that every element of A has a finite or countable spectrum. Then A has non-zero socle.*

It is even enough to suppose that the spectrum is finite or countable on an absorbing subset of A.

Extending the corresponding associative concept which has mainly been studied by B.A. Barnes, we can define a *modular annihilator Jordan-Banach algebra A* as a semisimple Jordan-Banach algebra such that $A/\text{Soc}\, A$ is radical. M. Benslimane and A. Rodríguez Palacios, using the ideas contained in [Au6] or in [Au7], Chapter V, §7, extended previous results of B.A. Barnes. A. Fernández López obtained similar results but his arguments are very technical.

Theorem 8.5.3 (M. Benslimane–A. Rodríguez Palacios [BeR], A. Fernández López [F1]) *Let A be a semisimple Jordan-Banach algebra such that the spectrum of every element has at most zero as a limit point. Then A is modular annihilator.*

From Theorems 8.4.2 and 8.5.3 we get the next structure theorem, similar to Theorem 3.4.3.

Theorem 8.5.4 (B. Aupetit–L. Baribeau [AB]) *Let A be a separable Jordan-Banach algebra. Suppose that every element of A has a finite or countable spectrum. Then there exists an ordinal number $\alpha_0 \in \Omega$ and a composition sequence $(I_\alpha)_{\alpha \leq \alpha_0}$ of closed ideals such that $I_0 = \text{Rad}\, A$, $I_{\alpha_0+1} = A$, and $I_{\alpha+1}/I_\alpha$ is modular annihilator for $\alpha \leq \alpha_0$.*

As in the associative case the I_α are in fact the transfinite hypersocles obtained from the successive α-Calkin algebras. Here it is not true that codim $I_{\alpha_0} < +\infty$.

A. Fernández López and A. Rodríguez Palacios [FR1] proved for a semisimple Jordan-Banach algebra A that the socle coincides with the largest von Neumann regular ideal. Given a semisimple Jordan-Banach algebra, A. Fernández López [F2] proved that its socle is an algebraic ideal, and conversely, if I is any algebraic ideal, then every element of I can be written as the sum of an element of the socle and a nilpotent element. This result was improved by M. Benslimane, O. Jaa and A. Kaïdi in [BeJK] where they proved that every element of a spectrally finite ideal can be written as the sum of an element of the socle and an element whose square is zero. Using recent results of O. Loos [L] it is even possible to prove much more.

Theorem 8.5.5 (M. Benslimane–O. Jaa–A. Kaïdi–O. Loos) *Let A be a semisimple Jordan-Banach algebra and let I be a spectrum finite ideal of A. Then I is included in the socle of A.*

For more comments of this question see [Ro] p. 40 and pp. 100–101 and [Lo].

As seen in section 3.5, B. Aupetit and H. du T. Mouton gave a purely spectral characterization of the socle in the case of Banach algebras which has very interesting consequences. Using sophisticated results on analytic multifunctions it is possible to prove the following five results.

Theorem 8.5.6 *Let A be a semisimple Jordan-Banach algebra, let $a \in A$ and let n be a* *non-negative integer. The following conditions are equivalent:*

(i) $Sp\,U_x(a)\backslash\{0\}$ *has at most* n *points for every* x *in* A;

(ii) $\bigcap_{t\in F} Sp(y+ta) \subset Sp(y)$, *for every* y *in* A *and* F *a finite subset of* $\mathbb{C}$ *having at most* $n+1$ *non-zero points.*

We denote by $\mathcal{F}_n$ the set of $a \in A$ satisfying one of the two previous properties. By Corollary 8.3.5, $\mathcal{F}_0$ is the radical of A, hence $\{0\}$. We shall say that $\mathcal{F}_n$ is the set of elements of A with rank less than or equal to n. The main result (Theorem 8.5.10) says that $\bigcup_{n\geq 0}\mathcal{F}_n$ coincides with the socle of A.

Theorem 8.5.7 *If* $a \in \mathcal{F}_n$, *then* $U_x(a) \in \mathcal{F}_n$ *for every* $x \in A$.

The proof of Theorem 8.5.9 needs a beautiful geometrical characterization of algebraic varieties of $\mathbb{C}^2$ we recently proved and which apparently was unknown until today [Au10].

Lemma 8.5.8 *Let* $V \neq \mathbb{C}^2$ *be a closed subset of* $\mathbb{C}^2$ *above* $\mathbb{C}$ *which has the following properties:*

(a) $\mathbb{C}^2\backslash V$ *is a pseudoconvex open set;*

(b) *for every* $\lambda \in \mathbb{C}$ *we have* $\#((\{\lambda\}\times\mathbb{C})\cap V) \leq m$ *or* $\{\lambda\}\times\mathbb{C}\subset V$;

(c) *for every* $\mu \in \mathbb{C}$ *we have* $\#((\mathbb{C}\times\{\mu\})\cap V) \leq n$ *or* $\mathbb{C}\times\{\mu\}\subset V$.

Then V *is a complex algebraic subvariety of* $\mathbb{C}^2$ *of degree at most* $m+n$.

From that we get the following

Theorem 8.5.9 *If* $a \in \mathcal{F}_m, b \in \mathcal{F}_n$, *then* $a+b \in \mathcal{F}_{m+n}$.

Using Theorems 8.5.5, 8.5.7 and 8.5.9 we immediately get the next result.

Theorem 8.5.10 $\bigcup_{n\geq 0} \mathcal{F}_n$ *coincides with the socle of* A.

All these theorems, which will appear in [Au9,10], imply the existence of an additive trace on the socle whose restriction to $\mathcal{F}_n$ is continuous. They also imply that every element of the socle of A is a finite sum of elements of $\mathcal{F}_1$. Another consequence is that a linear mapping from a semisimple Jordan-Banach algebra onto itself which preverves the spectrum leaves the $\mathcal{F}_n$ invariant and consequently also the socle.

To finish this paper we would say that there remains the fascinating task to try to understand more clearly the intimate correlation between Jordan-Banach algebras, homogeneous symmetric spaces and analytic multifunctions.

The reader intending to know more about the recent developments of the algebraic theory of Jordan algebras has to read the recent big survey by A. Rodríguez Palacios [Ro].

Chapter 9
Applications to complex dynamics

9.1 Introduction

Very recently the theory of analytic multifunctions – more precisely meromorphic multifunctions – has given interesting applications for Julia sets corresponding to the iteration of rational maps in the complex plane.

For a good introduction to the theory of iteration of rational functions see [B]. Let D be a domain in $\mathbb{C}$ and let $\{R_\lambda\}$ be a family of rational functions from $D \times \mathbb{C}_\infty$ into $\mathbb{C}_\infty$ (where $\mathbb{C}_\infty$ denotes the Riemann sphere) depending analytically on the parameter $\lambda \in D$. Suppose moreover that the degree of R_λ is greater than or equal to 2. If $J(\lambda)$ is the Julia set corresponding to R_λ the question is to know how behaves the multifunction J. It is well known from several examples that J is not upper semicontinuous in general. From famous results of R. Mañé, P. Sad and D. Sullivan ([MSS], Theorems A and B) there is a dense open subset D' of D on which $J(\lambda)$ is a holomorphic motion, which means that the graph of J restricted to D' is the disjoint union of graphs of meromorphic functions on D'. Is it possible to improve this result?

9.2 The results

The following results were obtained by L. Baribeau and T.J. Ransford [BR]. They complement former results of H. Kriete [Kr].

As in section 7.2 we shall denote the by $J^*(\lambda)$ the upper semicontinuous regularization of $J(\lambda)$, consequently J^* is upper semicontinuous on D and $J(\lambda) = J^*(\lambda)$ on $D' \subset D$ as defined previously.

We slightly extend the class of analytic multifunctions considering a *meromorphic multifunction* on D as being an upper semicontinuous multifunction with compact values in $\mathbb{C}_\infty$ such that if the set of values at $\lambda_0 \in D$ is not all the Riemann sphere, there exists a neighourhood of λ_0 on which the composition of the previous multifunction with some Möbius map is analytic multivalued.

Theorem 9.2.1 *Let D be a domain in $\mathbb{C}$ and let $\{R_\lambda\}$ be a family of rational functions depending analytically on the parameter $\lambda \in D$ such that* deg $R_\lambda \geq 2$. *Then*

(i) $\lambda \mapsto J^*(\lambda)$ *is a meromorphic multifunction on D;*

(ii) *there exists a polar subset E of D such that $\partial J^*(\lambda) \subset J(\lambda)$ for $\lambda \in D \backslash E$.*

Corollary 9.2.2 *If R_λ is a polynomial for every λ in D, then $\lambda \mapsto \hat{J}(\lambda)$ is an analytic multifunction, where $\hat{J}(\lambda)$ denotes the polynomially convex hull of $J(\lambda)$.*

Property (ii) says that for $\lambda \in D\backslash E$, $J^*(\lambda)$ is the union of $J(\lambda)$ with a possibly empty collection of its holes, that is to say, components of the Fatou set of R_λ. Fix $\lambda_0 \in D$. Because

$J^*(\lambda_0)$ is invariant under R_{λ_0}, we can suppose that every component F of the Fatou set is forward-invariant, that is $R_{\lambda_0}(F) = F$. There are only five categories of forward-invariant domains: attracting, super-attracting, parabolic domains, Siegel disks and Herman rings.

For the first two cases we have :

Theorem 9.2.3 *With the previous hypotheses, suppose that* $R_{\lambda_0}(z_0) = z_0$ *and* $|R'_{\lambda_0}(z_0)| < 1$ *and set* $A = \{z \in \mathbb{C}_\infty : R^n_{\lambda_0}(z) \to z_0 \text{ when } n \to \infty\}$. *Then* $J^*(\lambda_0)$ *and* A *are disjoint.*

The three last cases are very 'unstable' (see the examples given in [BR]). Nevertheless some partial results can be given.

Theorem 9.2.4 *With the previous hypotheses, suppose that* $R_\lambda(z_0) = z_0$ *and* $R'_\lambda(z_0) \in T$ *for each* $\lambda \in D$, *where* T *denotes the set of* $e^{2\pi i\alpha}$ *such that* α *is a real number with the property that* $\left|\alpha - \frac{p}{q}\right| > c/q^2$ $(p, q \in \mathbb{Z}, q \neq 0)$ *for some* $c > 0$. *Let* S_λ *be the Siegel disk of* R_λ *containing* z_0. *Then there exists a polar subset* E *of* D *such that for* $\lambda \in D \backslash E$ *the two sets* $J^*(\lambda)$ *and* S_λ *are disjoint.*

Theorem 9.2.5 *With the previous hypotheses, suppose that* $R_\lambda(z_0) = z_0$ *and* $R'_\lambda(z_0) = 1$ *for each* $\lambda \in D$. *Suppose moreover that the order of the fixed point* z_0 *is independent of* λ. *Setting* $P_\lambda = \{z \in \mathbb{C}_\infty : R^n_\lambda(z) \longrightarrow z_0 \text{ when } n \to \infty\}$, *then* $J^*(\lambda)$ *and* P_λ *are disjoint.*

Theorem 9.2.6 *With the previous hypotheses we have the following:*

(i) *if* S *is a Siegel disk of* R_{λ_0} *and* $S \not\subset J^*(\lambda_0)$, *then* $S \backslash J^*(\lambda_0)$ *is homeomorphic to a disk;*

(ii) *if* H *is a Herman ring of* R_{λ_0} *and* $H \not\subset J^*(\lambda_0)$, *then* $H \backslash J^*(\lambda_0)$ *is homeomorphic to an annulus.*

Setting

$$C_n(\lambda) = \{R^n_\lambda(z) : z \text{ a critical point of } R_\lambda\} \quad \text{and} \quad C(\lambda) = \overline{\bigcup_{n \geq 0} C_n(\lambda)},$$

then for $C(\lambda)$ the analogue of Theorem 9.2.1 is true. Using this result and Theorem 2.2.8 it is possible to prove that for $D = \mathbb{C}$, $R_\lambda(z) = \lambda + z^2$, we have $C^*(\lambda)$ countable for $|\lambda| = 1$ and $C^*(\lambda) = \mathbb{C}_\infty$ for $|\lambda| = 1$.

Considering the iteration of entire functions, the analogue of Theorem 9.2.1 is true for $\bar{J}(\lambda)$, the closure in $\mathbb{C}_\infty$ of $J(\lambda)$, which is a subset of $\mathbb{C}$.

For more details and also for the very interesting examples it contains, we advise the reader to have a look at [BR].

References

[A1] Albert, A.A., On Jordan algebras of linear transformations, *Trans. Amer. Math. Soc.* **59** (1946), 524–555.

[A2] Albert, A.A., A structure theory for Jordan algebras, *Ann. of Math.* **48** (1947), 446–467.

[AlW1] Alexander, H., Wermer, J., On the approximation of singularity sets by analytic varieties, *Pacific J. Math.* **104** (1983), 263–268.

[AlW2] Alexander, H., Wermer, J., Polynomial hulls with convex fibers, *Math. Ann.* **271** (1985), 99–109.

[AlW3] Alexander, H., Wermer, J., On the approximation of singularity sets by analytic varieties II, *Michigan Math. J.* **32** (1985), 227–235.

[AlW4] Alexander, H., Wermer, J., Envelopes of holomorphy and polynomial hulls, *Math. Ann.* **281** (1988), 13–22.

[ASS] Alfsen, E.M., Shultz, F.W., Størmer, E., A Gelfand-Neumark theorem for Jordan algebras, *Adv. Math.* **28** (1978), 11–56.

[Au1] Aupetit, B., Caractérisation spectrale des algèbres de dimension finie, *J. Funct. Anal.* **26** (1977), 232–250.

[Au2] Aupetit, B., *Propriétés spectrales des algèbres de Banach,* Lecture Notes in Math. **735**, Springer-Verlag, Berlin – Heidelberg – New York, 1979.

[Au3] Aupetit, B., Some applications of analytic multifunctions to Banach algebras, *Proc. Roy. Irish Acad. Sect. A* **81** (1981), 37–42.

[Au4] Aupetit, B., The uniqueness of the complete norm topology in Banach algebras and Banach-Jordan algebras, *J. Funct. Anal.* **47** (1982), 1–6.

[Au5] Aupetit, B., Analytic multivalued functions in Banach algebras and uniform algebras, *Adv. Math.* **44** (1982), 18–60.

[Au6] Aupetit, B., Inessential elements in Banach algebras, *Bull. London Math. Soc.* **18** (1986), 493–497.

[Au7] Aupetit, B., *A Primer on Spectral Theory,* Universitext, Springer-Verlag, New-York, 1991.

[Au8] Aupetit, B., Spectral characterization of the radical in Banach and Jordan-Banach algebras, *Math. Proc. Cambridge Philos. Soc.* **114** (1993), 31–35.

[Au9] Aupetit, B., Spectral characterization of the socle in Jordan-Banach algebras, to appear.

[Au10] Aupetit, B., A geometric characterization of algebraic varieties, to appear.

[AB] Aupetit, B., Baribeau, L., Sur le socle dans les algèbres de Jordan-Banach, *Canad. Math. J.* **41** (1989), 1090–1100.

[AD] Aupetit, B., Drissi, D., Local spectrum theory revisited, to appear.

[AM] Aupetit, B., Mouton, H. du T., Spectrum-preserving linear mappings in Banach algebras, to appear in *Studia Math.*

[AW] Aupetit, B., Wermer, J., Capacity and uniform algebras, *J. Funct. Anal.* **28** (1978), 386–400.

[AY] Aupetit, B, Youngson, M.A., On symmetry of Banach-Jordan algebras, *Proc. Amer. Math. Soc.* **91** (1984), 364–366.

[AZ] Aupetit, B., Zemánek, J., On zeros of analytic multivalued functions, *Acta Sci. Math. (Szeged)* **46** (1983), 311–316.

[AZr] Aupetit, A., Zraïbi, A., Propriétés analytiques du spectre dans les algèbres de Jordan-Banach, *Manuscripta Math.* **38** (1982), 381–386.

[Ba1] Baribeau, L., Sur les fonctions analytiques multiformes dont les valeurs sont des segments, *Canad. Math. Bull.* **33** (1989), 100–105.

[Ba2] Baribeau, L., Multifonctions analytiques polygonales, *Studia Math.* **96** (1990), 167–173.

[BaH] Baribeau, L., Harbottle, S., Two new open mapping theorems for analytic multivalued functions, *Proc. Amer. Math. Soc.* **115** (1992), 1009–1012.

[BaR] Baribeau, L., Ransford, T.J., Meromorphic multifunctions in complex dynamics, *Ergodic Theory Dynamical Systems* **12** (1992), 39–52.

[Bar1] Barnes, B.A., On the existence of minimal ideals in a Banach algebra, *Trans. Amer. Math. Soc.* **133** (1968), 511–517.

[Bar2] Barnes, B.A., A generalized Fredholm theory for certain maps in the regular representations of an algebra, *Canad. Math. J.* **20** (1968), 495–504.

[Bar3] Barnes, B.A., When it the spectrum of a convolution operator on L^p independent of p? *Proc. Edinburgh Math. Soc.* **33** (1990), 327–332.

[B] Beardon, A.F., *Iteration of Rational Functions*, Springer-Verlag, New York, 1991.

[Be] Behncke, H., Hermitian Jordan-Banach algebras, *J. London Math. Soc. (2)* **20** (1979), 327–333.

[Beh] Behrends, E., Points of symmetry of convex sets in the two-dimensional space, a counterexample to D. Yost's problem, *Math. Ann.* **290** (1991), 463–471.

[BeFK] Benslimane, M., Fernández López, A., Kaïdi, A., Caractérisation des algèbres de Jordan-Banach de capacité finie, *Bull. Sci. Math. (2)* **112** (1988), 473–480.

[BeJK] Benslimane, M., Jaa, O., Kaïdi, A., The socle and the largest spectrum finite ideal, *Quart. J. Math. Oxford Ser. (2)* **42** (1991), 1–7.

[BeK] Benslimane, M., Kaïdi, A., Structure des algèbres de Jordan-Banach non commutatives complexes régulières ou semisimples à spectre fini, *J. Algebra* **113** (1988), 201–206.

[BeR] Benslimane, M., Rodríguez Palacios, A., Caractérisation spectrale des algèbres de Jordan-Banach non commutatives complexes modulaires annihilatrices, *J. Algebra* **140** (1991), 344–354.

[BR] Berndtsson, B., Ransford, R.J., Analytic multifunctions, the $\overline{\partial}$-equation, and a proof of the corona theorem, *Pacific J. Math.* **124** (1986), 57–72.

[Bi] Bishop, E., Holomorphic completions, analytic continuation, and the interpolation of semi-norms, *Ann. of Math.* **78** (1963), 468–500.

[Bj1] Björk, J.-E., Analytic structures in the maximal ideal space of a uniform algebra, *Ark. Mat.* **8** (1971), 239–244.

[Bj2] Björk, J.-E., Holomorphic convexity and analytic structures in Banach algebras, *Ark. Mat.* **9** (1971), 39–54.

[BD] Bonsall, F.F., Duncan, J., *Complete Normed Algebras*, Ergebnisse der Mathematik und ihrer Grenzgebiete **80**, Springer-Verlag, Berlin – Heidelberg – New York, 1973.

[Br] Brelot, M., *Eléments de la théorie classique du potentiel*, Centre de documentation universitaire, Paris, 1965.

[Ca] Calderón, A.P., Intermediate spaces and interpolation, the complex method, *Studia Math.* **24** (1964), 113–190.

[CF] Colojoară, I., Foiaş, C., *Theory of Generalized Spectral Operators*, Gordon & Breach, New York, 1968.

[Da] Daneš, J., On local spectral radius, *Časopis Pěst. Mat.* **112** (1987), 177–187.

[D1] Dieudonné, J., *Calcul infinitésimal*, Hermann, Paris, 1968.

[D2] Dieudonné, J., (Ed.). *Abrégé d'histoire des mathématiques 1700–1900*, Vol.1 et 2, Hermann, Paris, 1978.

[D3] Dieudonné, J., *History of Functional Analysis*, North-Holland, Amsterdam, 1981.

[Du] Dunford, N., A survey of the theory of spectral operators, *Bull. Amer. Math. Soc.* **64** (1958), 217–274.

[EL] Erdelyi, I., Lange, R., *Spectral Decompositions on Banach Spaces*, Lecture Notes in Math. **623**, Springer-Verlag, Berlin – Heidelberg – New York, 1977.

[F1] Fernández López, A., Modular annihilator Jordan algebras, *Comm. Algebra* **13** (1985), 2597–2613.

[F2] Fernández López, A., Noncommutative Jordan Riesz algebras, *Quart. J. Math. Oxford Ser. (2)* **39** (1988), 67–80.

[FR1] Fernández López, A., Rodríguez Palacios, A., On the socle of a noncommutative Jordan algebra, *Manuscripta Math.* **56** (1986), 269–278.

[FR2] Fernández López, A., Rodríguez Palacios, A., Primitive noncommutative Jordan algebras with nonzero socle, *Proc. Amer. Math. Soc.* **96** (2) (1986), 199–206.

[Fo] Foiaş, C., Une application des distributions vectorielles à la théorie spectrale, *Bull. Sci. Math.* **84** (1960),147–158.

[FoV] Foiaş, C., Vasilescu, F.-H., On the spectral theory of commutators, *J. Math. Anal. Appl.* **31** (1970), 473–486.

[FS] Fong, C.-K., Sołtysiak, A., Existence of multiplicative functional and joint-spectra, *Studia Math.* **81** (1985), 213–220.

[For] Forstnerič, F., Polynomial hulls of sets fibered over the circle, *Indiana Univ. Math. J.* **37** (1988), 869–889.

[G] Gamelin, T.W., Polynomial approximation on thin sets, in: *Symposium on Several Complex Variables, Park City, Utah, 1970* (R.M. Brooks, ed.), Lecture Notes in Math. **184**, Springer-Verlag, Berlin - Heidelberg - New York, 1971; 50–78.

[GRC] Gelfand, I.M., Raĭkov, D.A., Chilov, G.E., *Les anneaux normés commutatifs*, Gauthier-Villars, Paris, 1964.

[GoK] Gohberg, I.C., Krejn, M.G., *Introduction à la théorie des opérateurs linéaires non auto-adjoints dans un espace hilbertien*, Dunod, Paris, 1971.

[H] Halmos, P.R., *A Hilbert Space Problem Book*, D. Van Nostrand, Princeton, 1967.

[HS] Hanche-Olsen, H., Størmer, E., *Jordan Operator Algebras*, Pitman, New York, 1984.

[Ha] Harte, R.E., Spectral mapping theorems, *Proc. Roy. Irish Acad. Sect. A* **72** (1972), 89–107.

[Har] Hartogs, F., Zur Theorie der analytischen Funktionen mehrerer unabhängiger Veränderlicher, insbesondere über die Darstellung derselben durch Reihen, welche nach Potenzen einer Veränderlichen fortschreiten, *Math. Ann.* **63** (1906), 1–88.

[Haw] Hawkins, T., Cauchy and the spectral theory of matrices, *Historia Math.* **2** (1975), 1–29.

[HK] Hayman, W.K., Kennedy, P.B., *Subharmonic Functions*, Vol. 1, Academic Press, New York - London, 1976.

[He] Helms, L. L., *Introduction to Potential Theory*, Robert E. Krieger, New York, 1975.

[HM] Helton, J.W., Marshall, D.E., Frequency domain design and analytic selections, *Indiana Univ. Math. J.* **39** (1990), 157–184.

[Her] Herstein, I.N., *Topics in Ring Theory*, Univ. of Chicago Press, Chicago, 1969.

[Hö] Hörmander, L., *An Introduction to Complex Analysis in Several Variables*, North-Holland, Amsterdam, 1973.

[Hu] Huruya, T., A spectral characterization of a class of C^*-algebras, *Sci. Rep. Niigata Univ. Ser. A* **15** (1978), 21–24.

[I] Istrăţescu, I., *Introduction to Linear Operator Theory*, Marcel Dekker, New York, 1981.

[J1] Jacobson, N., *Structure and Representations of Jordan Algebras,* Amer. Math. Soc. Colloq. Publ. **39**, Providence, RI, 1968.

[J2] Jacobson, N., *Lectures on Quadratic Jordan Algebras*, Tata Institute of Fundamental Research, Bombay, 1969.

[J3] Jacobson, N., *Structure Theory of Jordan Algebras*, Lecture Notes **5**, University of Arkansas, Fayettesville, 1981.

[JS] Jafarian, A.A., Sourour, A.R., Spectrum-preserving linear maps, *J. Funct. Anal.* **66** (1986), 255-261.

[JNW] Jordan, P., von Neumann, J., Wigner, E., On an algebraic generalization of the quantum mechanical formalism, *Ann. of Math.* **35** (1934), 29–64.

[K] Kaplansky, I., *Infinite Abelian Groups*, University of Michigan Press, 1969.

[Ka] Kato, T., *Perturbation Theory for Linear Operators*, Springer-Verlag, Heidelberg, 1966.

[Kat] Katznelson, Y., *An Introduction to Harmonic Analysis*, John Wiley, New York, 1968.

[Ki] Kirchberg, E., Banach algebras whose elements have at most countable spectra, I and II, submitted to *Studia Math.* 1979 or 1980, it never appeared.

[Kl] Klimek, M., Joint spectra and analytic set-valued functions, *Trans. Amer. Math. Soc.*, **294** (1986), 187–196.

[Kr] Kriete, H., The stability of Julia sets, *Math. Göttingensis* **22** (1988), 1–16.

[Ku1] Kuratowski, K., Les fonctions semi-continues dans l'espace des ensembles fermés, *Fund. Math.* **18** (1932), 148–159.

[Ku2] Kuratowski, K., Operations on semi-continuous set-valued mappings, in: *Seminari 1962–1963, Ist. Naz. Alta Mat. Roma*, Vol. II, Ediz. Cremonese, Roma, 1965; 449–461.

[L] Loos, O., Properly algebraic and spectrum-finite ideals in Jordan systems, *Math. Proc. Cambridge Philos. Soc.* **114** (1993), 149–161.

[Lü] Lützen, J., *Joseph Liouville (1809–1882). Master of Pure and Applied Mathematics,* Springer-Verlag, New York, 1990.

[MSS] Mañé, R., Sad, P., Sullivan, D., On the dynamics of rational maps, *Ann. École Norm. Sup. (4)* **16** (1983), 193–217.

[M] Martínez Moreno, J., *Sobre álgebras de Jordan normadas completas,* Tesis doctoral, Universidad de Granada, Granada, 1977.

[Mc1] McCrimmon, K., The radical of a Jordan algebra, *Proc. Nat. Acad. Sci. U.S.A.* **62** (1969), 671–678.

[Mc2] McCrimmon, K., A characterization of the radical of a Jordan algebra, *J. Algebra* **18** (1971), 103–111.

[Mc3] McCrimmon, K., Jordan algebras and their applications, *Bull. Amer. Math. Soc.* **84** (1978), 612–627.

[Mo] Monna, A.F., *Functional Analysis in Historical Perspective*, Oosthoek Publishing Co., Utrecht, 1973.

[Na] Narasimhan, R., *Several Complex Variables,* Univ. of Chicago Press, Chicago, 1971.

[Ne] Neuenschwander, E., Studies in the history of complex function theory II: Interactions among the French school, Riemann, and Weierstrass, *Bull. Amer. Math. Soc.* **5** (1981), 87–105.

[Ni] Nishino, T., Sur les ensembles pseudoconvexes, *J. Math. Kyoto Univ.* **1-2** (1962), 225–245.

[O] Oka, K., Note sur les familles de fonctions analytiques multiformes etc., *J. Sci. Hiroshima Univ.* 4 (1934), 93–98.

[Os] Osborn, J.M., Representations and radicals of Jordan algebras, *Scripta Math.* **29** (1973), 297–329.

[OsR] Osborn, J.M., Racine, M.L., Jordan rings with nonzero socle, *Trans. Amer. Math. Soc.* **251** (1979), 375–387.

[PS] Pełczyński, A., Semadeni, Z., Spaces of continuous fonctions III. Spaces $C(\Omega)$ for Ω without perfect subsets, *Studia Math.* **18** (1959), 211–222.

[Pu] Puiseux, V., Recherches sur les fonctions algébriques, *Journal de Mathématiques* **15** (1850), 365–480.

[PY] Putter, P.S., Yood, B., Banach-Jordan $*$-algebras, *Proc. London Math. Soc. (3)* **41** (1980), 21–44.

[RR] Radjavi, H., Rosenthal, P., *Invariant Subspaces*, Springer-Verlag, New York, 1973.

[Ra1] Ransford, T.J., *Analytic Multivalued Functions*, Doctoral Thesis, University of Cambridge, Cambridge, 1983.

[Ra2] Ransford, T.J., Open mapping, inversion and implicit function theorems for analytic multivalued functions, *Proc. London Math. Soc. (3)* **49** (1984), 537–562.

[Ra3] Ransford, T.J., On the range of an analytic multivalued function, *Pacific J. Math.* **123** (1986), 421–439.

[Ra4] Ransford, T.J., The spectrum of an interpolated operator and analytic multifunctions, *Pacific J. Math.* **121** (1986), 445–466.

[Ra5] Ransford, T.J., *Potential Theory in the Complex Plane*, book to appear.

[Re] Rellich, F., *Perturbation Theory of Eigenvalue Problems*, Gordon & Breach, New York, 1969.

[Ri] Rickart, C.E., *General Theory of Banach Algebras*, Van Nostrand, Princeton, 1966.

[Ro] Rodríguez Palacios, A., Jordan structures in analysis, preprint.

[Ru] Rudin, W., *Real and Complex Analysis*, McGraw-Hill, New York, 1974.

[Run] Runde, V., Intertwinning operators over $L^1(G)$ for $G \in [PG] \cap [SIN]$, to appear.

[Sa] Sadullaev, A., Pseudoconcave sets and algebraic lemniscates (in Russian), preprint.

[Se] Senichkin, V.N., Subharmonic functions and analytic structure in the maximal ideal space of a uniform algebra, *Math. USSSR Sb.* **36** (1980), 111–126.

[Sh1] Shcherbina, N.V., The Levi form for C^1-smooth hypersurfaces, and the complex structure on the boundary of domains of holomorphy (English translation), *Math. USSSR Izv.* **19** (1982), 874–895.

[Sh2] Shcherbina, N.V., On the fibering into analytic curves of the common boundary of two domains of holomorphy (English translation), *Math. USSSR Izv.* **21** (1983), 399–413.

[Si] Sibony, N., Multi-dimensional analytic structure in the spectrum of a uniform algebra, in: *Spaces of Analytic Functions, Kristiansand, Norway, 1975* (O.B. Bekken et al., eds.), Lecture Notes in Math. **512**, Springer-Verlag, Berlin – Heidelberg – New York, 1976; 139–165.

[Sic] Siciak, J., On some extremal functions and their applications in the theory of analytic functions of several complex variables, *Trans. Amer. Math. Soc.* **105** (1962), 322–357.

[S] Sierpiński, W., *Cardinal and Ordinal Numbers*, Polish Scientific Publishers, Warsaw, 1965.

[Sł1] Słodkowski, Z., Analytic set-valued functions and spectra, *Math. Ann.* **256** (1981), 363–386.

[Sł2] Słodkowski, Z., On subharmonicity of the capacity of the spectrum, *Proc. Amer. Math. Soc.* **81** (1981), 243–249.

[Sł3] Słodkowski, Z., Analytic families of operators: variation of the spectrum, *Proc. Roy. Irish Acad. Sect. A* **81** (1981), 121–126.

[Sł4] Słodkowski, Z., Uniform algebras and analytic multifunctions, *Atti Accad. Naz. Lincei Rend. Cl. Sci. Fis. Mat. Natur.* **75** (1983), 9–18.

[Sł5] Słodkowski, Z., Analytic perturbations of Taylor spectrum, *Trans. Amer. Math. Soc.* **297** (1986), 319–336.

[Sł6] Słodkowski, Z., An analytic set-valued selection and an application to the corona theorem, *Trans. Amer. Math. Soc.* **294** (1986), 367–377.

[Sł7] Słodkowski, Z., A generalization of Vesentini and Wermer's theorems, *Rend. Sem. Mat. Univ. Padova* **75** (1986), 157–171.

[Sł8] Słodkowski, Z., Polynomial hulls with convex sections and interpolating spaces, *Proc. Amer. Math. Soc.* **96** (1986), 255–260.

[Sł9] Słodkowski, Z., On bounded analytic functions in finitely connected domains, *Trans. Amer. Math. Soc.* **300** (1987), 721–736.

[Sł10] Słodkowski, Z., Polynomial hulls in $\mathbb{C}^2$ and quasi-circles, *Ann. Scuola Norm. Sup. Pisa* **16** (1989), 367–391.

[Sł11] Słodkowski, Z., Complex interpolation to normed and quasinormed spaces in several dimension II: Properties of harmonic interpolation, *Trans. Amer. Math.Soc.* **317** (1990), 255–285.

[Sł12] Słodkowski, Z., Polynomial hulls with convex fibers and complex geodesics, *J. Funct. Anal.* **94** (1990), 156–176.

[Sn] Sneiberg, I.Ya., Spectral properties of linear operators in interpolation families of Banach spaces (Russian), *Mat. Issled.* **9** (1974), 214–229.

[St] Stout, E.L., *The Theory of Uniform Algebras*, Bodgen & Quigley, Tarrytown-on-Hudson, 1971.

[T] Taylor, A.E., Notes on the history of the uses of analyticity in operator theory, *Amer. Math. Monthly* **78** (1971), 331–342.

[Ta] Taylor, J.L., A joint spectrum for several commuting operators, *J. Funct. Anal.* **6** (1970), 172–191.

[Ts] Tsuji, M., *Potential Theory in Modern Function Theory*, 2nd ed., Chelsea, New York, 1975.

[U] Upmeier, H., *Jordan Algebras in Analysis, Operator Theory, and Quantum Mechanics,* CBMS Regional Conf. Ser. in Math. **67**, American Mathematical Society, Providence, RI, 1987.

[Ve1] Vesentini, E., On the subharmonicity of the spectral radius, *Boll. Un. Mat. Ital.* **4** (1968), 427–429.

[Ve2] Vesentini, E., Maximum theorems for spectra, in: *Essays on Topology and Related Topics* (*Mémoires dédiés à Georges de Rham*), Springer-Verlag, Berlin – Heidelberg – New York, 1970; 111–117.

[Ve3] Vesentini, E., Carathéodory distances and Banach algebras, *Adv. Math.* **47** (1983), 50–73.

[Vl] Vladimirov V.S., *Methods of the Theory of Functions of Several Complex Variables,* MIT Press, Cambridge, MA, 1966.

[Vr] Vrbová, P., On local spectral properties of operators in Banach spaces, *Czechoslovak Math. J.* **23** (1973), 483–492.

[We1] Wermer, J., Subharmonicity and hulls, *Pacific J. Math.* **58** (1975), 283–290.

[We2] Wermer, J., *Banach Algebras and Several Complex Variables*, 2nd ed., Springer-Verlag, New York, 1976.

[We3] Wermer, J., Polynomially convex hulls and analyticity, *Ark. Math.* **20** (1982), 129–135.

[We4] Wermer, J., Maximum modulus algebras, *Contemp. Math.* **137** (1992), 469–478.

[Wr] Wright, J.D.M., Jordan C^*-algebras, *Michigan Math. J.* **24** (1977), 291–302.

[Ya] Yamaguchi, H., Sur une uniformité des surfaces constantes d'une fonction entière de deux variables complexes, *J. Math. Kyoto Univ.* **13** (1973), 417–433.

[Za] Zafran, M., Spectral theory and interpolation of operators, *J. Funct. Anal.* **36** (1980), 185–204.

[Zah] Zaharjuta, V., Transfinite diameter, Čebysev constants and capacity for compact in $\mathbb{C}^n$, *Math. USSR Sb.* **25** (1975), 350–364.

[ZKKP] Zaidenburg, M.G., Krein, S.G., Kuchment, P.A., Pankov, A.A., Banach bundles and linear operators, *Russian Math. Surveys* **30** (1975), 115–175.

[Zh] Zhevlakov, K.A., Slin'ko, A.M., Shestakov, I.P., Shirshov, A.I., *Rings That Are Nearly Associative*, Academic Press, New York, 1982.

[Zr] Zraïbi, A., *Sur les fonctions analytiques multiformes*, Thèse de doctorat, Université Laval, Québec, 1983.

Harmonic approximation on closed subsets of Riemannian manifolds

Thomas BAGBY
Department of Mathematics, Rawles Hall
Indiana University
Bloomington, IN 47405
U.S.A.

and

Paul M. GAUTHIER
Département de mathématiques et de statistique
and
Centre de recherches mathématiques
Université de Montréal
C.P. 6128-A, Montréal, Qué., H3C 3J7
Canada

Abstract

We discuss the problem of approximating functions on a closed subset of a noncompact Riemannian manifold by functions which are harmonic on the entire manifold.

1 Introduction

In this paper we study approximation by harmonic functions on Riemannian manifolds. We let Ω be a (connected, oriented, C^∞) noncompact Riemannian manifold, and F a closed subset of Ω. Let $A(F)$ denote the set of all continuous real-valued functions on F which are harmonic on the interior int F. We say that F is a *Runge set* provided that every function which is harmonic on an open neighborhood of F can be uniformly approximated on F by functions harmonic on all of Ω. We say that F is a *Mergelyan set* provided every function in $A(F)$ can be uniformly approximated by functions harmonic on all of Ω.

Characterizations of Mergelyan sets have been given in various settings in earlier papers [BG1], [BG2], [GO]. More recently Gardiner [G1] has given a characterization of Runge sets

Research supported in part by NSERC-Canada and FCAR-Québec.

P. M. Gauthier (ed.) and G. Sabidussi (techn. ed.), Complex Potential Theory, 75–87.

in open subsets of Euclidean space, from which he obtains a very explicit characterization of Mergelyan sets in open subsets of Euclidean space; in the present paper we will give analogous results for arbitrary Riemannian manifolds, summarized in the following three theorems. We refer to [BB] for a discussion of harmonic functions on Riemannian manifolds and the concept of thinness. We shall follow the usual practice of referring to a relatively compact subset of Ω as a *bounded* set. A *hole* of the closed set $F \subset \Omega$ is a component of $\Omega \backslash F$ which is bounded. We let $\hat{F}$ denote the union of F and all of its holes; then $\hat{F}$ is a closed subset of Ω with no holes. We say that a family of subsets of Ω satisfies *the long islands condition* provided that for each bounded set $B \subset \Omega$, the union of all members of the family which intersect B is bounded. We use the notation $\Omega^* = \Omega \cup \{*\}$ for the Alexandroff one-point compactification of Ω.

Theorem 1 *Let F be a closed subset of a noncompact Riemannian manifold Ω. Then the following are equivalent.*

(a) *F is a Runge set.*

(b) *All three of the following conditions hold:*

(i) *$\Omega \backslash \hat{F}$ and $\Omega \backslash F$ are thin at the same points of F.*

(ii) *The holes of F satisfy the long islands condition.*

(iii) *$\Omega^* \backslash \hat{F}$ is locally connected.*

Although Runge sets are characterized in Theorem 1, they have other important topological properties given in the following result.

Theorem 2 *Let F be a Runge set in a noncompact Riemannian manifold Ω.*

(a) $\partial \hat{F} = \partial F$.

(b) *If W is the union of the holes of F, then* int $\hat{F}$ *is the (disjoint) union of W and* int F.

Theorem 3 *Let F be a closed subset of a noncompact Riemannian manifold Ω. Then the following are equivalent.*

(a) *F is a Mergelyan set.*

(b) *All three of the following conditions hold:*

(i) *$\Omega \backslash \hat{F}$ and $\Omega \backslash$*int *F are thin at the same points of F.*

(ii) *The holes of F satisfy the long islands condition.*

(iii) *$\Omega^* \backslash \hat{F}$ is locally connected.*

The proofs of these three theorems will be given in sections 2, 3, 4, 5, 6, and 7. It is also interesting to consider approximation by harmonic functions at arbitrarily fast rates of speed. We say that the closed set $F \subset \Omega$ is a *Carleman set* provided that for every $u \in A(F)$ and every continuous function $\varepsilon : F \to (0,1]$, there exists a harmonic function $h : \Omega \to \mathbf{R}$ such that $|u - h| < \varepsilon$ on F. In the following theorem we limit ourselves to the case of a noncompact Riemann surface Ω; we recall that a Riemann surface can always be regarded as a two-dimensional Riemannian manifold with the same class of harmonic functions [BG2, section 3].

Theorem 4 *Let F be a closed subset of a noncompact Riemann surface Ω. Then the following are equivalent.*

(a) *F is a Carleman set.*

(b) *F is a Mergelyan set, and the components of* int F *satisfy the long islands condition.*

We remark that in view of condition (b)(ii) of Theorem 1, and Theorem 2(b), we obtain an equivalent statement of Theorem 4 if we replace int F by int $\hat{F}$. Theorem 4 was first stated in the latter form in [BG1]; the proof of the implication (b) $\Rightarrow$ (a) was given in [BG1], and will not be repeated here. The proof of the implication (a) $\Rightarrow$ (b) will be given in section 8 below, and our proof of this implication is valid when Ω is a noncompact Riemannian manifold of any dimension. Theorem 4 has been proved in [G2] and [GG] when Ω is an open subset of Euclidean space of any dimension.

In case the closed set F is compact and the Riemannian manifold Ω is hyperbolic, our results are contained in results on harmonic spaces given in [GGG].

Our discussion is largely self-contained, but will be logically dependent on the study of harmonic approximation on manifolds in [BB], [BG1] and [BG2]. In some of the proofs we will make use of the fine topology and finely harmonic functions on Riemannian manifolds [F], [BB, section 8].

2 Notation and preliminary results

For the rest of this paper we let Ω be a noncompact Riemannian manifold. We will repeatedly use the elementary fact that the holes of a closed set $F \subset \Omega$ are precisely the nonempty bounded regions W in $\Omega \backslash F$ such that $\partial W \subset F$. For each set $E \subset \Omega$ we let $H(E)$ denote the set of functions harmonic on (neighborhoods of) E. We will often use implicitly the fact that for any open set $W \subset \Omega$, the uniform limit of a sequence in $H(W)$ is again in $H(W)$ (see [BB, Theorem 3.1(e)]). In the following theorem we summarize some of the additional facts about harmonic functions which we will need.

Theorem 2.1

(a) (*Maximum principle*) *If u is a harmonic function on Ω which has a local maximum or a local minimum at a point of Ω, then u is a constant function on Ω.*

(b) (*Aronszajn-Cordes theorem*) *Let u be a harmonic function on Ω. Suppose that a is a point of a parametric ball $B \subset\subset \Omega$, and that in the local coordinates of B we have the partial derivatives $D^\alpha u(a) = 0$ for $|\alpha| \geq 0$. Then $u \equiv 0$ on Ω.*

(c) (*Lax-Malgrange theorem*) *Any compact set $K \subset \Omega$ with no holes is a Runge set.*

(d) *A subset E of a parametric ball $B \subset\subset \Omega$ is thin at a point $a \in B$ if and only if it is thin at a with respect to the usual flat metric on B.*

(e) *A parametric ball $B \subset\subset \Omega$ is nonthin at each point of ∂B.*

(f) *Let W be a bounded open subset of Ω. If $\Omega \backslash W$ is nonthin at each point of ∂W, then the Dirichlet problem for W has a unique solution; that is, for every continuous function f on ∂W there is a unique continuous function on $\overline{W}$ whose restriction to W is harmonic.*

Theorem 2.1 is well known. For part (a) we refer to [BB, Theorem 3.1(c)] and the reference given there. Part (b) follows from [Hö, Theorem 17.2.6] and a standard connectedness argument. A simple proof of part (c) is given in [BB, Theorem 3.10]. Part (d), on the equivalence of thinness with respect to various differential operators, is in [He, Theorem 2]. The property in part (e) is well known when the ball B carries the usual flat metric, and in the general case it then follows from part (d). Part (f) follows from the general theory of the Dirichlet problem in harmonic spaces [dP, Chapter 4], since W is a relatively compact subset of a regular subregion of Ω. (Note that any regular subregion of Ω satisfies Axiom P since it has a Green function [BB]. Any parametric ball in Ω satisfies Axiom D as explained in [BB, section 8], and hence any regular subregion of Ω satisfies Axiom D by the localization principle [CC, Corollary 9.2.2].)

We next recall some terminology associated with spaces of continuous functions. We let $C(X)$ denote the Banach space of all continuous real-valued functions on a compact Hausdorff space X, with the usual uniform norm. We say that a function $f \in C(X)$ *peaks* at the point $a \in X$ provided

$$f(a) = 1, \qquad f(x) < 1 \text{ for all } x \in X \backslash \{a\}.$$

If L is a closed subspace of $C(X)$, we say that a point $a \in X$ is a *peak point* for L provided that there is a function $f \in L$ which peaks at a; and the *Choquet boundary* of L, denoted by bL, is the set of all peak points of L.

Now let K be a compact subset of Ω. Note that $A(K)$ is a closed subspace of $C(K)$. We also wish to study $\overline{H}(K)$, the closure of $H(K)$ in $C(K)$; and $\overline{H}_K(\Omega)$, the closure of $H(\Omega)$ in $C(K)$.

Theorem 2.2 *Let K be a compact subset of Ω. Let $a \in K$.*

(a) *Let $L = A(K)$. Then $a \in bL$ if and only if $\Omega \setminus \operatorname{int} K$ is nonthin at a.*

(b) *Let $L = \overline{H}(K)$. Then $a \in bL$ if and only if $\Omega \setminus K$ is nonthin at a.*

(c) *Let $L = \overline{H}_K(\Omega)$. Then $a \in bL$ if and only if $\Omega \setminus \hat{K}$ is nonthin at a.*

Proof Part (a), which is known as the Keldysh lemma, was proved by Keldysh in a more classical setting. In the present generality, it follows, at least in one direction, from work of Brelot [B, Lemme 2], applied to the hyperbolic Riemannian manifold obtained by removing from Ω a closed parametric disk disjoint from K; proofs in both directions have been given in later papers by various authors (see for example [G, Lemma 6]).

We turn next to the proof of (b), and suppose first that a is a peak point for $\overline{H}(K)$. Then there exists $u \in \overline{H}(K)$ which peaks at a. From [F, 9.6] or [F, 11.9] we see that u is finely harmonic on the fine interior of K. Thus if a were a point of the fine interior of K, we could apply the mean-value property in the definition of finely harmonic functions [F, 8.3] to obtain a contradiction to the fact that u peaks at a. We conclude that a is not in the fine interior of K, which means that $\Omega \setminus K$ is nonthin at a.

To prove the other implication in (b), we suppose that $\Omega \setminus K$ is nonthin at a. It follows from part (d) of Theorem 2.1, the Wiener criterion, and the proof of the Keldysh lemma in [G] (replacing the appeal to Ancona's theorem by the more elementary regularity property of capacity), that there is a compact set $Q = \{a\} \cup e$, with $e \subset \Omega \setminus K$, which is nonthin at a. We can now obtain a bounded open set V, containing e and disjoint from K, such that V is nonthin at each of its boundary points; in fact, we may construct V as the union of a sequence of parametric balls in $\Omega \setminus K$ which cover $Q \setminus \{a\}$ and converge to the point a.

Now let U be any bounded open set in Ω which is regular for the Dirichlet problem and contains both K and V. Set $W = U \setminus \overline{V}$. Since W is regular for the Dirichlet problem, it is easy to construct a function $u \in A(\overline{W})$ which peaks at a. It follows that u is in the space $\overline{H}(\overline{W})$ (see [BB, Theorem 8.3] and the references given there). We conclude that the restriction of u to K is in $\overline{H}(K)$, so a is a peak point for $\overline{H}(K)$.

To prove (c), we suppose first that $\Omega \setminus \hat{K}$ is nonthin at a. From (b) we conclude that a is a peak point for $\overline{H}(\hat{K})$. This means that there exists a function $f \in C(\hat{K})$ which peaks at a, and a sequence of functions $f_n \in H(\hat{K})$, such that $|f - f_n| < 1/n$ on $\hat{K}$. According to the Lax-Malgrange theorem 2.1(c), there exists for each n a function $g_n \in H(\Omega)$ such that $|g_n - f_n| < 1/n$ on $\hat{K}$. We conclude that $f \in \overline{H}_K(\Omega)$, so a is a peak point for $\overline{H}_K(\Omega)$.

Finally, we suppose that a is a peak point for $\overline{H}_K(\Omega)$. This means that there is a function $f \in C(K)$ which peaks at a, and a sequence of functions $f_n \in H(\Omega)$ which converges to f uniformly on K. Applying the maximum principle 2.1(a) to each hole of K, we see that the sequence $\{f_n\}$ is uniformly Cauchy on $\hat{K}$, and hence converges uniformly on $\hat{K}$ to a function $\hat{f} \in \overline{H}_{\hat{K}}(\Omega)$ such that $\hat{f}|_K = f$. We then must have $\hat{f} < 1$ on any hole V of K; this follows from the maximum principle 2.1(a), since we have $\partial V \subset K$ and we know that $f(x) < 1$ for at least one point $x \in \partial V$. It follows that a is a peak point for $\overline{H}(\hat{K})$. From (b) we conclude that $\Omega \setminus \hat{K}$ is nonthin at a. This completes the proof of (c), and the proof of Theorem 2.2.

3 Proof of necessity in Theorem 1

Remark 3.1 *If F is a compact Runge set in Ω, then F has all the properties in part* (b) *of Theorem* 1. (In fact, condition (b)(i) follows from Theorem 2.2, since identical closed subspaces of $C(F)$ must have identical Choquet boundaries, and conditions (b)(ii) and (b)(iii) hold for any compact subset F of a noncompact Riemannian manifold (see [BG2, Section 2].)

The following lemma shows that *any* Runge set F in a Riemannian manifold must satisfy conditions (b)(ii) and (b)(iii) of Theorem 1.

Lemma 3.2 *Let F be a Runge set in Ω. Then*

(a) *for every nonempty compact set $K \subset \Omega$, the holes of $K \cup F$ satisfy the long islands condition.*

(b) *the holes of F satisfy the long islands condition.*

(c) *$\Omega^* \backslash \hat{F}$ is locally connected.*

In the proof of Lemma 3.2 we will need the following two lemmas.

Lemma 3.3 *Let F be a closed subset of Ω. If F is a Runge set, so is $\hat{F}$.*

Proof Let $\hat{u}$ be a function harmonic on an open neighborhood of $\hat{F}$, and let $\varepsilon > 0$. Since F is a Runge set, there exists a harmonic function v on Ω such that $\sup_F |\hat{u} - v| < \varepsilon$. Applying the maximum principle 2.1(a) to each hole of F, we conclude that $\sup_{\hat{F}} |\hat{u} - v| = \sup_F |\hat{u} - v| = \varepsilon$, which completes the proof of the lemma.

If W is any bounded subregion of Ω, and $q \in W$, then the *generalized Green function* $G_W(\cdot, q)$ is defined, and the following result is proved, in [BG2, Lemma 4.3].

Lemma 3.4 *Let W be a bounded region in Ω, and $q \in W$. Let $U : \overline{W} \backslash \{q\} \to \mathbf{R}$ be a continuous function which is harmonic on $W \backslash \{q\}$, and suppose that $U - G_W(\cdot, q)$ has a removable singularity at q. If $U \geq 0$ on ∂W, then $U - G_W \geq 0$ on W.*

We turn now to the proof of Lemma 3.2. If condition (a) of Lemma 3.2 fails, then there are distinct holes $W_1, W_2, \ldots$ of $K \cup F$ such that each W_j contains distinct points a_j and p_j, where the sequence $\{a_j\}$ is bounded and $\lim_{j \to \infty} p_j = *$. By passing to a subsequence (if necessary), we may assume that the holes W_j are mutually disjoint, and that the sequence $\{a_j\}$ converges to some point $a \in \Omega$. Since the points a_j lie in distinct components of $\Omega \backslash (K \cup F)$, we have $a \in K \cup F$. Now for each index j we define $G_j \equiv G_{W_j}(\cdot, p_j)$, and then select a finite constant $c_j j / G_j(a_j)$. We then apply a Mittag-Leffler theorem due to the authors and Blanchet [BG2, 4.2] to obtain a harmonic function u on all of $\Omega \backslash \bigcup_j \{p_j\}$ such that, for each j, the harmonic function $u - c_j G_j$ has a removable singularity at p_j. Using the fact that F is a Runge set, we see that there exists a harmonic function v on Ω such that $0 \leq u - v \leq 1$ on F, and we define $m = \min_K (u - v)$. For each j the hole W_j

must satisfy $\partial W_j \subset K \cup F$, so from Lemma 3.4 we obtain $\min\{0,m\} \le u - v - c_j G_j$ on W_j, and evaluating at a_j then gives $\min\{0,m\} + j \le u(a_j) - v(a_j)$. We now let $j \to \infty$ to obtain $(u-v)(a) = \lim_{j\to\infty}(u-v)(a_j) = +\infty$, and this contradiction completes the proof of condition (a).

Condition 3.2(b) follows easily from 3.2(a).

To prove 3.2(c), it suffices by [BG2, Theorem 2.3] to fix a compact set $K \subset \Omega$, and to prove that *the union of the holes of* $K \cup \hat{F}$ *is bounded.* This is easy to prove using the fact that *the holes of* $K \cup \hat{F}$ *satisfy the long islands condition* (by 3.2(a) and 3.3); and the fact that *each hole* W *of* $K \cup \hat{F}$ *must have a boundary which meets* K (since otherwise we would have $\partial W \subset \hat{F}$, so W would be a hole of $\hat{F}$). This completes the proof of Lemma 3.2, and the proof of the necessity of conditions (b)(ii) and (b)(iii) of Theorem 1.

In proving the necessity of condition (b)(i) of Theorem 1, we need the following remark.

Remark 3.5 *Let* F *be a closed subset of* Ω.

(a) *Let* W *be a hole of* F. *If* A *is a closed subset of* Ω *containing* W, *then* W *is a hole of* $A \cap F$.

(b) *Suppose the holes of* F *satisfy the long islands condition. If* $B \subset\subset \Omega$ *is a parametric ball, then there exists a compact set* A *with no holes such that* $A \supset \overline{B}$ *and*

$$B \cap \hat{F} = B \cap (A \cap F)^{\wedge}.$$

Proof (a) Since W is a hole of F, we have $\partial W \subset F$. Thus W is a nonempty bounded region in Ω satisfying $\partial W \subset A \cap F$, so W is a hole of $A \cap F$.

(b) Let V be the union of B and all holes of F which intersect B. By hypothesis the set $\overline{V}$ is compact, and hence $A \equiv (\overline{V})^{\wedge}$ is compact. Using (a), we see that A has the properties stated in (b), and 3.5 is proved.

For the rest of this section we suppose that F is a Runge set in Ω, and we prove that F has property (b)(i) of Theorem 1.

Claim: If A *is any compact subset of* Ω *with no holes, then* $K = A \cap F$ *is a Runge set in* Ω.

To prove this claim, it suffices to fix a function $u \in C_0^\infty(\Omega)$ which is harmonic on an open neighborhood N of K, and to prove that u can be uniformly approximated on K by functions harmonic on Ω. There is a function $\varphi \in C_0^\infty(N \cup (\Omega \backslash F))$ which is identically equal to one on an open neighborhood of A. We define $v = \mathcal{V}_\varphi[u]$, where $\mathcal{V}$ is the localization operator on the manifold Ω defined in [BB, Section 6]. From [BB, Lemma 6.2] we see that v has the following properties:

1° v is harmonic on N (since u is).

2° v is harmonic on an open neighborhood of $F \backslash N$ (since $\varphi \equiv 0$ on an open neighborhood of $F \backslash N$.)

3° $u - v$ is harmonic on an open neighborhood of A (since $\varphi \equiv 1$ on an open neighborhood of A.)

From 1° and 2° we see that v is harmonic on an open neighborhood of F. If $\varepsilon > 0$, then by hypothesis there is a harmonic function h on Ω such that

$$\sup_F |h - v| < \varepsilon/2.$$

From 3° and the Lax-Malgrange theorem 2.1(c) there is a harmonic function g on Ω such that

$$\sup_A |g - (u - v)| < \varepsilon/2.$$

Since K lies inside F and inside A, we conclude from the last two inequalities that

$$\begin{aligned} \sup_K |u - (h + g)| &= \sup_K |u - g - v + v - h| \\ &\leq \sup_K |u - g - v| + \sup_K |v - h| \\ &< \frac{\varepsilon}{2} + \frac{\varepsilon}{2} = \varepsilon. \end{aligned}$$

This proves the claim.

Finally, we fix an arbitrary parametric ball $B \subset\subset \Omega$. Using Remark 3.1, Lemma 3.2(b), Remark 3.5(b) and the preceding claim, we see that $\Omega \backslash \hat{F}$ and $\Omega \backslash F$ are thin at the same points of $B \cap F$. Thus F has property (b)(i) of Theorem 1, which completes the proof of necessity in Theorem 1.

4 Proof of sufficiency in Theorem 1

Lemma 4.1 *Let W be an open subset of Ω such that*

(a) *$\Omega \backslash W$ is nonthin at each point of ∂W, and*

(b) *the components of W satisfy the long islands condition.*

Then the Dirichlet problem for W has a unique solution; that is, for every continuous function f on ∂W there is a unique continuous function u on $\overline{W}$ whose restriction to W is harmonic.

Proof The uniqueness of the solution follows from applying the maximum principle 2.1(a) to each component of W.

To prove the existence of a solution, we write $\Omega = \bigcup_j K_j$, where each set K_j is compact and $K_j \subset \text{int } K_{j+1}$; and for each j we let W_j be the union of all components of W which meet

K_j. Then the complement of W_j is nonthin at each point of ∂W_j, so from Theorem 2.1(f) there exists a unique continuous function u_j on $\overline{W_j}$ which coincides with f on ∂W_j and is harmonic on W_j. If $j < k$, the functions u_j and u_k must coincide on $\overline{W_j}$, as we see from the uniqueness of u_j. Thus the union function u, defined to be equal to u_j on W_j, is well defined on $\overline{W} = \bigcup_j \overline{W_j}$, and satisfies the conditions of the lemma.

The rest of this section is devoted to the proof of sufficiency in Theorem 1. Throughout this discussion we assume that all three conditions in part (b) of Theorem 1 hold, and we let W be the union of the holes of F.

Claim 1: *W satisfies conditions* (a) *and* (b) *of Lemma* 4.1.

To prove Claim 1, we note that condition 4.1(b) is clear from hypothesis (b)(ii) of Theorem 1. To prove condition 4.1(a), we let $a \in \partial W$ be arbitrary, and consider two cases:

Case 1° : *$\Omega\backslash\hat{F}$ is nonthin at a.* In this case, $\Omega\backslash W \supset \Omega\backslash\hat{F}$ is nonthin at a.

Case 2° : *$\Omega\backslash\hat{F}$ is thin at a.* In this case, we know from hypothesis (b)(i) that $\Omega\backslash F$ is thin at a. Thus $W \subset \Omega\backslash F$ is thin at a, so $\Omega\backslash W$ is nonthin at a.

This completes the proof of Claim 1.

Now let f be harmonic on an open neighborhood of F. Then $f \in C(F)$, and f is finely harmonic on the fine interior of F. From Claim 1 and Lemma 4.1 we may find a function $h \in C(\overline{W})$ such that $h|_{\partial W} = f|_{\partial W}$, and h is harmonic on W. Then the function

$$w = \begin{cases} f & \text{on } F \\ h & \text{on } \overline{W} \end{cases}$$

is continuous on $\hat{F} = F \cup \overline{W}$, since its restriction to each of the closed sets F and $\overline{W}$ is continuous.

Claim 2: *The function w is finely harmonic on the fine interior of $\hat{F}$.*

To prove Claim 2 we note that in view of the sheaf property of finely harmonic functions [F, 8.6], it suffices to select any point a in the fine interior of $\hat{F}$, and to find a finely open set $N_a \subset \hat{F}$, containing a, on which w is finely harmonic. We consider two cases:

Case 1° : $a \in W$. In this case we may take $N_a = W$.

Case 2° : $a \in F$. Since a is in the fine interior of $\hat{F}$, we know that $\Omega\backslash\hat{F}$ is thin at a. From hypothesis b(i) we see that $\Omega\backslash F$ is thin at a, which means that a is in the fine interior of F. Therefore in this case we may take N_a to be the fine interior of F. This completes the proof of Claim 2.

We now complete the proof of sufficiency in Theorem 1. Let $\varepsilon > 0$ be arbitrary. From Claim 2 and [BB, Theorem 8.3] we see that there is a harmonic function $\hat{u}$ on an open neighborhood of $\hat{F}$ such that $\sup_{\hat{F}} |\hat{u} - w| < \varepsilon/2$. Since $\Omega^*\backslash\hat{F}$ is connected by [BG2, Theorem 2.3] and is locally connected by hypothesis, we conclude from [BB, Theorem 9.3] that there is a harmonic function u on Ω such that $\sup_{\hat{F}} |u - \hat{u}| < \varepsilon/2$. It follows that $\sup_F |u - f| < \varepsilon$, and the proof of sufficiency in Theorem 1 is complete.

5 Proof of Theorem 2

We devote this section to a proof of part (a) of Theorem 2. Part (b) of the theorem then follows at once, since the sets int F and W are clearly disjoint, and from part (a) we may write int $\hat{F} = \hat{F}\backslash\partial\hat{F} = \hat{F}\backslash\partial F = (\hat{F}\backslash F) \cup$ int $F = W \cup$ int F.

We next note that for any closed set $F \subset \Omega$, the inclusions $\partial\hat{F} \subset \partial F \subset F \subset \hat{F}$ are known [BG2, Remarks 2.2]. Thus to prove part (a) of Theorem 2 we suppose that $F \subset \Omega$ is a Runge set, and that there is a point $p \in \partial F\backslash\partial\hat{F} = \partial F \cap$ int $\hat{F}$, and we will obtain a contradiction.

We first claim that *there exists a parametric ball $B \subset \hat{F}\backslash F$ such that $\partial B \cap \partial F \cap$ int $\hat{F}$ contains a point q.* To prove this claim, note that since $p \in$ int $\hat{F}$, there is an open set $U \subset \hat{F}$ which contains p, and a local coordinate function which carries U into the open unit ball of $\mathbf{R}^n$ and p into the origin; using this local coordinate to transfer to U the usual Euclidean metric of the open unit ball of $\mathbf{R}^n$, we see that there must be a point $a \in U\backslash F$ whose distance to p is less than 1/2, and hence the claim is true if we take B to be the largest open Euclidean ball centered at a which is disjoint from F.

We now see that $\Omega\backslash F \supset B$ is nonthin at q, by Theorem 2.1(e); but $\Omega\backslash\hat{F}$ is thin at q, since $q \in$ int $\hat{F}$. This contradicts condition (b)(i) of Theorem 1, so Theorem 2 is proved.

6 Proof of necessity in Theorem 3

Any Mergelyan set is a Runge set, and hence by Theorem 1 must have properties (b)(ii) and (b)(iii) of Theorem 3.

We now assume that F is a Mergelyan set in Ω, and we will prove property (b)(i) in Theorem 3. Suppose that $\Omega\backslash\hat{F}$ is thin at a point $a \in F$. Since F must be a Runge set, we conclude from Theorem 1 that $\Omega\backslash F$ is thin at a. Since F has the property that every function in $A(F)$ can be uniformly approximated on F by functions harmonic on neighborhoods of F, we conclude from [BB, Theorem 8.1] that $\Omega\backslash$int F is thin at a. Property (b)(i) of Theorem 3 is now clear, and we have proved the necessity in Theorem 3.

7 Proof of sufficiency in Theorem 3

Let F be a closed subset of Ω which satisfies all three conditions in part (b) of Theorem 3. It follows from (b)(i) that the three sets

$$\Omega\backslash\hat{F}, \quad \Omega\backslash F, \quad \Omega\backslash\text{int } F$$

are thin at the same points of F.

Now let $f \in A(F)$ and $\varepsilon > 0$. Since $\Omega\backslash F$ and $\Omega\backslash\text{int } F$ are thin at the same points of Ω, we conclude from [BB, Theorem 8.1] that there is a harmonic function h on an open neighborhood of F such that $\sup_F |h - f| < \varepsilon/2$. Using the fact that $\Omega\backslash\hat{F}$ and $\Omega\backslash F$ are thin at the same points of F, together with hypotheses (b)(ii) and (b)(iii) of Theorem 3, we conclude from Theorem 1 that there is a harmonic function u on Ω such that $\sup_F |u - h| < \varepsilon/2$. It follows that $\sup_F |u - f| < \varepsilon$, which completes the proof of sufficiency in Theorem 3.

8 Proof of necessity in Theorem 4

To prove the necessity in Theorem 4 it suffices to prove the following necessary condition for Carleman approximation in manifolds of any dimension.

Theorem 8.1 *Let F be a Carleman set in a noncompact Riemannian manifold Ω. Then the components of* int F *satisfy the long islands condition.*

For the proof of Theorem 8.1 we need the following three lemmas concerning a noncompact Riemannian manifold Ω.

Lemma 8.2 *Let $B \subset\subset \Omega$ be a parametric ball, and K a compact subset of B. For each multi-index α there exists a constant $C = C(\Omega, K, \alpha)$ such that*

$$\sup_K |D^\alpha u| \leq C \sup_B |u|$$

for every bounded harmonic function u on B.

Proof Let $|\alpha| = k$ and let $B' \subset\subset B$ be a smaller parametric ball containing K. We fix a function $\varphi \in C_0^\infty(B')$ satisfying $\varphi \equiv 1$ on an open neighborhood N of K. If u is a bounded harmonic function on B, we define $U = \varphi u$. If we fix an integer $k' > n/2$, then according to the Sobolev lemma we can estimate $\sup_K |D^\alpha U|$ by (a constant times the sum of) the norms $\|D^\beta U\|_{L_2(B')}$ for $|\beta| \leq k + k'$, and each of the latter norms can be estimated by the norms $\|D^\beta u\|_{L_2(B')}$ for $|\beta| \leq k + k'$. (Here we have applied the product rule repeatedly to $U = \varphi u$.) Applying [N, 3.6.11] to the Laplace operator on B, we may estimate each of the latter norms by $\sup_B |u|$. This completes the proof.

Lemma 8.3 (Theorem of Montel type) *The set of harmonic functions on Ω, which satisfy $|u| \leq 1$ there, is compact with respect to the topology of uniform convergence on compact subsets of Ω.*

Lemma 8.3 follows from [C, Theorem 3]. The axioms assumed in [C] are satisfied here, as noted in [BB].

Lemma 8.4 (Transfer of smallness) *Let K be a compact subset of Ω, and $\mathcal{O}$ a nonempty open subset of Ω. For each positive constant ε there exists a positive constant δ with the following property. If u is a harmonic function on Ω satisfying $\sup_\Omega |u| \leq 1$ and $\sup_{\mathcal{O}} |u| \leq \delta$, then $\sup_K |u| \leq \varepsilon$.*

Proof We may assume that $\mathcal{O} \subset\subset \Omega$. If the lemma is not true, there exist a constant $\epsilon > 0$, and harmonic functions $u_1, u_2, \ldots$ on Ω, such that $\sup_\Omega |u_j| = 1$, and $\sup_K |u_j| > \epsilon$, for each positive integer j, but $\lim_{j\to\infty}(\sup_{\mathcal{O}} |u_j|) = 0$. We next use Lemma 8.3 to obtain a subsequence $\{u_{j_i}\}$ of $\{u_j\}$ such that $v(x) = \lim_{i\to\infty} u_{j_i}(x)$ is defined for each $x \in \Omega$, and the convergence is uniform on each compact subset of Ω. It follows that $\sup_K |v| \geq \epsilon$, $v \equiv 0$ on $\mathcal{O}$, and v is harmonic on Ω. This is impossible in view of the Aronszajn-Cordes theorem 2.1(b), so the proof of Lemma 8.4 is complete.

We now give the proof of Theorem 8.1. In view of [BG1, Theorem 3.1.3], it suffices to show that if F is any closed subset of Ω such that the family of components of int F does not satisfy the long islands condition, then there must exist a continuous function $\lambda : F \to (0, 1]$ such that any harmonic function $h : \Omega \to \mathbf{R}$ satisfying $|h| < \lambda$ on F must be identically equal to zero. To prove this we note that by assumption there exists a sequence of components W_j of int F, a sequence of points $a_j \in W_j$, and a sequence of parametric balls $B_j \subset W_j$, such that the sequence of points $\{a_j\}$ is bounded in Ω and the sequence of balls B_j converges to the ideal point $*$. We may assume that the sequence a_j converges to a point $a \in \Omega$; it is then clear that in fact $a \in F$. We also may assume that a and all the points a_j lie in a single parametric ball, and we will now use coordinates in this parametric ball.

We next apply Lemmas 8.2 and 8.4 to conclude that for each positive integer j there exists a positive constant δ_j with the following property: any harmonic function $h : \Omega \to \mathbf{R}$ such that $\sup_{W_j} |h| \leq 1$ and $\sup_{B_j} |h| \leq \delta_j$ must satisfy $|D^\alpha h(a_j)| \leq 1/j$ for every multi-index α with $|\alpha| \leq j$; it follows that such a function h must have all its derivatives at a equal to zero, and from the Aronszajn-Cordes theorem 2.1(b) we conclude that $h \equiv 0$ on Ω. Since we may construct a continuous function $\lambda : F \to (0, 1]$ such that $\lambda < \delta_j$ on B_j, it follows that every harmonic function $h : \Omega \to \mathbf{R}$ satisfying $|h| < \lambda$ on F must be identically zero, as required. This completes the proof of Theorem 8.1, and the proof of Theorem 4.

References

[B] Brelot, M., Sur un théorème de prolongement fonctionnel de Keldych concernant le problème de Dirichlet, *J. Analyse Math.* **8** (1961), 273–288.

[BB] Bagby, T. and Blanchet, P., Uniform harmonic approximation on Riemannian manifolds, to appear in *J. Analyse Math.*

[BG1] Bagby, T. and Gauthier, P.M., Approximation by harmonic functions on closed subsets of Riemann surfaces, *J. Analyse Math.* **51** (1988), 259–284.

[BG2] Bagby, T. and Gauthier, P.M., Uniform approximation by global harmonic functions, in: *Approximation by Solutions of Partial Differential Equations* (B. Fuglede et al., eds.), NATO ASI Ser. C 365, Kluwer Academic Publishers, Dordrecht, 1992; 15–26.

[BH] Bliedtner, J. and Hansen, W., *Potential Theory: An Analytic and Probabilistic Approach*, Springer-Verlag, Berlin – New York, 1986.

[C] Constantinescu, C., Equicontinuity in harmonic spaces, *Nagoya Math. J.* **29** (1967), 1–6.

[CC] Constantinescu, C. and Cornea, A., *Potential Theory on Harmonic Spaces*, Springer-Verlag, Berlin, 1972.

[dP] du Plessis, N., *An Introduction to Potential Theory*, Hafner, Darien, CT, 1970.

[F] Fuglede, B., *Finely Harmonic Functions*, Lecture Notes in Math. **289**, Springer-Verlag, Berlin – Heidelberg – New York, 1972.

[G] Gauthier, P.M., Uniform approximation, *this volume*, 235–271.

[G1] Gardiner, S., Superharmonic extensions and harmonic approximation, preprint.

[G2] Gardiner, S., Tangential harmonic approximation on relatively closed sets, to appear in *Illinois J. Math.*

[GG] Gardiner, S. and Goldstein, M., Carleman approximation by harmonic functions, to appear in *Amer. J. Math.*

[GGG] Gardiner, S., Goldstein, M. and GowriSankaran, K., Global approximation in harmonic spaces, to appear in *Proc. Amer. Math. Soc.*

[He] Hervé, R.-M., Quelques propriétés des fonctions surharmoniques associées à une équation uniformément elliptique de la forme $Lu = -\sum_i \frac{\partial}{\partial x_i}\left(\sum_j a_{ij}\frac{\partial u}{\partial x_j}\right) = 0$, *Ann. Inst. Fourier (Grenoble)* **15** (1965), 215–224.

[Hö] Hörmander, L., *The Analysis of Linear Partial Differential Operators III*, Springer-Verlag, New York, 1985.

[N] Narasimhan, R., *Analysis on Real and Complex Manifolds*, 3rd printing, North-Holland, Amsterdam, 1985.

Pick interpolation, Von Neumann inequalities, and hyperconvex sets

Brian J. COLE and John WERMER

Department of Mathematics
Brown University
Providence, RI 02912
USA

Abstract

In 1916 Georg Pick started the study of interpolation by bounded analytic functions in the unit disk. In 1951 John Von Neumann proved an inequality for contractions on a Hilbert space. These two subjects are in fact closely connected, as was shown by Donald Sarason in 1967. We continue the study of this relationship, from the point of view of representations of uniform algebras by operators on a Hilbert space. Our work leads us to define and investigate certain convex bodies in $\mathbf{C}^n$ which we call hyperconvex sets.

0 Introduction

These lectures treat problems arising from interpolation by bounded analytic functions in the unit disk.

The classical problem we start with is this: let $\alpha = (\alpha_1, \alpha_2, \ldots, \alpha_n)$ be an n-tuple of distinct points in the open unit disk $|\zeta| < 1$. For which n-tuples $w = (w_1, \ldots, w_n)$ of complex numbers does there exist a bounded analytic function f on $|\zeta| < 1$ with $\sup_{|\zeta|<1} |f(\zeta)| \le 1$, such that

$$f(\alpha_j) = w_j, \quad 1 \le j \le n?$$

We denote by $\mathcal{D}(\alpha)$ the set of points w in $\mathbf{C}^n$ which can be obtained in this way. $\mathcal{D}(\alpha)$ is a compact convex body. We call $\mathcal{D}(\alpha)$ the *Pick body corresponding to* α, after Georg Pick whose paper [Pi] in 1916 began this study. (See the book of Garnett, [Gar] pp. 1-10, for an exposition of Pick's results.)

Pick bodies have certain geometrical properties as subsets of $\mathbf{C}^n$. In particular, if the point $w = (w_1, \ldots, w_n)$ belongs to $\mathcal{D}(\alpha)$ and if P is a polynomial with $\max_{|\zeta|\le 1} |P(\zeta)| \le 1$, then the point $(P(w_1), \ldots, P(w_n))$ also lies in $\mathcal{D}(\alpha)$. A generalization of this property is taken as the basis of definition for a class of sets in $\mathbf{C}^n$ which we call *hyperconvex* sets. Such hyperconvex sets arise naturally in many other interpolation problems and also arise in problems in operator theory on Hilbert space.

P. M. Gauthier (ed.) and G. Sabidussi (techn. ed.), Complex Potential Theory, 89–129.

In 1967 Donald Sarason in [Sa1] proved a theorem which implies the following result: let $\alpha_1, \ldots, \alpha_n$ be n points as above, let $A(\Gamma)$ denote the disk algebra, and let I be the ideal of all functions g in $A(\Gamma)$ such that $g(\alpha_j) = 0$, $1 \leq j \leq n$. Then the quotient algebra $A(\Gamma)/I$ is isometrically isomorphic to an operator algebra on a certain n-dimensional Hilbert space, given explicitly.

A few years later, Brian Cole proved a generalization of this result, where the disk algebra is replaced by an arbitrary uniform algebra. We shall define uniform algebras in Chapter 2. An exposition of Cole's representation theorem is given in the book [B-D] of Bonsall and Duncan, pp. 270-273.

In Chapter 1 we study a class of operator algebras on an n-dimensional Hilbert space which will play a role in our work. In Chapter 3 we study representations of a given uniform algebra by operators on a Hilbert space. As an application, we give a "Pick interpolation theorem" for an arbitrary uniform algebra.

In Chapter 4 we apply the results of Chapter 3 to derive Pick's Theorem for the disk and Agler's Theorem, [Ag], for the bidisk.

Chapter 5 is concerned with hyperconvex sets.

In Chapter 6 we obtain more detailed information on representations of the disk algebra, and give applications.

Chapter 7 gives references to some of the related literature.

Much of the material in these lectures is a further development of our papers [C-L-W 1] and [C-L-W 2], which are joint work with Keith Lewis.

We wish to thank Federica Andreano for her help with proofreading and to thank Natalie Johnson for her work on the typing.

1 A class of operator algebras

Let H be an n-dimensional Hilbert space. We choose a basis $\Psi_1, \Psi_2, \ldots, \Psi_n$ of H, where we do not assume that the Ψ_j are mutually orthogonal. For each $\alpha = (\alpha_1, \ldots, \alpha_n)$ in $\mathbf{C}^n$ we denote by S_α the operator on H given by

$$S_\alpha\Big(\sum_{j=1}^{n} t_j \Psi_j\Big) = \sum_{j=1}^{n} t_j \alpha_j \Psi_j. \tag{1.1}$$

Definition 1.1 $\mathcal{A} = \{S_\alpha \,|\, \alpha \in \mathbf{C}^n\}$.

$\mathcal{A}$ is an n-dimensional commutative operator algebra on H. The basis vectors Ψ_j, $1 \leq j \leq n$, are simultaneous eigenvectors of the operators S_α, $\alpha \in \mathbf{C}^n$. Under the map: $\alpha \to S_\alpha$, the vector space $\mathbf{C}^n$ is mapped isomorphically on $\mathcal{A}$. $\mathbf{C}^n$ is an algebra under coordinatewise multiplication, and the map: $\alpha \to S_\alpha$ is an algebraic isomorphism of $\mathbf{C}^n$ on $\mathcal{A}$. We give $\mathcal{A}$ the operator norm.

Definition 1.2 $\mathcal{D}_{\mathcal{A}} = \{\alpha \in \mathbf{C}^n \mid ||S_\alpha|| \le 1\}$.

We denote by $\phi_1, \phi_2, \ldots, \phi_n$ the dual basis of $\Psi_1, \Psi_2, \ldots, \Psi_n$, i.e.

$$(\Psi_i, \phi_j) = \delta_i^j \quad 1 \le i, j \le n.$$

Then for each $\alpha \in \mathbf{C}^n$,

$$S_\alpha^* \phi_j = \overline{\alpha}_j \phi_j, \quad 1 \le j \le n.$$

It is clear that $\mathcal{D}_{\mathcal{A}}$ is a compact convex set in $\mathbf{C}^n$. The set $\mathcal{D}_{\mathcal{A}}$ also has some less obvious properties, which follow from operators theoretic results of Von Neumann, [Vo], and Ando, [An]. Using coordinatewise multiplication in $\mathbf{C}^n$ we have for $\alpha = (\alpha_1, \alpha_2, \ldots, \alpha_n)$ and P a polynomial in z,

$$P(\alpha) = (P(\alpha_1), \ldots, P(\alpha_n)).$$

We denote by Δ the closed unit disk $|z| \le 1$ and by $||P||_\Delta$ the supremum of $|P(z)|$ over Δ.

Theorem 1.1 *Let $\alpha \in \mathcal{D}_{\mathcal{A}}$. If P is a polynomial with $||P||_\Delta \le 1$, then $P(\alpha) \in \mathcal{D}_{\mathcal{A}}$.*

Proof Von Neumann in [Vo] proved the following inequality: if T is a linear transformation on a Hilbert space with $||T|| \le 1$, and if P is a polynomial in z, then

$$||P(T)|| \le ||P||_\Delta. \tag{1.2}$$

To prove Theorem 1.1 we observe that, for each j,

$$P(S_\alpha)\Psi_j = P(\alpha_j)\Psi_j = S_{P(\alpha)}\Psi_j$$

and so $P(S_\alpha) = S_{P(\alpha)}$. Since $\alpha \in \mathcal{D}_{\mathcal{A}}$, $||S_\alpha|| \le 1$. By (1.2), then $||S_{P(\alpha)}|| = ||P(S_\alpha)|| \le 1$, and so $P(\alpha) \in \mathcal{D}_{\mathcal{A}}$, as desired. □

We shall give a direct proof of Theorem 1.1 in our special situation, obtaining some intermediate results which we shall use later on, in Section 6.

Fix $\alpha = (\alpha_1, \alpha_2, \ldots, \alpha_n) \in \mathbf{C}^n$ with $|\alpha_j| < 1$ for all j. Put

$$l_j(\theta) = \frac{1}{1 - \overline{\alpha}_j e^{i\theta}}, \qquad 0 \le \theta \le 2\pi, \quad j = 1, \ldots, n.$$

Definition 1.3 For $t = \sum_j t_j \phi_j$ and $s = \sum_k s_k \phi_k$ in H and $0 \le \theta \le 2\pi$, put

$$K_\alpha(\theta, s, t) = \sum_{j,k=1}^{n} t_j \overline{s}_k l_j(\theta) \overline{l_k(\theta)} (1 - \overline{\alpha}_j \alpha_k)(\phi_j, \phi_k).$$

Lemma 1.1 *For all $t, s \in H$ and for all polynomials P,*

$$(P(S_\alpha)t, s) = \frac{1}{2\pi} \int_0^{2\pi} K_\alpha(\theta, s, t) P(e^{i\theta})\, d\theta. \tag{1.3}$$

Proof Fix $p \geq 0$ and fix j, k. Then

$$\begin{aligned}\frac{1}{2\pi}\int_0^{2\pi} e^{ip\theta} l_j(\theta)\overline{l_k(\theta)}\, d\theta &= \frac{1}{2\pi}\int_0^{2\pi} \frac{e^{ip\theta}}{(1-\overline{\alpha}_j e^{i\theta})(1-\alpha_k e^{-i\theta})}\, d\theta \\ = \frac{1}{2\pi}\int_0^{2\pi} \frac{e^{ip\theta}e^{i\theta}\, d\theta}{(1-\overline{\alpha}_j e^{i\theta})(e^{i\theta}-\alpha_k)} &= \frac{1}{2\pi i}\int_{|z|=1} \frac{z^p\, dz}{(1-\overline{\alpha}_j z)(z-\alpha_k)} \\ &= \frac{\alpha_k^p}{1-\overline{\alpha}_j\alpha_k}\end{aligned}$$

or, in other words,

$$\frac{1}{2\pi}\int_0^{2\pi} e^{ip\theta} l_j(\theta)\overline{l_k(\theta)}\, d\theta = \frac{\alpha_k^p}{1-\overline{\alpha}_j\alpha_k}. \tag{1.4}$$

It follows that for s, t in H,

$$\begin{aligned}\frac{1}{2\pi}\int_0^{2\pi} K_\alpha(\theta, s, t)e^{ip\theta}\, d\theta &= \sum_{j,k=1}^{n} t_j\overline{s}_k(1-\overline{\alpha}_j\alpha_k)(\phi_j, \phi_k)\frac{1}{2\pi}\int_0^{2\pi} l_j(\theta)\overline{l_k(\theta)}e^{ip\theta}\, d\theta \\ &= \sum_{j,k=1}^{n} t_j\overline{s}_k(\phi_j, \phi_k)\alpha_k^p.\end{aligned} \tag{1.5}$$

On the other hand,

$$\begin{aligned}(S_\alpha^p t, s) &= (t, S_\alpha^{*p}s) = \left(\sum_j t_j\phi_j, \sum_k \overline{\alpha}_k^p s_k\phi_k\right) \\ &= \sum_{j,k} t_j\overline{s}_k(\phi_j, \phi_k)\alpha_k^p.\end{aligned} \tag{1.6}$$

So we get

$$(S_\alpha^p t, s) = \frac{1}{2\pi}\int_0^{2\pi} K_\alpha(\theta, s, t)e^{ip\theta}\, d\theta, \quad p = 0, 1, 2, \ldots. \tag{1.7}$$

Since $P(z)$ is a finite linear combination of powers of z, (1.3) follows. □

Lemma 1.2 *Fix α. Suppose $K_\alpha(\theta, t, t) \geq 0$ for all t in H and all θ in $[0, 2\pi]$. Then*

$$\frac{1}{2\pi}\int_0^{2\pi} |K_\alpha(\theta, s, t)|\, d\theta \leq ||s||\, ||t||, \quad \textit{for all } s, t \in H. \tag{1.8}$$

Proof Fix θ in $[0, 2\pi]$ and define

$$[s, t]_\theta = K_\alpha(\theta, s, t) \quad \text{for } s,\ t \text{ in } H.$$

In view of our hypothesis, $[\ ,\ ]_\theta$ is a (semi-definite) inner product on H. The Schwarz inequality in this inner product gives

$$|K_\alpha(\theta, s, t)| \leq \sqrt{K_\alpha(\theta, s, s)}\,\sqrt{K_\alpha(\theta, t, t)}.$$

It follows that

$$\int_0^{2\pi} |K_\alpha(\theta,s,t)|\,d\theta \le \sqrt{\int_0^{2\pi} K_\alpha(\theta,s,s,)\,d\theta}\sqrt{\int_0^{2\pi} K_\alpha(\theta,t,t)\,d\theta}. \tag{1.9}$$

From (1.3) with $P=1$ we have that

$$\int_0^{2\pi} K_\alpha(\theta,t,t)\,d\theta = 2\pi(t,t),$$

and similarly for s. So (1.9) gives

$$\int_0^{2\pi} |K_\alpha(\theta,s,t)|\,d\theta \le 2\pi\|s\|\,\|t\|,$$

and so (1.8). □

Proof of Theorem 1.1 Fix $\alpha \in \mathcal{D}_A$ with $|\alpha_j|<1$ for all j. Then $\|S_\alpha^*\| = \|S_\alpha\| \le 1$. Fix $c_1,\ldots,c_n \in \mathbf{C}$ and put $c = \sum_j c_j\phi_j$. Then $\|S_\alpha^* c\|^2 \le \|c\|^2$, or

$$\left(\sum_j c_j\overline{\alpha}_j\phi_j, \sum_k c_k\overline{\alpha}_k\phi_k\right) \le \left(\sum_j c_j\phi_j \sum_k c_k\phi_k\right)$$

and so

$$\sum_{j,k} c_j\overline{c}_k(1-\overline{\alpha}_j\alpha_k)(\phi_j,\phi_k) \ge 0. \tag{1.10}$$

Fix $\theta \in [0,2\pi]$, fix s, t in H, and put $c_j = t_j l_j(\theta)$, $1 \le j \le n$. Then $\overline{c}_k = \overline{t}_k\overline{l_k(\theta)}$. So (1.10) gives

$$\sum_{j,k} t_j\overline{t}_k l_j(\theta)\overline{l_k(\theta)}(1-\overline{\alpha}_j\alpha_k)(\phi_j,\phi_k) \ge 0.$$

Thus $K_\alpha(\theta,t,t) \ge 0$ for all t in H.

Let P be a polynomial in z. By Lemma 1.1,

$$\begin{aligned}
|(P(S_\alpha)t,s)| &\le \frac{1}{2\pi}\int_0^{2\pi} |K_\alpha(\theta,s,t)|\;|P(e^{i\theta})|\,d\theta \\
&\le (\|P\|_\Delta)\frac{1}{2\pi}\int_0^{2\pi} |K_\alpha(\theta,s,t)|\,d\theta \\
&\le \|P\|_\Delta\|s\|\,\|t\|,
\end{aligned}$$

where we have used Lemma 1.2 in the last inequality. Since this holds for all s,t in H, we conclude that

$$\|P(S_\alpha)\| \le \|P\|_\Delta. \tag{1.11}$$

We now choose α in $\mathcal{D}_A$ and remove the restriction that $|\alpha_j|<1$ for all j. Fix u in $0<u<1$ and consider the point $u\alpha$. This point lies in $\mathcal{D}_A$ and satisfies $|(u\alpha)_j|<1$ for all j. Inequality (1.11) yields $\|P(S_{u\alpha})\| \le 1$ if $\|P\|_\Delta \le 1$. Letting $u \to 1$, we get $\|P(S_\alpha)\| \le 1$ and so $P(\alpha) \in \mathcal{D}_A$, as desired. □

Notation For each integer $k \ge 1$ we write Δ^k for the closed unit polydisk in $\mathbf{C}^n$ and for each polynomial P in k variables we write $\|P\|_{\Delta^k}$ for the maximum of $|P|$ over Δ^k.

Theorem 1.2 *Let $\mathcal{D}_A$ be as earlier. If α', α'' is a pair of points in $\mathcal{D}_A$ and if P is a polynomial in two variables with $||P||_{\Delta^2} \leq 1$, then $P(\alpha', \alpha'') \in \mathcal{D}_A$.*

Proof If $\alpha' = (\alpha'_1, \ldots, \alpha'_n)$ and $\alpha'' = (\alpha''_1, \ldots, \alpha''_n)$, then

$$P(\alpha', \alpha'') = (P(\alpha'_1, \alpha''_1), P(\alpha'_2, \alpha''_2), \ldots, P(\alpha'_n, \alpha''_n)) \text{ in } \mathbf{C}^n.$$

Theorem 1.2 follows from Ando's Theorem in [An]: Let T_1 and T_2 be commuting contractions on a Hilbert space and let P be a polynomial in two variables. Then

$$||P(T_1, T_2)|| \leq \max |P| \text{ over } \Delta^2. \tag{1.12}$$

Theorem 1.2 follows from Ando's inequality (1.12) in the same way that Theorem 1.1 follows from Von Neumann's inequality (1.2). □

We now show that the general case of Von Neumann's inequality may be reduced to the special case we considered in Theorem 1.1.

Let H be a Hilbert space and let T be a linear operator of norm ≤ 1 on H. We call such an operator a *contraction*.

Assume now that $\dim H < \infty$ and let T be a contraction on H. By elementary linear algebra we can approximate T arbitrarily closely by a contraction S on H such that S has distinct eigenvalues. Then the eigenvectors of S form a basis of H. Our proof of Theorem 1.1 shows that Von Neumann's inequality (1.2) holds for S. By approximation then, (1.2) holds for T.

Let now H be a Hilbert space with dimension $\leq \infty$ and let T be a contraction on H. Let P be a polynomial in one variable, of degree N. Fix a vector x in H with $||x|| \leq 1$. Let $\mathcal{W}$ denote the subspace of H spanned by the vectors $x, Tx, T^2x, \ldots, T^N x$, and denote by E the orthogonal projection of H on $\mathcal{W}$. Then $\mathcal{W}$ is finite dimensional. Define the operator T' on $\mathcal{W}$ by

$$T'y = ETy, \quad y \in \mathcal{W}.$$

Then T' is a contraction on $\mathcal{W}$. By the preceding, inequality (1.2) holds for T' on $\mathcal{W}$. $T'x = ETx = Tx$, since $Tx \in \mathcal{W}$. $(T')^2 x = T'(Tx) = ET^2x = T^2x$, since $T^2x \in \mathcal{W}$. Continuing in this way, we get that $(T')^k x = T^k x, 0 \leq k \leq N$. Hence $P(T')x = P(T)x$. It follows that $||P(T)x|| = ||P(T')x|| \leq ||P||_\Delta$. Since this holds for all x in H, $||P(T)|| \leq 1$. Thus (1.2) holds for T. So, this shows that Von Neumann's inequality holds in the general case.

2 Interpolation problems

H^∞ denotes the space of bounded analytic functions on the disk $|\zeta| < 1$; for g in H^∞,

$$||g||_\infty = \sup_{|\zeta|<1} |g(\zeta)|.$$

Fix an n-tuple of points $\alpha_1, \ldots, \alpha_n$ in $|\zeta| < 1$. We define a function N on $\mathbf{C}^n$ by

$$N(w) = \inf\{||g||_\infty \,|\, g \in H^\infty,\ g(\alpha_j) = w_j,\ 1 \le j \le n\}, \tag{2.1}$$

where $w = (w_1, \ldots, w_n) \in \mathbf{C}^n$. Since every bounded sequence of elements of H^∞ has a pointwise converging subsequence, this infimum is always attained. Hence, given w in $\mathbf{C}^n$, there exists an $f \in H^\infty$ with $||f||_\infty \le 1$ satisfying

$$f(\alpha_j) = w_j, \quad 1 \le j \le n, \tag{2.2}$$

if and only if $N(w) \le 1$.

Claim *The function N is a norm on $\mathbf{C}^n$.*

Proof of Claim Clearly $N(w) \ge 0$ for all w and $= 0$ only for $w = 0$. Fix a, b in $\mathbf{C}^n$. Fix $\epsilon > 0$. Choose $g \in H^\infty$ with $g(\alpha_j) = a_j$ for all j and $||g||_\infty \le N(a) + \epsilon$ and choose h similarly for b. Then $(g + h)(\alpha_j) = a_j + b_j = (a + b)_j$ for all j. Hence $N(a + b) \le ||g + h||_\infty \le ||g||_\infty + ||h||_\infty \le N(a) + N(b) + 2\epsilon$. Since this holds for all ϵ, we conclude $N(a + b) \le N(a) + N(b)$.

The reader can also check that $N(ta) = |t|N(a)$ for each a in $\mathbf{C}^n$ and $t \in \mathbf{C}$. The Claim is proved.

So far we lack a formula for $N(w)$ in terms of w. A result of Pick in the paper, [Pi]: *Über die Beschränkungen analytischer Funktionen, welche durch vorgegebene Funktionswerte bewirkt werden*, Math. Ann. 77 (1916), 7–23, will allow us to give such a formula. The result is *Pick's Theorem*:

Theorem 2.1 *Fix $w \in \mathbf{C}^n$. Then there is an f in H^∞ with $||f||_\infty \le 1$, such that $f(\alpha_j) = w_j$, $1 \le j \le n$, if and only if the inequality*

$$\sum_{j,k=1}^{n} t_j \bar{t}_k \frac{w_j \overline{w}_k}{1 - \alpha_j \overline{\alpha}_k} \le \sum_{j,k=1}^{n} t_j \bar{t}_k \frac{1}{1 - \alpha_j \overline{\alpha}_k} \tag{2.3}$$

holds for every $(t_1, \ldots, t_n)$ in $\mathbf{C}^n$.

We put

$$(t, s)_\alpha = \sum_{j,k=1}^{n} t_j \bar{s}_k \frac{1}{1 - \alpha_j \overline{\alpha}_k}, \quad t = (t_1, \ldots, t_n), \quad s = (s_1, \ldots, s_n). \tag{2.4}$$

Lemma 2.1 *For every $t = (t_1, \ldots, t_n)$ we have $(t, t)_\alpha \ge 0$. If $t \ne 0$, $(t, t)_\alpha > 0$.*

Proof For all j, k

$$\frac{1}{1 - \alpha_j \overline{\alpha}_k} = \sum_{p=0}^{\infty} (\alpha_j)^p (\overline{\alpha}_k)^p.$$

Hence

$$\begin{aligned}(t,t)_\alpha &= \sum_{j,k=1}^{n} t_j\bar{t}_k \frac{1}{1-\alpha_j\overline{\alpha}_k} = \sum_{j,k=1}^{n}\sum_{p=0}^{\infty} t_j\bar{t}_k(\alpha_j)^p(\overline{\alpha}_k)^p \\ &= \sum_{p=0}^{\infty}\left(\sum_{j,k=1}^{n}(t_j\alpha_j^p)\overline{(t_k\alpha_k^p)}\right) = \sum_{p=0}^{\infty}\left|\sum_{j=1}^{n} t_j\alpha_j^p\right|^2.\end{aligned}$$

Hence for all t, $(t,t)_\alpha \geq 0$. Also $(t,t)_\alpha = 0$ implies

$$\sum_{j=1}^{n} t_j(\alpha_j)^p = 0, \quad p = 0, 1, \ldots, n-1.$$

Since the Vandermonde determinant $\det(((\alpha_j)^p)) \neq 0$, it follows that $t = 0$, proving the Lemma. □

The function $(\ ,\ )_\alpha$ is a Hermitian bilinear form on $\mathbf{C}^n$ and in view of the Lemma, it is an inner product on $\mathbf{C}^n$ making $\mathbf{C}^n$ into an n-dimensional Hilbert space H_α. We define for each $w = (w_1, \ldots, w_n) \in \mathbf{C}^n$ an operator T_w on H_α by

$$T_w(t_1, \ldots, t_n) = (w_1t_1, \ldots, w_nt_n), \quad t \in H_\alpha.$$

Claim *Fix w in $\mathbf{C}^n$. Then $||T_w|| \leq 1$ in operator norm on H_α if and only if (2.3) holds for all t in $\mathbf{C}^n$.*

Proof of Claim Fix $t = (t_1, \ldots, t_n)$ in H_α. Then

$$\begin{aligned}||T_wt||^2 &= \sum_{j,k=1}^{n}(w_jt_j)(\overline{w}_k\bar{t}_k)\frac{1}{1-\alpha_j\overline{\alpha}_k} \quad \text{and} \\ ||t||^2 &= \sum_{j,k=1}^{n} t_j\bar{t}_k\frac{1}{1-\alpha_j\overline{\alpha}_k},\end{aligned}$$

in the norm of H_α. So $||T_w|| \leq 1$ if and only if for all $t \in H_\alpha$ $||T_wt|| \leq ||t||$ if and only if (2.3) holds. The Claim is proved.

This Claim, together with Theorem 2.1 gives that, if $w \in \mathbf{C}^n$, then there is an $f \in H^\infty$ with $||f||_\infty \leq 1$ such that $f(\alpha_j) = w_j$, $1 \leq j \leq n$, if and only if $||T_w|| \leq 1$.

We also know that such an f exists if and only if $N(w) \leq 1$. Thus $N(w) \leq 1$ and $||T_w|| \leq 1$ are equivalent. But $w \to ||T_w||$ evidently is a norm on $\mathbf{C}^n$ and $w \to N(w)$ is likewise a norm. These two norms have the same closed unit ball, hence must coincide. Thus we have

Theorem 2.2 *Let N be defined for w in $\mathbf{C}^n$ by*

$$N(w) = \inf\{||f||_\infty \mid f(\alpha_j) = w_j,\ 1 \leq j \leq n,\ \textit{where } f \in H^\infty\}.$$

Then $N(w) = ||T_w||$, $w \in \mathbf{C}^n$.

This gives a formula for $N(w)$ in terms of w, as desired. A more explicit, but also more complicated, formula for $N(w)$ can be obtained by expressing the operator norm $||T_w||$ by means of linear algebra.

We next shall generalize Theorem 2.2 by replacing the unit disk and the space H^∞ by a class of sets and algebras of functions on them. Let X be a compact Hausdorff space and A an algebra of continuous complex-valued functions defined on X. A is called a *uniform algebra* on X if A is closed under uniform convergence on X, contains the constants and separates the points of X. If A is such an algebra, it has a natural norm:

$$||f|| = \max_{x \in X} |f(x)|, \quad f \in A.$$

A is a commutative Banach algebra in this norm. By Gelfand's theory, X lies embedded as a closed subset of a certain compact Hausdorff space $\mathcal{M}$, the maximal ideal space of A, in such a way that each f in A has a canonical continuous extension to $\mathcal{M}$ which is again denoted f. A separates the points of $\mathcal{M}$. Also, if $M \in \mathcal{M}$, then $|f(M)| \leq ||f||$, $f \in A$.

We fix n points $M_1, \ldots, M_n$ in $\mathcal{M}$. Since A is an algebra and separates the points M_j, given $w = (w_1, \ldots, w_n)$ in $\mathbf{C}^n$ we can always choose elements $g \in A$ such that $g(M_j) = w_j$, $1 \leq j \leq n$.

Definition 2.1 The *interpolation set* $\mathcal{D}$ consists of all points $w = (w_1, \ldots, w_n)$ in $\mathbf{C}^n$ such that for each $\epsilon > 0$ there is an f in A with $||f|| \leq 1 + \epsilon$ and $f(M_j) = w_j$, $1 \leq j \leq n$.

Definition 2.2 The function N is defined on $\mathbf{C}^n$ by

$$N(w) = \inf\{||g|| \mid g \in A, \; g(M_j) = w_j, \; 1 \leq j \leq n\}, \quad w \in \mathbf{C}^n.$$

Clearly, $\mathcal{D} = \{w \mid N(w) \leq 1\}$. By the same argument as we used in the proof of the Claim earlier, we see that N is a norm on $\mathbf{C}^n$ and that $\mathcal{D}$ is the closed unit ball of $\mathbf{C}^n$ in this norm. We wish to find an analogue, for a given uniform algebra A and given points $M_1, \ldots M_n$, for Theorem 2.2 which expressed $N(w)$ in terms of w in the case of the unit disk. We shall give such a formula for $N(w)$ in the next chapter, in Theorem 3.2. This will then allow us, in Theorem 3.3, to give necessary and sufficient conditions on a point w in order that w belong to the interpolation set $\mathcal{D}$. These conditions will be inequalities generalizing (2.3).

We next give some examples of uniform algebras and their maximal ideal spaces.

Example 1 Γ is the unit circle $|\zeta| = 1$. $A(\Gamma)$ is the *disk algebra*, consisting of all continuous functions on Γ which admit analytic extensions to $|\zeta| < 1$. $\mathcal{M}$ is the closed unit disk $|\zeta| \leq 1$.

Example 2 T^2 is the torus $|z_1| = 1, |z_2| = 1$ in $\mathbf{C}^2$. $A(T^2)$ is the *bidisk algebra*, consisting of all continuous functions on T^2 which admit an analytic extension to the open bidisk: $|z_1| < 1$, $|z_2| < 1$ in $\mathbf{C}^2$ which is continuous on the closed bidisk. $A(T^2)$ is a uniform algebra on T^2. Here $\mathcal{M}$ is the closed bidisk.

Example 3 B^n is the closed unit ball in $\mathbf{C}^n$, S^{2n-1} is its boundary. $A(B^n)$ is the *ball algebra* consisting of all continuous functions on S^{2n-1} which admit analytic extensions to the open ball $\text{int}(B^n)$. $A(B^n)$ is a uniform algebra on the sphere S^{2n-1} and $\mathcal{M} = B^n$.

Example 4 Let K be a compact set in $\mathbf{C}^n$. $A = \mathcal{P}(K)$, the uniform closure on K of the restrictions to K of all polynomials in the coordinate functions $z_1, \ldots, z_n$. $\mathcal{P}(K)$ is a uniform algebra on K. Here $\mathcal{M}$ is the polynomially convex hull of K consisting of all points $\lambda \in \mathbf{C}^n$ such that $|Q(\lambda)| \leq \max_K |Q|$ for every polynomial Q in n variables.

The reader can extend this list of examples at will.

A general result about interpolation in a uniform algebra thus yields information in many different situations which arise in complex analysis. Background concerning uniform algebras can be found in particular in the books by A. Browder, [Br], T. W. Gamelin, [Gam], and E. L. Stout, [St].

3 Representations of a uniform algebra

Let A denote a uniform algebra on a space X and let $\mathcal{M}$ be the maximal ideal space of A. As earlier, we fix n points M_j, $1 \leq j \leq n$, in $\mathcal{M}$. We define

$$I = \{f \in A \mid f(M_j) = 0, \ \ 1 \leq j \leq n\}.$$

Then I is a closed ideal in A. We form the quotient algebra A/I of A by the ideal I, consisting of all the cosets $[f]$ relative to I, where $f \in A$. The norm of $[f]$ in A/I is given by

$$\|[f]\| = \inf \|g\|, \quad \text{taken over all } g \text{ in } [f].$$

In this norm A/I is a commutative Banach algebra. We wish to construct representations of the algebra A/I by operators on certain finite-dimensional Hilbert spaces. In greater generality, these representations were introduced by Brian Cole in 1969, and this work is covered in the book, [B-D], by Bonsall and Duncan, in Chapter 7, §50.

We fix a probability measure μ on X. Let $L^2(\mu)$ be the corresponding L^2-space and define $H^2(\mu)$ as the closure of A in $L^2(\mu)$. Then $I \subseteq H^2(\mu)$. We denote by $\overline{I}$ the closure of I in $H^2(\mu)$ and we denote by $I^\perp$ the orthogonal complement of $\overline{I}$ in $H^2(\mu)$. Then

$$H^2(\mu) = I^\perp \oplus \overline{I}.$$

Let E denote the orthogonal projection of $H^2(\mu)$ on $I^\perp$.

Definition 3.1 For each f in A, we put

$$S_f\phi = E(f\phi), \quad \phi \text{ in } H^2(\mu).$$

Then S_f is a bounded linear operator on $H^2(\mu)$. The assertions of the following Lemma may be verified immediately.

Lemma 3.1 *For all f, g in A we have:* (i) $S_{f+g} = S_f + S_g$, (ii) $S_{fg} = S_f S_g$, (iii) $S_1 =$ *the identity, and* (iv) $||S_f|| \leq ||f||$ *where the norm on the left is the operator norm of S_f on $H^2(\mu)$ and the norm on the right is the norm in A.*

Fix now f and g in A with $[f] = [g]$. Then $f - g \in \bar{I}$. Fix $\phi \in H^2(\mu)$. Then $(f-g)\phi \in \bar{I}$, and so $S_f(\phi) - S_g(\phi) = E(f\phi) - E(g\phi) = E((f-g)\phi) = 0$. Thus $S_f\phi = S_g\phi$. This holds for all ϕ and so $S_f = S_g$. Thus the operator S_f depends only on the coset of f in A/I.

Furthermore, for each $f \in A$ the operator S_f maps $I^\perp$ onto itself. We may hence regard S_f as a linear operator on $I^\perp$, and we shall do so.

Definition 3.2 For μ a probability measure on X, S^μ is the map of A/I on operators on $I^\perp$ which sends

$$[f] \to S_f \quad \text{for each} \quad f \quad \text{in} \quad A.$$

It follows directly from Lemma 3.1 that S^μ is a norm-decreasing homomorphism of A/I into the operators on $I^\perp$. Among all probability measures on X we shall now single out a class of measures μ for which the representation S^μ has special properties. We fix an n-tuple of points $M_1, \ldots, M_n$ in $\mathcal{M}$.

Definition 3.3 A probability measure μ on X is *dominating* if for each j, $1 \leq j \leq n$, the functional: $f \to f(M_j)$, $f \in A$, is bounded in $L^2(\mu)$-norm.

As an example, we may consider the disk algebra $A(\Gamma)$ and choose the points $M_1, \ldots, M_n$ in the open unit disk. Then the measure $d\theta/2\pi$ on Γ is a dominating measure, where we identify Γ with $[0, 2\pi]$ in the usual way.

Let now A, X, $M_1, \ldots, M_n$ be as above and fix a dominating measure μ. We define $I = \{g \in A \,|\, g(M_j) = 0,\ 1 \leq j \leq n\}$. Fix j. Since $f \to f(M_j)$ is a functional on A bounded in $L^2(\mu)$-norm there exists a unique element l_j in $H^2(\mu)$ such that

$$f(M_j) = (f, l_j) = \int_X f\bar{l}_j \, d\mu, \quad f \in A. \tag{3.1}$$

Here the inner product (,) is the inner product in $L^2(\mu)$. We shall call l_j the *Szegö functional* for M_j, relative to μ.

If $g \in I$, $0 = g(M_j) = (g, l_j)$ and so $l_j \perp g$. Thus $l_j \in I^\perp$ for each j. The set $l_1, \ldots, l_n$ is linearly independent. Hence $\dim(I^\perp) \geq n$. Also, I has codimension n in A. Hence $\dim(I^\perp) \leq n$. Thus $\dim(I^\perp) = n$. Hence $l_1, l_2, \ldots, l_n$ is a basis of $I^\perp$.

Lemma 3.2 *For any $f \in A$, $S_f^* l_j = \overline{f(M_j)}\, l_j$, $j = 1, \ldots, n$.*

Proof Fix $\phi \in I^\perp$. Choose a sequence $\{g_n\} \in A$ with $g_n \to \phi$ in $L^2(\mu)$-norm. Then, for all $f \in A$ and for all j,

$$(fg_n, l_j) = f(M_j)g_n(M_j) = f(M_j)(g_n, l_j).$$

Letting $n \to \infty$, we get

$$(f\phi, l_j) = f(M_j)(\phi, l_j)$$

and so

$$\begin{aligned}(\phi, S_f^* l_j) &= (S_f\phi, l_j) = (E(f\phi), l_j) = (f\phi, l_j) \\ &= f(M_j)(\phi, l_j) = (\phi, \overline{f(M_j)} l_j).\end{aligned}$$

Since this holds for all $\phi \in I^\perp$, we get the assertion of the Lemma. □

Theorem 3.1 *Let μ be a probability measure on X. The representation S^μ is an isomorphism if and only if μ is a dominating measure.*

Proof Suppose that μ is dominating. Fix f in A such that S^μ sends $[f]$ into 0. Then $S_f = 0$ and so $S_f^* = 0$. By the last Lemma, this gives that $f(M_j) = 0$ for all j, and so $[f] = 0$. Thus S^μ is an isomorphism.

Conversely, assume S^μ is an isomorphism. Let $\mathcal{A}$ denote the image of S^μ which consists of all operators S_f with $f \in A$. Since $\mathcal{A}$ is n-dimensional and semi-simple, and since $I^\perp$ is an n-dimensional Hilbert space, $I^\perp$ contains a basis of simultaneous eigenvectors of the S_f, $f \in A$. Let $\phi_1, \ldots, \phi_n$ be the dual basis. Then $S_f^*\phi_j = \overline{f(M_j)}\phi_j$, for each j, after a suitable relabeling of the ϕ_j. We can write

$$1 = e_1 + e_2, \quad \text{with } e_1 \in I^\perp \text{ and } e_2 \in \overline{I}.$$

Fix j. For each f in A, then, we have

$$\begin{aligned}(f, \phi_j) &= (Ef, \phi_j) = (E(fe_1 + fe_2), \phi_j) = (S_f e_1, \phi_j) + (E(fe_2), \phi_j) \\ &= (e_1, \overline{f(M_j)}\phi_j) = f(M_j)(e_1, \phi_j),\end{aligned}$$

where we have used that $fe_2 \in \overline{I}$. If $(e_1, \phi_j) = 0$, then $(f, \phi_j) = 0$ for all f in A and so $\phi_j = 0$, contrary to choice of ϕ_j. Hence $(e_1, \phi_j) \neq 0$, and so

$$f(M_j) = \frac{(f, \phi_j)}{(e_1, \phi_j)} \quad \text{for all } f \text{ in } A.$$

Hence the functional $f \to f(M_j)$ is bounded in $L^2(\mu)$-norm. This holds for each j, and so μ is dominating. We are done. □

Fix w in $\mathbf{C}^n$. Choose f in A with $f(M_j) = w_j$, $1 \le j \le n$. The quotient norm $\|[f]\|$ of $[f]$ in A/I is defined by $\|[f]\| = \inf\{\|g\| \,|\, g \in A,\ g \in [f]\}$. A given $g \in A$ belongs to $[f]$ if and only if $g(M_j) = w_j$, $1 \le j \le n$. Hence $\|[f]\| = \inf\{\|g\| \,|\, g \in A,\ g(M_j) = w_j,\ 1 \le j \le n\}$. But this coincides with $N(w)$ as defined in Definition 2.2. Hence

$$N(w) = \|[f]\|, \quad f \in A, \quad f(M_j) = w_j, \quad 1 \le j \le n.$$

Our objective is to give a formula for $N(w)$ in terms of w. By what we just saw, this is equivalent to giving a formula for $\|[f]\|$.

The following theorem gives such a formula.

Theorem 3.2 *For each f in A, we have*

$$|||[f]||| = \sup ||S^{\mu}([f])|| \tag{3.2}$$

taken over all dominating measures μ.

Lemma 3.3 *Fix f in A such that $|||[f]||| = 1$. Then there exists a probability measure λ on X such that if we form $L^2(\lambda)$, $H^2(\lambda)$, $\overline{I}$, $I^{\perp}$, and E, corresponding to λ, as above, then*

$$|Ef| = 1 \quad a.e.\text{-}d\lambda, \tag{3.3}$$

and

$$1 \in I^{\perp}. \tag{3.4}$$

Proof The quotient norm $|||[f]|||$ is the distance in A from f to I, and so this distance $= 1$. By Banach space theory, it follows that there exists a bounded linear functional $L \in A^*$ such that

$$L(j) = 0, \quad j \in I, \quad ||L||_{A^*} = 1, \quad L(f) = 1.$$

Further, L has a norm preserving extension to a functional on $C(X)$ which is given by a complex measure ν on X. Then

$$\int j\,d\nu = 0, \quad j \in I, \quad ||\nu|| = 1, \quad \int f\,d\nu = 1.$$

We put $\lambda = |\nu|$, the total variation measure of ν, and we form $L^{\infty}(\lambda)$. We may write

$$\nu = \Phi d\lambda, \quad \text{where} \quad |\Phi| = 1 \quad \text{a.e.-}d\lambda.$$

Choose a sequence $\{j_n\}$ in I such that $||f - j_n|| \to 1$ as $n \to \infty$. Then the sequence $\{f - j_n\}$ has a subnet converging weak-$*$ in $L^{\infty}(\lambda)$ to a function $g \in L^{\infty}(\lambda)$. Also $||g||_{\infty} \le 1$, and $g \in H^2(\lambda)$. Then

$$1 = \int (f - j_n)\,d\nu = \int (f - j_n)\Phi\,d\lambda \to \int g\Phi\,d\lambda.$$

So $\int g\Phi\,d\lambda = 1$. Since $||g||_{\infty} \le 1$ and $|\Phi| = 1$ a.e.-$d\lambda$, we must have $g\Phi = 1$ a.e.-$d\lambda$, and $|g| = 1$ a.e.-$d\lambda$. So

$$\Phi = \overline{g} \quad \text{and so} \quad \nu = \overline{g}\,d\lambda.$$

For $j \in I$,

$$\int j\overline{g}\,d\lambda = \int j\,d\nu = 0 \quad \text{and so} \quad g \in I^{\perp}.$$

We thus have

$$\Phi = \overline{g}, \quad |g| = 1 \quad \text{a.e.,} \quad g \in I^{\perp}. \tag{3.5}$$

Claim $f - g \in \overline{I}$.

Proof of Claim Fix $h \in I^{\perp}$. Choose a net $\{j_n\}$ in I such that $f - j_n \to g$ weak-$*$ in $L^{\infty}(\lambda)$. Then $f - g - j_n \to 0$ weak-$*$. Hence, letting (,) denote the inner product in

$L^2(\lambda)$, $(f-g,h) = (f-g-j_n,h) \to 0$ as $n \to \infty$ and so $h \perp f-g$. Since this holds for all $h \in I^\perp$, $f - g \in \bar{I}$, as claimed.

Hence $E(f-g) = 0$. Since $f = g + (f-g)$, we have

$$Ef = Eg = g \quad \text{and so} \quad |Ef| = 1 \quad \text{a.e.,}$$

giving (3.3). Finally, for $j \in I$,

$$(j,1) = \int j\,d\lambda = \int jg\,d\nu = \lim_{n\to\infty} \int j(f-j_n)\,d\nu = 0,$$

since $j(f-j_n) \in I$ for all n. So $1 \in I^\perp$, i.e. (3.4). We are done with the proof of the Lemma. □

Proof of Theorem 3.2 Without loss of generality, $||[f]|| = 1$. For each probability measure μ, then $||S^\mu([f])|| \le 1$. Hence the righthand side in (3.2) ≤ 1. To prove (3.2) it thus suffices to construct, for each t with $0 \le t < 1$, a dominating measure μ_t such that

$$||S^{\mu_t}([f])|| \ge t.$$

For each j, $1 \le j \le n$, we choose a representing measure σ_j for M_j, in the sense that σ_j is a probability measure on X such that

$$\int g\,d\sigma_j = g(M_j), \quad g \in A.$$

We now put $\sigma = \frac{1}{n}\sum_{j=1}^{n} \sigma_j$. Then σ is a dominating measure, as follows by use of Schwarz's inequality.

Using Lemma 3.3 we choose a probability measure λ on X such that (3.3) and (3.4) hold. Let $\bar{I}$, $I^\perp$, and E be taken with respect to λ. Fix $g \in I$. Then $f = Ef + b$, where $b \in \bar{I}$ and so

$$f - g = Ef + (b - g),$$

and so

$$\int |f-g|^2\,d\lambda = \int |Ef|^2\,d\lambda + \int |b-g|^2\,d\lambda.$$

In view of (3.3), $\int |Ef|^2\,d\lambda = 1$ and so we have

$$\int |f-g|^2\,d\lambda \ge 1, \quad g \in I. \tag{3.6}$$

For each t, $0 \le t \le 1$, we define the measure

$$\mu_t = t\lambda + (1-t)\sigma.$$

Since σ is dominating, there is a constant C with

$$|g(M_j)|^2 \le C \int |g|^2\,d\sigma, \quad g \in A, \quad 1 \le j \le n.$$

Then

$$|g(M_j)|^2 \le \frac{C}{1-t}\int |g|^2(1-t)\,d\sigma \le \frac{C}{1-t}\int |g|^2\,d\mu_t.$$

So μ_t is a dominating measure. We form the corresponding spaces $H^2(\mu_t)$ and the decomposition with respect to μ_t:

$$H^2(\mu_t) = I_t^{\perp} + \overline{I}_t,$$

and we denote by E^t the orthogonal projection of $H^2(\mu_t)$ on $I_t^{\perp}$. So,

$$\int |E^t f|^2\,d\mu_t = \inf_{g\in I}\int |f-g|^2\,d\mu_t.$$

Define for fixed $g \in I$,

$$\phi_g(t) = t\int |f-g|^2\,d\lambda + (1-t)\int |f-g|^2\,d\sigma.$$

Then $\int |f-g|^2\,d\mu_t = \phi_g(t)$, and so

$$\int |E^t f|^2\,d\mu_t = \inf_{g\in I}\phi_g(t). \tag{3.7}$$

By (3.6), $\phi_g(1) \ge 1$. Also $\phi_g(0) \ge 0$. Since ϕ_g is a linear function of t on $0 \le t \le 1$, it follows that

$$\phi_g(t) \ge t, \quad 0 \le t \le 1$$

and so

$$\int |E^t f|^2\,d\mu_t \ge t. \tag{3.8}$$

Also, if $\beta \in I$, then

$$\int \beta\,d\mu_t = t\int \beta\,d\lambda + (1-t)\int \beta\,d\sigma = 0,$$

where we have used (3.4) as well as the fact that

$$\int \beta\,d\sigma = \frac{1}{n}\sum_{j=1}^{n}\beta(M_j) = 0.$$

Thus $1 \in I_t^{\perp}$ and so

$$||S^{\mu_t}([f])(1)|| = ||E^t f|| \ge \sqrt{t},$$

and so

$$\text{the operator norm} \quad ||S^{\mu_t}([f])|| \ge \sqrt{t} \ge t.$$

For $0 \le t < 1$, μ_t is a dominating measure. So our construction is finished and the Theorem is proved. □

Using Theorem 3.2, we shall next give conditions on a point $w = (w_1, \ldots, w_n)$ in $\mathbf{C}^n$ in order that w belong to the interpolation set $\mathcal{D}$ defined in Definition 2.1. Observe that $w \in \mathcal{D}$ if and only if $f \in A$ and $f(M_j) = w_j$, $1 \le j \le n$, implies that $||[f]|| \le 1$.

Fix a dominating measure μ. Form $H^2(\mu)$ and let $l_1, \ldots, l_n$ be the Szegö functionals corresponding to μ, i.e. each $l_j \in H^2(\mu)$ and satisfies (3.1). We form the $n \times n$ matrix of inner products

$$(l_j, l_k)_\mu = \int l_j \bar{l}_k \, d\mu$$

and shall consider the corresponding Hermitian form on $\mathbf{C}^n$. We note that the matrix $(((l_j, l_k)_\mu))$ is positive definite, since if $t_1, \ldots, t_n \in \mathbf{C}$, then

$$\sum_{j,k=1}^{n} t_j \bar{t}_k (l_j, l_k)_\mu = \int \left| \sum_{j=1}^{n} t_j l_j \right|^2 d\mu \geq 0$$

and $= 0$ only if the $t_j = 0$, since the l_j are linearly independent.

Theorem 3.3 *Fix $w = (w_1, \ldots, w_n)$ in $\mathbf{C}^n$. Then $w \in \mathcal{D}$ if and only if the condition*

$$\sum_{j,k=1}^{n} t_j \bar{t}_k (1 - \overline{w}_j w_k)(l_j, l_k)_\mu \geq 0, \quad (t_1, \ldots, t_n) \in \mathbf{C}^n, \tag{3.9}$$

holds for each dominating measure μ.

We note that by taking complex conjugates and replacing t_j by $\bar{t}_j$ for each j, (3.9) is shown to be equivalent to

$$\sum_{j,k=1}^{n} t_j \bar{t}_k (1 - w_j \overline{w}_k) \overline{(l_j, l_k)_\mu} \geq 0, \quad (t_1, \ldots, t_n) \in \mathbf{C}^n. \tag{3.10}$$

Lemma 3.4 *Choose a dominating measure μ, choose $f \in A$, and put $w_j = f(M_j)$, $1 \leq j \leq n$. Condition (3.9), resp. (3.10), is satisfied if and only if $||S^\mu([f])|| \leq 1$.*

Proof Choose $t \in I^\perp$. Then $t = \sum_{j=1}^n t_j l_j$, $t_j \in \mathbf{C}$ for all j. Put $w_j = f(M_j)$ for all j. Then $S_f^* l_j = \overline{w}_j l_j$, where $S_f = S^\mu([f])$. So,

$$\begin{aligned} ||t||^2 &= \sum_{j,k} t_j \bar{t}_k (l_j, l_k)_\mu, \\ ||S_f^* t||^2 &= \sum_{j,k} t_j \bar{t}_k \overline{w}_j w_k (l_j, l_k)_\mu. \end{aligned}$$

Condition (3.9) thus says that $||S_f^* t||^2 \leq ||t||^2$ for each $t \in I^\perp$. This is equivalent to $||S_f^*|| \leq 1$ and so to $||S_f|| \leq 1$, as desired. □

Proof of Theorem 3.3 Fix $w \in \mathbf{C}^n$. Choose f in A with $f(M_j) = w_j$, $1 \leq j \leq n$. By Lemma 3.4, (3.9) holds for a dominating measure μ if and only if $||S^\mu([f])|| \leq 1$. Hence $\sup ||S^\mu([f])||$ over all dominating measures μ is ≤ 1 if and only if (3.9) holds for all such μ. By Theorem 3.2, then, $||[f]|| \leq 1$ if and only if (3.9) holds for every dominating measure. This proves the Theorem. □

4 Pick's Theorem and Agler's Theorem

Fix n points $\alpha_1, \ldots, \alpha_n$ in the open unit disk. There are two interpolation sets naturally associated with this n-tuple. One is the Pick body

$$\mathcal{D}(\alpha) = \{w \mid \exists f \in H^\infty,\ \|f\|_\infty \le 1,\ f(\alpha_j) = w_j,\ 1 \le j \le n\}.$$

The other is the set

$$\mathcal{D} = \{w \mid \forall \epsilon > 0\ \exists f \in A(\Gamma),\ \|f\| \le 1 + \epsilon,\ f(\alpha_j) = w_j,\ 1 \le j \le n\},$$

which is the interpolation set associated to the disk algebra $A(\Gamma)$.

Claim $\mathcal{D} = \mathcal{D}(\alpha)$.

Proof of Claim If $w = (w_1, \ldots, w_n) \in \mathcal{D}(\alpha)$, there exists an $f \in H^\infty$ with $f(\alpha_j) = w_j$, $1 \le j \le n$, and $\|f\|_\infty \le 1$. For each r, $0 < r < 1$, the function $\zeta \to f(r\zeta)$ belongs to $A(\Gamma)$ and has norm ≤ 1. Hence the point $(f(r\alpha_1), \ldots, f(r\alpha_n)) \in \mathcal{D}$. As $r \to 1$, this point $\to w$. Since $\mathcal{D}$ is closed, $w \in \mathcal{D}$. So $\mathcal{D}(\alpha) \subseteq \mathcal{D}$.

Conversely, fix $w \in \mathcal{D}$. Choose, for each integer $n \ge 1$, $f_n \in A(\Gamma)$ with $f_n(\alpha_j) = w_j$ and $\|f_n\| \le 1 + (1/n)$. Without loss of generality, the sequence f_n converges pointwise on $|\zeta| < 1$ to $f \in H^\infty$. Then $\|f\|_\infty \le 1$ and $f(\alpha_j) = w_j$ for all j. So $w \in \mathcal{D}(\alpha)$. Thus $\mathcal{D} \subseteq \mathcal{D}(\alpha)$. The Claim is proved.

Pick's Theorem (Theorem 2.1) can thus be stated as follows: For each w in $\mathbf{C}^n$, $w \in \mathcal{D}$ if and only if Condition (2.3) holds. We further recall that $w \in \mathcal{D}$ if and only if $f \in A(\Gamma)$ and $f(\alpha_j) = w_j$, $1 \le j \le n$, implies $\|[f]\| \le 1$. Condition (2.3) can be expressed in the language of matrices. Condition (2.3) can be written

$$\sum_{j,k=1}^{n} t_j \bar{t}_k \frac{(1 - w_j \overline{w}_k)}{(1 - \alpha_j \overline{\alpha}_k)} \ge 0 \quad \text{for all } t_1, \ldots, t_n,$$

or

$$\left(\left(\frac{1 - w_j \overline{w}_k}{1 - \alpha_j \overline{\alpha}_k}\right)\right) \ge 0.$$

We put $A_{jk} = \dfrac{1 - w_j \overline{w}_k}{1 - \alpha_j \overline{\alpha}_k}$, $1 \le j,\ k \le n$. Inequality (2.3) is then equivalent to

$$1 - w_j \overline{w}_k = (1 - \alpha_j \overline{\alpha}_k) A_{jk}, \quad 1 \le j, k \le n, \tag{4.1}$$

where the matrix $((A_{jk})) \ge 0$ (is positive semi-definite).

Lemma 4.1 *Let* $((A_{\alpha\beta}))$, $((K_{\alpha\beta}))$ *be the two* $n \times n$ *matrices such that* $((A_{\alpha\beta})) \ge 0$ *and* $((K_{\alpha\beta})) \ge 0$. *Then the matrix* $((A_{\alpha\beta} K_{\alpha\beta})) \ge 0$.

Proof Let $\lambda_1, \ldots, \lambda_n$ be the eigenvalues of $((A_{\alpha\beta}))$. In suitable new coordinates t'_s, $s = 1, \ldots, n$, with $t'_s = \sum_{j=1}^n C_j^{(s)} t_j$, we have for all $t_1, \ldots, t_n$:

$$\begin{aligned}\sum_{\alpha,\beta} A_{\alpha\beta} t_\alpha \bar{t}_\beta &= \sum_{s=1}^n \lambda_s \left| \sum_{j=1}^n C_j^{(s)} t_j \right|^2 = \sum_{s=1}^n \lambda_s \Big(\sum_\alpha C_\alpha^{(s)} t_\alpha \Big) \Big(\sum_\beta \overline{C_\beta^{(s)} t_\beta} \Big) \\ &= \sum_{s=1}^n \lambda_s \sum_{\alpha,\beta} C_\alpha^{(s)} \overline{C_\beta^{(s)}} t_\alpha \bar{t}_\beta = \sum_{\alpha,\beta} \Big(\sum_{s=1}^n \lambda_s C_\alpha^{(s)} \overline{C_\beta^{(s)}} \Big) t_\alpha \bar{t}_\beta,\end{aligned}$$

and so

$$A_{\alpha\beta} = \sum_s \lambda_s C_\alpha^{(s)} \overline{C_\beta^{(s)}} \quad \text{for all } \alpha, \beta.$$

Fix $t_1, \ldots, t_n$. The above equation gives

$$\begin{aligned}\sum_{\alpha,\beta} K_{\alpha\beta} A_{\alpha\beta} t_\alpha \bar{t}_\beta &= \sum_{\alpha,\beta} K_{\alpha\beta} \Big(\sum_{s=1}^n \lambda_s C_\alpha^{(s)} \overline{C_\beta^{(s)}} \Big) t_\alpha \bar{t}_\beta \\ &= \sum_{s=1}^n \lambda_s \Big(\sum_{\alpha,\beta} K_{\alpha\beta} (C_\alpha^{(s)} t_\alpha) \overline{(C_\beta^{(s)} t_\beta)} \Big) \geq 0\end{aligned}$$

since $((K_{\alpha\beta})) \geq 0$ and $\lambda_s \geq 0$ for all s. So the Lemma is proved. □

Proof of Theorem 2.1 We are given points $\alpha_1, \ldots, \alpha_n$ in $|\zeta| < 1$. Choose w in $\mathbf{C}^n$ such that (2.3) is satisfied. As we saw, then (4.1) holds, i.e.

$$1 - w_j \overline{w}_k = (1 - \alpha_j \overline{\alpha}_k) A_{jk} \text{ with } ((A_{jk})) \geq 0.$$

Fix a dominating measure μ on Γ. Let S^μ be the corresponding representation of $A(\Gamma)/I$. For f in $A(\Gamma)$, write $S_f = S^\mu([f])$. Let f_0 be the identity function in $A(\Gamma)$. Then $||f_0|| \leq 1$, so $||S_{f_0}|| \leq 1$. Also $f_0(\alpha_j) = \alpha_j$, for all j.

By Lemma 3.4, then for all $t = (t_1, \ldots, t_n)$

$$\sum_{j,k} t_j \bar{t}_k (1 - \alpha_j \overline{\alpha}_k) \overline{(l_j, l_k)_\mu} \geq 0.$$

Since $((A_{jk})) \geq 0$, Lemma 4.1 allows us to conclude from this that

$$\sum_{j,k} t_j \bar{t}_k (1 - \alpha_j \overline{\alpha}_k) \overline{(l_j, l_k)_\mu} A_{jk} \geq 0 \quad \text{for all } t.$$

Hence we have

$$\sum_{j,k} t_j \bar{t}_k (1 - w_j \overline{w}_k) \overline{(l_j, l_k)_\mu} \geq 0 \quad \text{for all } t.$$

By Theorem 3.3 it follows that $w \in \mathcal{D}$. Thus (2.3) is a sufficient condition for w to belong to $\mathcal{D}$.

Conversely, suppose $w \in \mathcal{D}$. By Theorem 3.3, then, (3.9) holds for each dominating measure μ and so in particular for $\mu = d\theta/2\pi$. The corresponding Szegö functionals l_j are given by $l_j(\zeta) = \dfrac{1}{1-\overline{\alpha}_j\zeta}$, since if $f \in A(\Gamma)$, then

$$\int_\Gamma f\,\overline{l}_j\,\frac{d\theta}{2\pi} = \frac{1}{2\pi}\int_\Gamma f(\zeta)\frac{1}{1-\alpha_j\overline{\zeta}}\,d\theta = \frac{1}{2\pi i}\int_\Gamma f(\zeta)\,\frac{d\zeta}{\zeta-\alpha_j} = f(\alpha_j).$$

Then $(l_j, l_k)_\mu = l_j(\alpha_k) = \dfrac{1}{1-\overline{\alpha_j}\alpha_k}$, so $\overline{(l_j,l_k)}_\mu = \dfrac{1}{1-\alpha_j\overline{\alpha}_k}$. (Compare the proof of Lemma 1.1 in Chapter 1.) Inequality (3.9) thus gives

$$\sum_{j,k} t_j\bar{t}_k(1-w_j\overline{w}_k)\frac{1}{1-\alpha_j\overline{\alpha}_k} \geq 0, \quad (t_1,\ldots,t_n) \in \mathbf{C}^n$$

which is just (2.3). Thus (2.3) is a necessary condition for $w \in \mathcal{D}$. Theorem 2.1 is proved. □

Recall the bidisk algebra $A(T^2)$, discussed as Example 2 in Chapter 2. Jim Agler, [Ag], has proved an analogue of Pick's Theorem for this algebra. We next state Agler's Theorem and prove it making use of Theorem 3.3.

We fix n points $M_j = (\lambda_1^j, \lambda_2^j)$, $1 \leq j \leq n$, in the interior of Δ^2 and we put

$$I = \{g \in A(T^2) \,|\, g(M_j) = 0,\ 1 \leq j \leq n\}$$

We now form the quotient algebra $A(T^2)/I$. We denote by $\mathcal{D}$ the corresponding interpolation set.

Theorem 4.1 (Agler's Theorem) *Fix $w = (w_1,\ldots,w_n)$ in $\mathbf{C}^n$. Choose $f \in A(T^2)$ with $f(M_j) = w_j$, $1 \leq j \leq n$. Then $\|[f]\| \leq 1$ if and only if the following condition holds:*

$$\begin{cases} \text{there exist } n\times n \text{ matrices } ((A^1_{jk})) \geq 0,\ ((A^2_{jk})) \geq 0 \text{ such that} \\ 1 - w_j\overline{w}_k = (1-\lambda_1^j\overline{\lambda_1^k})A^1_{jk} + (1-\lambda_2^j\overline{\lambda_2^k})A^2_{jk},\ 1 \leq j,k \leq n. \end{cases} \tag{4.2}$$

We note that (4.2) generalizes the condition (4.1).

Proof of Sufficiency We fix w in $\mathbf{C}^n$ such that (4.2) holds. Let f_1 be the first coordinate function on $\mathbf{C}^2$: $f_1(\lambda_1,\lambda_2) = \lambda_1$. Then $f_1(M_j) = \lambda_1^j$, $1 \leq j \leq n$, and

$$\|[f_1]\| \leq \|f_1\| = 1.$$

Let μ be a dominating measure on T^2 for the points M_j, $1 \leq j \leq n$. Choose Szegö functionals $l_1,\ldots,l_n$ in $I^\perp$ corresponding to μ. Then $\|S^\mu([f_1])\| \leq 1$. By Lemma 3.4, then, (3.10) holds for $w = (f_1(M_1),\ldots,f_1(M_n)) = (\lambda_1^1,\lambda_1^2,\ldots,\lambda_1^n)$ and so

$$\sum_{j,k} t_j\bar{t}_k(1-\lambda_1^j\overline{\lambda_1^k})\overline{(l_j,l_k)}_\mu \geq 0, \quad t \in \mathbf{C}^n. \tag{4.3}$$

Since the matrix $((A^1_{jk})) \geq 0$, Lemma 4.1 together with (4.3) gives

$$\sum_{j,k} t_j\bar{t}_k(1-\lambda_1^j\overline{\lambda_1^k})\overline{(l_j,l_k)}_\mu A^1_{jk} \geq 0. \tag{4.4}$$

Similarly, we have

$$\sum_{j,k} t_j\bar{t}_k(1-\lambda_2^j\overline{\lambda_2^k})\overline{(l_j,l_k)}_\mu A^2_{jk} \geq 0. \tag{4.5}$$

Adding the last two inequalities, we get

$$\sum_{j,k} t_j\bar{t}_k[(1-\lambda_1^j\overline{\lambda_1^k})A^1_{jk} + (1-\lambda_2^j\overline{\lambda_2^k})A^2_{jk}]\overline{(l_j,l_k)}_\mu \geq 0. \tag{4.6}$$

Since (4.2) holds, this gives

$$\sum_{j,k} t_j\bar{t}_k(1-w_j\overline{w}_k)\overline{(l_j,l_k)}_\mu \geq 0.$$

Since this holds for every dominating measure μ, Theorem 3.3 gives that $w \in \mathcal{D}$. Sufficiency is proved. □

In order to prove Necessity, we need some preliminaries. We denote by $\mathcal{H}^n$ the real vector space of $n \times n$ Hermitian matrices. For each $g = ((g_{jk})) \in \mathcal{H}^n$ we denote by L_g the linear functional on $\mathcal{H}^n$ given by

$$L_g(A) = \sum_{j,k} g_{jk}A_{jk}, \quad A = ((A_{jk})). \tag{4.7}$$

The map: $g \to L_g$ from $\mathcal{H}^n$ to $(\mathcal{H}^n)^*$ is linear. The map is one-one, since

$$L_g = 0 \quad \text{implies} \quad L_g(\overline{g}) = 0 \quad \text{implies} \quad \sum_{j,k} |g_{jk}|^2 = 0$$

and hence $g = 0$. Since $\dim(\mathcal{H}^n) = \dim[(\mathcal{H}^n)^*]$, the map is onto $(\mathcal{H}^n)^*$. So we have: if $L \in (\mathcal{H}^n)^*$, then there is a $g \in \mathcal{H}^n$ such that $L = L_g$.

With the points $M_j = (\lambda_1^j, \lambda_2^j)$ as above, we define the set $\mathcal{K}$ of matrices by

$$\mathcal{K} = \{(((1-\lambda_1^j\overline{\lambda_1^k})A^1_{jk} + (1-\lambda_2^j\overline{\lambda_2^k})A^2_{jk})) \mid ((A^1_{jk})) \geq 0, ((A^2_{jk})) \geq 0\}.$$

It is clear that $\mathcal{K}$ is a convex cone, with vertex at the origin, contained in $\mathcal{H}^n$. We claim

$$\mathcal{K} \text{ is closed in } \mathcal{H}^n. \tag{4.8}$$

To prove (4.8), we consider a sequence $\{B^r\}$, $r = 1, 2, \ldots$ of elements of $\mathcal{K}$ converging to a matrix B in $\mathcal{H}^n$. Then there exist matrices $\{A^{r1}\} \geq 0$, $\{A^{r2}\} \geq 0$, $r = 1, 2, \ldots$ such that

$$B^r_{jk} = (1-\lambda_1^j\overline{\lambda_1^k})A^{r1}_{jk} + (1-\lambda_2^j\overline{\lambda_2^k})A^{r2}_{jk} \quad \text{for all } j, k, r. \tag{4.9}$$

In particular

$$B^r_{jj} = (1 - |\lambda^j_1|^2)A^{r1}_{jj} + (1 - |\lambda^j_2|^2)A^{r2}_{jj} \quad \text{for all } j, r.$$

Since the sequence of numbers $\{B^r_{jj}\}_r$ is bounded for each j, it follows that the sequence $\{A^{r1}_{jj}\}_r$ is bounded and so the sequence $\{\text{trace}(A^{r1})\}_r$ is bounded. Hence the sequence of matrices $\{A^{r1}\}_r$ is bounded, and so we can choose a convergent subsequence with limit A^1, where $A^1 \geq 0$. Similarly $\{A^{r2}\}_r$ has a convergent subsequence with limit $A^2 \geq 0$.

It follows from (4.9) that

$$B_{jk} = (1 - \lambda^j_1\overline{\lambda^k_1})A^1_{jk} + (1 - \lambda^j_2\overline{\lambda^k_2})A^2_{jk} \quad \text{for all } j, k$$

and so $B \in \mathcal{K}$. Hence $\mathcal{K}$ is closed, i.e. (4.8) holds.

Lemma 4.2 *Given a positive definite matrix* $((g_{jk}))$ *such that*

$$\left.\begin{aligned} (((1 - \lambda^j_1\overline{\lambda^k_1})g_{jk})) &\geq 0 \\ (((1 - \lambda^j_2\overline{\lambda^k_2})g_{jk})) &\geq 0. \end{aligned}\right\} \tag{4.10}$$

Fix $w^0 = (w^0_1, \ldots, w^0_n)$ *such that* $w^0 \in \mathcal{D}$. *Then*

$$(((1 - w^0_j\overline{w^0_k})g_{jk})) \geq 0. \tag{4.11}$$

Proof We define an inner product $(\ ,\ )$ on $\mathbf{C}^n$ by

$$(t, s) = \sum_{j,k=1}^{n} t_j\bar{s}_k g_{jk}, \quad t, s \in \mathbf{C}^n.$$

We choose the standard basis $\{\Psi_j\}$ in $\mathbf{C}^n$, where Ψ_j has the jth entry $= 1$ and all other entries $= 0$. For each $\alpha = (\alpha_1, \ldots, \alpha_n)$ in $\mathbf{C}^n$ we define the linear operator T_α on $\mathbf{C}^n$ by

$$T_\alpha\Psi_j = \alpha_j\Psi_j, \quad 1 \leq j \leq n.$$

For $\alpha \in \mathbf{C}^n$, $||T_\alpha|| \leq 1$ if and only if $||T_\alpha t||^2 \leq ||t||^2$ for all t, and this is equivalent to

$$\sum_{j,k}(\alpha_j t_j)\overline{(\alpha_k t_k)}g_{jk} \leq \sum_{j,k} t_j\bar{t}_k g_{jk}, \quad \text{or}$$

$$\sum_{j,k}(1 - \alpha_j\bar{\alpha}_k)g_{jk}t_j\bar{t}_k \geq 0 \quad \text{for all } t. \tag{4.12}$$

The results of Chapter 1 apply to this situation. As in Chapter 1, we denote by $\mathcal{A}$ the algebra consisting of all operators T_α, $\alpha \in \mathbf{C}^n$, and by $\mathcal{D}_\mathcal{A}$, as in Definition 1.2, we denote the set of all α in $\mathbf{C}^n$ with $||T_\alpha|| \leq 1$. Then $\alpha \in \mathcal{D}_\mathcal{A}$, if and only if (4.12) holds.

We put

$$\alpha' = (\lambda^1_1, \lambda^2_1, \ldots, \lambda^n_1), \quad \alpha'' = (\lambda^1_2, \lambda^2_2, \ldots, \lambda^n_2).$$

Then $\alpha'_j\overline{\alpha'_k} = \lambda_1^j\overline{\lambda_1^k}$ and $\alpha''_j\overline{\alpha''_k} = \lambda_2^j\overline{\lambda_2^k}$ and so hypothesis (4.10) gives that

$$(((1-\alpha'_j\overline{\alpha'_k})g_{jk})) \geq 0 \quad \text{and} \quad (((1-\alpha''_j\overline{\alpha''_k})g_{jk})) \geq 0$$

and so α' and α'' satisfy (4.12). Hence α' and $\alpha'' \in \mathcal{D}_{\mathcal{A}}$. The statement (4.11) which is to be proved is equivalent to showing that $w^0 \in \mathcal{D}_{\mathcal{A}}$. Fix $\epsilon > 0$. By choice of w^0, there exists a $g \in A(T^2)$ with $g(M_j) = w_j^0$, $1 \leq j \leq n$, and $\|g\| \leq 1+\epsilon$. Without loss of generality we may assume that g is a polynomial. Put $P = g/(1+\epsilon)$. Then P is a polynomial in two variables with $\|P\|_{\Delta^2} \leq 1$. By Theorem 1.2 in Chapter 1, it follows that $P(\alpha',\alpha'') \in \mathcal{D}_{\mathcal{A}}$. Then

$$(P(\lambda_1^1,\lambda_2^1), P(\lambda_1^2,\lambda_2^2), \ldots, P(\lambda_1^n,\lambda_2^n)) = \left(\frac{g(M_1)}{1+\epsilon}, \ldots, \frac{g(M_n)}{1+\epsilon}\right) \in \mathcal{D}_{\mathcal{A}}.$$

This holds for each $\epsilon > 0$. Letting $\epsilon \to 0$ and using the fact that $\mathcal{D}_{\mathcal{A}}$ is closed, we conclude that $(g(M_1), \ldots, g(M_n)) = w^0$ is in $\mathcal{D}_{\mathcal{A}}$. As we noted above, this gives (4.11), and we are done. □

Proof of Necessity of (4.2) We are given $w \in \mathbf{C}^n$ such that if $f \in A(T^2)$ and $f(M_j) = w_j$, $1 \leq j \leq n$, then $|||[f]||| \leq 1$. Suppose (4.2) does not hold. Then, with $\mathcal{K}$ defined as above, $((1 - w_j\overline{w}_k)) \notin \mathcal{K}$. By the separation theorem for closed convex sets, there exists a linear functional L on $\mathcal{H}^n$ such that

$$L(((1-w_j\overline{w}_k))) < 0, \quad L(K) \geq 0 \quad \text{for} \quad K \in \mathcal{K}.$$

As we showed, there is a $g = ((g_{jk})) \in \mathcal{H}^n$ such that g represents L in the sense that

$$L(A) = \sum_{j,k} g_{jk}A_{jk}$$

for all A in $\mathcal{H}^n$. Then

$$\sum_{j,k} g_{jk}(1-w_j\overline{w}_k) < 0, \quad \text{and} \tag{4.13}$$

$$\sum_{j,k} g_{jk}(1-\lambda_1^j\overline{\lambda_1^k})A_{jk} \geq 0 \quad \text{if } ((A_{jk})) \geq 0, \tag{4.14}$$

and

$$\sum_{j,k} g_{jk}(1-\lambda_2^j\overline{\lambda_2^k})B_{jk} \geq 0 \quad \text{if } ((B_{jk})) \geq 0. \tag{4.15}$$

Claim *Fix $z_1, \ldots, z_n$ in $|z| < 1$. Then for each $(t_1, \ldots, t_n) \in \mathbf{C}^n$*

$$\left(\left(\frac{t_j\bar{t}_k}{1-z_j\bar{z}_k}\right)\right) \geq 0. \tag{4.16}$$

The Claim is proved in the same way as Lemma 2.1 in Chapter 2.

Fix $(t_1, \ldots, t_n) \in \mathbf{C}^n$ and put

$$B_{jk} = \frac{t_j\bar{t}_k}{1-\lambda_2^j\overline{\lambda_2^k}}, \quad 1 \leq j,k \leq n.$$

By the above Claim, $((B_{jk})) \geq 0$. So (4.15) yields

$$\sum_{j,k} g_{jk} t_j \bar{t}_k \geq 0. \tag{4.17}$$

Thus $((g_{jk})) \geq 0$.

Fix $\epsilon > 0$ and put

$$h_{jk} = g_{jk} + \epsilon \delta_j^k \quad \text{for all } j, k.$$

Then $((h_{jk})) > 0$. Also, for ϵ small, (4.13) implies

$$\sum_{j,k} h_{jk}(1 - w_j \overline{w}_k) < 0. \tag{4.18}$$

Fix $t \in \mathbf{C}^n$. Then

$$\sum_{j,k} h_{jk}(1 - \lambda_1^j \overline{\lambda_1^k}) t_j \bar{t}_k = \sum_{j,k} g_{jk}(1 - \lambda_1^j \overline{\lambda_1^k}) t_j \bar{t}_k + \epsilon \sum_j (1 - |\lambda_1^j|^2)|t_j|^2 \geq 0$$

where we have used (4.14). Thus

$$((h_{jk}(1 - \lambda_1^j \overline{\lambda_1^k}))) \geq 0.$$

Similarly

$$((h_{jk}(1 - \lambda_2^j \overline{\lambda_2^k}))) \geq 0.$$

By Lemma 4.2, it follows that

$$((h_{jk}(1 - w_j \overline{w}_k))) \geq 0.$$

Hence

$$\sum_{j,k} h_{jk}(1 - w_j \overline{w}_k) \geq 0.$$

This contradicts (4.18). So (4.2) holds. Necessity is proved. □

A Corollary of Agler's Theorem is the following:

Theorem 4.2 *Let P be a polynomial in two variables with $||P||_{\Delta^2} \leq 1$. Let (α_1, β_1), ..., (α_n, β_n) be an n-tuple of points in* $\operatorname{int} \Delta^2$. *Then there exist $n \times n$ matrices A, B with $A \geq 0$, $B \geq 0$ such that for all j, k*

$$1 - P(\alpha_j, \beta_j)\overline{P(\alpha_k, \beta_k)} = (1 - \alpha_j \overline{\alpha}_k) A_{jk} + (1 - \beta_j \overline{\beta}_k) B_{jk}. \tag{4.19}$$

Proof We apply the necessity of (4.2) in Agler's Theorem to the case $w_j = P(\alpha_j, \beta_j)$. □

Remark Suppose that we can find a direct proof of Theorem 4.2. We can then derive Theorem 1.2 in Chapter 1 (a finite-dimensional version of Ando's Theorem), as follows:

We are given a Hilbert space H and a basis $\Psi_1, \ldots, \Psi_n$ of H. Then for all $\alpha \in \mathbf{C}^n$, S_α is the operator on H with $S_\alpha \Psi_j = \alpha_j \Psi_j$, $1 \leq j \leq n$, and $\mathcal{D}_A = \{\alpha \mid ||S_\alpha|| \leq 1\}$. The assertion of Theorem 1.2 is:

Proposition *Let $\alpha = (\alpha_1, \ldots, \alpha_n)$ and $\beta = (\beta_1, \ldots, \beta_n)$ be points in $\mathcal{D}_A$ and let P be a polynomial in two variables with $||P||_{\Delta^2} \leq 1$. Then $P(\alpha, \beta) \in \mathcal{D}_A$.*

Proof We may assume without loss of generality that $|\alpha_j| < 1$, $|\beta_j| < 1$ for each j. Fix $w \in \mathbf{C}^n$. Then, by a familiar calculation, $||S_w|| \leq 1$ if and only if

$$\sum_{j,k} (1 - w_j \overline{w}_k) t_j \bar{t}_k (\Psi_j, \Psi_k) \geq 0, \quad t \in \mathbf{C}^n. \tag{4.20}$$

Our hypothesis then gives

$$\sum_{j,k} (1 - \alpha_j \overline{\alpha}_k) t_j \bar{t}_k (\Psi_j, \Psi_k) \geq 0, \quad \text{and} \tag{4.21}$$

$$\sum_{j,k} (1 - \beta_j \overline{\beta}_k) t_j \bar{t}_k (\Psi_j, \Psi_k) \geq 0, \quad t \in \mathbf{C}^n. \tag{4.22}$$

Since $||P||_{\Delta^2} \leq 1$ and (4.19) holds, there exist matrices $((A_{jk})) \geq 0$ and $((B_{jk})) \geq 0$ such that for all j, k

$$1 - P(\alpha_j, \beta_j)\overline{P(\alpha_k, \beta_k)} = (1 - \alpha_j \overline{\alpha}_k) A_{jk} + (1 - \beta_j \overline{\beta}_k) B_{jk}.$$

Lemma 4.1 applied to (4.21) and (4.22), gives

$$\begin{aligned} \sum_{j,k} (1 - \alpha_j \overline{\alpha}_k) t_j \bar{t}_k (\Psi_j, \Psi_k) A_{jk} &\geq 0, \quad \text{and} \\ \sum_{j,k} (1 - \beta_j \overline{\beta}_k) t_j \bar{t}_k (\Psi_j, \Psi_k) B_{jk} &\geq 0, \quad t \in \mathbf{C}^n. \end{aligned}$$

Adding gives

$$\sum_{j,k} ((1 - \alpha_j \overline{\alpha}_k) A_{jk} + (1 - \beta_j \overline{\beta}_k) B_{jk})(\Psi_j, \Psi_k) t_j \bar{t}_k \geq 0 \quad \text{for all } t$$

and so

$$\sum_{j,k} ((1 - P(\alpha_j, \beta_j)\overline{P(\alpha_k, \beta_k)})(\Psi_j, \Psi_k) t_j \bar{t}_k \geq 0 \quad \text{for all } t.$$

Also $P(\alpha, \beta) = (P(\alpha_1, \beta_1), \ldots, P(\alpha_n, \beta_n))$. By (4.20) we conclude that $||S_{P(\alpha,\beta)}|| \leq 1$. Hence $P(\alpha, \beta) \in \mathcal{D}_A$. So Theorem 1.2 is proved. □

A direct proof of Theorem 4.2 remains to be found.

In a similar way we may prove Theorem 1.1, a finite-dimensional version of Von Neumann's inequality, starting with Pick's Theorem. We use the preceding notations.

Proof of Theorem 1.1 We are given $\alpha = (\alpha_1, \ldots, \alpha_n) \in \mathcal{D}_A$ and a polynomial Q in one variable with $||Q||_\Delta \leq 1$. Without loss of generality, we assume $|\alpha_j| < 1$ for each j. We must show that $Q(\alpha) \in \mathcal{D}_A$. By Pick's Theorem, there exists a matrix $((A_{jk})) \geq 0$ such that for all j, k,

$$1 - Q(\alpha_j)\overline{Q(\alpha_k)} = (1 - \alpha_j\overline{\alpha}_k)A_{jk}.$$

Since $\alpha \in \mathcal{D}_A$, (4.20) gives

$$\sum_{j,k}(1 - \alpha_j\overline{\alpha}_k)t_j\overline{t}_k(\Psi_j, \Psi_k) \geq 0 \quad \text{for all } t.$$

Hence by Lemma 4.1,

$$\sum_{j,k}(1 - \alpha_j\overline{\alpha}_k)t_j\overline{t}_k(\Psi_j, \Psi_k)A_{jk} \geq 0 \quad \text{for all } t$$

and so

$$\sum_{j,k}(1 - Q(\alpha_j)\overline{Q(\alpha_k)})t_j\overline{t}_k(\Psi_j, \Psi_k) \geq 0 \quad \text{for all } t.$$

Then $||S_{Q(\alpha)}|| \leq 1$, and so $Q(\alpha) \in \mathcal{D}_A$. Theorem 1.1 is proved. □

5 Hyperconvex sets

We consider a uniform algebra A with maximal ideal space $\mathcal{M}$ and we fix an n-tuple of points $M_1, \ldots, M_n$ in $\mathcal{M}$. We then denote by $\mathcal{D}$ the interpolation set corresponding to this choice, as defined in Definition 2.1.

Lemma 5.1 *Let P be a polynomial in k variables with $||P||_{\Delta^k} \leq 1$. For each set of k points $\alpha', \alpha'', \ldots, \alpha^{(k)}$ in $\mathcal{D}$ we have that $P(\alpha', \ldots, \alpha^{(k)})$ again belongs to $\mathcal{D}$.*

Here $P(\alpha', \alpha'', \ldots, \alpha^{(k)})$ is calculated in the algebra $\mathbf{C}^n$ under coordinatewise multiplication. Hence

$$P(\alpha', \ldots, \alpha^{(k)}) = (w_1, \ldots, w_n) \text{ in } \mathbf{C}^n, \quad \text{where}$$
$$w_j = P(\alpha'_j, \ldots, \alpha_j^{(k)}), \quad 1 \leq j \leq n.$$

Proof Fix $\epsilon > 0$. Choose $\delta > 0$ such that $|\zeta_j| \leq 1 + \delta$, $1 \leq j \leq n$, implies $|P(\zeta)| \leq 1 + \epsilon$. Choose $f' \in A$, $f'(M_j) = \alpha'_j$, $1 \leq j \leq n$, with $||f'|| \leq 1 + \delta$; choose $f'', \ldots, f^{(k)}$ similarly for $\alpha'', \ldots, \alpha^{(k)}$. Then $P(f', \ldots, f^{(k)}) \in A$ and $||P(f', \ldots, f^{(k)})|| \leq 1 + \epsilon$. Also

$$P(f', \ldots, f^{(k)})(M_j) = P(\alpha'_j, \alpha''_j, \ldots, \alpha_j^{(k)}), \quad 1 \leq j \leq n.$$

Since ϵ is arbitrary, $P(\alpha', \ldots, \alpha^{(k)}) \in \mathcal{D}$. We are done. □

This Lemma suggests the following definition:

Definition 5.1 Fix $n \geq 1$. Let Y be a compact set in $\mathbf{C}^n$ with non-void interior. Assume that for any integer $k \geq 1$, whenever $\alpha', \alpha'', \ldots, \alpha^{(k)}$ is a k-tuple of points in Y and P is a polynomial in k variables with $||P||_{\Delta^k} \leq 1$, then the point $P(\alpha', \ldots, \alpha^{(k)})$ again lies in Y. We then say that Y is *hyperconvex*.

Theorem 5.1 *Let Y be a hyperconvex set in $\mathbf{C}^n$. Then Y has the following properties:* (i) *Y is convex.* (ii) *Y is circled, i.e. if $(\zeta_1, \ldots, \zeta_n) \in Y$ and $\theta \in \mathbf{R}$, then $(e^{i\theta}\zeta_1, \ldots, e^{i\theta}\zeta_n) \in Y$.* (iii) *$Y \subseteq \Delta^n$.* (iv) *$Y$ contains the diagonal of $\Delta^n = \{(\lambda, \ldots, \lambda) \mid |\lambda| \leq 1\}$.* (v) *$Y$ is closed under multiplication in $\mathbf{C}^n$.*

The proof of assertions (i) to (v) follows at once from the definition of hyperconvexity.

Lemma 5.1 tells us that each interpolation set $\mathcal{D}$ is a hyperconvex set. The converse is true. We have

Theorem 5.2 *Let Y be a hyperconvex set in $\mathbf{C}^n$. Then there exists a uniform algebra A with maximal ideal space $\mathcal{M}$ and an n-tuple $M_1, \ldots, M_n$ of points in $\mathcal{M}$ such that the corresponding interpolation set is Y.*

Proof By Theorem 5.1, Y is convex and circled. Also Y is compact and contains a neighborhood of 0 in $\mathbf{C}^n$. Hence there exists a norm $|||\ \ |||$ on $\mathbf{C}^n$ such that Y is the closed unit ball in this norm.

Let $w, w' \in \mathbf{C}^n$. Then $\dfrac{w}{|||w|||}, \dfrac{w'}{|||w'|||} \in Y$. By Theorem 5.1 (v), Y is closed under multiplication in $\mathbf{C}^n$, and so $\dfrac{|||ww'|||}{|||w|||\ |||w'|||} \leq 1$. Hence $|||ww'||| \leq |||w||| \cdot |||w'|||$. So $|||\ \ |||$ makes $\mathbf{C}^n$ into a normed algebra $\mathcal{L}$. Clearly, $\mathcal{L}$ is a commutative Banach algebra with unit. In addition, since Y is hyperconvex, the following implication holds:

$$\left.\begin{array}{l} \text{If } w', w'', \ldots, w^{(k)} \text{ are elements in } \mathcal{L} \text{ with } |||\ \ ||| \leq 1, \\ \text{and if } P \text{ is a polynomial in k variables with } ||P||_{\Delta^k} \leq 1, \\ \text{then } |||P(w', w'', \ldots, w^{(k)})||| \leq 1. \end{array}\right\} \tag{5.1}$$

By a result due independently to B. Cole and I. Craw (see [B-D], §50, p. 271, Proposition 5), (5.1) is a necessary and and sufficient condition for the existence of a uniform algebra A and a closed ideal J in A such that $\mathcal{L}$ is isometrically isomorphic to the quotient algebra A/J.

Let $\mathcal{M}$ denote the maximal ideal space of A. Since A/J is algebraically isomorphic to $\mathcal{L}$, hence to the algebra $\mathbf{C}^n$, there exist n distinct homomorphisms of A/J into $\mathbf{C}$. We call them $\phi_1, \ldots, \phi_n$. Each ϕ_j induces a point M_j in $\mathcal{M}$ under the composed map: $A \to A/J \to \mathbf{C}$. The M_j, $1 \leq j \leq n$, are distinct points in $\mathcal{M}$. We put

$$I = \{f \in A \mid f(M_j) = 0,\ \ 1 \leq j \leq n\}.$$

Then $J \subseteq I$. Also I and J each have codimension n in A. Hence $I = J$.

Let τ denote the isometric isomorphism of A/I onto $\mathcal{L}$ which was constructed above. Then τ preserves idempotents in the two algebras. It follows that for each $f \in A$, if we put $w_j = f(M_j)$, $1 \leq j \leq n$, then

$$\tau([f]) = (w_1, \ldots, w_n) \in \mathcal{L}.$$

Fix w^0 in $\mathbf{C}^n$ and choose f in A such that $\tau([f]) = w^0$. Then w^0 is in Y if and only if $|||w^0||| \leq 1$ if and only if $|||[f]||| \leq 1$ if and only if w^0 belongs to the interpolation set $\mathcal{D}$ which corresponds to A and $M_1, \ldots, M_n$. Hence $Y = \mathcal{D}$. We are done. □

Theorem 5.2 together with Lemma 5.1 implies that the problem of classifying interpolation sets of uniform algebras and the problem of classifying hyperconvex sets are the same.

In trying to classify hyperconvex sets we begin with the fact that each Pick body $\mathcal{D}(\alpha)$ and each polydisk Δ^n are hyperconvex sets in $\mathbf{C}^n$.

Theorem 5.3 *Put $n = 2$. Every hyperconvex set Y in $\mathbf{C}^2$ is either a Pick body or is the bidisk. Further, there exists a λ, $0 < \lambda \leq 1$, such that*

$$Y = \{(w_1, w_2) \,|\, |w_1| \leq 1, \ |w_2| \leq 1, \ \left|\frac{w_1 - w_2}{1 - \overline{w}_1 w_2}\right| \leq \lambda\}.$$

Proof See [C-L-W 1], Theorem 5. □

It follows that hyperconvex sets in $\mathbf{C}^2$ are simply ordered under inclusion. For $n \geq 2$, if Y_1 and Y_2 are two hyperconvex sets in $\mathbf{C}^n$, then their intersection $Y_1 \cap Y_2$ is again hyperconvex. This follows at once from the definition. Theorem 5.3 does not extend to $\mathbf{C}^3$. We showed in [C-L-W 1], Section 3.3, that there exist two Pick bodies $\mathcal{D}(z')$ and $\mathcal{D}(z'')$ in $\mathbf{C}^3$ such that the hyperconvex set $\mathcal{D}(z') \cap \mathcal{D}(z'')$ is not a Pick body, and of course $\neq \Delta^3$.

It is possible to characterize Pick bodies in $\mathbf{C}^n$.

Theorem 5.4 *Fix an integer $n \geq 1$ and let Y be a compact set in $\mathbf{C}^n$ with non-void interior. Then Y is a Pick body if and only if*

(i) *Y is hyperconvex, and*
(ii) *Y contains a point $z = (z_1, \ldots, z_n)$ with $|z_j| < 1$, $1 \leq j \leq n$, and $z_i \neq z_j$, if $i \neq j$, such that the points*

$$z, z^2, \ldots, z^{n-1}$$

all lie on the boundary of Y.

This result is proved as Theorem 1 in [C-L-W 2] and an improved version is given in Theorem 1′ in [C-L-W 2].

Next we observe that an arbitrary hyperconvex set can be defined by certain inequalities, as follows at once from Theorem 3.3.

Theorem 5.5 *Let Y be an arbitrary hyperconvex set in $\mathbf{C}^n$. Then there exists a family of positive definite $n \times n$ matrices $((L_{jk}^{(\lambda)}))$, where λ runs over a label set Λ, such that Y is defined by the family of inequalities:*

$$\sum_{j,k=1}^{n} t_j \bar{t}_k (1 - \overline{w}_j w_k) L_{jk}^{(\lambda)} \geq 0, \quad t \in \mathbf{C}^n, \ \lambda \in \Lambda. \tag{5.2}$$

Proof By Theorem 5.2, Y is the interpolation set for a certain uniform algebra A and points $M_1, \ldots, M_n$ in $\mathcal{M}$. By Theorem 3.3, a point w in $\mathbf{C}^n$ lies in Y if and only if the inequalities

$$\sum_{j,k=1}^{n} t_j \bar{t}_k (1 - w_j \bar{w}_k) \overline{(l_j, l_k)_\mu} \geq 0, \quad t \in \mathbf{C}^n, \tag{5.3}$$

are satisfied. We take for our label set Λ the set of all dominating measures μ corresponding to our data, and put $L_{jk}^{(\mu)} = \overline{(l_j, l_k)_\mu}$ for all μ in Λ. Theorem 3.3 then gives that $w \in Y$ if and only if (5.2) holds. We are done. □

Finally, we point out a property of hyperconvex sets which is related to holomorphic maps in $\mathbf{C}^n$. Let Y be a hyperconvex set in $\mathbf{C}^n$ and let Ω denote the interior of Y. Then Ω is a convex, circled domain in $\mathbf{C}^n$.

Fix a Moebius transformation of the complex plane:

$$\phi(\zeta) = e^{it} \frac{\zeta - \zeta_0}{1 - \bar{\zeta}_0 \zeta},$$

where $t \in \mathbf{R}$, $|\zeta_0| < 1$. For $w = (w_1, \ldots, w_n)$ in $\mathbf{C}^n$, define

$$\Phi(w) = (\phi(w_1), \ldots, \phi(w_n)).$$

Theorem 5.6 *For each Moebius transformation ϕ, the map: $w \to \Phi(w)$ is a biholomorphic self-map of Ω.*

Proof $|\phi(\zeta)| \leq 1$ on $|\zeta| \leq 1$. Fix $\epsilon > 0$ and choose a polynomial P_ϵ in ζ such that $|\phi(\zeta) - P_\epsilon(\zeta)| \leq \epsilon$ on $|\zeta| \leq 1$. Then $\|P_\epsilon\|_\Delta \leq 1 + \epsilon$, and so $\|\frac{P_\epsilon}{1+\epsilon}\|_\Delta \leq 1$. Fix $w \in Y$. Since Y is hyperconvex,

$$\frac{P_\epsilon}{1+\epsilon}(w) = \left(\frac{P_\epsilon(w_1)}{1+\epsilon}, \ldots, \frac{P_\epsilon(w_n)}{1+\epsilon} \right) \in Y.$$

As $\epsilon \to 0$, $\frac{P_\epsilon(w)}{1+\epsilon} \to (\phi(w_1), \ldots, \phi(w_n)) = \Phi(w)$. Since Y is closed, $\Phi(w) \in Y$. Thus Φ maps Y into Y. Similarly, Φ^{-1} maps Y into Y. Thus Φ is a homeomorphism of Y onto Y, and so induces a homeomorphism of Ω onto Ω. Also Φ and Φ^{-1} are holomorphic maps on Ω. Thus the assertion of the Theorem holds. □

Note Domains in $\mathbf{C}^n$ which admit a large group of holomorphic self-mappings, such as the ball and the polydisk, are quite special among domains in $\mathbf{C}^n$, and have been much studied. The interior of each hyperconvex set belongs in this class of domains, in view of Theorem 5.6.

6 Representations of the disk algebra

We shall assume basic facts about the disk algebra. These facts are covered, e.g. in Hoffman's book [Ho].

We fix points $\alpha_1, \ldots, \alpha_n$ in $|\zeta| < 1$ as our points $M_1, \ldots, M_n$ in $\mathcal{M}$, where we recall that the maximal ideal space $\mathcal{M}$ of $A(\Gamma)$ is the closed unit disk. We put

$$I = \{f \in A(\Gamma) \,|\, f(\alpha_j) = 0, \;\; 1 \leq j \leq n\}.$$

Let μ be any probability measure on Γ and form the spaces $L^2(\mu)$, $H^2(\mu)$, $\bar{I}$, and $I^\perp$, as defined in Chapter 3.

Claim *Either $I^\perp = \{0\}$ or μ is dominating. (If $I^\perp = \{0\}$, then the representation S^μ is trivial.)*

Proof of Claim Suppose $I^\perp \neq \{0\}$. Choose $k \in I^\perp$ with $k \neq 0$. Let B be the Blaschke product formed with the points $\alpha_1, \ldots, \alpha_n$. If $g \in A(\Gamma)$, then $\int_\Gamma gB\bar{k}\, d\mu = 0$. Fix $f \in A(\Gamma)$ and fix α, $|\alpha| < 1$. Then

$$\frac{f(\zeta) - f(\alpha)}{\zeta - \alpha} \in A(\Gamma), \text{ so } \int_\Gamma \frac{f(\zeta) - f(\alpha)}{\zeta - \alpha} B\bar{k}\, d\mu = 0.$$

Thus

$$\int_\Gamma \frac{f(\zeta)}{\zeta - \alpha} B\bar{k}\, d\mu = \int_\Gamma \frac{f(\alpha) B\bar{k}\, d\mu}{\zeta - \alpha}.$$

Define

$$\Phi(\alpha) = \int_\Gamma \frac{B\bar{k}}{\zeta - \alpha}\, d\mu, \quad \alpha \in \mathbf{C} \setminus \Gamma.$$

Then

$$\int_\Gamma \frac{f(\zeta)}{\zeta - \alpha} B\bar{k}\, d\mu = f(\alpha)\Phi(\alpha), \quad |\alpha| < 1.$$

Using the geometric series expansion of $(\zeta - \alpha)^{-1}$, we see that

$$\begin{aligned} \Phi(\alpha) &= -\sum_{n=0}^{\infty} \frac{1}{\alpha^{n+1}} \int_\Gamma \zeta^n B\bar{k}\, d\mu, \quad \text{if } |\alpha| > 1, \text{ and} \\ \Phi(\alpha) &= \sum_{n=0}^{\infty} \alpha^n \int_\Gamma \frac{1}{\zeta^{n+1}} B\bar{k}\, d\mu, \quad \text{if } |\alpha| < 1. \end{aligned}$$

For $|\alpha| > 1$, $\Phi(\alpha) = 0$, since $(\zeta - \alpha)^{-1} \in A(\Gamma)$. Suppose $\Phi(\alpha)$ is identically 0 for $|\alpha| < 1$. Then we have

$$\int_\Gamma \zeta^n B\bar{k}\, d\mu = 0, \quad n = 0, 1, 2, \ldots \text{ and}$$

$$\int_\Gamma \frac{1}{\zeta^{n+1}} B\bar{k}\, d\mu = 0, \quad n = 0, 1, 2, \ldots.$$

Thus the measure $B\bar{k}d\mu$ has all its Fourier coefficients $= 0$, and so $B\bar{k} = 0$ a.e.-$d\mu$. This contradicts the choice of k. So $\Phi(\alpha)$ is not identically 0 on $|\alpha| < 1$. Also Φ is analytic in $|\alpha| < 1$. Thus the zeros of Φ in $|\alpha| < 1$ form a discrete set. We choose a circle $\gamma : |\zeta| = r$ in $|\zeta| < 1$ which avoids all the zeros of Φ and such that each α_j, $1 \leq j \leq n$, lies inside γ.

Since $\Phi \neq 0$ on γ, there is a $\delta > 0$ with $|\Phi| \geq \delta$ at all points of γ. Fix α on Γ. Then

$$f(\alpha) = \frac{1}{\Phi(\alpha)} \int_\Gamma \frac{f(\zeta)}{\zeta - \alpha} B\overline{k}\, d\mu.$$

Hence

$$\begin{aligned} |f(\alpha)| &\leq \frac{1}{|\Phi(\alpha)|} \int_\Gamma \frac{|f(\zeta)|}{(1-r)} |k|\, d\mu \\ &\leq \frac{1}{(1-r)\delta} \sqrt{\int_\Gamma |f|^2\, d\mu} \sqrt{\int_\Gamma |k|^2\, d\mu} \\ &= C\sqrt{\int_\Gamma |f|^2\, d\mu}, \end{aligned}$$

where C is a constant independent of f and of α on γ. If $1 \leq j \leq n$, the maximum principle gives for all j:

$$|f(\alpha_j)| \leq \max_{\alpha\in\gamma} |f(\alpha)| \leq C\sqrt{\int_\Gamma |f|^2\, d\mu}.$$

So μ is dominating, proving the Claim.

We write μ_0 for the measure $d\theta/2\pi$ on Γ. Clearly μ_0 is dominating, for the chosen points. Also $H^2(\mu_0)$ is the usual Hardy space H^2. The functionals $l_1, \ldots, l_n$, as in Chapter 1, are the Szegö functionals for μ_0.

Theorem 6.1 *Let μ be a dominating measure on Γ, and let m_j be the corresponding j-th Szegö functional in $H^2(\mu)$. Then there exist constants $d_j \neq 0$, $1 \leq j \leq n$, such that*

$$\int_\Gamma m_j \overline{m}_k\, d\mu = (m_j, m_k) = \frac{d_j \overline{d}_k}{1 - \overline{\alpha}_j \alpha_k}, \quad 1 \leq j, k \leq n. \tag{6.1}$$

Proof Let S^μ be the representation of $A(\Gamma)/I$ corresponding to μ. Let f_0 be the identity function: $f_0(\zeta) = \zeta$ in $A(\Gamma)$. Then $S^\mu([f_0])$ is an operator on $I^\perp$.

We take H to be the n-dimensional Hilbert space $I^\perp$. The vectors $m_1, \ldots, m_n$ are a basis of H, and we take this to be the basis $\phi_1, \ldots, \phi_n$ as in Chapter 1. By Lemma 3.2,

$$(S^\mu([f_0]))^* m_j = \overline{\alpha}_j m_j, \quad 1 \leq j \leq n.$$

Hence in the notation of Chapter 1, $S^\mu([f_0]) = S_\alpha$.

Fix $t = \sum_{j=1}^n t_j m_j$ in H. Then for $p = 0, 1, 2, \ldots$,

$$\begin{aligned} (S_\alpha^p t, t) = (t, S_\alpha^{*p} t) &= \left(\sum_j t_j m_j, \sum_k t_k \overline{\alpha}_k^p m_k\right) \\ &= \sum_{j,k} t_j \overline{t}_k \alpha_k^p (m_j, m_k). \end{aligned}$$

Also

$$\begin{aligned}\int_\Gamma \Big|\sum_j t_j m_j\Big|^2 e^{ip\theta}\,d\mu(\theta) &= \sum_{j,k} t_j\bar{t}_k \int_\Gamma m_j\overline{m}_k e^{ip\theta}\,d\mu(\theta)\\ &= \sum_{j,k} t_j\bar{t}_k g_j(\alpha_k), \quad \text{where } g_j = m_j e^{ip\theta},\\ &= \sum_{j,k} t_j\bar{t}_k (m_j, m_k)\alpha_k^p.\end{aligned}$$

Thus

$$(S_\alpha^p t, t) = \int_\Gamma \Big|\sum_j t_j m_j\Big|^2 e^{ip\theta}\,d\mu(\theta). \tag{6.2}$$

By Lemma 1.1,

$$(S_\alpha^p t, t) = \frac{1}{2\pi}\int_\Gamma K_\alpha(\theta, t, t)e^{ip\theta}\,d\theta \tag{6.3}$$

where

$$K_\alpha(\theta, t, t) = \sum_{j,k} t_j\bar{t}_k l_j(\theta)\overline{l_k(\theta)}(1-\overline{\alpha}_j\alpha_k)(m_j, m_k)$$

with

$$l_j(\theta) = \frac{1}{1-\overline{\alpha}_j e^{i\theta}}, \quad 1 \le j \le n.$$

We have

$$||S_\alpha|| \le ||[f_0]|| \le ||f_0|| = 1,$$

and so as in the proof of Theorem 1.1, we get

$$K_\alpha(\theta, t, t) \ge 0 \quad \text{for all } \theta \in [0, 2\pi].$$

Equations (6.2) and (6.3) imply that

$$\Big|\sum_j t_j m_j\Big|^2 d\mu(\theta) = K_\alpha(\theta, t, t)\frac{d\theta}{2\pi}$$

as measures on Γ. We write

$$d\mu = p(\theta)\frac{d\theta}{2\pi} + d\mu_s(\theta)$$

where $p \in L^1(0, 2\pi)$, $p(\theta) \ge 0$ and μ_s is a singular measure on Γ. We thus have

$$\Big|\sum_j t_j m_j\Big|^2 p(\theta) = K_\alpha(\theta, t, t) \quad \text{a.e.-}d\theta,\ t \in H \tag{6.4}$$

and

$$\int_\Gamma m_j(\theta)\overline{m_k(\theta)}\,d\mu_s(\theta) = 0, \quad 1 \le j, k \le n. \tag{6.5}$$

Formula (6.4) gives, for fixed t in H and for a.a. θ,

$$\sum_{j,k} t_j\bar{t}_k m_j(\theta)\overline{m_k(\theta)}p(\theta) = \sum_{j,k} t_j\bar{t}_k l_j(\theta)\overline{l_k(\theta)}(1-\overline{\alpha}_j\alpha_k)(m_j, m_k). \tag{6.6}$$

This implies for $1 \le j,\ k \le n$ and a.a. θ,

$$m_j(\theta)\overline{m_k(\theta)}p(\theta) = l_j(\theta)\overline{l_k(\theta)}(1-\overline{\alpha}_j\alpha_k)(m_j, m_k). \tag{6.7}$$

Put $C_{jk} = (1-\overline{\alpha}_j\alpha_k)(m_j, m_k)$. Then (6.7) gives

$$\left(\frac{\sqrt{p(\theta)}m_j(\theta)}{l_j(\theta)}\right)\overline{\left(\frac{\sqrt{p(\theta)}m_k(\theta)}{l_k(\theta)}\right)} = C_{jk} \quad \text{for a.a. } \theta. \tag{6.8}$$

Fix θ_1 such that (6.8) holds for all j, k. Put

$$d_j = \frac{\sqrt{p(\theta_1)}m_j(\theta_1)}{l_j(\theta_1)}, \quad 1 \le j \le n.$$

Then $d_j\overline{d}_k = C_{jk}$, so by (6.8)

$$m_j(\theta)\overline{m_k(\theta)}p(\theta) = d_j\overline{d}_k l_j(\theta)\overline{l_k(\theta)}, \quad \text{a.a. } \theta.$$

Integration over $[0, 2\pi]$ gives for all j, k

$$\frac{1}{2\pi}\int_\Gamma m_j(\theta)\overline{m_k(\theta)}p(\theta)\,d\theta = d_j\overline{d}_k\frac{1}{1-\overline{\alpha}_j\alpha_k}.$$

Also, by (6.5)

$$\frac{1}{2\pi}\int_\Gamma m_j(\theta)\overline{m_k(\theta)}\,d\mu_s(\theta) = 0.$$

So

$$\frac{1}{2\pi}\int_\Gamma m_j(\theta)\overline{m_k(\theta)}\,d\mu(\theta) = \frac{d_j\overline{d}_k}{1-\overline{\alpha}_j\alpha_k}.$$

Thus (6.1) holds. If $d_j = 0$ for some j, then $(m_j, m_k) = 0$ for all k and so $m_j = 0$. This is a contradiction, and so $d_j \ne 0$ for all j. Theorem 6.1 is proved. □

Corollary *Let μ be a dominating measure on Γ, $I^\perp$ the corresponding space on which S^μ acts, and let $I_0^\perp$ denote the space on which S^{μ_0} acts. Then the representations S^μ and S^{μ_0} are unitarily equivalent, in the sense that there exists a unitary operator U of $I^\perp$ on $I_0^\perp$ such that*

$$S_f^{\mu_0} = US_f^\mu U^{-1} \quad \textit{for all } f \textit{ in } A(\Gamma), \tag{6.9}$$

where we write $S_f^\mu = S^\mu([f])$ and $S_f^{\mu_0} = S^{\mu_0}([f])$.

Proof Let m_j be the j-th Szegö functional for μ, as above. We denote by U the linear operator from $I^\perp$ to $I_0^\perp$ with

$$Um_j = d_jl_j, \quad \text{where the } d_j \text{ satisfy (6.1).}$$

Then for all j, k

$$(Um_j, Um_k) = (d_jl_j, d_kl_k) = \frac{d_j\overline{d}_k}{1-\overline{\alpha}_j\alpha_k} = (m_j, m_k).$$

Hence U is unitary. Also, writing S_f for S_f^{μ},

$$\begin{aligned}(US_f^*U^{-1})l_j &= US_f^*\left(\frac{1}{d_j}m_j\right) = U\overline{f(\alpha_j)}\frac{1}{d_j}m_j \\ &= \overline{f(\alpha_j)}\frac{d_jl_j}{d_j} = \overline{f(\alpha_j)}l_j = (S_f^{\mu_0})^*l_j.\end{aligned}$$

Since this holds for each j, $US_f^*U^{-1} = (S_f^{\mu_0})^*$. Taking adjoints, we have $US_fU^{-1} = S_f^{\mu_0}$, which is (6.9). □

Definition 6.1 Let $\mathcal{L}$ be a Banach algebra with unit. A *state* of $\mathcal{L}$ is a bounded linear functional L on $\mathcal{L}$ such that

$$L(1) = 1 \quad \text{and} \quad ||L|| \leq 1.$$

Proposition *Let A be a uniform algebra on X, I a closed ideal of A. Let L be a state of A/I. Then there is a probability measure μ on X such that*

$$\int_X j\,d\mu = 0, \qquad j \in I \tag{6.10}$$

and

$$L([f]) = \int_X f\,d\mu, \quad f \text{ in } A. \tag{6.11}$$

Proof We define a functional L_1 on A by $L_1(f) = L([f])$, $f \in A$. Then $|L_1(f)| \leq ||f||$, $f \in A$. By Banach space theory there exists a complex measure μ on X with $||\mu|| \leq 1$ such that

$$L_1(f) = \int_X f\,d\mu, \quad f \in A.$$

Since $L([1]) = 1$, it follows that μ is a probability measure. If $j \in I$, $[j] = 0$, so

$$\int_X j\,d\mu = L_1(j) = L([j]) = 0,$$

i.e. (6.10) holds. We are done. □

The converse clearly holds as well: each probability measure on X satisfying (6.10) induces a state L on A/I by

$$L([f]) = \int_X f\,d\mu.$$

So states on A/I and probability measures on X satisfying (6.10) are in one-to-one correspondence.

We can obtain examples of states on A/I as follows: let σ be a dominating measure on X. Choose $\phi \in I^{\perp}$ with $||\phi||_2 = 1$, where $||\ ||_2$ denotes the L^2-norm with respect to σ. Put

$$L(f) = \int_X f|\phi|^2\,d\sigma, \quad f \in A. \tag{6.12}$$

Fix $j \in I$. Then $j\phi \in \bar{I}$. Hence $j\phi \perp \phi$ in $L^2(\sigma)$. Thus $\int j|\phi|^2\, d\sigma = 0$. Hence $L = 0$ on I, and so L is defined on A/I. The choice of ϕ yields that the norm of L as a functional on A/I is ≤ 1, and that $L(1) = 1$. It follows that L is a state on A/I.

We next show that in the case that A is the disk algebra $A(\Gamma)$, every state of $A(\Gamma)/I$ can be written in the form (6.12), where the dominating measure σ is taken to be $d\theta/2\pi = \mu_0$.

Theorem 6.2 *Fix points $\alpha_1, \ldots, \alpha_n$ in the open unit disk and let*

$$I = \{g \in A(\Gamma) \,|\, g(\alpha_j) = 0,\ \ 1 \leq j \leq n\}.$$

Let L be a state of $A(\Gamma)/I$. Then there exist constants c_j, $1 \leq j \leq n$, such that if we put

$$\phi(\zeta) = \sum_{j=1}^{n} c_j \frac{1}{1 - \bar{\alpha}_j \zeta} \quad \textit{for } \zeta \in \Gamma,$$

then

$$L([f]) = \int_0^{2\pi} f(e^{i\theta})\, |\phi(e^{i\theta})|^2 \frac{d\theta}{2\pi}, \quad f \in A(\Gamma). \tag{6.13}$$

Proof By the Proposition, there is a probability measure μ on Γ such that (6.10) holds and

$$L(f) = L([f]) = \int_\Gamma f\, d\mu, \quad f \in A(\Gamma).$$

Claim *μ is a dominating measure.*

Proof of Claim Let B denote the Blaschke product with zeros at the points α_j, $1 \leq j \leq n$. Then for all $g \in A(\Gamma)$,

$$\int_\Gamma gB\, d\mu = 0.$$

Hence there is a $k \in H_0^1$ such that

$$B d\mu = k\frac{d\theta}{2\pi}, \quad \text{or} \quad d\mu = \frac{k}{B}\frac{d\theta}{2\pi}.$$

Fix j_0, $1 \leq j_0 \leq n$. Choose a rational function g with $g(0) = 0$ whose only possible pole is at α_{j_0} such that the residue of $\left(\frac{g}{z}\right)\left(\frac{k}{B}\right)$ at α_{j_0} is 1 and $\left(\frac{g}{z}\right)\left(\frac{k}{B}\right)$ is analytic at α_j if $j \neq j_0$. (Here we have assumed that $\alpha_{j_0} \neq 0$. If $\alpha_{j_0} = 0$, a similar argument applies.) By the Residue Theorem, then for all $f \in A(\Gamma)$,

$$\begin{aligned} \int_\Gamma fg\, d\mu &= \int_\Gamma fg\frac{k}{B}\frac{d\theta}{2\pi} = \int_\Gamma f\left(\frac{g}{z}\right)\frac{k}{B}\frac{dz}{2\pi i} \\ &= f(\alpha_{j_0}) \end{aligned}$$

and so

$$|f(\alpha_{j_0})| \leq \sqrt{\int_\Gamma |f|^2\, d\mu}\sqrt{\int_\Gamma |g|^2\, d\mu}.$$

Hence μ is dominating, as claimed.

Since μ is dominating, we know by the Corollary of Theorem 6.1 that S^{μ} and S^{μ_0} are related by

$$S_f^{\mu_0} = US_f^{\mu}U^{-1}, \quad f \in A(\Gamma),$$

where U is a unitary transformation of $I^{\perp}$ on $I_0^{\perp}$.

Since μ satisfies (6.10), $1 \in I^{\perp}$. If $f \in A(\Gamma)$, then

$$L(f) = \int f\, d\mu = (S_f^{\mu}1, 1) = (U^{-1}S_f^{\mu_0}U1, 1) = (S_f^{\mu_0}(U1), U1)_{L^2(\mu_0)}.$$

Put $\phi = U(1)$. Then $\phi \in I_0^{\perp}$ and hence

$$\phi = \sum_{j=1}^{n} c_j \frac{1}{1-\overline{\alpha}_j\zeta},$$

for suitable constants c_j. Then for every $f \in A(\Gamma)$, we have

$$L(f) = \int_{\Gamma} (f\phi)\overline{\phi}\frac{d\theta}{2\pi} = \int_{\Gamma} f|\phi|^2\frac{d\theta}{2\pi},$$

i.e. (6.13). We are done. □

As we know, the space $I_0^{\perp}$ corresponding to μ_0 is spanned by the Szegö functionals

$$l_j(\zeta) = \frac{1}{1-\overline{\alpha}_j\zeta}, \quad \zeta \in \Gamma, \quad 1 \leq j \leq n.$$

Put $\Pi(\zeta) = \prod_{j=1}^{n}(1-\overline{\alpha}_j\zeta)$. Then $I_0^{\perp}$ consists of all functions of the form

$$\frac{P(\zeta)}{\Pi(\zeta)}$$

where P is a polynomial of degree $p \leq n-1$.

Lemma 6.1 *Fix $\phi \in I_0^{\perp}$. Let $w_1, \ldots, w_k$ be the zeros of ϕ in $|\zeta| < 1$. We define g by*

$$\phi(\zeta) = \frac{P(\zeta)}{\Pi(\zeta)} = \left(\prod_{j=1}^{k}\frac{\zeta - w_j}{1-\zeta\overline{w}_j}\right)g(\zeta).$$

Then g is a rational function analytic in $|\zeta| \leq 1$ and free of zeros in $|\zeta| < 1$. Also $g \in I_0^{\perp}$.

Proof Since

$$\frac{P(\zeta)}{\prod_{j=1}^{n}(1-\overline{\alpha}_j\zeta)} = \left(\prod_{j=1}^{k}\frac{\zeta - w_j}{1-\zeta\overline{w}_j}\right)g(\zeta),$$

each w_j is a zero of P, and so $P(\zeta)/\prod_{j=1}^{k}(\zeta - w_j)$ is a polynomial of degree $p-k$. Then

$$g(\zeta) = \frac{\left(\prod_{j=1}^{k}(1-\overline{w}_j\zeta)\right)\left(P(\zeta)/\prod_{j=1}^{k}(\zeta - w_j)\right)}{\Pi(\zeta)} = \frac{Q(\zeta)}{\Pi(\zeta)},$$

where Q is a polynomial of degree $k+(p-k) = p \leq n-1$. Hence $g \in I_0^{\perp}$, and we are done. □

The following theorem is closely related to the work of Sarason in [Sa 1] which we discussed in the Introduction.

Theorem 6.3 *Let $f \in A(\Gamma)$ with $|||[f]||| = 1$. Then there exists a Blaschke product χ of order $\leq n-1$ and a $\phi \in I_0^{\perp}$ with $\int_\Gamma |\phi|^2\, d\mu_0 = 1$ such that*

$$\chi\phi \in I_0^{\perp} \tag{6.14}$$

and

$$\chi(\alpha_j) = f(\alpha_j), \quad 1 \leq j \leq n. \tag{6.15}$$

Proof $|||[f]||| = \inf_{j\in I}||f - j||$, and so by hypothesis this inf $= 1$. By the distance formula in the Banach space $A(\Gamma)$, this gives that there is a linear functional Λ on $A(\Gamma)$ with

$$\Lambda = 0 \text{ on } I, \quad ||\Lambda|| = 1, \quad \text{and } \Lambda(f) = 1.$$

Hence there exists a complex measure ν on Γ such that

$$\int j\, d\nu = 0, \quad j \in I, \quad ||\nu|| = 1, \quad \int f\, d\nu = 1.$$

Let B denote the Blaschke product with zeros at the α_j, $1 \leq j \leq n$. Then for all $g \in A(\Gamma)$, $\int Bg\, d\nu = 0$, whence $B\, d\nu$ is an annihilating measure for $A(\Gamma)$. Hence there is an $F \in H_0^1$ such that

$$B d\nu = F\frac{d\theta}{2\pi}, \quad \text{or } d\nu = \frac{F}{B}\frac{d\theta}{2\pi}.$$

In particular, $d\nu$ and $d\theta$ are mutually absolutely continuous measures on Γ.

Since $|||[f]||| = 1$, we can choose a sequence $\{f_n\}$ in the coset $[f]$ such that $||f_n|| \to 1$. Then there exists a subsequence, again denoted $\{f_n\}$, which converges weak-$*$ in L^∞ to some function χ in L^∞ with $||\chi||_\infty \leq 1$. Since each $f_n \in H^\infty$ it follows that $\chi \in H^\infty$. Further,

$$\int f_n\, d\nu = \int f_n \frac{F}{B}\frac{d\theta}{2\pi} \to \int \chi \frac{F}{B}\frac{d\theta}{2\pi} = \int \chi\, d\nu,$$

and so $1 = \int \chi\, d\nu$. Since $|\chi| \leq 1$ a.e. and $||\nu|| \leq 1$, it follows that $|\chi| = 1$ a.e. and $d\nu = \overline{\chi} d|\nu|$, where $|\nu|$ denotes the total variation of ν.

Fix $j \in I$. Then

$$\int j\, d|\nu| = \int j\chi\, d\nu = \lim_{n\to\infty} \int j f_n\, d\nu = 0$$

since $jf_n \in I$ for all n.

We define $L(g) = \int g\, d|\nu|$ for $g \in A(\Gamma)$. Then L is a state of $A(\Gamma)/I$, and hence by Theorem 6.2, there is a $\phi \in I_0^\perp$ such that

$$\int g\, d|\nu| = \int g|\phi|^2 \frac{d\theta}{2\pi}.$$

This holds for all $g \in A(\Gamma)$ and so $d|\nu| = |\phi|^2\, d\theta/2\pi$, as measures. By Lemma 6.1, we may write $\phi = hg$, where h is a finite Blaschke product and g is an outer function in H^2 which belongs to $I_0^\perp$. Then $|\phi| = |g|$, so without loss of generality, we may assume

$$\text{there exists a } \phi \in I_0^\perp, \quad \phi \text{ outer}, \quad \text{such that } d|\nu| = |\phi|^2 \frac{d\theta}{2\pi}.$$

Fix $k \in A(\Gamma)$. Then $Bk \in I$. Hence

$$\begin{aligned} 0 &= \int Bk\, d\nu = \int Bk\overline{\chi}\, d|\nu| = \int Bk\overline{\chi}|\phi|^2 \frac{d\theta}{2\pi} \\ &= \int (Bk\phi)(\overline{\chi\phi}) \frac{d\theta}{2\pi}. \end{aligned}$$

Thus

$$\chi\phi \perp \{Bk\phi \,|\, k \in A(\Gamma)\}.$$

Since ϕ is outer, $\{\phi k \,|\, k \in A(\Gamma)\}$ is dense in H^2. Hence $\chi\phi \perp BH^2$. In particular, $\chi\phi \in I_0^\perp$, i.e. (6.14) holds.

Fix $g \in I_0^\perp$. For each n, $f_n - f \in I$. Hence

$$\int_\Gamma (\chi - f)\overline{g} \frac{d\theta}{2\pi} = \lim_{n\to\infty} \int_\Gamma (f_n - f)\overline{g} \frac{d\theta}{2\pi} = 0.$$

So $g \perp \chi - f$. It follows that

$$\int_\Gamma (\chi - f)\overline{l}_j \frac{d\theta}{2\pi} = 0, \quad 1 \le j \le n,$$

and so (6.15) holds.

It remains to show that χ is a Blaschke product of order $\le n-1$. Put $\Psi = \chi\phi$. By (6.14), $\Psi \in I_0^\perp$. Hence there exist polynomials R and S of degree $\le n-1$ such that

$$\phi = \frac{R}{\Pi}, \qquad \Psi = \frac{S}{\Pi},$$

where $\Pi(\zeta) = \prod_{j=1}^n (1 - \overline{\alpha}_j\zeta)$. Hence $\chi = \dfrac{\Psi}{\phi} = \dfrac{S}{R}$. Thus χ is a rational function of degree $\le n-1$. Also $\chi \in H^\infty$ and $|\chi| = 1$ a.e. on Γ. Hence χ is a Blaschke product of order $\le n-1$. Theorem 6.3 is proved. □

Corollary *For each f in $A(\Gamma)$, we write S_f for $S^{\mu_0}([f])$. If $f \in A(\Gamma)$, then $\|S_f\| = \|[f]\|$, where the norm on the left is operator norm on $I_0^\perp$ and the norm on the right is the quotient norm in $A(\Gamma)/I$.*

(This is Sarason's result in [Sa 1]).

Proof Without loss of generality, $|||[f]||| = 1$. We choose χ and ϕ by Theorem 6.3. Since $\chi(\alpha_j) = f(\alpha_j)$, $f - \chi \in I$, and hence $(f - \chi)\phi$ is in the closure of I in $H^2(\mu_0)$. It follows that

$$E[(f - \chi)\phi] = 0, \quad \text{and so} \quad E(f\phi) = E(\chi\phi),$$

where E is the orthogonal projection of $H^2(\mu_0)$ on $I_0^\perp$. So

$$S_f\phi = E(f\phi) = E(\chi\phi) = \chi\phi,$$

since $\chi\phi \in I_0^\perp$. So

$$||S_f\phi||^2 = \int_\Gamma |S_f\phi|^2 \, d\mu_0 = \int_\Gamma |\chi\phi|^2 \, d\mu_0 = \int |\phi|^2 \, d\mu_0.$$

Thus $||S_f|| \geq 1$. Also $||S_f|| \leq |||[f]||| = 1$. So $||S_f|| = 1$. We are done. □

As a final remark, we give a well-known characterization of analytic functions, based on Pick's Theorem.

Theorem 6.4 *Let F be a complex-valued function defined on the open unit disk $|\zeta| < 1$, such that $|F| < 1$ for all ζ in $|\zeta| < 1$. Then F is analytic on $|\zeta| < 1$ if and only if for each integer $N \geq 1$ and each N-tuple of points $\alpha_1, \alpha_2, \ldots, \alpha_N$ in $|\zeta| < 1$ we have*

$$\sum_{j,k=1}^{N} t_j \bar{t}_k F(\alpha_j)\overline{F(\alpha_k)} \frac{1}{1 - \alpha_j \bar{\alpha}_k} \leq \sum_{j,k=1}^{N} t_j \bar{t}_k \frac{1}{1 - \alpha_j \bar{\alpha}_k} \tag{6.16}$$

for all $(t_1, \ldots, t_N)$ in $\mathbf{C}^N$.

Proof The Necessity of Condition (6.16) follows from Pick's Theorem, Theorem 2.1.

Suppose now that F satisfies (6.16). Let $Q = \{\alpha_j\}_{j=1}^{\infty}$ be any fixed countable dense set in $|\zeta| < 1$. For each integer N, the N-tuple of values $w_1 = F(\alpha_1), \ldots, w_N = F(\alpha_N)$ satisfies Condition (2.3) for the points $\alpha_1, \alpha_2, \ldots, \alpha_N$. By Theorem 2.1, there exists f_N in H^∞ with $||f_N||_\infty \leq 1$ such that

$$f_N(\alpha_j) = F(\alpha_j), \quad 1 \leq j \leq N.$$

The sequence $\{f_N\}$ is a normal family on $|\zeta| < 1$. Hence we can choose a subsequence $\{f_{N_j}\}_j$ converging pointwise on $|\zeta| < 1$ to an analytic limit function f. Then $f(\alpha_j) = F(\alpha_j)$, $j = 1, 2, \ldots$, i.e. $f = F$ on Q.

We claim that in addition $f = F$ on $|\zeta| < 1$. Fix z with $|z| < 1$. By repeating the preceding argument, there exists f' analytic on $|\zeta| < 1$ so that $f' = F$ on $Q' = Q \cup \{z\}$. Since $f' = f$ on Q, this yields $f' = f$ on the open disk. In particular, $f(z) = F(z)$. Since z is arbitrary, it follows that $f = F$ on $|\zeta| < 1$. But f is analytic by construction. So F is analytic, and we are done. □

7 Appendix

The problem of interpolation by bounded analytic functions in the disk is often called the "Nevanlinna-Pick problem". As we have noted, Pick wrote his paper [Pi] in 1916. Rolf Nevanlinna made two major contributions to the problem, in 1919, [Ne 1], and in 1929, [Ne 2]. In his 1929 paper, Nevanlinna wrote "The general interpolation problem, in the case of finitely many assigned values, was first posed and solved by Herr Pick in [Pi]" (translation from the German).

The study of the quotient of a uniform algebra by a closed ideal was initiated by N. Varopoulos in the late 1960's. Such quotient algebras were called "Q-algebras". B. Cole showed that each Q-algebra has an isometric representation by operators on a Hilbert space. This result and other results on Q-algebras are given in the book [B-D] by Bonsall and Duncan, §50. It was natural to ask whether every commutative norm-closed algebra of operators on a Hilbert space, with identity, is a Q-algebra. This question was settled in the negative by N. Varopoulos in [Va 1], Theorem 2. This work showed at the same time that Von Neumann's inequality, ((1.11) in Chapter 1 above), cannot be generalized to an arbitrary k-tuple of commuting contractions on a Hilbert space, for $k > 2$. See also [Va 2]. A related counterexample is due to Crabb and Davie, [C-D]. The generalization of Von Neumann's inequality for two commuting contractions is true, as was shown by Ando in [An]. Recent results on generalizations of Von Neumann's inequality were given by Lotto, [Lo], and Lotto and Steger, [L-S]. See also earlier work by Drury in [Dr].

Sarason's paper [Sa 1], mentioned in the Introduction, established a connection between representations of a Q-algebra by operators on a Hilbert space, in the special case of the disk algebra and Pick interpolation. This connection, generally valid for Q-algebras, is the basis of the present paper. The connection was used in a related manner by Eric Amar in 1977, in [Am]. In particular, Amar gave a result closely related to our Theorem 3.3. Another related result is due to Nakazi, [Na].

Much of the work reported on in the present paper is joint with Keith Lewis and is contained in our joint papers, [C-L-W 1] and [C-L-W 2]. Agler's Theorem is in [Ag], and much of Chapter 4 was stimulated by Agler's work. Pick interpolation for finitely connected plane domains was studied by Abrahamse in [Ab]. Interpolation problems for the polydisk have been investigated recently by Cotlar and Sadosky in [C-S].

The literature on interpolation by bounded analytic functions as well as the literature on generalizations of Von Neumann's inequality for operators on a Hilbert space is extremely rich. The references we have just given are therefore of necessity a very partial list.

References

[Ab] Abrahamse, M., The Pick interpolation theorem for finitely connected domains, *Michigan Math. J.* **26** (1979), 195–203.

[Ag] Agler, J., Interpolation, to appear in *J. Funct. Anal.*

[Am] Amar, E., Ensembles d'interpolation dans le spectre d'une algèbre d'opérateurs, Thesis, University of Paris XI, 1977.

[An] Ando, T., On a pair of commutative contractions, *Ann. of Math.* **24** (1963), 88–90.

[B-D] Bonsall, F. and Duncan, J., *Complete Normed Algebras*, Ergebnisse der Mathematik **80**, Springer-Verlag, Berlin-Heidelberg-New York, 1973.

[Br] Browder, A., *Introduction to Function Algebras*, W. A. Benjamin, New York, 1969.

[C-L-W 1] Cole, B., Lewis, K. and Wermer, J., Pick conditions on a uniform algebra and von Neumann inequalities, *J. Funct. Anal.* **107** (1992), 235–254.

[C-L-W 2] Cole, B., Lewis, K. and Wermer, J., A characterization of Pick bodies, *J. London Math. Soc.* **48** (1993), 316–328.

[C-S] Cotlar, M. and Sadosky, C., Nehari and Nevanlinna-Pick problems and holomorphic extensions in the polydisk in terms of restricted BMO, to appear.

[C-D] Crabb, M. and Davie, A., Von Neumann's inequality for Hilbert space operators, *Bull. London Math. Soc.* **7** (1975), 49–50.

[Dr] Drury, S.W., Remarks on von Neumann's inequality, in: *Banach Spaces, Harmonic Analysis and Probability Theory (Storrs, Conn., 1980-1981)* (R. Blei and S. Sidney, eds.) Lecture Notes in Math. **995**, Springer-Verlag, Berlin-Heidelberg-New York, 1983, 14–32.

[Ga] Gamelin, T., *Uniform Algebras*, Prentice-Hall, 1969.

[Gar] Garnett, J., *Bounded Analytic Functions*, Academic Press, New York, 1981.

[Ho] Hoffman, K., *Banach Spaces of Analytic Functions*, Prentice-Hall, 1962.

[Lo] Lotto, B., Von Neumann's inequality for commuting, diagonalizable contractions, I, to appear in *Proc. Amer. Math. Soc.*

[L-S] Lotto, B. and Steger, T., Von Neumann's inequality for commuting diagonalizable contractions, II, to appear in *Proc. Amer. Math. Soc.*

[Na] Nakazi, T., Commuting dilations and uniform algebras, *Canad. J. Math.* **42** (1990), 776–789.

[Ne 1] Nevanlinna, R., Über beschränkte Funktionen die in gegebenen Punkten vorgeschriebene Werte annehmen, *Ann. Acad. Sci. Fenn. Ser. A* **13** (1919), 1–71.

[Ne 2] Nevanlinna, R., Über beschränkte analytische Funktionen, *Ann. Acad. Sci. Fenn. Ser. A* **32** (1929), 1–75.

[Pi] Pick, G., Über die Beschränkungen analytischer Funktionen, welche durch vorgegebene Funktionswerte bewirkt werden, *Math. Ann.* **77** (1916), 7–23.

[Sa 1] Sarason, D., Generalized interpolation in H^∞, *Trans. Amer. Math. Soc.* **127** (1967), 179–203.

[Sa 2] Sarason, D., Operator-theoretic aspects of the Nevanlinna-Pick interpolation problem, in: *Operators and Function Theory* (S.C. Power, ed.), NATO Adv. Sci. Inst. Ser. C **153**, Reidel, Dordrecht, 1985, 279–314.

[St] Stout, E., *The Theory of Uniform Algebras*, Bogden and Quigley, 1971.

[Va 1] Varopoulos, N., Sur une inégalité de Von Neumann, *C.R. Acad. Sci. Paris Sér. A* **277** (1973), 19–22.

[Va 2] Varopoulos, N., On an inequality of von Neumann and an application of the metric theory of tensor products to operator theory, *J. Funct. Anal.* **16** (1974), 83–100.

[Vo] Von Neumann, J., Eine Spektraltheorie für allgemeine Operatoren eines unitären Raumes, *Math. Nachr.* **4** (1951), 258–281.

Complex dynamics in higher dimensions

John Erik FORNÆSS
Mathematics Department
The University of Michigan
Ann Arbor, MI 48109
U.S.A.

Nessim SIBONY
Bât. 425, Mathématiques
Université de Paris-Sud
F-91405 Orsay Cédex
France

Notes partially written by
Estela A. GAVOSTO

Abstract

The field of complex dynamics in higher dimension was initiated in the 1920's by Fatou. It was motivated by studies in Newton's method, celestial mechanics and functional equations. Recently, new methods from pluripotential theory have been introduced to the subject. These techniques have produced many new interesting results. We give an introduction to this subject and a summary of the most relevant developments.

1 Introduction to complex dynamics

The topic of these notes is Complex Dynamics in Higher Dimensions, that is, the study of the iteration of holomorphic mappings in $\mathbf{C}^n$ or $\mathbf{P}^n$. Although complex dynamics is an old field, there has been so much recent progress that there is a need for an introductory text. (See [FS3] and [FS4] for references.) Our goal with these notes is to provide such an introduction. We will have in mind readers with varied backgrounds. For the readers with a several complex variables background we will state the basic concepts and facts from dynamics. For the experts in complex dynamics in one variable we will give a survey of basic tools from the theory of several complex variables. We also think that it is necessary to include the basic concepts and results from pluripotential theory and currents.

The theory of complex dynamics goes back to Schröder. He was the first one to use Newton's method to study complex roots in one complex variable (see [Sc1] and [Sc2]). The main problem is to describe the initial guesses leading to a root. In order to introduce some

P. M. Gauthier (ed.) and G. Sabidussi (techn. ed.), Complex Potential Theory, 131–186.

notation, let us consider the case of two polynomials of two complex variables,

$$\begin{cases} P(z,w) &= 0 \\ Q(z,w) &= 0 \end{cases}$$

For simplicity we write $F = (P,Q)$. F is a map from $\mathbf{C}^2$ to $\mathbf{C}^2$. We want to solve $F = 0$. Newton's method consists of guessing a root (z_1, w_1) and getting a better guess from a formula like

$$(z_2, w_2) = (z_1, w_1) - (F'(z_1, w_1))^{-1} F(z_1, w_1)$$

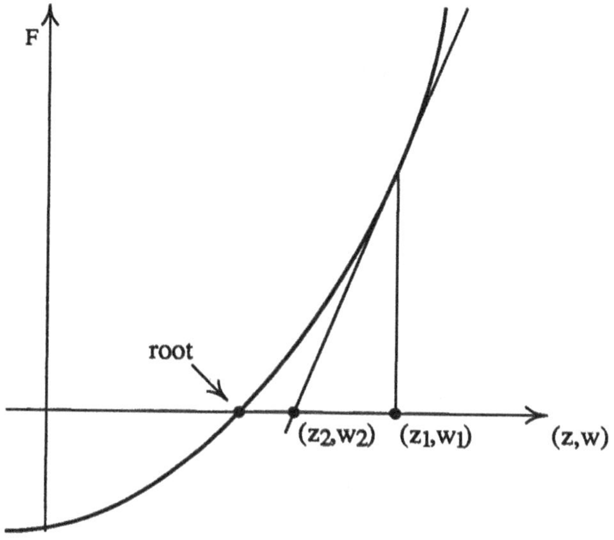

Figure 1

Here F' is a (2×2) matrix. See Figure 1. So we obtain a root of $(F = 0)$ by iteration of the map

$$R(z,w) = (z,w) - (F')^{-1} F(z,w).$$

We get inductively a sequence $\{(z_n, w_n)\}$,

$$(z_{n+1}, w_{n+1}) = R(z_n, w_n).$$

It turns out that for some initial values (z_1, w_1) the sequence $\{(z_n, w_n)\}$ does not converge to a root. Since R is a rational function, it is natural to work on $\mathbf{P}^n$ rather than $\mathbf{C}^n$, where

$$\mathbf{P}^n = \text{Space of complex lines through 0 in } \mathbf{C}^{n+1}.$$

To illustrate this, consider the equations:

$$\begin{cases} P(z,w) &= \frac{1}{2}z - \frac{\epsilon}{2}w - \frac{1}{2}z^2 &= 0 \\ Q(z,w) &= \frac{1}{2}w - \frac{\epsilon}{2}z &= 0. \end{cases}$$

Here $(0,0)$ is a common root. One can replace F' in Newton's method by the constant matrix

$$A = \begin{bmatrix} \frac{1}{2} & 0 \\ 0 & \frac{1}{2} \end{bmatrix},$$

assuming that ϵ is small. In that case Newton's method becomes

$$R(z,w) = (z,w) - A^{-1}F = (z^2 + \epsilon w, \epsilon z).$$

This map is an example of a complex Hénon map. The root $(0,0)$ is a fixed point for R. Moreover, it is an attracting fixed point, i.e. the eigenvalues of $R'(0,0)$ are both smaller than one. Hence, all points (z_1, w_1) in a small neighborhood of $(0,0)$ are attracted to $(0,0)$. The largest such set is the basin of attraction of $(0,0)$, and is an open set consisting of all initial guesses giving that root. This kind of set was studied already in the 20's and 30's by Fatou [Fa] and Bieberbach [Bi]. They discovered that these sets were quite large, holomorphically equivalent to the whole space $\mathbf{C}^2$, nevertheless they were also quite small. So both the set of initial guesses giving the root and initial guesses not giving the root are quite large. A more precise description of these sets was only obtained in the last few years. For more details, see section 7.

The iterates $\{R^n\}$ form a normal family on the basin of attraction of $(0,0)$. In general, the largest open set where $\{R^n\}$ is a normal family is called the Fatou set. But note that in the case where the limits of sequences of $\{R^n\}$ are infinite it will depend on the context whether the domain is considered as part of the Fatou set or not. The complement of the Fatou set is called the Julia set. A main problem then is to describe the Fatou set and the Julia set of a map R. The points on the Julia set are never roots of F. But even some connected components of the Fatou set might not give correct initial guesses of the roots. So natural questions are to ask whether the Julia set has zero volume and what are the possible kinds of Fatou components.

To discuss the behavior of R on the Julia set, one can introduce the concept of chaos. For example, it is known that rational maps on $\mathbf{P} = \mathbf{C} \cup \{\infty\}$ or polynomial maps on $\mathbf{C}$ are always chaotic on their Julia sets. We say that a continuous map $f : K \to K$ is chaotic on a compact metric space K (cf. [Dev]) with metric d if

(i) f is sensitive to initial conditions, i.e., there exists a $\delta > 0$ so that if $x \in K$ and $\epsilon > 0$, then there is a $y \in K$, $d(x,y) < \epsilon$ and an integer $n \geq 1$ so that $d(f^n(x), f^n(y)) > \delta$.

(ii) f is topologically transitive, i.e., if $x, y \in K$, $\delta > 0$, then there exists a $z \in K$ and an integer $n \geq 1$ so that if $d(x,z) < \delta$, $d(f^n(z), y) < \delta$.

(iii) Periodic orbits are dense in K.

We will see that in higher complex dimensions the Julia set might be too big. We will need to restrict our study to the nonwandering part. Let $f : M \to M$ be a continuous map on a manifold M. A point $p \in M$ is said to be nonwandering if for every open set U, $p \in M$, there is an integer $n \geq 1$ so that $f^n(U) \cap U \neq \emptyset$. Otherwise the point p is said to be wandering.

Concepts related to chaos are ergodicity and mixing. Let $f : K \to K$ be a continuous map on a compact metric space. We say that a Borel measure μ is invariant if $\mu(f^{-1}(E)) = \mu(E)$ for all Borel sets $E \subset K$. Let μ be an invariant probability measure, $\mu(K) = 1$. Then μ is ergodic if whenever E is an invariant Borel set, i.e., $F(E) = F^{-1}(E) = E$, then $\mu(E) = 0$ or 1. A stronger condition than ergodicity is that of (strong) mixing. The invariant probability measure μ is (strong) mixing if whenever E and F are Borel measurable subsets of K, then $\mu(f^{-n}(E) \cap F) \to \mu(E) \cdot \mu(F)$ when $n \to \infty$. Later in these notes we will show that holomorphic maps in higher dimension carry natural measures which are ergodic.

Fixed points usually play a big role in dynamics. Let F be a holomorphic map $F : M \to M$, M a complex manifold. Let $p \in M$, be a fixed point of F. Consider the eigenvalues of the matrix $F'(p)$. If all the eigenvalues are of modulus strictly less than one, then p is an attracting fixed point and there is an open set, the attracting basin of p, consisting of points z for which $F^n(z) \to p$, as we have seen above. If all the eigenvalues are strictly larger than one, the point p is repelling, and there exists an open neighborhood U of p such that whenever $q \in U \setminus \{p\}$, then there exists an integer $n \geq 1$ so that $F^n(q) \notin U$.

Another case is when some eigenvalues $\{\lambda_i^s\}_{i=1}^k$, but not all, are strictly less than one and all the others, $\{\lambda_j^u\}_{j=1}^l$ are strictly larger than one. In this case p is said to be a saddle point. Then, there are arbitrarily small neighborhoods U_p of p containing also complex submanifolds $W_{U_p}^s$, $W_{U_p}^u$; the stable and unstable manifolds of complex dimension k and l, respectively. The fixed point $p \in W_{U_p}^s \cap W_{U_p}^u$, and the tangent space of $W_{U_p}^u$ at p consists of the unstable eigenspace. The tangent space of $W_{U_p}^s$ at p contains the stable eigenspace and is transverse to the tangent space of $W_{U_p}^u$, but note that some of the stable eigenvalues might vanish.

The local stable manifold $W_{U_p}^s$ consists of those points $q \in U_p$ so that $F^n(q) \to p$ as $n \to \infty$ and $\{F^n(q)\}_{n=1}^\infty \subset U_p$. The local unstable manifold $W_{U_p}^u$ consists of those points $q \in U_p$ for which there exists a sequence $\{q_n\}_{n=1}^\infty \subset U_p$, $q_n \to p$, $F(q_{n+1}) = q_n$, $q_1 = q$. See Figure 2.

Moreover, if q is any point in $U_p \setminus (W_{U_p}^s \cap W_{U_p}^u)$ then there exists an integer $n \geq 1$ so that $F^n(q) \in U_p$ while $F^{n+1}(q) \notin U_p$. Also, inductively define $S^1 = \{q\}$ and $S^{n+1} = F^{-1}(S^n) \cap U_p$. Then $S^n = \emptyset$ for all large enough n.

The stable set, W_p^s, of p consists of all points $q \in M$ so that $F^n(q) \to p$, as $n \to \infty$. The unstable set, W_p^u, consists of all points $q \in M$ so that there exists a sequence $\{q_n\}_{n=1}^\infty \subset M$, $q_1 = q$, $F(q_{n+1}) = q_n$, $q_n \to p$.

The three kinds of points considered above, attracting fixed points, repelling points, and saddle points are said to be hyperbolic fixed points. These are fixed points where no eigenvalue of the derivative is on the unit circle. Similarly, one says that a periodic orbit $\{z_k\}_{k=0}^l$, $z_l = z_0$, is hyperbolic if $(F^l)'(z_0)$ has no eigenvalue on the unit circle.

The local dynamics near hyperbolic periodic orbits is simple to describe, as we have just seen. Furthermore this local dynamics is stable, i.e. maps close to F also have hyperbolic periodic points close to p.

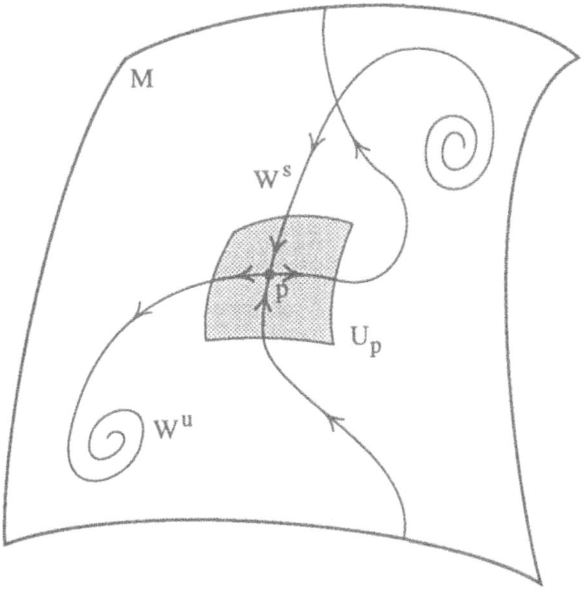

Figure 2

One can generalize the concept of hyperbolicity to more general compact sets [Ru]. So let again $F : M \to M$ be a holomorphic map on a complex manifold M and let K be a compact set. We assume that K is surjectively forward invariant, i.e., $F(K) = K$. The space $\hat{K} = K^{\mathbf{N}}$ of orbits $\{p_n\}_{n=-\infty}^{-1}$, $F(P_n) = p_{n+1}$, is compact in the product topology. By the tangent bundle T_K of $\hat{K}$ we mean the space (p, ξ) where $p = \{p_n\} \in \hat{K}$ and $\xi \in T_M(p_1)$ is a tangent vector. We give this tangent bundle the obvious topology. Then F lifts to a homeomorphism $\hat{F} : \hat{K} \to \hat{K}$,

$$\hat{F}(\{\cdots, p_{-2}, p_{-1}\}) = (\{\cdots, p_{-2}, p_{-1}, F(p_{-1})\}),$$

and F' lifts to a map $\hat{F}'$ on T_K in the obvious way.

Let $K \subset M$ be a compact surjectively forward invariant set. Then F is said to be hyperbolic on K if there exists a continuous splitting $E^u \oplus E^s$ of the tangent bundle of $\hat{K}$ such that the subbundles E^u, E^s are preserved by $\hat{F}'$, i.e., $\hat{F}'_{E_p^s} \subset E^s_{\hat{F}(p)}$ and $\hat{F}'_{E_p^u} \subset E^u_{\hat{F}(p)}$, and for some constants $C, c > 0$, $\lambda > 1$, $\mu < 1$, and some Hermitian metric on M,

$$\begin{aligned} |(\hat{F}^n)'(\xi)| &\geq c\lambda^n|\xi|, \quad \xi \in E^u, \\ |(\hat{F}^n)'(\xi)| &\geq C\mu^n|\xi|, \quad \xi \in E^s, \end{aligned}$$

$n = 1, 2, \cdots$.

One of the main questions for rational maps on $\mathbf{P}^1$ is whether the maps which are hyperbolic on their Julia set are dense in the rational maps.

A weaker property is that of stability. A family of maps $\{f : M \to M\}_{f \in A}$ is stable at f_0 if there exists an open neighborhood U of f_0 in A such that all maps $g \in U$ are topologically conjugate to f_0, i.e., there exists a homeomorphism $h : M \to M$ so that $g \circ h = h \circ f$.

It is known that the space of rational maps on $\mathbf{P}^1$ is stable on an open dense set [MSS].

2 Introduction to several complex variables

The purpose of this section is to collect some basic facts from the theory of Several Complex Variables.

One of the main concepts in the theory of several complex variables is that of pseudoconvexity. This is a holomorphically invariant version of convexity; see below for a precise definition. Pseudoconvex domains are the natural domains of definition for holomorphic functions. If a domain fails to be pseudoconvex, all holomorphic functions extend to a fixed strictly larger domain. It might therefore not be surprising that Fatou components are pseudoconvex. This will be explored and proved in later sections.

To define pseudoconvexity, define at first the class F of domains which are locally biholomorphic to convex domains. An open set $U \subset M^n$, M a complex manifold of dimension n, belongs to F if for every $p \in \partial U$ there exists an open neighborhood V of p and a biholomorphic map $\phi : V \to B$, the open unit ball in $\mathbf{C}^n$, so that $\phi(p) = 0$ and $\phi(V \cap U)$ is convex. See Figure 3.

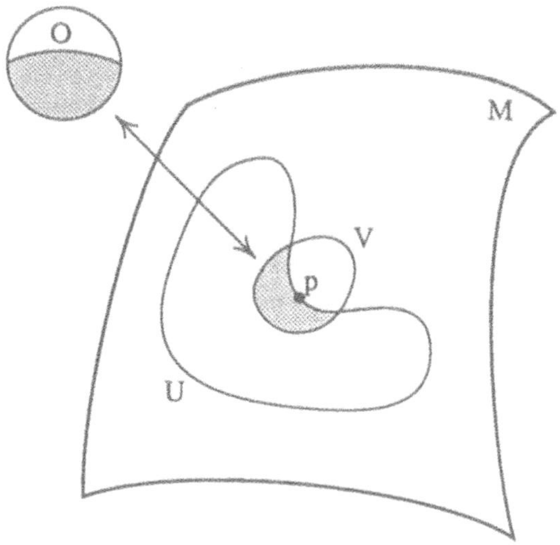

Figure 3

An open set $U \subset M^n$ is pseudoconvex if for every $p \in \partial U$ there exists a neighborhood W of p such that for every compact set $K \subset W \cap U$, there exists a set $O \in F$, $K \subset O \subset W \cap U$. See Figure 4.

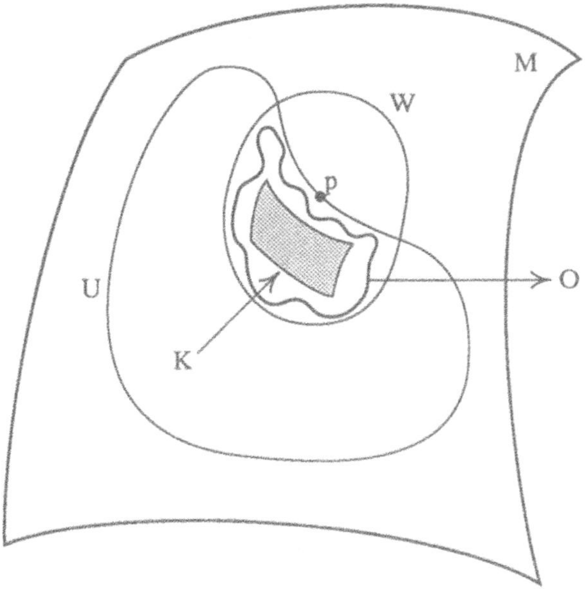

Figure 4

All open sets in $\mathbf{C}$ are pseudoconvex. All convex domains in $\mathbf{C}^n$ are pseudoconvex. A basic model of a non-pseudoconvex domain is a Hartogs figure. Pick numbers $0 < r_1, r < 1$. Then let

$$\begin{aligned} H := & \{(z,w) \in \mathbf{C}^n : z \in \mathbf{C}, w \in \mathbf{C}^{n-1}, |z| < 1, |w| < r\} \\ & \cup \{(z,w) : r_1 < |z| < 1, |w| < 1\}. \end{aligned}$$

Let $\hat{H}$ denote the convex hull of H, $\hat{H} := \{(z,w) : |z| < 1, |w| < 1\}$. See Figure 5.

The basic fact is that all holomorphic functions defined on H extend as holomorphic functions to $\hat{H}$. In fact, an equivalent definition of pseudoconvexity can be given using Hartogs figures. A domain $U \subset M^n$ is pseudoconvex if whenever $\phi : \hat{H} \to M$ is a one to one holomorphic map and $\phi(H) \subset U$, then $\phi(\hat{H}) \subset U$. See Figure 6.

In later sections we will see that in higher dimension it is natural to introduce different Fatou sets. The ordinary Fatou set will be shown to be pseudoconvex, as described using the Hartogs figures. To distinguish them, we will call these domains $(n-1)$ pseudoconvex and the Hartogs figure a $(1, n-1)$ Hartogs figure. In general, we have for $0 < k < n$ an $(n-k, k)$ Hartogs figure,

$$\begin{aligned} H_k := & \{(z,w) \in \mathbf{C}^n : z \in \mathbf{C}^{n-k}, w \in \mathbf{C}^k, |z| < 1, |w| < r\} \\ & \cup \{(z,w) : r_1 < |z| < 1, |w| < 1\}. \end{aligned}$$

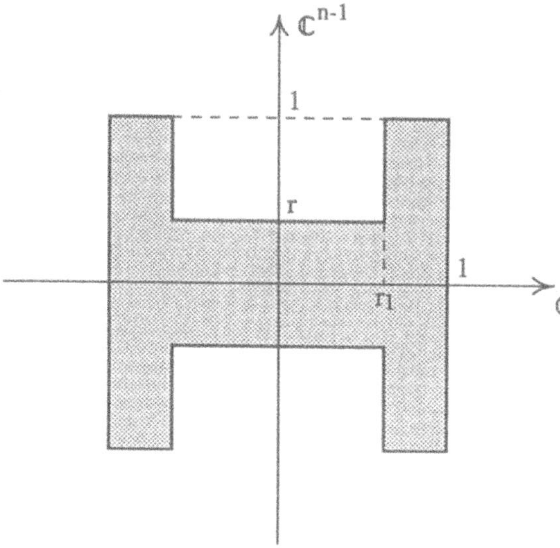

Figure 5

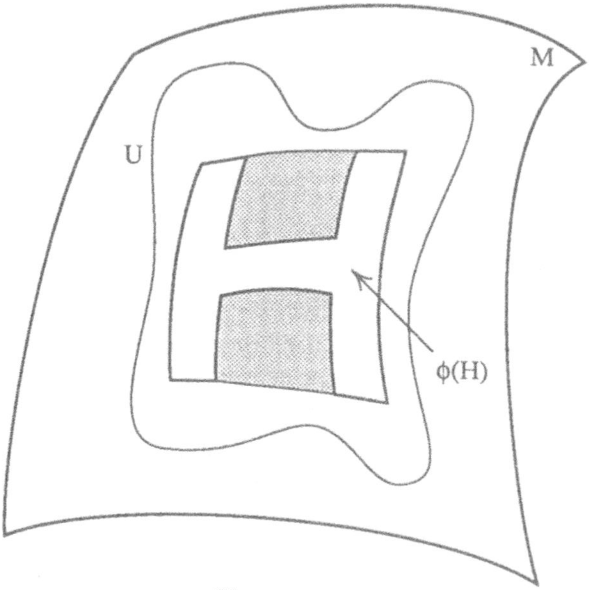

Figure 6

A domain $U \subset M^n$ is k-pseudoconvex if whenever $\phi : \hat{H}_k \to M$ is a one-one holomorphic map and $\phi(H_k) \subset U$, then $\phi(\hat{H}_k) \subset U$. One of the main questions in the theory of several complex variables in the period 1910–1950 was whether pseudoconvex domains were domains of holomorphy. A domain $U \subset \mathbf{C}^n$ or $\mathbf{P}^n$ ($U = \mathbf{P}^n$ is the only exclusion) is a domain of holomorphy if there is no point $p \in \partial U$, no connected neighborhood V of p and no nonempty connected component W of $V \cap U$ so that whenever f is holomorphic on U, there exists a holomorphic function g on V so that $f = g$ on W. See Figure 7.

The above is a precise way to say, even when the boundary is bad, that holomorphic functions cannot be extended to a fixed larger domain.

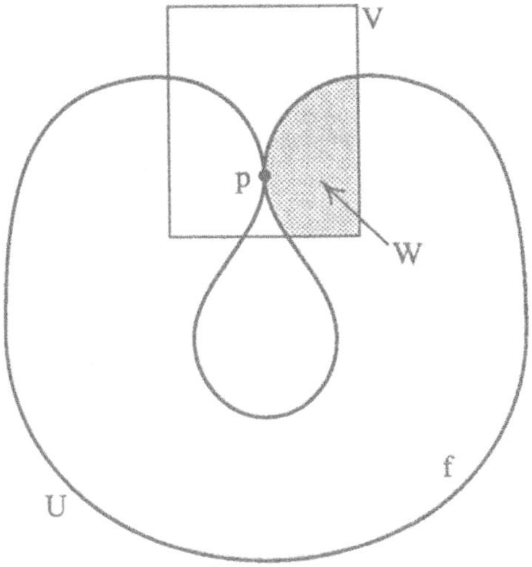

Figure 7

Theorem 2.1 [Ta] *If $U \subset \mathbf{C}^n$ or $\mathbf{P}^n$, $U \neq \mathbf{P}^n$ and U is pseudoconvex, then U is a domain of holomorphy.*

To mention a couple of consequences of dynamical interest, keep in mind that we will show in later sections that the Fatou set is pseudoconvex. So the Julia set is the complement of a domain of holomorphy.

Theorem 2.2 [Ta] *The complement of a domain of holomorphy in $\mathbf{P}^n$, $n \geq 2$, is connected.*

The above theorem fails in $\mathbf{P}^1$, and indeed there are rational maps with totally disconnected Julia sets in $\mathbf{P}^1$ (see [CG, p. 67]). There are many rational maps on $\mathbf{P}^1$ for which the set of critical points, and even the closure of their critical orbit, is disjoint from the Julia sets. In fact, these maps are precisely the hyperbolic rational maps.

Theorem 2.3 [Ta] *A domain of holomorphy in $\mathbf{P}^n$, $n \geq 2$, cannot contain a compact complex variety of positive dimension.*

A consequence of this theorem is that the critical set of a holomorphic map $R : \mathbf{P}^n \to \mathbf{P}^n$, $n \geq 2$, of degree $d \geq 2$, $\{\det R' = 0\}$ must intersect the Julia set.

A useful fact is that the holomorphic maps f on $\mathbf{P}^k$ can be lifted to homogeneous polynomial F on $\mathbf{C}^{k+1}$ in such a way that the following diagram commutes:

$$\begin{array}{ccc} \mathbf{C}^{k+1} & \xrightarrow{F} & \mathbf{C}^{k+1} \\ \downarrow \pi & & \downarrow \pi \\ \mathbf{P}^k & \xrightarrow{f} & \mathbf{P}^k \end{array}$$

where $\pi : \mathbf{C}^{k+1} \to \mathbf{P}^k$ is the canonical projection sending lines through zero to points. The map F has the crucial property that F only vanishes at the origin. It is clear that any such F gives rise to a holomorphic map on $\mathbf{P}^k$. We need to show that we can find an F given f. To do this, we will lift the map $g := f \circ \pi : \mathbf{C}^{k+1} \to \mathbf{P}^k$. We may assume that $f \circ \pi(\mathbf{C}^{k+1})$ is not contained in any coordinate plane $\{z_j = 0\}$. The quotients $\{z_j \circ g / z_0 \circ g\}$ are then well defined meromorphic functions on $\mathbf{C}^{k+1} \setminus \{0\}$ and they are constant on complex lines through zero. It follows from the Weierstrass-Hurwitz Theorem that $z_j \circ g / z_0 \circ g$ extend to meromorphic functions on $\mathbf{C}^{k+1}$ which are given as quotients of homogeneous polynomials of the same degree $z_j \circ g / z_0 \circ g = F_j / G_j$. Next consider the mapping

$$\tilde{F} : z \mapsto (\pi G_j, (F_1/G_1)\pi G_j, \ldots, (F_k/G_k)\pi G_j).$$

$\tilde{F}$ is a homogeneous polynomial mapping which lifts f outside the zero set $\{z_0(g) = 0\}$. Let F denote the homogeneous polynomial map obtained from cancelling all common homogeneous factors from the components of $\tilde{F}$. Then F still lifts f outside $\{z_0(g) = 0\}$.

Consider next a point $p \in \{z_0(g) = 0\} \setminus \{0\}$. We may assume that $f \circ \pi(p) = [0 : 1 : \cdots]$. Consider the germs $\{z_j(g) = 0\}$ at p. Notice that $z_i(g(p)) \neq 0$. Also note that no germ $\{z_j(g) = 0\}$, $j \geq 2$, can contain a whole irreducible germ of $z_0(g) = 0$, because then, by dimension counting, the algebraic hypersurfaces $\{z_j = 0\}_{j=0}^{k}$ would have a point in common in $\mathbf{P}^k$. At such a point, the map f would be undefined, i.e., f would not be holomorphic there, the point would be a point of indeterminacy. Hence all the G_j have the same zero set (with multiplicity) at p. It follows that $F(p) \neq 0$.

A main goal in higher dimensional complex dynamics is to characterize the nature of the Fatou components. So we need a tool to detect whether a domain is large, in the sense of containing arbitrarily large discs or even a copy of $\mathbf{C}$. The Kobayashi metric serves for this purpose. We will give the precise definition and state some basic properties. See [La] for a systematic study.

Let M denote a complex manifold. Pick a point $p \in M$ and a tangent vector ξ at p. We will search for the largest disc in M through p in the direction of ξ. Let Δ denote the unit disc in the complex plane.

Consider any holomorphic map $f : \Delta \to M$, $f(0) = p$, $f'_*(\partial/\partial z) = c\xi$. We define the infinitesimal Kobayashi pseudo-metric $ds(p, \xi) = \inf_f \{1/|c|\}$. See Figure 8.

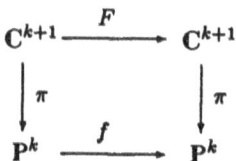

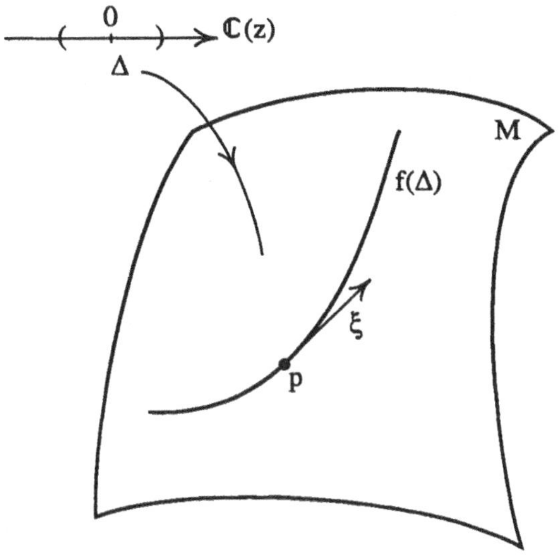

Figure 8

If $M = \mathbf{C}^n$, c can be arbitrarily large, so $ds \equiv 0$. If $M = \Delta$, ds is just the Poincaré metric.

A complex manifold M is said to be Kobayashi hyperbolic if $ds > 0$ when $\xi \neq 0$ uniformly and locally on the tangent bundle of M, TM. We will mention some key facts which follow easily from the definition.

Theorem 2.4 *If $F : M \to N$ is a holomorphic covering map between complex manifolds, then F is distance decreasing, i.e.,*

$$ds(F(p), F'_*(\xi)) \leq ds(p, \xi).$$

In particular, if F has a holomorphic inverse, then F is an isometry.

Theorem 2.5 *If $F : M \to N$ is a holomorphic covering map between complex manifolds, then F is pointwise an isometry in ds.*

Since there is a covering between Δ and $\mathbf{C} - \{0, 1\}$ it follows that $\mathbf{C} - \{0, 1\}$ is Kobayashi hyperbolic. Observe also that a bounded open set in $\mathbf{C}^n$ is Kobayashi hyperbolic. This is an easy consequence of Schwarz's lemma. We need another criterion to determine when a domain is Kobayashi hyperbolic. A complex hypersurface $D \subset \mathbf{P}^n$ is a set which locally can be described by an equation $f = 0$ where f, $f \not\equiv 0$, is a holomorphic function.

Green ([Gr1] and [Gr2]) has proved the following theorems, generalizing the fact that $\mathbf{C} \setminus \{0, 1\} = \mathbf{P}^1 \setminus \{0, 1, \infty\}$ is Kobayashi hyperbolic.

Theorem 2.6 *Let $D_1, \ldots, D_m$ be compact complex hypersurfaces in $\mathbf{P}^n$. Then $\mathbf{P}^n \setminus (\bigcup_j D_j)$ is Kobayashi hyperbolic if*

(i) *there is no non-constant holomorphic map from $\mathbf{C} \to \mathbf{P}^n \setminus (\bigcup_j D_j)$,*

(ii) *there is no non-constant holomorphic map $\mathbf{C} \to (D_{i_1} \cap \ldots \cap D_{i_k}) \setminus (D_{j_1} \cup \ldots \cup D_{j_l})$ for any $\{i_1, \ldots, i_k, j_1, \ldots, j_l\} = \{1, \ldots, m\}$.*

Theorem 2.7 *Suppose $f : \mathbf{C} \to \mathbf{P}^k$ omits $k+2$ distinct complex hypersurfaces. Then $f(\mathbf{C})$ is contained in a compact complex hypersurface.*

Let $P(z) = z^k + \cdots$ be a polynomial in one complex variable, vanishing to order k at the origin. If $k = 1$, P has a zero of order 1, i.e., P gives a local holomorphic coordinate. If $k > 1$, then for arbitrarily small ϵ, $\epsilon \neq 0$, $P - \epsilon$ has k distinct zeroes of order 1 arbitrarily close to 0. Then we say that P has a zero of multiplicity k. This terminology can be carried over to common roots of several polynomials.

Theorem 2.8 (Bezout's Theorem) *Let $P_1(z_1, \ldots, z_{n+1}), \ldots, P_n(z_1, \ldots, z_{n+1})$ be homogeneous polynomials of degree $d_1, \ldots, d_n$. Their zero sets can be interpreted as complex hypersurfaces $X_1, \ldots, X_n$ in $\mathbf{P}^n$. Suppose $X_1 \cap X_2 \cap \ldots \cap X_n$ contains only finitely many points. Then the number of points, counted with multiplicity, is the product of the degrees, $d_1 d_2 \ldots d_n$.*

Bezout's Theorem can be used to count periodic points with multiplicity.

Theorem 2.9 *Let $f : \mathbf{P}^k \to \mathbf{P}^k$ be a holomorphic map of degree $d \geq 2$. The number of fixed points of f counted with multiplicity is equal to $(d^{k+1} - 1)/(d-1)$.*

To prove the above theorem, one lifts f to a homogeneous map

$$F = (F_1, \ldots, F_{k+1}) : \mathbf{C}^{k+1} \to \mathbf{C}^{k+1}(z_1, \ldots, z_{k+1})$$

and considers the common solutions of $\{F_j - z_j = 0\}_{j=1}^{k+1}$ after introducing an extra variable t to get homogeneous equations $\{F_j - t^{d-1} z_j = 0\}_{j=1}^{k+1}$. Then Bezout's Theorem applies.

3 Critically finite maps

Consider the quadratic family $\{z^2 + c\}$ parametrized by c. The Mandelbrot set $M(c)$ in the parameter plane $\mathbf{C}(c)$ can be defined as $\{c : J_c \text{ is connected }\}$, where J_c is the Julia set of $P_c(z) = z^2 + c$. The interior components of the Mandelbrot set consist of stable components for the dynamics. If the conjecture that hyperbolic maps are dense were true, then each interior component of the Mandelbrot set would contain a unique map with a particularly nice dynamics, a critically finite map. A critically finite map is a map for which a critical point has a periodic orbit.

We give a more general definition that will apply to higher dimensions. For simplicity, we will consider two complex dimensions.

Let $R : \mathbf{P}^2 \to \mathbf{P}^2$ be a holomorphic map of degree $d \geq 2$. Denote by C the critical set, and let $\{C_i\}$ denote the irreducible components of C. We say that R is critically finite if each C_i is periodic or preperiodic, i.e. there exists integers $l_i \geq 0, n_i \geq 10$ (minimally chosen) so that

$$C_i \to R(C_i) \to \cdots \to R^{l_i}(C_i) \to \cdots \to R^{l_i+n_i}(C_i) = R^{l_i}(C_i).$$

We say that R is strictly critically finite if in addition the maps $R^{n_i} : R^{l_i}(C_i) \to R^{l_i}(C_i)$ are also critically finite maps, i.e. if all their critical points are periodic of preperiodic. Since the compact Riemann surfaces might be singular, that is, they might have self-crossings of cusp singularities, one must be careful with the definition of critical points. One possibility is to include all the singular points in the critical set. Another possibility is to consider the normalizations $\tilde{C}_i$ of $F^{l_i}(C_i)$ and lift R^{n_i} to a holomorphic map $S^i : \tilde{C}_i \to \tilde{C}_i$ between smooth, nonsingular, Riemann surfaces and restrict to the critical points of S^i. This definition is wider than the one indicated above for the quadratic family z^2+c. It includes also the maps $P_c(z)$ for which the critical point is preperiodic. These are the so-called Misiurewicz points and they belong to the boundary of the Mandelbrot set. (See [CG, p. 133].)

If all the critical points of a rational map on $\mathbf{P}^1$ are preperiodic, then it is known that the Julia set is all of $\mathbf{P}^1$ [Mil]. McMullen [Bie] asked whether there is a similar theory in $\mathbf{P}^2$.

We will see in section 4 that there is a way to find examples of maps on $\mathbf{P}^2$ which are critically finite using maps on $\mathbf{P}^1$ and a suspension argument [U], but it is trickier to find a "genuine" 2-dimensional example.

It is shown in [FS2] that the map $g : \mathbf{P}^2 \to \mathbf{P}^2$, $g([z : w : t]) = [(z-2w)^2 : (z-2t)^2 : z^2]$ is strictly critically finite and that the Julia set is all of $\mathbf{P}^2$. To prove this result, one first computes the critical set and shows that the forward orbit of the critical set lands on a cycle of three complex lines. Direct computation shows that g^3 is a critically finite map on each of these lines. Moreover, all critical points are strictly preperiodic on each of these lines. Hence the Julia set contains these three lines and therefore their preimages. In particular, the critical orbit of g on $\mathbf{P}^2$ is contained in the Julia set.

Next one can show that the complement of the critical orbit is Kobayashi hyperbolic. This implies that the map g is strictly volume increasing in the Kobayashi metric on the complement of the critical orbit. With this one can show that any Fatou component must converge under iteration to the critical orbit. But then one gets a contradiction to the expansiveness of g^3 in the above three lines.

This result generalizes to other strictly critically finite maps, but one must add the technical hypothesis that the complement of the critical orbit, V, has to be Kobayashi hyperbolic. An attracting orbit $z_0 \mapsto z_1 \mapsto \cdots \mapsto z_k = z_0$ is superattracting if $(f^k)'(z_0) = 0$.

Theorem 3.1 *Let $f : \mathbf{P}^2 \to \mathbf{P}^2$ be a holomorphic map with critical set C. Assume that f is strictly critically finite and that $\mathbf{P}^2 \setminus V$ is Kobayashi hyperbolic. Then the only Fatou* *components of f are preperiodic to or equal to superattracting components. In particular if no critical point is periodic, then the Julia set of f is $\mathbf{P}^2$.*

The proof is similar to the proof for the above example g except that one takes into account the possibility that f might have nonempty Fatou components when restricted to V. Such Fatou components are necessarily superattracting on V. One needs an additional computation to show that both eigenvalues of $(f^k)'(z_0)$ vanish on attracting orbits inside V (see [FS4]).

Until now we have been studying complex dynamics from an analytic point of view. Hubbard and Oberste-Vorth [HOV] have made a topological study of Hénon mappings on $\mathbf{C}^2$. Let us recall some notation. A complex Hénon map of degree d is a map of the form $F(z,w) = (z^d + q(z) - aw, z)$. Define the set K^+ as

$$K^+ = \{(z,w) : (\|F^n(z,w)\|) \text{ is bounded}\}$$

and the set $U^+ = \mathbf{C}^2 \setminus K^+$. A main question is to study the dynamics of F on the invariant closed set K^+. The open set U^+ is foliated by level sets of a pluriharmonic function G^+. (Recall that $G^+(z,w) = \lim_{n\to\infty} \frac{1}{d^n} \log_+ \|F^n(z,w)\|$.) The function G^+ vanishes precisely on K^+ and is strictly positive on U^+. Also the function G^+ satisfies a functional equation $G^+(F(z,w)) = dG^+(z,w)$. Hence the map F sends level sets $\{G^+ = c\}$ to $\{G^+ = dc\}$. It is therefore natural to study F on ∂K^+ by considering the topological nature of the level sets $\{G^+ = c\}$ and the nature of the maps between the level sets, especially the limiting behavior when $c \to 0$.

Actually, they showed that the map $G^+ : U^+ \mapsto \mathbf{R}^+ \setminus \{0\}$ is a trivial fibration. They also discovered a solenoidal nature of the level sets. Let Π denote the solid torus, $\Pi = \{(\xi, z) : |\xi| = 1, |z| \leq 2\}$. Consider the solenoidal mapping $\tau_d : \Pi \to \Pi$, $\tau_d(\xi, z) = (\xi^d, \xi + \epsilon z \xi^{1-d})$ for $0 < \epsilon \ll 1$. Define next the solenoid as the intersection $\Sigma_d = \cap_{m=1}^{\infty} \tau_d^m(\Pi)$. Then Σ_d is a compact invariant subset of the torus Π. We can consider $\Sigma_d \subset \mathbf{R}^3 \subset \mathbf{R}^3 \cup \{\infty\} = S^3$.

A main result of Hubbard and Oberste-Vorth is that the level sets of G^+ are homeomorphic to $S^3 \setminus \Sigma_d$. Moreover, the map $h_d : \Pi \to \Pi$ extends to a homeomorphism $h_d : S^3 \to S^3$. This map has two invariant solenoids $\Sigma_d = \Sigma_+$, which is attracting, and Σ_-, which is repelling. Analytically, the natural compactifications of $\mathbf{C}^2$ are obtained by adding one or more compact complex curves at infinity. Hubbard and Oberste-Vorth have a topological compactification of $\mathbf{C}^2$, $\overline{\overline{\mathbf{C}}}^2$, obtained by adding a copy of S^3 at infinity. They can do this compactification in such a way that the Hénon map extends to a homeomorphism $\tilde{F}$ of $\overline{\overline{\mathbf{C}}}^2$ and $\tilde{F}$ is the solenoidal mapping h_d on S^3.

Our final topic in this section is an investigation of Kobayashi hyperbolicity of Fatou components. A linear map in $\mathbf{C}$ such as $z \to \frac{1}{2}z$ has a Fatou component equivalent to $\mathbf{C}$ which is not Kobayashi hyperbolic. However, for rational maps of degree d, $d \geq 2$, on $\mathbf{P}^1$, all Fatou components are hyperbolic. Passing to two dimensions, we have seen with Fatou-Bieberbach domains that Fatou components of (non-linear) Hénon maps can be biholomorphic to $\mathbf{C}^2$. These maps extend to $\mathbf{P}^2$ as meromorphic maps. However, if we restrict to holomorphic maps of $\mathbf{P}^k$ of degree $d \geq 2$, the next theorem shows that a Fatou component can never be biholomorphic to $\mathbf{C}^k$.

Theorem 3.2 [U] *If $f : \mathbf{P}^k \to \mathbf{P}^k$ is a holomorphic map of degree $d \geq 2$, then all Fatou components are Kobayashi hyperbolic. In particular, no Fatou component can be holomorphically equivalent to $\mathbf{C}^k$.*

Proof We can lift f to a homogeneous polynomial map $F : \mathbf{C}^{k+1} \to \mathbf{C}^{k+1}$. Since f is holomorphic there is a constant C so that $\frac{1}{C}\|z\|^d \leq \|F(z)\| \leq C\|z\|^d$.

The idea of the proof is to lift the Fatou components to the boundary of the basin of attraction of 0 for F. Then Kobayashi hyperbolicity follows from the fact that this basin of attraction is a bounded set. Consider the Green function, or escape function,

$$G(z) = \lim_{n\to\infty} \frac{1}{d^n} \log \|F^n(z)\|.$$

Recall that

$$G(\lambda z) = G(z) + \log|\lambda| \qquad (*)$$

and that $G(z)$ is continuous on $\mathbf{C}^{k+1} \setminus \{0\}$. It follows that $U = \{z : G(z) < 0\}$ is the basin of attraction of $\{0\}$ and that its boundary is given by $\{G = 0\}$. Also it follows that U is a bounded set.

To lift a Fatou component V to ∂U, we explore the fact that G is pluriharmonic on $\Pi^{-1}(V)$ (see section 6). Pick a point $p \in V$ and let $q \in \mathbf{C}^{k+1} \setminus \{0\}$, $\Pi(q) = p$. Then in a small neighborhood of q, there is a function H so that $G + iH$ is holomorphic. Let $\Phi = e^{G+iH}$. Then by $(*)$, $\Phi(cz) = |c|\Phi(z)$. Assume for example that we have a set $V' = \{(q_1, \ldots, q_k, 1)\}$ in this neighborhood, covering a neighborhood of V. This set, V' is a local lifting of a neighborhood of p. We find another lifting V'' by the formula $\{\frac{1}{\Phi(q_1,\ldots,q_k,1)}\}$. Note that on V'', $|\Phi| \equiv 1$, hence $G \equiv 0$ there, so $V'' \subset \partial U$. We get another local lifting V''' by multiplying with $e^{i\theta}$, but these are the only liftings to ∂U. This means that local liftings patch together. Hence there is a complex submanifold $\tilde{V}$ in ∂U which is a covering of V. But since U is bounded, this means that $\tilde{V}$ and hence V is Kobayashi hyperbolic.

4 Fatou components

In this section we will discuss what is known about the nature of Fatou components in higher dimensions. Recall that for rational maps on $\mathbf{P}^1$ in general and for polynomial maps on $\mathbf{C}$ in particular, there is a very satisfactory classification. There are four types of Fatou components in one dimension:

(i) Attracting basins

(ii) Parabolic basins

(iii) Siegel discs

(iv) Herman rings

We will define them in general. Let M be a complex manifold and let $f : M \to M$ be a holomorphic map. A periodic orbit $z_0 \to z_1 = f(z_0) \to \cdots \to z_k = f(z_{k-1}) = z_0$ is attracting if all the eigenvalues of $(f^k)'(z_0)$ have norm strictly less than one. In this case there is an open set U, the attracting basin of z_0, consisting of the points $z \in M$ such that $f^{kn}(z) \to z_0$, $n \to \infty$. The connected component of U containing z_0 is the immediate basin of attraction of z_0. A periodic orbit is called parabolic if at least one eigenvalue has norm one and the others, if any, have norm strictly less than one. In this situation, there might be some nonempty open set U, the parabolic basin of z_0, consisting of the largest open set on which $f^{kn} \to z_0$ uniformly on compact sets. The Siegel domains are Fatou components on which there is a subsequence of iterates converging to the identity map uniformly on compact sets. In one variable, these domains are divided into two types, Siegel discs, on which f^k is conjugate to the rotation $z \to e^{i\theta}z$ on the unit discs, and Herman rings, on which f^k is conjugate to the rotation $z \to e^{i\theta}z$ on an annulus.

In addition, there are components that are called preperiodic. They include the parts of the basin of attraction which are not in the immediate basin of attraction.

Let $f : \mathbf{P}^2 \mapsto \mathbf{P}^2$ be a holomorphic map of degree $d \geq 2$. We want to classify periodic Fatou components Ω for f, so $f^k(\Omega) = \Omega$ for some $k \geq 1$. Without loss of generality we can assume that $f(\Omega) = \Omega$, replacing f by an iterate if necessary.

A Fatou component Ω is recurrent if for some $p_0 \in \Omega$ the ω-limit set of p_0 intersects Ω. More precisely, there exists $p_0 \in \Omega$ such that $f^{n_i}(p_0)$ is relatively compact in Ω for some subsequence n_i.

In $\mathbf{P}^1$, the recurrent Fatou components are attracting basins, Siegel discs and Herman rings. We say that a recurrent Fatou component is oscillatory if there exists some $p_0 \in \Omega$ such that the orbit $f^{n_i}(p_0)$ clusters both at some interior point and at some boundary point. In $\mathbf{P}^1$ there are no oscillatory recurrent Fatou components. Fix a smooth metric on $\mathbf{P}^2$.

Theorem 4.1 *Let $f \in H_d$, $d \geq 2$. Let Ω be a recurrent Fatou component such that $f(\Omega) = \Omega$. Then one of the following happens:*

(i) *There is a fixed attracting point $p \in \Omega$, the eigenvalues of f' at p are $|\lambda_1| \cdot |\lambda_2| < 1$.*

(ii) *There exists a Riemann surface $\tilde{\Sigma}$ which is a closed complex submanifold of Ω and $f|\tilde{\Sigma} \mapsto \tilde{\Sigma}$ is an automorphism; moreover $d(f^n(K), \tilde{\Sigma}) \mapsto 0$ for any compact set K in Ω. The Riemann surface $\tilde{\Sigma}$ is biholomorphic to a disc, a punctured disc or an annulus, and $f|\tilde{\Sigma}$ is conjugate to a rotation. The limit h of any convergent subsequence, f^{n_i}, has the same image. Any two limits h_1, h_2 differ only by a rotation in $\tilde{\Sigma}$. The domain Ω is not oscillatory.*

(iii) *The domain Ω is a Siegel domain, i.e. there is a sequence $\{n_i\} \mapsto \infty$ such that $f^{n_i} \mapsto \mathrm{Id}$ uniformly on compact subsets of Ω.*

We refer the reader to [FS5] for the proof of this theorem.

Remarks

(1) It is known that a rational map F of degree d in the complex plane can have at most $d+1$ attracting orbits (see [CG, p. 58]). The proof goes by showing that each attracting orbit attracts a critical point, and then one can show that a rational map of degree d can have

at most $d+1$ critical points. Gavosto [Ga] has shown recently that there are holomorphic maps on $\mathbf{P}^2$ with infinitely many attracting periodic points.

(2) Ueda [U] has a method to carry over examples from maps on $\mathbf{P}^1$ to $\mathbf{P}^2$. It is based on the following fact. There is a 2-1 holomorphic map $\Phi : \mathbf{P}^1 \times \mathbf{P}^1 \to \mathbf{P}^2$. The fibers of the map consist of pairs $\{(z,w),(w,z)\}$. Moreover, the map is branched on the diagonal $\{(z,z)\}$. The main idea is that whenever $f : \mathbf{P}^1 \to \mathbf{P}^1$ is a rational map, we get a holomorphic map $F : \mathbf{P}^2 \to \mathbf{P}^2$ with similar dynamical properties from the commutative diagram,

$$\begin{array}{ccc} \mathbf{P}^1 \times \mathbf{P}^1 & \xrightarrow{(f(z),f(w))} & \mathbf{P}^1 \times \mathbf{P}^1 \\ \downarrow \phi & & \downarrow \phi \\ \mathbf{P}^2 & \xrightarrow{F} & \mathbf{P}^2 \end{array}$$

The map Φ can be described explicitly by

$$\Phi([z_1 : z_2],[w_1, w_2]) = [z_1 w_1 : z_2 w_2 : z_1 w_2 + z_2 w_1].$$

This method can be used to find an example of a holomorphic map F on $\mathbf{P}^2$ with a Siegel domain of the form an annulus cross an annulus. Simply pick a rational map f on $\mathbf{P}^1$ with two Herman rings A_1 and A_2. Then F contains a Siegel domain of the form $A_1 \times A_2$. In general, if f has two distinct Fatou components U and V, F will have $U \times V$ as a Fatou component. Another use of this tool is to find a holomorphic map on $\mathbf{P}^2$ whose repelling periodic points are dense by choosing a function f with Julia set all of $\mathbf{P}^1$.

We will conclude these notes with a list of open problems.

Problem 1 Does every holomorphic map $F : \mathbf{P}^k \to \mathbf{P}^k$ of degree at least two have a repelling periodic point? Do meromorphic maps have periodic orbits?

Problem 2 Is the support of μ equal to the closure of the repelling periodic orbits?

Problem 3 Does there exist a holomorphic map $F : \mathbf{P}^k \to \mathbf{P}^k$ with a wandering Fatou component U, i.e., $F^n(U) \bigcap F^m(U) = \emptyset$ for all $n \neq m$?

Problem 4 Classify the dynamics around a fixed point of a holomorphic map on $\mathbf{P}^k$ or even just defined in a neighborhood of p.

Problem 5 (Herman) Let $F_c : (z,w) \to (z^2+c-w, z), c \in \mathbf{C}$, a symplectic automorphism of $\mathbf{C}^2$, i.e., F_c is biholomorphic and $F_c^* s(dz \wedge dw) = dz \wedge dw$. Can there exist a $c \in \mathbf{C}$ and a periodic orbit $\{z_i\}_{i=0}^k$ for F_c such that z_0 belongs to a Siegel domain?

Problem 6 (Herman) Let G denote the group of all symplectic automorphisms of $\mathbf{C}^2$ with the topology of uniform convergence on compact subsets. Does G contain a dense G_δ subset of maps F for which most orbits go to infinity, i.e., $\{z \in \mathbf{C}^2;\ \{\|F^n(z)\|\}_{n=1}^\infty$ is bounded$\}$ has no interior?

Problem 7 Let $f : \mathbf{P}^2 \to \mathbf{P}^2$ be a holomorphic map of degree $d \geq 2$. Assume K is a totally invariant set. Let C denote the critical set of f. Assume $\overline{\bigcup_{n=1}^{\infty} f^n(C)} \bigcap K = \emptyset$. Is f hyperbolic on K?

Problem 8 Let H_d denote the space of holomorphic maps $f : \mathbf{P}^2 \to \mathbf{P}^2$ of degree d. This is a finite dimensional space parametrized by the coefficients. Does the set of $f \in H_d$ with infinitely many attracting basins have measure zero?

Problem 9 Let $f : \mathbf{P}^2 \to \mathbf{P}^2$ be a holomorphic map of degree $d \geq 2$ and let $\lambda_1 \leq \lambda_2$ be the Lyapunov exponents for the ergodic measure $\mu = T \wedge T$. Is $\lambda_1 > 0$?

Problem 10 Classify critically finite maps on $\mathbf{P}^3$.

5 Currents and p.s.h. functions

5.1 Subharmonic and p.s.h. functions, a convergence theorem

Let Ω be an open set in $\mathbf{R}^n$. Let $u : \Omega \to [-\infty, +\infty[$ an upper semicontinuous function ($u.s.c.$) which is not identically $-\infty$ on any component of Ω. The function u is *subharmonic* in Ω if for all r such that $0 < r < d(x_0, \Omega)$ we have

$$u(x_0) \leq \frac{1}{c_n} \int_{|y|=1} u(x + ry)dy.$$

Here dy denotes the Lebesgue measure in S_{n-1}, the unit sphere of $\mathbf{R}^n$, and c_n is the area of S_{n-1}.

It follows that $u \in L^1_{\text{loc}}$, hence u defines a distribution in Ω. We then have that $\Delta u \geq 0$ in Ω, where Δ denotes the Laplacian in $\mathbf{R}^n$.

Conversely, if u is a distribution in Ω, $u \in \mathcal{D}'(\Omega)$, such that $\Delta u \geq 0$ then, as a distribution, u coincides with a subharmonic function u_0.

The following result is classical, see [Hö], Theorem 4.1.9.

Theorem 5.1 *Let (v_j) be a sequence of subharmonic functions in Ω with a uniform upper bound on compact sets. Then:*

(a) *if (v_j) does not converge to $-\infty$ uniformly on compact sets, then there exists a subsequence (v_{j_k}) which is convergent in $L^1_{\text{loc}}(\Omega)$;*

(b) *if (v_j) is a sequence of subharmonic functions which is convergent to v in $\mathcal{D}'(\Omega)$, then v is subharmonic and $v_j \to v$ in L^1_{loc}. Moreover,*

$$\overline{\lim}\, v_j(x) \leq v(x), \quad x \in \Omega$$

with the two sides equal and finite a.e. More generally, for every compact $K \subset \Omega$ and every continuous function f on K,

$$\overline{\lim_{j \to \infty}} \sup_K (v_j - f) \leq \sup_K (v - f).$$

Examples

(1) Let (a_j) be a dense sequence in the unit disc $D \subset \mathbf{C}$. Choose $\lambda_j > 0$ such that $\sum \lambda_j \log \frac{1}{|a_j|} < \infty$. Define

$$v(z) = \sum \lambda_j \log \left| \frac{z - a_j}{2} \right|.$$

The function v is subharmonic in D; however $(v = -\infty)$ is a G_δ dense set. So $(v = -\infty)$ contains other points than the sequence (a_j). We have $\Delta v = 2\pi \sum_j \lambda_j \delta_{a_j}$, where δ_a is the Dirac mass at a.

(2) Let v be a continuous subharmonic function in $\Omega \subset \mathbf{R}^n$. Assume $v_j \leq v$, v_j subharmonic and $v_j(x) \to v(x)$ for x in a dense set $E \subset \Omega$. Then $v_j \to v$ in L^1_{loc}. It is enough to show that if ψ is a limit of (v_{j_k}) then $\psi = v$.

Let Ω be an open set in $\mathbf{C}^n$. Let $u : \Omega \to [-\infty, +\infty[$ be a u.s.c. function which is not identically $-\infty$ on any component of Ω. The function u is *plurisubharmonic* (p.s.h. for short) if $\tau \to u(z + \tau w)$ is subharmonic (or identically $-\infty$) where defined, where $z \in \Omega$, $w \in \mathbf{C}^n$ and $\tau \in \mathbf{C}$. Every p.s.h. function is in $L^1_{\text{loc}}(\Omega)$, and hence defines a distribution. Moreover,

$$\langle (Lu)w, w \rangle := \sum_{i,j} \frac{\partial^2 u}{\partial z_j \partial \bar{z}_k} w_j \bar{w}_k \geq 0 \quad \text{if } w \in \mathbf{C}^n.$$

This means that $\langle (Lu)w, w \rangle$ is a positive measure for each w. Conversely, if $u \in \mathcal{D}'(\Omega)$ is such that $\langle (Lu)w, w \rangle$ is a positive measure for every $w \in \mathbf{C}^n$, then u is defined by a p.s.h. function u_0. The function u_0 is uniquely determined.

The p.s.h. functions are clearly subharmonic for the underlying real structure. So Theorem 5.1 holds for p.s.h. functions also.

Examples

(1) If f is holomorphic in Ω and not identically zero, then $\log |f|$ is p.s.h. in Ω.

(2) If $g : \Omega \to \Omega'$ is holomorphic and u is p.s.h. in Ω', then $u \circ g$ is p.s.h. in Ω or identically $-\infty$. This allows one to define p.s.h. functions on complex manifolds.

Let B denote the unit ball in $\mathbf{C}^n$. We have the following approximation result.

Proposition 5.2 *Let $\alpha \in C_0^\infty(B)$, $\alpha \geq 0$, $\int \alpha = 1$, and $\alpha_\epsilon(w) = \epsilon^{-2n}\alpha(w/\epsilon)$. Assume that α depends only on $|w|$. Then for u p.s.h. in Ω the functions $u_\epsilon = u * \alpha_\epsilon$ are C^∞ p.s.h. in $\Omega_\epsilon = \{z : d(z, \Omega) > \epsilon\}$. Moreover, $u_\epsilon \downarrow u$ and the u_ϵ converge to u in L^1_{loc}.*

5.2 Currents

Let M be a real smooth manifold of dimension m. Let $\mathcal{D}^p(M)$ denote the smooth differential forms of degree p with compact support. A current T of dimension p (and degree $m - p$) on M is a continuous linear form on $\mathcal{D}^p(M)$. More precisely,

$$\varphi \to \langle T, \varphi \rangle := \int T \wedge \varphi$$

is linear and satisfies the following continuity property. If (φ_n) is a sequence of forms of degree p supported on a fixed compact K, and if $\varphi_n \to 0$ in the C^∞ topology, then

$$\langle T, \varphi_n \rangle \to 0.$$

A current can be restricted to an open set $U \subset M$; this means that we consider T as acting on $\mathcal{D}^p(U)$. If we have local coordinates $x_1, \ldots, x_m$ in U, then the current $T_{|U}$ (the restriction to U) can be expressed as a differential form of degree $m-p$ with distributions as coefficients:

$$T = \sum_{|I|=m-p} T_I dx^I$$

where $I = (i_1, \ldots, i_{m-p})$, $i_1 < i_2 < \cdots < i_{m-p}$, and $dx^I := dx_{i_1} \wedge \cdots \wedge dx_{i_{m-p}}$.

If $J = (j_1, \ldots, j_p)$, $j_1 < j_2 < \cdots < j_p$, then we have

$$\begin{aligned} \langle T_I dx^I, \varphi_J dx^J \rangle &= 0 \qquad \text{if } I \cup J \neq \{1, \cdots, m\} \\ &= (-1)^{\sigma(I,J)} \langle T_I, \varphi_J \rangle, \end{aligned}$$

where $\sigma(I, J)$ is the signature of the permutation $i_1, \ldots, i_{m-p}, j_1, \ldots, j_p$ and $\langle T_I, \varphi_J \rangle$ is the value of the distribution T_I acting on the test function φ_J.

The space of currents of dimension p is denoted by $\mathcal{D}'_p$. It is identified with the space of currents of degree $m-p$, i.e. $\mathcal{D}'^{m-p}(M)$. We will always consider the usual topology on currents, i.e. if $T_j \in \mathcal{D}'_p(M)$, then $T_j \to 0$ if and only if $\langle T_j, \varphi \rangle \to 0$ for every $\varphi \in \mathcal{D}^p(M)$.

Examples

(1) $[M]$ the current of integration on M is a current of dimension m; it acts on $\varphi \in \mathcal{D}^m(M)$ as follows

$$\langle [M], \varphi \rangle = \int_M \varphi.$$

(2) Let $f : Y \to M$ be a *proper* smooth map from a smooth manifold Y of dimension p to the manifold M. We do not require that f is an embedding. Then

$$\varphi \to \int_Y f^* \varphi$$

is a current of dimension p on M. We will denote this current by $f_*[Y]$. Recall that since f is proper, f^{-1}(compact) is compact. The notion of support of a current is quite clear. The current $f_*[Y]$ is supported on $f(Y)$, which is closed since f is proper.

(3) If T is a current of dimension p on M, and α a smooth form of degree k we define

$$\langle T \wedge \alpha, \varphi \rangle := \langle T, \alpha \wedge \varphi \rangle.$$

$T \wedge \alpha$ is a current of dimension $p-k$.

Basic operations on currents

(i) *Exterior derivative of a current*

There is a natural inclusion of $\mathcal{E}^{m-p}(M)$, the space of smooth forms of degree $m-p$ on M, into the space of currents of degree $m-p$. If $\alpha \in \mathcal{E}^{m-p}(M)$ and $\varphi \in \mathcal{D}^p(M)$ we define

$$\langle \alpha, \varphi \rangle := \int_M \alpha \wedge \varphi.$$

The exterior differentiation operator defined on $\mathcal{E}^q(M)$ extends to currents.

Given a current T of degree q we define the degree $q+1$ current dT by the formula

$$\langle dT, \varphi \rangle = (-1)^{q+1} \langle T, d\varphi \rangle, \qquad \varphi \in \mathcal{D}^{m-q-1}(M).$$

The current T is closed if $dT = 0$. We define the current bT the boundary of the current T by the formula :

$$\langle bT, \varphi \rangle := \langle T, d\varphi \rangle.$$

(ii) *Push-forward of a current*

We generalize the construction in Example 2. Let $f : M \to N$ be a smooth map from a manifold M of dimension m, to a smooth manifold N of dimension n.

Let $T \in \mathcal{D}'_p(M)$. We assume that f restricted to $\operatorname{supp} T$ is proper, i.e. if X is compact in N, then $f^{-1}(X) \cap \operatorname{supp} T$ is compact. Then we define the direct image f_*T of T as follows

$$\langle f_*T, \varphi \rangle = \langle T, f^*\varphi \rangle.$$

Then $f_*T \in \mathcal{D}'_p(N)$, i.e., f_* preserves the dimension of currents. The definition makes sense, since f is proper on $\operatorname{supp} T$. The following properties are easy to check.

(a) We have that $\operatorname{supp} f_*T \subset f(\operatorname{supp} T)$.

(b) If $\psi \in \mathcal{E}^k(N)$, then

$$f_*(T \wedge f^*\psi) = f_*T \wedge \psi.$$

(c) $b(f_*T) = f_*bT$ and $d(f_*T) = (-1)^{m+n} f_*(dT)$. So when $m+n$ is even, d commutes with f_*.

(iii) *Pull-back of a current*

Let f $M \to N$ be a smooth *submersion*. Let ψ be a $\mathcal{C}^k$ form (resp. a form in L^1_{loc}) on M. Assume f is proper on $\operatorname{supp} \psi$. Then the form $f_*\psi$ is of class $\mathcal{C}^k$ (resp. in L^1_{loc}), and it is obtained by integrating ψ on the fibres of f. It is important to observe that $f_*\mathcal{D}^{m-p}(M) \to \mathcal{D}^{n-p}(N)$ is continuous. When ψ is a test form in $\mathcal{D}^{m-p}(M)$, we consider it as a current of dimension p and we compute $f_*\psi$ thanks to the relation

$$\langle f_*\psi, \varphi \rangle = \langle \psi, f^*\varphi \rangle,$$

it is not necessary to assume that f is proper when ψ has compact support.

If T is a current on N and $f : M \to N$ is a submersion , then we define f^*T, the pull-back of T, by the formula

$$\langle f^*T, \varphi \rangle = \langle T, f_*\varphi \rangle.$$

Observe that when T is a smooth form the f^*T is the usual pull-back of the smooth form T.

The following properties are easy to establish:

(a) $\deg f^*T = \deg T$;

(b) If ψ is a smooth form, then $f^*(T \wedge \psi) = f^*T \wedge f^*\psi$;

(c) $d(f^*T) = f^*(dT)$;

(d) $\operatorname{supp} f^*T \subset f^{-1}(\operatorname{supp} T)$;

(e) If $T_j \to T$, then $f^*T_j \to f^*T$.

(iv) *Smoothing*

We recall the following result of smoothing of currents, [deRh].

Theorem 5.3 *Let M be a smooth manifold of dimension m. For $\epsilon > 0$ there is an operator $R_\epsilon : \mathcal{D}'_p \to \mathcal{E}^{m-p} \subset \mathcal{D}'_p$ and an operator $A_\epsilon : \mathcal{D}'_p \to \mathcal{D}'_{p+1}$ such that*

$$R_\epsilon - \mathrm{Id} = b \circ A_\epsilon - A_\epsilon \circ b. \tag{5.1}$$

Moreover, if $T \in \mathcal{E}^{m-p}$, then $A_\epsilon(T) \in \mathcal{E}^{m-p-1}$ and for every T, $R_\epsilon(T) \to T$ as $\epsilon \to 0$.

When $M = \mathbf{R}^m$, the regularization R_ϵ is just a convolution. But on a manifold there is no convolution, so the result is quite tricky. Identity (5.1) implies that R_ϵ commutes with d. If $bT = 0$ then $R_\epsilon T - T = b(A_\epsilon T)$, so $b(R_\epsilon T) = 0$. If $T = bV$, then $R_\epsilon T = bV + b(A_\epsilon V)$. Hence the regularization process preserves boundaries and closed currents.

Currents representable by integration

A current $T = \sum T_I dx^I$ is representable by integration if each T_I is a regular Borel measure on M. The space $\mathcal{M}_p(M)$ of currents of dimension p representable by integration is the dual of continuous forms with compact support $\mathcal{C}^p(M)$. If T is representable by integration and f is proper on $\operatorname{supp} T$, then f_*T is representable by integration. We want to introduce some semi-norms on currents representable by integration.

Let $\alpha = \Sigma' c_I dx^I$ be a p vector in $\mathbf{R}^m$, where Σ' means that the summation is performed over strictly increasing multi-indices. Define $|\alpha| := \left(\sum' |c_I|^2\right)^{1/2}$. This is the standard Hilbert norm on p covectors in $\mathbf{R}^m$. For a current T of dimension p, representable by integration on an open set Ω in $\mathbf{R}^m$, we define

$$M_U(T) = \sup\{|\langle T, \varphi\rangle| : \varphi \in \mathcal{D}_p(U), \ |\varphi(x)| \le 1, \ x \in U\}$$

for each $U \subset\subset \Omega$. For K compact in Ω we define

$$M_K(T) = \inf\{M_U(T) : \ U \text{ open } U \supset K\}.$$

$M_K(T)$ is the mass of T on K. By choosing charts on the manifold N it is easy to generalize this notion for currents representable by integration on a manifold N.

5.3 Currents on complex manifolds

When Ω is an open set in $\mathbf{C}^m$, let $\mathcal{D}^{p,q}(\Omega)$ denote the smooth forms φ with compact support that can be written

$$\varphi = \sum_{|I|=p,|J|=q} \varphi_{IJ} dz_I \wedge d\bar{z}_J,$$

where $dz_I = dz_{i_1} \wedge \cdots \wedge dz_{i_p}$, and $d\bar{z}_J = dz_{j_1} \wedge \cdots \wedge dz_{j_p}$.

$\mathcal{D}'_{p,q}(\Omega)$ will denote the space of currents of bidimension (p,q), i.e. the dual space of $\mathcal{D}_{p,q}(\Omega)$. As in the previous paragraph, it can be shown that T can be written as a differential form of bidegree $(m-p, m-q)$ with distributions as coefficients:

$$T = \sum_{\substack{|I'|=m-p \\ |J'|=m-q}} T_{I'J'} dz_{I'} \wedge d\bar{z}_{J'}.$$

These notions generalize easily to complex manifolds.

Recall that on a complex manifold $d = \partial + \bar{\partial}$, where

$$\partial\varphi = \sum \frac{\partial \varphi_{I,J}}{\partial z_k} dz_k \wedge dz_I \wedge d\bar{z}_J \quad \text{and} \quad \bar{\partial}\varphi = \sum \frac{\partial \varphi_{I,J}}{\partial \bar{z}_k} d\bar{z}_k \wedge dz_I \wedge d\bar{z}_J.$$

The operator d^c is defined as $d^c := \frac{i}{2\pi}(\bar{\partial} - \partial)$; it is a real operator in the sense that $\overline{d^c u} = d^c \bar{u}$. We also have that $dd^c = \frac{i}{\pi}\partial\bar{\partial}$. The operations ∂ and $\bar{\partial}$ extend to currents.

So if T is of bidegree (p,q),

$$\langle \partial T, \varphi \rangle = (-1)^{p+q+1} \langle T, \partial\varphi \rangle \quad \text{and} \quad \langle \bar{\partial} T, \varphi \rangle = (-1)^{p+q+1} \langle T, \bar{\partial}\varphi \rangle.$$

So we get if T is of bidimension (p,p)

$$\langle dT, \varphi \rangle = -\langle T, d\varphi \rangle$$

$$\langle d^c T, \varphi \rangle = -\langle T, d^c\varphi \rangle$$

and

$$\langle dd^c T, \varphi \rangle = \langle T, dd^c\varphi \rangle.$$

We now introduce the notion of positivity. A current T of bidimension (p,p) is *positive* if $\langle T, \varphi \rangle \geq 0$ for all forms

$$\varphi = i\alpha_1 \wedge \bar{\alpha}_1 \wedge \cdots i\alpha_p \wedge \bar{\alpha}_p, \qquad \alpha_j \in \mathcal{D}^{1,0}(M).$$

It is quite easy to show that a $(1,1)$ current $T = i\sum T_{jk} dz_j \wedge d\bar{z}_k$ is positive if and only if for all $(w_1, \cdots, w_n) \in \mathbf{C}^m$ the distribution

$$\sum_{j,k} T_{jk} w_j \bar{w}_k \geq 0.$$

So, it is a positive measure and hence the coefficients T_{ij} are measures. It follows that a function u is p.s.h. on M if and only if

$$dd^c u = \frac{i}{\pi}\sum \frac{\partial^2 u}{\partial z_i \partial \bar{z}_j} dz_i \wedge d\bar{z}_j \geq 0.$$

Hence a positive $(1,1)$ current is representable by integration. One shows similarly that a positive (p,p) current is representable by integration, so locally it has measure coefficients. We give now an important example of a positive closed current. Let X be a complex analytic subvariety of pure $\dim p$ in M. We define

$$\langle [X], \varphi \rangle = \int_{\operatorname{Reg} X} \varphi.$$

Here $\operatorname{Reg} X$ denotes the manifold points of X. Then $[X]$ is a positive closed current of bidimension (p,p). The fact that the integral converges for $\varphi \in \mathcal{D}_{p,p}(M)$ is non-trivial. It is a consequence of the fact that $\operatorname{Reg}(X)$ has locally bounded volume near the singular points.

Assume M is a Kähler manifold with Kähler form ω. Let S be a positive current of bidimension (p,p) we define for a compact $K \subset M$

$$|S|_K := \int_K S \wedge \frac{\omega^p}{p!};$$

we call $|S|$ the volume measure of S.

It is easy to check that for a compact $X \subset M$ there exists a constant C such that for every compact $K \subset X$ we have

$$\frac{1}{C} M_K(S) \leq |S|_K \leq C\, M_K(S).$$

When S has compact support we denote

$$||S|| := \int S \wedge \frac{\omega^p}{p!}.$$

We now wish to examine the influence of holomorphicity on the basic operations.

Let $f : M \to N$ be a holomorphic map between complex manifolds. Then f^* preserves the bidegree of smooth forms. Consequently if T is a current of bidimension (p,p) and f is proper on support T, then $f_* T$ is a current of bidimension (p,p).

Similarly, if f is a submersion and T is a current of bidegree (q,q) on N, then f^*T is a current of bidegree (q,q) on M.

Extension of currents

Let A be a closed set in a complex manifold Ω, and let T be a positive current on $\Omega \backslash A$. Assume that T is closed in $\Omega \backslash A$. A natural question is to find a positive closed current T_1 in Ω such that $T_{1|\Omega\backslash A} = T$. This problem contains in particular the question of the extension of analytic varieties in $\Omega \backslash A$ through A.

The positive current $\tilde{T}$ in Ω obtained by extending the coefficients of T by zero on A is called the *simple extension* of T. It is a positive current on Ω provided the mass of T is locally bounded near A. If $\chi_\nu \in C_0^\infty(\Omega\backslash A)$, $0 \leq \chi_\nu \leq 1$, and $\chi_\nu \nearrow \chi_{\Omega\backslash A}$, the characteristic function of $\Omega\backslash A$, then clearly $\tilde{T} = \lim_\nu \chi_\nu T$. If T is closed in $\Omega\backslash A$ it is not true in general that $\tilde{T}$ is closed. It is however the case in some important situations.

Theorem 5.4 (Harvey [HaP]) *Let Ω be an open set in $\mathbf{C}^n$ and A a closed set in Ω. Let T be a positive closed current of bidimension (p,p) on $\Omega\backslash A$. Assume that $\Lambda_{2p-1}(A) = 0$, where Λ_{2p-1} denotes the Hausdorff measure of dimension $2p-1$. Then $\tilde{T}$, the simple extension of T, has locally bounded mass near A and is a closed positive current on Ω.*

Theorem 5.5 (Skoda [Sk]) *Let Ω be an open set in $\mathbf{C}^n$, and let A be an analytic variety in Ω. Let T be a positive closed current of bidimension (p,p) in $\Omega\backslash A$. Assume that T has locally bounded mass near A. Then $\tilde{T}$ is a positive closed current on Ω.*

Remark 5.6 It follows from the previous result that if T is a positive current of bidimension (p,p) in Ω, and if A is an analytic variety of dimension p, then $T_{|A} = c[A]$, where c is locally constant. Results of this type have been proved first by Siu [Siu].

A simple proof of Theorem 5.5 is given in [Si].

Theorem 5.7 *Let T be a positive closed current of bidegree $(1,1)$ in $\mathbf{C}^n$. Then there exists a function u such that $dd^c u = T$. It follows that u is p.s.h. in $\mathbf{C}^n$.*

Remark 5.8 The theorem is also true for a ball B instead of $\mathbf{C}^n$, or more generally for a domain of holomorphy Ω such that $H^2_{\text{de Rham}}(\Omega, \mathbf{R}) = 0$.

5.4 Positive closed currents of bidegree (1,1) on $\mathbf{P}^k$

Let $\mathbf{P}^k$ be the complex projective space of dimension k, and $\pi : \mathbf{C}^{k+1}\backslash(0) \to \mathbf{P}^k$ the canonical projection. We consider the cone $\mathcal{P}$ of p.s.h. functions u in $\mathbf{C}^{k+1}$, satisfying the following homogeneity property. There exists $c > 0$ such that if $\lambda \in \mathbf{C}$, then

$$u(\lambda z) = c \log|\lambda| + u(z) \quad \text{for} \quad z \in \mathbf{C}^{k+1}.$$

The functions in $\mathcal{P}$ are normalized so that $\sup_B u = 0$, where B denotes the unit ball in $\mathbf{C}^{k+1}$.

With a function u in $\mathcal{P}$ we are going to associate a positive closed current T of bidimension $(k-1, k-1)$ in $\mathbf{P}^k$. Let U be an open set in $\mathbf{P}^k$ such that there is a holomorphic inverse $s : U \to \mathbf{C}^k\backslash\{0\}$ of π, i.e., $\pi \circ s = \text{Id}$. Define T_s on U by $T_s = dd^c(u \circ s)$. Then T_s is independent of s. If s' is another section on U, then $s' = \varphi s$ for some holomorphic function φ on U. Hence

$$\begin{aligned} T_{s'} &= dd^c(u \circ s') = dd^c(u(\varphi s)) \\ &= dd^c c \log|\varphi| + dd^c u(s) = T_s. \end{aligned}$$

So, with this local definition, we have an operator $L : \mathcal{P} \to \mathcal{D}'_{k-1,k-1}(\mathbf{P}^k)$, $L(u) = T$, where T coincides with T_s on U if s is a holomorphic section of π. Since π is a submersion, it follows that $\pi^*T = dd^c u$ is well-defined on $\mathbf{C}^{k+1}\setminus\{0\}$. Since $dd^c u$ is a current on $\mathbf{C}^{k+1}$, we can consider π^*T as a positive closed current on $\mathbf{C}^{k+1}$. For example, the Kähler form ω on $\mathbf{P}^k$ is the $(1,1)$ current such that $\pi^*\omega = dd^c \log|z|$, where $|z|$ denotes the Euclidean norm of z.

Theorem 5.9 *The operator L is an isomorphism between $\mathcal{P}$ and $\mathcal{C}'_{k-1,k-1}(\mathbf{P}^k)$, the cone of positive closed currents of bidimension $(k-1, k-1)$ in $\mathbf{P}^k$. Moreover, if $T = L(u)$ and $u(\lambda z) = c \log|\lambda| + u(z)$, then $||T|| := \int T \wedge \omega^{k-1} = c$.*

Proof If $L(u_1) = L(u_2)$, then $dd^c(u_1 - u_2) = 0$ in $\mathbf{C}^{k+1}$, so $u_1 - u_2$ is pluriharmonic of logarithmic growth, hence constant. Therefore $u_1 = u_2 + c'$; the normalization of the functions in $\mathcal{P}$ implies that $c' = 0$.

We now show that L is surjective. Let S be a positive closed current of bidegree $(1,1)$ on $\mathbf{P}^k$. Then π^*S is positive closed of bidegree $(1,1)$ in $\mathbf{C}^{k+1}\setminus\{0\}$. Theorem 5.4 implies that it has a positive closed extension to $\mathbf{C}^{k+1}$. So using Theorem 5.5 there is a p.s.h. function v in $\mathbf{C}^{k+1}$ such that $\pi^*S = dd^c v$. Define

$$u(z) = \frac{1}{2\pi}\int_0^{2\pi} v(e^{i\theta}z)d\theta.$$

Let $h_\theta(z) = e^{i\theta}z$. Since $h_\theta^*(\pi^*S) = (\pi \circ h_\theta)^*S = \pi^*S$. We still have $\pi^*S = dd^c u$ and u is radially invariant. For $\alpha \in \mathbf{C}$, define $u_\alpha(z) = u(\alpha z)$. Since $dd^c u_\alpha = dd^c u$ and both functions are radial subharmonic functions on any complex line through zero, there is a constant $C(\alpha)$ such that $u_\alpha(z) = C(\alpha) + u(z)$. Define $u'(z) = u(z) - \frac{C(2)}{\log 2}\log|z|$. Then

$$u'(2z) = u(2z) - \frac{C(2)}{\log 2}\log|2z| = u(z) - \frac{C(2)}{\log 2}\log|z| = u'(z).$$

So u' is constant on complex lines and

$$\begin{aligned} u(\lambda z) &= u'(\lambda z) + \frac{C(2)}{\log 2}\log|\lambda z| \\ &= u'(z) + \frac{C(2)}{\log 2}\log|z| + \frac{C(2)}{\log 2}\log|\lambda| \\ &= u(z) + \frac{C(2)}{\log 2}\log|\lambda|. \end{aligned}$$

This shows that L is surjective.

We need a smoothing lemma for functions in $\mathcal{P}$. Let $\mathcal{P}_c$ denote the functions in $\mathcal{P}$ such that $u(\lambda z) = c\log|\lambda| + u(z)$.

Lemma 5.10 *Let $u \in \mathcal{P}_c$. There exists $u_\epsilon \downarrow u$, u_ϵ smooth in $\mathbf{C}^{k+1}\setminus\{0\}$ and $u_\epsilon \in \mathcal{P}_c$ for all ϵ.*

Proof Let α be a test function in $\mathbf{C}^{k+1}$ depending only on $|z|$, i.e., $\alpha \in C_0^\infty$, $\alpha \geq 0$, $\int \alpha = 1$. Define $\alpha_\epsilon(z) = \epsilon^{-(2k+2)}\alpha(z/\epsilon)$. Set

$$\begin{aligned} u_\epsilon(z) &= \frac{1}{|z|^{2(k+1)}} \int u(w)\alpha_\epsilon\left(\frac{z-w}{|z|}\right) d\lambda(w) \\ &= \int u(z - |z|w)\alpha_\epsilon(w)d\lambda(w). \end{aligned}$$

Clearly u_ϵ is smooth out of 0. Using the second expression one checks easily that $u_\epsilon(\lambda z) = c\log|\lambda| + u_\epsilon(z)$. The fact that u_ϵ is decreasing can be checked on the unit sphere, using the first formula. To check that u_ϵ is p.s.h. it is enough to show that for $r > 0$, $\ell_r(z) = \int_{|w|=r} u(z - |z|w)d\lambda(w)$ is p.s.h.. But one checks easily that

$$\ell_r(z) = \int_\Gamma u(z - r\gamma(z))d\gamma,$$

where Γ is the unitary group and $d\gamma$ is the Haar measure on Γ. It is then clear that ℓ_r is p.s.h.

End of proof of Theorem 5.9 Let S be a positive closed $(1,1)$ current in $\mathbf{P}^k$. We know there is $u \in \mathcal{P}$ such that $L(u) = S$, i.e., $\pi^*S = dd^c u$. Let $S_\epsilon = L(u_\epsilon)$, where u_ϵ is the smoothing introduced in Lemma 5.10. Clearly $S_\epsilon \to S$ in the sense of currents, and S_ϵ are smooth forms. If $u \in \mathcal{P}_1$ then $u_\epsilon \in \mathcal{P}_1$. So the function $v_\epsilon([z]) = u_\epsilon(z) - \log|z|$, where $[z] = \pi(z)$, is a smooth function on $\mathbf{P}^k$. Hence on $\mathbf{P}^k$ we have $S_\epsilon - \omega = dd^c v_\epsilon$. Using Stokes' Theorem we get:

$$\begin{aligned} ||S|| &= \int S \wedge \omega^{k-1} = \lim_{\epsilon\to 0} \int S_\epsilon \wedge \omega^{k-1} \\ &= \lim_{\epsilon \to 0} \int (\omega + dd^c v_\epsilon) \wedge \omega^{k-1} = \int \omega^k = 1. \end{aligned}$$

Now let us prove the continuity of L when $\mathcal{P}$ carries the L^1_{loc} topology and the topology on currents is the usual weak topology. It is clear that L is continuous on $\mathcal{P}$ since d and d^c are continuous. Assume conversely that $S_n \to S$ in the sense of currents. Then $||S_n|| \to ||S||$. Let $u_n \in \mathcal{P}$ such that $\pi^*S_n = dd^c u_n$. We have $u_n(z) = u_n(z/|z|) + c_n \log|z|$. Since $\sup_B u_n = 0$, the sequence u_n does not converge uniformly to $-\infty$. If $u_{n_i} \to v$ in L^1_{loc}, then $v \in \mathcal{P}$ and $L(v) = S$. So $v = u$, where u is the unique function in $\mathcal{P}$ such that $dd^c u = \pi^*S$.

Example Let V be an algebraic hypersurface of degree d in $\mathbf{P}^k$. More precisely, let P be a homogeneous irreducible polynomial of degree d and

$$V = \{z = [z_0, \cdots, z_k] \in \mathbf{P}^k : \quad P(z_0, \ldots, z_k) = 0\}.$$

The current $[V]$ is by definition the current of integration on the regular points of V.

As follows from the Lelong-Poincaré formula

$$\pi^*[V] = dd^c \log|P|.$$

The formula can be proved using local coordinates.

The above theorem implies that $||[V]|| = d$.

5.5 Wedge product of currents

Let S be a positive closed current of bidimension (p,p), $p > 0$, in Ω, an open set in $\mathbf{C}^k$. Let $|S|$ denote the volume measure of S. Recall $|S| = S \wedge \beta^p$ where $\beta = dd^c|z|^2$. Let u be a p.s.h. function in Ω such that $u \in L^1_{\mathrm{loc}}(|S|)$. We define $dd^c u \wedge S := dd^c(uS)$. The definition makes sense because uS is a current. It coincides with the usual one when u is smooth since S is closed.

Proposition 5.11 *The wedge product $dd^c u \wedge S$ is a closed positive current.*

Proof Without loss of generality we can assume $u < 0$ on Ω. Let $u_\epsilon \downarrow u$ and u_ϵ p.s.h. smooth as in Proposition 5.2. We have $|u_\epsilon| \leq |u|$, so $u_\epsilon \to u$ in $L^1_{\mathrm{loc}}(|S|)$ and $u_\epsilon S \to uS$ in the sense of currents. So $dd^c(u_\epsilon S) \to dd^c(uS)$. Since u_ϵ is smooth $dd^c(u_\epsilon S) = dd^c u_\epsilon \wedge S \geq 0$ so the weak limit $dd^c u \wedge S$ is also positive. □

The current $dd^c u \wedge S$ can be defined for every closed current S when u is locally bounded, see [BT]. In particular, if $u_1, \ldots, u_q$ are locally bounded p.s.h. functions, then one can define inductively $dd^c u_1 \wedge \cdots \wedge dd^c u_q \wedge S$. It is shown in [BT] that this is a positive closed current, and the wedge product is symmetric with respect to $u_1, \ldots, u_q$.

The following inequality, basically due to Chern-Levine-Nirenberg, is quite crucial in the theory. See [De1] or [Si].

Theorem 5.12 *Let M be a complex manifold and let $L \subset\subset K$ be two compacta in M. Let $V, u_1, \cdots, u_q$ be p.s.h. functions in M such that $u_1, \cdots, u_q$ are locally bounded. Then*

$$|V\, dd^c u_1 \wedge \cdots \wedge dd^c u_q|_L \leq C_{K,L} ||V||_{L^1(K)} ||u_1||_{L^\infty(K)} \cdots ||u_q||_{L^\infty(K)}.$$

So, in particular, if $u_1, \cdots u_n$ are locally bounded p.s.h. in $\Omega \subset \mathbf{C}^n$, then the measure $\nu = dd^c u_1 \wedge \cdots \wedge dd^c u_n$ is a positive measure and every p.s.h. function V in Ω is in $L^1_{\mathrm{loc}}(\nu)$. In particular, ν has no mass on pluripolar sets, i.e. sets contained in $\{V = -\infty\}$, where V is p.s.h.

Let M be a complex manifold. A set E is *locally pluripolar* if for every $p \in M$, there is a neighborhood $U(p)$ of p, and a p.s.h. function u in $U(p)$ such that $E \cap U(p) \subset \{z \in U(p); u(z) = -\infty\}$. For example analytic varieties in $\mathbf{P}^k$ are locally pluripolar.

The following result due to Bedford-Taylor [BT] is quite useful.

Theorem 5.13 *Given a positive closed current T of bidimension (p,p) in Ω, let $u_1, \ldots, u_q$ be locally bounded p.s.h. functions in Ω and let $u_1^n, \cdots, u_q^n$ be decreasing sequences of p.s.h. functions converging pointwise to $u_1, \cdots, u_q$. Then*

(a) $u_1^n dd^c u_2^n \wedge \cdots \wedge dd^c u_q^n \wedge T \to u_1 dd^c u_2 \wedge \cdots \wedge dd^c u_q \wedge T$ *weakly;*

(b) $dd^c u_1^n \wedge \cdots \wedge dd^c u_q^n \wedge T \to dd^c u_1 \wedge \cdots \wedge dd^c u_q \wedge T$ *weakly.*

Some extensions to the case where (u_j) are not locally bounded are in [Si] [De1] [FS5].

We will need the following consequence of the Chern-Levine-Nirenberg inequality. Let $f : \Omega_1 \to \Omega_2$ be a holomorphic map between two equidimensional complex manifolds. Assume the critical set C is a proper subvariety. Then $f : \Omega_1 \backslash f^{-1}(f(C)) \to \Omega_2 \backslash f(C)$ is a submersion. Let $u_1, \cdots, u_q$ be locally bounded p.s.h. functions on Ω_2. Then $dd^c u_1 \wedge \cdots \wedge dd^c u_q$ is, as we have seen, a positive closed current on Ω_2. Let $f^*[dd^c u_1 \wedge \cdots \wedge dd^c u_q]$ denote the simple extension of the pull-back which is a priori defined on $\Omega_1 \backslash f^{-1}(f(C))$. Since on $\Omega_1 \backslash f^{-1}(f(C))$ the current is equal to $dd^c(u_1 \circ f) \wedge \cdots \wedge dd^c(u_q \circ f)$, and since by C.L.N. inequality this current has no mass on $f^{-1}(f(C))$, we then have

$$f^*[dd^c u_1 \wedge \cdots \wedge dd^c u_q] = dd^c(u_1 \circ f) \wedge \cdots \wedge dd^c(u_q \circ f).$$

The right hand side will be the closed extension of $f^*[dd^c u_1 \wedge \cdots \wedge dd^c u_q]$ to Ω_1.

We will need the following special case of a result from [FS5].

Proposition 5.14 *Let $u_1, \cdots, u_q \in \mathcal{P}$. Assume they are locally bounded in $\mathbf{C}^{k+1} \backslash \{0\}$. Define $T_1, \cdots, T_q$ such that $\pi^*(T_j) = dd^c u_j$. Then*

$$||T_1 \wedge \cdots \wedge T_q|| = ||T_1|| \cdots ||T_q||.$$

Proof Let $S = T_2 \wedge \cdots \wedge T_q \wedge \omega^{k-q}$. Take $u_\epsilon \downarrow u$, $u_\epsilon \in \mathcal{P}$ and u_ϵ smooth in $\mathbf{C}^{k+1} \backslash \{0\}$ as in Lemma 5.10. Let T_ϵ be such that $\pi^* T_\epsilon = dd^c u_\epsilon$. If $c = ||T_1||$, it follows from Theorem 5.9 that $T_\epsilon - c\omega = dd^c v_\epsilon$, where v_ϵ is a smooth function on $\mathbf{P}^k$. Using Stokes' Theorem we have:

$$\begin{aligned} ||T_1 \wedge \cdots \wedge T_q|| &= ||T_1 \wedge S|| = \lim_{\epsilon \to 0} ||T_\epsilon \wedge S|| \\ &= \lim_{\epsilon \to 0} c \int S \wedge \omega + \int S \wedge dd^c v_\epsilon \\ &= c \int S \wedge \omega \\ &= ||T_1|| \, ||S||. \end{aligned}$$

5.6 A capacity for p.s.h. functions and extremal functions

The following useful capacity has been introduced by Bedford and Taylor [BT].

Let Ω be a connected bounded open set in $\mathbf{C}^k$, and let $P(\Omega)$ denote the set of p.s.h. functions in Ω. For every Borel set $E \subset \Omega$ we define

$$c(E, \Omega) = \sup \left\{ \int_E (dd^c u)^k \ : \ u \in P(\Omega), \ \ 0 \le u \le 1 \right\}.$$

Bedford and Taylor proved that a Borel set is pluripolar in Ω if and only if $C(E, \Omega) = 0$.

Let $K \subset\subset B$, where B denotes the unit ball in $\mathbf{C}^k$. Define

$$u_{K,B} = \sup\{v \ : \ v \in P(B), \ v \le 0 \text{ on } K, \ 0 \le v \le 1\}.$$

Let $u^*_{K,B}$ be the upper semicontinuous regularization of U_K. It is also a result due to Bedford and Taylor that

$$C(K, B) = \int_B (dd^c u^*_{K,B})^k.$$

The function $u^*_{K,B}$ is the extremal function of K with respect to B. If $k = 1$, then $u^*_{K,B}$ is just the harmonic measure, i.e. if K is regular in the sense of potential theory, then $u^*_{K,B}$ is harmonic on $B \backslash K$ with value zero on K and 1 on ∂B.

We now introduce the extremal function with respect to $\mathbf{C}^k$.

Let $\mathcal{L}$ denote the family of p.s.h. functions u in $\mathbf{C}^k$ such that $u(z) \leq \log^+|z| + O(1)$ at infinity. Given a bounded set E, let

$$U_E(z) = \sup\{u(z) \ : \ u \in \mathcal{L}, \ u \leq 0 \text{ on } E\}.$$

U^*_E denotes the upper semicontinuous regularization of U_E. It is quite easy to show that if U^*_E is not identically $+\infty$, then $U^*_E \in \mathcal{L}$ and satisfies $U^*_E(z) = \log^+|z| + O(1)$ at infinity. It satisfies also $(dd^c u^*_E)^k = 0$ in $\mathbf{C}^k \backslash \bar{E}$.

If $E = B(0,r)$ denotes the ball with center 0 and radius r, then $U^*_E = \log^+|z|/r$. We will denote by B the unit ball and we will assume that $E \subset B$. Let $\mathcal{M}(E) = \sup_B U^*_E$. Using that $U^*_B = \log^+|z|$, it follows that

$$\log^+|z| \leq U^*_E(z) \leq m(E) + \log^+|z|. \tag{5.1}$$

We need an inequality due to Alexander and Taylor [AT] comparing $m(E)$ and $C(E,B)$.

Theorem 5.15 *For $r < 1$ there is a constant $A(r)$ such that for any compact $K \subset B(0,r)$ then*

$$C(K,B)^{-1/k} \leq m(K) \leq \frac{A(r)}{c(K,B)}.$$

This result is a consequence of the Chern-Levine-Nirenberg inequality and of the fact that $C(K,B) = \int_B (dd^c u^*_{K,B})^k$.

Observe that when $k = 1$, then U^*_E is just the Green function with pole at infinity. In which case $U^*_E(z) = \log|z| + \gamma + o(1)$ at ∞, and $\operatorname{cap}(E) := e^{-\gamma}$ is the logarithmic capacity. The following proposition is quite straightforward, see Tsuji [Ts].

Proposition 5.16 *Let v be a subharmonic function in $\mathcal{L}(\mathbf{C})$. Assume*

$$\limsup_{z \to \infty} (v(z) - \log|z|) = A > -\infty.$$

Let $E_m = \{z : |z| \leq 1, \ v(z) < -m\}$. Then

$$\operatorname{cap}(E_m) \leq e^{-A} e^{-m} \quad \textit{and} \quad \operatorname{area}(E_m) \leq \pi e(\gamma(E))^2.$$

Proof We have $v + m \leq 0$ on E_m, so $v(z) + m \leq U^*_{E_m}(z) = \log|z| + \gamma_m + o(1)$. Consequently $A + m \leq \gamma_m$ and $\operatorname{cap}(E_m) \leq e^{-A}e^{-m}$. The estimate on the area follows, see Tsuji [Ts], p. 58. □

Notes The book [Hö] by Hörmander contains the proofs of the basic results on subharmonic and p.s.h. functions. The classical reference on currents is the book by de Rham [deRh].

The book by Lelong-Gruman [LG] has several chapters on positive currents in Complex Analysis.

The Bedford-Taylor capacity is studied in their article [BT]. Several extensions for the definition of the wedge product of a positive closed current T and $dd^c u$, where u is p.s.h., are considered in Demailly [De2], Sibony [Si] and Fornæss and Sibony [FS3].

6 Holomorphic maps in $\mathbf{P}^k$

6.1 Holomorphic self-maps of $\mathbf{P}^k$

Theorem 6.1 *Let f be a non-constant holomorphic map from $\mathbf{P}^k$ to $\mathbf{P}^k$. Then f is given in homogeneous coordinates by $[f_0 : f_1 : \cdots : f_k]$, where each f_j is a homogeneous polynomial of degree r, and the f_j have no common zero in $\mathbf{C}^{k+1}$ except the origin.*

This result follows from standard fact, see [FS1] for a proof. We will denote by $F = (f_0, \cdots, f_k)$ the corresponding map in $\mathbf{C}^{k+1}$. Observe that F is defined up to a non-zero multiplicative constant. In what follows we will always assume that the common degree r of the f_j's is ≥ 2. And we will denote by $\mathcal{H}_r$ the space of holomorphic self-maps of $\mathbf{P}^k$ of degree r. If we parametrize such maps it is easy to see that they form a Zariski dense open set in some $\mathbf{P}^N$, hence $\mathcal{H}_r$ is connected [FS1]. Observe however that, as follows from Bezout's Theorem, $f^{-1}(a)$ consists of r^k points taking into account the multiplicity. The mapping f has always a non-empty critical set which is an algebraic variety of degree $(k+1)(d-1)$.

As for rational functions in $\mathbf{P}^1$, we define the Fatou set of $f \in \mathcal{H}_r$ as follows. A point $p \in \mathbf{P}^k$ is in the *Fatou set* if there is a neighborhood $U(p)$ such that (f^n) is equicontinuous on $U(p)$. We call *Fatou components* the components of the Fatou set. The complement of the Fatou set is called the *Julia set* and is denoted by J_0.

6.2 The Green current

Given a self-map $f : \mathbf{P}^k \to \mathbf{P}^k$, we are going to associate with f a current T (the Green current) which carries interesting information on the dynamics of f.

We denote by $|\ |$ the euclidean norm in $\mathbf{C}^p$, for any p. Let $f \in \mathcal{H}_r$ and $F = (f_0, \cdots, f_k)$ a lifting. We know that $F^{-1}(0) = 0$, and that $\pi \circ F = f \circ \pi$. For each n we define $G_n(z) = \frac{1}{r^n} \mathrm{Log}\,|F^n(z)|$. Here $F^h := F \circ F \circ \cdots \circ F$. Recall that $\mathcal{P}_1$ denotes the p.s.h. functions u in $\mathbf{C}^{k+1}$ such that

$$u(\lambda z) = \mathrm{Log}\,|\lambda| + u(z).$$

It is clear that $G_n \in \mathcal{P}_1$. We define

$$G(z) = \lim G_n(z) = \lim_n \frac{1}{r^n} \mathrm{Log}\,|F^n(z)|.$$

Proposition 6.2 *The function G is continuous p.s.h. in $\mathbf{C}^{k+1}\setminus\{0\}$. Moreover,*

$$G(\lambda z) = \operatorname{Log}|\lambda| + G(z), \qquad \lambda \in \mathbf{C},$$

and

$$G(F(z)) = rG(z). \tag{6.1}$$

Proof Since $F^{-1}(0) = 0$, there is a constant $C > 0$ such that

$$\frac{1}{C}|z|^r \leq |F(z)| \leq C|z|^r.$$

So

$$\frac{1}{C}|F^n(z)|^r \leq |F^{n+1}(z)| \leq C|F^n(z)|^r.$$

Taking Log,

$$|G_{n+1}(z) - G_n(z)| \leq \frac{\operatorname{Log} C}{r^n}.$$

Therefore

$$|G_{n+q}(z) - G_n(z)| \leq 2\frac{\operatorname{Log} C}{r^n}.$$

It follows that $G = \lim_n G_n$ is well defined and that

$$|G - G_n| \leq \frac{C}{r^n}. \tag{6.2}$$

All the other properties are clear. □

Remark 6.3 Let $a \to f_a$ be a continuous family of maps in $\mathcal{H}_r$, where a is a parameter in the unit disc D. We can choose a lifting F_a that varies continuously. It follows that $(a, z) \to G_a(z)$ is continuous in $D \times \mathbf{C}^{k+1}\setminus\{0\}$. If f_a varies holomorphically then $(a, z) \to G_a(z)$ is p.s.h. in $D \times \mathbf{C}^{k+1}$.

Since $G \in \mathcal{P}_1$, there is a positive closed current T in $\mathbf{P}^{k+1}$ such that $L(G) = T$ or equivalently such that $\pi^*T = dd^cG$. Moreover, it follows from Theorem 5.9 that $||T|| = \int T \wedge \omega^{k-1} = 1$.

Observe that although F is defined up to a multiple constant the current T depends only on f. We call T the *Green current* associated with f.

Theorem 6.4 *The current T depends continuously on $f \in \mathcal{H}_r$. It has no mass on locally pluripolar sets and satisfies the equation*

$$f^*T = rT.$$

The support of T is equal to the Julia set J_0 of f.

Proof The fact that T depends continuously on f is a consequence of Remark 6.3. Let C be the critical set of f. Since T has locally a bounded potential $T = dd^cG \circ s$, where $\pi \circ s = \mathrm{Id}$, it follows for example from the Chern-Levine-Nirenberg inequality (Theorem 5.12) that T has no mass on locally pluripolar sets and in particular no mass on analytic varieties.

The map $f : \mathbf{P}^k \backslash f^{-1}(f(C)) \to \mathbf{P}^k \backslash f(C)$ is a submersion, and f^*T is well defined as a current on $\mathbf{P}^k \backslash f^{-1}(f(C))$. Moreover,

$$\begin{aligned} \pi^* f^* T &= F^* \pi^* T \\ &= F^* dd^c G = dd^c G \circ F \\ &= r\pi^* T, \end{aligned}$$

as follows from relation (6.1).

So $f^*T = rT$ on $\mathbf{P}^k \backslash f^{-1}(f(C))$. The trivial extension of f^*T, which we still denote by f^*T, is closed and satisfies the same functional equation. Observe that $f^*T = L(G \circ F)$.

We show first that $\operatorname{supp} T$ is contained in J_0. Let $U \subset\subset \Omega$, where Ω is a Fatou component. Let f^{n_i} be a subsequence which converges uniformly on U to a holomorphic map h. Shrinking U if necessary we may assume that

$$f^{n_i}(U) \subset \left\{ \left|\frac{z_1}{z_0}\right| < 2, \cdots, \left|\frac{z_k}{z_0}\right| < 2 \right\}.$$

We can then write

$$F^{n_i} = F_0^{n_i}(1, A_1^i, \cdots, A_k^i),$$

where $A_1, \cdots, A_k$ are bounded over U in $\mathbf{C}^{k+1}$. Hence

$$G_{n_i} = \frac{1}{r^{n_i}} \operatorname{Log} |F_0^{n_i}| + \frac{1}{r^{n_i}} \operatorname{Log} |(1, A_1^i, \cdots, A_k^i)|.$$

The last term converges uniformly to zero and the first term is pluriharmonic, hence π^*T has no mass on $\pi^{-1}(U)$, so T has no mass on U.

Conversely, assume T vanishes on an open set U. It follows that G is pluriharmonic on $\pi^{-1}(U)$. Shrinking U, we may assume that there is a holomorphic function h on $\pi^{-1}(U)$ such that $G = \operatorname{Log}|h|$. It follows from relation (6.2) that

$$\left| \frac{1}{r^n} \operatorname{Log} \frac{|F^n|}{|h|^{r^n}} \right| \le \frac{C}{r^n} \quad \text{on } \pi^{-1}(U), \qquad \text{i.e.,} \qquad e^{-c} \le \frac{|F^n|}{|h|^{r^n}} \le e^c,$$

so the sequence (f^n) is normal. □

6.3 Green currents of higher degree

We want to define $T^\ell = T \wedge \cdots \wedge T$, $\ell \le k$. Let s be a holomorphic section of π on an open set U. We define $T^\ell = T \wedge T \wedge \cdots \wedge T := (dd^c(G \circ s))^\ell$. The definition makes sense since G is p.s.h. and continuous on $\mathbf{C}^{k+1} \backslash \{0\}$. As in section 5.4 one checks easily that the current obtained is independent of the holomorphic section s.

Theorem 6.5 *The support of T^ℓ is connected if $2\ell \le k$. $\mathbf{P}^k \backslash (\operatorname{supp} T^\ell)$ is $(k-\ell)$-pseudoconvex.*

For $\ell = 1$ the theorem is proved in [FS2]; it is proved in [FS3] in the general case; $(k-1)$-pseudoconvexity corresponds to the usual pseudoconvexity, and hence the Fatou set is a domain of holomorphy as a consequence of the solution of the Levi problem in $\mathbf{P}^k$. A domain is $(k-\ell)$-pseudoconvex if it satisfies the *Kontinuitätssatz* with respect to ℓ-dimensional polydiscs.

We give a proof of the connectedness of $J_0 = \operatorname{supp} T$, which can be generalized to prove the connectedness of $\operatorname{supp} T^\ell$, $2\ell \le k$.

Proof of connectedness of J_0 Suppose J_0 is not connected. Let U be an open set such that $\partial U \cap J_0 = \emptyset$ but both U and $U^c = \mathbf{P}^k \backslash \bar{U}$ intersect support T. Let $T_1 := T_{|U}$, $T_2 := T_{|U^c}$. The currents T_1, T_2 are both positive and closed. Let (S_n) be a sequence of smooth closed currents whose limit is T_1, $\pi^* S_n = dd^c v_n$, with $v_n \downarrow$. Observe also that there exists a continuous function u_2 on $\mathbf{P}^k$ such that $T_2 = c_2\omega + dd^c u_2$, where $c_2 = ||T_2||$ and ω is the Kähler form on $\mathbf{P}^k$, as follows from Theorem 5.9. We then have:

$$
\begin{aligned}
0 = \int T_1 \wedge T_2 \wedge \omega^{k-2} &= \lim \int S_n \wedge T_2 \wedge \omega^{k-2} \\
&= \lim \left[c_2 \int S_n \wedge \omega^{k-1} + \int S_n \wedge dd^c u \wedge \omega^{k-2} \right] \\
&= c_2 \int T_1 \wedge \omega^{k-1}.
\end{aligned}
$$

So either $||T_2|| = 0$ or $||T_1|| = 0$. □

Corollary 6.6 *If (f^n) has a convergent subsequence on U, an open set in $\mathbf{P}^k$, then U is contained in a Fatou component. The Julia set J_0 intersects the support of any positive closed current S of bidimension (p,p), $p > 0$. In particular, J_0 intersects the critical set. The Julia set does not have a Stein neighborhood.*

Proof If (f^n) has a convergent subsequence on U, then as in the proof of Theorem 6.5, we have that G is pluriharmonic on $\pi^{-1}(U)$. So T has no mass on U and hence U is in the Fatou set.

We can assume that S is of bidimension $(1,1)$. Let $T_n = L(G_n)$, i.e., $\pi^* T_n = dd^c G_n$. We know that $T = \lim T_n$, and since G_n converges uniformly to G, then $S \wedge T = \lim S \wedge T_n$. We also have that $T_n = \omega + dd^c u_n$, where u_n is a smooth function on $\mathbf{P}^k$. So

$$\int S \wedge T = \lim \int S \wedge T_n = \int S \wedge \omega + \lim_n \int S \wedge dd^c u_n = \int S \wedge \omega$$

i.e.,

$$||S \wedge T|| = ||S||\, ||T||.$$

So the support of S intersects the support of T, i.e. the Julia set.

We show next that there is no smooth strictly p.s.h. function in a neighborhood of J_0. So in particular J_0 does not have a Stein neighborhood. Let ρ be p.s.h. in a neighborhood of J_0, and bounded in a neighborhood V of J_0. Assume there exists $c > 0$, $dd^c\rho \geq c\omega$ on V. Let $\theta \in C_0^\infty(V)$, $0 \leq \theta \leq 1$, $\theta = 1$ in a neighborhood of J_0. Then on V, $dd^c\rho \wedge T \wedge \omega^{k-2} \geq c\omega^{k-1} \wedge T > 0$. But, on V,

$$\begin{aligned} dd^c\rho \wedge T \wedge \omega^{k-2} &= dd^c(\rho T) \wedge \omega^{k-2} \\ &= dd^c(\rho\theta T) \wedge \omega^{k-2} \\ &= \lim_n dd^c(\rho_n T) \wedge \omega^{k-2} = 0, \end{aligned}$$

where the ρ_n are smooth decreasing to $\rho\theta$ and with compact support in V. The last equality is a consequence of Stokes' Theorem. □

6.4 A convergence result

Theorem 6.7 *Assume $f \in \mathcal{H}_r$ is a holomorphic map on $\mathbf{P}^2$ with $r \geq 3$. Suppose that the local multiplicity of f is at most $(r-1)$ except on a finite set E which contains no periodic point. Let S be a bidegree $(1,1)$ positive closed current on $\mathbf{P}^2$, $||S|| = 1$. Then $\frac{(f^n)^*S}{r^n} \to T$ in the sense of currents.*

Proof (sketch) As we already observed, f is not a submersion, so we have to define $\frac{(f^{n*})S}{r^n}$. Let $u \in \mathcal{P}_1$ such that $\pi^*S = dd^c u$. Then

$$\frac{(f^n)^*S}{r^n} := L\left(\frac{u \circ F^n}{r^n}\right), \quad \text{i.e.,} \quad \pi^*\frac{(f^n)^*S}{r^n} = \frac{u \circ F^n}{r^n}.$$

As we have seen this definition coincides with the usual one on $\mathbf{P}^2 \backslash f^{-1}(f(C))$ where C is the critical set for f, see section 5.5.

The sequence $u_n := \frac{u(F^n)}{r^n}$ is uniformly bounded above on $\{||z|| \leq 1\}$. We show first that no subsequence converges uniformly on compact sets to $-\infty$. Indeed,

$$\frac{1}{r^n}u(F^n) - G_n = \frac{1}{r^n}u\left(\frac{F^n}{|F^n|}\right).$$

If u_{n_i} converges uniformly to $-\infty$ on $|z| = 1$, then $F^{n_i}/|F^{n_i}|$ will not be surjective on $|z| = 1$, a contradiction. So assume $u_{n_i} \to v$ in L^1_{loc} (Theorem 5.1). We want to show $v = G$. Since $u(z) \leq \text{Log}||z|| + c$ then clearly $v \leq G$. Since v is upper semicontinuous and G is continuous, $\{v < G\}$ is open. Let $\omega \subset \mathbf{P}^2$ such that $v < G - 2\delta$, $\delta > 0$, on $\pi^{-1}(\omega)$. By Theorem 5.1, it follows that for n_i large enough,

$$\frac{1}{r^{n_i}}u\left(\frac{F^{n_i}}{|F^{n_i}|}\right) < -\delta$$

on $\pi^{-1}(\omega)$. So $\pi^{-1}(F^{n_i}(\omega)) \cap \{\frac{1}{2} \leq ||z|| \leq 1\}$ is contained in $X := \{u < -\delta r^{n_i}\}$.

The hypothesis on f implies that $X \cap \{\frac{1}{2} \leq |z| \leq 1\}$ contains balls of radius of magnitude $\epsilon^{(r-1/2)^{n_i}}$ for some $\epsilon > 0$. See Theorem 4.15 in [FS2] for the proof of this fact which relies on a Lojasiewicz type inequality. Fix (z_0, w_0, t_0) such that $u(z_0, w_0, t_0) = A > -\infty$. Let (a, b, c) be a center of a ball of radius $\epsilon^{(r-1/2)^{n_i}}$ contained in $X \cap \{\frac{1}{2} \leq |z| \leq 1\}$. Define $v(\lambda) = u(a + \lambda z_0, b + \lambda w_0, c + \lambda t_0)$. Then $\limsup_{\lambda \to \infty} v(\lambda) - \operatorname{Log}|\lambda| = A$. So any disc contained in $X \cap \left\{\frac{1}{2} < |z| \leq 1\right\}$ has radius of order of magnitude at most $e^{-\delta r^{n_i}}$, as follows from Proposition 5.16, a contradiction. □

Remark 6.8 It is shown in [FS2] that the hypothesis in Theorem is satisfied in the complement of a countable union of closed proper subvarieties of $\mathcal{H}_r$, $r \geq 3$.

Example Let $f(z, w, t) = [z^d : w^d : t^d]$, $d \geq 2$. Then

$$G(z, w, t) = \sup(\operatorname{Log}|z|, \operatorname{Log}|w|, \operatorname{Log}|t|).$$

The Julia set

$$J_0 = \{(|z| = |w| \geq |t|) \cup (|z| = |t| \geq |w|) \cup (|t| = |w| \geq |z|)\}.$$

The Fatou set consists of three components Ω_1, Ω_2, Ω_3,

$$\Omega_1 = \{|z| < |t|, |w| < |t|\}, \quad \Omega_2 = \{|z| < |w|, |t| < |w|\}, \quad \Omega_3 = \{|w| < |z|, |t| < |z|\};$$

for example, Ω_1 is the domain of attraction of $[0 : 0 : 1]$. In this case it is possible to describe the current T explicitly; for example in the chart $t = 1$. The current T corresponds to integration on the analytic varieties on J_0 and then averaging. $T_{(|z|=1, |w|<1)} = \int_0^{2\pi} [z = e^{i\theta}, |w| < 1] d\theta$, where $[z = e^{i\theta}, |w| < 1]$ denotes the current of integration on the disc $z = e^{i\theta}$, $|w| < 1$.

6.5 Hyperbolicity of Fatou components

As we have observed in Theorem 6.5, $\mathbf{P}^k \backslash (\operatorname{supp} T)$ is $(k-1)$-pseudoconvex. It follows from the solution of the Levi problem that $\mathbf{P}^k \backslash (\operatorname{supp} T)$, i.e., $\mathbf{P}^k \backslash J_0$ is a domain of holomorphy, so every Fatou component is a domain of holomorphy.

In [U] Ueda proved that Fatou components are Kobayashi hyperbolic. We give a slight refinement of his result. Denote by Δ^ℓ the unit polydisc in $\mathbf{C}^\ell$, ℓ fixed, $1 \leq \ell \leq k$.

Theorem 6.9 *Let $f \in \mathcal{H}_r$, $f : \mathbf{P}^k \to \mathbf{P}^k$, let Ω be a Fatou component for f. Then Ω is Kobayashi hyperbolic. More generally, let Σ be a parametrized variety in $\bar{\Omega}$. Assume that for every $p \in \Sigma$ there exists a neighborhood V_p of p in Σ and parametrized varieties $\varphi_i : \Delta^\ell \to D_i \subset U_i$, where U_i is open in Ω, such that φ_i converges to φ, a parametrization of $\Sigma \cap V_p$. Then Σ is Kobayashi hyperbolic.*

Proof Let U be a simply connected domain in Ω, and let s be a holomorphic section of π over U. Since $G \circ s$ is pluriharmonic on U, there exists a holomorphic function h such that $G \circ s = \operatorname{Log}|h|$ on U. So $(s(z)/h(z))$ is a holomorphic section of π whose image is in $\mathcal{A} := \{z; G(z) = 1\}$. The set $\mathcal{A}$ is a compact set in $\mathbf{C}^{k+1}$, since G grows like $\operatorname{Log}\|z\|$ at infinity. Observe that sections of π with values in $\mathcal{A}$ are unique up to a constant factor. So if Ω is a Fatou component we get by analytic continuation a covering manifold of Ω which is contained in the compact set $\mathcal{A}$. Therefore Ω is Kobayashi hyperbolic, see [Ko].

In the more general case, let $\tilde{\varphi}_i$ be liftings of φ_i with values on $\mathcal{A}$, $\pi \circ \tilde{\varphi}_i = \varphi_i$. The total collection of liftings is obtained from one of them by multiplication by $e^{i\theta}$. So we can assume $\tilde{\varphi}_i \to \tilde{\varphi}$ and $\pi \circ \tilde{\varphi} = \varphi$. We then construct a covering manifold $\tilde{\Sigma}$ of Σ which is contained in $\mathcal{A}$. So Σ is Kobayashi hyperbolic. □

Corollary 6.10 *Let $f \in \mathcal{H}_r$, $f : \mathbf{P}^k \to \mathbf{P}^k$. Assume that (f^{n_i}) is a subsequence which converges on Ω to a function h. Then $\Sigma = h(\Omega)$ is a hyperbolic abstract Kobayashi hyperbolic complex manifold.*

Proof As we noticed in Corollary 6.6, Ω is then a Fatou component. Observe that $\Sigma = h(\Omega)$ is not a priori closed and can intersect the Fatou set as well as the Julia set. It follows from Theorem 6.9 that Σ is Kobayashi hyperbolic.

If we assume that $f(\Omega) = \Omega$, then it is then easy to show that f induces a surjective map from Σ to Σ. □

Notes The material in this chapter is mostly taken from [FS2] and [FS3]. The idea to use potential theory in one-variable complex iteration was first considered in Brolin [Br]. It was developed more systematically by Sibony (unpublished course notes) and Tortrat [T]. Hubbard and Papadopol [HP] used the function G on $\mathbf{C}^{k+1}$ to study the domain of attraction of 0 for the lifted map F in $\mathbf{C}^{k+1}$.

Ueda [U] proved that Fatou components in $\mathbf{P}^k$ are Kobayashi hyperbolic.

7 Hénon mappings

7.1 Polynomial automorphisms of $\mathbf{C}^2$

Let us consider first some examples of holomorphic automorphisms of $\mathbf{C}^2$. An elementary map is a map e of the form:

$$e(z, w) = (\alpha z + p(w), \beta w + \gamma)$$

for constants α, β, γ with $\alpha\beta \neq 0$ and some polynomial p of degree $d > 1$.

If we consider the meromorphic extension to $\mathbf{P}^2$ we get

$$e[z : w : t] = [\alpha z t^{d-1} + t^d p(\frac{w}{t}) : \beta w t^{d-1} + \gamma t^d : t^d].$$

There is one point of indeterminacy $p = [1 : 0 : 0]$ for e, and $(t = 0)$ is mapped to the point of indeterminacy. We will see that this implies that the degree of $e \circ e$ is strictly less than d^2.

The dynamics of elementary maps is quite simple, since $w = w_0$ is mapped to $w = w_1$:

If $\beta = 1$, then $e^n(z,w) \to \infty$ and indeed in $\mathbf{P}^2$, $e^n \to p$.

If $\beta \neq 1$, then the map is conjugate to

$$e_1(z,w) = (\alpha z + p_1(w), \beta w).$$

If $|\beta| > 1$, $e_1^n \to p$.

For $|\beta| < 1$ all limits are in $z = 0$ and we let the reader discuss the various cases according to the modulus of α. The case $|\beta| = 1$ is also of interest.

Consider now a map g of the following form

$$g(z,w) = (P(z) - aw, z)$$

with degree $P = d \geq 2$.

The map g is an automorphism of $\mathbf{C}^2$ of Jacobian a. If we consider the extension to $\mathbf{P}^2$ we get

$$g[z : w : t] = [t^d P(\frac{z}{t}) - awt^{d-1} : zt^{d-1} : t^d].$$

The map g has exactly one point of indeterminacy $p_- = [0 : 1 : 0]$. The hyperplane $(t = 0)$ is mapped to $p_+ = [1 : 0 : 0]$, and it is easy to check that p_+ is a fixed point for g and that both eigenvalues of $g'(p_+)$ vanish.

Let

$$\mathcal{G} := \{g; g = g_m \circ \cdots \circ g_1,\ g_j(z,w) = (p_j(z) - a_j w, z),\ \ a_j \neq 0\}.$$

$\mathcal{G}$ is by definition the semigroup of Hénon mappings. We always assume that $\deg p_j = d_j \geq 2$. All Hénon mappings have the same point of indeterminacy p_- and the same superattractive point p_+.

In [FM], Friedland and Milnor proved the following result.

Theorem 7.1 [FM] *Let f be a polynomial automorphism of $\mathbf{C}^2$. Then either f is affinely conjugate to an elementary mapping or f is affinely conjugate to a Hénon mapping.*

It follows from this result that the only polynomial automorphisms of $\mathbf{C}^2$ which are of interest from the dynamical point of view are the Hénon mappings.

If g is a polynomial automorphism we call the degree of g, the maximum of the degrees of the components. It is of interest to understand the growth of the degrees of the iterates of a map f.

Proposition 7.2 *Let f_1, f_2 be meromorphic maps in $\mathbf{P}^2$ of maximal rank 2. Let I_1, I_2 be their sets of indeterminacy. Then*

$$\deg(f_2 \circ f_1) < (\deg f_1) \cdot (\deg f_2)$$

if and only if there is a variety C of dimension 1 such that $f_1(C) \in I_2$.

Proof Observe first that I_2 is a finite set of points. So if $f_1(C) \in I_2$ it implies that f_1 is constant on C, we can assume that C is irreducible. Let $\{h = 0\}$ be an equation for C. Then all components of $f_2 \circ f_1$ vanish on C, hence we can factor out a power of h in order to write the map $f_2 \circ f_1$ properly. Hence $\deg(f_2 \circ f_1) < (\deg f_1)(\deg f_2)$. It is clear that if the components of $f_2 \circ f_1$ have no common factor then degree of $f_2 \circ f_1$ equals $(\deg f_2)(\deg f_1)$.

Corollary 7.3 *If g is a Hénon mapping of degree d, then g^n is of degree d^n.*

Proof Indeed the only variety which is mapped to a constant is $(t = 0)$, which is mapped to the point $p_+ \notin I$. □

For another approach see Theorem 2.1 in [FM].

In what follows we will study the dynamics of Hénon mappings. For the sake of simplicity we will sometimes restrict ourselves to maps of the form

$$g(z, w) = (P(z) - aw, z), \qquad \deg P \geq 2,$$

but the results are valid for all Hénon maps.

Following Hubbard [Hu] define

$$K^{\pm} = \{p \in \mathbf{C}^2 : g^n(p),\ n \geq 0, \text{ is bounded}\}.$$

Observe that

$$g^{-1}(z, w) = \left(w, \frac{-z + P(w)}{a}\right).$$

Similarly, define

$$K^{-} = \{p \in \mathbf{C}^2 : g^{-n}(p),\ n \geq 0 \text{ is bounded}\}.$$

As we have seen, $p_+ = [1 : 0 : 0]$ is a fixed attractive point for g. Let U^+ denote the domain of attraction of p_+, and U^- the domain of attraction of p_- for the map g^{-1}.

The point $p_- = [0 : 1 : 0]$ is an attractive fixed point for g^{-1}, and p_+ is the point of indeterminacy for the extension of g^{-1} to $\mathbf{P}^2$.

We also define $J^{\pm} = (\partial K^{\pm})$ and $K := K^+ \cap K^-$, $J := J^+ \cap J^-$.

Following [FM] we introduce the following domains. Fix $R > 0$ such that

$$\{|z| > R,\ \ |w| \leq |z|\} \Rightarrow |P(z) - aw| > |z|.$$

Such an R exists since $\deg p \geq 2$. Now define

$$\begin{aligned} V^+ &= \{(z, w) : |z| > R \text{ and } |z| > |w|\} \\ V^- &= \{(z, w) : |w| > R \text{ and } |w| > |z|\} \\ V &= \{(z, w) : |z| \leq R \text{ and } |w| \leq R\}. \end{aligned}$$

We will also denote by g the extension of g to $\mathbf{P}^2$.

Proposition 7.4 *Let U^+ denote the domain of attraction of p_+. Then*

$$U^+ = \bigcup_{n\geq 0} g^{-n}(V^+)$$

and $K^+ = \mathbf{C}^2 \backslash U^+$. The set K is compact.

Proof If $(z_n, w_n) = g^n(z_0, w_0)$, then $(z_n, w_n) = (z_n, z_{n-1})$. The choice of R shows that if $(z_0, w_0) \in V^+$, then $(z_1, z_0) \in V^+$ and $z_n \to \infty$. Moreover it is clear that for n large enough, $|z| > c|z_{n-1}|^d$ so $(z_n, z_{n-1}) \to p_+$. Hence $\bigcup_{n\geq 0} g^{-n}(V^+) \subset U^+$. Since $V^+ \cup (t = 0)$ is a neighborhood of p_+ in $\mathbf{P}^2$ we then have that $\bar{U}^+ \subset \bigcup_{n\geq 0} g^{-n}(V^+)$.

From the definition of U^+ it follows that $K^+ \subset \mathbf{C}^2 \backslash U^+$. Fix $(z_0, w_0) \in K^+$. Choose n such that $|z| > R$ and $|z_n| > |z_0|$. We can choose n as small as possible. Then$|z_n| \geq |z_{n-1}|$ and hence $z_n \to \infty$. So $K^+ = \mathbf{C}^2 \backslash U^+$. We can show similarly that $\bigcup_{n\geq 0} g^n(V^-) = U^-$, so $K = K^- \cap K^+$ is compact. □

Remark 7.5 Let $\Phi : \mathbf{C}^2 \to \mathbf{P}^2$, $\Phi(z, w) = [z : w : 1]$. Define $\tilde{U}^+$ as the domain of attraction of p_+ for the holomorphic map $g : \mathbf{P}^2 \backslash (p_-) \to \mathbf{P}^2 \backslash (p_-)$. Then

$$U^+ = \Phi^{-1}(\Phi(\mathbf{C}^2) \cap \tilde{U}^+), \quad \text{and} \quad \tilde{U}^+ = \Phi(U^+) \cup \{(t = 0) \backslash p_-\}.$$

Definition 7.6 A domain Ω in $\mathbf{C}^k$ is a *Fatou-Bieberbach domain* if Ω is biholomorphic to $\mathbf{C}^k$ and $\bar{\Omega} \neq \mathbf{C}^k$.

Theorem 7.7 *Let g be a polynomial automorphism of $\mathbf{C}^2$ which is neither affine nor affinely conjugate to an elementary mapping. Let p_0 be a fixed attractive point for g and*

$$\Omega = \{p : \ p \in \mathbf{C}^2 \quad g^n(p) \to p_0\}.$$

Then Ω is a Fatou-Bieberbach domain.

Proof Theorem 7.1 implies that g is affinely conjugate to a Hénon mapping. On the other hand it is classical that if g is an automorphism of $\mathbf{C}^2$ with $g(p_0) = p_0$, and if the eigenvalues λ_1, λ_2 of $g'(p_0)$ satisfy $|\lambda_1| < 1$, $|\lambda_2| < 1$, then Ω, the domain of attraction of p_0, is biholomorphic to $\mathbf{C}^2$. See Rosay and Rudin [RR] for a proof.

We can assume that g is a Hénon mapping, hence U^+ is non-empty. So $\bar{\Omega} \neq \mathbf{C}^2$. □

7.2 Green's function for Hénon mappings

Let f be a Hénon mapping in $\mathbf{C}^2$ of degree d. We denote by $\tilde{f}$ the extension of f to $\mathbf{P}^2$. There exists a map $F : \mathbf{C}^3 \to \mathbf{C}^3$ such that $\pi \circ F = \tilde{f} \circ \pi$, where π as usual denotes the canonical map from $\mathbf{C}^3 \backslash \{0\}$ onto $\mathbf{P}^2$. For example, if $f(z, w) = (z^d + c + aw, z)$, then $F(z, w, t) = (z^d + ct^d - awt^{d-1}, zt^{d-1}, t^d)$.

We have observed that $p_+ = [1:0:0]$ is attractive for $\tilde{f}$ and a point of indeterminacy for $\tilde{f}^{-1}$, and similarly $p_- = [0:1:0]$ is attractive for $\tilde{f}^{-1}$ and a point of indeterminacy for $\tilde{f}$.

It is clear that there is a constant $C > 0$ such that

$$|F(Z)| \leq C|Z|^d,$$

where $Z = (z, w, t)$. Let B denote the unit ball in $\mathbf{C}^3$. Let ω be a small conic neighborhood in $\mathbf{C}^3$ of the line $\{(0, \lambda, 0),\ \lambda \in \mathbf{C}\}$. Let $C' = \min\{|F(Z)|,\ Z \in \partial B \backslash \omega\}$; then $C' > 0$. Since p_- is repelling for f we have that $\frac{F^n(Z)}{|F^n(Z)|} \notin \omega$ if $Z \notin \omega$. So for $Z \notin \omega$,

$$C'|F^n(Z)|^d \leq |F^{n+1}(Z)| \leq C|F^n(Z)|^d. \tag{7.1}$$

Hence we define

$$\tilde{G}^+(Z) = \lim_{n\to\infty} \frac{1}{d^n} \mathrm{Log}|F^n(Z)|.$$

Proposition 7.8 *The function $\tilde{G}^+$ is p.s.h. on $\mathbf{C}^3$. It is continuous on $\mathbf{C}^3 \setminus \{(0, w, 0) : w \in \mathbf{C}\}$ and continuous with values on $[-\infty, +\infty[$ on $\mathbf{C}^3$. Moreover, $\tilde{G}$ satisfies the following relations:*

$$\tilde{G}^+(\lambda Z) = \mathrm{Log}\,|\lambda| + \tilde{G}^+(Z), \tag{7.2}$$

$$\tilde{G}^+(F(Z)) = d\tilde{G}^+(Z). \tag{7.3}$$

Proof Relation (7.1) implies that out of $\{\lambda Z,\ \lambda > 0,\ Z \in \omega\}$ we have for $n \geq 1$

$$\left| \frac{1}{d^{n+1}} \mathrm{Log}\,|F^{n+1}(Z)| - \frac{1}{d^n} \mathrm{Log}\,|F^n(Z)| \right| \leq \frac{M}{d^n},$$

where M is a constant depending of ω. It follows that the convergence is uniform on compact sets disjoint from $\{(0, w, 0),\ w \in \mathbf{C}\}$.

Observe that $u_n(Z) := \frac{1}{d^n}\mathrm{Log}\,|F^n|(Z) + \sum_n^\infty (\mathrm{Log}\, C)/d^k$ is decreasing, so $\tilde{G}(Z)$ is p.s.h. on $\mathbf{C}^3$. We have $\tilde{G}(0, w, 0) = -\infty$. Hence $\exp(\tilde{G})$ is continuous.

The other properties are clear. □

We define

$$\begin{aligned} G^+(z, w) = \tilde{G}^+(z, w, 1) &= \lim_n \frac{1}{d^n} \mathrm{Log}\left(|f^n(z, w)|^2 + 1\right)^{1/2} \\ &= \lim_n \frac{1}{d^n} \mathrm{Log}^+ |f^n(z, w)|. \end{aligned}$$

The function G^+ is continuous p.s.h. in $\mathbf{C}^2$, and satisfies

$$G^+(f(z, w)) = dG^+(z, w).$$

We also have that

$$\tilde{G}^+(z, w, t) = \mathrm{Log}\,|t| + G^+\left(\frac{z}{t}, \frac{w}{t}\right),$$

as follows from Proposition 7.8.

Let $F^{-1}(z,w,t) = t^d f^{-1}\left(\frac{z}{t},\frac{w}{t}\right)$ be associated to f^{-1} as F is to f. We define

$$\begin{aligned}\tilde{G}^-(Z) &= \lim_n \frac{1}{d^n}\mathrm{Log}\,|F^{-n}(Z)|,\\ G^-(z,w) &= \lim \frac{1}{d^n}\mathrm{Log}^+|f^{-n}(z,w)|.\end{aligned}$$

We also have that

$$G^-(f^{-1}(z,w)) = d\,G^-(z,w).$$

Hence

$$G^-(f(z,w)) = \frac{1}{d}G^-(z,w). \tag{7.4}$$

The functional relations for F are however different.

We can easily check that

$$F\circ F^{-1}(z,w,t) = F^{-1}\circ F(z,w,t) = t^{d^2-1}(z,w,t).$$

Therefore

$$d\tilde{G}^-(F(z,w,t)) = \tilde{G}^-(t^{d^2-1}(z,w,t)) = (d^2-1)\mathrm{Log}\,|t| + \tilde{G}^-(z,w,t). \tag{7.5}$$

This relation will be of interest to us.

We will also consider the continuous p.s.h. function

$$G(z,w) := \sup\{G^+(z,w), G^-(z,w)\}.$$

Proposition 7.9 *Let $f:\mathbf{C}^2\to\mathbf{C}^2$ be a Hénon mapping of degree d. The function $G^\pm$ is pluriharmonic and positive on $U^\pm$. Put $\{G^\pm = 0\} = K^\pm$. On K^+,* $\mathrm{dist}(f^n(z,w),K)$ *converges uniformly to zero on compact sets. In particular, (f^n) is a normal family on* $\mathrm{int}\,K^+$.

Proof If $(z_n, z_{n-1}) = f^n(z,w)$ and $(z,w)\in U^+$, we have seen that $z_{n-1} = O(z_n)$ hence, on U^+

$$G^+(z,w) = \lim_n \frac{1}{d^n}\mathrm{Log}\,|f_1^n(z,w)|.$$

Hence G^+, being a limit of pluriharmonic functions, is a pluriharmonic function; therefore G^+ is strictly positive on U^+, by the minimum principle. It is also clear that $(G^+>0)\subset U^+$, so $K^+ = \{G^+=0\}$. We have for $|w|\le C$,

$$G^+(z,w) = \mathrm{Log}^+|z| + O(1)$$

uniformly on w.

Observe that since $K = K^+\cap K^-$, $\{G=0\} = K$. Hence $\{G<\epsilon\}$ is a neighborhood basis of K.

Let X be a compact set on K^+. Assume $G^- \leq C$ on X. Then

$$G^-(f^n(z,w)) \leq \frac{C}{d^n} \quad \text{on } X.$$

Therefore

$$G(f^n(z,w)) \leq \frac{C}{d^n},$$

which gives the uniform convergence on compact sets towards K.

7.3 The currents μ^+ and μ^- and the measure μ

Since $\tilde{G}^\pm$ is p.s.h. on $\mathbf{C}^3$ and satisfies the homogeneity condition $\tilde{G}^\pm(\lambda Z) = \operatorname{Log}|\lambda| + \tilde{G}^\pm(Z)$. We know by Theorem 5.9 that there are positive closed $(1,1)$ currents $\mu^\pm$ on $\mathbf{P}^2$ such that $\pi^*\mu^\pm = dd^c\tilde{G}^\pm$. If we consider the restriction of $\mu^\pm$ to $\mathbf{C}^2$, i.e. $(t \neq 0)$, we find that

$$\mu^\pm = dd^cG^\pm.$$

Since $\{(t = 0\backslash p_-\}$ is mapped by $\tilde{f}$ to p_+ it follows that μ^+ has no mass near $\{(t = 0)\backslash p_-\}$. So we can identify $\mu^+ = dd^cG^+$ with its trivial extension to $\mathbf{P}^2$. It is however of interest to consider $\mu^\pm$ as positive closed currents on $\mathbf{P}^2$.

Theorem 7.10 *The currents $\mu^\pm$ are of mass one, and* $\operatorname{supp}\mu^+ = J^+ \cup \{p^-\}$, $\operatorname{supp}\mu^- = J^- \cup \{p^+\}$. *We also have:*

$$f^*\mu^+ = d\mu^+, \qquad f^*\mu^- = \frac{1}{d}\mu^-,$$

and

$$\mu^+ \wedge \mu^+ = \delta_{p_-}, \qquad \mu^- \wedge \mu^- = \delta_{p_+},$$

where δ_{p_-} denotes the Dirac mass at p_- and δ_{p_+} the Dirac mass at p_+.

Proof Since $\tilde{G}^+(\lambda Z) = \operatorname{Log}|\lambda| + \tilde{G}^+(Z)$, we know by Theorem 5.9 that μ^+ has mass one on $\mathbf{P}^2$, similarly for μ^-. The function G^+ is pluriharmonic on U^+, so μ^+ is supported on $\partial K^+ = J^+$. If a point q of $\partial K^+ = J^+$ were not in the support of μ^+, then G^+ would be pluriharmonic in a neighborhood of q, but $G^+(q) = 0$, so this will contradict the minimum principle for G^+. Since $G^+(f) = dG^+$ we get $f^*\mu^+ = d\mu^+$; this functional equation is to be interpreted in $(t \neq 0)$. We can extend it to $\mathbf{P}^2$ using the observations after Theorem 5.13. For $\epsilon > 0$ define $G^+_\epsilon = \max\{G^+, \epsilon\}$. Since G^+ is pluriharmonic on U^+, through every point in $\mathbf{C}^2$ there is an analytic disc on which G^+_ϵ is constant, hence $dd^cG^+_\epsilon \wedge dd^cG^+_\epsilon = 0$. Letting $\epsilon \to 0$ we find that $\mu^+ \wedge \mu^+ = 0$ on $(t \neq 0)$. Since μ^+ vanishes near $\{(t = 0)\backslash p_-\}$ it is also clear that $\mu^+ \wedge \mu^+ = 0$ on $\mathbf{P}^2\backslash p_-$. Fix a neighborhood ω of p_- in $\mathbf{P}^2$. In ω, μ^+ has a potential bounded near $\partial\omega$, so μ^+ is wedgeable with itself, see [FS3], and the mass of $\mu^+ \wedge \mu^+$ is one, hence $\mu^+ \wedge \mu^+ = \delta_{p_-}$. □

We define $\mu := \mu^+ \wedge \mu^-$. The wedge product is well defined, since local potentials of μ^+ and μ^- are continuous.

Theorem 7.11 *The measure μ is a probability measure. It satisfies*

$$f^*\mu = \mu, \quad \operatorname{supp}\mu := J_1 \subset J.$$

Moreover, $\mu = dd^cG \wedge dd^cG$.

Proof That μ is a probability measure follows from the fact that $||\mu^+|| = ||\mu^-|| = 1$ and $\mu = \mu^+ \wedge \mu^-$. See [FS3] for a general Bezout Theorem. We have:

$$f^*\mu = f^*\mu^+ \wedge f^*\mu^- = d\mu^+ \wedge \frac{1}{d}\mu^- = \mu^+ \wedge \mu^- = \mu.$$

Since μ^+ is supported on J^+, and μ^- is supported on J^-, the support J_1 of μ is contained in J.

We now prove the last relation. We have

$$G = \lim_{\epsilon \to 0} \max(G_\epsilon^+, G_\epsilon^-).$$

So $(dd^cG)^2 = \lim_{\epsilon\to 0}(dd^c \max(G_\epsilon^+, G_\epsilon^-))^2$. We show then that $(dd^c \max(G_\epsilon^+, G_\epsilon^-))^2 = dd^cG_\epsilon^+ \wedge dd^cG_\epsilon^-$. Indeed one checks by computation that

$$(dd^c\max(Re\, z)^+, (Re\, w)^+)^2 = dd^c(Re\, z)^+ \wedge dd^c(Re\, w)^+.$$

Choose ϵ to be a regular value for G^+ and G^-, and change coordinates, since near $G^+ = \epsilon$, $G^- = \epsilon$ the functions G^+ and G^- are pluriharmonic. This gives the formula. Since $G_\epsilon^\pm$ converges uniformly to $G^\pm$ we get that

$$(dd^cG)^2 = \mu^+ \wedge \mu^-. \qquad \square$$

Theorem 7.12 *If S is a positive closed current on $\mathbf{P}^2$ supported on $K^+ \cup p_-$, then $S = c\mu^+$ with $c = \int S \wedge \mu^-$.*

Lemma 7.13 *Let $\mathcal{S}$ be the set of positive closed currents S, $||S|| = 1$, supported on $K^+ \cup p_-$. Then $\frac{1}{d^n}(f^n)^*S$ converges uniformly on $\mathcal{S}$ to μ^+.*

Proof The currents in $\mathcal{S}$ have no mass on $t = 0$. Let (S_k) be a sequence in $\mathcal{S}$. Then $S_k = dd^c u_k$ in $t \neq 0$, which is identified with $\mathbf{C}^2$. Moreover, $u_k(z, w) \leq \mathrm{Log}^+|(z, w)| + C$, where C is independent of k, as follows from Theorem 5.9. Recall also that $u_k(z, w) = \tilde{u}_k(z, w, 1)$, where $\tilde{u}_k \in \mathcal{P}_1$. Let v be a limit in L^1_{loc} of a subsequence $v_k := \left(\frac{u_k \circ f^{n_k}}{d^{n_k}}\right)$, with $n_k \to \infty$. We have to show that $v = G^+$.

Since the u_k's are pluriharmonic on V^+ we have that for $|w| \leq R$, $u_k(z, w) = \log|z| + O(1)$ at infinity with $O(1)$ bounded independently of w. So $v = G^+$ on U^+. Since v is upper semicontinuous we have $v = G^+$ on $\overline{U^+}$.

We show next that $v = 0$ on $\operatorname{int} K^+$. Assume $v < -2\alpha$, $\alpha > 0$, on $\Omega \subset\subset \operatorname{int} K^+$. Then, using Theorem 5.1, $v_k < -\alpha$ on Ω for k large enough. Define

$$E_k = \{(z, w) \in V : \quad u_k \leq -\alpha d^{n_k}\}.$$

Since u_k is of logarithmic growth we can estimate the logarithmic capacity of $E_k(w)$ for each fixed $|w| < R$, $\text{cap}(E_k(w)) \leq Ce^{-\alpha d^{n_k}}$. So using Proposition 5.16 and Fubini's Theorem we get that $\text{vol}(E_k) \leq C_1 e^{-\alpha d^{n_k}}$. But $f^{n_k}(\Omega) \subset E_k$ for k large enough, hence $\text{vol}(E_n) \geq \text{vol}\, f^{n_k}(\Omega) = |a|^{2n_k}\text{vol}(\Omega)$, a contradiction. So $v = G^+$.

Proof of Theorem 7.12 Let S be a positive closed current supported on $K^+ \cup p_-$. Then S has no mass on $t = 0$, see Remark 5.6. For $n > 0$ the current $(f^{-n})^*S$ is well defined on $t \neq 0$, it is positive and closed and has bounded mass near $t = 0$, so the trivial extension $\widetilde{(f^{-n})^*S}$ is positive and closed on $\mathbf{P}^2$ (see [Sk] or [Si]). We assume $||S|| = 1$ and we want to show that $||(f^{-n})^*S|| = \frac{1}{d^n}$.

Let u be a p.s.h. function on $\mathbf{C}^2$ of logarithmic growth such that $dd^cu = S$. Define $v := u \circ f^{-1}$; v is a potential for $(f^{-1})^*S$. Let $\tilde{v}$ be a p.s.h. function in $\mathbf{C}^3$ such that

$$\tilde{v}(\lambda Z) = c \log|\lambda| + \tilde{v}(Z) \quad \text{and} \quad \pi^*[\widetilde{(f^{-1})^*S}] = dd^c\tilde{v}.$$

As we have seen in Theorem 5.9, $||\widetilde{(f^{-1})^*S}|| = c$. Let $L_1 = \{(0, w, 0) : \ w \in \mathbf{C}\}$ be the complex line through $(0, 1, 0)$ in $\mathbf{C}^3$. The current $\widetilde{(f^{-1})^*S}$ vanishes near $(t = 0)\backslash p_-$, so $\tilde{v}$ is pluriharmonic near $(t = 0)\backslash L_1$. Since $(t = 0)\backslash p_-$ is mapped to $[1 : 0 : 0]$, and since $\tilde{v}$ is pluriharmonic near $(z, 0, 0)$, $z \neq 0$, we get that $\tilde{v} \circ F$ is pluriharmonic near $(t = 0)\backslash L_1$. We also have

$$(\tilde{v} \circ F)(\lambda Z) = dc\text{Log}\,|\lambda| + (\tilde{v} \circ F)(z).$$

But

$$\pi^*S = \pi^*[f^*((f^{-1})^*S)] = F^*[\pi^*(f^{-1})^*S] = F^*dd^c\tilde{v} = dd^c(\tilde{v} \circ F).$$

Consequently $\tilde{v} \circ F$ is a potential for S, and hence $c = \frac{1}{d}$. This finally shows that $||(f^{-n})^*S|| = \frac{1}{d^n}$, proving the claim.

One should observe that if $\tilde{u} \in \mathcal{P}_1$ is such that $\pi^*S = dd^c\tilde{u}$, then $\tilde{v} \neq \tilde{u} \circ F^{-1}$. For example, if $\tilde{u} = \tilde{G}^+$ then since $(F^{-1} \circ F)(z, w, t) = t^{d^2-1}(z, w, t)$, we get

$$\tilde{G}^+ \circ F^{-1} \circ F = (d^2 - 1)\text{Log}\,|t| + \tilde{G}^+.$$

So when we consider $\tilde{v}$ we forget the "polar part" of $\tilde{u} \circ F^{-1}$ on $t = 0$.

We write

$$S = \frac{1}{d^n}(f^n)^*[d^{+n}(\widetilde{f^{n*}S})].$$

Let $\sigma_n = d^n\widetilde{(f^{-n})^*S}$. So $S = \frac{1}{d^n}(f^n)^*\sigma_n$. We know that $||\sigma_n|| = 1$. Lemma 7.11 implies that $\frac{1}{d^n}(f^n)^*\sigma_n$ converges to μ^+. So $S = \mu^+$.

Corollary 7.14 *Let S be a positive closed $(1, 1)$ current on $\mathbf{P}^2$. Assume $p_+ \notin \text{supp}\, S$. Then $\frac{1}{d^n}(f^n)^*S$ converges to a constant multiple of μ^+.*

Proof Since $\text{supp}(f^n)^*S \subset f^{-n}(\text{supp}\, S)$ and since p_+ is attracting for f, it follows that any cluster point of $\frac{1}{d^n}(f^n)^*S$ is supported on K^+, hence it is equal to $c\mu^+$,

$$c = \lim \int \frac{(f^{n_i})^*S}{d^{n_i}} \wedge \mu^- = \lim \int S \wedge \frac{(f^{-n_i})^*\mu^-}{d^{n_i}} = \int S \wedge \mu^-.$$

So c is independent of the subsequence.

Corollary 7.15 *There is no non-zero positive closed* $(1,1)$ *current on* $\mathbf{P}^2$ *with support* $X \subset \overline{K^+}$ *and such that* $X \cap \operatorname{int} K^+$ *is non-empty. In particular, the closure of a domain of attraction of an attractive fixed point contains no algebraic curve.*

Proof This is clear since the only positive closed $(1,1)$ currents on K^+ are multiples of μ^+, which is supported on ∂K^+. □

Remark 7.16 Theorem 7.12 implies that the current μ^+ is extremal in the convex cone of positive closed $(1,1)$ currents in $\mathbf{P}^2$. It is also not the current of integration over an analytic curve. Such currents were first constructed by J. P. Demailly [De2].

7.4 Density of stable manifolds in J^+

In this section we prove an interesting result due to Bedford and Smillie [BS2]. Given a saddle point p in J^+, the stable manifold through p is contained and dense in J^+.

We have seen in Corollary 7.14 that if S is a positive closed current with $||S|| = 1$, and such that $p^+ \notin \operatorname{supp} S$, then $\frac{1}{d^n}(f^n)^* S$ converges to μ^+. We first deduce from that result a convergence theorem, towards μ^+, for some non-closed positive currents.

Theorem 7.17 [BS2] *Given a positive closed current* R *in an open set* $\Omega \subset \mathbf{C}^2$, *let* ψ *be a test function supported in* Ω. *Define* $R_n = \frac{1}{d^n}(f^n)^*(\psi R)$. *Then* $R_n \to c\mu^+$ *with* $c = \int \psi R \wedge \mu^-$.

Proof Let ω be the standard Kähler form on $\mathbf{P}^2$. Then

$$\langle R_n, \omega \rangle = \langle \psi R, \frac{(f^{-n})^* \omega}{d^n} \rangle \to \int \psi R \wedge \mu^-,$$

since by definition of μ^-, the $\frac{(f^{-n})^*\omega}{d^n}$ converge to μ^- because the corresponding potentials converge uniformly to G^-.

So the sequence of currents R_n has bounded mass. Let S be a weak limit of R_{n_i}; we show that S is closed.

Let Φ be a smooth $(0,1)$ form of compact support in $\mathbf{C}^2$. Then

$$\left|\int \partial R_n \wedge \Phi\right| = \frac{1}{d^n}\left|\int (f^n)^* \partial(\psi R) \wedge \Phi\right| = \frac{1}{d^n}\left|\int \partial\psi \wedge R \wedge (f^{-n})^*\Phi\right|.$$

Here we have used that $\partial(\psi R) = \partial\psi \wedge R$, since R is closed near $\operatorname{supp}\psi$. Now we use Schwarz's inequality:

$$\begin{aligned}
\left|\int \partial R_n \wedge \Phi\right| &\leq \frac{1}{d^n}\left|\int R \wedge \partial\psi \wedge i\bar{\partial}\psi\right|^{1/2} \left|\int_{(\operatorname{Supp}\psi)} R \wedge (f^{-n})^*\Phi \wedge i\overline{(f^{-n})^*\Phi}\right|^{1/2} \\
&\leq \frac{C}{d^{n/2}}\left|\int_{(\operatorname{Supp}\psi)} \frac{(f^n)^* R}{d^n} \Phi \wedge i\bar{\Phi}\right|^{1/2} \\
&= O\left(\frac{1}{d^{n/2}}\right).
\end{aligned}$$

The last integral is bounded since Φ is smooth and the mass of the current $\frac{(f^n)^*R}{d^n}$ is bounded, as we have seen in the first part of the proof.

So all limits of (R_n) are closed positive currents supported on K^+. Theorem 7.12 shows that they are equal to $c\mu^+$. We have

$$c = \lim \int \frac{(f^n)^* \psi R}{d^n} \wedge \omega = \int \psi R \wedge \mu^-.$$

So c is independent of the subsequence.

Corollary 7.18 [BS2] *Let f be a Hénon map. Suppose $f(p) = p$ and let λ_1, λ_2 be the eigenvalues of $f'(p)$. Assume $|\lambda_1| < 1$, $|\lambda_2| > 1$, i.e. that p is hyperbolic. Let*

$$W^s(p) = \{q : \lim_{n\to\infty} f^n(q) = p\}.$$

Then $W^s(p)$ is dense in J^+. The boundary of any basin of attraction is J^+.

Proof Let M be an open disc in $W^s(p)$ containing the saddle point p, and locally closed in an open set Ω. If $[M]$ denotes the current of integration on M, then $[M] \wedge \mu^- \neq 0$; indeed $[M] \wedge \mu^- = dd^c G^-_{|M}$. It is then enough to prove that G^- is not harmonic on M. Assume G^- is harmonic on M. Since $G^-(p) = 0$ and $G^- \geq 0$ we would have that G^- is identically zero on M. Hence $G^- = 0$ on $W^s(p)$ and $W^s(p)$ would be contained in K, $f^{-n}{}_{|M}$ would be a normal family contradicting that $|\lambda_1| < 1$. So for any test function ψ non-identically zero and supported in Ω we have

$$\lim_{n\to\infty} \frac{(f^n)^* \psi [M]}{d^n} = c\mu^+, \qquad c \neq 0.$$

Hence J^+ the support of μ^+ is contained in $\overline{\cup f^{-n}(M)} \subset \overline{W^s(p)}$.

It is quite clear that M cannot intersect $\operatorname{int} K^+$. Let v_2 be a unit vector such that $f'(p)v_2 = \lambda_2 v_2$, $|\lambda_2| > 1$. f' is expanding on vectors on M parallel to v_2, hence M cannot intersect $\operatorname{int} K^+$. So $M \subset J^+$, hence $W^s(p) \subset J^+$ and therefore $\overline{W^s(p)} = J^+$.

Assume $f(p) = p$ and that p is attractive, i.e. the two eigenvalues of $f'(p)$, λ_1, λ_2 are both of modulus strictly less than one. Since $p \in K^-$, we can find a disc D through p such that D is not contained in K^-. Hence by the maximum principle $G^-{}_{|D}$ is not harmonic, so if χ is a non-zero positive test function supported near p, then $\int \chi[D] \wedge \mu^- = c \neq 0$. It follows that $d^{-n} f^{n*}(\chi[D]) \to c\mu^+$. Let $\Omega = \{q : \lim f^n(q) = p\}$. It is then clear that $\operatorname{supp} \mu^+ \subset \partial\Omega$. Since on the other hand $\partial\Omega \subset J^+$ we get that $\partial\Omega = J^+$. □

Notes Hénon maps where considered first by Hénon as automorphisms on $\mathbf{R}^2$. Benedicks and Carleson [BC] did significant work for real Hénon mappings, proving the existence of strange attractors.

Following the approach from one complex variable, Hubbard [Hu] introduced the functions G^+ and G^-. He proved that G^+ is pluriharmonic on U^+, and studied the level sets of G^+ in U^+. Theorem 7.1 is in Friedland and Milnor [FM] who discussed the dynamics of Hénon mappings. Corollary 7.3 and Proposition 7.4 are also in [FM].

The idea to study the currents μ^+, μ^- and $\mu = (dd^cG)^2$ is due to Sibony. The first results in this direction were obtained by Bedford and Sibony; some of them appear in section 3 of Bedford and Smillie [BS1]. Theorem 7.10 in $\mathbf{C}^2$ and Theorem 7.11 can be found there.

The idea to extend Hénon mappings to $\mathbf{P}^2$ is in Fornæss and Sibony [FS4] which contains a version of Theorem 7.12 weaker versions of Theorem 7.3.3 appear in [BS2] and [FS4]. The density of stable manifolds in J^+ is due to Bedford and Smillie [BS2].

8 Mixing for holomorphic mappings in $\mathbf{P}^k$ and applications

8.1 Mixing

Let $f : \mathbf{P}^k \to \mathbf{P}^k$ be a holomorphic map in $\mathcal{H}_d$. We have defined a p.s.h. function $G : \mathbf{C}^{k+1} \to \mathbf{R}$, $G \in \mathcal{P}$, which satisfies the functional equation

$$G(f(z)) = dG(z).$$

We have also studied some properties related to the $(1,1)$ positive closed current T defined by the relation $\pi^*T = dd^cG$.

Since the function G is continuous we can define for $\ell \leq k$,

$$T^\ell := T \wedge \cdots \wedge T = (dd^c(G \circ s))^\ell,$$

where s is a holomorphic section of π. We have that $\pi^*T^\ell = (dd^cG)^\ell$. Using the observations of section 5.5 we also have

$$f^*(T^\ell) = d^\ell T^\ell.$$

So if $\ell = k$, T^ℓ can be identified with a measure which we denote by μ. Proposition 5.14 shows that μ is a probability measure. We summarize these observations.

Proposition 8.1 *The measure $\mu := T^k$ is a probability measure which satisfies the functional equation*

$$f^*\mu = d^k\mu.$$

Observe that as a consequence of Bezout's Theorem f is a d^k-to-1 map.

We want to study some dynamical properties of μ. We recall the following definition. Let $(X, \mathcal{B}, m)$ be a probability space and $f : X \to X$ a measure preserving transformation, i.e. for all $A \in \mathcal{B}$, $m(f^{-1}(A)) = m(A)$. Then f is (*strongly*) *mixing* if for all $A, B \in \mathcal{B}$,

$$\lim_{n\to\infty} m(f^{-n}A \cap B) = m(A)m(B).$$

Equivalently

$$\lim_{n\to\infty} \int \psi(f^n)\varphi \, dm = \left(\int \psi \, dm\right)\left(\int \varphi \, dm\right) \tag{8.1}$$

for all $\varphi, \psi \in L^2(m)$.

It is also clear that it is enough to prove (8.1) for ψ, φ in a dense subspace of $L^2(m)$. See Walters [Wa] for background on ergodic theory.

We want to show that if $f \in \mathcal{H}_d$, $f : \mathbf{P}^k \to \mathbf{P}^k$, then f is mixing for the measure μ introduced above.

Theorem 8.2 *The map $f \in \mathcal{H}_d$ is mixing for $\mu = T^k$.*

Proof Let φ, ψ be two smooth non-negative test functions on $\mathbf{P}^k$. Define

$$\lambda_n(a, \varphi) := \frac{(f_*^n \varphi)(a)}{d^{kn}} = \frac{1}{d^{kn}} \sum_i \varphi(f_i^{-n}(a)),$$

where $f_i^{-n}(a)$ are the solutions of $f^n(z) = a$, counted with multiplicity. This means that if z_0 is a multiple root of $f(z) = a$, then z_0 is repeated, according to multiplicity, in the previous sum.

Lemma 8.3 *There exists a constant M such that*

$$\mu(|\lambda_n(a,\varphi) - c| \geq s) \leq \frac{M|\varphi|_2}{sd^n},$$
$$C((|\lambda_n(a,\varphi) - c| \geq s), B) \leq \frac{M|\varphi|_2}{sd^n},$$

where $c := \int \varphi \, d\mu$ and $|\varphi|_2$ denotes the C^2 norm of φ. Here C denotes the Bedford-Taylor capacity introduced in section 5.6.

Proof It is enough to prove the estimate locally in $\mathbf{P}^k$. So we can assume we are in a chart, for example $z_0 \neq 0$. Define

$$K_s = \{a : \ a \in B(0, \frac{1}{2}) \subset \mathbf{C}^k, \lambda_n(a, \varphi) - c \geq s\}.$$

Let u_s be the Siciak extremal function of K_s in $\mathbf{C}^k$ as defined in section 5.6.

We will assume that K_s is non-pluripolar, otherwise $\mu(K_s) = 0$ as follows from the Chern-Levine-Nirenberg inequality in Theorem 5.12.

Let $z = (z_1, \cdots, z_k)$. Define the function

$$v_s(z_0, z) = u_s(\frac{z}{z_0}) + \log|z_0|.$$

This is a p.s.h. function in $\mathcal{P}$.

Let S be a positive closed $(1,1)$ current on $\mathbf{P}^k$ such that $\pi^* S = dd^c v_s$.

As we have seen (Proposition 5.14), since v_s is locally bounded in $\mathbf{C}^{k+1} \backslash \{0\}$, $\nu_s = S^k$ is a probability measure.

The measure ν_s being supported on $\lambda_n(a,\varphi)-c \geq s$, we have

$$\begin{aligned} s &\leq \int (\lambda_n(a,\varphi)-c)d\nu_s \\ &= \int \lambda_n(a,\varphi)d\nu_s - c \\ &= \int \lambda_n(a,\varphi)d\nu_s - \int \lambda_n(a,\varphi)d\mu. \end{aligned}$$

We have used that $\frac{(f^n)^*\mu}{d^{kn}} = \mu$, hence $c = \langle \mu, \varphi\rangle = \langle \mu, \frac{f^n_*\varphi}{d^{kn}}\rangle$. So

$$s \leq \int \lambda_n(a,\varphi)[S^k - T^k] = \int \lambda_n(a,\varphi)[S-T]\wedge \left[\sum_{i=0}^{k-1} S^i \wedge T^{k-1-i}\right],$$

and $s \leq I$, where

$$I := \int \frac{\varphi}{d^n}(f^n)^*[S-T]\wedge \frac{1}{d^{(k-1)n}}\sum_{i=0}^{k-1}(f^n)^*S^i \wedge (f^n)^*(T^{k-1-i}).$$

Here we have used the remarks after Theorem 5.13 about the pull-back of the wedge product. Now we have

$$\pi^*(f^n)^*(S-T) = (f^n)^*\pi^*(S-T) = dd^c[(v_s - G)\circ F^n].$$

The function $v_s - G$ is well-defined on $\mathbf{P}^k$, hence

$$(f^n)^*(S-T) = dd^c[(v_s-G)\circ f^n].$$

Proposition 5.14 shows that the current $(f^n)^*S^i\wedge(f^n)^*(T^{k-1-i})$ has mass equal to $(d^n)^{k-1}$. So

$$I = \int \frac{dd^c\varphi}{d^n}(v_s - G)\circ f^n \wedge \frac{1}{d^{(k-1)n}}\sum_{i=0}^{k-1}(f^n)^*S^i \wedge (f^n)^*(T^{k-1-i}).$$

Hence

$$I \leq \frac{|\varphi|_2}{d^n}\sup|v_s - G|\int \omega \wedge \frac{1}{d^{(k-1)n}}\sum_{i=0}^{k-1}(f^n)^*S^i \wedge (f^n)^*(T^{k-1-i})$$

and

$$I \leq \frac{|\varphi|_2}{d^n}\sup|v_s - G|\cdot k. \tag{8.2}$$

Now we estimate $\sup|v_s - G|$. We know from (5.1) that

$$\mathrm{Log}^+|z| \leq u_s \leq m(s) + \log^+|z|,$$

where $m(s) = \sup_B u_s$, B being the unit ball in $\mathbf{C}^k$. So

$$\begin{aligned} u_s(\frac{z}{z_0}) - G(1,\frac{z}{z_0}) &\leq m(s) - G\left(\frac{1,\frac{z}{z_0}}{|\frac{z}{z_0}|\vee 1}\right) \\ &\leq m(s) + M, \end{aligned}$$

where M is a constant independent of s.

Similarly,

$$G\left(1, \frac{z}{z_0}\right) - u_s\left(\frac{z}{z_0}\right) \leq G\left(\frac{1, \frac{z}{z_0}}{|\frac{z}{z_0}| \vee 1}\right) \leq M,$$

whence $|v_s - G| \leq m(s) + M$.

The estimate (8.2) gives

$$s \leq I \leq \frac{|\varphi|_2}{d^n} k(m(s) + M). \tag{8.3}$$

Using the Alexander-Taylor inequality (Theorem 5.15), we get

$$m(s) + M \leq \frac{A(\frac{1}{2})}{C(K_s, B)} + M \leq \frac{M'}{C(K_s, B)}; \tag{8.4}$$

for the last inequality we have used that $C(K_s, B) \leq C(B(\frac{1}{2}), B)$. Inequalities (8.3) and (8.4) imply that

$$C(K_s, B) \leq \frac{1}{s} |\varphi|_2 \frac{M'}{d^n}.$$

In the chart $z_0 \neq 0$, $\mu = (dd^c G(1, z))^k$. The function $\lambda = \frac{1}{2}\left(\frac{G(1,z)}{\sup_B |G(1,z)|} + 1\right)$ is p.s.h. on B, and satisfies $0 \leq \lambda \leq 1$. Hence by the very definition of $C(K_s, B)$ we have

$$\int_{K_s} (dd^c\lambda)^k \leq C(K_s, B).$$

From the definition of λ, there exists also a constant α independent of s, such that

$$\alpha^{-1}\mu(K_s) \leq \int_{K_s} (dd^c\lambda)^k.$$

Finally,

$$\mu(K_s) \leq \alpha C(K_s, B) \leq \frac{\alpha}{s} |\varphi|_2 \frac{M'}{d^n}.$$

A similar computation with the set $H_s = \{a : \ c - \lambda_n(a, \varphi) \geq s\}$ finishes the proof of the lemma.

End of proof of Theorem 8.2 Define

$$I_n := \langle \mu, \psi(f^n)\varphi \rangle - \langle \mu, \psi \rangle \langle \mu, \varphi \rangle.$$

We have

$$\begin{aligned} \langle \mu, \psi(f^n)\varphi \rangle &= \langle \frac{(f^n)^*\mu}{d^{kn}}, \psi(f^n)\varphi \rangle \\ &= \langle \mu, \frac{(f^n)_*}{d^{kn}} \psi(f^n)\varphi \rangle \\ &= \langle \mu, \psi \frac{(f^n)^*\varphi}{d^{kn}} \rangle. \end{aligned}$$

So

$$I_n = \langle \mu, \psi[\lambda_n(a,\varphi) - c] \rangle.$$

Let $L = 2\sup|\varphi|$. Fix $q > 1$ and let p be such that $\frac{1}{p} + \frac{1}{q} = 1$. Using Hölder's inequality we have

$$\begin{aligned}
|I_n| &\le \left(\int \psi^q d\mu\right)^{1/q} \left(\int |\lambda_n(a,\varphi) - c|^p d\mu\right)^{1/p} \\
&\le ||\psi||_q \left(\int_0^L p\, s^{p-1} \mu(|\lambda_n(a,\varphi) - c| \ge s) ds\right)^{1/p} \\
&\le ||\psi||_q \left(\int_0^L p\, s^{p-2} M \frac{|\varphi|_2}{d^n} ds\right)^{1/p} \\
&\le ||\psi||_q \left(\frac{p}{p-1}\right)^{1/p} (2||\varphi||_\infty)^{\frac{p-1}{p}} |\varphi|_2^{1/p} d^{-n/p} M^{1/p} \\
&\le C_p ||\psi||_q ||\varphi||_\infty^{(1-1/p)} |\varphi_2|^{1/p} d^{-n/p}.
\end{aligned}$$

So $\lim I_n = 0$ and μ is mixing.

Remark 8.4 We have given an estimate of the decay of the coefficient of correlation. Indeed we have shown that if φ is C^2 and ψ is bounded then for any $\epsilon > 0$ there exists C_ϵ such that

$$\left|\int \psi(f^n)\varphi\, d\mu - \left(\int \psi\, d\mu\right)\left(\int \varphi\, d\mu\right)\right| \le C_\epsilon |\varphi|_2 ||\psi||_\infty d^{-n(1-\epsilon)}.$$

8.2 Properties of $J_k := \operatorname{supp}\mu$

In this section we show that the support of μ enjoys some of the properties of the Julia set of rational functions on $\mathbf{P}^1$. In particular, if $\operatorname{supp}\mu$ contains an open set, then it is equal to $\mathbf{P}^k$.

We first recall a few facts from pluri-potential theory. Let E be a Borel set in the unit ball of $\mathbf{C}^k$. We have defined in section 5.6 the capacity of E relative to B

$$C(E,B) = \sup\left\{\int_E (dd^c u)^k \ : \ u \in P(B),\ 0 \le u \le 1\right\}.$$

It is quite clear that if $E = \bigcup_{j\ge1} E_j$, where the E_j's are Borel sets, then

$$C(E,B) \le \sum_{j\ge1} C(E_j, B). \tag{8.4}$$

Proposition 8.5 *Given $f \in \mathcal{H}_d$, let U be an open set intersecting the support of μ. Define $E_U = \mathbf{P}^k \setminus \bigcup_N \left(\bigcap_{n\ge N} f^n(U)\right)$. Then E_U is locally pluripolar in $\mathbf{P}^k$.*

Proof Observe that E_U is the set of points a, for which there exists a sequence $n_i \to \infty$ such that $a \notin f^{n_i}(U)$. Let φ be a positive test function in U such that $c := \langle \mu, \varphi \rangle \neq 0$. As in Theorem 8.2 we can assume that we are in the chart $z_0 \neq 0$, and we want to show that $E'_U := E_U \cap B(0, \frac{1}{2})$ is pluripolar.

From our interpretation of E_U, $a \in E_U$ if and only if $\lambda_{n_i}(a, \varphi) := ((f_*^{n_i}\varphi)(a))/d^{kn_i} = 0$ for a sequence $n_i \to \infty$.

Let $E_n := \{a \in B(0, \frac{1}{2}) : |\lambda_n(a, \varphi) - c| \geq \frac{c}{2}\}$. Hence $E'_U \subset \bigcap_N \bigcup_{n \geq N} E_n$.

Using inequality (8.4) and Lemma 8.3, we get for every N,

$$C(E'_U, B) \leq \sum_{n \geq N} C(E_n, B) \leq \sum_{n \geq N} \frac{2}{c} M |\varphi|_2 d^{-n}.$$

So $C(E'_U, B) = 0$. Consequently the Bedford-Taylor Theorem (see section 5.6) shows that E'_U is pluripolar in B. □

Corollary 8.6 *If* $J_k := \operatorname{supp} \mu$ *contains an open set, then* $J_k = \mathbf{P}^k$.

Proof Since $f^*\mu = d^k\mu$, then if f is injective on U, $\mu(f(U)) = d^k\mu(U)$. Since the critical set of f is a variety, and since μ gives no mass to pluripolar sets, J_k is forward (and backward) invariant.

If J_k contains an open set U it follows from Proposition 8.5 that the complement of J_k is locally pluripolar. But J_k is closed, so $J_k = \mathbf{P}^k$. □

Notes The results of this chapter were proved for holomorphic maps in $\mathbf{P}^2$ in [FS2]. The situation in $\mathbf{P}^1$ was studied for polynomials by Brolin [Br] and for rational maps by Freire-Lopez-Mañe [FLM] and Lyubich [Ly].

References

[AT] Alexander, H., Taylor, B.A., Comparison of two capacities in $\mathbf{C}^n$, *Mat. Z.* **186** (1984), 407–417.

[BC] Benedicks, M., Carleson, L., The dynamics of the Hénon map, *Ann. of Math.* **133** (1991), 73–169.

[Be] Beardon, A., *Iteration of Rational Functions*, Springer-Verlag, Berlin - Heidelberg - New York, 1991.

[Bi] Bieberbach, L., Beispiel zweier ganzer Funktionen zweier komplexer Variablen, welche eine schlicht volumentreue Abbildung des $\mathbf{R}_4$ auf einen Teil seiner selbst vermitteln, *Preuss. Akad. Wiss. Sitzungsber.* **14/15** (1933), 476–479.

[Bie] Bielefeld, B., *Conformal Dynamics Problem List*, Preprint 1, SUNY Stony Brook, 1990.

[Br] Brolin, H., Invariant sets under iteration of rational functions, *Ark. Mat.* **6** (1965), 103–144.

[BS1] Bedford, E., Smillie, J., Polynomial diffeomorphisms of $\mathbf{C}^2$, *Invent. Math.* **87** (1990), 69–99.

[BS2] Bedford, E., Smillie, J., Polynomial diffeomorphisms of $\mathbf{C}^2$. II, *J. Amer. Math. Soc.* **4** (1991), 657–679.

[BT] Bedford, E., Taylor, B.A., A new capacity for p.s.h. functions, *Acta Math.* **149** (1982), 1–39.

[CG] Carleson, L., Gamelin, T., *Complex Dynamics*, Springer-Verlag, Berlin – Heidelberg – New York, 1993.

[Ch] Chirka, E.M., *Complex Analytic Sets*, Kluwer, Dordrecht, 1989.

[CLN] Chern, S.S., Levine, H. and Nirenberg, L., Intrinsic norms on a complex manifold, in: *Global Analysis (Papers in Honor of K. Kodaira)*, Univ. Tokyo Press, 1969; 119–139.

[De1] Demailly, J.-P., Monge-Ampère operators, Lelong numbers, and intersection theory, in: *Complex Analysis and Geometry* (V. Ancona and A. Silva, eds.). Plenum, New York – London, 1993; 115–193.

[De2] Demailly, J.-P., Courants positifs extrémaux et conjecture de Hodge, *Invent. Math.* **69** (1982), 347–374.

[Dev] Devaney, R.L., *An Introduction to Chaotic Dynamical Systems*, Addison–Wesley, 1989.

[Fa] Fatou, P., Sur les équations fonctionnelles, *Bull. Soc. Math. France* **47** (1919), 161–271.

[FLM] Freire, A., Lopez, A. and Mañe, R., An invariant measure for rational maps, *Bol. Soc. Brasil. Mat.* **6** (1983), 45–62.

[FM] Friedland, S., Milnor, J., Dynamical properties of plane polynomial automorphisms, *Ergodic Theory Dynamical Systems* **9** (1989), 67–99.

[FS1] Fornæss, J.E., Sibony, N., Complex Hénon mappings in $\mathbf{C}^2$ and Fatou-Bieberbach domains, *Duke Math. J.* **65** (1992), 345–380.

[FS2] Fornæss, J.E., Sibony, N., Critically finite rational maps on $\mathbf{P}^2$, *Contemp. Math.* **137** (1992), 245–260.

[FS3] Fornæss, J.E., Sibony, N., Complex dynamics in higher dimension I, to appear in *Astérisque*.

[FS4] Fornæss, J.E., Sibony, N., Complex dynamics in higher dimension II, to appear in *Ann. of Math. Studies.*

[FS5] Fornæss, J.E., Sibony, N., Oka's inequality for currents and applications, to appear in *Math. Ann.*

[FS6] Fornæss, J.E., Sibony, N., Classification of recurrent domains for some holomorphic mappings, preprint.

[Ga] Gavosto, E.A., Attracting basins in $\mathbf{P}^2$, preprint.

[Gr1] Green, M., The hyperbolicity of the complement of $2n+1$ hyperplanes in general position in $\mathbf{P}_n$, and related results, *Proc. Amer. Math. Soc.* **66** (1977), 109–113.

[Gr2] Green, M., Some Picard theorems for holomorphic maps to algebraic varieties, *Amer. J. Math.* **97** (1975), 43–75.

[HaP] Harvey, R., Polking, J., Extending analytic objects, *Comm. Pure Appl. Math.* **28** (1975), 701–727.

[Hö] Hörmander, L., *The Analysis of Linear Partial Differential Operators*, vol. I., Springer-Verlag, Berlin – Heidelberg – New York, 1983.

[HOV] Hubbard, J., Oberste-Vorth, W., Hénon mappings in the complex domain I. The global topology of dynamical space, preprint.

[HP] Hubbard, J.H., Papadopol, P., Superattractive fixed points in $\mathbf{C}^n$, preprint.

[Hu] Hubbard, J.H., The Hénon mapping in the complex domain, in: *Chaotic Dynamics and Fractals* (M.F. Barnsley and S.G. Demko, eds.), Academic Press, New York, 1986; 101–111.

[Kl] Klimek, M.K., *Pluripotential Theory*, Oxford Univ. Press, 1991.

[Ko] Kobayashi, S., *Hyperbolic Manifolds and Holomorphic Mappings*, Marcel Dekker, New York, 1970.

[La] Lang, S., *Introduction to Complex Hyperbolic Spaces*, Springer-Verlag, Berlin – Heidelberg – New York, 1987.

[Le] Lelong, P., *Fonctions plurisousharmoniques et formes différentielles positives*, Gordon & Breach, Dunod, Paris, Londres, New York, 1968.

[LG] Lelong, P., Gruman, L., *Entire Functions of Several Complex Variables*, Springer-Verlag, Berlin – Heidelberg – New York, 1986.

[Ly] Lyubich, M., Entropy properties of rational endomorphisms of the Riemann sphere, *Ergodic Theory Dynamical Systems* **3** (1983), 351–385.

[Mi1] Milnor, J., *Dynamics in One Complex Variable: Introductory Lecture*, Institute for Mathematical Sciences, SUNY Stony Brook, 1990.

[Mi2] Milnor, J., Notes on complex dynamics, preprint, SUNY Stony Brook.

[MSS] Mañé, R., Sad P. and Sullivan, D., On the dynamics of rational maps, *Ann. Sci. École Norm. Sup. (4)* **16** (1982), 193–217.

[deRh] de Rham, G., *Variétés différentiables*, Hermann, Paris, 1955.

[RR] Rosay, J.-P., Rudin, W., Holomorphic maps from $\mathbf{C}^n$ to $\mathbf{C}^n$, *Trans. Amer. Math. Soc.* **10** (1988), 47–86.

[Ru] Ruelle, D., *Elements of Differentiable Dynamics and Bifurcation Theory*, Academic Press, New York, 1989.

[Sc1] Schröder, E., Ueber unendlich viele Algorithmen zur Auflösung der Gleichungen, *Math. Ann.* **2** (1870), 317–365.

[Sc2] Schröder, E., Ueber iterierte Functionen, *Math. Ann.* **3** (1871), 296–322.

[Sch] Schwartz, L., *Théorie des distributions*, Hermann, Paris, 1966.

[Si] Sibony, N., Unpublished manuscript, course at UCLA, 1984.

[Siu] Siu, Y.T., Analyticity of sets associated to Lelong numbers and extension of closed positive currents, *Invent. Math.* **27** (1974), 53–156.

[Sk] Skoda, H., Prolongement des courants positifs, fermés, de masse finie, *Invent. Math.* **66** (1982), 361–376.

[T] Tortrat, P., Aspects potentialistes de l'itération des polynômes. in: *Séminaire de Théorie du Potentiel Paris, No. 8*, Lecture Notes in Math. **1235**, Springer-Verlag, Berlin – Heidelberg – New York, 1987; 195–209.

[Ta] Takeuchi, A., Domaines pseudoconvexes infinis et la métrique riemannienne dans un espace projectif, *J. Math. Soc. Japan* **16** (1964), 159–181.

[Ts] Tsuji, M., *Potential Theory in Modern Function Theory*, Mazuren, Tokyo, 1959.

[U] Ueda, T., Fatou set in complex dynamics in projective spaces, preprint.

[Wa] Walters, P., *An Introduction to Ergodic Theory*, Springer-Verlag, Berlin – Heidelberg – New York, 1981.

Analytic functions on Banach spaces

Theodore W. GAMELIN
Mathematics Department
University of California at Los Angeles
Los Angeles, CA 90024
USA

Abstract

In these lectures we discuss various topics related to algebras of analytic functions on Banach spaces. We begin in Chapter 1 with a brief development of function theory on a Banach space, and we prove Ryan's theorem that the Dunford-Pettis property implies the polynomial Dunford-Pettis property. Chapter 2 is devoted to extensions of analytic functions to the bidual. It includes a proof of the Aron-Hervés-Valdivia theorem. Chapter 3 is devoted to approximation by finite-type polynomials, including theorems of Littlewood and Pitt. In Chapter 4 we study the algebra of entire functions that are bounded on bounded sets, and in Chapter 5 we take a brief look at the algebra of bounded analytic functions on the open unit ball of a Banach space.

Contents

P. M. Gauthier (ed.) and G. Sabidussi (techn. ed.), Complex Potential Theory, 187–233.

Introduction

In recent years the study of polynomials on Banach spaces has been quite active. This study finds its roots in the work of M. Frechet and the early Polish school of functional analysts. The fundamental contributions of A. Grothendieck to topological tensor products were decisive, and A. Pelczynski played an important role in the early development of the area. A typical problem is to relate the properties of the space of analytic m-homogeneous analytic functions on a Banach space to properties of the underlying Banach space. A more specific problem is to determine which analytic functions are approximable by polynomials of finite type, that is, polynomials in continuous linear functionals. Natural problems in polynomial approximation that are trivial in the plane become difficult in the setting of Banach spaces.

Let Δ denote the open unit disk $\{|\lambda| < 1\}$ in the complex plane. The ***disk algebra*** $A(\Delta)$ consists of the analytic functions on Δ that extend continuously to the closed disk $\bar{\Delta}$. Any such function can be approximated uniformly on the disk by (analytic) polynomials in λ. There are two standard approximation procedures. One is to approximate the dilated function $f_r(\lambda) = f(r\lambda)$ by its power series expansion, and then let r increase to 1. The other is to approximate f by the Cesàro means of the partial sums of its power series. Either way we see that the polynomials are dense in $A(\Delta)$, in the topology of uniform convergence. The same idea works in finite dimensions. What happens in an infinite-dimensional setting?

Consider a dual Banach space $\mathcal{Z}$. Let U denote the open unit ball of $\mathcal{Z}$, and define $A(U)$ to be the algebra of analytic functions on U that extend to be weak-star continuous on the closed unit ball $\bar{U}$. Then $A(U)$ is a uniform algebra on the weak-star compact space $\bar{U}$, that is, $A(U)$ is a unital point-separating closed subalgebra of $C(\bar{U})$, with the usual supremum norm. The role of the polynomials is played by the weak-star continuous finite-type polynomials, that is, the algebra generated by the weak-star continuous linear functionals on $\mathcal{Z}$. We can ask some very basic questions about approximation. Are these polynomials dense in $A(U)$? Can any bounded analytic function on U be approximated pointwise by a net of polynomials uniformly bounded on U? We will see that the answer to both these questions is "no" in general. Thus we have the problem of characterizing those Banach spaces for which approximation is always possible, and here there are no completely satisfactory answers. Chapter 1 includes background material on analytic functions on Banach spaces and their Taylor expansions. In Chapter 2 we study the natural spaces of analytic functions that arise in this setting. The approximation problem is treated in Chapter 3.

The *spectrum* of a uniform algebra consists of the nonzero complex-valued homomorphisms of the algebra. Any such homomorphism φ is continuous and satisfies $\varphi(1) = 1 = ||\varphi||$, so that the spectrum is a subset of the unit sphere of the dual space of the algebra. The spectrum inherits the weak-star topology from the dual, and equipped with this topology it is compact.

The spectrum of a uniform algebra is an abstract entity that usually does not give much hard information on any given concrete algebra under study. Nevertheless it has played an important role in certain problems in analysis. Gelfand's striking proof of Wiener's theorem on absolutely convergent Fourier series drew a lot of attention to Banach algebras when they were introduced. Gelfand's proof depended on the correspondence between maximal ideals and points of the spectrum. Wermer's theorem on approximation of functions on curves in $\mathbf{C}^n$ depended on the spectrum of the algebras at hand. One of the key technical tools in Wermer's work involved the introduction of analytic structure in the spectrum, and this aspect was developed successfully especially by E. Bishop. The corona problem arose as a problem about the spectrum of $H^\infty(\Delta)$, the algebra of bounded analytic functions on Δ. While the answer to the corona question has not had any far-reaching consequences by itself, Carleson's solution to the corona problem did provide a number of powerful new ideas and techniques, which have had a profound effect on complex analysis.

In the case of the disk algebra $A(\Delta)$, each homomorphism φ is determined by its value on the generating function λ. The correspondence $\varphi \to \varphi(\lambda)$ maps the spectrum homeomorphically onto the closed unit disk $\bar{\Delta}$, so that each such φ is the evaluation homomorphism at the point $\varphi(\lambda) \in \bar{\Delta}$. There is a similar theorem, due to Arens, for the algebra $A(D)$ associated with an arbitrary bounded open subset of the complex plane, even though the polynomials are not dense. Now the argument above shows that the spectrum of the uniform algebra on $\bar{U}$ generated by the weak-star continuous linear functionals coincides with the closed unit ball $\bar{U}$, in the sense that each homomorphism is the evaluation homomorphism at some point of $\bar{U}$. Meanwhile it remains an open problem to determine the spectrum of $A(U)$. We will discuss the spectra of various algebras of entire functions in Chapter 4, and we will focus in Chapter 5 on the spectrum of the algebra $H^\infty(B)$ of bounded analytic functions on the open unit ball of a Banach space.

Our notation is as follows. We will denote by $\mathfrak{X}$ a complex Banach space, and by B the open unit ball of $\mathfrak{X}$. The open ball centered at x_0 of radius r is denoted by $B(x_0, r)$, and $B_r = B(0, r)$ is the dilate of B by r. The open unit ball of the bidual $\mathfrak{X}^{**}$ of $\mathfrak{X}$ is denoted by B^{**}, and the closed unit balls of $\mathfrak{X}$ and $\mathfrak{X}^{**}$ are denoted respectively by $\bar{B}$ and $\bar{B}^{**}$. We use the notation ℓ^p for the classical Banach space of p-summable sequences, and c_0 for the null sequences. We denote by $H^\infty(D)$ the algebra of bounded analytic functions on D, with the usual supremum norm.

Each of the five chapters is an amplified version of the corresponding lecture given at the NATO summer school in Montreal in the summer of 1993. I would like to thank the organizers of the summer school, and in particular Paul Gauthier, for their abundant hospitality, and for all the hard work that went into making the project a success.

Chapter 1
Analytic functions and Taylor series

This lecture is devoted to developing basic background material on analytic functions defined on a domain in a Banach space. Analytic functions have Taylor series expansions, and in this setting the Cauchy-Hadamard formula expresses the radius of bounded convergence of the series. The Taylor coefficients are restrictions to the diagonal of multilinear functionals. The space of m-linear functionals is representable as the dual of an m-fold projective tensor product.

1.1 Analytic functions

A complex-valued function on an open subset of a Banach space $\mathfrak{X}$ is *analytic* if it is locally bounded and its restriction to every complex one-dimensional affine subspace of $\mathfrak{X}$ is analytic. In other words, f is analytic on D if f is locally bounded, and if for every $x_0 \in D$ and direction $x \in \mathfrak{X}$, the function $\lambda \to f(x_0 + \lambda x)$ depends analytically on λ for λ near 0. From the corresponding facts in one complex variable it follows that sums, products, and uniform limits of analytic functions are analytic.

A locally bounded function f on D is analytic if and only if the restriction of f to the intersection of D and any finite-dimensional subspace of $\mathcal{X}$ is analytic. Thus statements about analytic functions that involve only a finite number of points of $\mathcal{X}$ will hold in general once they hold for analytic functions of several complex variables.

Let $\mathcal{P}(^m\mathcal{X})$ denote the space of analytic functions on $\mathcal{X}$ that are m-homogeneous, that is, that satisfy $f(\lambda x) = \lambda^m f(x)$, $x \in \mathcal{X}$. For example, the formula $f(x) = \sum x_j^m$ defines an m-homogeneous analytic function on ℓ^p whenever $m \geq p$. We endow $\mathcal{P}(^m\mathcal{X})$ with the supremum norm over the unit ball B of $\mathcal{X}$,

$$||f|| = \sup\{|f(x)| : x \in B\}, \qquad f \in \mathcal{P}(^m\mathcal{X}),$$

and then $\mathcal{P}(^m\mathcal{X})$ becomes a Banach space.

Let f be analytic on D, and let $x_0 \in D$. Suppose $|f| \leq C$ on the ball $B(x_0, r)$. For fixed $y \in \mathcal{X}$, the function $\zeta \to f(x_0 + \zeta y)$ has a Taylor series expansion

$$f(x_0 + \zeta y) = \sum_{m=0}^{\infty} A_m(y)\zeta^m,$$

which converges for $|\zeta| < r/||y||$. One checks easily that each $A_m(y)$ is m-homogeneous. Since $A_m(y)$ depends analytically on y in any finite-dimensional subspace of $\mathcal{X}$, and since it is locally bounded, it is analytic, hence in $\mathcal{P}(^m\mathcal{X})$. From the Cauchy estimates we have $|A_m(y)| \leq C||y||^m/r^m$, and consequently $||A_m|| \leq C/r^m$. If we set $\zeta = 1$ and $x = x_0 + y$, we have the following.

Theorem 1.1.1 *If f is analytic on D and $x_0 \in D$, then the Taylor series expansion*

$$f(x) = \sum_{m=0}^{\infty} A_m(x - x_0)$$

of f at x_0 converges uniformly in a neighborhood of x_0. The summands $A_m(y)$ are m-homogeneous analytic functions on $\mathcal{X}$, that is, $A_m \in \mathcal{P}(^m\mathcal{X})$. If $|f| \leq C$ on $B(x_0, r)$, then the A_m's satisfy the Cauchy estimates

$$||A_m|| \leq C/r^m, \qquad m \geq 0. \tag{1}$$

Now consider a Taylor series $\sum A_m(x)$ centered at 0. We define its *radius of bounded convergence* to be the supremum of all $r \geq 0$ such that $\sum A_m(x)$ extends to be a bounded analytic function on the open ball $B(0, r)$. In the case at hand, the Cauchy-Hadamard formula for the radius of convergence of a power series expresses the radius of bounded convergence.

Theorem 1.1.2 *The radius of bounded convergence of a Taylor series $\sum A_m(x)$ is given by*

$$R = 1/\limsup ||A_m||^{1/m}.$$

Proof Define R by the formula above. The Cauchy estimates show that if $f(x) = \sum A_m(x)$ is bounded for $||x|| < r$, then $\limsup ||A_m||^{1/m} \leq 1/r$, so that $r \leq R$. Thus the radius of bounded convergence is at most R. Let $r < s < R$. The defining formula for R shows that $||A_m||^{1/m} \leq 1/s$ for m large. Hence $r^m||A_m|| \leq r^m/s^m$ for m large, and consequently $\sum A_m(x)$ converges uniformly for $||x|| < r$, by the Weierstrass M-test, to a bounded analytic function. Thus the radius of bounded convergence coincides with R. □

There is a striking difference between analytic functions on an infinite-dimensional Banach space and those on $\mathbf{C}^n$. In a Banach space even an entire function can have a Taylor series with a finite radius of bounded convergence. As an example, consider the m-homogeneous polynomial on c_0 defined by $A_m(t_1, t_2, \ldots) = t_m^m$. Evidently $||A_m|| = 1$, so the analytic function

$$f(t_1, t_2, \ldots) = \sum_{m=1}^{\infty} t_m^m, \qquad (t_1, t_2, \ldots) \in c_0,$$

has radius of bounded convergence $R = 1$. However, this series converges pointwise on the entire Banach space c_0, and the convergence is uniform on any ball of radius strictly less than 1. Thus f is an entire analytic function on c_0. The power series extends to an entire function on ℓ^p for $1 \leq p < \infty$, though not to an entire function on ℓ^∞.

We mention in passing a problem, which is to find an analog of the theorem that the radius of convergence of a power series in one complex variable is the distance to the nearest singularity. Can the radius of bounded convergence of a Taylor series be explained in terms of the nonextendability of the analytic function to a ball of larger radius in some natural ambient space?

Our analytic functions will usually be complex-valued. Occasionally it will be useful to consider functions with values in a normed space. We define such a function to be *analytic* if it is locally bounded and its composition with every continuous linear functional is an analytic complex-valued function. The discussion above goes through for analytic functions with values in another Banach space $\mathcal{Y}$, with absolute values replaced by the norm of $\mathcal{Y}$ where appropriate. In this case, the Taylor coefficient A_m is an m-homogeneous analytic function with values in $\mathcal{Y}$. Thus the first Taylor coefficient in the expansion at x_0 of an analytic function f from $\mathcal{X}$ to $\mathcal{Y}$ is a linear operator from $\mathcal{X}$ to $\mathcal{Y}$. It coincides with the usual Frechet derivative $f'(x_0)$ of f at x_0, whose defining property is that $f(x) = f(x_0) + f'(x_0)(x - x_0) + o(||x - x_0||)$ as $x \to x_0$.

1.2 Multilinear functionals

A complex-valued m-linear functional F on the Banach space $\mathcal{X}$ is continuous if and only if F is bounded on B^m. The space of continuous m-linear functionals forms a Banach space, with the norm

$$||F|| = \sup\{|F(x_1, \ldots, x_m)| : ||x_1|| \leq 1, \ldots, ||x_m|| \leq 1\}.$$

We will denote this Banach space by $\mathcal{L}(^m\mathcal{X})$. In the case $m = 1$ we have the dual space $\mathcal{X}^*$ of $\mathcal{X}$.

The symmetric m-linear functionals form a closed subspace of $\mathcal{L}(^m\mathcal{X})$, which will be denoted by $\mathcal{L}_s(^m\mathcal{X})$. There is a natural symmetrization projection from $\mathcal{L}(^m\mathcal{X})$ onto $\mathcal{L}_s(^m\mathcal{X})$, obtained by permuting the variables and averaging,

$$F_s(x_1,\dots,x_m) = \frac{1}{m!}\sum_\pi F(x_{\pi(1)},\dots,x_{\pi(m)}).$$

Evidently $||F_s|| \leq ||F||$.

We define a *continuous m-homogeneous polynomial* f on $\mathcal{X}$ to be the restriction to the diagonal of a continuous m-linear functional F,

$$f(x) = F(x,\dots,x), \qquad x \in \mathcal{X}.$$

Evidently such an f is an m-homogeneous analytic function, and its norm in $\mathcal{P}(^m\mathcal{X})$ is estimated by $||f|| \leq ||F||$. If F is symmetric, we can recover it from its restriction to the diagonal by the polarization formula,

$$F(x_1,\dots,x_m) = \frac{1}{m!2^m}\sum \varepsilon_1\cdots\varepsilon_m f(\varepsilon_1 x_1 + \cdots + \varepsilon_m x_m),$$

where the summation is extended over the 2^m independent choices of $\varepsilon_j = \pm 1$. In particular, a symmetric m-linear functional on $\mathcal{X}$ is uniquely determined by its restriction to the diagonal. If F is arbitrary, the polarization formula gives the symmetrization F_s of F.

Theorem 1.2.1 *Every continuous m-homogeneous polynomial f on $\mathcal{X}$ is the restriction to the diagonal of a unique continuous symmetric m-linear functional F on $\mathcal{X}$, and the norms are equivalent,*

$$||f|| \leq ||F|| \leq \frac{m^m}{m!}||f||.$$

For $\mathcal{X} = \ell^1$, this estimate is sharp.

Proof The estimate is obtained by making the obvious estimate in the polarization formula, with $||\varepsilon_1 x_1 + \cdots + \varepsilon_m x_m|| \leq m$. To see that the upper estimate is sharp in ℓ^1, consider the m-linear functional

$$G(x_1,\dots,x_m) = x_{11}\cdots x_{mm}, \qquad x_j = (x_{j1},\dots,x_{jm}) \in \ell^1, \quad 1 \leq j \leq m,$$

and let f be the restriction of G to the diagonal. In this case, the polarization formula expresses the symmetrization F of G. It shows that $F(e_1,\dots,e_m) = 1/m!$, where $e_1, e_2, \dots$ is the standard basis for ℓ^1. On the other hand, we calculate using calculus that the maximum of the product $t_1\cdots t_m$ subject to the conditions $t_j \geq 0$, $\sum t_j = 1$, is $1/m^m$, and this implies that $||f|| = 1/m^m$. Thus the estimate is sharp. □

For most Banach spaces, the constant $m^m/m!$ can be improved upon. The best constant for a fixed Banach space $\mathcal{X}$ is called the *polarization constant* of $\mathcal{X}$. In the case of Hilbert space it can be shown that $||f|| = ||F||$, so that the polarization constant of Hilbert space is 1.

Theorem 1.2.2 *Let $m \geq 0$. Any m-homogeneous analytic function $f \in \mathcal{P}(^m\mathcal{X})$ is an m-homogeneous polynomial, that is, f is the restriction to the diagonal of a unique continuous symmetric m-linear functional $F \in \mathcal{L}_s(^m\mathcal{X})$.*

Proof On account of the uniqueness of symmetric extensions, it suffices to check this on finite-dimensional subspaces. On an N-dimensional subspace with basis $e_1, \ldots, e_N$, such an f is a finite linear combination of functions of the form $g(\lambda_1 e_1 + \cdots + \lambda_N e_N) = \lambda_1^{m_1} \cdots \lambda_N^{m_N}$, where $m_1 + \cdots + m_N = m$. Such a g is the restriction to the diagonal of the product G of m coordinate functionals,

$$G(x_1, \ldots, x_m) = x_{11} \cdots x_{m_1 1} x_{m_1+1,2} \cdots x_{mN},$$

where the jth coordinate of exactly m_j of the variables $x_1, \ldots, x_m$ is inserted into the product. Evidently G is m-linear, and the symmetrization of G is the unique symmetric extension of g. □

1.3 Projective tensor products

We wish to represent the space of symmetric m-linear functionals on $\mathcal{X}$ as the dual of a tensor product space. To simplify the discussion, we discuss first the projective tensor product of two fixed Banach spaces $\mathcal{X}$ and $\mathcal{Y}$.

The algebraic tensor product $\mathcal{X} \otimes \mathcal{Y}$ consists of finite sums $w = \sum x_j \otimes y_j$, with the usual rules for algebraic manipulation. The dual of $\mathcal{X} \otimes \mathcal{Y}$ is the space of all bilinear functionals on $\mathcal{X} \times \mathcal{Y}$, where the linear functional $\tilde{F}$ corresponding to the bilinear functional F is given by $\tilde{F}(\sum x_j \otimes y_j) = \sum F(x_j, y_j)$.

There are a number of useful norms that can be introduced into $\mathcal{X} \otimes \mathcal{Y}$. The *projective tensor product norm* is given by

$$||w|| = \inf \sum ||x_j|| \, ||y_j||,$$

where the infimum is taken over all representations of w as a finite sum $\sum x_j \otimes y_j$. This is evidently the largest norm with the property that $||x \otimes y|| \leq ||x|| \, ||y||$ for all $x \in \mathcal{X}$, $y \in \mathcal{Y}$. The completion of $\mathcal{X} \otimes \mathcal{Y}$ with respect to this norm is the *projective tensor product* of $\mathcal{X}$ and $\mathcal{Y}$, denoted by $\mathcal{X} \hat{\otimes} \mathcal{Y}$.

As an example, if $\mathcal{X} = L^1(\mu)$, then $\mathcal{X} \hat{\otimes} \mathcal{Y}$ can be identified with the space $L^1(\mu, \mathcal{Y})$ of integrable $\mathcal{Y}$-valued functions on the measure space. If further $\mathcal{Y} = L^1(\nu)$, then $\mathcal{X} \hat{\otimes} \mathcal{Y}$ is isometric to $L^1(\mu \times \nu)$.

Lemma 1.3.1 *The projective tensor product norm satisfies $||x \otimes y|| = ||x|| \, ||y||$. The dual of $\mathcal{X} \hat{\otimes} \mathcal{Y}$ is isometrically isomorphic to the space of continuous bilinear functionals on $\mathcal{X} \times \mathcal{Y}$.*

Proof Let F be a continuous bilinear functional on $\mathcal{X} \times \mathcal{Y}$, with corresponding linear functional $\tilde{F}$ on $\mathcal{X} \otimes \mathcal{Y}$. If $w = \sum x_j \otimes y_j$, then $|\tilde{F}(w)| = |\sum F(x_j, y_j)| \leq ||F|| \sum ||x_j|| \, ||y_j||$. Taking an infimum, we obtain $||\tilde{F}|| \leq ||F||$. Consider the bilinear functional $F(u, v) =$

$L(u)M(v)$, where L and M are functionals of unit norm on $\mathcal{X}$ and $\mathcal{Y}$ respectively that satisfy $L(x) = ||x||$ and $M(y) = ||y||$. Then $||\tilde{F}|| \leq ||F|| = 1$, while $\tilde{F}(x \otimes y) = ||x|| \, ||y||$. Thus $||x \otimes y|| = ||x|| \, ||y||$. Finally, taking x and y of unit norm such that $F(x, y) \approx ||F||$, we obtain $||x \otimes y|| = 1$ while $\tilde{F}(x \otimes y) \approx ||F||$. Hence $||\tilde{F}|| = ||F||$. □

In a similar manner we can define the projective tensor product $\mathcal{X}_1 \hat{\otimes} \cdots \hat{\otimes} \mathcal{X}_m$ and show that the dual is the space of continuous m-linear functionals on $\mathcal{X}_1 \times \cdots \times \mathcal{X}_m$.

We can also consider multilinear functions on $\mathcal{X}_1 \times \cdots \times \mathcal{X}_m$, with values in a Banach space $\mathcal{Y}$. The space of continuous multilinear functions from $\mathcal{X}_1 \times \cdots \times \mathcal{X}_m$ to $\mathcal{Y}$ will be denoted by $\mathcal{L}(\mathcal{X}_1, \ldots, \mathcal{X}_m; \mathcal{Y})$. It becomes a Banach space, with the usual supremum norm over the product of the unit balls. With this notation, the space of bounded linear operators from $\mathcal{X}$ to $\mathcal{Y}$ is denoted by $\mathcal{L}(\mathcal{X}; \mathcal{Y})$. The discussion of projective tensor products applies to vector-valued multilinear functions, and we obtain the following.

Theorem 1.3.2 *There is a natural isometric isomorphism of the space of continuous multilinear functions from $\mathcal{X}_1 \times \cdots \times \mathcal{X}_m$ to $\mathcal{Y}$ and the space of continuous linear operators from the projective tensor product $\mathcal{X}_1 \hat{\otimes} \cdots \hat{\otimes} \mathcal{X}_m$ to $\mathcal{Y}$,*

$$\mathcal{L}(\mathcal{X}_1, \ldots, \mathcal{X}_m; \mathcal{Y}) \cong \mathcal{L}(\mathcal{X}_1 \hat{\otimes} \cdots \hat{\otimes} \mathcal{X}_m; \mathcal{Y}).$$

The m-fold projective tensor product of $\mathcal{X}$ with itself is denoted by $\hat{\otimes}^m \mathcal{X}$. Its dual is isometric to the continuous m-linear functionals on $\mathcal{X}$,

$$(\hat{\otimes}^m \mathcal{X})^* \cong \mathcal{L}(^m \mathcal{X}).$$

A reordering of the coordinates by a permutation π induces a linear operator on $\otimes^m \mathcal{X}$, which is an isometry with respect to the projective tensor product norm. The tensors that are fixed by these operators, for all permutations π, are called *symmetric tensors*. The symmetric tensors form a closed subspace of $\hat{\otimes}^m \mathcal{X}$, which will be denoted by $\hat{\otimes}_s^m \mathcal{X}$. There is a norm-one projection of $\hat{\otimes}^m \mathcal{X}$ onto $\hat{\otimes}_s^m \mathcal{X}$, which is determined by the symmetrization operator

$$x_1 \otimes \cdots \otimes x_m \to \frac{1}{m!} \sum x_{\pi(1)} \otimes \cdots \otimes x_{\pi(m)},$$

summed over all permutations π. The dual of this projection operator, regarded as an operator on m-linear functionals, is the symmetrization projection $F \to F_s$ introduced above. Now if P is a norm-one projection on a Banach space, then P^* is a norm-one projection on the dual space, and the image of P^* is isometric to the dual of the image of P. Thus we obtain the following.

Theorem 1.3.3 *The dual of the space $\hat{\otimes}_s^m \mathcal{X}$ of symmetric tensors is isometric to the space $\mathcal{L}_s(^m \mathcal{X})$ of symmetric m-linear functionals on $\mathcal{X}$.*

We mention in passing how to manufacture various other tensor product norms. Let V be any linear subspace of continuous bilinear functionals on $\mathcal{X} \times \mathcal{Y}$ containing the functionals $x \otimes y \to L(x)M(y)$, $L \in \mathcal{X}^*$, $M \in \mathcal{Y}^*$. Then

$$||w||_V = \sup\{|\Lambda(w)| : \Lambda \in V, ||\Lambda|| \leq 1\}$$

is a norm on $\mathcal{X}\otimes\mathcal{Y}$ for which $\|x\otimes y\| = \|x\|\,\|y\|$. The corresponding dual space is isometric to a closed subspace of the continuous bilinear functionals containing V. If we take for V the smallest possible subspace, namely the finite linear combinations of the bilinear functionals $L(x)M(y)$, we obtain the ***injective tensor product norm*** on $\mathcal{X}\otimes\mathcal{Y}$. This can be defined on $w = \sum x_j \otimes y_j$ by

$$\|w\| = \sup\left\{\left|\sum L(x_j)M(y_j)\right| : L\in\mathcal{X}^*, M\in\mathcal{Y}\right\},$$

and interpreted as the norm of the operator $T_w : \mathcal{X}^* \to \mathcal{Y}$ defined by $T_w(L) = \sum L(x_j)y_j$. Thus the injective tensor product $\mathcal{X}\overset{\vee}{\otimes}\mathcal{Y}$ can be interpreted as the closure in $\mathcal{L}(\mathcal{X}^*;\mathcal{Y})$ of the finite-rank operators that are weak-star continuous.

There are other useful norms on $\mathcal{X}\otimes\mathcal{Y}$ that satisfy $\|x\otimes y\| = \|x\|\,\|y\|$, which lie between the projective and the injective tensor product norms, such as the nuclear norm and the integral norm. For each of these various tensor product norms, we obtain a linear space of polynomials of the corresponding type, generated by the m-homogeneous polynomials for which the corresponding symmetric functionals are continuous with respect to the tensor product norm.

1.4 The predual of $\mathcal{P}(^m\mathcal{X})$

Since $\mathcal{P}(^1\mathcal{X}) \cong \mathcal{X}^*$, in particular $\mathcal{P}(^1\mathcal{X})$ is a dual Banach space. It turns out that each of the Banach spaces $\mathcal{P}(^m\mathcal{X})$ is a dual space. The easiest way to see this is to regard $\mathcal{P}(^m\mathcal{X})$ as a subspace of $\ell^\infty(B)$, the space of all bounded functions on the unit ball B of $\mathcal{X}$, which is the dual of $\ell^1(B)$.

Lemma 1.4.1 *The Banach space $\mathcal{P}(^m\mathcal{X})$ of m-homogeneous polynomials is a weak-star closed subspace of $\ell^\infty(B)$, consequently the dual space of $\ell^1(B)/\mathcal{P}(^m\mathcal{X})^\perp$. A bounded net converges weak-star in $\mathcal{P}(^m\mathcal{X})$ if and only if it converges pointwise on $\mathcal{X}$.*

Proof By the Krein-Schmulian theorem, it suffices to show that the limit of any bounded net in $\mathcal{P}(^m\mathcal{X})$ that converges pointwise on B again belongs to $\mathcal{P}(^m\mathcal{X})$. This latter statement is easy to check, using the fact that any bounded family of analytic functions is normal on any finite-dimensional subspace. □

Theorem 1.4.2 *The natural map $(\hat{\otimes}_s^m\mathcal{X})^* \to \mathcal{P}(^m\mathcal{X})$, which restricts a symmetric m-linear functional to the diagonal, is the dual of an isomorphism of $\ell^1(B)/\mathcal{P}(^m\mathcal{X})^\perp$ onto $\hat{\otimes}_s^m\mathcal{X}$, under which the coset of $\sum a_j\delta_{x_j}$ is mapped to $\sum a_j x_j\otimes\cdots\otimes x_j$.*

Proof The rule $\sum a_j\delta_{x_j} \to \sum a_j x_j\otimes\cdots\otimes x_j$ defines a continuous linear operator from $\ell^1(B)$ into $\hat{\otimes}_s^m\mathcal{X}$, whose dual is the restriction operator of the theorem. Since the dual operator is a (Banach space) isomorphism onto $\mathcal{P}(^m\mathcal{X})$, the predual has nullspace $\mathcal{P}(^m\mathcal{X})^\perp$, and furthermore the quotient map on $\ell^1(B)/\mathcal{P}(^m\mathcal{X})^\perp$ is a (Banach space) isomorphism. □

This shows in particular that every element of the projective tensor product has a representation of the form $\sum a_j x_j \otimes \cdots \otimes x_j$, where the a_j's are summable and the x_j's are bounded. It is easy to readjust this representation so that in fact $x_j \to 0$.

A host of natural Banach space problems now arises. What are the dual and bidual of $\mathcal{P}(^m\mathcal{X})$? When is $\mathcal{P}(^m\mathcal{X})$ reflexive? We will return briefly to some of these questions at the end of Chapter 3.

1.5 The Dunford-Pettis property

An operator from $\mathcal{X}$ to $\mathcal{Y}$ is *completely continuous* if it takes weakly convergent sequences to norm convergent sequences. We say that $\mathcal{X}$ has the *Dunford-Pettis property* if every weakly compact operator from $\mathcal{X}$ to another Banach space is completely continuous. It is straightforward to check that $\mathcal{X}$ has the Dunford-Pettis property if and only if whenever $\{x_n\}$ is a sequence in $\mathcal{X}$ converging weakly to 0, and $\{L_n\}$ is a sequence in $\mathcal{X}^*$ converging weakly to 0 in $\mathcal{X}^*$, then $L_n(x_n) \to 0$. From this characterization it is clear that if a dual Banach space has the Dunford-Pettis property, then so does its predual. A reflexive Banach space has the Dunford-Pettis property if and only if it is finite-dimensional.

From the characterization of the weakly compact subsets of $L^1(\mu)$ in terms of uniform integrability, it is easy to deduce that $L^1(\mu)$ has the Dunford-Pettis property, as does the space $M(X)$ of finite regular Borel measures on a compact space X. Thus $C(X)$ also has the Dunford-Pettis property.

An m-linear function F from $\mathcal{X}$ to $\mathcal{Y}$ is *weakly compact* if the image under F of the m-fold product of the unit ball B^m is weakly precompact in $\mathcal{Y}$. Since the unit ball of $\widehat{\otimes}^m\mathcal{X}$ is the closed convex hull of the tensors corresponding to elements of B^m, and since the closed convex hull af a weakly compact set is weakly compact, the function F is weakly compact if and only if the associated linear operator $\tilde{F}$ from $\widehat{\otimes}^m\mathcal{X}$ to $\mathcal{Y}$ is weakly compact.

Theorem 1.5.1 *Suppose that $\mathcal{X}$ has the Dunford-Pettis property. Let F be a weakly compact m-linear function on $\mathcal{X}$ with values in a Banach space $\mathcal{Y}$. If $\{x_k^{(j)}\}$ is a weak Cauchy sequence in $\mathcal{X}$ for $1 \le k \le m$, then $F(x_1^{(j)}, \ldots, x_m^{(j)})$ converges in norm. If further $x_k^{(j)}$ converges weakly to 0 for some index k, then $F(x_1^{(j)}, \ldots, x_m^{(j)})$ converges to 0 in norm.*

Proof The first statement follows from the second, by applying it with $x_k^{(j)}$ replaced by $x_k^{(i_j)} - x_k^{(j)}$ for subsequences i_j of the integers. Thus we can assume that $x_m^{(j)}$ tends weakly to 0 as $j \to \infty$. The case $m = 1$ is the Dunford-Pettis property for $\mathcal{X}$, so we can assume that $m \ge 2$, and we make the induction hypothesis that the theorem is true for $(m-1)$-linear functions.

Let $c_0(\mathcal{Y})$ denote the Banach space of sequences in $\mathcal{Y}$ that converge to 0 in norm. Define $G(x_1, \ldots, x_{m-1})$ to be the sequence whose kth component is $F(x_1, \ldots, x_{m-1}, x_m^{(k)})$. On account of the Dunford-Pettis property, this sequence belongs to $c_0(\mathcal{Y})$, and G is a continuous $(m-1)$-linear function on $\mathcal{X}$ with values in $c_0(\mathcal{Y})$. Let $\tilde{G} : \widehat{\otimes}^{m-1}\mathcal{X} \to c_0(\mathcal{Y})$ be the corresponding linear operator. We claim that $\tilde{G}$ is weakly compact. Once this is established, the proof is concluded rapidly, as follows. The induction hypothesis shows that

$\widetilde{G}(x_1^{(j)},\ldots,x_{m-1}^{(j)})$ converges in $c_0(\mathcal{Y})$ as $j\to\infty$, say to $\{y_k\}_{k=1}^{\infty}\in c_0(\mathcal{Y})$. In other words, $F(x_1^{(j)},\ldots,x_{m-1}^{(j)},x_m^{(k)})-y_k$ tends to 0 as $j\to\infty$, uniformly in k. Setting $k=j$ and noting that $y_j\to 0$, we obtain $F(x_1^{(j)},\ldots,x_m^{(j)})\to 0$ as $j\to\infty$, which was to be proved.

Now an operator from a Banach space $\mathcal{X}_1$ to another $\mathcal{X}_2$ is weakly compact if and only if the range of the second adjoint operator from $\mathcal{X}_1^{**}$ to $\mathcal{X}_2^{**}$ is contained in (the canonically embedded image of) $\mathcal{X}_2$. In the case at hand, the bidual of $c_0(\mathcal{Y})$ is $\ell^\infty(\mathcal{Y}^{**})$, and we must show that the range of $\widetilde{G}^{**}:\mathcal{L}(^{m-1}\mathcal{X})^*\to\ell^\infty(\mathcal{Y}^{**})$ is contained in $c_0(\mathcal{Y})$.

Fix $\varphi\in\mathcal{L}(^{m-1}\mathcal{X})^*$. Define $\Phi:\mathcal{X}\to\mathcal{L}(^m\mathcal{X})^*$ by

$$\langle\Phi(x),H\rangle=\langle\varphi,H_x\rangle,\qquad H\in\mathcal{L}(^m\mathcal{X}),$$

where $H_x(x_1,\ldots,x_{m-1})=H(x_1,\ldots,x_{m-1},x)$. Let $\alpha\in\mathcal{Y}^*$. Define $\alpha_k\in\ell^1(\mathcal{Y}^*)$ to have α in the kth place and 0 elsewhere. The operator $\widetilde{G}^*:\ell^1(\mathcal{Y}^*)\to\mathcal{L}(^{m-1}\mathcal{X})$ is determined by

$$(\widetilde{G}^*\alpha_k)(x_1,\ldots,x_{m-1})=\alpha_k(\widetilde{G}(x_1,\otimes\cdots\otimes x_{m-1}))=\alpha(F(x_1,\ldots,x_{m-1},x_m^{(k)})).$$

Now $\widetilde{F}:\widehat{\otimes}^m\mathcal{X}\to\mathcal{Y}$ has dual $\widetilde{F}^*:\mathcal{Y}^*\to\mathcal{L}(^m\mathcal{X})$ given by

$$\widetilde{F}^*(\alpha)(x_1,\ldots,x_m)=\alpha(\widetilde{F}(x_1\otimes\cdots\otimes x_m))=\alpha(F(x_1,\ldots,x_m)).$$

The definition of Φ then yields $\langle\varphi,\widetilde{G}^*\alpha_k\rangle=\langle\Phi(x_m^{(k)}),\widetilde{F}^*(\alpha)\rangle$. Thus for all $\alpha\in\mathcal{Y}^*$ we have $\langle\widetilde{G}^{**}\varphi,\alpha_k\rangle=\langle\widetilde{F}^{**}(\Phi(x_m^{(k)})),\alpha\rangle$, hence the kth component of $\widetilde{G}^{**}\varphi$ coincides with $\widetilde{F}^{**}(\Phi(x_m^{(k)}))$. Since $\widetilde{F}$ is weakly compact, the image of $\widetilde{F}^{**}$ is contained in $\mathcal{Y}$, and consequently $\widetilde{G}^{**}\varphi\in\ell^\infty(\mathcal{Y})$. Since $x_m^{(k)}\to 0$ weakly in $\mathcal{X}$ as $k\to\infty$, and since the composition of $\widetilde{F}^{**}$ and Φ is weakly compact, $\widetilde{F}^{**}(\Phi(x_m^{(k)}))\to 0$ in norm as $k\to\infty$. Thus $\widetilde{G}^{**}\varphi\in c_0(\mathcal{Y})$, and $\widetilde{G}$ is weakly compact. □

Note in Theorem 1.5.1 that if the weak Cauchy sequences are actually weakly convergent, then the values of F on the sequence converge to its value at the limit, that is, F is weakly sequentially continuous. Note also that the theorem automatically applies to scalar-valued m-linear functionals. By summing a uniformly convergent Taylor series, we then obtain the following corollary.

Corollary 1.5.2 *If $\mathcal{X}$ has the Dunford-Pettis property, then a bounded (complex-valued) analytic function on the open unit ball B of $\mathcal{X}$ is weakly sequentially continuous on any subball B_r, $r<1$.*

Notes Good sources for the basic material on analytic functions on a Banach space are [Na], [Din] and [Mu]. For basic properties of Banach spaces, including the weak topology and weakly compact operators, and also the circle of ideas involving uniform integrability, see [DS]. For tensor products we refer to [Pie], and also to the short resumé in [Pis]. For information on the polarization constants and more references, see [LR]. The survey article [Die1] is devoted to the Dunford-Pettis property. Theorem 1.5.1 is due to R. Ryan [Ry2], and we have followed his proof.

Chapter 2
Extension of analytic functions to the bidual

Continuous linear functionals on $\mathcal{X}$ extend in a natural way to weak-star continuous functionals on the bidual $\mathcal{X}^{**}$. Our aim here is to develop a canonical extension procedure for extending arbitrary analytic functions on $\mathcal{X}$ to be analytic on $\mathcal{X}^{**}$. The procedure is simply to expand the analytic function in a Taylor series, extend the associated m-linear functionals to the bidual, and sum the resulting Taylor series on the bidual. We will also relate continuity properties of analytic functions to those of their extensions.

2.1 The canonical extension

We begin with an m-homogeneous polynomial $P \in \mathcal{P}_m$. Let $F(x_1, \ldots, x_m)$ be the symmetric m-linear functional corresponding to P. If we fix the first $m-1$ variables, we have a continuous linear functional in the mth variable, which extends to $\mathcal{X}^{m-1} \times \mathcal{X}^{**}$ to be weak-star continuous in the last variable. We continue this procedure, extending by weak-star continuity one variable at a time, and after m steps we obtain an m-linear functional $\hat{F}$ on $\mathcal{X}^{**}$ such that for $1 \leq k \leq m$, for fixed $x_1, \ldots, x_{k-1} \in \mathcal{X}$, and for fixed $z_{k+1}, \ldots, z_m \in \mathcal{X}^{**}$, the functional $z \rightarrow \hat{F}(x_1, \ldots, x_{k-1}, z, z_{k+1}, \ldots, z_m)$ is weak-star continuous on $\mathcal{X}^{**}$. If in this procedure we take the variables in a different order, we arrive in general at a different extension. (The bidual of a commutative Banach algebra, with Arens multiplication, need not be commutative.) However, from the symmetry of F it follows immediately that the restriction $\hat{P}$ of $\hat{F}$ to the diagonal does not depend on the order of the variables. The function $\hat{P}$ is an m-homogeneous analytic function on $\mathcal{X}^{**}$, and it coincides with P on $\mathcal{X}$. Using Theorem 1.1.1, we obtain

$$||\hat{P}|| \leq ||\hat{F}|| = ||F|| \leq \frac{m^m}{m!}||P||.$$

Since $m! \sim (m/e)^m$, this estimate shows that if the Taylor series $\sum P_m$ has radius of bounded convergence R, then the Taylor series $\sum \hat{P}_m$ in $\mathcal{X}^{**}$ has radius of bounded convergence at least R/e. We aim to show that the radii of bounded convergence are equal.

Theorem 2.1.1 *The canonical extension $\hat{P}$ of an m-homogeneous polynomial P to the bidual $\mathcal{X}^{**}$ satisfies $||\hat{P}|| = ||P||$, that is, the extension operator is an isometry.*

Proof Recall that B^{**} is the open unit ball of $\mathcal{X}^{**}$, and note that B is weak-star dense in $\bar{B}^{**}$. Let $z \in \bar{B}^{**}$. With F as above, choose $x_1 \in B$ such that $|\hat{F}(x_1, z, \ldots, z) - \hat{F}(z, z, \ldots, z)| < \varepsilon$. Then choose $x_2 \in B$ to satisfy the same condition as x_1, and additionally to satisfy $|\hat{F}(x_1, x_2, z, \ldots, z) - \hat{F}(x_1, z, \ldots, z)| < \varepsilon$. Proceeding in this fashion, we construct a sequence $\{x_j\}$ in B so that

$$|\hat{F}(x_{i_1}, x_{i_2}, \ldots, x_{i_{r-1}}, x_{i_r}, z, \ldots, z) - \hat{F}(x_{i_1}, x_{i_2}, \ldots, x_{i_{r-1}}, z, z, \ldots, z)| < \varepsilon.$$

whenever $1 \leq i_1 < i_2 < \cdots < i_r$. By means of a telescoping sum we then estimate

$$|F(x_{i_1}, \ldots, x_{i_m}) - \hat{F}(z, \ldots, z)| < m\varepsilon, \qquad 1 \leq i_1 < i_2 < \cdots < i_m.$$

Now fix N large, and set $y_N = (x_1 + \cdots + x_N)/N$. Then

$$P(y_N) - \hat{P}(z) = \frac{1}{N^m} \sum_{i_1,\dots,i_m=1}^{N} \left[F(x_{i_1},\dots,x_{i_m}) - \hat{F}(z,\dots,z)\right].$$

Each summand for which all the indices are distinct is estimated by $m\varepsilon$, so this part of the normalized sum is majorized by $m\varepsilon$. We count at most $N^{m-1}m(m-1)/2$ summands for which the indices are not distinct, so this part of the normalized sum is estimated by $m(m-1)||F||/N$, which tends to 0 as $N \to \infty$. Thus $\limsup |P(y_N) - \hat{P}(z)| \leq m\varepsilon$. Since $y_N \in B$, and since $\varepsilon > 0$ is arbitrary, we obtain $||\hat{P}|| \leq ||P||$. □

The above approximation procedure is of interest in its own right. From the proof we extract the following theorem, which we record for later use.

Theorem 2.1.2 *Let $\{x_\alpha\}$ be a net in B that converges weak-star to $z \in \bar{B}^{**}$. Then there is a net $\{y_\beta\}$ in B such that each y_β is an arithmetic mean of a finite number of x_α's, and $P(y_\beta) \to \hat{P}(z)$ for every polynomial P on $\mathcal{X}$.*

Proof Simply observe that the estimates in the preceding proof can be arranged to hold simultaneously for a fixed finite number of homogeneous polynomials. □

We say that a net $\{z_\alpha\}$ in $\mathcal{X}^{**}$ converges to $z \in \mathcal{X}^{**}$ in the *polynomial-star topology* if $\hat{P}(z_\alpha) \to \hat{P}(z)$ for all polynomials P on $\mathcal{X}$. The preceding theorem asserts that B is polynomial-star dense in B^{**}. This extends Goldstine's theorem, that B is weak-star dense in B^{**}.

Now suppose for convenience that f is analytic at $0 \in \mathcal{X}$, and expand f in a Taylor series

$$f(x) = \sum_{m=0}^{\infty} P_m(x), \qquad ||x|| < R.$$

Define an extension $\hat{f}$ of f to $\mathcal{X}^{**}$ by

$$\hat{f}(z) = \sum_{m=0}^{\infty} \hat{P}_m(z), \qquad z \in \mathcal{X}^{**}.$$

On account of Theorem 2.1.1 and the Cauchy-Hadamard formula, the radius of boundedness of the series for $\hat{f}$ is the same as that of f, call it R, so that $\hat{f}$ is analytic on B_R^{**}. If $r < R$, and $\{x_\alpha\}$ is a net in B_r converging polynomial-star to z, then since the Taylor series for $\hat{f}$ converges uniformly on B_r^{**}, $f(x_\alpha)$ converges to $\hat{f}(z)$. This makes it clear that the extension operator $f \to \hat{f}$ is not only linear but also multiplicative, an algebra isomorphism. We have proved the following.

Theorem 2.1.3 *The canonical extension operator $f \to \hat{f}$ is an isometric algebra isomorphism of $H^\infty(B)$ onto a closed subalgebra of $H^\infty(B^{**})$.*

Finally we observe that the canonical extension does not depend on the point at which we center the Taylor expansion. Indeed, suppose f has Taylor series expansions $\sum P_m(x-x_0)$ and $\sum Q_n(x-x_1)$, uniformly convergent for $||x-x_j|| \leq r_j$, and that $z \in \mathcal{X}^{**}$ satisfies $||z-x_j|| < r_j$, $j=1,2$. Starting with a weak-star convergent net and passing to a net of convex combinations, we obtain a net in $B(x_0,r_0) \cap B(x_1,r_1)$ that converges polynomial-star to z. Then we may pass to the limit in $\sum P_m(x-x_0) = \sum Q_n(x-x_1)$ to obtain $\sum \hat{P}_m(z-x_0) = \sum \hat{Q}_n(x-x_1)$, and the two defining series for $\hat{f}(z)$ give the same result. Thus any bounded analytic function f on a domain D in $\mathcal{X}$ has a canonical extension to a bounded analytic function $\hat{f}$ on some ambient domain in $\mathcal{X}^{**}$ containing D, and the correspondence is an isometric algebra isomorphism.

2.2 The Aron-Hervés-Valdivia theorem

We are concerned now with giving conditions under which the canonical extension $\hat{f}$ of f is (locally) weak-star continuous. If $\hat{f}$ is weak-star continuous on bounded subsets of $\mathcal{X}^{**}$ (that is, its restriction to any bounded subset of $\mathcal{X}^{**}$ is weak-star continuous), then its restriction f to $\mathcal{X}$ is weakly continuous on bounded subsets of $\mathcal{X}$. We wish to establish the converse. For linear functionals, the converse assertion is trivial. Any discontinuous linear functional on $\mathcal{X}$ is zero on a dense subset of $\mathcal{X}$. Consequently a linear functional that is continuous on bounded subsets of $\mathcal{X}$ is continuous, and its canonical extension to $\mathcal{X}^{**}$ is weak-star continuous. For multilinear functionals, the converse assertion remains true, but it is not at all obvious.

Theorem 2.2.1 *Let $m \geq 2$, and let F be a continuous m-linear functional on $\mathcal{X}$. For $1 \leq j \leq m$, define a linear operator T_j from $\mathcal{X}$ to the space $\mathcal{L}(^{m-1}\mathcal{X})$ of $(m-1)$-linear functionals on $\mathcal{X}$ by*

$$T_j(x)(x_1,\ldots,x_{m-1}) = F(x_1,\ldots,x_{j-1},x,x_j,\ldots,x_{m-1}), \qquad x \in \mathcal{X}.$$

The following are equivalent.

(1) *F is weakly continuous on bounded sets.*

(2) *Each T_j is a compact operator.*

(3) *F is uniformly continuous on bounded sets, with respect to the weak topology.*

(4) *F extends to an m-linear functional $\hat{F}$ on $\mathcal{X}^{**}$ that is weak-star continuous on bounded sets.*

Proof The key implication is (1) $\Rightarrow$ (2). In order to prove this, we establish some lemmas. Let Q be a subset of $\mathcal{X}^*$. We say that F is *Q-continuous on bounded sets* if whenever the bounded nets $x_k^{(\alpha)}$ in $\mathcal{X}$ and points $x_k \in \mathcal{X}$ satisfy $L(x_k^{(\alpha)}) \to L(x_k)$ for $L \in Q$ and $1 \leq k \leq m$, then $F(x_1^{(\alpha)},\ldots,x_m^{(\alpha)}) \to F(x_1,\ldots,x_m)$.

Lemma 2.2.2 *Let Q be a subset of $\mathfrak{X}^*$, and let F be an m-linear functional on $\mathfrak{X}$ that is Q-continuous on bounded sets. For $1 \le k \le m$, let $\{x_k^{(j)}\}_{j=1}^{\infty}$ be a bounded sequence in $\mathfrak{X}$ such that $L(x_k^{(j)})$ converges for all $L \in Q$. Suppose furthermore that for one of the indices k we have $L(x_k^{(j)}) \to 0$ as $j \to \infty$ for all $L \in Q$. Then $F(x_1^{(j)}, \ldots, x_m^{(j)}) \to 0$ as $j \to \infty$.*

Proof We may assume the lemma is true for $(m-1)$-linear functionals. By the continuity hypothesis on F, the lemma is true if each of the m sequences Q-converges to 0. We proceed backwards by induction on an index ℓ with $1 \le \ell \le m-1$. We assume the lemma is true if $\ell+1$ of the sequences Q-converge to 0, and we must prove it if ℓ of the sequences Q-converge to 0.

So suppose that $x_k^{(j)}$ Q-converges to 0 for $1 \le k \le \ell$. By the induction hypothesis with respect to m, for each fixed i the sequence $F(x_1^{(j)}, \ldots, x_{m-1}^{(j)}, x_m^{(i)})$ tends to 0 as $j \to \infty$. Hence we can choose an increasing sequence j_i such that $F(x_1^{(j_i)}, \ldots, x_{m-1}^{(j_i)}, x_m^{(i)})$ tends to 0 as $i \to \infty$. Now $x_m^{(j_i)} - x_m^{(i)}$ Q-converges to 0 as $i \to \infty$. By the induction hypothesis with respect to ℓ, the sequence $F(x_1^{(j_i)}, \ldots, x_{m-1}^{(j_i)}, x_m^{(j_i)} - x_m^{(i)})$ tends to 0. Hence $F(x_1^{(j_i)}, \ldots, x_m^{(j_i)})$ tends to 0, and in particular the original sequence has a subsequence on which F tends to 0. Since this is also true for every subsequence of the original sequence, we conclude that $F(x_1^{(j)}, \ldots, x_m^{(j)})$ tends to 0. □

Lemma 2.2.3 *Suppose $\mathfrak{X}$ is separable. If F is weakly continuous on bounded sets, then there is a countable subset Q of $\mathfrak{X}^*$ such that F is Q-continuous on bounded sets.*

Proof Let E be a countable dense subset of $\mathfrak{X}$. For $N \ge 1$ and $y_1, \ldots, y_m \in E \cap B_N$, choose a finite set $Q_0 = Q_0(N, y_1, \ldots, y_m)$ of functionals in $\mathfrak{X}^*$ such that if $x_1, \ldots, x_m \in B_N$ satisfy $|L(x_j) - L(y_j)| < 1$ for $1 \le j \le m$ and all $L \in Q_0$, then $|F(x_1, \ldots, x_m) - F(y_1, \ldots, y_m)| < 1/N$. The aggregate Q of the Q_0's does the trick. □

Now we prove that (1) implies (2). A linear operator is compact if and only if its restriction to every separable subspace is compact. Thus we may assume that $\mathfrak{X}$ is separable, and we choose Q as in Lemma 2.2.3.

Arguing by contradiction, we suppose T_m is not compact. Then there exist $\varepsilon > 0$ and $w_m^{(j)} \in B$ such that $\|T_m(w_m^{(j)}) - T_m(w_m^{(i)})\| > \varepsilon$ for $i \ne j$. Passing to a subsequence, we can assume that $L(w_m^{(j)})$ converges for all $L \in Q$. For $j \ge 1$ and $1 \le k \le m-1$, choose $x_k^{(j)} \in B$ such that

$$|F(x_1^{(j)}, \ldots, x_{m-1}^{(j)}, w_m^{(j+1)}) - F(x_1^{(j)}, \ldots, x_{m-1}^{(j)}, w_m^{(j)})| > \varepsilon.$$

Passing to a subsequence again, we can assume for $1 \le k \le m-1$ that $L(x_k^{(j)})$ converges for all $L \in Q$. Set $x_m^{(j)} = w_m^{(j+1)} - w_m^{(j)}$. Then $L(x_m^{(j)}) \to 0$ as $j \to \infty$ for all $L \in Q$. However, the estimate above shows that $F(x_1^{(j)}, \ldots, x_{m-1}^{(j)}, x_m^{(j)})$ does not tend to 0, and this contradicts Lemma 2.2.2.

The proof that (2) implies (3) depends upon the following lemma.

Lemma 2.2.4 *A linear operator $A : \mathfrak{X} \to \mathfrak{Y}$ is compact if and only if A is continuous from the weak topology on bounded subsets of $\mathfrak{X}$ to the norm topology of $\mathfrak{Y}$.*

Proof Suppose A is compact. Let x_α be a bounded net in $\mathfrak{X}$ converging weakly to $x \in \mathfrak{X}$. If $L \in \mathfrak{Y}^*$, then $L(A(x)) = (A^*L)(x) = \lim(A^*L)(x_\alpha) = \lim L(A(x_\alpha))$. Consequently $L(Ax) = L(y)$ for any point adherent to the net $A(x_\alpha)$ in the norm topology of $\mathfrak{Y}$. Since A is compact, $A(x_\alpha)$ must converge in norm to $A(x)$. The converse, which we do not need here, is proved as follows. If A is not compact, there are $\varepsilon > 0$ and a sequence x_j in $\mathfrak{X}$ such that $||x_j|| = 1$ and $||Ax_j - Ax_k|| \geq \varepsilon$ whenever $j \neq k$. The x_j's have a subnet, call it x_α, that converges weak-star in $\mathfrak{X}^{**}$. For each α, choose any $\beta > \alpha$ such that $x_\beta \neq x_\alpha$, and set $y_\alpha = x_\alpha - x_\beta$. Then $y_\alpha \to 0$ weakly in $\mathfrak{X}$, but $||Ay_\alpha|| \geq \varepsilon > 0$. □

Now suppose that each T_j is compact. For $1 \leq j \leq m$, let $x_j^{(\alpha)}$ and $y_j^{(\alpha)}$ be bounded nets in $\mathfrak{X}$ such that $x_j^{(\alpha)} - y_j^{(\alpha)} \to 0$ weakly. We must show that

$$F(x_1^{(\alpha)}, \ldots, x_m^{(\alpha)}) - F(y_1^{(\alpha)}, \ldots, y_m^{(\alpha)}) \to 0.$$

Write this difference as a sum of m terms of the form

$$F(x_1^{(\alpha)}, \ldots, x_{j-1}^{(\alpha)}, x_j^{(\alpha} - y_j^{(\alpha)}, y_{j+1}^{(\alpha)}, \ldots, y_m^{(\alpha)}) = T_j(x_j^{(\alpha)} - y_j^{(\alpha)})(x_1^{(\alpha)}, \ldots, x_{j-1}^{(\alpha)}, y_{j+1}^{(\alpha)}, \ldots, y_m^{(\alpha)}).$$

Since T_j maps bounded weakly convergent nets to norm convergent nets, $T_j(x_j^{(\alpha)} - y_j^{(\alpha)})$ tends to 0 in norm. Thus (2) implies (3).

The equivalence of (3) and (4) is a consequence of the fact that the completion of the unit ball B, with respect to the uniform structure determined by the weak topology, is the closed unit ball $\bar{B}^{**}$, with the weak-star topology. Since a uniformly continuous function extends continuously to the completion, (3) implies (4). This can also be established by a simple direct argument. Since (3) and (4) trivially imply (1), the proof of Theorem 2.2.1 is complete. □

Theorem 2.2.5 *Let $m \geq 2$, and let F be a continuous m-linear functional on $\mathfrak{X}$. For $1 \leq j \leq m$, define an $(m-1)$-linear function S_j on $\mathfrak{X}$ with values in $\mathfrak{X}^*$ by*

$$S_j(x_1, \ldots, x_{m-1})(x) = F(x_1, \ldots, x_{j-1}, x, x_j, \ldots, x_{m-1}), \qquad x \in \mathfrak{X}.$$

The following are equivalent.

(1) *F is weakly continuous on bounded sets.*

(2) *Each S_j is compact, that is, $S_j(B^{m-1})$ is precompact in $\mathfrak{X}^*$.*

(3) *S_m extends to an $(m-1)$-linear function on $\mathfrak{X}^{**}$ with values in $\mathfrak{X}^*$ that is continuous from the product weak-star topology of $(B^{**})^{m-1}$ to the norm topology of $\mathfrak{X}^*$.*

(4) *S_m is continuous from the product weak topology on bounded sets in $\mathfrak{X}^{m-1}$ to the norm topology of $\mathfrak{X}^*$.*

Proof Suppose (1) holds. Let $\hat{F}$ denote the extension of F to $\mathcal{X}^{**}$ given by Theorem 2.2.1, which is weak-star continuous on bounded sets. We claim that the operator $\hat{S}_m(z_1,\ldots,z_{m-1})(x_m) = \hat{F}(z_1,\ldots,z_{m-1},x_m)$ is continuous, from the weak-star topology on bounded subsets of $(\mathcal{X}^{**})^{m-1}$to the norm topology of $\mathcal{X}^*$. Indeed, otherwise there are bounded nets $z_j^{(\alpha)}$ in $\mathcal{X}^{**}$, $1 \le j \le m-1$, and corresponding $x_m^{(\alpha)}$ in $\mathcal{X}$, such that $z_j^{(\alpha)} \to 0$ weak-star for each j, $||x_m^{(\alpha)}|| = 1$, and $|F(z_1^{(\alpha)},\ldots,z_{m-1}^{(\alpha)},x_m^{(\alpha)})| \ge \varepsilon > 0$. Passing to a subnet, we can assume that $x_m^{(\alpha)} \to z_m$ weak-star. Then $F(z_1^{(\alpha)},\ldots,z_{m-1}^{(\alpha)},x_m^{(\alpha)})$ tends to $F(0,\ldots,0,z_m) = 0$, a contradiction. Thus (1) implies (3), and (3) trivially implies (4).

Suppose that (4) holds, and let $x_k^{(\alpha)}$ be a bounded net in $\mathcal{X}$ converging weakly to x_k, $1 \le k \le m$. Then

$$||S_m(x_1^{(\alpha)},\ldots,x_{m-1}^{(\alpha)}) - S_m(x_1,\ldots,x_{m-1})|| \to 0,$$

from which it follows that $F(x_1^{(\alpha)},\ldots,x_m^{(\alpha)}) = S_m(x_1^{(\alpha)},\ldots,x_{m-1}^{(\alpha)})(x_m^{(\alpha)})$ has the same limit as $S_m(x_1,\ldots,x_{m-1})(x_m^{(\alpha)})$, which is $S_m(x_1,\ldots,x_{m-1})(x_m) = F(x_1,\ldots,x_m)$. Hence (1) holds, and (1), (3) and (4) are equivalent.

Since the condition (1) is not affected by permuting the variables, each of the operators S_j has the properties (3) and (4) just as soon as one of them does. If S_j has the continuity property in (3), then $S_j(\bar{B}^{m-1})$ is compact in $\mathcal{X}^*$. Hence (3) implies (2).

Suppose that (2) holds. For $1 \le k \le m-1$, let $x_k^{(\alpha)}$ be a bounded net in $\mathcal{X}$ converging weakly to x_k. For $x_m \in \mathcal{X}$ fixed, define $G(x_1,\ldots,x_{m-1}) = F(x_1,\ldots,x_{m-1},x_m)$. We may assume the theorem is true for $(m-1)$-linear functionals. Since each of the slice operators S_j associated with G is compact, G is weakly continuous on bounded sets. Hence $S_m(x_1^{(\alpha)},\ldots,x_{m-1}^{(\alpha)})(x_m) = F(x_1^{(\alpha)},\ldots,x_{m-1}^{(\alpha)},x_m)$ has a limit $F(x_1,\ldots,x_{m-1},x_m) = S_m(x_1,\ldots,x_{m-1})(x_m)$. This limit coincides with $L(x_m)$ for any adherent point L of the net $S_m(x_1^{(\alpha)},\ldots,x_{m-1}^{(\alpha)})$ in $\mathcal{X}^*$. This shows that $S_m(x_1,\ldots,x_{m-1})$ is the unique adherent point of the net in $\mathcal{X}^*$, and so the net converges in $\mathcal{X}^*$, and (4) holds. □

An inspection of the proof reveals that in (2) we need require only $m-1$ of the S_j's to be compact. Likewise in Theorem 2.2.1 we need require only $m-1$ of the T_j's to be compact. However, compactness of $m-2$ of the S_j's does not necessarily imply the compactness of the others. Consider $F(x_1,x_2,x_3) = G(x_1,x_2)L(x_3)$, where $L \in \mathcal{X}^*$ and G is a continuous bilinear functional that is not weakly continuous on bounded sets. The operator S_3 has one-dimensional range, hence it is compact, though neither S_1 nor S_2 is compact.

If F is symmetric, then all of the S_j's coincide, as do all of the T_j's, so we have to check compactness for only one function.

Theorems 2.2.1 and 2.2.5 are also valid for m-linear functions on $\mathcal{X}$ with values in a Banach space $\mathcal{Y}$. In this case, the T_j's are operators from $\mathcal{X}$ to $\mathcal{L}(^{m-1}\mathcal{X};\mathcal{Y})$, and the S_j's are $(m-1)$-linear functions on $\mathcal{X}$ with values in $\mathcal{L}(\mathcal{X};\mathcal{Y})$. Compactness in $\mathcal{X}^*$ is replaced by compactness in $\mathcal{L}(\mathcal{X};\mathcal{Y})$. The proofs above go through virtually verbatim, with absolute values replacing norms where appropriate.

2.3 Weak-star continuous extensions

We wish to deduce continuity properties of f from the corresponding continuity properties of the Taylor coefficients of f. The proof of the following lemma is typical of how this is done.

Lemma 2.3.1 *Let f be bounded and analytic on the open unit ball B in $\mathfrak{X}$, with Taylor series $\sum P_m$. If f is weakly continuous on B, then each Taylor coefficient P_m is weakly continuous on bounded sets. Conversely, if each P_m is weakly continuous on bounded sets, and if f is uniformly continuous on B with respect to the norm, then f extends to be weakly continuous on the closed unit ball $\bar{B}$.*

Proof Suppose f is weakly continuous on B. To show that P_m is weakly continuous on bounded sets, it suffices by homogeneity to show that P_m is weakly continuous on some ball rB. So fix $0 < r < 1$, and let $\{x_\alpha\}$ be a net in rB converging weakly to x. Then λx_α converges weakly to λx for all complex λ. By the weak continuity of f, $g_\alpha(\lambda) = f(\lambda x_\alpha)$ converges pointwise to $g(\lambda) = f(\lambda x)$ for $|\lambda| < 1/r$. Furthermore, the g_α's are uniformly bounded by $||f||_B$ for $|\lambda| \le 1/r$. Hence for $|\lambda| \le 1$,

$$g_\alpha(\lambda) = f(\lambda x_\alpha) = \sum P_m(\lambda x_\alpha) = \sum \lambda^m P_m(x_\alpha)$$

converges uniformly to

$$g(\lambda) = \sum \lambda^m P_m(x).$$

It follows that for each fixed $m \ge 0$, $P_m(x_\alpha)$ converges to $P_m(x)$, and P_m is weakly continuous on rB.

For the converse, suppose that the Taylor coefficients of f are weakly continuous on bounded sets, and that f is uniformly continuous on B. Then the dilates $f_r(x) = f(rx)$ converge uniformly to f as r increases to 1. The Taylor coefficients of f_r are also weakly continuous on bounded sets, and the Taylor series of f_r converges uniformly on B. Consequently f_r is weakly continuous on B, as is f. □

Let now f be analytic in a neighborhood of x_0, and expand f in a Taylor series,

$$f(x) = \sum_{m=0}^{\infty} P_m(x - x_0)$$

centered at x_0. Suppose P_m is the restriction to the diagonal of a symmetric m-linear functional F_m. Define the $(m-1)$-linear function C_m from $\mathfrak{X}$ to $\mathfrak{X}^*$ by

$$C_m(x_1, \ldots, x_{m-1})(x) = F_m(x_1, \ldots, x_{m-1}, x), \qquad x \in \mathfrak{X}.$$

We say that *the mth Taylor coefficient of f is compact* if C_m is compact, that is, if the image under C_m of B^{m-1} is a precompact subset of $\mathfrak{X}^*$. By Theorem 2.2.5, this is equivalent to the weak continuity of F_m on bounded sets. In turn, this is equivalent to the weak continuity of P_m on bounded sets, by the polarization formula.

Note that $C_1 = F_1 \in \mathcal{X}^*$ is constant hence trivially compact. The form C_2 is a linear operator from $\mathcal{X}$ to $\mathcal{X}^*$. The compactness of the second Taylor coefficient means simply that C_2 is a compact operator.

The following theorem provides a local characterization of the functions whose canonical extensions are weak-star continuous on bounded sets. We scale to the unit ball.

Theorem 2.3.2 *The following are equivalent, for a bounded analytic function f on the open unit ball B of $\mathcal{X}$.*

(1) *For some $r > 0$, f is weakly continuous on the subball B_r.*

(2) *f extends to an analytic function on the open unit ball B^{**} of $\mathcal{X}^{**}$ that is weak-star continuous on the subball B_r^{**}, for any $r < 1$.*

(3) *All the Taylor coefficients of f at 0 are compact.*

(4) *For each $r < 1$, the Frechet derivatives $f'(x)$ for $x \in B_r$ form a precompact subset of $\mathcal{X}^*$.*

Proof Suppose (1) holds. By the lemma, each P_m is weakly continuous on bounded sets. As remarked above, each C_m is then compact, so (3) holds. In turn, if (3) holds, then by Theorem 2.2.1 each $\hat{P}_m$ is weak-star continuous on bounded sets. Since the Taylor series of $\hat{f}$ converges uniformly on B_r^{**}, (2) holds. That (2) implies (1) is trivial, so (1), (2) and (3) are equivalent.

Suppose (3) holds. The Frechet derivative $x \to f'(x)$ is analytic as a function on B with values in $\mathcal{X}^*$. Its Taylor series expansion at 0 is

$$f'(x) = \sum C_m(x, \ldots, x) = f'(0) + C_2(x) + C_3(x, x) + \cdots,$$

which converges uniformly for $x \in B_r$. Since the C_m's are compact, for each fixed N the image of B under $\sum_{m=1}^{N} C_m(x, \ldots, x)$ is precompact in $\mathcal{X}^*$. Since the series converges uniformly on B_r, the image of B_r under the limit function $f'(x)$ is totally bounded in $\mathcal{X}^*$, hence precompact. Thus (3) implies (4).

Conversely, suppose that (4) holds, and let $0 < r < 1$. Let E denote the closed circled convex hull of the $f'(x)$, $||x|| \leq r$. Then E is compact in $\mathcal{X}^*$. We claim that if $||x|| < r$, then $C_m(x, \ldots, x) \in E$, and in view of the polarization formula this will show that C_m is compact. So fix such an x, and let $\Lambda \in \mathcal{X}^{**}$. The scalar-valued function $(\Lambda \circ f')(\lambda x)$ depends analytically on the complex parameter λ and has Taylor series

$$(\Lambda \circ f')(\lambda x) = \sum_{m=1}^{\infty} \Lambda(C_m(x, \ldots, x))\lambda^{m-1}.$$

The Cauchy estimates give

$$|\Lambda(C_m(x, \ldots, x))| \leq \sup\{|\Lambda(f'(y))| : ||y|| \leq r\} = \sup\{|\Lambda(w)| : w \in E\}.$$

Since this holds for all $\Lambda \in \mathcal{X}^{**}$, the strict separation theorem for convex sets shows that $C_m(x, \ldots, x) \in E$. Thus (4) implies (3), and the proof is complete. □

Note that the above properties propagate. If f is analytic on a domain D in $\mathcal{X}$, and if the Taylor coefficients at one point are all compact, then the Taylor coefficients are compact at all points of D, and f is locally weakly continuous on D.

Theorem 2.3.3 *If f is a bounded analytic function on the open unit ball B of X, then f extends to be weak-star continuous on the closed unit ball $\bar{B}^{**}$ of X^{**} if and only if f is weakly continuous on B and f is uniformly continuous with respect to the norm.*

Proof Suppose first that f has an extension $\hat{f}$ that is weak-star continuous on $\bar{B}^{**}$. In particular, f is weakly continuous on B. Suppose that $\{z_j\}$ and $\{w_j\}$ are sequences in B such that $\|z_j - w_j\| \to 0$. Let $\{z_{j(\alpha)}\}$ be a subnet that converges weak-star to $z \in \bar{B}^{**}$. Then also $\{w_{j(\alpha)}\}$ converges weak-star to z. Hence $f(z_{j(\alpha)}) \to \hat{f}(z)$ and $f(w_{j(\alpha)}) \to \hat{f}(z)$, so that $|f(z_{j(\alpha)}) - f(w_{j(\alpha)})| \to 0$. Since this holds for any weak-star convergent subnet of $\{z_j\}$, in fact $|f(z_j) - f(w_j)| \to 0$. It follows that f is uniformly continuous on B.

The converse follows immediately from Theorem 2.3.2, once we note that on account of the uniform continuity, the dilates f_r of f converge uniformly to f on $\bar{B}$ as r increases to 1. □

The theorems, appropriately amended, also hold for analytic functions with values in a Banach space $\mathcal{Y}$. In this case, the $(m-1)$-form S_m has values in $\mathcal{L}(\mathcal{X};\mathcal{Y})$ instead of $\mathcal{X}^*$, and we say that the mth Taylor coefficient of f is compact if the image of B^{m-1} is a precompact subset of $\mathcal{L}(\mathcal{X};\mathcal{Y})$. The Frechet derivative $f'(x)$ also belongs to $\mathcal{L}(\mathcal{X};\mathcal{Y})$, and the condition (4) of Theorem 2.3.2 must be modified by replacing $\mathcal{X}^*$ by $\mathcal{L}(\mathcal{X};\mathcal{Y})$.

2.4 Banach spaces not containing ℓ^1

We say that a sequence $\{x_j\}$ in a Banach space $\mathcal{X}$ is *equivalent to the standard basis of* ℓ^1 if the operator assigning x_j to the jth standard basis element of ℓ^1 extends to a Banach space isomorphism of ℓ^1 and the closed linear span of the x_j's. In this case we say that ℓ^1 *embeds in* $\mathcal{X}$. A proof of the following beautiful result can be found in [LT].

Rosenthal's dichotomy *If S is a bounded subset of a Banach space, then either S contains a sequence equivalent to the standard basis of ℓ^1, or every sequence in S has a weak Cauchy subsequence.*

Many Banach spaces, such as reflexive Banach spaces and Banach spaces with separable duals, do not have subspaces isomorphic to ℓ^1. For these, only the second alternative of Rosenthal's dichotomy can hold, and we can sharpen the characterizations given in the preceding section. The upshot is that if ℓ^1 does not embed in $\mathcal{X}$, then weak sequential continuity implies weak continuity on bounded sets.

Theorem 2.4.1 *Suppose ℓ^1 does not embed in $\mathfrak{X}$. Let F be a continuous m-linear functional on $\mathfrak{X}$. For $1 \leq j \leq m$, define the T_j's as in Theorem 2.2.1 and the S_j's as in Theorem 2.2.5. Then the following are equivalent.*

(1) *F is weakly continuous on bounded sets.*

(2) *F is weakly sequentially continuous.*

(3) *Each T_j is a completely continuous operator from $\mathfrak{X}$ to $\mathcal{L}(^{m-1}\mathfrak{X})$.*

(4) *S_m is sequentially continuous, from the product weak topology of B^{m-1} to the norm topology of $\mathfrak{X}^*$.*

Proof The continuity property (1) implies the others, so we must get back from each of (2), (3), (4) to (1). The only tricky part is to get from (2) to (1). For this, we proceed as in the proof that (1) implies (2) in Theorem 2.2.1, except that we invoke Lemma 2.2.2 with $Q = \mathfrak{X}^*$ (so that Lemma 2.2.3 is not required). In view of Rosenthal's dichotomy, we can pass to weak Cauchy sequences as in that proof, thereby obtaining a contradiction to Lemma 2.2.2 that establishes the compactness of the T_j's.

That (3) implies (1) follows from Theorem 2.2.5 and the observation that if ℓ^1 does not embed in $\mathfrak{X}$, then any completely continuous operator from $\mathfrak{X}$ to another Banach space is compact. Indeed, according to Rosenthal's dichotomy any bounded sequence in $\mathfrak{X}$ has a weak Cauchy subsequence, and completely continuous operators map weak Cauchy sequences to norm convergent sequences. Finally, we get easily from (4) to (2) in the same way as in the corresponding implication of Theorem 2.2.1. □

The proof of Lemma 2.3.1 shows that if an analytic function is weakly sequentially continuous, then its Taylor coefficients are also weakly sequentially continuous. As a corollary to Theorems 2.4.1 and 1.5.2, we then obtain the following addendum to Theorem 2.3.2.

Theorem 2.4.2 *Suppose ℓ^1 does not embed in $\mathfrak{X}$. Then any weakly sequentially continuous analytic function on the open unit ball B of $\mathfrak{X}$ is weakly continuous on any subball B_r, $r < 1$. If further $\mathfrak{X}$ has the Dunford-Pettis property, then any bounded analytic function on the open unit ball B of $\mathfrak{X}$ is weakly continuous on any subball B_r, $r < 1$.*

The first statement of the preceding theorem actually characterizes spaces that do not contain ℓ^1. In fact, by appealing to some name theorems, we can prove the following converse statement.

Theorem 2.4.3 *Suppose ℓ^1 embeds in $\mathfrak{X}$. Then there is a 2-homogeneous analytic function P on $\mathfrak{X}$ that is weakly sequentially continuous, but that is not weakly continuous on the unit ball B of $\mathfrak{X}$.*

Proof We will use the theorem that any operator from an L^1-space to Hilbert space factors through an L^∞-space. This comes from a circle of deep results stemming from Grothendieck.

In fact, any such operator is two-summing (Theorem 5.10 of [Pis]), and any two-summing operator factors (Corollary 1.8 of [Pis]). We will also use the injectivity of $L^\infty(\mu)$, that any operator from a subspace of a Banach space to $L^\infty(\mu)$ can be extended to the Banach space. This important result is due to Nachbin, and the proof boils down to the observation that the classical proof of the Hahn-Banach theorem goes through when the real scalars are replaced by a real L^∞-space with its natural order.

Let now S be any operator from ℓ^1 to ℓ^2 that is not compact, say the inclusion operator. We factor S through $L^\infty(\mu)$ for some probability measure μ, and we extend the factor going from ℓ^1 to $L^\infty(\mu)$ to an operator from $\mathfrak{X}$ to $L^\infty(\mu)$. This yields an extension R of S, which goes from $\mathfrak{X}$ to ℓ^2 and which factors through $L^\infty(\mu)$. In view of the natural self-duality of ℓ^2, we can define the operator $T = R^*R : \mathfrak{X} \to \mathfrak{X}^*$. Let P be the 2-homogeneous analytic function corresponding to T, so that

$$P(x) = \langle Tx, x\rangle = \sum (Rx)_n^2, \qquad x \in \mathfrak{X}.$$

The factor of S from $L^\infty(\mu)$ to ℓ^2 is weakly compact, and $L^\infty(\mu)$ has the Dunford-Pettis property, so this factor is completely continuous. Hence T is completely continuous, and P is weakly sequentially continuous. However, since R is not compact, neither is T, and consequently by Theorem 2.2.1 (or 2.2.5), P is not weakly continuous on the unit ball of $\mathfrak{X}$. □

Notes The extension operator of section 2.1 was introduced for entire functions in [AB]. Our discussion is based on [DG], where the isometric extension for balls is obtained. For more on extensions of analytic functions, see [LR] and [G²M²]. Section 2.2 is based on [AHV]. Extension theorems for entire functions that are weakly continuous on bounded sets were obtained in [Mo], in the more general setting of linear topological vector spaces. Our discussion in section 2.3 follows [ACG2], though the equivalence of the conditions (3) and (4) in Theorem 2.3.2 comes from [AS]. Section 2.4 is based on [AHV]. Rosenthal's dichotomy was proved originally for real Banach spaces, and the proof was extended to complex Banach spaces by L. Dor. See [LT]. The example at the end of section 2.4 comes from [Gu]. It is a variant of an example of J. Diestel given in [AHV].

Chapter 3
Approximation by polynomials of finite type

We are concerned in this chapter with the uniform approximation and also the pointwise bounded approximation of analytic functions by finite-type polynomials.

3.1 Polynomials of finite type

A *polynomial of finite type* on $\mathfrak{X}$ is a finite linear combination of finite products of functionals in $\mathfrak{X}^*$, plus the constants. For instance, the finite sums $\sum_{j=1}^N x_j^2$ are finite-type polynomials on ℓ^2. The infinite series $f(x) = \sum_{j=1}^\infty x_j^2$ defines a 2-homogeneous polynomial on ℓ^2, but it is not of finite type.

An m-linear functional is *of finite type* if it is a finite linear combination of functionals of the form $G(x_1,\ldots,x_m) = L_1(x_1)\cdots L_m(x_m)$, where the L_j's are in $\mathfrak{X}^*$. The restrictions of these to the diagonal are m-homogeneous polynomials of finite type, and conversely the polarization formula shows that the symmetric extension of an m-homogeneous polynomial of finite type is an m-linear functional of finite type.

The finite-type polynomials form a unital algebra, which is generated by $\mathfrak{X}^*$. The Taylor coefficients of a finite-type polynomial are again polynomials of finite type. Any m-homogeneous finite-type polynomial is a finite linear combination of m-homogeneous monomials of the form $g(x) = L_1(x)\cdots L_m(x)$, where the L_j's are fixed in $\mathfrak{X}^*$. Each such function is analytic on $\mathfrak{X}$, and furthermore it is weakly continuous.

Lemma 3.1.1 *If f is an entire function on $\mathfrak{X}$ that is weakly continuous, then there is a closed subspace $\mathcal{Y}$ of $\mathfrak{X}$ of finite codimension, and an entire function g on $\mathfrak{X}/\mathcal{Y}$, such that f is the composition of g and the quotient map of $\mathfrak{X}$ onto $\mathfrak{X}/\mathcal{Y}$.*

Proof Choose $L_1,\ldots,L_n \in \mathfrak{X}^*$ such that f is bounded on the weak neighborhood $U = \{x : |L_j(x)| < 1, 1 \le j \le n\}$ of 0. Let $\mathcal{Y} = \{x : L_j(x) = 0, 1 \le j \le n\}$. If $x \in U$ and $y \in \mathcal{Y}$, then $x + y \in U$, so that $y \to f(x+y)$ is a bounded entire function on $\mathcal{Y}$ hence constant. Thus $f(x+y) = f(x)$ for all $x \in U$ and $y \in \mathcal{Y}$. If $f = \sum f_m$ is the Taylor series of f, then also $f_m(x+y) = f_m(x)$ for all $x \in U$, $y \in \mathcal{Y}$, hence by homogeneity for all $x \in \mathfrak{X}$, $y \in \mathcal{Y}$. Let F_m be the symmetric m-linear functional corresponding to f_m. Noting the pairwise cancellation of terms corresponding to $\varepsilon_m = \pm 1$ in the polarization formula (section 1.1), we obtain $F_m(x_1,\ldots,x_{m-1},y) = 0$ whenever $x_1,\ldots,x_{m-1} \in \mathfrak{X}$ and $y \in \mathcal{Y}$. It follows that $F_m(x_1,\ldots,x_m) = G_m(x_1+\mathcal{Y},\ldots,x_m+\mathcal{Y})$ for some symmetric m-linear functional G_m on $\mathfrak{X}/\mathcal{Y}$. Let g_m be the restriction of G_m to the diagonal, and set $g = \sum g_m$. Then $f(x) = g(x+\mathcal{Y})$ for all $x \in \mathfrak{X}$. □

Corollary 3.1.2 *Any entire function on $\mathfrak{X}$ that is weakly continuous is uniformly approximable on bounded sets by polynomials of finite type. Moreover, any m-homogeneous analytic function that is weakly continuous is a polynomial of finite type.*

3.2 The approximation property

Recall that the Banach space $\mathfrak{X}$ has the *approximation property* if for every $\varepsilon > 0$ and compact subset E of $\mathfrak{X}$, there is a finite-dimensional operator Q on $\mathfrak{X}$ such that $\|Qy - y\| < \varepsilon$ for all $y \in E$.

The approximation property guarantees that every compact operator can be approximated in operator norm by finite-dimensional operators. To see this, one simply composes the compact operator with a finite-dimensional Q that is close to the identity on the image of the unit ball under the compact operator. This simple line of proof yields also approximation theorems for multilinear functionals.

Theorem 3.2.1 *Suppose that the dual $\mathfrak{X}^*$ of $\mathfrak{X}$ has the approximation property. Then the m-linear functionals that are weakly continuous on bounded sets are precisely the limits in the norm of $\mathcal{L}(^m\mathfrak{X})$ of m-linear functionals of finite type.*

Proof The lemma is true if $m = 1$, and we make the induction hypothesis that it holds for $(m-1)$-linear functionals. Let F be an m-linear functional that is weakly continuous on bounded sets, and let S be the $(m-1)$-linear operator from $\mathfrak{X}^{m-1}$ to $\mathfrak{X}^*$ that represents F as in Theorem 2.2.1. By that theorem, the image under S of B^{m-1} is precompact in $\mathfrak{X}^*$. Hence there is a finite-dimensional operator Q on $\mathfrak{X}^*$ such that $\|QS(x_1,\ldots,x_{m-1}) - S(x_1,\ldots,x_{m-1})\| \leq \varepsilon$ whenever $x_j \in B$, $1 \leq j \leq m-1$. Choose L_j's in $\mathfrak{X}^*$ and Λ_j's in $\mathfrak{X}^{**}$ such that $Q(L) = \sum \Lambda_j(L)L_j$ for $L \in \mathfrak{X}^*$. Since S is weakly continuous on bounded sets, each $(m-1)$-linear functional $(x_1,\ldots,x_{m-1}) \rightarrow \Lambda_j(S(x_1,\ldots,x_{m-1}))$ is weakly continuous on bounded sets. By the induction hypothesis, each of these $(m-1)$-linear functionals is approximable in norm by $(m-1)$-linear functionals of finite type, and consequently F is approximable in norm by m-linear functionals of finite type. □

Theorem 3.2.2 *Suppose that the dual $\mathfrak{X}^*$ of $\mathfrak{X}$ has the approximation property. Then an analytic function f on B extends to be weak-star continuous on the closed unit ball $\bar{B}^{**}$ if and only if f is uniformly approximable on B by polynomials of finite type.*

Proof Suppose f extends weak-star continuously to $\bar{B}^{**}$. By Lemma 2.3.1, the Taylor coefficients of f are weakly continuous on bounded sets. By Theorem 3.2.1, these are then uniformly approximable by polynomials of finite type. Thus each dilate f_r is approximable, for $r < 1$, and these dilates tend uniformly to f as r increases to 1. The converse implication is trivial. □

3.3 Failure of approximation

We have seen that, roughly speaking, a weakly continuous analytic function on B is approximable by finite-type polynomials if and only if (i) its Taylor coefficients are approximable, and (ii) some condition holds that guarantees convergence of the dilates of the function back to itself. In turn, an m-homogeneous analytic function that is weakly continuous on B is uniformly approximable if and only if the corresponding symmetric m-linear functional is uniformly approximable by functionals of finite type. This implies that the corresponding compact linear operator from $\widehat{\otimes}_s^{m-1}\mathfrak{X}$ to $\mathfrak{X}^*$ is approximable in operator norm by finite dimensional operators. Conversely, approximability of certain compact operators by finite dimensional operators implies approximability of m-linear functionals. To clarify these ideas, we specialize to the case $m = 2$.

An operator $T : \mathfrak{X} \rightarrow \mathfrak{X}^*$ is *symmetric* if the corresponding bilinear form on $\mathfrak{X}$ is symmetric, that is, $(Tx)(y) = (Ty)(x)$ for all $x, y \in \mathfrak{X}$. Thus the space of 2-homogeneous polynomials is isomorphic (though not isometric) to the space of continuous symmetric linear operators from $\mathfrak{X}$ to $\mathfrak{X}^*$, where the operator T corresponds to the function $f(x) = (Tx)(x)$. Under this correspondence, the 2-homogeneous polynomials of finite type correspond to the symmetric finite-dimensional operators from $\mathfrak{X}$ to $\mathfrak{X}^*$. Since the supremum norm of the polynomial over the unit ball B is equivalent to the operator norm of the corresponding symmetric operator, we obtain immediately the following.

Lemma 3.3.1 *A 2-homogeneous analytic function is uniformly approximable on B by polynomials of finite type if and only if the corresponding symmetric operator from $\mathcal{X}$ to $\mathcal{X}^*$ is approximable in operator norm by finite-dimensional operators.*

Theorem 3.3.2 *Let V be a continuous linear operator on a reflexive Banach space $\mathcal{Y}$, and let f be the 2-homogeneous polynomial on $\mathcal{Y} \oplus \mathcal{Y}^*$ defined by $f(y \oplus L) = L(Vy)$. Then f is uniformly approximable on the unit ball of $\mathcal{Y} \oplus \mathcal{Y}^*$ by polynomials of finite type if and only if V is approximable in operator norm by finite-dimensional operators.*

Proof The symmetric operator T from $\mathcal{X} = \mathcal{Y} \oplus \mathcal{Y}^*$ to $\mathcal{X}^* = \mathcal{Y}^* \oplus \mathcal{Y}$ corresponding to $2f$ is given by $T(x \oplus L) = V^*(L) \oplus V(x)$. Now T is approximable by finite-dimensional operators if and only if both V and V^* are, and this occurs if and only if V is approximable. Lemma 2.3.1 then yields the conclusion. □

According to a theorem of P. Enflo, there is a compact operator on some Banach space that is not approximable by finite-dimensional operators. Moreover, we can take the Banach space to be reflexive (see [LT]). Thus the corresponding function f of the preceding theorem is not uniformly approximable on the unit ball by polynomials of finite type, though by Theorem 2.2.1, f is weakly continuous on bounded sets.

3.4 Littlewood's theorem

J.E. Littlewood proved in 1930 that every bilinear functional on c_0 can be approximated by functionals of finite type. The theorem was extended to m-linear functionals by W. Bogdanowicz and A. Pelczynski. We are in a position to prove a theorem that generalizes this result considerably. First we give an elementary proof along the lines of Littlewood. We will use c_{00} to denote the space of sequences in ℓ^∞ with only finitely many nonzero entries. Thus c_{00} is a dense linear subset of c_0.

Theorem 3.4.1 *Let $a_{i_1 \cdots i_m}$ be complex numbers, $1 \le i_j < \infty$, $1 \le j < \infty$, and define*

$$Q(z_1, \ldots, z_m) = \sum_{i_1, \cdots, i_m \ge 1} a_{i_1 \cdots i_m} z_1^{(i_1)} \cdots z_m^{(i_m)}$$

for z_j's in c_{00}. Suppose that Q is bounded, that is,

$$\sup\{|Q(z_1, \ldots, z_m)| \,:\, z_j \in c_{00}, ||z_j|| \le 1\} = C < \infty.$$

Then

$$\hat{Q}(z_1, \ldots, z_m) = \lim_{N_1, \ldots, N_m \to \infty} \sum_{i_1=1}^{N_1} \cdots \sum_{i_m=1}^{N_m} a_{i_1 \cdots i_m} z_1^{(i_1)} \cdots z_m^{(i_m)}$$

exists uniformly for z_j's in the unit ball of ℓ^∞ and defines a continuous m-linear functional on ℓ^∞ of norm C. In particular, every continuous m-linear functional on c_0 is approximable in norm by m-linear functionals of finite type.

Proof The proof is by induction on m. We assume that $m \geq 2$ and that the theorem is true for $m-1$. For the proof, we denote by B_{00} the unit ball of c_{00}. It suffices to show that

$$\sup\left\{\left|\sum_{\max(i_1,\ldots,i_m)>N} a_{i_1\cdots i_m} z_1^{(i_1)}\cdots z_m^{(i_m)}\right| : z_1,\ldots,z_m \in B_{00}\right\} \tag{1}$$

tends to 0 as $N \to \infty$.

Let $\varepsilon > 0$. Choose $\zeta_1,\ldots,\zeta_m \in B_{00}$ such that $|Q(\zeta_1,\ldots,\zeta_m)| > C-\varepsilon$, and choose $L \geq 1$ so that $\zeta_j^{(i)} = 0$ for $i > L$ and $1 \leq j \leq m$. Suppose $w_j \in c_{00}$ satisfies $w_j^{(i)} = 0$ for $1 \leq i \leq L$. Using the multilinearity to expand

$$Q(\zeta_1 + e^{i\theta}w_1,\ldots,\zeta_{m-1}+e^{i\theta}w_{m-1},\zeta_m + e^{i(1-m)\theta}w_m), \tag{2}$$

we obtain a trigonometric polynomial in θ with constant coefficient equal to

$$Q(\zeta_1,\ldots,\zeta_m) + Q(w_1,\ldots,w_m). \tag{3}$$

The entries in (2) have at most unit norm, so that the modulus of (2) is bounded by C. Thus the modulus of (3) is also bounded by C, and since we are free to multiply each w_j by a unimodular constant, we conclude that $|Q(w_1,\ldots,w_m)| < \varepsilon$. Thus we obtain

$$\left|\sum_{i_1>L}\cdots\sum_{i_m>L} a_{i_1\cdots i_m} z_1^{(i_1)}\cdots z_m^{(i_m)}\right| < \varepsilon, \qquad z_1,\ldots,z_m \in B_{00}. \tag{4}$$

By the inductive hypothesis, we can find N_r such that for $N \geq N_r$ and $1 \leq i_r \leq L$,

$$\left|\sum_{\max(i_1,\ldots,i_{r-1},i_{r+1},\ldots,i_m)>N} a_{i_1\cdots i_m} z_1^{(i_1)}\cdots z_{r-1}^{(i_{r-1})} z_{r+1}^{(i_{r+1})}\cdots z_m^{(i_m)}\right| < \varepsilon/L \tag{5}$$

for all $z_1,\ldots,z_m \in B_{00}$. Let $N = \max(L,N_1,\ldots,N_m)$. Denote by $\mathfrak{I}$ the set of multiindices $i = (i_1,\ldots,i_m)$ for which $\max(i_1,\ldots,i_m) > N$. We partition $\mathfrak{I}$ into a disjoint union of $\mathfrak{I}_k$'s, $\mathfrak{J}_k$'s, and $\mathcal{K}$, as follows. We put i in $\mathcal{K}$ if $i_k > N$ for all k; otherwise if $i_j > N$ for $1 \leq j \leq k$ and $i_k \leq N$, we put i in $\mathfrak{I}_k$ if $1 \leq i_k \leq L$, and we put i in $\mathfrak{J}_k$ if $L+1 \leq i_k \leq N$. By applying (4) with appropriate coefficients of the z_j's replaced by 0, we bound each of the sums over the $\mathfrak{J}_k$'s and over $\mathcal{K}$ by ε. Similarly, the subsum over the indices in $\mathfrak{I}_k$ with i_k fixed is estimated by ε/L, by applying (5) with appropriate coefficients of the z_j's replaced by 0, so that the sum over each $\mathfrak{I}_k$ is estimated by ε. Thus the grand sum (1) is estimated by $(2m+1)\varepsilon$, and the result is proved. □

Since bilinear functionals on $\mathcal{X}$ correspond to linear operators from $\mathcal{X}$ to $\mathcal{X}^*$, Littlewood's theorem can be rephrased as saying that every continuous linear operator from c_0 to ℓ^1 is compact. If we are permitted to use several "name" theorems, we can prove this directly as follows. By Grothendieck's theorem, any operator from c_0 to ℓ^1 factors through ℓ^2, and in particular is weakly compact. On account of the Eberlein-Schmulian theorem (weak compactness = weak sequential compactness) and Schur's theorem (weakly convergent sequences in ℓ^1 are norm convergent), any weakly compact subset of ℓ^1 is norm compact. Consequently any weakly compact operator into ℓ^1 is norm compact.

Now we give an abstract version of Littlewood's theorem. It is obtained immediately upon combining Theorems 3.2.1 and 2.4.2. The proof is not self-contained, in that it depends on Rosenthal's dichotomy, which we did not prove.

Theorem 3.4.2 *Suppose that ℓ^1 does not embed in $\mathfrak{X}$, that $\mathfrak{X}$ has the Dunford-Pettis property, and that $\mathfrak{X}^*$ has the approximation property. Let f be a bounded analytic function on the open unit ball B of $\mathfrak{X}$. Then there is a sequence of polynomials of finite type that is uniformly bounded on B and that converges uniformly to f on each subball B_r, $r < 1$.*

3.5 Pitt's theorem

There is a version of Littlewood's theorem that holds for the spaces ℓ^p. The proof, which follows the lines of Pitt's original proof for bilinear forms, shows clearly how the modulus of uniform continuity enters.

Theorem 3.5.1 *Let $1 < p < \infty$. If $m < p$, then every continuous m-linear functional on ℓ^p is approximable in norm by functionals of finite type.*

Proof We will actually prove the statement in Theorem 3.4.1, with the norm of ℓ^∞ replaced by that of ℓ^p. Thus we let Q and C be as in Theorem 3.4.1, and we use B_{00} to denote the sequences in the unit ball of ℓ^p that have only finitely many nonzero entries.

Let $\varepsilon > 0$. We claim that there exists $L \geq 1$ such that the norm of the functional Q on sequences in ℓ^p with the first L entries equal to 0 is less than ε, that is, such that the estimate (4) in the proof of Theorem 3.4.1 holds. Once this is done, the remainder of the proof is exactly the same as that of Theorem 3.4.1.

Let $\delta > 0$ be small. Choose $\zeta_1, \ldots, \zeta_m \in B_{00}$ so that $Q(\zeta_1, \ldots, \zeta_m) > C - C\delta$. Suppose that coefficients of ζ_j are 0 beyond the L_0th entry. We proceed as before to estimate the norm of Q on w's with first L_0 coefficients equal to 0. This time, since $||\zeta_j + w_j|| \leq 2^{1/p}$ we estimate (2) and hence (3) by $C2^{m/p}$. This leads to the estimate $|Q(w_1, \ldots, w_m)| < C2^{m/p} - (C - C\delta) = \sigma\delta$, where $\sigma = 2^{m/p} + \delta - 1$. We choose δ so small that $\sigma < 1$, and then we find that the norm of Q on w's with first L_0 coefficients equal to 0 is at most $C\sigma$. Now we repeat this procedure and find $L_1 > L_0$ so that the norm of Q on w's with first L_1 coefficients equal to 0 is at most $C\sigma^2$. After a finite number of repetitions, we arrive at an index L such that the norm of Q on w's with first L coefficients equal to 0 is at most ε. □

The condition $m < p$ in Pitt's theorem is sharp. If $1 < p < \infty$ and $m \geq p$, the function $f(x) = \sum x_j^m$ is an m-homogeneous analytic function on ℓ^p that is not approximable by finite-type polynomials. Indeed, the standard basis elements e_j tend to 0 weakly in ℓ^p, while $f(e_j) = 1$, so f is not weakly continuous on bounded sets, hence not approximable. In the case $p = 2$, there is a more striking result. A.S. Nemirovskii [Ne] has shown that for any fixed $N \geq 1$, the closed subalgebra of $H_b(\ell^2)$ generated by the homogeneous polynomials of degree at most N is proper.

There is a corresponding theorem about operators being compact. It is fairly easy to show (see [LT]) that if $1 \leq r < p < \infty$, then any continuous operator from ℓ^p to ℓ^r is compact, as is any operator from c_0 to ℓ^r. (This provides another proof of Littlewood's theorem.) By sharpening the argument used for Theorem 3.5.1 above, it is possible to establish the following more general result, which we state without proof.

Theorem 3.5.2 *If $1 \le r < \infty$, and if $(1/p_1) + \cdots + (1/p_m) < 1/r$, then any continuous linear operator from $\ell^{p_1} \hat{\otimes} \cdots \hat{\otimes} \ell^{p_m}$ to ℓ^r is compact.*

The statement remains true when we replace some of the ℓ^{p_j}'s by c_0 and the corresponding $1/p_j$'s by 0. If we replace all of the ℓ^{p_j}'s by c_0, we obtain the version of Littlewood's theorem for multilinear functionals. If we take all of the p_j's to be p and r to be the conjugate index of p, we obtain Pitt's theorem for approximation of $(m+1)$-linear functionals.

3.6 The principle of local reflexivity

The principle local reflexivity states that any finite-dimensional subspace of $\mathcal{X}^{**}$ is close to being isometric to a finite-dimensional subspace of $\mathcal{X}$. One consequence of this principle is that properties of Banach spaces that can be formulated in terms of finite-dimensional subspaces (so-called "local" properties) are inherited by the bidual. We will not use the principle explicitly, but rather a related result, from which local reflexivity can be easily derived.

Let $\mathcal{B}(\mathcal{X}, \mathcal{Y})$ denote the space of continuous linear operators from $\mathcal{X}$ to $\mathcal{Y}$, with the uniform operator norm. The following theorem is a generalization of Goldstine's theorem on the weak-star density of a Banach space in its bidual.

Theorem 3.6.1 *Let V be a finite-dimensional Banach space. Then:*

(1) *$\mathcal{B}(V, \mathcal{X})^{**}$ is isometrically isomorphic to $\mathcal{B}(V, \mathcal{X}^{**})$.*

(2) *If T is any linear operator from V to $\mathcal{X}^{**}$, there is a net of linear operators $\{T_\alpha\}$ from V to $\mathcal{X}$ such that $||T_\alpha|| \le ||T||$ and $T_\alpha v$ converges weak-star to Tv for all $v \in V$.*

Proof It is easy to see that the spaces in (1) are isomorphic. The crux of the matter is to prove isometry. The statement (2) follows immediately from (1) and Goldstine's theorem. Conversely, from (2) it is easy to deduce isometry in (1).

If V is the n-dimensional sequence space ℓ^1_n, then $\mathcal{B}(V, \mathcal{X})^*$ is the n-fold direct sum of $\mathcal{X}^*$ with the ℓ^∞-norm, and $\mathcal{B}(V, \mathcal{X})^{**}$ is easily seen to be isometric to $\mathcal{B}(V, \mathcal{X}^{**})$. If V is a quotient space of ℓ^1_n modulo a closed subspace W, then $\mathcal{B}(V, \mathcal{X})$ is isometric to the subspace of $\mathcal{B}(\ell^1_n, \mathcal{X})$ consisting of the operators that vanish on W. The bidual is then isometric to a closed subspace of $\mathcal{B}(\ell^1_n, \mathcal{X})^{**} \cong \mathcal{B}(\ell^1_n, \mathcal{X}^{**})$, which is easily identified as the subspace of operators vanishing on W, so that again the theorem holds for V.

Since an arbitrary norm on V can be approximated by the norm arising as a quotient space of an ℓ^1-space, the approximation statement (2) is easily seen to hold for arbitrary norms on V, and then so does (1). □

We will actually make use of the following dual version of the preceding theorem. Recall that $\bar{B}^*$ is the closed unit ball of $\mathcal{X}^*$.

Theorem 3.6.2 *Let T be a continuous linear operator from $\mathcal{X}^*$ to a finite-dimensional Banach space V. Then there is a net $\{T_\alpha\}$ of continuous linear operators from $\mathcal{X}^*$ to V such that*

(1) $T_\alpha(z) \to T(z)$ *for all* $z \in \mathcal{X}^*$,

(2) $T_\alpha(\bar{B}^*) \subseteq T(\bar{B}^*)$ *for all a,*

(3) *each T_α is continuous with respect to the weak-star topology of $\mathcal{X}^*$.*

Proof Replacing V by the range of T, we can assume that $T(\mathcal{X}^*) = V$. Renorming V, we can assume that $T(\bar{B}^*)$ coincides with the closed unit ball of V. Then $||T|| = 1$, and (2) is equivalent to $||T_\alpha|| \leq 1$. By the preceding theorem, there is a net $\{S_\alpha\}$ in $\mathcal{B}(V^*, \mathcal{X})$ such that $||S_\alpha|| \leq 1$ and $S_\alpha(w) \to T^*(w)$ weak-star for all $w \in V^*$. Then $T_\alpha = S_\alpha^*$ has the properties of the theorem. □

Now we are in a position to obtain another generalization of Goldstine's theorem for polynomials, in a different direction as the generalization given in the preceding chapter.

Theorem 3.6.3 *Let f be a finite-type polynomial on $\mathcal{X}^*$. Then there is a net $\{f_\alpha\}$ of weak-star continuous finite-type polynomials on $\mathcal{X}^*$ such that $||f_\alpha||_{B^*} \leq ||f||_{B^*}$ for all α, and $f_\alpha(z) \to f(z)$ for all $z \in \mathcal{X}^*$.*

Proof Let $\varphi_1, \ldots, \varphi_n$ be the aggregate of functionals in $\mathcal{X}^{**}$ used to express f as a finite-type polynomial. Then $f(z) = p(T(z))$, where T is the operator from $\mathcal{X}^*$ to $\mathbf{C}^n$ given by $T(z) = (\varphi_1(z), \ldots, \varphi_n(z))$, and p is a polynomial on $\mathbf{C}^n$. Evidently $|p(w)| \leq ||f||_{B^*}$ for all $w \in T(B^*)$. Let $\{T_\alpha\}$ be the net from Theorem 3.6.2 above. Since T_α is continuous with respect to the weak-star topology of $\mathcal{X}^*$, there are $x_\alpha^1, \ldots, x_\alpha^n$ in $\mathcal{X}$ such that $T_\alpha(z) = (z(x_\alpha^1), \ldots, z(x_\alpha^n))$, $z \in \mathcal{X}^*$. Thus $f_\alpha = p o T_\alpha$ defines a net of weak-star continuous finite-type polynomials, which has the desired properties. □

3.7 Pointwise bounded approximation

Let f be a bounded analytic function on the open unit ball B of the Banach space $\mathcal{X}$. We are concerned with the following questions. When is there a net (or sequence) $\{f_\alpha\}$ of finite-type polynomials that is uniformly bounded in norm and that converges pointwise to f on B? When can the approximators be taken to satisfy $||f_\alpha||_B \leq ||f||_B$?

There are functions f that cannot be approximated pointwise boundedly. This is seen by modifying the example given in section 3.2 for which uniform approximation fails. We start with a separable reflexive Banach space $\mathcal{X}$ on which there is a compact operator that is not approximable by finite-dimensional operators. From such an operator we construct a 2-homogeneous polynomial f on $\mathcal{X} \oplus \mathcal{X}^*$ that is weakly continuous on $\bar{B} \oplus \bar{B}^*$ but that is not uniformly approximable there by finite-type polynomials. We claim that f is not even approximable pointwise boundedly by finite-type polynomials. Indeed, if it were, then on account of the separability, f would be approximable pointwise boundedly by a *sequence* of finite-type polynomials. On account of the reflexivity, $\bar{B} \oplus \bar{B}^*$ is weak-star compact, and f is weak-star continuous. Consequently the approximating sequence converges weakly to f in the Banach space of weak-star continuous functions on $\bar{B} \oplus \bar{B}^*$. A sequence of convex combinations then converges uniformly to f, and this contradicts the choice of f.

A Banach space $\mathcal{X}$ has the *bounded approximation property* if there are $\lambda \geq 1$ and a net $\{Q_\alpha\}$ of finite-dimensional operators on $\mathcal{X}$ such that $||Q_\alpha|| \leq \lambda$ for all α, and $Q_\alpha(x) \to x$ for all $x \in \mathcal{X}$. In this case Q_α converges uniformly to the identity on any compact subset of $\mathcal{X}$, so that $\mathcal{X}$ has the approximation property. If we can choose the Q_α's above to satisfy $||Q_\alpha|| \leq 1$, we say that $\mathcal{X}$ has the *metric approximation property.*

If $\mathcal{X}$ has the metric approximation property, then any bounded analytic function f on B is approximable pointwise on B by a net $\{f_\alpha\}$ of finite-type polynomials satisfying $||f_\alpha||_B \leq ||f||_B$. Simply choose a polynomial g_α on the range of Q_α that approximates f there and that is bounded on the unit ball of the range of Q_α by $||f||$, and set $f_\alpha = g_\alpha \circ Q_\alpha$.

If $\mathcal{X}$ has only the bounded approximation property, we can still approximate polynomials pointwise boundedly by polynomials of finite type. However, the obvious estimates grow with the degree of the polynomial, and it is not clear whether arbitrary bounded analytic functions on B are pointwise boundedly approximable.

If we combine the observations above with Theorem 3.6.3, we obtain the following result.

Theorem 3.7.1 *Suppose that the dual space $\mathcal{X}^*$ has the metric approximation property. Then any bounded analytic function f on the open unit ball B^* of $\mathcal{X}^*$ can be approximated pointwise on B^* by a net $\{f_\alpha\}$ of weak-star continuous polynomials of finite type with supremum norms over B^* satisfying $||f_\alpha|| \leq ||f||$.*

3.8 Reflexivity of $\mathcal{P}(^m\mathcal{X})$

There has been some continuing interest in the spaces $\mathcal{P}(^m\mathcal{X})$ as Banach spaces. We present one elementary result, which provides a point of entrance for this area.

First observe that for $1 \leq k \leq m-1$, each $\mathcal{P}(^k\mathcal{X})$ embeds in $\mathcal{P}(^m\mathcal{X})$. Indeed, fix $L \in \mathcal{X}^*$ such that $||L|| = 1$. The map $f \to fL^{m-k}$ maps $\mathcal{P}(^k\mathcal{X})$ continuously into $\mathcal{P}(^m\mathcal{X})$. To show that it is an embedding, it suffices to obtain a lower bound $||fL^{m-k}|| \geq c||f||$. We claim in fact that $|f(x)| \leq 4^{m-k}||fL^{m-k}||$ for $x \in B$. This is clear if $|L(x)| \geq 1/4$. If $|L(x)| < 1/4$, take $x_0 \in B$ such that $|L(x_0)| > 3/4$, and consider $y_\lambda = (x + \lambda x_0)/2$. For $|\lambda| = 1$ we have $y_\lambda \in B$ and $|L(y_\lambda)| \geq 1/4$, so that $|f(y_\lambda)| \leq 4^{m-k}||fL^{m-k}||$. By the maximum principle, the same estimate holds for $y_0 = x$, which gives the lower bound in all cases.

In particular, $\mathcal{X}^*$ embeds in $\mathcal{P}(^m\mathcal{X})$. Thus if $\mathcal{P}(^m\mathcal{X})$ is reflexive, then so is $\mathcal{X}$, as are all the spaces $\mathcal{P}(^k\mathcal{X})$ for $1 \leq k \leq m$.

A natural problem is to find necessary and sufficient conditions for $\mathcal{P}(^m\mathcal{X})$ to be reflexive.

The following theorem is a variant of a theorem that appears in [Ry1]. Recall that $\mathcal{P}(^m\mathcal{X})$ is a dual Banach space, and the predual can be taken to be $\ell^1(\bar{B})/\mathcal{P}(^m\mathcal{X})^\perp$.

Theorem 3.8.1 *Let $\mathcal{Q}_m$ be the weak-star closure in $\mathcal{P}(^m\mathcal{X})$ of the polynomials of finite type. Then $\mathcal{Q}_m$ is reflexive if and only if $\mathcal{X}$ is reflexive and the functions in $\mathcal{Q}_m$ are weakly continuous on bounded subsets of $\mathcal{X}$.*

Proof Suppose first that $\mathcal{Q}_m$ is reflexive. The preliminary remarks show that $\mathcal{X}^*$ embeds in $\mathcal{Q}_m$, so that $\mathcal{X}$ is reflexive. Let $x_\alpha \in \bar{B}$ converge weakly (=weak-star) to x. We claim that $f(x_\alpha) \to f(x)$ for all $f \in \mathcal{Q}_m$, that is, that x_α converges to x in the $\mathcal{Q}_m$-topology. Now the predual for $\mathcal{Q}_m$ is $\ell^1(\bar{B})/\mathcal{Q}_m^\perp$, whose unit ball is compact in the $\mathcal{Q}_m$-topology. Passing to a subnet, we can then assume that $\delta_{x_\alpha} + \mathcal{Q}_m^\perp$ converges to $a + \mathcal{Q}_m^\perp$ in the $\mathcal{Q}_m$-topology, for some $a \in \ell^1(\bar{B})$. If $L \in \mathcal{X}^*$, then $L^m \in \mathcal{Q}_m$, and $L(x_\alpha)^m \to \langle L^m, a\rangle$. On the other hand, $L(x_\alpha)^m \to L(x)^m = \langle L^m, \delta_x\rangle$. Thus $\delta_x - a \perp L^m$ for all $L \in \mathcal{X}^*$. Consequently $\delta_x - a$ is orthogonal to all m-homogeneous polynomials of finite type, hence by weak-star density to $\mathcal{Q}_m$. It follows that $x_\alpha \to x$ in the $\mathcal{Q}_m$-topology.

For the converse, suppose that the two conditions hold. Then $\bar{B}$ is weakly compact in $\mathcal{X}$. Since the functions in $\mathcal{Q}_m$ are weakly continuous on $\bar{B}$, it is compact in the $\mathcal{Q}_m$-topology. Consequently the subset $\{\delta_x + \mathcal{Q}_m^\perp : x \in \bar{B}\}$ is weakly compact in $\ell^1(\bar{B})/\mathcal{Q}_m^\perp$. Since the closed convex hull of a weakly compact set is weakly compact, we conclude that the closed unit ball of $\ell^1(\bar{B})/\mathcal{Q}_m^\perp$ is weakly compact, hence the space is reflexive, as is its dual $\mathcal{Q}_m$. □

If $\mathcal{X}^*$ has the bounded approximation property, then the polynomials of finite type are pointwise boundedly dense in $\mathcal{P}(^m\mathcal{X})$, so that $\mathcal{P}(^m\mathcal{X}) = \mathcal{Q}_m$, and the preceding theorem applies. Even without this hypothesis, most of the preceding proof goes through for $\mathcal{P}(^m\mathcal{X})$. It is not known whether the reflexivity of $\mathcal{P}(^m\mathcal{X})$ implies bounded-weak continuity. It does in the presence of the approximation property for $\mathcal{X}^*$; see [Al2].

Fix $1 < p < \infty$, and consider the space $\mathcal{X} = \ell^p$. Pitt's theorem (Theorem 3.5.1) shows that $\mathcal{P}(^m\mathcal{X})$ is reflexive if $m < p$. On the other hand, if $m \geq p$, then the m-homogeneous analytic function $f(x) = \sum x_j^m$ is not weakly continuous on $\bar{B}$, so that $\mathcal{P}(^m\mathcal{X})$ is not reflexive. Another interesting example is provided by the original Tsirelson space, which is a reflexive Banach space denoted by T^* in the literature, its dual being denoted by T. It is proved in [AAD] that the spaces $\mathcal{P}(^mT^*)$ are reflexive for all $m \geq 2$, and this leads to an interesting example of a reflexive space of analytic functions. The space $\mathcal{P}(^2T)$ already fails to be reflexive.

Notes For background on the approximation property, see [LT]. The example of nonapproximability by polynomials of finite type is in [ACG2]. Littlewood's theorem is in [Li], and the papers of Bogdanowicz and Pelczynski are [Bo] and [Pe1]. Pelczynski actually proved (see [Pe3]) that the theorem on uniform approximation by finite-type polynomials holds for the Banach space $C(X)$ if and only if the compact set X is scattered (every nonempty closed subset has an isolated point). This topological condition is equivalent to every measure on X being atomic, and also to the condition that ℓ^1 does not embed in $C(X)$; see [Sa]. Pitt's theorem was proved for bilinear forms in [Pit]. The version given in Theorem 3.5.2 was established, in a more general setting, by Pelczynski [Pe1]. For an interesting related paper, see [BF]. Theorem 3.6.1 already appears in [Sch]. It was used by Dean [De] to give a brief proof of the principle of local reflexivity, which is due to Lindenstrauss and Rosenthal. The applications to pointwise bounded approximation are in [CCG]. For further information and references on the duals and biduals of the spaces $\mathcal{P}(^m\mathcal{X})$, see also [Fa] and [Va].

Chapter 4
The algebra of entire functions

Let $H_b = H_b(\mathcal{X})$ be the algebra of entire functions on the Banach space $\mathcal{X}$. Endowed with the topology of uniform convergence on bounded sets, H_b becomes a Frechet algebra. The translation on $\mathcal{X}$ induces a convolution operation on the dual H_b^* of H_b. The spectrum of H_b is invariant under the convolution, and with convolution it becomes an (associative) semigroup with identity.

4.1 The radius function

Each $f \in H_b$ has a Taylor series expansion $f = \sum f_m$, which converges uniformly on bounded sets, that is, which converges in H_b. Since the radius of bounded convergence of f is infinite, we obtain from section 1.3 that

$$\limsup_{m\to\infty} ||f_m||^{1/m} = 0. \tag{1}$$

Conversely, if functions $f_m \in \mathcal{P}_m$ satisfy (1), then $\sum f_m$ is the Taylor series of a function $f \in H_b$. Note that the condition (1) is equivalent to the existence for each (large) $r > 0$ of $C = C(r)$ such that

$$||f_m|| \le Cr^{-m}, \qquad m \ge 0. \tag{2}$$

There is a similar description for the dual space H_b^*. Before discussing this, we introduce some notation.

Recall that B_r the open ball rB in $\mathcal{X}$ of radius r centered at 0. We denote by $||f||_r$ the supremum norm of a function f over B_r :

$$||f||_r = \sup\{|f(x)| : ||x|| < r\}.$$

Thus $||\cdot||_1$ coincides with the norm on $\mathcal{P}_m$.

For $\varphi \in H_b^*$, let φ_m denote the restriction of φ to $\mathcal{P}_m$, and let $||\varphi_m||$ denote the norm of φ_m in $\mathcal{P}_m^*$. Suppose φ is continuous with respect to the norm of uniform convergence on the ball B_r. Then $|\varphi_m(f)| \le C||f||_r = Cr^m||f||$ for $f \in \mathcal{P}_m$, hence

$$||\varphi_m|| \le Cr^m, \qquad m \ge 0. \tag{3}$$

Conversely, suppose functionals $\varphi_m \in \mathcal{P}_m^*$ satisfy (3) for some $r > 0$. Let $s > r$. If $f = \sum f_m \in H_b$, then $|\varphi_m(f_m)| \le Cr^m||f_m|| \le C(r/s)^m||f_m||_s$. Consequently $\varphi(f) = \sum \varphi_m(f_m)$ converges absolutely and satisfies $|\varphi(f)| \le C(1-r/s)^{-1}||f||_s$. Thus $\varphi \in H_b^*$, and in fact φ is continuous with respect to the norm $||\cdot||_s$ for any $s > r$.

This leads us to define the *radius function* $R(\varphi)$, for $\varphi \in H_b^*$, to be the infimum of all $r > 0$ such that φ is continuous with respect to the norm of uniform convergence on B_r. Thus $0 \leq R(\varphi) < \infty$. The following theorem is dual to the formula for the radius of bounded convergence.

Theorem 4.1.1 *Functionals $\varphi_m \in \mathcal{P}_m^*$, for $m \geq 0$, arise from a functional $\varphi \in H_b^*$ if and only if* $\limsup ||\varphi_m||^{1/m}$ *is finite. Moreover,*

$$\limsup_{m\to\infty} ||\varphi_m||^{1/m} = R(\varphi), \qquad \varphi \in H_b^*.$$

Proof This follows immediately from the preceding discussion and the fact that the limes supremum coincides with the infimum of $r > 0$ for which there is $C > 0$ satisfying (3). □

Recall from Chapter 2 that every $f \in H_b(\mathcal{X})$ has a canonical extension $\hat{f}$ to $\mathcal{X}^{**}$. The correspondence $f \to \hat{f}$ is an algebra isomorphism, which is continuous from $H_b(\mathcal{X})$ into $H_b(\mathcal{X}^{**})$. For $z \in \mathcal{X}^{**}$, we denote by δ_z the evaluation functional of the canonical extension at z:

$$\delta_z(f) = \hat{f}(z), \qquad f \in H_b.$$

Lemma 4.1.2 $R(\delta_z) = ||z||$ *for* $z \in \mathcal{X}^{**}$.

Proof From the density theorem for the polynomial-star topology (Theorem 2.1.2), scaled to a ball of radius $||z||$, we obtain $|\hat{P}(z)| \leq ||z||^m ||P||$ for all polynomials P in $\mathcal{P}_m$. Consequently the norm of δ_z on $\mathcal{P}_m$ is at most $||z||^m$. If $L \in \mathcal{X}^*$ satisfies $||L|| = 1$ and $L(z) = ||z||$, then from $\delta_z(L^m) = ||z||^m$ we see that the norm of δ_z on $\mathcal{P}_m$ is exactly $||z||^m$. Thus Theorem 4.1.1 gives the result. □

4.2 Convolution on the dual of H_b

The translation operator T_x on H_b is defined by

$$(T_x f)(y) = f(x+y), \qquad x, y \in \mathcal{X},\ f \in H_b.$$

It induces a dual action on H_b^*,

$$(T_x^* \varphi)(f) = \varphi(T_x f), \qquad f \in H_b,\ \varphi \in H_b^*.$$

It is straightforward to check that

$$R(T_x^* \varphi) \leq R(\varphi) + ||x||, \qquad x \in \mathcal{X},\ \varphi \in H_b^*.$$

The action of translation on H_b is regular enough to induce a convolution operation on H_b^*. The key point is the following lemma.

Lemma 4.2.1 *If $f \in H_b$ and $\varphi \in H_b^*$, then the function $x \to \varphi(T_x f)$ belongs to H_b.*

Proof The estimate for $R(T_x^*\varphi)$ gives an estimate of the form

$$|\varphi(T_x f)| \le c\|f\|_{R(\varphi)+r+\epsilon}, \qquad \|x\| \le r,$$

so that $\varphi(T_x f)$ is locally bounded. For analyticity, we can assume that f is an m-homogeneous analytic function, say the restriction to the diagonal of the m-linear functional F. Then $(T_x f)(y) = F(x+y, \ldots, x+y)$. If we apply φ to this, as a function of y, we have an analytic function of x. □

Now for $\varphi, \theta \in H_b^*$, we define the convolution $\varphi * \theta$ by

$$(\varphi * \theta)(f) = \varphi(\theta(T_x f)), \qquad f \in H_b.$$

In other words, if g is the auxiliary function defined by $g(x) = \theta(T_x f)$, then since $g \in H_b$ we may define $(\varphi * \theta)(f) = \varphi(g)$.

It is easy to check that for $x, y \in \mathcal{X}$, the convolution of the evaluation functionals δ_x and δ_y corresponds to addition of x and y, that is, $\delta_x * \delta_y = \delta_{x+y}$. More generally, convolution with δ_x coincides with the dual translation action:

$$\delta_x * \varphi = \varphi * \delta_x = T_x^*\varphi, \qquad x \in \mathcal{X},\ \varphi \in H_b^*.$$

Two basic properties of convolution are contained in the following lemma. We omit the proof, which is a straightforward argument based on the definitions.

Lemma 4.2.2 *If $\varphi, \theta \in H_b^*$, then $\varphi * \theta \in H_b^*$, and $R(\varphi * \theta) \le R(\varphi) + R(\theta)$. The convolution operation is associative, that is, $\varphi * (\theta * \xi) = (\varphi * \theta) * \xi$.*

We have seen that H_b^* can be regarded as the sum of the various dual spaces $\mathcal{P}_m^*$, $m \ge 0$. We aim to show that convolution respects the natural grading, and to investigate the action of convolution on these spaces. For this purpose, we identify $\mathcal{P}_m^*$ with the subspace of functionals in H_b^* that annihilate $\mathcal{P}_k$ for all $k \ne m$. Thus $\mathcal{P}_0^* \cong \mathbf{C}$ is the one-dimensional subspace spanned by δ_0, and $\mathcal{P}_1^* \cong \mathcal{X}^{**}$. For $\varphi \in H_b$, we regard the restriction φ_m of φ to $\mathcal{P}_m$ as an element of H_b^* by declaring it to be 0 on $\mathcal{P}_k$ for $k \ne m$. With this convention, note that $\varphi_0 = \varphi(1)\delta_0$.

From $(\varphi * \theta)(1) = \varphi(1)\theta(1)$ we obtain $(\varphi * \theta)_0 = \varphi_0\theta_0$. The calculation of the component $(\varphi * \theta)_1$ of $\varphi * \theta$ in $\mathcal{X}^{**}$ is almost as easy. If $L \in \mathcal{X}^*$, then $(T_x L)(y) = L(x) + L(y)$, so the auxiliary function g above is $g(x) = \theta(1)L(x) + \theta(L)$, and $(\varphi * \theta)(L) = \varphi(g) = \theta(1)\varphi(L) + \varphi(1)\theta(L)$. Thus

$$(\varphi * \theta)_1 = \theta(1)\varphi_1 + \varphi(1)\theta_1, \qquad \varphi, \theta \in H_b^*.$$

In general, we have

$$(\varphi * \theta)_m = \sum_{k=0}^{m} \varphi_k * \theta_{m-k}, \qquad m \ge 0,\ \varphi, \theta \in H_b^*.$$

This is an immediate consequence of the first statement in the following theorem.

Theorem 4.2.3 *If $\varphi \in \mathcal{P}_j^*$ and $\theta \in \mathcal{P}_k^*$, then $\varphi * \theta \in \mathcal{P}_{j+k}^*$. Furthermore, if $P \in \mathcal{P}_{j+k}$ has associated symmetric $(j+k)$-linear functional F, then*

$$(\varphi * \theta)(P) = \frac{(j+k)!}{j!k!}\varphi^{(x)}(\theta^{(y)}(F(x,\ldots,x,y,\ldots,y))),$$

where there are j entries x and k entries y, and where $\theta^{(y)} = \theta$ operates with respect to the y variables and $\varphi^{(x)} = \varphi$ with respect to the x variables.

Proof Let $P \in \mathcal{P}_m$ have corresponding symmetric m-form F. The function $(T_xP)(y)$ then has the form

$$F(x+y,\ldots,x+y) = \sum_{j=0}^{m} \frac{m!}{j!(m-j)!}F(x,\ldots,x,y,\ldots,y),$$

where x appears j times in the jth summand. From this, the formula follows easily. □

For $z_1,\ldots,z_m \in \mathcal{X}^{**} \cong \mathcal{P}_1^*$, the convolution $z_1 * \cdots * z_m$ belongs to $\mathcal{P}_m^*$. It is identified explicitly in the following theorem.

Theorem 4.2.4 *Let $P \in \mathcal{P}_m$ have associated symmetric m-linear functional F, and let $\hat{F}$ be the extension of F to $\mathcal{X}^{**}$ obtained by extending by weak-star continuity, one variable at a time, from last to first. Then*

$$(z_1 * \cdots * z_m)(P) = m!\hat{F}(z_1,\ldots,z_m), \qquad z_1,\ldots,z_m \in \mathcal{X}^{**}.$$

Proof In this case the auxiliary function g is given by

$$g(x) = (z_2 * \cdots * z_m)(F(x+y,\ldots,x+y)).$$

Expanding the expression on the right and noting that $z_2 * \cdots * z_m$ annihilates all terms except the m terms that are $(m-1)$-homogeneous in y, we obtain $g(x) = m(z_2 * \cdots * z_m)(F(x,y,\ldots,y))$. We make the induction assumption that $(z_2 * \cdots * z_m)(F(x,y,\ldots,y)) = (m-1)!\hat{F}(x,z_2,\ldots,z_m)$, and then we simply apply z_m to obtain the result. □

In view of the definition of the canonical extension $\hat{P}$, we obtain from Theorem 4.2.4 that the component $(\delta_z)_m$ of δ_z in $\mathcal{P}_m^*$ is $(z * \cdots * z)/m!$, where z appears m times. Thus

$$\delta_z = \exp(*z) = 1 + z + \frac{z*z}{2!} + \frac{z*z*z}{3!} + \cdots, \qquad z \in \mathcal{X}^{**}.$$

Our next goal is to determine when the evaluation functionals δ_z commute under convolution.

Theorem 4.2.5 *The following are equivalent, for a Banach space $\mathcal{X}$.*

(1) *$z * w = w * z$ for all $z, w \in \mathcal{X}^{**}$.*

(2) $\delta_z * \delta_w = \delta_w * \delta_z$ *for all* $z, w \in \mathcal{X}^{**}$.

(3) $\delta_{z+w} = \delta_z * \delta_w$ *for all* $z, w \in \mathcal{X}^{**}$.

(4) *Every continuous symmetric bilinear functional on* $\mathcal{X}$ *extends to be a separately weak-star continuous bilinear functional on* $\mathcal{X}^{**}$.

(5) *For all* $m \geq 1$, *each continuous symmetric* m*-linear functional on* $\mathcal{X}$ *extends to be a separately weak-star continuous symmetric* m*-linear functional on* $\mathcal{X}^{**}$.

(6) *Every continuous symmetric linear operator from* $\mathcal{X}$ *to* $\mathcal{X}^*$ *is weakly compact.*

Proof The exponential formula for δ_z shows that (1) implies (2) and (3). The $\mathcal{P}_2$-term of $\delta_z * \delta_w - \delta_w * \delta_z$ is $z * w - w * z$, while that of $\delta_{z+w} - \delta_z * \delta_w$ is $(z * w - w * z)/2$, so that each of (2) and (3) implies (1).

If (1) holds, then Theorem 4.2.4 shows that every symmetric bilinear functional F satisfies $\hat{F}(z, w) = \hat{F}(w, z)$, that is, $\hat{F}$ is symmetric. Since $\hat{F}$ is always weak-star continuous with respect to the first variable, (4) holds. Conversely, if F has a separately weak-star continuous extension, then $\hat{F}$ is symmetric, and from Theorem 4.2.4 we conclude that (1) holds. Thus (1) through (4) are equivalent.

If (4) holds, then in the process of extending F one variable at a time by weak-star continuity, if we interchange the order of two consecutive variables, we arrive at the same extension, since each of the partial extensions is symmetric in the untreated variables. By making a finite number of such switches, we can reach any ordering of the variables, and hence the extension is independent of the order in which we take the variables. This implies the extension is separately weak-star continuous, and hence also symmetric. Thus (4) and (5) are equivalent.

For (6), we use again the fact that a continuous linear operator from $\mathcal{X}$ to $\mathcal{Y}$ is weakly compact if and only if the image of $\mathcal{X}^{**}$ under the second adjoint is contained in $\mathcal{Y}$. Let $T : \mathcal{X} \to \mathcal{X}^*$ be the symmetric operator associated to the symmetric bilinear functional F. Then the bilinear functional $\langle T^{**}z, w\rangle = \langle z, T^*w\rangle$ is an extension of F to $\mathcal{X}^{**}$ that is weak-star continuous in w for each fixed $z = x$ in $\mathcal{X}$, and weak-star continuous in z for each fixed $w \in \mathcal{X}^{**}$, hence coincides with $\hat{F}$. This functional is weak-star continuous in w for each fixed $z \in \mathcal{X}^{**}$ if and only if the image of T^{**} is contained in $\mathcal{X}^*$, and this occurs if and only if T is weakly compact. □

Theorem 4.2.5 applies for instance to the spaces $C(X)$, where X is a compact Hausdorff space. According to a theorem of Grothendieck ([Pis]), every continuous linear operator from $C(X)$ to $C(X)^*$ factors through Hilbert space and hence is weakly compact.

The conditions in Theorem 4.2.5 fail for ℓ^1. One way to see this is as follows. Let $\{y_n\}$ be any bounded sequence in ℓ^∞ with no weakly convergent subsequence; for instance, set $y_{nj} = 1$ for j odd, $j > n$, and $y_{nj} = 0$ otherwise. Define a continuous linear operator S from ℓ^1 to ℓ^∞ by specifying $S(e_n) = y_n$ on the standard basis elements e_n of ℓ^1. Then S is not weakly compact. To manufacture from S a symmetric operator T that is not weakly compact, simply define $T(x \oplus y) = S(y) \oplus S^*(x)$ on $\ell^1 \oplus \ell^1 \cong \ell^1$.

A striking difference between linear functionals and polynomials is that the Hahn-Banach theorem, one of the most powerful tools of linear analysis, fails for polynomials. In fact, the preceding example can be used to illustrate the failure of the Hahn-Banach theorem for bilinear functionals hence also for polynomials of degree two. Take X so that ℓ^1 embeds in $C(X)$, say $X = [0,1]$, and transfer the bilinear functional determined by T above to the embedded image of ℓ^1, a closed subspace of $C(X)$. Since the bilinear functional has no separately weak-star continuous extension to the bidual of the subspace, which can be regarded as a closed subspace of $C(X)^{**}$, and since all continuous bilinear functionals on $C(X)$ extend to be separately weak-star continuous on $C(X)^{**}$, the bilinear functional cannot be extended continuously to $C(X)$.

The following question arises in connection with Theorem 4.2.5. If every symmetric operator from $\mathfrak{X}$ to $\mathfrak{X}^*$ is weakly compact, then is *every* operator from $\mathfrak{X}$ to $\mathfrak{X}^*$ weakly compact?

4.3 The spectrum of the algebra H_b

The ***spectrum of*** H_b is the set of nonzero continuous complex-valued homomorphisms of H_b. The spectrum will be denoted by M_b, or by $M_b(\mathfrak{X})$. Thus M_b is a subset of H_b^*. We endow M_b with the weak topology determined by H_b, so that a net $\{\varphi_\alpha\}$ converges to φ in M_b if and only if $\varphi_\alpha(f) \to \varphi(f)$ for all $f \in H_b$.

Theorem 4.3.1 *The spectrum M_b of H_b is closed under convolution, hence it forms a semigroup with identity element δ_0.*

Proof Let $\varphi, \theta \in M_b$, and let $f, h \in H_b$. Since T_x and θ are multiplicative, $\theta(T_x(fh)) = \theta(T_x(f)T_x(h)) = \theta(T_x(f))\theta(T_x(h))$. Applying φ and using its multiplicativity, we see that $\varphi * \theta$ is multiplicative. □

Any nonzero homomorphism satisfies $\varphi(1) = 1$, so that $\varphi_0 = 1$ for all $\varphi \in M_b$. We denote by $\pi(\varphi)$ the restriction of $\varphi \in M_b$ to $\mathfrak{X}^*$, so that $\pi(\varphi) = \varphi_1 \in \mathfrak{X}^{**}$. From the preceding section we then have

$$\pi(\varphi * \theta) = \pi(\varphi) + \pi(\theta), \qquad \varphi, \theta \in M_b.$$

Since $\pi(\delta_z) = z$ for $z \in \mathfrak{X}^{**}$, the projection π maps M_b onto $\mathfrak{X}^{**}$. The fibers of this projection are generally quite large. However, if the finite-type polynomials are dense in H_b, then each $\varphi \in M_b$ is uniquely determined by its behavior on $\mathfrak{X}^*$, the projection π is one-to-one, and M_b coincides with $\mathfrak{X}^{**}$ as a point set. The following theorem provides a partial converse.

Theorem 4.3.2 *If $M_b(\mathfrak{X})$ coincides with $\mathfrak{X}^{**}$ (as a point set), then the canonical extensions of the functions in H_b are weak-star continuous on bounded subsets of $\mathfrak{X}^{**}$.*

Proof The set of $\varphi \in M_b$ satisfying $R(\varphi) \leq 1$ is H_b-compact. From our hypothesis, we see that the closed ball $\bar{B}_r^{**}$ of X^{**} is H_b-compact. Since the H_b-topology is finer than the weak-star topology, and since the ball is weak-star compact, the two topologies coincide on $\bar{B}_r^{**}$. Thus the functions in H_b are weak-star continuous on any ball in X^{**}. □

If now $\mathfrak{X}^*$ has the approximation property, then the analytic functions that are weak-star continuous on bounded sets are approximable by finite-type polynomials, by Theorem 3.2.1. In this case, $M_b(\mathfrak{X})$ coincides with $\mathfrak{X}^{**}$ (as a point set) if and only if the finite-type polynomials are dense in H_b.

There is an abundance of analytic structure in M_b. Convolution with the δ_z's allows us to pass an analytic copy of $\mathfrak{X}^{**}$ through any $\theta \in M_b$. The following theorem is proved by checking it first on polynomials and then passing to the sum of a Taylor series.

Theorem 4.3.3 *Let $\theta \in M_b$. The correspondence $z \to \delta_z * \theta$ is a one-to-one map of $\mathfrak{X}^{**}$ into M_b. If $f \in H_b$, then the function $z \to (\delta_z * \theta)(f)$ depends analytically on z, and is the canonical extension to $\mathfrak{X}^{**}$ of the function $x \to \theta(T_x f)$ in $H_b(\mathfrak{X})$.*

If $\delta_z * \delta_w \neq \delta_z + \delta_w$, then the embedded image of $\mathfrak{X}^{**}$ passing through δ_w, produced above, does not coincide with $\mathfrak{X}^{**}$. This gives two distinct analytic copies of $\mathfrak{X}^{**}$ passing through δ_w.

There is another way in which analytic structure can be embedded in M_b. Fix $\varphi \in H_b^*$. For each complex number λ, define $\varphi^\lambda \in H_b^*$ by $\varphi^\lambda = \sum \lambda^m \varphi_m$. In other words, if $f \in H_b$ has Taylor series $f = \sum f_m$, then $\varphi^\lambda(f) = \sum \lambda^m \varphi(f_m)$. Evidently φ^λ depends analytically on λ, and $R(\varphi^\lambda) = |\lambda| R(\varphi)$. The map $\lambda \to \varphi^\lambda$ determines a one-dimensional complex analytic set passing through $\varphi^1 = \varphi$ and $\varphi^0 = \delta_0$.

Theorem 4.3.4 *If $\varphi \in M_b$, then also $\varphi^\lambda \in M_b$ for all complex λ. Furthermore, if $\varphi_1, \ldots, \varphi_N \in M_b$, the correspondence $(\lambda_1, \ldots, \lambda_N) \to \varphi_1^{\lambda_1} * \cdots * \varphi_N^{\lambda_N}$ is an analytic map of $\mathbf{C}^N$ into M_b whose image passes through the φ_j's and also through δ_0.*

Proof If $f, h \in H_b$, then $\varphi^\lambda(fh) = \sum_m \lambda^m (\sum_j f_j h_{m-j}) = (\sum \lambda^j f_j)(\sum \lambda^k h_k) = \varphi^\lambda(f)\varphi^\lambda(h)$. Hence $\varphi^\lambda \in M_b$. The analyticity is proved by first checking it on polynomials, then passing to a limit. □

In the same manner we see that if φ, θ are arbitrary points of M_b, there is an analytic disk in M_b passing through the two points. The analytic map $\lambda \to \varphi^\lambda * \theta^{1-\lambda}$ determines a one-dimensional analytic set that contains both points.

4.4 An example: $L^1(\mu)$

One of the few concrete spaces $\mathfrak{X}$ for which the polynomial spaces $\mathcal{P}_m(\mathfrak{X})$ can be easily described is $\mathfrak{X} = L^1(\mu)$, where μ is a measure that is (say) sigma-finite. Let μ_n denote the n-fold product measure of μ with itself. As mentioned in section 1.3, the n-linear functionals on $L^1(\mu)$ are isometric to $L^\infty(\mu_n)$. The n-linear functional corresponding to $h \in L^\infty(\mu_n)$ is given explicitly by

$$F(f_1, \ldots, f_n) = \int f_1(t_1) \cdots f_n(t_n) h(t_1, \ldots, t_n) d\mu_n(t_1, \ldots, t_n), \qquad f_1, \ldots, f_n \in L^1(\mu_n).$$

Symmetric functions h correspond to symmetric functionals F.

Now $L^\infty(\mu_n)$ is isometric to the continuous functions $C(X_n)$ on a certain compact space X_n, namely, the spectrum of $L^\infty(\mu_n)$. The dual of $L^\infty(\mu_n)$ is thus the space of finite regular Borel measures on X_n. The dual of the subspace of symmetric functions in $L^\infty(\mu_n)$ is the subspace of symmetric measures, that is, measures invariant under the homeomorphisms of X_n induced by interchanging variables. The dual space of $H_b(L^1(\mu_n))$ is then the space of *measure chains*, defined to be sequences $\{\beta_n\}_{n=0}^\infty$, where each β_n is a symmetric measure on X_n, and $||\beta_n|| \le Cr^n$, $n \ge 0$, for some $C, r > 0$. Here we agree that X_0 is a one-point space.

There is a natural projection $X_{j+k} \to X_j \times X_k$, whose components are obtained respectively by restricting a functional in $L^\infty(\mu_n)^*$ to functions of the first j and last k variables. The spectrum $M_b(L^1(\mu))$ consists of the measure chains with the property that for all $j, k > 0$, the projection of β_{j+k} onto $X_j \times X_k$ is $\beta_j \times \beta_k$. The term β_0 is irrelevant, since $\varphi(1) = 1$ for all $\varphi \in M_b$. The term β_1 is simply the projection $\pi(\varphi)$ of the functional into $L^1(\mu)^{**}$. The convolution operation can be described in terms of measure chains, and the description can be used to show that convolution inverses do not always exist in M_b, so that M_b is not in general a group. For more details, see [ACG1].

4.5 Another example: ℓ^2

A homomorphism $\varphi \in M_b$ can annihilate infinitely many of the spaces $\mathcal{P}_m(\ell^2)$ and still be distinct from the evaluation functional at 0. This result is due to D. Deghoul [Deg]. The proof depends on Borsuk's theorem, that a continuous antipodal self-map of the n-sphere S^n is onto. In fact, such a map has odd degree, hence cannot be homotopic to a point.

Theorem 4.5.1 *There exists $\varphi \in M_b(\ell^2)$, $\varphi \ne \delta_0$, such that $\varphi = 0$ on $\mathcal{P}_m(\ell^2)$ for all odd integers m.*

Proof Let $P_1, \ldots, P_n$ be any finite collection of homogeneous polynomials of odd degree. Choose a point $x = (x_1, \ldots, x_{n+1})$ with real coordinates such that $P_j(x) = 0$ for $1 \le j \le n$, while $\sum x_j^2 = 1$. Such a point exists, or else the map $x \to (P_1(x), \ldots, P_n(x), 0)/\sum P_j(x)^2$ would be an antipodal self-map of S^n that is not onto, in contradiction to Borsuk's theorem. We order the finite sets of odd polynomials by inclusion, and then the corresponding x's determine a net on the unit sphere of ℓ^2. Any adherent point φ of the net in M_b satisfies $R(\varphi) \le 1$ and is orthogonal to homogeneous polynomials of odd degree. Since $\varphi(\sum x_j^2) = 1$, φ is not the evaluation functional at 0. □

The proof shows in particular that no power $(\sum x_j^2)^N$ belongs to the closed subalgebra generated by the spaces $\mathcal{P}_m(\ell^2)$ for m odd.

Notes Most of the material in this chapter comes from [ACG1]. Theorem 4.3.2 and the consequent characterization of approximability by finite-type polynomials is in [ACG2].

Chapter 5
The algebra of bounded analytic functions on the unit ball

We turn to $H^\infty(B)$, the algebra of bounded analytic functions on the the unit ball B of the Banach space $\mathcal{X}$. Our aim is to give a brief glimpse into the spectrum of $H^\infty(B)$. We discuss the natural fibering of the spectrum over the closed unit ball $\bar{B}^{**}$ of $\mathcal{X}^{**}$, and we indicate various methods for embedding analytic structure into the spectrum. $\bar{B}^{**}$ of $\mathcal{X}^{**}$. Details of proofs are omitted.

5.1 The spectrum of $H^\infty(B)$

We denote the spectrum of $H^\infty(B)$ by $\mathcal{M}$. It is a compact space, to which the functions in $H^\infty(B)$ extend continuously. We may regard B^{**} as a subset of $\mathcal{M}$, by identifying $z \in B^{**}$ with the evaluation homomorphism δ_z. The closure of B^{**} in $\mathcal{M}$ is then a compactification of B, with the $H^\infty(B)$-topology. It is characterized as the smallest compactification to which the functions in $H^\infty(B)$ extend continuously.

If we regard $\mathcal{X}^*$ as a subspace of $H^\infty(B)$, we obtain a natural projection π of $\mathcal{M}$ into $\mathcal{X}^{**}$, defined so that $\pi(\varphi)$ is simply the restriction of φ to $\mathcal{X}^*$. Since $||\varphi|| = 1$, we have $||\pi(\varphi)|| \le 1$, and π projects $\mathcal{M}$ into the closed unit ball $\bar{B}^{**}$ of $\mathcal{X}^{**}$. Clearly $\pi(\delta_z) = z$ for $z \in B^{**}$. Since π is continuous, from $\mathcal{M}$ to the weak-star topology of B^{**}, the image of π coincides with $\bar{B}^{**}$. We define the *fiber of* $\mathcal{M}$ *over* $z \in \bar{B}^{**}$ to be $\mathcal{M}_z = \pi^{-1}(z)$.

In the case that $\mathcal{X}$ is one-dimensional, we obtain a fibering of the spectrum of $H^\infty(\Delta)$ over the closed unit disk $\bar{\Delta}$ in the complex plane. In this case, the fiber $\mathcal{M}_\lambda$ over any fixed $\lambda \in \Delta$ consists solely of the evaluation functional δ_λ at λ. This is a simple consequence of the fact that any $f \in H^\infty(\Delta)$ has a decomposition $f(w) = f(\lambda) + (w - \lambda)g(w)$, where $g \in H^\infty(\Delta)$. Thus we have a picture of $\mathcal{M} = \mathcal{M}(\Delta)$ as the open unit disk Δ with complicated fibers $\mathcal{M}_\lambda$ attached over every boundary point $\lambda \in \partial\Delta$. The fiber $\mathcal{M}_\lambda$ at $\lambda \in \partial\Delta$ reflects the cluster behavior of functions in $H^\infty(\Delta)$ at λ. For a detailed discussion of the space $\mathcal{M}(\Delta)$, see [Ho].

In the cases that $\mathcal{X}$ is ℓ^2_n or ℓ^∞_n, something similar occurs. The ideal of functions in $H^\infty(B)$ vanishing at a fixed $\lambda \in B$ is the principal ideal generated by the coordinate functions $w_j - \lambda_j$, $1 \le j \le n$. This ensures that the fibering of $\mathcal{M}(B)$ over $\bar{B}$ is one-to-one over B. Again the fibers over points of the boundary of B are quite large. The *corona problem* for B is to determine whether B is dense in $\mathcal{M}(B)$. For $n > 1$, this problem remains unsolved.

In the infinite-dimensional case, it turns out that all of the fibers $\mathcal{M}_z$ are quite large. This can be seen as follows. By the Josefson-Nissenzweig theorem, there is a sequence of points $z_j \in B^{**}$ such that $||z_j|| \to 1$, while $z_j \to 0$ in the weak-star topology. The latter condition implies that any accumulation point of the sequence in $\mathcal{M}$ lies in the fiber $\mathcal{M}_0$ over 0. On the other hand, it is easy to construct functions in $H^\infty(B)$ that do not have limits along the sequence, so that $\mathcal{M}_0$ has more than one point. In fact, it is possible to extract an interpolating subsequence from the z_j's, and to construct Blaschke-type products in $H^\infty(B)$, to show that $\mathcal{M}_0$ contains a copy of $\beta(N)\backslash N$. For details, see [ACG1]. With more effort, one can sharpen the Josefson-Nissenzweig theorem, and use it to prove the following theorem from [CGJ].

Theorem 5.1.1 *If $\mathcal{X}$ is infinite-dimensional, then each fiber $\mathcal{M}_z$ over $z \in \bar{B}^{**}$ contains a copy of $(\beta(N)\backslash N) \times \Delta$, so that the functions in $H^\infty(B)$ extend to be analytic on the disks $\{s\} \times \Delta$, $s \in \beta(N)\backslash N$.*

Proof The idea is to construct a sequence z_j as in the Josefson-Nissenzweig theorem, so that each z_j is metrically far from the other points of the sequence. Then disks can be found centered at z_j, so that the Blaschke product construction provides an ample supply of functions, and the restriction of $H^\infty(B)$ to the sequence of disks is close to the ℓ^∞-direct sum of $H^\infty(\Delta)$. □

5.2 The radius function on the spectrum

There is another way of picturing the spectrum of $H^\infty(B)$, which we describe briefly. We regard the entire functions H_b as a subalgebra of $H^\infty(B)$, and we restrict the homomorphisms in $\mathcal{M}$ to H_b. This yields a projection ρ of $\mathcal{M}$ into the spectrum M_b of H_b, which is continuous with respect to the weak topologies. Suppose that $\varphi \in M_b$ satisfies $R(\varphi) < 1$, that is, there is some $r < 1$ such that φ is continuous on H_b with respect to the norm of uniform convergence on B_r. Since any function in $H^\infty(B)$ can be approximated uniformly on B_r by polynomials, the homomorphism φ extends uniquely to a homomorphism of $H^\infty(B)$. Thus the projection ρ maps the homomorphisms in $\mathcal{M}$ that are continuous with the norm of uniform convergence on B_r for some $r < 1$ one-to-one onto the subset $\{R < 1\}$ of M_b. On the other hand, any $\psi \in \mathcal{M}$ is continuous with respect to the norm of uniform convergence on B, so that the restriction $\varphi = \rho(\psi)$ satisfies $R(\varphi) \le 1$. Fix $\varphi \in M_b$ such that $R(\varphi) < 1$. Since $R(\varphi^\zeta) = |\zeta|$, the map $\zeta \to \rho(\varphi^\zeta)$ maps the disk $\{|\zeta| < 1\}$ analytically into $\mathcal{M}$. By continuity of ρ, the points $\rho(\varphi^\zeta)$ cluster on the fiber $\rho^{-1}(\varphi)$ as $\zeta \to 1$. In particular, ρ maps $\mathcal{M}$ onto the subset $\{R \le 1\}$ of M_b.

Thus we can view $\mathcal{M}$ as a ball $\{R < 1\}$ in M_b, with fibers attached over points of the closure $\{R \le 1\}$ of the ball. The "interior" of the ball is filled out by analytic structure, and as we have seen, any two points of the set lie on an analytic disk. For more details, see [ACG1].

5.3 Embedding inseparable Hilbert balls in the spectrum

Now we give another method for embedding large analytic structures in the spectrum. Instead of using Blaschke products, we pass to a limit through appropriate subspaces.

Let Δ^∞ denote the infinite unit polydisk, which can be viewed as the open unit ball of ℓ^∞. Suppose that $\{x_j\}$ is a basic sequence in $\mathcal{X}$, normalized so that $||x_j|| = 1$. We define operators T_k from ℓ^∞ to $\mathcal{X}$ by

$$T_k(s) = \sum_{j=1}^{\infty} \frac{1}{2^j} s_j x_{k+j}, \qquad s \in \ell^\infty,\ k \ge 1.$$

Then T_k maps Δ^∞ into B. We regard T_k as a map from Δ^∞ into $\mathcal{M}$, and we let k tend to ∞ through an ultrafilter. Since $\mathcal{M}$ is compact, T_k converges pointwise to a map T from Δ^∞ into $\mathcal{M}$. The limit map T is easily seen to be analytic, in the sense that $f \circ T$ is analytic on Δ^∞ for each (extended) $f \in H^\infty(B)$.

Theorem 5.3.1 *Let the analytic map $T : \Delta^\infty \to \mathcal{M}$ be constructed as above. Let $\{L_k\}$ be a bounded sequence of functionals in $\mathcal{X}^*$ such that $L_k(x_k) = 1$, while $L_k(x_j) = 0$ if $j \neq k$. Suppose that there is $m \geq 1$ such that $\sum |L_j(x)|^m < \infty$ for all $x \in \mathcal{X}$. Then T is one-to-one. If further the x_k's form a basis for $\mathcal{X}$, and if the linear span of the L_k's is dense in $\mathcal{X}^*$, then T embeds Δ^∞ into the fiber $\mathcal{M}_0$ over 0.*

Now we generalize this process. Instead of taking limits along a basic sequence, we take a limit of a sequence unit balls of finite-dimensional subspaces that are almost orthogonal. Let $\{\mathcal{E}_k\}$ be a sequence of finite-dimensional subspaces of $\mathcal{X}$, and let $\{\mathcal{W}_k\}$ be a sequence of finite-dimensional subspaces of $\mathcal{X}^*$, such that $\mathcal{W}_j \perp \mathcal{E}_k$ for $k \neq j$. We assume that there is a constant $c > 0$ such that

$$\sup\{|L(x)| : L \in \mathcal{W}_j\} \geq c||x||, \qquad x \in \mathcal{E}_j,\ j \geq 1,$$

that is, $\mathcal{W}_j$ c-norms $\mathcal{E}_j$. We also assume that for some $p > 1$ and $C \geq 1$, the sequence of subspaces $\{\mathcal{W}_j\}$ satisfies

$$||\sum L_j|| \leq C(\sum ||L_j||^p)^{1/p}, \qquad L_j \in \mathcal{W}_j,\ j \geq 1.$$

This condition is called an *upper p-estimate*. It guarantees that $f(x) = \sum L_j(x)^m$ is an m-homogeneous polynomial of norm $||f|| \leq C^q$ whenever $m \geq p$, where q is the conjugate index of p.

Fix a free ultrafilter $\mathcal{U}$ on the positive integers (a point of $\beta(N)\backslash N$), and let $\tilde{\mathcal{E}}$ be the limit of the $\mathcal{E}_j$'s along the ultrafilter $\mathcal{U}$. In other words, $\tilde{\mathcal{E}}$ is the quotient Banach space obtained from the ℓ^∞-direct sum of the $\mathcal{E}_j$'s by dividing by the subspace of sequences whose norms tend to 0 along the ultrafilter $\mathcal{U}$.

Theorem 5.3.2 *Under the conditions above, there is an analytic embedding of the unit ball $\tilde{B}$ of the Banach space $\tilde{\mathcal{E}}$ into the spectrum $\mathcal{M}$ of $H^\infty(B)$, which is uniformly bicontinuous from the norm metric of $\tilde{\mathcal{E}}$ to the Gleason metric of $\mathcal{M}$.*

Here the Gleason metric of $\mathcal{M}$ is the metric induced by the norm of the dual space of $H^\infty(B)$.

As an example, take $\mathcal{X}$ to be ℓ^2, and index an orthonormal basis to have the form $\{e_{kj} : 1 \leq j \leq k,\ 1 \leq k < \infty\}$. Let $\mathcal{E}_k$ be the linear span of the e_{kj}'s for $1 \leq j \leq k$, and let $\mathcal{W}_k = \mathcal{E}_k$. Then $\mathcal{W}_k$ 1-norms $\mathcal{E}_k$, and the $\mathcal{E}_k$'s satisfy an upper 2-estimate. In this case there is a natural scalar product on $\tilde{\mathcal{E}}$, defined to be the limit through $\mathcal{U}$ of the scalar products of the components of the representing sequences. Thus $\tilde{\mathcal{E}}$ is a Hilbert space. We claim that $\tilde{\mathcal{E}}$ is not separable. To see this, consider for each $0 < \theta < 1$ the sequence $x_\theta = \{x_{\theta,k}\}$

whose kth component is the basis element e_{kj}, where $j = j(k)$ is the first integer such that $j/k \geq \theta$. For different θ's, the kth components of x_θ are eventually orthogonal, hence the corresponding elements $\widetilde{x}_\theta \in \widetilde{\mathcal{E}}$ form an uncountable orthonormal set.

Theorem 5.3.2 can be applied to superreflexive Banach spaces. See [Die2] for a discussion of superreflexivity. The superreflexive Banach spaces include the L^p-spaces for $1 < p < \infty$, and they can be characterized as the Banach spaces that admit a uniformly convex norm. For superreflexive Banach spaces, the Gurarii-Gurarii-James theorem provides upper p-estimates for large p. With the aid of Dvoretsky's spherical sections theorem, the spaces $\mathcal{E}_k$ and $\mathcal{W}_k$ can be chosen by induction so that the $\mathcal{E}_k$'s are almost euclidean. The procedure above for constructing embeddings then yields the following.

Theorem 5.3.3 *Let $\mathcal{X}$ be a superreflexive Banach space. Then there is an analytic embedding of the unit ball of an inseparable Hilbert space into the fiber $\mathcal{M}_0$ over 0, which is uniformly bicontinuous from the norm metric of the unit ball of the Hilbert space to the Gleason metric of $\mathcal{M}$.*

For details and references, see [CGJ].

References

[Al1] Alencar, R., Multilinear mappings of nuclear and integral type, *Proc. Amer. Math. Soc.* **94** (1985), 33–38.

[Al2] Alencar, R., On reflexivity and bases for $\mathcal{P}(^m E)$, *Proc. Royal Irish Acad. Sect. A* **85** (1985), 131–138.

[AAD] Alencar, R., Aron, R.M. and Dineen, S., A reflexive space of holomorphic functions in infinitely many variables, *Proc. Amer. Math. Soc.* **90** (1984), 407–411.

[AB] Aron, R.M. and Berner, P.D., A Hahn-Banach extension theorem for analytic mappings, *Bull. Soc. Math. France* **106** (1978), 3–24.

[ACG1] Aron, R.M., Cole, B.J. and Gamelin, T.W., Spectra of algebras of analytic functions on a Banach space, *J. Reine Angew. Math.* **415** (1991), 51–93.

[ACG2] Aron, R.M., Cole, B.J. and Gamelin, T.W., Weak-star continuous analytic functions, preprint.

[AHV] Aron, R.M., Hervés, C. and Valdivia, M., Weakly continuous mappings on Banach spaces, *J. Funct. Anal.* **52** (1983), 189–204.

[AS] Aron, R.M. and Schottenloher, M., Compact holomorphic mappings on Banach spaces and the approximation property, *J. Funct. Anal.* **21** (1976), 7–30.

[Bo] Bogdanowicz, W., On the weak continuity of the polynomial functionals on the space c_0, *Bull. Acad. Polon. Sci. Cl. III.* **5** (1957), 243–246.

[BF] Bonic, R. and Frampton, J., Smooth functions on Banach manifolds, *J. Math. Mech.* **15** (1966), 877–898.

[CCG] Carne, T.K., Cole, B.J. and Gamelin, T.W., A uniform algebra of analytic functions on a Banach space, *Trans. Amer. Math. Soc.* **314** (1989), 639–659.

[CGJ] Cole, B.J., Gamelin, T.W. and Johnson, W.B., Analytic disks in fibers over the unit ball of a Banach space, *Michigan Math. J.* **39** (1992), 551–569.

[DG] Davie, A.M. and Gamelin, T.W., A theorem on polynomial-star approximation, *Proc. Amer. Math. Soc.* **106** (1989), 351–356.

[Dea] Dean, D.W., The equation $L(E,X^{**}) = L(E,X)^{**}$ and the principle of local reflexivity, *Proc. Amer. Math. Soc.* **40** (1973), 146–148.

[Deg] Deghoul, D., Construction de caractères exceptionnels sur une algèbra de Fréchet, *C. R. Acad. Sci. Paris, Sér.I Math.* **312** (1991), 579–580.

[Die1] Diestel, J., A survey of results related to the Dunford-Pettis property, in: *Proceedings of the Conference on Integration, Topology and Geometry in Linear Spaces* (W. Graves, ed.), Contemp. Math. **2**, American Mathematical Society, Providence, RI, 1980; 15–60.

[Die2] Diestel, J., *Geometry of Banach Spaces – Selected Topics*, Lecture Notes in Math. **485**, Springer-Verlag, Berlin – Heidelberg – New York, 1975.

[Die3] Diestel, J., *Sequences and Series in Banach Spaces*, Springer-Verlag, Berlin – Heidelberg – New York, 1984.

[Din] Dineen, S., *Complex Analysis in Locally Convex Spaces*, North-Holland, Amsterdam, 1981.

[DS] Dunford, N. and Schwartz, J., *Linear Operators, Part I*, Interscience, 1959.

[Fa] Farmer, J., Polynomial reflexivity in Banach spaces, to appear.

[FJ] Farmer, J. and Johnson, W.B., Polynomial Schur and polynomial Dunford-Pettis properties, in: *Banach Spaces* (Bor-Luh Lin and W.B. Johnson, eds.), Contemp. Math. **144**, American Mathematical Society, Providence, RI, 1993.

[Ga] Gamelin, T.W., *Uniform Algebras*, 2nd ed., Chelsea, 1984.

[G^2M^2] Galindo, P., Garcia, D., Maestre, M. and Mujica, J., Extension of multilinear mappings on Banach spaces, to appear in *Studia Math.*

[Gu] Gutiérrez, J., Weakly continuous functions on Banach spaces not containing ℓ_1, *Proc. Amer. Math. Soc.* **119** (1993), 147–152.

[JP] Jaramillo, J.A. and Prieto, A., Weak-polynomial convergence on a Banach space, to appear in *Proc. Amer. Math. Soc.*

[Jo] Josefson, B., Weak sequential convergence in the dual of a Banach space does not imply norm convergence, *Ark. Mat.* **13** (1975), 79–89.

[LT] Lindenstrauss, J. and Tsafriri, L., *Classical Banach Spaces, Vol. 1*, Springer-Verlag, Berlin – Heidelberg – New York, 1977.

[LR] Lindström, M. and Ryan, R.A., Applications of ultraproducts to infinite dimensional holomorphy, to appear.

[Li] Littlewood, J.E., On bounded bilinear forms in an infinite number of variables, *Quart. J. Math. Oxford* **1** (1930), 164–174.

[Mo] Moraes, L., Extension of holomorphic mappings from E to E'', to appear in *Proc. Amer. Math. Soc.*

[Mu] Mujica, J., *Complex Analysis in Banach Spaces*, Math. Studies **120**, North-Holland, Amsterdam, 1986.

[Na] Nachbin, L., *Topology on Spaces of Holomorphic Mappings*, Springer-Verlag, Berlin – Heidelberg – New York, 1969.

[Ne] Nemirovskii, A.S., On a certain chain of algebras on a Hilbert space, *Functional Anal. Appl.* **5** (1971), 85–86.

[Ni] Nissenzweig, A., w^* Sequential convergence, *Israel J. Math.* **22** (1975), 266–272.

[Pe1] Pelczynski, A., A property of multilinear operations, *Studia Math.* **16** (1957), 173–182.

[Pe2] Pelczynski, A., On weakly compact polynomial operators on B-spaces with Dunford-Pettis property, *Bull. Acad. Polon. Sci. Sér. Sci. Math. Astronom. Phys.* **11** (1963), 371–378.

[Pe3] Pelczynski, A., A theorem of Dunford-Pettis type for polynomial operators, *Bull. Acad. Polon. Sci. Sér. Sci. Math. Astronom. Phys.* **11** (1963), 379–386.

[Pie] Pietsch, A., *Nuclear Locally Convex Spaces*, Springer-Verlag, Berlin – Heidelberg – New York, 1972.

[Pis] Pisier, G., *Factorization of Linear Operators and Geometry of Banach Spaces*, CBMS Regional Conf. Series in Math. **60**, Amer. Math. Soc., 1986.

[Pit] Pitt, H.R., A note on bilinear forms, *J. London Math. Soc.* **11** (1936), 174–180.

[Ry1] Ryan, R.A., *Application of Topological Tensor Products to Infinite Dimensional Holomorphy*, Ph.D. thesis, Trinity College, Dublin, 1980.

[Ry2] Ryan, R.A., Dunford-Pettis properties, *Bull. Acad. Polon. Sci. Sér. Sci. Math. Astronom. Phys.* **27** (1979), 373–379.

[Ry3] Ryan, R.A., Weakly compact holomorphic mappings on Banach spaces, *Pacific J. Math.* **131** (1988), 179–190.

[Sa] Samedani, Z., *Banach Spaces of Continuous Functions*, Monograf. Mat. **55**, PWN Warszawa, 1971.

[Sch] Schatten, R., *A Theory of Cross-Spaces*, Ann. of Math. Stud. **26**, Princeton Univ. Press, 1950.

[Va] Valdivia, M., Banach spaces of polynomials without copies of ℓ^1, to appear in *Proc. Amer. Math. Soc.*

Uniform approximation

Paul M. GAUTHIER

Département de mathématiques et de statistique
and
Centre de recherches mathématiques
Université de Montréal
C.P. 6128-A, Montréal, Qué., H3C 3J7
Canada

Abstract

In this course we give a partial survey of the theory of uniform approximation on unbounded sets by holomorphic, harmonic, or subharmonic functions. We show, by several examples, that this theory is convenient for the construction of functions having various interesting properties.

1 Introduction

Approximation theory can be approached from two viewpoints: the qualitative or the quantitative viewpoint. Qualitative approximation is more abstract. It is concerned with density theorems, that is, with the mere *possibility* of approximation. Quantitative approximation takes a more practical stance. It is concerned with actually finding efficient algorithms for *performing* the approximations. Because of these two aspects of their subject, approximation theorists have long been sensitive to the distinction between non-constructive and constructive mathematics. It is interesting to note that, before devoting himself to constructivism, Errett Bishop had obtained fundamental results in qualitative approximation, some of which will be discussed in the present survey.

There is another way of classifying approximation theory from two viewpoints: smoothing-type approximation and extension-type approximation. The first type of approximation seeks to approximate a given function by a nicer function defined on (more or less) the same domain whereas extension-type approximation seeks to approximate a given function by a function (of the same sort but) defined on a larger domain.

The present notes deal only with qualitative extension-type approximation. For an overview of the state of the subject in the early 1980's, the reader may consult, for example, [14], [20] and [30]. However, there have been many subsequent developments. We shall

This research was supported by NSERC (Canada) and FCAR (Québec).

P. M. Gauthier (ed.) and G. Sabidussi (techn. ed.), Complex Potential Theory, 235–271.

discuss a (very) few of these as well as a selection of classical results. Many better known theorems concerning approximation of functions defined on a compact set have been extended to the less known context of functions defined on closed (possibly unbounded) sets. We point out two difficulties in passing from approximation on compact sets to approximation on closed sets. First of all, approximation theory often makes use of functional analysis. The space $C(F)$ is a topological space (with the topology of uniform convergence) and it is a vector space. However, if F is unbounded, it is not a topological vector space. Indeed, scalar multiplication is not continuous: If λ_n are non-zero scalars with $\lambda_n \to 0$, and f is unbounded, then $\lambda_n \cdot f \not\to 0$. A second difficulty is that proofs in complex approximation often use integral representations. On unbounded sets, integration is of course possible, but more subtle.

In order to motivate the study of approximation on unbounded sets, we shall present a number of striking applications thereof. Many of these applications are known results, that is, known to some senior members of my audience, but probably not to the "real" students for whom this course is, after all, intended. The proofs we shall present in these applications are extremely short and elegant. The classical proofs, when they exist, are more elementary and more tedious. In this respect, I like to paraphrase one of my friends and favourite mathematical authors, Larry Zalcman [51]: "Approximation on compacta (ingeniously applied) often provides a useful tool in the construction of analytic functions having a certain prescribed boundary behaviour. The powerful generalization of this theory, allowing approximation on unbounded sets, often applies directly to such situations, rendering ingenuity superfluous." I have never been quite able to figure out whether this was intended as a compliment or not.

Some of the applications are very recent. One of these, due to Gardiner, and not yet published, I consider to be among the most interesting mathematical achievements of the past few years. Namely, Gardiner has completely characterized those unbounded domains on which the Dirichlet problem is *classically* solvable.

These notes are an expanded version of lectures given during a special year on complex analysis at the Nankai Institute [26] and which are being published by the International Press. I extend my thanks to the International Press for granting me permission to draw from the material of those lectures, to Stephen Gardiner for allowing me to mention some new results from his forthcoming lecture notes [21] on harmonic approximation, to Myron Goldstein for allowing me to present an unpublished idea of his for deriving complex approximation from harmonic approximation, and to Thomas Bagby, Stephen Gardiner, Louis-Philippe Giroux, and Larry Zalcman, who examined various versions of this manuscript and provided helpful comments.

2 Weierstrass

The year 1885 was a very special year for the qualitative theory of approximation, for that was the year that both Weierstrass and Runge proved their famous approximation theorems. In the present section we discuss various extensions of the Weierstrass theorem; in the following section, we shall do the same for Runge's theorem.

To state these and related results, we introduce some notation. If X is a topological space, we denote by $C(X)$ the space of continuous complex valued functions on X, and if X is a subset of a complex manifold Y, we denote by $H(X)$ the space of holomorphic functions on (open sets containing) X.

Theorem 1 (Weierstrass) *Let I be a closed interval on the real axis. Then, for each $f \in C(I)$ and for each $\epsilon > 0$, there is a polynomial p such that*

$$|f - p| < \epsilon.$$

The following beautiful result was proved by Carleman [15] in 1927 and deserves to be much better known than it is.

Theorem 2 (Carleman) *Suppose f is an arbitrary continuous functions on the real axis $\mathbf{R}$ and $\epsilon > 0$. Then, there is an entire function g on the finite complex plane $\mathbf{C}$ such that on $\mathbf{R}$,*

$$|f - g| < \epsilon.$$

Note that this gives a very nice generalization of the Weierstrass theorem since any entire function is the uniform limit, on compact subsets, of polynomials. Indeed, we may take the partial sums of the Maclaurin expansion.

This theorem was extended to functions of several variables by Scheinberg [48]. Scheinberg's proof is completely different from that of Weierstrass. Of course Carleman's theorem appears to be a generalization of the Weierstrass theorem. But Scheinberg has taken the original proof of Weierstrass himself and shown that it leads rapidly to Carleman's theorem. In my opinion, Scheinberg's proof has three merits: to advertise Weierstrass' own proof of his approximation theorem (most books on approximation give other proofs); to provide a new and interesting proof of Carleman's theorem; and to provide a tool for approximation in $\mathbf{C}^n$. Indeed the Weierstrass-Scheinberg approach has been successfully exploited in the very important paper of Baouendi and Trèves [8].

I now sketch very briefly Scheinberg's proof in the case $n = 1$. All details can be found in Scheinberg's paper.

Let $f \in L^1(\mathbf{R})$. Then,

$$g_n(z) = \frac{n}{\sqrt{2\pi}} \int_{-\infty}^{+\infty} f(t) e^{-[n(t-z)]^2/2} dt$$

defines an entire function of $z \in \mathbf{C}$. Note that for $x \in \mathbf{R}$, $g_n(x)$ is the expected value of f with respect to the normal probability law with mean x and standard deviation $1/n$. For large n, the deviation is small and so it is natural that the expected value $g_n(x)$ should be near $f(x)$.

In fact, Weierstrass approximated precisely by such functions g_n. Since g_n are entire, he could, of course, then approximate by polynomials.

Indeed, if f is not only integrable but also continuous at the point x, it is easy to see that $g_n(x)$ converges to $f(x)$. The same proof yields that if $f \in L^1(\mathbf{R})$ is uniformly continuous, then the convergence is uniform. In particular, we have the Weierstrass theorem.

Suppose now f is any continuous function on $\mathbf{R}$ (not necessarily in L^1) and ϵ is a positive constant. Let $\{\varphi_j\}$ be a partition of unity of $\mathbf{R}$ and write

$$f = \sum_{j=1}^{\infty} \varphi_j f.$$

Then, each $\varphi_j f$ is continuous and of compact support. Thus, by the previous paragraphs, we may choose n_j so large that

$$|\varphi_j f - g_{n_j}| < \frac{\epsilon}{2^{j+1}}.$$

Since $|g_n|$ depends on $\mathrm{Re}[t-z]^2$ and since the supports of $\varphi_j f$ are locally finite, it is not hard to choose n_j so large, for each j, that the series $\sum g_{n_j}$ is uniformly convergent on compact subsets of $\mathbf{C}$. Thus,

$$g = \sum_{j=1}^{\infty} g_{n_j}$$

is an entire function and performs the required approximation.

3 Runge

If Y is a subset of a topological space X, then we shall use the expression "hole of Y" (relative to X) to designate any relatively compact complementary component of Y.

Theorem 3 (Runge) *Let K be a compact subset of the Riemann sphere $\overline{\mathbf{C}}$, containing at least two points, and let P be a subset of $\overline{\mathbf{C}}$. A necessary and sufficient condition in order that for each $f \in H(K)$ and for each $\epsilon > 0$, there be a rational function r, whose poles lie in P, such that $|f - r| < \epsilon$ on K, is that P meet each hole of K. Thus, in particular, if $K \subset \mathbf{C}$, a necessary and sufficient condition in order that each $f \in H(K)$ be the uniform limit of polynomials is that $\mathbf{C} \setminus K$ be connected.*

One can prove the sufficiency in the Runge theorem, for example, using the following fundamental consequence of Green's theorem.

Theorem 4 (Cauchy-Green Formula) *If D is a smoothly bounded domain in $\mathbf{C}$ and $f \in C^1(\overline{D})$, then*

$$f(z) = \frac{1}{2\pi i} \int_{\partial D} \frac{f(\zeta)}{\zeta - z} d\zeta - \frac{1}{\pi} \iint_{D} \frac{\partial f}{\partial \overline{\zeta}}(\zeta) \frac{d\xi d\eta}{\zeta - z},$$

for $z \in D$.

Corollary 1 (Cauchy Formula) *If D is a smoothly bounded domain in* $\mathbf{C}$ *and* $f \in C^1(\overline{D}) \cap H(D)$, *then*

$$f(z) = \frac{1}{2\pi i} \int_{\partial D} \frac{f(\zeta)}{\zeta - z} d\zeta. \tag{1}$$

Corollary 2 *If* $f \in C_0^1(\mathbf{C})$, *then*

$$f(z) = -\frac{1}{\pi} \iint_{\mathbf{C}} \frac{\partial f}{\partial \overline{\zeta}}(\zeta) \frac{d\xi d\eta}{\zeta - z}. \tag{2}$$

The sufficiency in Runge's theorem can be easily deduced from either of these corollaries. First of all, we may assume that $K \subset \mathbf{C}$. Now, let $f \in H(K)$ and let $\epsilon > 0$. We may choose a smoothly bounded neighbourhood D of K such that $f \in C^1(\overline{D}) \cap H(D)$. Then from the first corollary, $f(z)$ can be expressed as a Cauchy integral over ∂D, for $z \in K$. If we approximate this integral (4) by a Riemann sum, we may obtain a rational function s, with poles on ∂D, which approximates f on K to within $\epsilon/2$. Let us decompose s into partial fractions: $s = s_1 + s_2 + \cdots + s_n$, where each s_j is a rational function having only one pole q_j. By assumption, to each such pole q_j, we may associate a point p_j of P which lies in the same hole of K as the point q_j. By invoking the following pole-pushing lemma, we may, for each j, approximate s_j on K to within $\epsilon/(2n)$ by a rational function r_j whose only pole is at the point p_j. Then, setting $r = r_1 + r_2 + \cdots + r_n$, we have that $|f - r| < \epsilon$ on K, which completes the proof of sufficiency in Runge's theorem.

Lemma 1 (Pole-pushing) *Let p and q be points lying in a domain U of the Riemann sphere $\overline{\mathbf{C}}$. If r_q is a rational function whose only pole is at the point q and $\epsilon > 0$, then there is a rational function r_p whose only pole is at the point p and such that $|r_p - r_q| < \epsilon$ on $\overline{\mathbf{C}} \setminus U$.*

In order to prove this lemma, we may, first of all, assume $\infty \notin U$. Let D_j, $j = 1, 2, \ldots, m$ be a chain of discs going from q to p in U. That is, if z_j denotes the center of the disc D_j, then $p = z_m$ and, setting $q = z_0$, we have that $z_j \in D_{j+1}$, for each $j = 0, 1, \ldots, m-1$. The pole-pushing lemma now follows by applying the following lemma finitely many times.

Lemma 2 *Let D be a disc in* $\mathbf{C}$, *centered at the point p, and let g be holomorphic on $\overline{\mathbf{C}} \setminus D$. Then g is the uniform limit of polynomials in $(z - p)^{-1}$.*

This is an immediate consequence of the representation of g on $\overline{\mathbf{C}} \setminus D$ as a Laurent series centered at p.

We can also prove the sufficiency in Runge's theorem using the second corollary. Namely, by modifying f outside of a neighbourhood of K, we may assume that $f \in C_0^1(\mathbf{C})$. Thus, we have the integral representation (5). Again, by taking Riemann sums and pole-pushing, we arrive at the desired approximation.

Looking at Corollary 1, we are reminded that the fundamental solution of the partial differential operator $\partial/\partial\overline{\zeta}$ is the function $-(\pi\zeta)^{-1}$. Thus, for example, if μ is any (complex) measure of compact support, then

$$F(z) = -\frac{1}{\pi}\int \frac{d\mu(\zeta)}{\zeta - z}$$

is a locally integrable function which is a (generalized) solution of the equation

$$\frac{\partial F}{\partial \overline{\zeta}} = \mu. \tag{3}$$

Moreover, it follows from the explicit form of the solution that if μ has a certain regularity, then so does the solution F. For example, if $\mu = \varphi dm(\zeta)$, where $dm(\zeta)$ is Lebesgue (planar) measure and $\varphi \in L^\infty$ is of compact support, then the solution F is continuous. In fact, for any $f \in L^1_{loc}$, the function

$$F(z) = \int f(\zeta - z)\varphi(\zeta)dm(\zeta)$$

is continuous. This is easy to see when the function f is continuous. In the general case, we may construct a sequence $\{f_n\}$ of continuous functions which converge locally-L^1 to f. The corresponding integrals, F_n, are then continuous and, since they converge locally uniformly to F, it follows that F is also continuous.

As an application of Runge's theorem, we now show that we can, in fact solve equation (6) in any domain Ω and for any measure μ (not necessarily of compact support) in Ω. Indeed, let $\{\Omega_n\}$, $n = 1, 2, \ldots$, be a normal exhaustion of Ω. Thus, for each n, Ω_n is smoothly bounded, each hole of Ω_n contains a hole of Ω, $\Omega_n \subset\subset \Omega_{n+1}$, and $\Omega = \bigcup \Omega_n$. Set $\Omega_0 = \phi$ and for $n \geq 1$, set $\mu_n = \mu\, |_{\Omega_n - \Omega_{n-1}}$. Let F_n be a continuous solution of the equation (6) for μ_n. By Runge's theorem, for each $n > 2$, there is a rational function g_n with poles outside of Ω such that $|F_n - g_n| < 2^{-n}$ on $\overline{\Omega}_{n-2}$. Set $g_1 = g_2 = 0$. The locally integrable function $F = \sum(F_n - g_n)$ is the desired solution of our problem in Ω.

Runge's theorem has been extended to Riemann surfaces. We begin by reformulating Runge's theorem.

Theorem 5 *Let $W \subset \overline{\mathbf{C}}$ be open and let P be a subset of $\overline{\mathbf{C}}$. Then, each function in $H(W)$ can be approximated uniformly on compact subsets of W by rational functions whose poles lie in P, provided that P meets each hole of W. If $W \subset \mathbf{C}$, then each function in $H(W)$ can be approximated uniformly on compact subsets of W by polynomials if and only if $\mathbf{C} \setminus W$ is connected.*

Notice that, in contrast to our initial version of Runge's theorem, we have only claimed the sufficiency, and not the necessity, of the condition that P meet each hole of W. Indeed, in the present formulation, this condition is *not* necessary. Although, in general, an open set W may have uncountably many holes, it turns out that there is always a countable set P from which we may choose our poles. To see this, let W_j be a regular exhaustion of W, so that, for each j, $W_j \subset W_{j+1}$ and W_j has finitely many holes, each of which contains a hole of W. Choose a finite subset P_j of the complement of W which meets each hole of W_j. Now set $P = \bigcup_j P_j$. It is not hard to see that each function in $H(W)$ can be uniformly approximated on compact subsets by rational functions whose poles lie in the set P. Indeed, let $f \in H(W)$ and let K be a compact subset of W. Then, for some j, $K \subset W_j$. By the first formulation of Runge's theorem, f can be uniformly approximated on K by rational functions whose poles lie in P_j and hence in P.

A consequence of the previous paragraph is that the space $H(W)$ of holomorphic functions on an open set W, endowed with the topology of uniform convergence on compact subsets, contains a countable dense subset of rational functions. Indeed, let P be a countable set of possible poles associated to the set W as in the previous paragraph. We may write any rational function by its partial fraction representation. If, in this representation, we allow only coefficients whose real and imaginary parts are rational and only poles from the countable set P, then we have a countable set of rational functions all of which are in $H(W)$ and which form a dense subset thereof.

Definition Let K be an open subset of a Riemann surface Ω. The pair (K, Ω) is said to be a *Runge pair* if for each $f \in H(K)$ and for each $\epsilon > 0$, there is a $g \in H(\Omega)$ such that $|f - g| < \epsilon$ on K. Let W be an open subset of a Riemann surface Ω. The pair (W, Ω) is said to be a *Runge pair* (for compact sets) if for each $f \in H(W)$, for each compact $K \subset W$ and for each $\epsilon > 0$, there is a $g \in H(\Omega)$ such that $|f - g| < \epsilon$ on K.

Notation Let us denote by $\Omega^* = \Omega \cup \{*\}$ the Alexandrov one-point compactification of Ω, where $*$ denotes the ideal Alexandrov point at infinity.

Riemann surface theory can be subdivided into two areas of study: the study of compact surfaces and the study of open (non compact) surfaces. Each of these has its principal theorem. For compact surfaces it is the Riemann-Roch theorem while for open surfaces it is the following theorem of Behnke and Stein [9] which is in fact the natural extension of Runge's theorem to Riemann surfaces. It may also be said that, to a great extent, this extension of Runge's theorem, which Behnke and Stein published in 1949, inspired the remarkable flourishment of several complex variables which took place in the Séminaire Henri Cartan in the succeeding decade.

Theorem 6 (Behnke-Stein) *The following equivalent statements hold:*

(a) *for each open Riemann surface Ω and each compact subset K of Ω, the pair (K, Ω) is a Runge pair if and only if $\Omega^* \setminus K$ is connected;*

(b) *for each open Riemann surface Ω and each open subset W of Ω, the pair (W, Ω) is a Runge pair (for compact sets) if and only if $\Omega^* \setminus W$ is connected.*

4 Mergelian

Runge's theorem assures us that if a compact subset K of the complex plane has connected complement, then each function f holomorphic on K can be uniformly approximated by polynomials. The hypothesis that f be holomorphic on K is unnecessarily strong. Of course, if f is approximable, then f is necessarily continuous on K and holomorphic on the interior of K. The following theorem extends simultaneously both the Runge-Behnke-Stein theorem as well as the Weierstrass theorem.

Theorem 7 (Mergelian-Bishop) *For a compact subset K of an open Riemann surface Ω, the following are equivalent:*

(a) *for each f continuous on K and holomorphic on K^0 and for each $\epsilon > 0$, there is a function g, holomorphic on Ω, such that*

$$|f - g| < \epsilon;$$

(b) *$\Omega^* \setminus K$ is connected.*

This was established in 1952 by Mergelian [40] in the case where Ω is the complex plane and extended in 1958 by Bishop [10] to Riemann surfaces.

5 Arakelyan

The Weierstrass and Runge theorems were extended to closed sets by Carleman [15] in 1927 and Roth [44] in 1938 respectively. Subsequent extensions were obtained by Keldysh and Lavrentiev. Finally, a complete solution to the problem of uniform approximation on compacts sets by polynomials was given by Mergelian [40]. Arakelyan eventually generalized Mergelian's theorem by giving a complete solution to the problem of uniform approximation on closed sets by entire functions [1]. Arakelyan used his approximation theorem to disprove a conjecture of Nevanlinna: namely, that an entire function of finite order ρ has at most 2ρ deficient values. Indeed, Arakelyan [2] showed that if $\{a_k\}$ is any sequence of distinct complex numbers, then, for each ρ, with $1/2 < \rho < +\infty$, there is an entire function f of order ρ, for which each of the values a_k is deficient. The restriction $1/2 < \rho$ is essential since it is known that an entire function of order $\leq 1/2$ has *no* finite deficient values.

The generalizations of the Weierstrass theorem due to Mergelian and Arakelyan are better known than the generalization of the Runge theorem due to Roth. We shall present first the generalization of Roth. Although the theorems of Mergelian and Arakelyan are more powerful than that of Roth, we shall see that Roth's theorem is nevertheless sufficiently strong to yield several surprising applications.

In the previous section, we stated the Behnke-Stein generalization of Runge's theorem which, in place of approximation by functions holomorphic on a plane domain Ω, dealt with approximation by functions holomorphic on a Riemann surface Ω. In 1938, Alice Roth [44] had already generalized the Runge theorem, but in a different direction. She proved an extension of Theorem 5 in which she considered approximation on *unbounded* closed subsets of the plane $\mathbf{C}$ rather than on compact subsets. Then, in 1973 [45] (note the time span!), she generalized her own result by approximating on closed subsets of an arbitrary plane open set Ω, where "closed" means "closed in the relative topology of Ω". Here, then, is the Runge theorem of Roth.

Theorem 8 (Roth) *Let F be a closed subset of a plane open set Ω such that $\Omega^* \setminus F$ is both connected and locally connected. Then, for each $f \in H(F)$ and for each $\epsilon > 0$, there is a function $g \in H(\Omega)$ such that*

$$|f - g| < \epsilon.$$

With the hope of making the reader more comfortable with the topological conditions in the above Runge-type theorems, we present some examples in which we take Ω to be the complex plane $\mathbf{C}$.

Example 1 $\mathbf{C}^* \setminus F$ *is connected if and only if* $\mathbf{C} \setminus F$ *has no bounded components. Thus, if* $F = \mathbf{R}$, *then* $\mathbf{C}^* \setminus \mathbf{R}$ *is connected.* (Note that $\mathbf{C} \setminus \mathbf{R}$ is not!)

Example 2 *If F is the unit circle ($|z| = 1$), then $\mathbf{C}^* \setminus F$ is not connected.*

Example 3 *Let*

$$F = [0, +\infty) \cup \{z = x + iy : x = (1/y)|sin(1/y)|, 0 < y \leq 1\}.$$

This set is known as Arakelyan's glove. It has the property that both $\mathbf{C} \setminus F$ and $\mathbf{C}^* \setminus F$ are connected but $\mathbf{C}^* \setminus F$ is not locally connected.

Definition A family $\mathcal{A}$ of sets in Ω is said to have *no long islands* if for each compact subset K of Ω there is a compact set Q in Ω such that any member of $\mathcal{A}$ which meets K is necessarily contained in Q.

Lemma 3 *The following conditions are equivalent:*

(1) *$\Omega^* \setminus F$ is locally connected;*

(2) *$\Omega^* \setminus F$ is locally connected at $*$;*

(3) *for each compact set K in Ω, the family of bounded components of $\Omega \setminus (F \cup K)$ has no long islands.*

If $\Omega = \mathbf{C}$, these conditions are also equivalent to the following:

(4) *for each $r > 0$, there exists $r' > r$ such that any two points a, b in $\overline{\mathbf{C}} \setminus F$ and outside of the disc $D_{r'}$, centered at 0 and of radius r', can be joined by a path in $\overline{\mathbf{C}} \setminus F$ and outside of D_r.*

Approximation on unbounded sets has many applications (to pure mathematics). We have already mentioned Arakelyan's disproof of the Nevanlinna conjecture. In the sequel, we shall present several more applications, the first of which is a "counterexample" to one of the best known theorems of function theory.

Application 1 ("Counterexample" to Liouville's theorem) *There exists a nonconstant entire function which is bounded on each line.*

Proof Let

$$F_1 = (y \geq 1/x) \cup (x \leq 0) \cup (y \leq 0)$$

and choose a point z_0 not in F_1. Set $F = F_1 \cup z_0$. Then $\mathbf{C}^* \setminus F$ is connected and locally connected. Define f to be 0 at the point z_0 and 1 on the set F_1. Then $f \in H(F)$ and so by Roth's theorem there is a $g \in H(\mathbf{C})$ such that $|f - g| < 1/2$ on F. Since each line is, except for a bounded portion, contained in F, it follows that g is bounded on each line. The function g cannot be constant since it approximates two different constants (0 and 1) too well. In fact we can do better (or worse, depending on how perverse we are). Set $g_0 = (g(z) - g(z_0))/(z - z_0)$. Then g_0 is also a nonconstant entire function and g_0 is not only bounded, but even tends to 0 on each line!

In order to give the next application we recall the beautiful approximation theorem which Whitney published in 1934. Let $C^\omega(\mathbf{R}^n)$ denote the set of analytic functions on $\mathbf{R}^n$.

Theorem 9 (Whitney) *For each $f \in C(\mathbf{R}^n)$ and for each $\epsilon > 0$, there is a function $g \in C^\omega(\mathbf{R}^n)$ such that*

$$|f - g| < \epsilon.$$

The classical Dirichlet problem for the half-plane was solved by R. Nevanlinna in 1925. As an application of approximation theorems on unbounded sets, we now present an extremely short and elegant solution to this problem which was introduced by W. Kaplan in 1955. Let $U = \{z = x + iy : y > 0\}$ denote the open upper half of the complex plane $\mathbf{C}$ and let $\mathbf{R}$ denote the real axis in $\mathbf{C}$. Also, for $\varphi \in C(\mathbf{R})$, we denote by P_φ the Poisson integral of φ for the half-plane U.

Application 2 (Dirichlet problem for half-plane) *Given $\varphi \in C(\mathbf{R})$ find $u \in C(\overline{U})$ such that u is harmonic on U and $u = \varphi$ on $\mathbf{R}$.*

Solution By Whitney's theorem there is a $g \in C^\omega(\mathbf{R})$ such that $|\varphi - g| < 1$. Extend g to $\tilde{g} \in H(\mathbf{R})$. By Roth's theorem, there is an $f \in H(\mathbf{C})$ such that $|f - \tilde{g}| < 1$. The function

$$u = \mathrm{Re} f + P_{\varphi - \mathrm{Re} f}$$

is then a solution of the given Dirichlet problem. One might ask why we did not simply take the Poisson integral of the function φ itself. But recall that the function φ is an *arbitrary* continuous function on $\mathbf{R}$ and so its Poisson integral may not even converge. However, the Poisson integral of any *bounded* continuous function does exist.

In the above proof, we only used Whitney's theorem for $n = 2$. In this case the Whitney theorem and the Roth theorem can be replaced by a single (earlier !) theorem of Carleman. We shall present Carleman's theorem further along.

Application 3 (Domains of holomorphy) *Every domain $\Omega \subset \mathbf{C}$ is a domain of holomorphy. That is, there exists $g \in H(\Omega)$ such that g does not extend holomorphically to a (strictly) larger domain.*

Proof Indeed, one can construct such a function g of the form $g(z) = \sum a_n(z - z_n)^{-1}$, where $\{z_n\}$ is a sequence dense on $\partial\Omega$ and the coefficients $\{a_n\}$ decrease sufficiently rapidly. However, it is also very simple to show the existence of a non-extendable g using Roth's theorem. Let $\{z_n\}$ be a sequence in Ω having no limit point in Ω and having every point of $\partial\Omega$ as limit point. Set

$$F = \bigcup_{n=1}^{\infty}\{z_n\}.$$

Then, $\Omega^* \setminus F$ is connected and locally connected. Also, the function

$$f(z_n) = n$$

is in $H(F)$. Thus by Roth's theorem, there is a $g \in H(\Omega)$ such that $|f - g| < 1$ on F. Since the function g is unbounded at each point of $\partial\Omega$, it cannot extend holomorphically to any domain strictly containing Ω.

Application 4 (Mittag-Leffler Theorem) *Let D be a plane domain, $\{z_n\}$ a sequence in D without limit points in D, and for each n, let s_n be a function holomorphic in a punctured neighbourhood of z_n. Then, there exists a function*

$$g \in H(D \setminus \bigcup_{n=1}^{\infty}\{z_n\})$$

such that, for all n, $g - s_n \in H(z_n)$.

Proof For each n we may choose a disc D_n centered at z_n such that the function s_n is holomorphic on $\overline{D}_n \setminus \{z_n\}$ and such that the family $\{\overline{D}_n\}$ is disjoint and locally finite. Set

$$\Omega = D \setminus \bigcup_{n=1}^{\infty}\{z_n\}, \quad F = \bigcup_{n=1}^{\infty}(\overline{D}_n \setminus \{z_n\}).$$

Then, $\Omega^* \setminus F$ is connected and locally connected. Moreover, the function f, defined on F by setting $f = s_n$ on $\overline{D}_n \setminus \{z_n\}$ for each n, is holomorphic on F. Thus, by Roth's theorem, there is a function $g \in H(\Omega)$ such that $|f - g| < 1$. The function g has the required properties. Notice that, whereas in the usual formulation of the Mittag-Leffler theorem the prescribed singularities are poles, in the present formulation they are *arbitrary* isolated singularities.

We now have two kinds of extensions of Runge's theorem, the theorem of Roth which extends Runge's theorem to closed subsets of *plane* domains, and the theorem of Mergelian which sharpens Runge's theorem so as to include the Weierstrass theorem. Perhaps the best known theorem concerning approximation on unbounded sets is the following theorem of Arakelyan ([1] [3]) which includes both the theorems of Roth and of Mergelian as special cases.

Theorem 10 (Arakelyan) *For a closed subset F of a plane domain Ω, the following are equivalent:*

(a) *for each f continuous on F and holomorphic on F^0 and for each $\epsilon > 0$, there is a function g, holomorphic on Ω, such that*

$$|f - g| < \epsilon;$$

(b) *$\Omega^* \setminus F$ is connected and locally connected.*

Rosay and Rudin [43] have given a short and elegant proof of Arakelyan's theorem (which implies Roth's theorem). Note that Arakelyan also showed that the condition in Roth's theorem is not only sufficient for approximation but also necessary.

The necessity remains true on open Riemann surfaces. That is, if F is a closed subset of an open Riemann surface Ω such that for each $f \in H(F)$ and for each $\epsilon > 0$, there is a function $g \in H(\Omega)$ such that $|f - g| < \epsilon$, then $\Omega^* \setminus F$ is both connected and locally connected. However, the sufficiency fails! That is, although the compact Runge theorem holds on Riemann surfaces (Theorem 7), the closed Runge theorem (Theorem 8) fails to extend. The construction of a counterexample [30] uses the existence of certain surfaces first constructed by Myrberg (see, e.g. [47, Chapter I, section 10]).

Example 4 *There exists an open Riemann surface X having a closed parametric disc K whose complement $X \setminus K$ admits no non-constant bounded holomorphic functions.*

Proof Let X_1 and X_2 be two copies of the complex plane $\mathbf{C}$ each having the following slits:

$$(1/(2k+1), 1/(2k)), \ k = 1, 2, \ldots.$$

The Riemann surface X is formed by joining X_1 and X_2 in the usual way along their slits. Let D be that portion of X which lies over the open unit disc in $\mathbf{C}$. Let K be a closed disc on X_1 which is disjoint from D and suppose $f \in H(X \setminus K)$. Let $p(z) = \tilde{z}$ be the idempotent automorphism of X which maps a point $z \in X$ to the corresponding point $\tilde{z}$ on the other sheet of X. Note that, at the branch points (which lie over the points $1/j$), $z = \tilde{z}$. Now set

$$\varphi(z) = [f(z) - f(\tilde{z})]^2, \ z \in D.$$

Then $\varphi \in H(D)$ and $\varphi(z) = \varphi(\tilde{z})$. Thus, φ is well defined on the *unslit* punctured unit disc $0 < |z| < 1$. If f is bounded, then the isolated singularity of φ at the origin is removable and so φ extends to a function holomorphic in the unit disc. Since $\varphi(1/j) = 0, \ j = 1, 2, \ldots,$ it follows that $\varphi = 0$. From the definition of φ, this means that on D, the function f takes the same value on both sheets. Thus, $f|X_2$ extends holomorphically to the slits of X_2. But X_2 with its slits closed up is just the plane $\mathbf{C}$ punctured at the origin. Since f is bounded, f extends to a bounded entire function, which by Liouville's theorem is necessarily constant. Since $X \setminus K$ is connected, f is constant on all of $X \setminus K$.

Example 5 **[30]** *Let X and K be as in the above theorem and let a and b be distinct points of K^0. Then, the analog of the closed Runge Theorem 8 fails on the Riemann surface $\Omega = X \setminus \{a\}$.*

Proof Set $F = (X \setminus K^0) \cup \{b\}$. Then F is a closed subset of Ω and $\Omega^* \setminus F$ is both connected and locally connected. Setting

$$f = \begin{cases} 0 & \text{on } X \setminus K^0, \\ 1 & \text{at } b, \end{cases}$$

we have that $f \in H(F)$. Suppose, now, that $g \in H(\Omega)$ and $|f - g| < 1/2$ on F. Then, by Theorem 9, the function $g\,|_{X \setminus K}$ is constant and hence g is constant. But this contradicts the fact that g approximates both 0 and 1 within 1/2.

The problem of characterizing those closed subsets F of a Riemann surface Ω for which every holomorphic function on F can be approximated by functions holomorphic on Ω remains open. However, in subsequent sections, we will see that analogous questions for harmonic approximation have recently found satisfactory answers.

6 Arbitrary speed

Thus far, we have been interested in uniform approximation. In the present section, we consider the possibility of doing much better than uniform approximation. In an earlier section we stated the Carleman theorem about uniform approximation on the whole real axis. In fact, we withheld an important aspect of Carleman's theorem, for, in fact, he obtained approximations which were much better than uniform. Indeed, the complete statement of Carleman's theorem is as follows.

Theorem 11 (Carleman) *Suppose f and ϵ are arbitrary continuous functions on the real axis $\mathbf{R}$ with ϵ positive. Then, there is an entire function g on $\mathbf{C}$ such that on $\mathbf{R}$,*

$$|f - g| < \epsilon.$$

Note that this allows one to approximate with an error ϵ which gets smaller and smaller for large values and, in fact, tends to zero with any preassigned speed.

Proof From a modern perspective, there are several ways to prove Carleman's theorem. Assume, first, that ϵ is a constant. The result then follows from the earlier cited theorems of Whitney and of Roth and the triangle inequality. To prove Carleman's theorem for a general ϵ, we note that there always exists an entire h such that, on the real axis, h is zero-free and $|h| < \epsilon$. Let g be an entire function such that $|g - f/h| < 1$ on $\mathbf{R}$. Then the entire function hg is the desired approximator of f.

Let F be a closed subset of a Riemann surface Ω. We shall say that the pair (F, Ω) is an *Arakelyan* (respectively, *Carleman*) *pair* if, for each function f continuous on F and holomorphic on F^0 and each positive constant (respectively, continuous function) ϵ, there is a function g, holomorphic on Ω such that $|f - g| < \epsilon$ on F. Arakelyan's theorem characterizes Arakelyan pairs, in case Ω is a plane domain and Carleman's theorem asserts that $(\mathbf{R}, \mathbf{C})$ is a Carleman pair. Of course any Carleman pair is *a fortiori* an Arakelyan pair. The following result gives a complete characterization of those Arakelyan pairs which are Carleman pairs, that is, pairs for which it is possible to approximate with arbitrary speed. Recall that a family $\mathcal{A}$ of sets in Ω is said to have *no long islands* if for each compact subset K of Ω there is a compact set Q in Ω such that any member of $\mathcal{A}$ which meets K is necessarily contained in Q.

Theorem 12 *Let F be closed subset of an open Riemann surface Ω. Then the following are equivalent:*

(a) (F, Ω) *is a Carleman pair;*

(b) $\Omega^* \setminus F$ *is connected and locally connected and the family of components of F^0 has no long islands;*

(c) (F, Ω) *is an Arakelyan pair and the family of components of F^0 has no long islands.*

In case Ω is a plane domain, this theorem is due to Nersesyan [41]. The long islands condition had been introduced in [23], where it had been shown to be necessary for Carleman type approximation. In the present form, that is, on Riemann surfaces, the above theorem is due to Boivin [11]. Recall that we saw earlier that there is no known characterization of sets F on Riemann surfaces for which uniform approximation is possible. That is, there is no characterization of Arakelyan pairs. It is thus all the more interesting that via condition (b) we *do* have a characterization for approximations which are in fact much better than uniform.

Our next application concerns so-called universal functions. In 1929 G.D. Birkhoff showed the following striking result.

Theorem 13 (Birkhoff) *There exists an entire function whose translates are dense among all entire functions.*

What this means is that there exists an entire function g with the following property. For each entire function h, there is a sequence of points $\{a_j\}$ such that $g(z + a_j) \to h(z)$ uniformly on compact subsets of $\mathbf{C}$. Such a function g is called a universal function. There is a certain resemblance between the notion of a universal function as above and that of a wavelet.

Proof Since the polynomials are dense in the enitre functions, it is sufficient to construct a g which approximates all polynomials. In fact, it is sufficient to approximate a dense set of polynomials. Let $\mathcal{P}$ be a dense set of polynomials which is countable, for example, the polynomials whose coefficients have both real and imaginary parts rational. Let $p_j, j = 1, 2, \ldots$ be a sequence of polynomials in which each polynomial in $\mathcal{P}$ recurs an infinite number of times. Let $\{a_j\}$ be a sequence of distinct points which diverges sufficiently rapidly to infinity that we may construct closed discs $\overline{D}_j, j = 1, 2, \ldots,$ centered at the points a_j respectively, which form a pairwise disjoint locally finite family and whose radii tend to infinity. Now let F be the closed set formed by the union of these closed discs. Then $\overline{\mathbf{C}} \setminus F$ is connected and locally connected. We define a holomorphic function f on F by setting $f(z) = p_j(z - a_j)$ on $\overline{D}_j, j = 1, 2, \ldots$. By the previous theorem, there is an entire function g such that $|f - g| < 1/j$ on $\overline{D}_j, j = 1, 2, \ldots$. This function g has the required properties. Of course it is possible (as Birkhoff did) to use less sophisticated machinery, but the proof would be a little longer.

Once we are over the shock of the existence of a universal entire function, it is not very surprising that, in fact, most entire functions are universal. However, as is often the case with such assertions that most functions are wild, it is not obvious how to come up with an explicit example of a universal function. It turns out that one of the most interesting functions of all, the Riemann zeta function, is "sort of" universal in its critical strip. Indeed, Voronin [50] showed that for any $0 < r < 1/4$, for any $f \in H(|z| \leq 1/4)$ which is zero free, and for any $\epsilon > 0$, there exists a real value a such that

$$\max_{|z|\leq r} |f(z) - \zeta(z + ia)| < \epsilon.$$

Note that, without the assumption that f is zero free, this would contradict the Riemann hypothesis.

Carleman approximation is very suitable for the study of asymptotic behaviour of functions and, indeed, Carleman himself made important contributions to this subject. Also, the best known theorem concerning approximation on unbounded sets, the theorem of Arakelyan, was developed in order to study asymptotic behaviour. By an asymptotic path we understand a continuous path $\sigma : [0, +\infty) \to \mathbf{C}$ such that $\sigma(t) \to \infty$ as $t \to \infty$. Let us say that an entire function g has the value w as asymptotic value if there is an asymptotic path σ such that $f(z) \to w$ as $z \to \infty$ along σ. It is possible to introduce a notion of multiplicity with which f has a value w as asymptotic value so that, for example, the function $\exp z$ has each of the values 0 and ∞ as asymptotic values with multiplicity 1, whereas the function $\exp z^2$ has each of the values 0 and ∞ as asymptotic values with multiplicity 2.

Application 5 *There exists an entire function having each value as asymptotic value continually many times.*

Proof Let us denote by E the Cantor set lying on the segment, $S = (\xi = 1, 0 \leq \eta \leq 1)$, of the complex $\zeta = \xi + i\eta$ plane. Let T be a tree lying in the strip $0 < \xi < 1$ which grows horizontally towards the right and whose branches approach each point of E by continually many different paths. Let φ be a continuous function defined on the segment S which maps the Cantor set E onto the Riemann sphere $\overline{\mathbf{C}}$. Such a function can be constructed by composing a Cantor function with a Peano curve. Now extend the function φ to a continuous function $\varphi : \overline{T} \to \overline{\mathbf{C}}$, which is finite valued on T. Let F_0 be the closed subset of $\mathbf{C}$ which is the image of T be the stretching

$$\zeta \longmapsto z = \frac{\xi}{1-\xi}\zeta.$$

Thus, F_0 is a tree which branches to ∞ towards the right. We define a continuous function f on F_0 by setting $f(z) = \varphi(\zeta)$. Now let $\epsilon(z)$ be any positive continuous function on $\mathbf{C}$ which decreases to zero as $z \to \infty$. By the above theorem on Carleman approximation, there is an entire function g_0 such that $|f - g_0| < \epsilon$ on F_0. Then g_0 has each value as asymptotic value along continually many asymptotic paths lying in F_0. However, we should modify this construction slightly to be sure that the continually many asymptotic paths associated to any given asymptotic value are pairwise "non-equivalent" so that the multiplicity of any asymptotic value is actually that of the continuum. Let $\{z_n\}$ be a sequence which tends to ∞ in the complement of F_0 and which is frequently "between" any two boundary paths contained in F_0. We extend f to the sequence $\{z_n\}$ in such a way, that for any two asymptotic paths in F_0, f restricted to the values of $\{z_n\}$ has no limit as $z \to \infty$ along those values of $\{z_n\}$ which lie "between" these two paths. Now let F be the union of F_0 with the values of the sequence $\{z_n\}$. Again, by the theorem of Carleman approximation, there is an entire function g such that $|f - g| < \epsilon$. Then, g has the same asymptotic values as g_0 along the asymptotic paths contained in F_0, but the behaviour of g along the sequence $\{z_n\}$ assures us that any two such asymptotic paths are, by any reasonable definition, non-equivalent for g.

7 Plurisubharmonic functions

If f is a holomorphic function, then $\ln |f|$ is a plurisubharmonic function. An attempt to determine whether general plurisubharmonic functions can be built up from functions of the form $\ln |f|$, with f holomorphic, leads naturally to the notion of a Hartogs function. Moreover, it turns out that Hartogs functions are somehow connected to our subject, holomorphic approximation (see [49]).

Definition A real-valued function φ, defined in a domain $\Omega \subset \mathbf{C}^n$ is called a Hartogs function if it belongs to the intersection

$$\bigcap_{G \subset\subset \Omega} \Phi_G,$$

where Φ_G denotes the smallest class of functions, containing all functions of the form $u = \ln |f|$, with f holomorphic in G, and closed with respect to the operations:

1)

$$\varphi = \lambda_1 \varphi_1 + \lambda_2 \varphi_2,$$

with $\lambda_1 \geq 0, \lambda_2 \geq 0$;

2)

$$\varphi = \sup_{\alpha} \varphi_\alpha,$$

where $\{\varphi_\alpha\}$ is any family which is uniformly upper-bounded on each compact set $K \subset G$;

3)

$$\varphi = \lim_{j \to \infty} \varphi_j,$$

where $\{\varphi_j\}$ is any decreasing sequence of functions.

4) Upper semicontinuous regularization $\varphi \longmapsto \varphi^*$:

$$\varphi^*(z) = \lim_{\delta \to 0+} \sup_{|\zeta - z| < \delta} \{\varphi(\zeta)\}.$$

We remark that condition 4) is, in fact, superfluous [34], at least for the definition of Hartogs functions on domains in $\mathbf{C}^n$. However, this is less clear if we wish to consider Hartogs functions on manifolds.

Any upper semicontinuous Hartogs function is plurisubharmonic. For domains of holomorphy, the following theorem of Bremermann [12] asserts the converse.

Theorem 14 (Bremermann) *Every function u, plurisubharmonic in a domain of holomorphy Ω, is a Hartogs function.*

Proof (n=1) For $n = 1$ the result is due to Lelong and asserts that any subharmonic function in any domain is a Hartogs function. We present a proof for the case $n = 1$ which is a (slight) simplification of a proof given by Ronkin [42]. Let $G \subset\subset \Omega$ and μ the Riesz measure of the function u. Let $\mathcal{P}$ be a (finite) partition of G and for each $R \in \mathcal{P}$, choose a point $\zeta_R \in R$ and set $\mu_{\mathcal{P}} = \sum_{\mathcal{P}} \mu(R)\delta_{\zeta_R}$.

Lemma 4 *For each uniformly continuous function f on G,*

$$\lim_{|\mathcal{P}|\to 0} \int_G f d\mu_{\mathcal{P}} = \int_G f d\mu.$$

Consider now the integral

$$v_\mu^t(z) = \int_G h_t(z-\zeta)d\mu_\zeta,$$

where

$$h_t(z) = \max\{\ln|z|, t\},$$

and the potential

$$v_\mu(z) = \int_G \ln|z-\zeta|d\mu_\zeta.$$

Analogously, we define the functions $v_{\mu_{\mathcal{P}}}^t$ and $v_{\mu_{\mathcal{P}}}$. We remark that by the above lemma, as $|\mathcal{P}| \to 0$,

$$v_{\mu_{\mathcal{P}}}^t(z) \to v_\mu^t(z).$$

From this, using the obvious inequality

$$v_{\mu_{\mathcal{P}}}^t(z) \geq v_{\mu_{\mathcal{P}}}(z),$$

we conclude that, as $|\mathcal{P}| \to 0$,

$$\limsup v_{\mu_{\mathcal{P}}}(z) \leq v_\mu^t(z).$$

Passing to the limit as $t \to -\infty$, we have, further, that for $z \in G$,

$$\limsup v_{\mu_{\mathcal{P}}}(z) \leq v_\mu(z),$$

as $|\mathcal{P}| \to 0$. At the same time, we will show that, for each $\epsilon > 0$, as $|\mathcal{P}| \to 0$,

$$\limsup v_{\mu_{\mathcal{P}}}(z) > v_\mu(z) - \epsilon \tag{4}$$

holds on a dense subset of G. Indeed, if for some point $z^0 \in G$, we have

$$\limsup v_{\mu_{\mathcal{P}}}(z) \leq v_\mu(z) - \epsilon, \tag{5}$$

as $|\mathcal{P}| \to 0$, in some disc $E_r(z^0) \subset\subset G$, then by Fatou's lemma, for any positive continuous function φ of compact support in $E_r(z^0)$, we have

$$\limsup \int_G \int_G \varphi(z) \ln|z-\zeta| d\mu_{\mathcal{P}}(\zeta) d\omega(z) =$$

$$\limsup \int_G \varphi(z) v_{\mu_{\mathcal{P}}}(z) d\omega(z) \leq \int_G \varphi(z)\{\limsup v_{\mu_{\mathcal{P}}}(z)\} d\omega(z) \leq$$

$$\int_G \varphi(z) v_\mu(z) d\omega(z) - \epsilon \int_G \varphi(z) d\omega(z) < \int_G \varphi(z) v_\mu(z) d\omega(z).$$

On the other hand, since the function

$$\int_G \ln|z-\zeta| \varphi(z) d\omega(z)$$

is clearly continuous in ζ, by the above lemma, we have

$$\lim \int_G \int_G \varphi(z) \ln|z-\zeta| d\mu_{\mathcal{P}}(\zeta) d\omega(z) =$$

$$\lim \int_G d\mu_{\mathcal{P}}(\zeta) \int_G \varphi(z) \ln|z-\zeta| d\omega(z) = \int_G \varphi(z) v_\mu(z) d\omega(z).$$

This contradiction shows that there does not exist a disc $E_r(z^0)$ on which (5) holds. Thus, for each $\epsilon > 0$, (4) holds on a dense subset of G, from which it immediately follows that the regularization v^* of the function

$$v = \limsup_{|\mathcal{P}| \to 0} v_{\mu_{\mathcal{P}}} \tag{6}$$

coincides with the function v_μ in G. Notice that each function $v_{\mu_{\mathcal{P}}}$ belongs to the class Φ_G. Indeed, from the definition of the measures $\mu_{\mathcal{P}}$, it follows that

$$v_{\mu_{\mathcal{P}}}(z) = \int_G \ln|z-\zeta| d\mu_{\mathcal{P}}(\zeta) = \sum_{\mathcal{P}} \mu(R) \ln|z-\zeta_R|.$$

Since the functions $v_{\mu_{\mathcal{P}}}$ belong to F_G, $v^* = v_\mu$, and the regularization is still a Hartogs function, we conclude that the potential v_μ also belongs to Φ_G.

We now show that any function h, harmonic in G, also belongs to Φ_G. To this end, since any domain Ω can be exhausted by finitely-connected domains, we restrict our attention only to such domains, i.e., finitely connected domains G. In this case, it is easy to see that there exist (positive) real numbers $\lambda_1, \ldots, \lambda_p$ and complex numbers $a_1 \notin G, \ldots, a_p \notin G$, such that the function

$$h_1(z) = h(z) - \sum_j \lambda_j \ln |z - a_j|$$

is the real part of a holomorphic function in G. It follows that $h_1 \in \Phi_G$ from which it is immediate that also $h \in \Phi_G$. Thus, both terms in the Riesz representation of u belong to Φ_G, and hence u itself also belongs to $\Phi_{G'}$. Now, if we exhaust the domain Ω by domains G', it follows that u is a Hartogs function. This completes the proof of the theorem.

The preceding theorem of Bremermann shows that in domains of holomorphy, all plurisubharmonic functions can be derived from holomorphic functions. For continuous plurisubharmonic functions, Bremermann [13] showed much more. In fact, continuous plurisubharmonic functions can be obtained from holomorphic functions via *uniform* approximation.

Theorem 15 (Bremermann) *If u is a function, continuous and plurisubharmonic in a domain of holomorphy Ω, then for each compact $K \subset \Omega$ and each $\epsilon > 0$, there exist functions $f_1, f_2, \cdots, f_N \in H(\Omega)$ and positive constants $\alpha_1, \alpha_2, \cdots, \alpha_N$ such that on K,*

$$|u - \max\{\alpha_1 \ln |f_1|, \alpha_2 \ln |f_2|, \cdots, \alpha_N \ln |f_N|\}| \leq \epsilon.$$

The approximation may even be chosen to be one-sided, that is, less than or equal to u or greater than or equal to u, as we wish.

8 Harmonic and subharmonic approximation

Throughout this section, Ω will denote an open Riemannian manifold (or more precisely, a "noncompact smooth orientable Riemannian manifold") and W will denote an open subset of Ω. We say that (W, Ω) is a subharmonic extension pair for *compact* sets if for every function u subharmonic on W, there is a sequence of functions u_n, subharmonic on Ω, such that for each compact subset E of W, $u_n = u$ on E, for all sufficiently large n. We say that (W, Ω) is a subharmonic (respectively harmonic) Runge pair for *compact* sets if for every function u subharmonic (respectively harmonic) on W, there is a sequence of functions u_n, subharmonic (respectively harmonic) on Ω which converges pointwise to u on W and such that for each compact subset E of W, u_n is decreasing on E, for sufficiently large n. Similarly, we define the respective notions of subharmonic extension pair, and subharmonic Runge pair, for *closed* sets by replacing compact subsets E of W, in the above definitions, by subsets E of W which are closed (in Ω, not just in W). We define the notions of *continuous* subharmonic extension pairs and *continuous* subharmonic Runge pairs for compact sets by replacing "subharmonic" by "continuous subharmonic" in the above definitions. It follows from Dini's theorem that in the case of continuous subharmonic (and *a fortiori* harmonic) Runge pairs the convergence is uniform on each compact set E. In fact, since constants are harmonic, the possibility of uniform approximation of a continuous function by a sequence of continuous subharmonic (harmonic) functions is equivalent to the possibility of such approximation by a decreasing such sequence, but since Runge approximation is usually considered for continuous functions, decreasing sequences of approximants are not usually required in the definition. Finally, we say that (W, Ω) is a continuous subharmonic (respectively harmonic) Runge pair for closed sets if for every function u continuous subharmonic (respectively harmonic) on W, there is a sequence of functions u_n, continuous subharmonic (respectively harmonic) on Ω which converges pointwise to u on W and such that the convergence is uniform on each closed subset E of W.

Theorem 16 *The following are equivalent:*

(a) (W, Ω) *is a harmonic Runge pair for compact sets;*

(b) (W, Ω) *is a harmonic Runge pair for closed sets;*

(c) (W, Ω) *is a (continuous) subharmonic Runge pair for compact sets;*

(d) (W, Ω) *is a (continuous) subharmonic Runge pair for closed sets;*

(e) (W, Ω) *is a (continuous) subharmonic extension pair for compact sets;*

(f) (W, Ω) *is a (continuous) subharmonic extension pair for closed sets;*

(g) $\Omega^* \setminus W$ *is connected.*

The preceding theorem is due to various authors (see [24] for more information on this). Notice that the equivalence of (a) and (g) is the harmonic analog of the Behnke-Stein theorem. Similarly, it can be shown that the holomorphic analog of (g) $\rightarrow$ (b) is equivalent to the statement in Roth's theorem.

As in the case of the classical complex Runge theorem, one can prove the implication (g) $\rightarrow$ (a) using Green's theorem. We indicate this route in the simple case when Ω is a domain in the complex plane $\mathbf{C}$.

Theorem 17 (Poisson-Green Formula) *If D is a smoothly bounded domain in $\mathbf{C}$ and $u \in C^2(\overline{D})$, then*

$$\begin{aligned} u(z) &= \frac{1}{2\pi}\int_{\partial D}[u(\zeta)\frac{\partial}{\partial n_\zeta}\cdot \ln|\zeta - z| - \ln|\zeta - z|\cdot\frac{\partial u}{\partial n_\zeta}(\zeta)]\,ds(\zeta) \\ &+ \frac{1}{2\pi}\iint_D \ln|\zeta - z|\cdot\Delta u(\zeta)dm(\zeta), \end{aligned}$$

for $z \in D$, where $\partial/\partial n_\zeta$ denotes the outward normal derivative, $ds(\zeta)$ denotes arc length, and $dm(\zeta)$ denotes Lebesgue planar measure, all with respect to the variable ζ.

Let us denote by $h(D)$ the class of functions harmonic on D.

Corollary 3 *If D is a smoothly bounded domain in $\mathbf{C}$ and $f \in C^2(||\overline{D}) \cap h(D)$, then, for all $z \in D$,*

$$u(z) = \frac{1}{2\pi}\int_{\partial D}[u(\zeta)\frac{\partial}{\partial n_\zeta}\cdot \ln|\zeta - z| - \ln|\zeta - z|\cdot\frac{\partial u}{\partial n_\zeta}(\zeta)]\,ds(\zeta). \tag{7}$$

Corollary 4 *If $u \in C_0^2(\mathbf{C})$, then*

$$u(z) = \frac{1}{2\pi} \iint\limits_{\mathbf{C}} \ln|\zeta - z| \cdot \Delta u(\zeta) dm(\zeta). \tag{8}$$

The implication (g) → (a) can be easily deduced from either of these corollaries. Indeed, let $u \in h(W)$ and let K be a compact subset of Ω. We may choose a smoothly bounded neighbourhood D of K such that $u \in C^2(\|\overline{D}) \cap h(D)$. Then from the first corollary, for $z \in K$, $u(z)$ can be expressed as an integral over ∂D. If we approximate this integral (7) by Riemann sums, we obtain linear combinations of logarithms having "poles" on ∂D, but we also obtain linear combinations of normal derivatives of logarithms. If we approximate the latter by differential quotients, then we obtain an approximation of u on K by linear combinations of logarithms only, whose "poles" lie on ∂D. As in the case of complex rational approximation, we may "push these poles" outside of Ω.

We may arrive at the same result using the second corollary. Namely, by modifying u outside of a neighbourhood of K, we may assume that $u \in C_0^2(\mathbf{C})$. Thus, we have the integral representation (8). Now, if we take Riemann sums, we have only logarithmic terms, so it is actually simpler to use the second corollary rather than the first. In the case of complex approximation, both analogous corollaries were equally efficient.

In the preceding paragraph, we have referred to pole-pushing for harmonic functions. In our earlier discussion of the classical Runge theorem for complex rational approximation, we pointed out that the method of pole-pushing is based on Laurent expansions. Since the harmonic analog of the Laurent expansion is not so widely taught, we shall say a few words about such expansions. For $a \in \mathbf{C}$ and $0 \leq R_1 < R_2 \leq +\infty$, we denote by $A_a(R_1, R_2)$ the annulus $R_1 < |z - a| < R_2$ in the complex plane $\mathbf{C}$. Also, we denote by $D_a(R)$ the disc $0 < |z - a| < R$ and by $A_a(R)$ the "outer disc" $R < |z - a| \leq +\infty$.

Lemma 5 (Laurent expansion) *Harmonic functions u have the following expansions:*

in a disc $D_a(R)$,

$$u(z) = \sum_{0}^{+\infty} r^n (\alpha_n \cos n\theta + \beta_n \sin n\theta); \tag{9}$$

in an outer disc $A_a(R)$,

$$u(z) = \sum_{0}^{+\infty} r^{-n} (\alpha_n \cos n\theta + \beta_n \sin n\theta); \tag{10}$$

in an annulus $A_a(R_1, R_2)$,

$$u(z) = \lambda \ln r + \sum_{-\infty}^{+\infty} r^n (\alpha_n \cos |n|\theta + \beta_n \sin |n|\theta); \tag{11}$$

where $z = a + re^{in\theta}$ and λ, the α_n's, and the β_n's are scalars.

In the case of the outer disc $A_a(R)$, we have abused the notation for the sake of elegance. Actually, the expansion we have given makes no sense for $z = +\infty$, but in this case, it is natural to define the value of $u(+\infty)$ as α_0.

To justify the above expansion in the case of a function u harmonic in a disc $D_a(R)$, fix $0 < r < R$. Then the Fourier series for the function $u(re^{i\theta})$ can indeed be written in the form (9). Our task is to verify that if we change the value of r, then the coefficients α_n and β_n do not change. From the formula of De Moivre, we see that the functions $r^n \cos n\theta$ and $r^n \sin n\theta$ are harmonic and by Parseval's theorem, we have that the series $\sum r^{2n}(|\alpha_n|^2 + |\beta_n|^2)$ converges. It follows that the series in (9) converges uniformly on any disc $D_a(\rho)$, with $0 < \rho < r$ and therefore represents a harmonic function on $D_a(r)$. Since this harmonic function coincides with the function u on the boundary of $D_a(r)$, it follows from the maximum principle, that it also coincides with u in the whole disc $D_a(r)$. Since r was any number between 0 and R, we have, in any such disc $D_a(r)$, a representation of the form (9). From the uniqueness of Fourier series, we have that all of these representations have the same coefficients, which proves the validity of the representation (9) in all of $D_a(R)$.

The justification of the expansion in an outer disc is almost identical and makes use of the fact that isolated singularities (in this case, $+\infty$), are removable for bounded harmonic functions.

Suppose, now, that u is harmonic in an annulus $A_a(R_1, R_2)$. Suppose $R_1 < r_1 < r_2 < R_2$. For $z \in A_a(r_1, r_2)$, $u(z)$ has a representation as an integral of the form (7). Thus, if $C_a(r)$ denotes the circle centered at a of radius r, $u(z)$ can be represented as an integral I_2 over the outer boundary $C_a(r_2)$ of $A_a(r_1, r_2)$ minus an integral I_1 over the inner boundary $C_a(r_1)$. As functions of z, I_1 and I_2 are harmonic on the complement of $C_a(r_1)$ and $C_a(r_2)$ respectively. In particular, I_2 is harmonic in the disc $D_a(r_2)$ and so has a series representation of the form (9). Since by (7), the value of I_2 is invariant if z is fixed while r_2 is replaced by any larger r less than R_2, we have that in $A_a(r_1, R_2)$, u is represented as a series of the form (9) minus the integral I_1. We now write the integral I_1 as a difference of two integrals:

$$I_1(z) = \frac{1}{2\pi} \int_{C_a(r_1)} u(\zeta) \frac{\partial}{\partial n_\zeta} \cdot \ln|\zeta - z| \, ds(\zeta) - \frac{1}{2\pi} \int_{C_a(r_1)} \ln|\zeta - z| \cdot \frac{\partial u}{\partial n_\zeta}(\zeta) \, ds(\zeta). \tag{12}$$

Both integrals on the right side of (12) are harmonic in the outer disc $A_a(r_1)$ and the first of these is bounded at infinity so it has a representation as a series of the form (10). The difference of the second integral and $\lambda \ln|z - a|$ is also bounded at infinity, where

$$\lambda = \frac{1}{2\pi} \int_{C_a(r_1)} \frac{\partial u}{\partial n_\zeta}(\zeta) \, ds(\zeta),$$

and so this difference also has a representation as a series of the form (10). We have shown that u has a Laurent expansion (11) in the annulus $A_a(r_1, R_2)$. Again, letting r_1 decrease to R_1 we find that this same series is valid in the entire annulus $A_a(R_1, R_2)$ which completes the proof of the Laurent expansion for harmonic functions.

We could also have used the fact that, for some λ, $u(z) - \lambda \ln|z - a|$ is the real part of a holomorphic function f in $A_a(R_1, R_2)$. A Laurent expansion for u can then be found by adding $\lambda \ln |z-a|$ to the real part of the Laurent expansion for f. However, we thought it of interest to also present a proof which was less dependent on function theory.

9 Iversen's maximum principle

As in the complex case, the theory of harmonic and subharmonic approximations and extensions on closed sets allows interesting applications.

Theorem 18 (Maximum Principle) *If* Ω *is a* bounded *open subset of the finite complex plane* $\mathbf{C}$*, then*

$$\sup_{\Omega} s = \sup_{\partial\Omega} s,$$

for all s *subharmonic in* Ω.

The right side of the above equality requires some explanation, since the function s may not be defined on the boundary of Ω. We adopt the following convention:

$$\sup_{\partial\Omega} s = \sup_{y\in\partial\Omega} \{\overline{\lim_{x\to y}}\, s(x)\}.$$

The following trivial examples show that the Maximum Principle is not in general true if Ω is not bounded.

Example 6 $\Omega = (|z| > 1)$, $s(z) = |z|$.

Example 7 $\Omega = (\mathrm{Re}\, z > 0)$, $s(z) = |e^z|$.

It may come as a surprise that the Maximum Principle, nevertheless, does hold for some unbounded sets. Moreover, we shall give a simple characterization of such sets.

We have stated the maximum principle for subharmonic functions, which implies the maximum principle for harmonic functions as well as for holomorphic functions (since the absolute value of a holomorphic function is subharmonic). Harmonic functions are classically defined on open subsets of $\mathbf{R}^n$, whereas the natural domain of definition for a holomorphic function is a Riemann surface. Riemannian manifolds generalize both kinds of domains. We shall say that $\tilde{\Omega}$ is a second countable compactification of a Riemannian manifold Ω if $\tilde{\Omega}$ is a compact Hausdorff space having a countable basis of open sets and containing (a homeomorphic copy of) Ω as an open (not necessarily dense) subset. We denote by $\tilde{\partial}\Omega = \tilde{\Omega} \setminus \Omega$ the ideal boundary of Ω in $\tilde{\Omega}$. A subset E of the ideal boundary $\tilde{\partial}\Omega = \tilde{\Omega} \setminus \Omega$ is said to be *accessible* (from Ω) if there exists a continuous path $\sigma : [0, +\infty) \to \Omega$ which is eventually in each neighbourhood of E. That is, for each neighbourhood V of E in $\tilde{\Omega}$, there is a t_V such that $\sigma(t) \in V$ for each $t > t_V$. An upper semicontinuous function $s : \Omega \to [-\infty, +\infty)$ is said to be *subharmonic* if for any relatively compact open set V in Ω, any harmonic function which dominates s on ∂V dominates s on V.

The following result, obtained jointly with Chen Huaihui [16], characterizes those subsets of the boundary which can be disregarded in the maximum principle. Earlier versions are due independently to Sahakian [46] and Gauthier-Grothmann-Hengartner [27], but especially (and I thank Alex Eremenko [17] for bringing this to my attention) to Iversen [38, p. 24].

Theorem 19 (Generalized Maximum Principle) *Let E be a closed subset of $\tilde{\partial}\Omega$. A necessary and sufficient condition in order that*

$$\sup_{\Omega} s = \sup_{\tilde{\partial}\Omega \setminus E} s,$$

for all s subharmonic on Ω, is that E be not accessible.

There are two instances which are of particular interest. Firstly, there is the case where Ω is a (not necessarily bounded) open subset of $\mathbf{R}^n$ and $\tilde{\Omega} = ||\overline{\Omega} \cup \{\infty\}$. Let us say that the maximum principle holds on such an open set Ω if the conclusion of Theorem 1 holds for Ω. We have the following generalization of Theorem 1.

Corollary 5 *The maximum principle holds on an open subset Ω of $\mathbf{R}^n$ if and only if ∞ is not accessible from Ω.*

Note that in Examples 1 and 2 above, ∞ is accessible, whereas if Ω is bounded, then ∞ is not accessible. A less trivial example is the following.

Example 8 *Let*

$$\Omega = \{(x, y) \in \mathbf{R}^2 : 0 < x < 1,\ 0 < y < x^{-1}|sin(x^{-1})| + 1\}.$$

Then Ω is an unbounded domain from which ∞ is not accessible.

Remark When the maximum principle for unbounded domains is mentioned, the first thought that usually comes to mind is the Phragmén-Lindelöf theorem which for nice domains may be paraphrased as follows:

(boundary estimate) + (global estimate) $\Rightarrow$ (better global estimate).

We wish to emphasize that we assume *no* global estimates. Thus, we do not assume that s is bounded; rather, we *infer* that s is bounded – and by the same bound as on the boundary!

The sufficiency in Corollary 1 follows from the contrapositive of the following.

Theorem 20 (Fuglede-Iversen [18]) *If s is a subharmonic function on an open subset Ω of $\mathbf{R}^n$ and*

$$\sup_{\Omega} s > \sup_{\partial\Omega} s,$$

then there is a continuous path $\sigma : [0, +\infty) \to \Omega$ along which s tends to $\sup_{\Omega} s$.

A second instance of Theorem 2 which is of particular interest is when Ω is a bounded subset of $\mathbf{R}^n$ and $\tilde{\Omega} = \|\overline{\Omega}$.

Corollary 6 *Let E be a closed subset of the boundary of a bounded open subset Ω of $\mathbf{R}^n$. Then, a necessary and sufficient condition in order that*

$$\sup_{\Omega} s = \sup_{\partial\Omega \setminus E} s,$$

for all s subharmonic on Ω, is that E be not accessible.

Example 9 *Let Ω be the square $\{0 < x < 1,\ 0 < y < 1\}$ with the following segments removed:*

$$x = 1/(2^j),\ 0 < y \leq 2/3, \textit{and } x = 1/(2^j + 1),\ 1/3 \leq y < 1; \quad j = 1, 2, \ldots.$$

Then, the set $E_1 = \{x = 0, 0 \leq y \leq 1\}$ is accessible whereas the set $E_2 = \{x = 0, 0 \leq y \leq 1/2\}$ is not.

Under the additional assumption that s is bounded (a Phragmén-Lindelöf type assumption), the sufficiency in Corollary 2 is an immediate consequence of the two-constants theorem of Nevanlinna and the fact that any non-accessible set is of harmonic measure zero. We present an amusing proof of the latter. Recall that the harmonic measure at a point $x \in \Omega$ of a Borel set $E \subset \partial\Omega$ is the probability that Brownian motion starting at x first leave Ω through E. Since Brownian motion is continuous, if E is not accessible, it is *impossible*, hence *improbable*, that Brownian motion leave through E! Again, we repeat that this proof works only under the additional assumption that s is bounded.

For a proof of the sufficiency in Theorem 2, we refer the reader to [16]. We now prove the necessity as an application of approximations and extensions.

Application 6 (Necessity in maximum principle) *Suppose E is an accessible closed subset of $\tilde{\partial}\Omega$. Then, there exists a function s subharmonic on Ω such that*

$$\sup_{\Omega} s > \sup_{\tilde{\partial}\Omega \setminus E} s.$$

Proof Let σ be a path in Ω which tends to E. We may assume that σ is simple and hence it is possible to construct open sets U and V such that

$$\sigma \subset U \subset \overline{U} \subset V \subset \overline{V} \subset \Omega,$$

where the closures are to be taken in the compactification $\tilde{\Omega}$ and, setting $W = U \cup (\Omega \setminus \overline{U})$, we have that $\Omega^* \setminus W$ is connected. If we define u to be 1 on U and 0 on $\Omega \setminus \overline{U}$, then, by the implication (g) $\to$ (f), there is a function s, subharmonic (and continuous) on all of Ω which is 1 on σ and 0 on the "neighbourhood" $\Omega \setminus V$ of $\tilde{\partial}\Omega \setminus E$. It follows that the function s has the required properties. In fact, by the implication (g) $\to$ (b) we even have a *harmonic* counterexample to the generalized maximum principle for this E. A detailed justification of the preceding topological assertions can be given following the methods of [5].

10 Local approximation

We have seen that, near the end of the 19th century, K. Runge considered (quite successfully) the problem of describing those sets K for which every holomorphic function on K can be approximated by polynomials. The problem, then, quite naturally arose of approximating functions, given on K, by functions holomorphic on K. Of course, any function which can be uniformly approximated on K by functions holomorphic on K is necessarily continuous on K and holomorphic in the interior of K, and early in the present century, J. L. Walsh [25] showed that in some very important cases, no more is required of the approximatees. The problem, then, quite naturally arose of characterizing those sets for which *every* function continuous on K and holomorphic on the interior of K can be approximated by functions holomorphic on K. In this section, we present the complete solution to this problem which was found, first for the analogous harmonic problem by M.V. Keldysh, and then in the complex case by A.G. Vitushkin, both of the Russian school. The solutions to these problems turn out to be local and can be described in terms of capacity and, in the harmonic case, the notion of thinness.

Recall that the *fine topology* on a Riemannian manifold Ω is the topology generated by the subharmonic functions on Ω.

Theorem 21 *For a subset E and a point x_0 of $\mathbf{R}^n$, the following are equivalent:*

(a) *E is thin at x_0;*

(b) $(\mathbf{R}^n \setminus E) \cup \{x_0\}$ *is a fine neighbourhood of* $\{x_0\}$;

(c) *Brownian motion starting from x_0 remains in $\mathbf{R}^n \setminus E$ for positive time.*

The equivalence of (a) and (b) persists on Riemannian manifolds. The notion of Brownian motion can also be introduced on Riemannian manifolds, but, in any case, we will not require the equivalence of the last condition to the first two.

Let us denote by $C(E)$, $C_a(E)$ and $C_{ac}(E)$ respectively the capacity, analytic capacity and continuous analytic capacity of a set E. Without stating the definitions of the various capacities, we recall that they have the following important property of characterizing removable sets for their associated classes of functions. Namely, a set is removable for the class of bounded harmonic, bounded holomorphic or (uniformly) continuous holomorphic functions if and only if it is respectively of harmonic, analytic or continuous analytic capacity zero.

We denote by $h(F)$ the class of functions harmonic on (a neighbourhood of) a set F and (as before), by $H(F)$ the class of functions holomorphic on F. Let $\overline{h}(F)$ (respectively, $\overline{H}(F)$) denote the functions on F which are uniform limits of functions in $h(F)$ (respectively, $H(F)$). Further, we denote by $a(F)$ (respectively, $A(F)$) the classes of functions continuous on F and harmonic (respectively, holomorphic) on the interior of F. Clearly, we have $\overline{h}(F) \subset a(F)$ and $\overline{H}(F) \subset A(F)$.

Theorem 22 (Keldysh, Deny, Labrèche) *Let F be a closed subset of a domain Ω of Euclidean space $\mathbf{R}^n$. The following are equivalent:*

(a) $\overline{h}(F) = a(F)$;

(b) *for each* $x \in F$, *there is a closed ball* $\overline{B}_x \subset \Omega$ *such that*

$$\overline{h}(F \cap \overline{B}_x) = a(F \cap \overline{B}_x);$$

(c) *for each open* U *in* Ω,

$$\mathcal{C}(U \setminus F) = \mathcal{C}(U \setminus F^0);$$

(d) $\Omega \setminus F$ *and* $\Omega \setminus F^0$ *are thin at the same points of* F.

This theorem was first shown for F compact by Keldysh and later, independently, by Deny. The closed case is due to Labrèche. Recently [4] this result has been extended to Riemannian manifolds (omitting condition (c)).

We now give two examples of compact sets K for which approximation fails, that is, $\overline{h}(K) \neq a(K)$, followed by an application involving a closed set F where approximation is possible.

Example 10 *There is a set* $K \subset \mathbf{R}^2$ *of the form*

$$K = \overline{D} \setminus \bigcup_{j=1}^{\infty} D_j$$

such that $\overline{h}(K) \neq a(K)$, *where* $\overline{D}$ *is the unit disc and* $\{D_j\}$ *is a sequence of discs having disjoint closures in* D.

Such an example can be constructed [31], for example, by choosing the sequence $\{D_j\}$ in the upper half-disc and clustering precisely to the whole interval $[0, 1]$ and nowhere else. It follows that $\mathbf{R}^2 \setminus K^0$ is not thin at 0 since it contains the interval $[0, 1]$. On the other hand, the discs $\{D_j\}$ can be chosen so small that, by the Wiener criterion, their union, which locally is $\mathbf{R}^2 \setminus K$, is thin at 0.

If a compact subset K of $\mathbf{R}^2$ has connected complement, then automatically the complement is thin at no point of the boundary of K and so $\overline{h}(K) = a(K)$. This situation does not persist in higher dimensions. In fact there is the following striking example.

Example 11 *There is a compact starlike set* $K \subset \mathbf{R}^3$ *for which* $\overline{h}(K) \neq a(K)$.

Such an example was constructed by Bagby and the author [7] in the form

$$K = \overline{B} \setminus \bigcup_{j=1}^{\infty} S_j,$$

where $\overline{B}$ is the unit ball and $\{S_j\}$ is a sequence of Lebesgue spines having disjoint closures in B and clustering precisely to all points of the disc $\overline{D} = \overline{B} \cap \{x_3 = 0\}$ and to no other points. The argument that K has the desired properties is similar to that in the preceding example.

Let us denote an arbitrary point x of Euclidean n-space $\mathbf{R}^n$, $n \geq 2$, by $x = (x', x_n)$, where $x' \in \mathbf{R}^{n-1}$ and $x_n \in \mathbf{R}$. Let U denote the open upper half-space $\{x : x_n > 0\}$. As an illustration of the power of the above theorem on harmonic approximation on unbounded sets, we will now give an elegant solution to the higher dimensional analog of the Dirichlet problem for the half-plane.

Application 7 (Dirichlet problem for half-space) *Given $\varphi \in C(\mathbf{R}^{n-1})$, find $u \in C(\overline{U})$ such that u is harmonic on U and $u(x', 0) = \varphi(x')$ for $x' \in \mathbf{R}^{n-1}$.*

Solution By Labrèche's theorem there is a $g \in h(\mathbf{R}^{n-1})$ such that $|\varphi - g| < 1$. By section 4, there is an $f \in h(\mathbf{R}^n)$ such that $|f - g| < 1$. The function

$$u = f + P_{\varphi - f}$$

is then a solution of the given Dirichlet problem.

In regard to the preceding application, the reader may want to look at Gardiner's paper [22], where a stronger result is proved.

The preceding theorem has a complex analog proved first by Vitushkin for compact F and later extended to closed sets by Nersesyan and Roth (see [30]).

Theorem 23 (Vitushkin, Nersesyan, Roth) *Let F be a closed subset of a domain Ω of the complex plane $\mathbf{C}$. The following are equivalent:*

(a) $\overline{H}(F) = A(F)$;

(b) *for each $z \in F$, there is a closed disc $\overline{D}_z \subset \Omega$ such that*

$$\overline{H}(F \cap \overline{D}_z) = A(F \cap \overline{D}_z);$$

(c) *for each open U in Ω,*

$$C_{ac}(U \setminus F) = C_{ac}(U \setminus F^0).$$

A particular, and most interesting, instance of this theorem is obtained by taking F to be the real axis. This yields Whitney's approximation theorem (as we have stated it) on uniform approximation of continuous functions on $\mathbf{R}$ by analytic ones. (Actually, Whitney obtained *better*-than-uniform approximation, and in $\mathbf{R}^n$, not just in $\mathbf{R}$.) Thus, we are able to "bypass" the use of Whitney's theorem in "our" earlier solution of the Dirichlet problem for the half-plane.

11 Individual functions

The theorems of the preceding section give necessary and sufficient conditions on a closed set F in order that each function belonging to the *class* of functions continuous on F and harmonic (respectively, holomorphic) on the interior of F can be approximated by functions harmonic (respectively, holomorphic) on F. Such theorems are called *class* theorems. Suppose now we consider an arbitrary closed set F, which may or may not satisfy the conditions of the class theorem, and hence there may be some functions in $a(F)$ (respectively, $A(F)$) which cannot be approximated. A more general question than that addressed by the class theorems is to ask just which functions on such an arbitrary F *can* be approximated. Theorems which deal with this latter question are called *individual function* theorems. In the harmonic case this problem has been solved on compact sets by Debiard and Gaveau in terms of fine potential theory.

Definition Let x be a point of a finely open set V and let $E \subset \partial V$. We define the (fine) *harmonic measure* $\omega_x^V(E)$ as the probability that Brownian motion starting at x first exit V through E.

Remarks

1. If V is an ordinary open set, then ω_x^V is ordinary harmonic measure.
2. ω_x^V is carried by the fine boundary of V.

Definition A finely continuous function u defined on a finely open set U is said to be *finely harmonic* if there is a base $\mathcal{B}$ for the induced fine topology on U such that for all $V \in \mathcal{B}$, the fine closure of V is contained in U and

$$u(x) = \int u d\omega_x^V, \text{ for all } x \in V.$$

We remark that on a usual open set U, the harmonic functions are the finely harmonic functions which are locally bounded.

We may now state the complete characterization of approximable functions, known as the Debiard and Gaveau theorem. This was originally proved for compact sets, later extended to closed sets of Euclidean space by Ladouceur and the author and then extended to closed subsets of Riemannian manifolds by Bagby and Blanchet [4].

Theorem 24 (Debiard-Gaveau theorem for closed sets) *Suppose F is a closed subset of an open Riemannian manifold, and let $u \in C(F)$. Then the following are equivalent.*

(a) $u \in \overline{h}(F)$.

(b) *u is finely harmonic on the fine interior of F.*

The above individual function theorem of Debiard and Gaveau is stronger than the class theorem of Keldysh. Let us verify this by deriving the equivalence of (a) and (d) in the Keldysh theorem from the Debiard and Gaveau theorem.

Suppose, then, that $\Omega \setminus F$ and $\Omega \setminus F^0$ are thin at the same points and let $u \in a(F)$. We claim that $u \in \overline{h}(F)$. By the Debiard and Gaveau theorem, it is sufficient to show that u is finely harmonic on the fine interior of F. Let x be a fine interior point of F. We may assume that x lies on the boundary of F, for otherwise u is by hypothesis harmonic on an open neighbourhood of x and *a fortiori* finely harmonic in a fine neighbourhood of x. To say that the boundary point x lies on the fine interior of F is to say that $\Omega \setminus F$ is thin at x, but then, by hypotheses, $\Omega \setminus F^0$ is also thin at x. Thus, F^0 is a fine deleted neighbourhood of x. Since u is continuous on F it is bounded and so by a removable singularity theorem for finely harmonic functions, u extends to be finely harmonic at x. But since u is already continuously defined at x, u is itself this fine harmonic extension. Thus, $u \in \overline{h}(F)$. This shows that (d) $\rightarrow$ (a) in the Keldysh theorem.

The following lemma, first proved by Keldysh for compact sets, is fundamental. A set is said to be *thick* at a point if it is not thin at that point.

Lemma 6 (Keldysh Lemma) *Let F be a closed subset of a Riemannian manifold Ω and $x \in F$. Then, $\Omega \setminus F^0$ is thick at x if and only if x is a peak point for the class $a(F)$.*

Proof Suppose, first, that x is a peak point for the class $a(F)$ and let u be a function in $a(F)$ which peaks at x. If $\Omega \setminus F^0$ were thin at x, then F^0 would be a deleted fine neighbourhood of x and, since u is bounded near x, by the theorem on removable singularities for finely harmonic functions, u would extend to be finely harmonic at x. But u is already continuously defined at x and so u itself would be finely harmonic at x. Since u peaks at x, this would violate the maximum principle for finely harmonic functions. Thus, $\Omega \setminus F^0$ is thick at x.

Suppose, conversely, that $\Omega \setminus F^0$ is thick at x. Fix a parametric ball B centered at x. By the Wiener criterion, the thickness of the set $\Omega \setminus F^0$ at x is equivalent to the divergence of a certain Wiener series $\sum \lambda_n C(A_n \setminus F^0)$, where the A_n are appropriate "annuli' centered at x and the λ_n are appropriate constants. We may assume that the series $\sum \lambda_{2n} C(A_{2n} \setminus F^0)$ also diverges. Now, by a famous result of Ancona, for any $\epsilon_n > 0$, there is a compact set K_n contained in $A_{2n} \setminus F^0$ such that

$$C(K_n) > C(A_{2n} \setminus F^0) - \epsilon_n$$

and K_n is thick at each of its points. Set

$$G = \Omega \setminus \bigcup_{n=1}^{\infty} K_n.$$

The open set G is regular for the Dirichlet problem. Hence, there is a function $u \in a(\overline{G})$ which peaks at the point $x \in \partial G$. Since $F^0 \subset G$ it follows that u can be extended to a function in $a(F)$ which peaks at x.

In the above proof of the Keldysh Lemma, we have used the concept of capacity and the Wiener criterion for thinness, on a parametric ball in a Riemannian manifold. Here our concept of thinness comes from regarding the parametric ball as a harmonic space ("harmonic functions" defined by the Riemannian metric). We know of no reference for the Wiener criterion in this generality, although the Wiener criterion is of course well known when harmonic functions are defined by the Euclidean metric. However, Hervé [37] has shown that these two concepts of thinness in the parametric ball are the same. (A similar problem and solution occurs in [4, 1, Remark 8.2 (e)].)

We now present a proof that (a) $\to$ (d) in the Keldysh theorem. Suppose, then, that $\Omega \setminus F$ and $\Omega \setminus F^0$ are *not* thin at the same points and let us construct a $u \in a(F)$ which is not in $\overline{h}(F)$. There exists, by hypothesis, some point $x \in \partial F$ such that $\Omega \setminus F$ is thin at x but $\Omega \setminus F^0$ is not. By the Keldysh lemma, x is a peak point for the class $a(F)$. Thus, there is a function $u \in a(F)$ which peaks at x. By the maximum principle for finely harmonic functions, this function cannot be finely harmonic at x. Thus, by the theorem of Debiard and Gaveau this function u is not in $\overline{h}(F)$ and this shows that (a) $\to$ (d) in the Keldysh theorem.

Let us now turn to the individual function problem for holomorphic functions. A function f is said to be *finely holomorphic* if f' exists (in the fine sense) and is finely continuous. Let us denote by $\tilde{A}(F)$ the set of functions continuous on F and finely holomorphic on the fine interior of F. Then, as in the harmonic situation, we have the following inclusions:

$$\overline{H}(F) \subset \tilde{A}(F) \subset A(F).$$

However, in contrast to the harmonic case, in which the Debiard and Gaveau theorem asserts that the harmonic analogs of the first two classes coincide, Fuglede [19] has shown that the inclusions in the holomorphic case can be strict. Thus, the complex analog of the Debiard and Gaveau theorem fails.

Nevertheless, another type of individual function theorem does hold for complex approximation, namely, the well known theorem of Vitushkin which, in terms of certain estimates involving continuous analytic capacity, gives a complete description of those functions on a compact set which can be approximated by holomorphic functions. This powerful result of Vitushkin (which we do not state) has been extended to closed sets by Hadjiiski [36].

12 Holomorphic vs. harmonic

A recurring theme in these lectures has been the resemblance between the theory of harmonic functions and that of holomorphic functions of *one* complex variable. This is mostly due to the fact that harmonic functions of two variables (x, y) are (locally) precisely the real parts of holomorphic functions in the variable $z = x + iy$. The similarities are also to some extent due to the fact that harmonic functions are the solutions of the homogeneous Laplace equation $\Delta u = 0$, holomorphic functions are the solutions of the Cauchy-Riemann equation $\partial f/\partial \| \overline{z} = 0$, and both the Laplacian and the Cauchy-Riemann operator are partial differential operators of elliptic type. The differences are to some extent due to the fact that the Laplacian is an operator of order two while the Cauchy-Riemann operator is of order one.

The relationship between harmonic functions and holomorphic functions of *several* variables is much more tenuous. While it is true that the real part of a holomorphic function is always harmonic, it is no longer true in higher dimensions that each harmonic function is locally the real part of a holomorphic function. Harmonic functions in any number of variables are the solutions of a single elliptic equation in one unknown: $\Delta u = 0$. However, holomorphic functions of n complex variables are the solutions of the *system* of n Cauchy-Riemann equations in one unknown: $\partial f/\partial \| \overline{z}_j = 0, j = 1, 2, \ldots, n$. If $n > 1$, this system is overdetermined which accounts to a great extent for some of the differences between, on the one hand, harmonic functions of one or several variables and holomorphic functions of a single complex variable, and on the other hand, holomorphic functions of more than one variable. In fact, from the Cauchy-Kovaleska theorem, it is not hard to see that, in the case of more than one complex variable, there is *no* elliptic differential equation, $Pf = 0$, with analytic coefficients, whose solutions coincide with the holomorphic functions.

The holomorphic Carleman theorem has been extended to several complex variables by Scheinberg [48]. Myron Goldstein has observed to us that the holomorphic Carleman theorem follows from the harmonic Carleman theorem. The proof of this observation uses the property (interesting in itself) that the Taylor series of a harmonic entire function has infinite radius of convergence. We see this as follows. For fixed n, there is a positive constant λ such that any harmonic function in a ball of radius R in $\mathbf{R}^n$ extends holomorphically to a ball of radius λR in $\mathbf{C}^n$. It follows that any entire harmonic function u in $\mathbf{R}^n$ extends to an entire (holomorphic) function $\tilde{u}$ in $\mathbf{C}^n$. Since the Taylor series of u and $\tilde{u}$ about any point of $\mathbf{R}^n$ are the same, we conclude that the radius of convergence of any Taylor series of u is infinite. Now, let f be a continuous function on $\mathbf{R}^n$. By the harmonic Carleman theorem, we can approximate f by an entire harmonic function h in $\mathbf{R}^{n+1}$. Then, the Taylor series of h has infinite radius. It follows that the Taylor series of the restriction u of h to $\mathbf{R}^n$ also has infinite radius. Let $\tilde{u}$ be the entire holomorphic function in $\mathbf{C}^n$ obtained by replacing x by z in the Taylor series for u. Then $\tilde{u}$ approximates f and this completes the proof of the holomorphic Carleman theorem.

The holomorphic and harmonic Runge theorems which we have given for approximation on compact subsets of an open set W can both be stated in the same way.

Theorem 25 *Let W be an open subset of an open Riemann surface (respectively, Riemannian manifold) Ω. Then, (W, Ω) is a holomorphic (respectively, harmonic) Runge pair if and only if $\Omega^* \setminus W$ is connected.*

We have thus seen that the same condition characterizes harmonic Runge pairs for approximation on *closed* subsets of W. It can be shown that the complex analog of this last statement is also true provided that Ω is a plane domain. However, this is not the case for Ω an arbitrary Riemann surface as our counterexample to the Roth theorem on Riemann surfaces shows.

Runge pairs (W, Ω) have been defined above, only when W is open in Ω. Replacing the open set W by a closed set in the definition of a Runge pair for closed sets, we shall say that for a closed subset F of a Riemann surface (respectively, Riemannian manifold) Ω, the pair (F, Ω) is a holomorphic (respectively, harmonic) Runge pair if each function holomorphic (respectively, harmonic) on F is the uniform limit of functions holomorphic (respectively, harmonic) on all of Ω. With this terminology, the theorem of Roth (with the necessity coming from Arakelyan) takes the following form.

Theorem 26 *Let F be a relatively closed subset of a plane domain Ω. Then, (F, Ω) is a holomorphic Runge pair if and only if $\Omega^* \setminus F$ is connected and locally connected.*

A partial analog of this result was found by Goldstein, Ow and myself for the harmonic situation.

Theorem 27 ([28], [29]) *Let F be a relatively closed subset of a domain Ω in $\mathbf{R}^n$ such that $\Omega^* \setminus F$ is connected and locally connected. Then (F, Ω) is a harmonic Runge pair.*

This result was extended to Riemannian manifolds by Bagby and Blanchet [4, Theorem 9.3].

The following elementary example shows that, in contrast to the holomorphic situation, the conditions in the last theorem, although sufficient, are no longer necessary in the harmonic situation.

Example 12 *Let $F = (|z| = 1)$ and $\Omega = \mathbf{C}$. Then (F, Ω) is a harmonic Runge pair but $\mathbf{C}^* \setminus F$ is not connected.*

Proof Suppose $u \in h(|z| = 1)$ and $\epsilon > 0$. Let $\tilde{u}$ be the solution of the Dirichlet problem on $(|z| \leq 1)$, with $\tilde{u}(e^{i\theta}) = u(e^{i\theta})$. For $\rho < 1$, set $u_\rho(z) \equiv \tilde{u}(\rho z)$. Then, $u_\rho \in h(|z| \leq 1)$ and $u_\rho \to u$. There exists a ρ such that $|u_\rho - u| < \epsilon/2$ on F and by the harmonic Runge theorem, there is a $v \in h(\mathbf{C})$ such that $|v - u_\rho| < \epsilon/2$ on F. Thus, $u \in \overline{h}(\mathbf{C})$.

A complete characterization of harmonic Runge pairs (F, Ω), when F is a closed subset of Ω has been obtained only recently by Gardiner. In order to state Gardiner's theorem we introduce some terminology. A subset of Ω is said to be *Ω-bounded* if its closure is compact in Ω. Let F be a subset of Ω. An *Ω-hole* of F is an Ω-bounded component of the complement $\Omega \setminus F$. We denote by $\hat{F}$ the union of F and its Ω-holes.

Theorem 28 (Gardiner [21]) *Let F be a relatively closed subset of a domain Ω in $\mathbf{R}^n$. The following are equivalent:*

(a) *(F, Ω) is a subharmonic Runge pair;*

(b) *(F, Ω) is a harmonic Runge pair;*

(c) *(F, Ω) satisfies*

  (i) *$\Omega \setminus \hat{F}$ and $\Omega \setminus F$ are thin at the same points of F,*

  (ii) *for each compact $K \subset \Omega$ there is a compact $Q \subset \Omega$ which contains every Ω-hole of $F \cup K$ whose closure intersects K.*

Of course, if F is compact, condition (ii) is superfluous. When $n = 2$, one can replace (i) by the condition that $\partial\hat{F} = \partial F$. It follows that, if we denote the boundary of the unbounded complementary component of a bounded subset K of the plane by $\partial_\infty K$, we have the following result.

Corollary 7 *Let K be a compact subset of the complex plane $\mathbf{C}$. A necessary and sufficient condition in order that every $u \in h(K)$ be uniformly approximable by harmonic polynomials is that $\partial K = \partial_\infty K$.*

It is remarkable that, prior to Gardiner's beautiful theorem, the answer to such a natural (and old) problem as the characterization of harmonic Runge pairs, for closed subsets F, was unknown, even when F is compact.

13 Other norms and applications

Throughout these lectures, we have considered approximation only in the uniform norm (which to me seems the most natural). However, it is also of importance to study approximation in other norms and, in fact, the most interesting remarks concerning the topic of the previous section "holomorphic uniform versus harmonic uniform approximation" involve other norms. Firstly, it follows from work of Khavin [39] that the problem of harmonic uniform approximation (for classes) in $\mathbf{R}^2$ is equivalent to that of holomorphic L^2-approximation in $\mathbf{C} = \mathbf{R}^2$. Moreover there are indications (for example, [32]) that the problem of holomorphic uniform approximation (for classes) is related (equivalent seems to be too strong a claim) to that of harmonic C^1-approximation

One can also seek to approximate in other norms, for example, in the C^m-norms or in the Sobolev $W^{m,p}$-norms. In the case of uniform approximation on a compact subset K of the complex plane $\mathbf{C}$, we are approximating in the space $C(K)$ of continuous functions on K endowed with the sup-norm. This space is quite familiar to us and it is interesting to note that, by the Tietze extension theorem, any continuous function on K can be considered to be the restriction of a continuous function on all of $\mathbf{C}$ which has the same norm on $\mathbf{C}$. It is less obvious how one should define the spaces $C^m(K)$ and its norm, especially if the set K is not smoothly bounded. It is even less clear how to define the restrictions to K of functions in the Sobolev space $W^{m,p}$. In fact, functions in $W^{m,p}$ (and their derivatives up to order m) are basically functions of class L_p. and hence, it is meaningless to talk about their values at individual points of K. However, according to the spectral synthesis theorem of Hedberg and Wolff, these can be considered as functions defined almost everywhere with respect to appropriate capacities. With an appropriate quotient topology, $W^{m,p}(K)$ is a Banach space. For a sample from this point of view and further references, see for example [33].

References

[1] Arakelyan, N.U., Uniform approximation on closed sets by entire functions (Russian), *Izv. Akad. Nauk SSSR Ser. Mat.* **28** (1964), 1187–1206.

[2] Arakelyan, N.U., Entire functions of finite order with an infinite set of deficient values, *Dokl. Akad. Nauk SSSR* **170** (1966), 999–1002; English translation *Soviet Math. Dokl.* **7** (1966), 1303–1306.

[3] Arakelyan, N.U., Uniform and tangential approximation with analytic functions (Russian), *Izv. Akad. Nauk Armyan. SSR Ser. Mat.* **3** (1968), 273–286.

[4] Bagby, T. and Blanchet, P., Uniform harmonic approximation on Riemannian manifolds, to appear in *J. Analyse Math.*

[5] Bagby, T., Cornea, A. and Gauthier, P.M., Harmonic approximation on arcs, *Constr. Approx.* **9** (1993), 501–507.

[6] Bagby, T. and Gauthier, P.M., Approximation by harmonic functions on closed subsets of Riemann surfaces, *J. Analyse Math.* **51** (1988), 259–284.

[7] Bagby, T. and Gauthier, P.M., Uniform approximation by global harmonic functions, in: *Approximation by Solutions of Partial Differential Equations* (B. Fuglede et al., eds.), NATO ASI Ser. C365, Kluwer Academic Publishers, Dordrecht, 1992; 15–26.

[8] Baouendi, M.S. and Trèves, F., A property of the functions and distributions annihilated by a locally integrable system of complex vector fields, *Ann. of Math.* **113** (1981), 387–421.

[9] Behnke, H. and Stein, K., Entwicklungen analytischer Funktionen auf Riemannschen Flächen, *Math. Ann.* **120** (1948), 430–461.

[10] Bishop, E., Subalgebras of functions on a Riemann surface, *Pacific J. Math.* **8** (1958), 29–50.

[11] Boivin, A., Carleman approximation on Riemann surfaces, *Math. Ann.* **276** (1986), 57–70.

[12] Bremermann, H.J., On the conjecture of the equivalence of the plurisubharmonic functions and the Hartogs functions, *Math. Ann.* **131** (1956), 76–86.

[13] Bremermann, H.J., Die Charakterisierung Rungescher Gebiete durch plurisubharmonische Funktionen, *Math. Ann.* **136** (1958), 173–186.

[14] Burckel, R.B., *An Introduction to Classical Complex Analysis I*, Academic Press, 1979.

[15] Carleman, T., Sur un théorème de Weierstrass, *Ark. Mat. Astronom. Fys.* **20B** (1927), 1–5.

[16] Chen H. and Gauthier, P.M., A maximum principle for subharmonic and plurisubharmonic functions, *Canad. Math. Bull.* **35** (1992), 34–39.

[17] Eremenko, A.E., private communication.

[18] Fuglede, B., Asymptotic paths for subharmonic functions and polygonal connectedness of fine domains, in: *Séminaire de théorie du potentiel Paris, No.5*, Lecture Notes in Math. **814**, Springer-Verlag, Berlin - Heidelberg - New York, 1980; 97–115.

[19] Fuglede, B., Sur les fonctions finement holomorphes, *Ann. Inst. Fourier (Grenoble)* **31** (1981), 57–88.

[20] Gaier, D., *Lectures on Complex Approximation*, Birkhäuser, Basel, 1987.

[21] Gardiner, S.J., *Harmonic Approximation*, in preparation.

[22] Gardiner, S.J., The Dirichlet and Neumann problems for harmonic functions in half-spaces, *J. London Math. Soc. (2)* **24** (1981), 502–512.

[23] Gauthier, P.M., Tangential approximation by entire functions and functions holomorphic in a disc, *Izv. Akad. Nauk Armyan. SSR Ser. Mat.* **4** (1969), 319–326.

[24] Gauthier, P.M., Subharmonic extensions and approximations, *Canad. Math. Bull.* **37** (1994), 46–53.

[25] Gauthier, P.M., J.L. Walsh and Qualitative Approximation, to appear in: *Walsh Selecta* (T.J. Rivlin and E.B. Saff, eds.).

[26] Gauthier, P.M., Uniform approximation: holomorphic, harmonic, subharmonic, to appear in: *Proceedings of 1991–1992 Special Academic Year on Complex Analysis & International Conference on Complex Analysis, Nankai Institute of Math.*, International Press, Hong Kong.

[27] Gauthier, P.M., Grothmann, R. and Hengartner, W., Asymptotic maximum principles for subharmonic and plurisubharmonic functions, *Canad. J. Math.* **40** (1988), 477–486.

[28] Gauthier, P.M., Goldstein, M, and Ow, W.H., Uniform approximation on closed sets by harmonic functions with logarithmic singularities, *Trans. Amer. Math. Soc.* **261** (1980), 169–183.

[29] Gauthier, P.M., Goldstein, M, and Ow, W.H., Uniform approximation on closed sets by harmonic functions with Newtonian singularities, *J. London Math. Soc. (2)* **28** (1983), 71–82.

[30] Gauthier, P.M. and Hengartner, W., Approximation uniforme qualitative sur des ensembles non bornés, Sém. Math. Sup. **82**, Les Presses de l' Université de Montréal, 1982.

[31] Gauthier, P.M., W. Hengartner and Labrèche, M., Approximation harmonique, approximation holomorphe et topologie, *Canad. J. Math.* **34** (1982), 216–219.

[32] Gauthier, P.M. and Paramonov, P.V., Approximation by harmonic functions in the C^1-norm and harmonic C^1-content of compact subsets in $\mathbf{R}^n$, *Mat. Zametki* **53** (1993), 21–30 (Russian).

[33] Gauthier, P.M. and Tarkhanov, N.N., Degenerate cases of uniform approximation by solutions of systems with surjective symbols, *Canad. J. Math.* **45** (1993), 740–757.

[34] Giroux, L.-P., Master's Thesis, Université de Montréal, in preparation.

[35] Gol'berg, A.A., private communication.

[36] Hadjiiski, V.H., Vitushkin's type theorems for meromorphic approximation on unbounded sets, in: *Proc. Conf. Complex Analysis and Applications '81-Varna*, Bulgarian Acad. Sci., Sofia, 1984; 229–238.

[37] Hervé, R.-M., Quelques propriétés des fonctions surharmoniques associées à une équation uniformément elliptique de la forme $Lu = -\sum_i \frac{\partial}{\partial x_i}\left(\sum_j a_{ij}\frac{\partial u}{\partial x_j}\right) = 0$, *Ann. Inst. Fourier (Grenoble)* **15**(2) (1965), 215–224.

[38] Iversen, F., Recherches sur les fonctions inverses des fonctions méromorphes, Thèse, Helsingfors, 1914.

[39] Khavin, V.P., Approximation in the mean by analytic functions, *Dokl. Akad. Nauk SSSR* **178** (1968), 1025–1028; English translation: *Soviet Math. Dokl.* **9** (1968), 245–248.

[40] Mergelyan, S.N., Uniform approximations to functions of a complex variable, English translation: Translations Amer. Math. Soc. **3** (1962), 294–391.

[41] Nersesyan, A.A., On Carleman sets, *Izv. Akad. Nauk Armyan. SSR Ser. Mat.* **6** (1971), 465–471; English tranlation: Amer. Math. Soc. Transl. (2) **122** (1984), 99–104.

[42] Ronkin, L.I., *Introduction to the Theory of Entire Functions of Several Variables*, Nauka, Moscow, 1971; English translation: Transl. Math. Monographs **44**, Amer. Math. Soc., Providence, RI, 1974.

[43] Rosay, J.-P. and Rudin, W., Arakelian's approximation theorem, *Amer. Math. Monthly* **96** (1989), 432–434.

[44] Roth, A., Approximationseigenschaften und Strahlengrenzwerte meromorpher und ganzer Funktionen, *Comment. Math. Helv.* **11** (1938), 77–125.

[45] Roth, A., Meromorphe Approximationen, *Comment. Math. Helv.* **48** (1973), 151–176.

[46] Sahakian, R.Sh., On a generalization of the maximum principle, *Izv. Akad. Nauk Armyan. SSR Ser. Mat.* **22** (1987), 94–101; English translation: *Soviet J. Contemporary Math. Anal.* **22** (1987), 94–102.

[47] Sario, L. and Nakai, M., *Classification Theory of Riemann Surfaces*, Springer-Verlag, Berlin, 1970.

[48] Scheinberg, S., Uniform approximation by entire functions, *J. Analyse Math.* **29** (1976), 16–19.

[49] Shirinbekov, M., Stability of pseudoconvex domains, *Dokl. Akad. Nauk SSSR* **287** (1986), 1177–1192; English translation: *Soviet Math. Dokl.* **33** (1986), 388–391.

[50] Voronin, S.M., Theorem on the "universality" of the Riemann zeta-function, *Izv. Akad. Nauk SSSR Ser. Mat.* **39** (1975); English translation: *Math USSR-Izv.* **9** (1975), 443–453.

[51] Zalcman, L., *Math. Reviews* **46**-2062.

Plurisubharmonic functions and their singularities

Christer O. KISELMAN
Department of Mathematics
Uppsala University
P. O. Box 480
S-751 06 Uppsala
Sweden

Abstract

The theme of these lectures is local and global properties of plurisubharmonic functions. First differential inequalities defining convex, subharmonic and plurisubharmonic functions are discussed. It is proved that the marginal function of a plurisubharmonic function is plurisubharmonic under certain hypotheses. We study the singularities of plurisubharmonic functions using methods from convexity theory. Then in the final chapter we generalize the classical notions of order and type of an entire function of finite order to functions of arbitrarily fast growth.

Contents

This work was partially supported by the Swedish Natural Science Research Council.

P. M. Gauthier (ed.) and G. Sabidussi (techn. ed.), Complex Potential Theory, 273–323.

Introduction

The plurisubharmonic functions appear in complex analysis as logarithms of moduli of holomorphic functions and as analogues of potentials. Their usefulness for many constructions is due to the fact that they are easier to manipulate than holomorphic functions—this is why Lelong [1985] includes them among "les objets souples de l'analyse complexe."

In these lectures we shall first consider analogies between the convex, subharmonic, and plurisubharmonic functions: these three classes can be defined using differential inequalities. We shall also study marginal functions of plurisubharmonic functions, i.e., functions of the form

$$g(x_1, ..., x_n) = \inf_{y_1, ..., y_m} f(x_1, ..., x_n, y_1, ..., y_m).$$

It is a known fact that marginal functions of convex functions are convex, but the corresponding result is not true for plurisubharmonic functions. However, it is true under some extra hypotheses, and we shall establish one such result, called the minimum principle, in Chapter 1.

In Chapter 2, we use the minimum principle to prove that sets related to plurisubharmonic functions are analytic varieties. The model result here is Siu's theorem, which says that the set of points where the Lelong number is larger than or equal to a certain number is an analytic variety. We shall see that the minimum principle provides us with a family of plurisubharmonic functions related to a given one, and that there are analyticity theorems for families of plurisubharmonic functions which are easy to deduce from the Hörmander–Bombieri theorem.

In the third chapter we shall take a look at the classical notions of order and type for entire functions. To every entire function F we can in a natural way associate a convex function f which describes its growth:

$$f(t) = \sup_{|z|=e^t} \log |F(z)|, \qquad t \in \mathbf{R}.$$

We call f the growth function of F. That f is convex is the content of Hadamard's three-circle theorem. These classical definitions can quite naturally be extended to plurisubharmonic functions; just replace $\log|F|$ by an arbitrary plurisubharmonic function. What we do in classical complex analysis is to compare the growth of two convex functions, the growth function f and the growth function $g(t) = e^t$ of the exponential function $G(z) = e^z$. The notion of relative order, the order of f relative to g, arises from such a comparison of two convex functions. The notion of relative type of one function with respect to another is the result of a slightly different comparison.

All classical results on order and type can now be considered in this more general setting, and many of them have very precise counterparts. It should be stressed that the functions we consider may grow arbitrarily fast, whereas classically one considers functions of finite order. We have adjusted the definitions so that order and type become dual in the sense of convexity theory. This fact is very useful, for we can often choose to do calculations either on the functions themselves or on their conjugate functions, their Fenchel transforms.

The relative order determines the maximal domain in which a solution to a natural extension problem exists. This extension problem can be formulated for convex, plurisubharmonic or entire functions—the resulting domain of existence is the same in all three cases.

Acknowledgments I am grateful to the *Séminaire de Mathématiques Supérieures* for the invitation to participate in this summer school. It was a great experience! It is also a pleasure to acknowledge the good help provided by Stefan Halvarsson, who typed Chapter 1 into TEX, made many useful suggestions, and proofread all the chapters. My thanks go also to Maciej Klimek for checking the manuscript and for valuable comments on the presentation. Chapter 3 is essentially taken from my paper [1993] (which contains four additional sections). The London Mathematical Society has kindly given its permission to include this material here.

Chapter 1
Convexity and plurisubharmonicity

1.1 Introduction

Let us first recall that the real-valued convex functions on the real line are those that satisfy the inequality

$$f((1-t)x+ty) \leqslant (1-t)f(x)+tf(y), \qquad 0 \leqslant t \leqslant 1, \ x, y \in \mathbf{R}. \tag{1.1.1}$$

In particular, for $t = 1/2$ they satisfy

$$f(c) \leqslant \tfrac{1}{2}f(c-r)+\tfrac{1}{2}f(c+r), \qquad c, r \in \mathbf{R}, \tag{1.1.2}$$

which can be written as

$$f(c) \leqslant M_{\partial I} f,$$

denoting by M the mean value over a set, in this case $\partial I = \{c-r, c+r\}$, which is the boundary of the one-dimensional ball $c + rB$.

Some regularity has to be imposed if we use (1.1.2) though, for while (1.1.1) implies that f is continuous (where it is real-valued), (1.1.2) does not:

Example Take a Hamel basis for the vector space of all real numbers over the rational numbers with 1 and $\sqrt{2}$ as basis elements. Define f to be a $\mathbf{Q}$-linear form $f: \mathbf{R} \to \mathbf{Q}$ such that $f(1) = 1$, $f(\sqrt{2}) = 0$. Then obviously f satisfies (1.1.1) for rational t (with equality), in particular (1.1.2), but it is not continuous (and we would not like to call it convex). Indeed, $f(s + t\sqrt{2}) = s$ for rational s, t, which shows that f is unbounded near any point.

However, (1.1.2) plus some mild regularity assumption (like semicontinuity or even measurability) is equivalent to (1.1.1) for real-valued functions.

The definition of a subharmonic function is a generalization of this: a function f is called *subharmonic* in an open subset Ω of $\mathbf{R}^n$ if it takes its values in $[-\infty, +\infty[$, is upper semicontinuous, and satisfies the mean-value inequality

$$f(c) \leqslant M_{\partial A} f$$

whenever A is a closed ball of center c contained in $\Omega \subset \mathbf{R}^n$. We shall write $f \in SH(\Omega)$. The constant $-\infty$ is allowed.

However, we can generalize the notion of a convex function of one variable in a different direction: we consider a function in $\mathbf{R}^n$ and look at its restrictions to real lines, in other words at its pull-backs $\varphi^* f = f \circ \varphi$ for an arbitrary affine function $\varphi: \mathbf{R} \to \mathbf{R}^n$. If this pull-back is always convex, then f is called convex in $\mathbf{R}^n$. (Actually such a function should be called "pluriconvex" if we were to follow the idea that has led to the word *plurisubharmonic*!) We shall write $f \in CVX(\Omega)$ if f is real-valued and convex in a convex open set Ω.

Remark In convexity theory one usually allows values in $[-\infty, +\infty]$. A function $f: \mathbf{R}^n \to [-\infty, +\infty]$ is defined to be convex if its *epigraph*

$$\operatorname{epi} f = \{(x, t) \in \mathbf{R}^n \times \mathbf{R};\ f(x) \leqslant t\} \tag{1.1.3}$$

is convex as a subset of $\mathbf{R}^n \times \mathbf{R}$. It is sometimes more convenient to use the *strict epigraph*

$$\operatorname{epi}_s f = \{(x, t) \in \mathbf{R}^n \times \mathbf{R};\ f(x) < t\}. \tag{1.1.4}$$

It is easy to see that the epigraph and the strict epigraph are convex simultaneously. For real-valued functions, the definition using the epigraph is equivalent to (1.1.1).

We can now generalize the subharmonic functions of one complex variable in the same way as we did when we defined convex functions in $\mathbf{R}^n$. If $\varphi^* f = f \circ \varphi$ is subharmonic for all complex affine mappings $\varphi: \mathbf{C} \to \mathbf{C}^n$ and has in addition some kind of regularity, then f is called plurisubharmonic. The additional regularity assumption is usually taken to be upper semicontinuity, which means the the strict epigraph $\operatorname{epi}_s f$ (cf. (1.1.4)) is assumed to be open.

Definition 1.1.1 We say that f is *plurisubharmonic* in an open set Ω in $\mathbf{C}^n$ if $f:\Omega \to [-\infty, +\infty[$ is upper semicontinuous in Ω and, for all $a, b \in \mathbf{C}^n$, $z \mapsto f(a+zb)$ is subharmonic as a function of the complex variable z in the open set where it is defined. Notation: $f \in PSH(\Omega)$.

The scheme of generalizations can be illustrated as follows:

$n = 1$ subharmonic = convex	$\longrightarrow$	$n > 1$ convex
$\downarrow$		$\downarrow$
$n = 2k = 1$ subharmonic = plurisubharmonic	$\longrightarrow$	$k > 1$ plurisubharmonic

In all cases, the mean-value inequality $f(c) \leqslant M_{\partial A} f$ is imposed, but with different balls A: they can be real one-dimensional or complex one-dimensional or full-dimensional. This will lead to important analogies between the different cones of functions: the cone PSH is sometimes analogous with SH, sometimes with CVX.

A very natural question is this: if the pull-back $\varphi^* f$ is subharmonic for all affine functions φ mapping the complex plane into $\mathbf{C}^n$, is f plurisubharmonic? In other words, is the assumption of upper semicontinuity superfluous? The answer seems to be unknown. There is a similar question whether separately subharmonic[1] functions are subharmonic: this is not true as shown by Wiegerinck [1988]. However, if we add some, even very weak, integrability condition, separately subharmonic functions are indeed subharmonic; see Riihentaus [1989].

It is not difficult to prove the following inclusions:

$$CVX_{loc}(\Omega) \subset SH(\Omega), \qquad \Omega \subset \mathbf{R}^n, \tag{1.1.5}$$

and

$$CVX_{loc}(\Omega) \subset PSH(\Omega) \subset SH(\Omega), \qquad \Omega \subset \mathbf{C}^n, \tag{1.1.6}$$

where $CVX_{loc}(\Omega)$ is the cone of functions which are locally convex, i.e., convex in some ball around an arbitrary point. They can be proved using the mean-value inequalities, but they will also follow from the differential inequalities to be presented in the next section.

For general information about plurisubharmonic functions see Hörmander [1990; forthc.], Klimek [1991], and Lelong [1969].

1.2 Conditions on the derivatives of convex and plurisubharmonic functions

We shall now take a look at various differential inequalities which are related to convexity, subharmonicity and plurisubharmonicity. The simplest is this:

[1]This means that the function is subharmonic in each variable when the others are kept fixed.

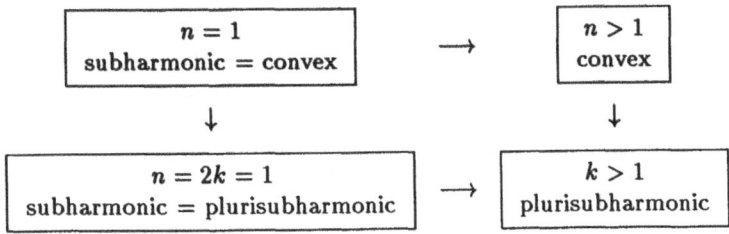

Proposition 1.2.1 *Let $f \in C^2(I)$, where $I \subset \mathbf{R}$ is an interval. Then f is convex if and only if $f'' \geqslant 0$.*

This can of course be proved directly, but since it is a special case of Proposition 1.2.3 below, we omit the proof.

We shall write $\mathcal{D}(\Omega)$ for the set of all test functions in an open set Ω and $\mathcal{D}'(\Omega)$ for the set of all distributions in Ω, the space dual to $\mathcal{D}(\Omega)$.

Proposition 1.2.2 *Let $f \in L^1_{\text{loc}}(I)$, I being an interval. Then f is equal to a convex function almost everywhere if and only if $f'' \geqslant 0$ in the sense of distributions, i.e., $\int_I \varphi'' f d\lambda \geqslant 0$ for all $\varphi \in \mathcal{D}(I)$ satisfying $\varphi \geqslant 0$. Moreover, if u is a distribution in I, $u \in \mathcal{D}'(I)$, then there exists a convex function f such that $\int_I f\varphi d\lambda = u(\varphi)$ for every test function $\varphi \in \mathcal{D}(I)$ if and only if $u'' \geqslant 0$.*

This result is a special case of Proposition 1.2.4 below.

Proposition 1.2.3 *Let $f \in C^2(\Omega)$, $\Omega \subset \mathbf{R}^n$. Then $f \in SH(\Omega)$ if and only if $\Delta f \geqslant 0$, where $\Delta = \partial^2/\partial x_1^2 + \cdots + \partial^2/\partial x_n^2$ is the Laplacian.*

Proof We shall write B for the closed unit ball and S for its boundary, the unit sphere, so that $c + rB$ is the closed ball of radius r and center at c, and $c + rS$ its boundary. Let E be the fundamental solution of the Laplacian such that $\Delta E = \delta_c$ and E vanishes on the sphere $c + rS$. Then Green's formula yields

$$f(c) - \fint_{c+rS} f = \int_{c+rB} E\Delta f d\lambda, \tag{1.2.1}$$

where $d\lambda$ denotes Lebesgue measure. We use a barred integral sign to denote mean value, thus

$$M_A(f) = \fint_A f d\lambda = \int_A f d\lambda \Big/ \int_A d\lambda \quad \text{provided} \quad 0 < \int_A d\lambda < +\infty. \tag{1.2.2}$$

Since $E \leqslant 0$ in the ball $x + rB$, $\Delta f \geqslant 0$ implies

$$f(c) - \fint_{c+rS} f \leqslant 0.$$

This holds for all $c \in \Omega$ and all r such that $c + rB \subset \Omega$. This is the mean-value inequality for f.

For the other direction, assume $\Delta f(c) < 0$ at some point c. Take r so small that $\Delta f < 0$ in $c + rB$. Then (1.2.1) shows that

$$f(c) - \fint_{c+rS} f > 0$$

for these r, so f does not satisfy the mean-value inequality.

Proposition 1.2.4 *Let $u \in \mathcal{D}'(\Omega)$, $\Omega \subset \mathbf{R}^n$. Then there exists $f \in SH(\Omega)$ such that $\int f\varphi d\lambda = u(\varphi)$ for all $\varphi \in \mathcal{D}(\Omega)$ if and only if $\Delta u \geqslant 0$ in the sense of distributions, i.e., $u(\Delta\varphi) \geqslant 0$ for all $\varphi \in \mathcal{D}(\Omega)$ satisfying $\varphi \geqslant 0$.*

Proof First let $f \in SH(\Omega)$. Form $f_\varepsilon \in C^\infty(\Omega_\varepsilon)$ by convolution:

$$f_\varepsilon(x) = (f * \psi_\varepsilon)(x) = \int f(y)\psi_\varepsilon(x-y)d\lambda(y) = \int f(x-\varepsilon y)\psi(y)d\lambda(y), \qquad x \in \mathbf{R}^n,$$

where ψ is a radial[2] C^∞ function with support in the unit ball and of integral one satisfying $\psi \geqslant 0$, and $\psi_\varepsilon(x) = \varepsilon^{-n}\psi(x/\varepsilon)$. Then f_ε is subharmonic in $\Omega_\varepsilon = \{x \in \Omega;\, x + \varepsilon B \subset \Omega\}$. Indeed, the integral $\int f(x-\varepsilon y)\psi(y)d\lambda(y)$ is a limit of finite sums $\sum f(x - \varepsilon y^j)c_j$ with positive c_j. Since f_ε is smooth, Proposition 1.2.3 implies that $\Delta f_\varepsilon \geqslant 0$. When $\varepsilon \to 0$, f_ε tends to f in $L^1_{loc}(\Omega)$ and the positivity in the sense of distributions is preserved: $\Delta f \geqslant 0$.

Conversely, if $u \in \mathcal{D}'(\Omega)$ with $\Delta u \geqslant 0$, form $u_\varepsilon = u * \psi_\varepsilon$. Then $u_\varepsilon \in C^\infty(\Omega_\varepsilon)$ and $\Delta u_\varepsilon \geqslant 0$. Hence by Proposition 1.2.3, $u_\varepsilon \in SH(\Omega_\varepsilon)$. I claim that u_ε is an increasing function of ε. To see this, note that the solution χ_ε of $\Delta\chi_\varepsilon = \psi_\varepsilon$ in $\mathbf{R}^n \setminus \{0\}$ which is zero for $|x| > \varepsilon$ can be written

$$\chi_\varepsilon(x) = \int_{|x|}^{\varepsilon} s^{-n+1}ds \int_{s/\varepsilon}^{1} t^{n-1}\Psi(t)dt, \qquad 0 < |x| \leqslant \varepsilon,$$

where $\Psi(|x|) = \psi(x)$. This formula shows that χ_ε is increasing in $\varepsilon > 0$, because the integrand is non-negative and the domain of integration increases with ε. Now if $\varepsilon \geqslant \delta > 0$, then $\chi_\varepsilon - \chi_\delta \in \mathcal{D}(\mathbf{R}^n)$ and $\psi_\varepsilon - \psi_\delta = \Delta(\chi_\varepsilon - \chi_\delta)$ in all of $\mathbf{R}^n$, not only in $\mathbf{R}^n \setminus \{0\}$. Moreover $\chi_\varepsilon - \chi_\delta \geqslant 0$, so that by the positivity of Δu, $(u * (\psi_\varepsilon - \psi_\delta))(0) = u(\psi_\varepsilon - \psi_\delta) \geqslant 0$. Translating this we get $(u_\varepsilon - u_\delta)(x) = (u * (\psi_\varepsilon - \psi_\delta))(x) \geqslant 0$ for all x such that this has a sense, i.e., for all $x \in \Omega_\varepsilon$. This proves the claim that u_ε is an increasing function of ε.

By known properties of subharmonic functions, the limit $f = \lim u_\varepsilon$ is subharmonic in Ω, and since the convergence holds in $L^1_{loc}(\Omega)$, f defines the distribution u. This proves the proposition.

If $f \in C^2(\Omega)$, $\Omega \subset \mathbf{R}^n$, then by definition f is convex if and only if the function $t \mapsto f(a+tb) = f_{a,b}(t)$ is convex for all $a \in \Omega$ and $b \in \mathbf{R}^n$ where it is defined. Hence by the chain rule

$$f''_{a,b}(t) = \sum \frac{\partial^2 f}{\partial x_j \partial x_k}(a+tb)b_j b_k \geqslant 0, \qquad a, b \in \mathbf{R}^n,\ t \in \mathbf{R},\ a + tb \in \Omega.$$

It suffices to take $t = 0$. We state the result as a proposition:

Proposition 1.2.5 *Let $\Omega \subset \mathbf{R}^n$ be convex and $f \in C^2(\Omega)$. Then f is convex if and only if*

$$\sum \frac{\partial^2 f}{\partial x_j \partial x_k}(a)b_j b_k \geqslant 0, \qquad a \in \Omega,\ b \in \mathbf{R}^n. \tag{1.2.3}$$

[2] A function is called *radial* if it is a function of the distance to the origin.

Proposition 1.2.6 *Let $u \in \mathcal{D}'(\Omega)$, where $\Omega \subset \mathbf{R}^n$ is convex. Then there exists $f \in CVX(\Omega)$ such that*

$$u(\varphi) = \int_\Omega f\varphi d\lambda, \qquad \varphi \in \mathcal{D}(\Omega),$$

if and only if

$$\sum \frac{\partial^2 u}{\partial x_j \partial x_k} b_j b_k \geqslant 0, \qquad b \in \mathbf{R}^n, \tag{1.2.4}$$

in the sense of distributions.

Proof If $f \in CVX(\Omega)$, form $f_\varepsilon = f * \psi_\varepsilon \in CVX(\Omega_\varepsilon)$ with ψ as in the proof of Proposition 1.2.4. Then $f_\varepsilon \to f$ in $\mathcal{D}'(\Omega)$, which implies

$$\sum \frac{\partial^2 f_\varepsilon}{\partial x_j \partial x_k} b_j b_k \to \sum \frac{\partial^2 f}{\partial x_j \partial x_k} b_j b_k$$

in $\mathcal{D}'(\Omega)$, since convergence there is stable under differentiation. (We use here the weak topology $\sigma(\mathcal{D}'(\Omega), \mathcal{D}(\Omega))$, meaning that $u_j \to u$ if $u_j(\varphi) \to u(\varphi)$ for every test function φ.) Positivity is preserved under passage to the limit, which means that (1.2.4) holds.

Conversely, if u satisfies the positivity condition (1.2.4), form $u_\varepsilon = u * \psi_\varepsilon \in C^\infty(\Omega_\varepsilon)$. Then also u_ε satisfies the positivity condition (1.2.4), which is the same as (1.2.3) since u_ε is a smooth function. Therefore u_ε is convex by Proposition 1.2.5. Moreover u_ε tends decreasingly (cf. the proof of Proposition 1.2.4) to some function f, which is then necessarily convex as a pointwise limit of convex functions. Since convergence holds in $L^1_{loc}(\Omega)$, f defines the given distribution u.

Proposition 1.2.7 *Let $f \in C^2(\Omega)$, $\Omega \subset \mathbf{C}^n$. Then f is plurisubharmonic if and only if*

$$\sum \frac{\partial^2 f}{\partial z_j \partial \bar{z}_k}(a) b_j \bar{b}_k \geqslant 0, \qquad a \in \Omega,\ b \in \mathbf{C}^n. \tag{1.2.5}$$

Proof This follows from the chain rule and Proposition 1.2.3.

Proposition 1.2.8 *Let $u \in \mathcal{D}'(\Omega)$, $\Omega \subset \mathbf{C}^n$. Then there exists $f \in PSH(\Omega)$ such that $u(\varphi) = \int_\Omega f\varphi d\lambda$ for every test function $\varphi \in \mathcal{D}(\Omega)$ if and only if*

$$\sum \frac{\partial^2 u}{\partial z_j \partial \bar{z}_k} b_j \bar{b}_k \geqslant 0, \qquad b \in \mathbf{C}^n, \tag{1.2.6}$$

in the sense of distributions.

Proof The proof is analogous to the convex case, Proposition 1.2.6.

It is now easy to prove the inclusions (1.1.5) and (1.1.6). The first follows from taking $b_j = \delta_j^k$ in (1.2.4) and then summing over k. In (1.1.6) the first inclusion follows from (1.1.5) and the second from (1.2.6): again take $b_j = \delta_j^k$ and sum over k.

Proposition 1.2.9 *Let $u \in PSH(\Omega)$ be locally independent of the imaginary part of z, i.e., for any $z \in \Omega$, $f(z') = f(z)$ if z' is sufficiently close to z and $\operatorname{Re} z' = \operatorname{Re} z$. Then f is locally convex in Ω (thus convex if Ω is convex).*

Proof If u is a plurisubharmonic function it satisfies (1.2.6), but if it is locally independent of the imaginary part of the variables z_j, that condition reduces to (1.2.4) for u regarded as a function of the $x_j = \operatorname{Re} z_j$. Thus by Proposition 1.2.6 there is a locally convex function f which defines the same distribution as u. The regularizations u_ε and f_ε are therefore equal, which implies that also their limits $\lim_{\varepsilon\to 0} u_\varepsilon = u$ and $\lim_{\varepsilon \to 0} f_\varepsilon = f$ are equal at every point.

Corollary 1.2.10 *If Ω is a pseudoconvex open set in $\mathbf{C}^n$ which is independent of the imaginary parts of the variables in the sense that $z \in \Omega$ and $\operatorname{Re} z' = \operatorname{Re} z$ implies $z \in \Omega$, then every component of Ω is convex.*

Proof Consider the function $u = -\log d$, where d is the distance to the complement of Ω. Thus u is plurisubharmonic if Ω is pseudoconvex—this is indeed one of the possible definitions of pseudoconvexity; see Hörmander [1990, Theorem 2.6.7]. By the proposition, u is locally convex. Therefore the restriction of u to any segment contained in Ω is convex. Now if a^0 and a^1 are two points which belong to the same component of Ω, there is a curve from one to the other, say $[0,1] \ni t \mapsto a^t \in \Omega$. We claim that the segment from a^0 to a^t must be contained in Ω for all t. Indeed the set T of all such t is open in $[0,1]$ by the openness of Ω, and it is closed by the definition of u, for the smallest distance from any point on the segment $[a^0, a^t]$ to the complement of Ω is never smaller than the distance from $\{a^0, a^t\}$ to $\mathbf{C}^n \setminus \Omega$ by the convexity of u on $[a^0, a^t]$. Moreover T is not empty, for $0 \in T$. This proves that T is equal to all of $[0,1]$. Thus the segment $[a^0, a^1]$ is contained in Ω.

These results illustrate some of the many analogies between the three cones CVX, SH and PSH. Let us mention one aspect where this analogy is not clear. Given any cone K in a vector space we may form the space $\delta K = K - K$ of all differences of elements of K. Thus we form three subspaces $\delta CVX(\Omega)$, $\delta SH(\Omega)$ and $\delta PSH(\Omega)$ of $L^1_{loc}(\Omega)$ (or $\mathcal{D}'(\Omega)$) consisting of all differences of functions that are, respectively, convex and finite-valued, subharmonic and finite almost everywhere, and plurisubharmonic and finite almost everywhere in Ω (Ω being convex and open in $\mathbf{R}^n$ in the first case, just open in the second, and open in $\mathbf{C}^n$ in the last case). Each of these spaces has a local variant consisting of those locally integrable functions that admit a representation $f = f_1 - f_2$ with $f_j \in K$ in a neighborhood of an arbitrary point. It is now easy to prove that $\delta SH_{loc}(\Omega) = \delta SH(\Omega)$ for all open sets (it is the space of all locally integrable functions f such that Δf is a measure). Also $\delta CVX_{loc}(\Omega) = \delta CVX(\Omega)$ if Ω is convex. But it seems not to be known whether $\delta PSH_{loc}(\Omega) = \delta PSH(\Omega)$ (for example in a pseudoconvex open set). See Kiselman [1977] for details.

1.3 The minimum principle

For any given function f defined in $\mathbf{R}^n \times \mathbf{R}^m$, we call

$$g(x) = \inf_{y \in \mathbf{R}^m} f(x,y), \qquad x \in \mathbf{R}^n, \tag{1.3.1}$$

the *marginal function* of f. (It defines a kind of margin of the epigraph of f.)

Theorem 1.3.1 *Let* $f: \mathbf{R}^n \times \mathbf{R}^m \to [-\infty, +\infty]$ *be convex. Then its marginal function* (1.3.1) *is convex.*

Proof The strict epigraph of f (cf. 1.1.4) is

$$\operatorname{epi}_s f = \{(x, y, t) \in \mathbf{R}^n \times \mathbf{R}^m \times \mathbf{R};\ f(x, y) < t\}.$$

We now observe that $\operatorname{epi}_s g = \pi(\operatorname{epi}_s f)$, where π is the projection $(x, y, t) \mapsto (x, t)$. If f is convex, then $\operatorname{epi}_s f$ is convex, and any linear image of a convex set is convex, so $\operatorname{epi}_s g = \pi(\operatorname{epi}_s f)$ is also convex. This means that the function g is convex.

Calculus proof (Not that it is necessary now—we shall do it only as a warm-up for the plurisubharmonic case.) Let us assume that the function is of class C^2 and that the infimum is attained at a point $y = w(x)$ for each x which depends in a C^1 fashion on x:

$$y = (w_1(x), \ldots, w_m(x))^{\mathrm{T}},$$

where the exponent means transpose, so that y is regarded as a column vector. Assume also $x \in \mathbf{R}$, i.e., $n = 1$. This is enough; in general we consider $g(x^0 + tx^1)$, $t \in \mathbf{R}$.

Thus $g(x) = f(x, w(x))$; the chain rule yields

$$\frac{\partial g}{\partial x} = \frac{\partial f}{\partial x} + \sum \frac{\partial f}{\partial y_k} \frac{\partial w_k}{\partial x}.$$

At a minimum point we have $\partial f / \partial y_k = 0$, so that $\partial g / \partial x = \partial f / \partial x$ when $y = w(x)$. By the chain rule again

$$\frac{\partial^2 g}{\partial x^2} = \frac{\partial^2 f}{\partial x^2} + \sum \frac{\partial^2 f}{\partial x \partial y_k} \frac{\partial w_k}{\partial x} = f_{xx} + A\alpha,$$

where A is the row matrix

$$A = (A_1, \ldots, A_m) \text{ with } A_k = \frac{\partial^2 f}{\partial x \partial y_k}$$

and α the column matrix

$$\alpha = \left(\frac{\partial w_1}{\partial x}, \ldots, \frac{\partial w_m}{\partial x} \right)^{\mathrm{T}}.$$

We now apply the chain rule to the equation $\partial f / \partial y_k(x, w(x)) = 0$, which gives

$$\frac{\partial^2 f}{\partial x \partial y_k} + \frac{\partial^2 f}{\partial y_j \partial y_k} \frac{\partial w_j}{\partial x} = 0, \text{ in other words } A + \alpha^{\mathrm{T}} H = 0,$$

where

$$H = \left(\frac{\partial^2 f}{\partial y_j \partial y_k} \right)$$

is the Hessian matrix of f with respect to y. Summing up:

$$g_{xx} = f_{xx} + A\alpha = f_{xx} - \alpha^{\mathrm{T}} H \alpha.$$

Now what do we know about H? The convexity of f in all variables $(x, y_1, ..., y_m)$ implies that

$$f_{xx} + \sum \frac{\partial^2 f}{\partial x \partial y_k} b_k + \sum \frac{\partial^2 f}{\partial y_j \partial x} b_j + \sum \frac{\partial^2 f}{\partial y_j \partial y_k} b_j b_k \geqslant 0,$$

for all b (column vectors) or

$$f_{xx} + Ab + b^{\mathrm{T}} A^{\mathrm{T}} + b^{\mathrm{T}} H b \geqslant 0.$$

Since Ab is a scalar, $Ab = b^{\mathrm{T}} A^{\mathrm{T}}$ and we have $f_{xx} + 2Ab + b^{\mathrm{T}} H b \geqslant 0$, and since $A = -\alpha^{\mathrm{T}} H$ this can be written as

$$f_{xx} - 2\alpha^{\mathrm{T}} H b + b^{\mathrm{T}} H b \geqslant 0$$

for any column vector b. Now choose $b = \alpha$. Then we finally obtain

$$g_{xx} = f_{xx} - \alpha^{\mathrm{T}} H \alpha \geqslant 0$$

and we are done.

During this calculation we needed that $w(x)$ is a C^1 function of x. It is the solution of the system $\partial f / \partial y_j = 0$, and it follows from the implicit function theorem that w is C^1 if the Hessian H is positive definite, for the Hessian is precisely the Jacobian matrix of this system and we need the Jacobian (determinant) to be non-zero. Hence $w \in C^1$ and the chain rule can be applied as above. Note as a matter of curiosity that $g(x) = f(x, w(x))$ is C^1 since $w \in C^1$, but since $g_x = f_x$ when $y = w(x)$ we see that g_x is also C^1, hence $g \in C^2$. This concludes our calculations on convex functions.

The condition that $f \in C^2$ and that the infimum is attained can be removed. Regularization and addition of a coercive function will help! We shall not show this now, since we shall do it soon in the plurisubharmonic case in detail.

We shall now investigate similarly the Levi form of a minimum of a plurisubharmonic function f. Thus as before $g(x) = f(x, w(x))$, where $y = w(x)$ defines a stationary point of $y \mapsto f(x, y)$. We let $x \in \mathbf{C}^n = \mathbf{C}$ and $y \in \mathbf{C}^m$. It is enough to consider $n = 1$, because for plurisubharmonicity in x we consider complex lines in $\mathbf{C}^n$.

We shall use the notation

$$A_k = \frac{\partial^2 f}{\partial x \partial y_k}, \quad B_k = \frac{\partial^2 f}{\partial \bar{x} \partial y_k}, \quad H_{jk} = \frac{\partial^2 f}{\partial y_j \partial y_k}, \quad L_{jk} = \frac{\partial^2 f}{\partial y_j \partial \bar{y}_k}, \tag{1.3.2}$$

and put $A = (A_1, ..., A_m)$, $B = (B_1, ..., B_m)$. Here $H = (H_{jk})$ is the complex Hessian matrix and $L = (L_{jk})$ is the Levi matrix with respect to the y variables. We write

$$\mathcal{H}(b) = \sum H_{jk} b_j b_k = b^{\mathrm{T}} H b \tag{1.3.3}$$

for the Hessian form, which is a symmetric quadratic form (thus $H^{\mathrm{T}} = H$, $H^* = \overline{H}$), and

$$\mathcal{L}(b) = \sum L_{jk} b_j \bar{b}_k = b^{\mathrm{T}} L \bar{b} \tag{1.3.4}$$

for the Levi form, which is an Hermitian form if f is real-valued; thus $L^{\mathrm{T}} = \overline{L}$ and $L^* = L$ in that case. We write $\alpha_j = \partial w_j / \partial x$, $\alpha = (\alpha_1, ..., \alpha_m)^{\mathrm{T}}$ and $\beta_j = \partial w_j / \partial \bar{x}$, $\beta = (\beta_1, ..., \beta_m)^{\mathrm{T}}$.

The result is this:

Proposition 1.3.2 *Let f be a real-valued C^2 function in some open set Ω in the space of $1+m$ complex variables, $(x,y)\in\mathbf{C}\times\mathbf{C}^m$. If $y=w(x)$ is a stationary point of $y\mapsto f(x,y)$ which depends in a C^1 fashion on x, then the Laplacian of $g(x)=f(x,w(x))$ satisfies*

$$\tfrac14\Delta g = g_{x\bar x} = f_{x\bar x} - 2\operatorname{Re}\mathcal{H}(\alpha,\beta) - \mathcal{L}(\alpha) - \mathcal{L}(\beta), \tag{1.3.5}$$

where $\mathcal{H}$ and $\mathcal{L}$ are given by (1.3.2–4) and $\mathcal{H}(\alpha,\beta)=\alpha^{\mathrm T}H\beta$ is obtained by polarization.

Proof If we differentiate $g(x)=f(x,w(x))$ once we get

$$g_x(x)=\frac{\partial g}{\partial x}(x)=f_x(x,w(x))+\sum\frac{\partial f}{\partial y_j}\alpha_j+\sum\frac{\partial f}{\partial \bar y_j}\beta_j=f_x(x,w(x)),$$

since $\partial f/\partial y_j$ and $\partial f/\partial\bar y_j$ both vanish at a stationary point. This shows that g_x is of class C^1. We can therefore apply $\partial/\partial\bar x$ to the equation $g_x=f_x$ and get

$$g_{x\bar x}=\frac{\partial^2 g}{\partial x\partial\bar x}=\frac{\partial^2 f}{\partial x\partial\bar x}+\sum\frac{\partial^2 f}{\partial x\partial y_k}\frac{\partial w_k}{\partial\bar x}+\sum\frac{\partial^2 f}{\partial x\partial\bar y_k}\frac{\partial\bar w_k}{\partial\bar x}.$$

Since f is real-valued it follows that $\partial^2 f/\partial x\partial\bar y_k=\overline{B}_k$, thus

$$g_{x\bar x}=f_{x\bar x}+\sum A_k\beta_k+\sum\overline{B}_k\bar\alpha_k=f_{x\bar x}+A\beta+\overline{B\alpha}. \tag{1.3.6}$$

To determine A and B we differentiate the equation $\partial f/\partial y_k=0$ with respect to x to get

$$\frac{\partial^2 f}{\partial x\partial y_k}+\sum\frac{\partial^2 f}{\partial y_j\partial y_k}\frac{\partial w_j}{\partial x}+\sum\frac{\partial^2 f}{\partial\bar y_j\partial y_k}\frac{\partial\bar w_j}{\partial x}=0,$$

or $A_k+\sum\alpha_jH_{jk}+\sum\bar\beta_jL^{\mathrm T}_{jk}=0$, which in matrix notation becomes

$$A=-\alpha^{\mathrm T}H-\beta^*L^{\mathrm T}. \tag{1.3.7}$$

Next we differentiate $\partial f/\partial y_k=0$ with respect to $\bar x$ and get

$$\frac{\partial^2 f}{\partial\bar x\partial y_k}+\sum\frac{\partial^2 f}{\partial y_j\partial y_k}\frac{\partial w_j}{\partial\bar x}+\sum\frac{\partial^2 f}{\partial\bar y_j\partial y_k}\frac{\partial\bar w_j}{\partial\bar x}=0,$$

or $B_k+\sum\beta_jH_{jk}+\sum\bar\alpha_jL_{kj}=0$, which in matrix notation gives

$$B=-\beta^{\mathrm T}H-\alpha^*L^{\mathrm T}. \tag{1.3.8}$$

Now insert the values $A=-\alpha^{\mathrm T}H-\beta^*L^{\mathrm T}$ and $B=-\beta^{\mathrm T}H-\alpha^*L^{\mathrm T}$ into (1.3.6). Then we get

$$g_{x\bar x}=f_{x\bar x}-\alpha^{\mathrm T}H\beta-\beta^*L^{\mathrm T}\beta-\overline{\beta^{\mathrm T}H\alpha}-\overline{\alpha^*L^{\mathrm T}\alpha}=f_{x\bar x}-2\operatorname{Re}(\alpha^{\mathrm T}H\beta)-\alpha^{\mathrm T}L\bar\alpha-\beta^{\mathrm T}L\bar\beta,$$

which in terms of $\mathcal{H}$ and $\mathcal{L}$ is just (1.3.5). This proves Proposition 1.3.2.

So far we have not assumed any plurisubharmonicity! We have just used the identity $g(x)=f(x,w(x))$, where $\partial f/\partial y_k(x,w(x))=0$ and $\partial f/\partial\bar y_k(x,w(x))=0$, equations which hold since $y=w(x)$ is a stationary point of $y\mapsto f(x,y)$. Note, by the way, that these two equations are equivalent if f is real-valued. We shall now assume that f is plurisubharmonic, and deduce a lower bound for its partial Laplacian with respect to x:

Proposition 1.3.3 *If f is plurisubharmonic and of class C^2 in an open set in $\mathbf{C} \times \mathbf{C}^m$, then*

$$f_{x\bar{x}} = \frac{\partial^2 f}{\partial x \partial \bar{x}} \geqslant \overline{B} M B^{\mathrm{T}}, \tag{1.3.9}$$

where

$$B = (B_1, ..., B_m), \; B_k = \frac{\partial^2 f}{\partial \bar{x} \partial y_k},$$

and M is an Hermitian quasi-inverse of the Levi matrix

$$L = \left(\frac{\partial^2 f}{\partial y_j \partial \bar{y}_k} \right),$$

i.e., $M^* = M$ and $LML = L$.

Remark In a nice coordinate system $L = L_1 \oplus 0$, where L_1 is positive definite. Any Hermitian quasi-inverse then has the form $M = M_1 \oplus M_2 = L_1^{-1} \oplus M_2$, where M_2 is Hermitian. We get $LM = ML = I \oplus 0$, so $LML = L$. Moreover $MLM = L_1^{-1} \oplus 0$ ($= M$ if $M_2 = 0$).

Proof of Proposition 1.3.3 What does it mean that f is plurisubharmonic? By Proposition 1.2.7 it means that

$$\frac{\partial^2 f}{\partial x \partial \bar{x}} s\bar{s} + \sum \frac{\partial^2 f}{\partial x \partial \bar{y}_k} s\bar{z}_k + \sum \frac{\partial^2 f}{\partial y_j \partial \bar{x}} z_j \bar{s} + \sum \frac{\partial^2 f}{\partial y_j \partial \bar{y}_k} z_j \bar{z}_k \geqslant 0,$$

for all $s \in \mathbf{C}$, $z \in \mathbf{C}^m$. It suffices to take $s = 1$:

$$f_{x\bar{x}} + \overline{Bz} + Bz + z^{\mathrm{T}} L \bar{z} \geqslant 0, \qquad z \in \mathbf{C}^n,$$

or equivalently

$$f_{x\bar{x}} \geqslant -\inf_z (z^{\mathrm{T}} L \bar{z} + Bz + \overline{Bz}). \tag{1.3.10}$$

To find the best possible use of the plurisubharmonicity we need to determine the infimum in terms of B. The result is this:

Lemma 1.3.4 *Suppose $F(z) = z^{\mathrm{T}} L \bar{z} + 2\,\mathrm{Re}\, Bz$ is bounded from below. Then its infimum is*

$$\inf_{z \in \mathbf{C}^m} (z^{\mathrm{T}} L \bar{z} + 2\,\mathrm{Re}\, Bz) = -\overline{B} M B^{\mathrm{T}},$$

and is attained at $z = -\overline{M} B^*$, where M is any Hermitian quasi-inverse of L. (Then $LMB^{\mathrm{T}} = B^{\mathrm{T}}$, and this property is sufficient for the formula above to hold.)

Proof of Lemma 1.3.4 If the infimum is attained at a point a, we must have

$$F(z) = (z - a)^{\mathrm{T}} L \overline{(z - a)} - a^{\mathrm{T}} L \bar{a},$$

for the linear terms must vanish in an expansion around a. Hence $\inf F(z) = -a^{\mathrm{T}} L \bar{a}$. Now assume M is such that $LMB^{\mathrm{T}} = B^{\mathrm{T}}$ and $M^* = M$. Then we just calculate:

$$F(z) = z^{\mathrm{T}} L \bar{z} + 2\,\mathrm{Re}\, Bz = (z + \overline{M} B^*)^{\mathrm{T}} L \overline{(z + \overline{M} B^*)} - \overline{B} MLM B^{\mathrm{T}}.$$

Thus $\inf F = -\overline{B}MLMB^{\mathrm{T}} = -\overline{B}MB^{\mathrm{T}}$ and it is attained at the point $z = -\overline{M}B^*$ (not necessarily unique, since it depends on the choice of quasi-inverse). Here we only used the fact that M satisfies $LMB^{\mathrm{T}} = B^{\mathrm{T}}$ and $M^* = M$.

For completeness we shall also show that if $LML = L$, $M^* = M$, then necessarily $LMB^{\mathrm{T}} = B^{\mathrm{T}}$. If this is not true there is a row-vector c such that $cB^{\mathrm{T}} \neq 0$ but $cL = 0$. (We have $LMx = x$ for all columns of L, hence for x in the linear span of those columns, so if B^{T} does not belong to this span, there is a linear form which annihilates the columns of L without annihilating B^{T}). Now consider

$$F(sc^{\mathrm{T}}) = (sc^{\mathrm{T}})^{\mathrm{T}}L\overline{(sc^{\mathrm{T}})} + Bsc^{\mathrm{T}} + \overline{Bsc^{\mathrm{T}}} = 2\,\mathrm{Re}(sBc^{\mathrm{T}}).$$

This real-linear form is not identically zero by hypothesis, and hence not bounded from below. But we assumed F to be bounded from below. The set of all column vectors x such that $LMx = x$ includes all columns of L and therefore also B^{T}.

Thus Lemma 1.3.4 and hence Proposition 1.3.3 are proved.

Theorem 1.3.5 *Let f be plurisubharmonic and of class C^2 in an open set in $\mathbf{C} \times \mathbf{C}^m$ and $y = w(x)$ a stationary point of $y \mapsto f(x, y)$ with w of class C^1. We write*

$$\beta = \Big(\frac{\partial w_1}{\partial \bar{x}}, \ldots, \frac{\partial w_m}{\partial \bar{x}}\Big)^{\mathrm{T}}, \qquad H = \Big(\frac{\partial^2 f}{\partial y_j \partial y_k}\Big), \qquad L = \Big(\frac{\partial^2 f}{\partial y_j \partial \bar{y}_k}\Big),$$

and let M be an arbitrary Hermitian quasi-inverse of L, i.e., $M = M^*$, $LML = L$. Define

$$N = HM^{\mathrm{T}}\overline{H} - L. \tag{1.3.11}$$

Then $g(x) = f(x, w(x))$ satisfies

$$g_{x\bar{x}} \geqslant \beta^{\mathrm{T}}(HM^{\mathrm{T}}\overline{H} - L)\overline{\beta} = \mathcal{M}(\overline{H\beta}) - \mathcal{L}(\beta) = \beta^{\mathrm{T}}N\overline{\beta} = \mathcal{N}(\beta), \tag{1.3.12}$$

where $\mathcal{M}(b) = b^{\mathrm{T}}M\bar{b}$ and $\mathcal{N}(b) = b^{\mathrm{T}}N\bar{b}$ denote the Hermitian forms defined by M and N (cf. (1.3.4)). In particular, g is subharmonic if $\mathcal{N}(\beta) \geqslant 0$.

Thus for every plurisubharmonic function f of class C^2 we have defined an Hermitian matrix $N = HM^{\mathrm{T}}\overline{H} - L$ which is of interest. It is highly non-linear in f.

Proof The criterion (1.3.9) of Proposition 1.3.3, $f_{x\bar{x}} \geqslant \overline{B}MB^{\mathrm{T}}$, takes the form $f_{x\bar{x}} \geqslant \mathcal{M}(\overline{H\beta}) + 2\,\mathrm{Re}\,\mathcal{H}(\alpha, \beta) + \mathcal{L}(\alpha)$ if we are at a critical point. Indeed, $B = -\beta^{\mathrm{T}}H - \alpha^* L^{\mathrm{T}}$ (see (1.3.8)), so

$$\overline{B}MB^{\mathrm{T}} = \beta^*\overline{H}MH\beta + \beta^*\overline{H}ML\overline{\alpha} + \alpha^{\mathrm{T}}LMH\beta + \alpha^{\mathrm{T}}LML\overline{\alpha}.$$

To simplify this expression we use the equations $LML = L$ and $LMB^{\mathrm{T}} = B^{\mathrm{T}}$, which give $LMH\beta = H\beta$ and $\beta^*\overline{H}ML = \beta^*\overline{H}$. Therefore

$$f_{x\bar{x}} \geqslant \overline{B}MB^{\mathrm{T}} = \beta^*\overline{H}MH\beta + \beta^*\overline{H}\overline{\alpha} + \alpha^{\mathrm{T}}H\beta + \alpha^{\mathrm{T}}L\overline{\alpha} = \mathcal{M}\left(\overline{H\beta}\right) + 2\,\mathrm{Re}\,\mathcal{H}(\alpha, \beta) + \mathcal{L}(\alpha).$$

On the other hand we calculated $g_{x\bar{x}}$ in Proposition 1.3.2:

$$g_{x\bar{x}} = f_{x\bar{x}} - 2\,\mathrm{Re}\,\mathcal{H}(\alpha, \beta) - \mathcal{L}(\alpha) - \mathcal{L}(\beta).$$

Using the estimate for $f_{x\bar{x}}$ we get $g_{x\bar{x}} \geqslant \mathcal{M}\left(\overline{H\beta}\right) - \mathcal{L}(\beta) = \mathcal{N}(\beta)$, which concludes the proof of the theorem.

Let us look at a few special cases of the theorem.

1. If w is a holomorphic function, then $\beta = 0$ so g is subharmonic. This is no surprise, $g(x) = f(x, w(x))$ being the composition of a plurisubharmonic function and a holomorphic mapping.
2. If $N = HM^{\mathsf{T}}\overline{H} - L \geqslant 0$ (positive semi-definite), then g is subharmonic.
3. The term $\beta^{\mathsf{T}} HM^{\mathsf{T}}\overline{H\beta}$ is equal to $x^{\mathsf{T}}M^{\mathsf{T}}\bar{x}$ with $x = H\beta$, so it is always greater than or equal to zero if $L \geqslant 0$. Therefore $g_{x\bar{x}} \geqslant -\mathcal{L}(\beta)$. Suppose we know that $L \leqslant aI$, $|\beta| \leqslant b$. Then $-\beta L\beta^* \geqslant -a|\beta|^2 \geqslant -ab^2$, so that $g(x) + ab^2|x|^2$ is subharmonic. This means that we have some control of the lack of subharmonicity.
4. If L is invertible, the condition $HM^{\mathsf{T}}\overline{H} \geqslant L$ means that $P = L^{-1}H$ satisfies $P\overline{P} \geqslant I$. Is there a nice interpretation of this inequality?
5. For $m = 1$ it is easy to analyze the condition. It becomes
$$g_{x\bar{x}} \geqslant (HM\overline{H} - L)|\beta|^2.$$
Hence g is subharmonic if either $\beta = 0$ or $|H| \geqslant L$. At a minimum we must have $|H| \leqslant L$, so the case $|H| \geqslant L$ is then equivalent to $|H| = L$, which means that there exists a direction where the second derivative is zero. (If $m > 1$ and L and H can be diagonalized simultaneously then we have more or less this case.)
6. Again for $m = 1$, the expression $N = HM\overline{H} - L$ is equal to
$$N = \frac{f_{y'y''}^2 - f_{y'y'}f_{y''y''}}{f_{y'y'} + f_{y''y''}} = -\frac{\text{real Monge–Ampère}(f)}{\text{Laplacian}(f)},$$
where $y = y' + iy''$, $y', y'' \in \mathbf{R}$. Same conclusion as in 5.
7. Consider the special case $L = 0$. Then f is plurisubharmonic if $B = 0$ and $f_{x\bar{x}} \geqslant 0$. Taking $M = 0$ in the theorem we see that $g_{x\bar{x}} \geqslant 0$, which is true, since in Proposition 1.3.2 we have $g_{x\bar{x}} = f_{x\bar{x}}$. Indeed, $0 = B = -\beta^{\mathsf{T}}H$, so $\mathcal{H}(\alpha, \beta) = \alpha^{\mathsf{T}}H\beta = 0$. The conclusion cannot be improved.
8. Consider now the special case $H = 0$. Then f is plurisubharmonic if and only if $f_{x\bar{x}} \geqslant \mathcal{L}(\alpha)$. In fact, the necessary and sufficient condition for plurisubharmonicity (see (1.3.10)) is
$$\begin{aligned} f_{x\bar{x}} &\geqslant -\inf_z(\mathcal{L}(z) + 2\operatorname{Re} Bz) \\ &= -\inf_z(\mathcal{L}(z) - 2\operatorname{Re} \alpha^* L^{\mathsf{T}} z) \\ &= -\inf_z(\mathcal{L}(\alpha - z) - \mathcal{L}(\alpha)) = \mathcal{L}(\alpha). \end{aligned}$$
In Proposition 1.3.2 we have $g_{x\bar{x}} = f_{x\bar{x}} - \mathcal{L}(\alpha) - \mathcal{L}(\beta)$. The theorem says that $g_{x\bar{x}} \geqslant -\mathcal{L}(\beta)$, which is true and cannot be improved.

We have thus seen in 7. and 8. that if either H or L vanishes, the conclusion of the theorem cannot be improved.

9. If f is independent of $\operatorname{Im} y$, then $H = L$, so

$$N = HM^{\mathrm{T}}\overline{H} - L = LM^{\mathrm{T}}\overline{L} - L = (LML)^{\mathrm{T}} - L^{\mathrm{T}} = 0^{\mathrm{T}} = 0,$$

for $L^* = L$, $H^{\mathrm{T}} = H$. So then the matrix N vanishes identically! Thus we have proved:

Corollary 1.3.6 *If $f \in C^2(\Omega) \cap PSH(\Omega)$ is locally independent of* $\operatorname{Im} y$, *then $g(x) = f(x, w(x))$ is plurisubharmonic if $y = w(x)$ is a stationary point (local minimum) of the function $y \mapsto f(x,y)$ which depends in a C^1 manner of x.*

It is now a matter of routine to eliminate the smoothness assumptions in Corollary 1.3.6. We then obtain the following theorem:

Theorem 1.3.7 (The Minimum Principle, Kiselman [1978]) *Let $\Omega \subset \mathbf{C}^n \times \mathbf{C}^m$ be pseudoconvex and $f \in PSH(\Omega)$. Assume that Ω and f are both independent of the imaginary part of $y \in \mathbf{C}^m$, i.e., if $(x,y) \in \Omega$ and y' is a point in $\mathbf{C}^m$ with* $\operatorname{Re} y'_j = \operatorname{Re} y_j$, *then $(x,y') \in \Omega$ and $f(x,y') = f(x,y)$. Assume also (now only for simplicity) that the fiber $\pi^{-1}(x) \cap \Omega$ is connected (thus a convex set according to Corollary* 1.2.10) *for each $x \in \mathbf{C}^n$, where π is the projection $\mathbf{C}^n \times \mathbf{C}^m \to \mathbf{C}^n$ defined by $\pi(x,y) = x$. Define*

$$g(x) = \inf_y f(x,y).$$

Then $\pi(\Omega)$ is pseudoconvex and $g \in PSH(\pi(\Omega))$.

Remarks If the fiber $\pi^{-1}(x)$ is not connected, it consists of several convex components, and the theorem makes sense in this case also; however, the function g will not be defined in a subset of $\mathbf{C}^n$ but on a Riemann domain over $\mathbf{C}^n$. See Kiselman [1978] for details. — If $m = 1$, then each component of a fiber $\pi^{-1}(x) \cap \Omega$ is a strip or a half-plane or the whole plane. In most of the applications that we are going to present we do have $m = 1$, and the fiber is a half-plane, in particular connected.

A special case of the theorem is when $f = 0$ in Ω and $g = 0$ in $\pi(\Omega)$. Then the theorem just says that the projection $\pi(\Omega)$ is pseudoconvex. This special case is equivalent to the whole theorem. Indeed, let

$$\Omega_f = \{(x,y,t) \in \Omega \times \mathbf{C};\ f(x,y) < \operatorname{Re} t\}.$$

Then

$$\pi(\Omega_f) = \{(x,t) \in \pi(\Omega) \times \mathbf{C};\ g(x) < \operatorname{Re} t\}.$$

It is known that Ω_f is pseudoconvex if and only if Ω is pseudoconvex and $f \in PSH(\Omega)$. Therefore, if we have proved the theorem in the special case of zero functions, it follows that $\pi(\Omega_f)$ is pseudoconvex, which is equivalent to g being plurisubharmonic.

Proof of Theorem 1.3.7 We shall successively reduce the theorem to Corollary 1.3.6.

First we shall show that if the result holds for a function f which tends to $+\infty$ at the boundary in the sense that the set

$$\Omega^a = \{(x,y) \in \Omega;\ f(x,y) < a\}$$

satisfies

$$\Omega^a \cap (\mathbf{C}^n \times \mathbf{R}^m) \Subset \Omega, \qquad a \in \mathbf{R}, \tag{1.3.13}$$

then it holds generally. To do this we form

$$f_j = \max(-j, f) + \frac{1}{j}\left(\max(0, -\log d) + |x|^2 + |\operatorname{Re} y|^2\right),$$

where d is the distance to the complement of Ω. The functions f_j satisfy (1.3.13), and if the result holds for them, so that $g_j = \inf_y f_j(x,y)$ is plurisubharmonic, then it follows that $\lim g_j = \inf_j g_j$ is plurisubharmonic. Clearly the decreasing limit $\inf_j g_j$ is precisely g. This means that the theorem holds for general f.

Next suppose that a function f satisfies (1.3.13). Then we form

$$f_\varepsilon = f * \psi_\varepsilon + \varepsilon |\operatorname{Re} y|^2$$

like in the proof of Proposition 1.2.4, but of course with $\psi_\varepsilon(x,y) = \varepsilon^{-n-m}\psi((x,y)/\varepsilon)$. This convolution is well-defined in the set Ω_ε of points of distance larger than ε to the complement of Ω. Given an arbitrary relatively compact subdomain ω of $\pi(\Omega)$ we shall prove that g is plurisubharmonic in ω. Now g is bounded from above in ω, say $g < a$ there. Pick ε with $0 < \varepsilon \leqslant 1$ and $b > a$ such that $\Omega^a + \varepsilon B \subset \Omega^b$. Then Ω_ε contains Ω^a, so that $f * \psi_\varepsilon$ is well-defined in Ω^a. Next let

$$c = b + \sup_{(x,y)\in\Omega^a} |\operatorname{Re} y|^2 < +\infty.$$

Then $\Omega^a \subset \Omega^c \subset \Omega_\delta$ for some small positive δ. For $x \in \omega$ we have

$$c > \inf_y (f_\varepsilon(x,y);\ (x,y) \in \Omega^a) \geqslant \inf_y (f_\varepsilon(x,y);\ (x,y) \in \Omega^c).$$

In $\Omega_\varepsilon \setminus \Omega^c$ we have $f_\varepsilon \geqslant f \geqslant c$, so the last infinimum is equal to $\inf_y(f_\varepsilon(x,y); (x,y) \in \Omega_\varepsilon)$; we denote this quantity by $g_\varepsilon(x)$.

Thus f_ε is a strongly convex[3] function of $\operatorname{Re} y$ and the infimum when y varies is attained at a unique real point $y = w_\varepsilon(x)$. Corollary 1.3.6 can be applied to such functions. To see this, we first have to prove that the function w_ε is well-defined and of class C^1. Now this follows from the implicit function theorem, for the point y is the solution of the system of equations $\partial f_\varepsilon / \partial y_j = 0$, whose Jacobian is

$$\det_{j,k}\left(\frac{\partial^2 f_\varepsilon}{\partial(\operatorname{Re} y_j)\partial(\operatorname{Re} y_k)}\right)(x, w(x)).$$

[3]This means that we can subtract a small positive multiple of $|\operatorname{Re} y|^2$ and still have a convex function.

But this determinant is also the determinant of the real Hessian matrix of f_ε as a function of $\operatorname{Re} y$, and is therefore non-zero in view of the strong convexity of f_ε as a function of $\operatorname{Re} y$. This proves that w_ε is of class C^∞.

We also have to ensure that the fibers $\pi^{-1}(x) \cap \Omega_\varepsilon$ are connected, even though the set Ω_ε itself need not be connected. To see this, define first

$$W_x(\varepsilon) = \{y \in \mathbf{C}^m;\ (x, y) + (\varepsilon B \cap (\{0\} \times \mathbf{C}^m)) \subset \Omega\} \subset \mathbf{C}^m, \qquad x \in \mathbf{C}^n,\ \varepsilon > 0.$$

Since $\pi^{-1}(x) \cap \Omega$ is connected, thus convex, the set $W_x(\varepsilon)$ is convex. Therefore $\{x\} \times W_{x'}(\varepsilon)$ is convex as well; it is a subset of $\pi^{-1}(x)$. But then also the intersection

$$\bigcap_{x' \in x + \varepsilon B} \{x\} \times W_{x'}\left(\sqrt{\varepsilon^2 - |x' - x|^2}\right) = \pi^{-1}(x) \cap \Omega_\varepsilon$$

is convex. Thus Corollary 1.3.6 can be applied, and we deduce that $g_\varepsilon(x) = \inf_y f_\varepsilon(x, y)$ is a plurisubharmonic function of x. Letting ε tend to 0, we conclude that $g = \lim_{\varepsilon \to 0} g_\varepsilon = \inf_\varepsilon g_\varepsilon$ is plurisubharmonic in ω. Since ω was an arbitrary relatively compact subdomain of $\pi(\Omega)$, this proves the theorem in general.

Chapter 2
The Lelong number and the integrability index

2.1 Introduction

In the present chapter we shall show how to construct in a straight-forward way new plurisubharmonic functions from old ones using standard methods of convex analysis. These new functions can then be used to find analytic varieties that are connected with the original function, or rather with its singularities. We shall therefore first describe how one can measure the singularity of a plurisubharmonic function: this is done using the Lelong number and the integrability index.

The Lelong number measures how big (or "heavy") the singularities of a plurisubharmonic function are. It generalizes the notion of multiplicity of a zero of a holomorphic function. To define it, we first form the measure $\mu = (2\pi)^{-1}\Delta f$, where Δ is the Laplacian in all $2n$ real variables $\operatorname{Re} z_j$, $\operatorname{Im} z_j$. Note that when $f = \log|h|$ is the logarithm of the absolute value of a holomorphic function of one variable, then μ is a sum of point masses, one at each zero of h and with weight equal to the multiplicity of the zero. The *Lelong number of f at a point x* is by definition the $(2n-2)$-dimensional density of the measure μ at x. More explicitly, it is the limit as $r \to 0$ of the mean density of μ in the ball of center x and radius r:

$$\nu_f(x) = \lim_{r \to 0} \frac{\mu(x + rB)}{\lambda_{2n-2}(rB \cap \mathbf{C}^{n-1})}, \tag{2.1.1}$$

where λ_k denotes k-dimensional Lebesgue measure. Note that we compare the mass of μ in the ball $x + rB$ with the volume of the ball of radius r in $\mathbf{C}^{n-1}$, i.e., of real dimension $2n-2$. This makes sense, because if $f = \log|h|$ with h holomorphic, then μ is a mass distribution on the $(2n-2)$-dimensional zero set of h. If $n = 1$, then $\lambda_{2n-2}(rB \cap \mathbf{C}^{n-1}) = \lambda_0(\{0\}) = 1$, and $\nu_f(x)$ is just the mass of μ at x.

One often approximates a plurisubharmonic function f by $f_j = \max(-j, f)$ or by smooth functions $f_j = f * \psi_j$ obtained by convolution. However, in these cases the functions f_j never take the value $-\infty$, so their Lelong numbers $\nu_{f_j}(x)$ are zero everywhere; their singularities as measured by the Lelong number do not approach those of f as $j \to +\infty$. Here we shall construct functions f_τ depending on a non-negative number τ such that $f_0 = f$ and f_τ has Lelong number $\nu_{f_\tau}(x) = (\nu_f(x) - \tau)^+$. It turns out that the family $(f_\tau)_\tau$ can be used in various constructions. The singularities of f_τ are the same as those of f but attenuated in a certain sense. More precisely, the important property is that $\nu_{f_\tau}(x) > 0$ if $\tau < \nu_f(x)$, whereas the singularity is completely killed, i.e., $\nu_{f_\tau}(x) = 0$, if $\tau > \nu_f(x)$. In this context it is convenient to define the Lelong number of a family of plurisubharmonic functions. We prove analyticity theorems for the superlevel sets of such numbers; see section 2.4.

If f is plurisubharmonic and t a positive number, the function $\exp(-f/t)$ may or may not be integrable. The set of all t such that this function is locally integrable in the neighborhood of a certain point is an interval, and its endpoint measures how singular f is. This is the reason behind the integrability index ι_f to be defined in section 2.3 (see (2.3.4)). From the Hörmander–Bombieri theorem we get analyticity theorems for the integrability index (see (2.3.4)). There is a relation between the integrability index and the Lelong number: $\iota_f \leqslant \nu_f \leqslant n\iota_f$, where n is the complex dimension of the space; see Theorem 2.3.5. This relation cannot be improved (see Example 2.3.6), but nevertheless it will suffice to yield analyticity theorems for the Lelong number. The reason for this is roughly speaking that if we subtract the same quantity τ from two numbers like $\nu_f(x)$ and $\nu_f(x') > \nu_f(x) > \tau$, then the quotient between $\nu_f(x') - \tau$ and $\nu_f(x) - \tau$ can be large, for instance larger than the dimension n. This is why analyticity theorems for sets of plurisubharmonic functions are useful when it comes to proving analyticity theorems for a single function. For other studies of Lelong numbers, see Abrahamsson [1988], Demailly [1987, 1989], and Wang [1991].

2.2 Spherical means and spherical suprema

Let f and q be two given plurisubharmonic functions in an open set Ω in $\mathbf{C}^n$, thus $f, q \in PSH(\Omega)$. We define an open set Ω_q in $\mathbf{C}^n \times \mathbf{C}$ as

$$\Omega_q = \{(x,t) \in \Omega \times \mathbf{C};\ q(x) + \operatorname{Re} t < 0\}, \tag{2.2.1}$$

and we note immediately that Ω_q is pseudoconvex if Ω is pseudoconvex, for the function $(x,t) \mapsto q(x) + \operatorname{Re} t$ is plurisubharmonic in $\Omega \times \mathbf{C}$. We shall assume that $q(x) \geqslant -\log d_\Omega(x)$ for all $x \in \Omega$, denoting by $d_\Omega(x)$ the distance from x to the complement of Ω, and we note that then $(x,t) \in \Omega_q$ implies that the closed ball of center x and radius $|e^t|$ is contained in Ω. We define two functions u and U in Ω_q by putting

$$u(x,t) = u_f(x,t) = u_{f,q}(x,t) = \fint_{z \in S} f(x + e^t z), \qquad (x,t) \in \Omega_q; \tag{2.2.2}$$

$$U(x,t) = U_f(x,t) = U_{f,q}(x,t) = \sup_{z \in S} f(x + e^t z), \qquad (x,t) \in \Omega_q. \tag{2.2.3}$$

Here S is the Euclidean unit sphere, and the barred integral sign indicates the mean value; see (1.2.2). So $u_f(x,t)$ is the mean value of f over the sphere $x+e^tS$, and $U_f(x,t)$ is the supremum of f over the same sphere. Since we usually keep q fixed, the dependence on that function need not always be shown. If $\Omega \neq \mathbf{C}^n$, the simplest choice of q is just $q = -\log d_\Omega$. Then $q > -\infty$ everywhere. However, if $\Omega = \mathbf{C}^n$, then it is usually not convenient to use $q = -\log d_\Omega = -\infty$, because with this choice of q, the behavior of f at infinity would influence the local properties of the functions we construct. In this case it is best just to take $q = 0$.

The functions $u_{f,q}$ and $U_{f,q}$ are well defined and $< +\infty$ in Ω_q, thanks to our assumption $\exp(-q(x)) \leqslant d_\Omega(x)$. We define them to be $+\infty$ outside Ω_q.

Clearly $u_f \leqslant U_f$, and we shall see that there are inequalities in the opposite direction. We can note quickly that $u_{af+bg} = au_f + bu_g$ for non-negative a,b, even for real a,b, which implies that the function u_f depends linearly of f in the linear space of all Borel measurable functions which are integrable on spheres, thus in particular on the space $\delta PSH(\Omega)$ of delta-plurisubharmonic functions, i.e., the vector space spanned by those plurisubharmonic functions which are not identically minus infinity in any open component of Ω (see the end of section 1.2). We shall see that this implies that the Lelong number is a linear function on $\delta PSH(\Omega)$. As to the function U_f we can only say that $U_{af+bg} \leqslant aU_f + bU_g$ for $a,b \geqslant 0$, which implies that U_f is a convex function of f, and the Lelong number a concave function of f. But when it comes to the maximum of two functions, we have $U_{\max(f,g)} = \max(U_f, U_g)$ which implies that $\nu_{\max(f,g)} = \min(\nu_f, \nu_g)$, whereas for the mean we can say only that $u_{\max(f,g)} \geqslant \max(u_f, u_g)$ which implies that $\nu_{\max(f,g)} \leqslant \min(\nu_f, \nu_g)$. It is therefore useful to know that the Lelong number can be defined by either u_f or U_f, because this enables us to use the best properties of either one.

We can define the Lelong number as the slope at minus infinity of the function $t \mapsto u(x,t)$. As a consequence of the maximum principle, $u(x,t)$ and $U(x,t)$ are increasing in t; by Hadamard's three-circle theorem, they are convex functions of t. Therefore their slopes at $-\infty$ exist:

$$\nu_f(x) = \lim_{t\to-\infty} \frac{u(x,t)}{t} \quad \text{and} \quad N_f(x) = \lim_{t\to-\infty} \frac{U(x,t)}{t} \tag{2.2.4}$$

both exist. This follows from the fact that the slopes

$$\frac{u(x,t)-u(x,t_0)}{t-t_0} \quad \text{and} \quad \frac{U(x,t)-U(x,t_0)}{t-t_0}$$

are increasing in t. The first limit $\nu_f(x)$ is the *Lelong number of f at x*, and the definition we shall use in this chapter. The definition (2.1.1) of the Lelong number as the density of a measure is equivalent to (2.2.4) as can be proved without difficulty using Stokes' theorem (Kiselman [1979]). To see this we shall calculate the mean density assuming that f is of class C^2. We first express the mass of μ in a ball in terms of the derivative of u:

$$\mu(x+rB) = \frac{1}{2\pi}\int_{x+rB} \Delta f = \frac{1}{2\pi}\int_{x+rS} \frac{\partial f}{\partial r} dS = \frac{1}{2\pi}\frac{\partial u}{\partial t}\frac{dt}{dr}\int_{rS} dS = \frac{1}{2\pi r}\frac{\partial u}{\partial t}\int_{rS} dS.$$

We now compare with the integral over a ball of lower dimension:

$$\int_{rS} dS = r^{2n-1}\int_S dS = 2\pi r^{2n-1}\int_{B^{2n-2}} d\lambda_{2n-2} = 2\pi r\int_{rB^{2n-2}} d\lambda_{2n-2} = 2\pi r\lambda_{2n-2}(rB^{2n-2}).$$

Note that we use the unit sphere of dimension $2n-1$ and the unit ball of dimension $2n-2$ here; the remarkable fact is that the quotient

$$\frac{\text{area}(S^{2n-1})}{\text{volume}(B^{2n-2})} = 2\pi$$

is independent of the dimension. The mean density $\mu(x+rB)/\lambda_{2n-2}(rB\cap\mathbf{C}^{n-1})$ is therefore equal to the slope $\partial u/\partial t$ at the point $t = \log r$, and the density at the point x is equal to the limit $\lim_{t\to-\infty} \partial u/\partial t(x,t)$. We can now get rid of the extra assumption that f is of class C^2, the derivative of u being replaced by the derivative from the right (we use closed balls).

Since $u_f \leqslant U_f$ we have $\nu_f(x) \geqslant N_f(x)$. We shall now see that the two numbers are equal. To this end we shall use Harnack's inequality, which takes the form

$$\frac{1+|x|/r}{(1-|x|/r)^{m-1}}h(0) \leqslant h(x) \leqslant \frac{1-|x|/r}{(1+|x|/r)^{m-1}}h(0) \tag{2.2.5}$$

for harmonic functions which satisfy $h \leqslant 0$ in the ball of radius r in $\mathbf{R}^m$. If f is subharmonic in a neighborhood of the closed ball e^sB in $\mathbf{C}^n$, we can consider its harmonic majorant h there, which satisfies $f(x) \leqslant h(x)$ and

$$h(0) = \fint_{z\in S} h(e^sz) = \fint_{z\in S} f(e^sz) = u(0,s).$$

Therefore

$$U(0,t) = \sup_{e^tS} f \leqslant \sup_{e^tS} h \leqslant \frac{1-e^{t-s}}{(1+e^{t-s})^{2n-1}}u(0,s), \qquad t < s,$$

provided only $f \leqslant 0$ in e^sB. If we apply this inequality to the function $f - U(0,s)$, which is non-positive in e^sB by definition, we get, writing $U(t)$ instead of $U(0,t)$ for simplicity:

$$U(t) - U(s) \leqslant \frac{1-e^{t-s}}{(1+e^{t-s})^{2n-1}}(u(s) - U(s)),$$

equivalently,

$$U(t) \leqslant (1-\lambda_{t-s})U(s) + \lambda_{t-s}u(s), \qquad t < s, \tag{2.2.6}$$

where λ_t is defined for $t < 0$ as

$$\lambda_t = \frac{1-e^t}{(1+e^t)^{2n-1}}.$$

We can now prove that the two limits in (2.2.4) are equal. As already noted, $\nu_f(x) \geqslant N_f(x)$. In the other direction we can take for instance $s = t+1$ in (2.2.6) to obtain the estimate

$$U(t) \leqslant (1-\lambda_{-1})U(t+1) + \lambda_{-1}u(t+1),$$

whence

$$\frac{U(t)}{t} \geqslant (1-\lambda_{-1})\frac{U(t+1)}{t} + \lambda_{-1}\frac{u(t+1)}{t}, \qquad t < 0.$$

Letting t tend to $-\infty$ we see that $N_f(x) \geqslant \nu_f(x)$.

To any given $f, q \in PSH(\Omega)$ we define

$$\varphi_\tau(x) = \inf_t [u_f(x,t) - \tau \operatorname{Re} t], \qquad x \in \Omega,\ \tau \geqslant 0. \tag{2.2.7}$$

In view of our convention that $u_f(x,t) = +\infty$ if $(x,t) \notin \Omega_q$, the infimum is effectively only over those t that satisfy $\operatorname{Re} t < -q(x)$. The function $\tau \mapsto -\varphi_\tau(x)$ is the Fenchel transform of $\mathbf{R} \ni t \mapsto u_f(x,t)$; cf. (3.4.1). We assume all the time that $e^{-q(x)}$ does not exceed the distance $d_\Omega(x)$ from x to the boundary of Ω, so that u_f is well defined. The function $(x,t) \mapsto u_f(x,t) - \tau \operatorname{Re} t$ is plurisubharmonic in Ω_q and independent of the imaginary part of t. Therefore the minimum principle, Theorem 1.3.7, can be applied and yields that φ_τ is plurisubharmonic in Ω.

Example Let us look at the simplest example: $f(x) = \log|x|$, $x \in \mathbf{C}^n$. We choose $q = 0$ and form $U_f(x,t) = \log(e^t + |x|)$ for $t < q(x) = 0$. Then $\varphi_\tau(x) = \inf_{t<0}(U_f(x,t) - \tau t)$ can be calculated explicitly: it is $\varphi_\tau(x) = (1-\tau)\log|x| + C_\tau$ for $0 \leqslant \tau < 1$, where C_τ is a constant which depends on the parameter τ, and $\varphi_\tau(x) = \log(1+|x|)$ for $\tau \geqslant 1$. Thus the Lelong number of φ_τ at the origin is $\max(1-\tau, 0)$ for all $\tau \geqslant 0$.

There is no apparent reason why the Lelong number of the plurisubharmonic function φ_τ at x should be a function of $\nu_f(x)$ and τ; it could as well depend in some other way on the behavior of f near x. However, it turns out that the simple formula for the Lelong number of φ_τ in the example holds quite generally:

Theorem 2.2.1 (Kiselman [1979]) *Let $f, q \in PSH(\Omega)$ with $q \geqslant -\log d_\Omega$. Define φ_τ by* (2.2.2) *and* (2.2.7)*. Then $\varphi_\tau \in PSH(\Omega)$. If $\nu_q(x) = 0$, then the Lelong number of φ_τ is*

$$\nu_{\varphi_\tau}(x) = \max(\nu_f(x) - \tau, 0) = (\nu_f(x) - \tau)^+, \qquad x \in \Omega,\ \tau \geqslant 0. \tag{2.2.8}$$

We can also use the function U_f instead of u_f in the construction; the proof is the same. If $\tau < 0$, then of course $\nu_{\varphi_\tau}(x) = +\infty$.

We shall give a simplified proof of Theorem 2.2.1 under the slightly stronger hypothesis that $q(x) > -\infty$. This is quite enough for the applications we have in mind. (As soon as $\Omega \neq \mathbf{C}^n$, we must indeed have $q > -\infty$ everywhere.)

Lemma 2.2.2 *With f and φ_τ as in Theorem* 2.2.1 *we have*

$$\nu_{\varphi_\tau}(x) \geqslant \nu_f(x) - \tau.$$

Proof We first note that by the definition of φ_τ, we have for any t'

$$\varphi_\tau(y) \leqslant u_f(y,t') - t'\tau.$$

Taking the mean over the sphere $x + e^t S$ then gives

$$\begin{aligned} u_{\varphi_\tau}(x,t) &= \fint_{z\in S} \varphi_\tau(x+ze^t) \leqslant \fint_{z\in S}\fint_{w\in S} f(x+ze^t+we^{t'}) - t'\tau \\ &\leqslant \fint_{z\in S} f(x+ze^{t''}) - t'\tau = u_f(x,t'') - t'\tau. \end{aligned}$$

Here t and t' are arbitrary and t'' is determined from them by the equation $e^{t''} = e^t + e^{t'}$. The only interesting choice is $t = t'$, so that $t'' = t + \log 2$. Thus

$$\frac{u_{\varphi_\tau}(x,t)}{t} \geqslant \frac{u_f(x,t+\log 2)}{t} - \tau, \qquad t < 0,$$

and letting t tend to $-\infty$ we get the desired conclusion.

Lemma 2.2.3 *With f and φ_τ as in Theorem 2.2.1, take a number $\tau > \nu_f(x)$. Assume that $q(x) > -\infty$. Then $\varphi_\tau(x) > -\infty$. In particular $\nu_{\varphi_\tau}(x) = 0$.*

Proof Since $\nu_f(x) < \tau < +\infty$, f is not equal to $-\infty$ identically in a neighborhood of x. The value $\varphi_\tau(x)$ is the infimum of the convex function $u_f(x,t) - t\tau$ of the real variable t when t varies in the interval $]-\infty, -q(x)[$. This interval is by hypothesis bounded from the right. Moreover by the choice of τ, the function is strictly decreasing when $t \ll 0$. Thus its infimum is finite.

Proof of Theorem 2.2.1, assuming that $q(x)$ **is finite** The proof consists of the following steps (cf. Kiselman [1992]). First we note that φ_τ is a concave function of τ with $\varphi_0 = f$. Therefore $\nu_{\varphi_\tau}(x)$ is a convex function of τ taking the value $\nu_f(x)$ for $\tau = 0$, for the Lelong number is as we have seen a linear function of f, the limit of $u(x,t)/t$. Second we see from Lemma 2.2.2 that $\nu_{\varphi_\tau}(x) \geqslant \nu_f(x) - \tau$. Third we know from Lemma 2.2.3 (if $\nu_f(x)$ is finite) that $\nu_{\varphi_\tau}(x) = 0$ if $\tau > \nu_f(x)$. Now the only convex function of τ which has these properties is $\tau \mapsto (\nu_f(x) - \tau)^+$.

2.3 The Hörmander–Bombieri theorem and the integrability index

The purpose of this chapter is to show how the singularities of plurisubharmonic functions give rise to, and can be described by, analytic varieties. To do so we shall of course need a method to construct varieties defined by a given plurisubharmonic function. This method is the technique of solving the $\bar\partial$ equation using plurisubharmonic functions as weights, most elegantly expressed by the Hörmander–Bombieri theorem:

Theorem 2.3.1 *Let Ω be a pseudoconvex open set in $\mathbf{C}^n$, and let $\varphi \in PSH(\Omega)$. For every $a \in \Omega$ such that $e^{-\varphi} \in L^2_{loc}(a)$ there exists a holomorphic function $h \in \mathcal{O}(\Omega)$ such that $h(a) = 1$ and*

$$\int_\Omega |h|^2 e^{-2\varphi}(1+|z|^2)^{-3n}\, d\lambda_{2n}(z) < +\infty. \tag{2.3.1}$$

Here $L^2_{loc}(a)$ denotes the set of all functions that are square integrable in some neighborhood of the point a.

For the proof see Hörmander [1990, Theorem 4.4.4]. (The exponent $-3n$ can be improved to $-n-\varepsilon$ for any positive ε; see Hörmander [forthc.]. This is, however, not important in a local study like ours.) Let us denote by $\mathcal{O}(\Omega,\varphi)$ the set of all holomorphic functions h in Ω which satisfy condition (2.3.1) for a given function φ. The intersection

$$V(\varphi)=\bigcap_h \left(h^{-1}(0);\, h\in \mathcal{O}(\Omega,\varphi)\right) \tag{2.3.2}$$

is an intersection of zero sets of holomorphic functions, and therefore itself an analytic set. Let us define

$$I(\varphi)=\{a\in\Omega;\, e^{-\varphi}\notin L^2_{loc}(a)\}.$$

With this notation the theorem says that $V(\varphi)\subset I(\varphi)$. It is however obvious that $I(\varphi)\subset V(\varphi)$.

In view of this theorem it is natural to measure the singularity of a plurisubharmonic function φ at a point a by its *integrability index* $\iota_\varphi(a)$:

$$\iota_\varphi(a)=\inf_{t>0}[t;\, e^{-\varphi/t}\in L^2_{loc}(a)]. \tag{2.3.3}$$

It is easy to see that if $e^{-\varphi/t}\in L^2_{loc}(a)$, then also $e^{-\varphi/s}\in L^2_{loc}(a)$ for every $s>t$. Thus the set of $t>0$ such that $e^{-\varphi/t}\in L^2_{loc}(a)$ is an interval, either $[\iota_\varphi(a),+\infty[$ or $]\iota_\varphi(a),+\infty[$. In all examples I have seen, it is an open interval. It seems to be unknown whether it is always open for plurisubharmonic φ.

Let Φ be an arbitrary subset of $PSH(\Omega)$ and κ a functional on Φ in the sense that there is given a function $\kappa_\varphi\colon\Omega\to[0,+\infty]$ for every $\varphi\in\Phi$. We introduce a notation for the superlevel sets of such functionals:

$$E^\kappa_c(\varphi)=\{a\in\Omega;\, \kappa_\varphi(a)\geqslant c\},\qquad c\geqslant 0.$$

The superlevel sets of the integrability index are analytic varieties. In fact, by the definition of ι

$$E^\iota_c(\varphi)\subset I(\varphi/t)\subset E^\iota_t(\varphi),\qquad 0<t<c.$$

Using the Hörmander–Bombieri theorem we see that

$$E^\iota_c(\varphi)\subset V(\varphi/t)\subset E^\iota_t(\varphi),\qquad 0<t<c.$$

We now note that by the definition of the superlevel set, the intersection of all $E^\iota_t(\varphi)$ when t varies in the interval $0<t<c$ is just $E^\iota_c(\varphi)$, so that

$$E^\iota_c(\varphi)=\bigcap_{0<t<c} V(\varphi/t),\qquad c\geqslant 0. \tag{2.3.4}$$

Suppose that κ is a functional which is comparable to the integrability index in the sense that the inequality

$$s\iota_\varphi(x)\leqslant\kappa_\varphi(x)\leqslant t\iota_\varphi(x),\qquad \varphi\in\Phi,\ x\in\Omega, \tag{2.3.5}$$

holds for some positive constants s and t. Then there is of course a relation between the superlevel sets of the two functionals:

$$E^{\kappa}_{tc}(\varphi) \subset E^{\iota}_{c}(\varphi) \subset E^{\kappa}_{sc}(\varphi), \qquad \varphi \in \Phi. \tag{2.3.6}$$

If we know that $E^{\kappa}_{tc}(\varphi) = E^{\kappa}_{sc}(\varphi)$, then $E^{\kappa}_{tc}(\varphi)$ equals $E^{\iota}_{c}(\varphi)$ and so is an analytic variety. Of course functions which admit such an interval of constancy in their superlevel sets are very special. But we shall see in the next section that for a set of plurisubharmonic functions such intervals of constancy can appear quite naturally.

We now ask whether the Lelong number is comparable to the integrability index in the sense of (2.3.5). The answer is well-known, but will be quoted here for convenience.

Theorem 2.3.2 *If $\varphi \in PSH(\Omega)$ where $\Omega \subset \mathbf{C}^n$, and $\nu_\varphi(a) \geqslant n$, then $e^{-\varphi} \notin L^2_{loc}(a)$. Thus $E^{\nu}_{n}(\varphi) \subset I(\varphi) \subset V(\varphi)$. In terms of the integrability index we have $\nu_\varphi(x) \leqslant n\iota_\varphi(x)$.*

Proof This result is contained in Skoda [1972, Proposition 7.1], but it is easy to give a proof using the function $U = U_\varphi$ defined by (2.2.3). If $\nu_\varphi(a) \geqslant n$, then the slope of $t \mapsto U(a,t)$ at minus infinity is at least n, and we have

$$U(a,t) \leqslant U(a,t_0) + n(t-t_0), \qquad t \leqslant t_0,$$

for some t_0. Rewriting this in terms of φ we see that

$$\varphi(z) \leqslant \varphi(z_0) + \log\left(\frac{|z-a|^n}{|z_0-a|^n}\right), \qquad |z-a| \leqslant |z_0-a|,$$

for a suitable point z_0 on the sphere $|z-a| = e^{t_0}$, or equivalently

$$e^{-2\varphi(z)} \geqslant e^{-2\varphi(z_0)}\frac{|z_0-a|^{2n}}{|z-a|^{2n}},$$

where the right-hand side is a non-integrable function near a.

In the other direction we have:

Theorem 2.3.3 *If $\varphi \in PSH(\Omega)$ has a finite value at a point $a \in \Omega$, then $e^{-\varphi} \in L^2_{loc}(a)$.*

For the proof of this result, see Hörmander [1990, Theorem 4.4.5]. The theorem says that $I(\varphi)$ is contained in the *polar set* $P(\varphi) = \varphi^{-1}(-\infty)$ of φ, thus $I(\varphi) \subset P(\varphi)$.

Combining Theorems 2.3.1, 2.3.2 and 2.3.3 we see that

$$E^{\nu}_{nc}(\varphi) \subset V(\varphi/c) \subset P(\varphi), \qquad c > 0. \tag{2.3.7}$$

A stronger result in the same direction is

Theorem 2.3.4 *If $\nu_\varphi(a) < 1$, then $e^{-\varphi} \in L^2_{loc}(a)$. Thus $I(\varphi) \subset E^{\nu}_{1}(\varphi)$. Also $\iota_\varphi(x) \leqslant \nu_\varphi(x)$.*

For the proof see Skoda [1972, Proposition 7.1].

Combining Theorems 2.3.1, 2.3.2 and 2.3.4 we get

Theorem 2.3.5 *The Lelong number ν is comparable to the integrability index ι in the sense of* (2.3.5); *more precisely,*

$$\iota_\varphi(x) \leqslant \nu_\varphi(x) \leqslant n\iota_\varphi(x), \qquad \varphi \in PSH(\Omega),\ x \in \Omega \subset \mathbf{C}^n, \tag{2.3.8}$$

and

$$E^\nu_{nc}(\varphi) \subset V(\varphi/c) \subset E^\nu_c(\varphi) \subset V(n\varphi/c), \qquad \varphi \in PSH(\Omega),\ c > 0. \tag{2.3.9}$$

These inequalities are sharp.

This result is the basis for the analyticity theorems that we shall state. However, the weaker result (2.3.7) is often sufficient.

That the comparison in (2.3.8) between the Lelong number and the integrability index cannot be improved follows from simple examples:

Example 2.3.6 The function

$$f(z) = \max(\log|z_1|^a, \log|z_2|^b), \qquad z \in \mathbf{C}^2,$$

has integrability index $\iota_f(0) = ab/(a+b)$ and Lelong number $\nu_f(0) = \min(a,b)$. Thus

$$\frac{\nu_f(0)}{\iota_f(0)} = \frac{a+b}{\max(a,b)} = \frac{a^{-1}+b^{-1}}{\max(a^{-1},b^{-1})} \in]1,2].$$

A little more generally, if we take $1 \leqslant k \leqslant n$ and positive numbers $a_1, ..., a_k$ and define

$$f(z) = \max_{1\leqslant j\leqslant k} \log|z_j|^{a_j}, \qquad z \in \mathbf{C}^n,$$

then $\iota_f(0) = (\sum a_j^{-1})^{-1}$ and $\nu_f(0) = \min a_j$, so that

$$\frac{\nu_f(0)}{\iota_f(0)} = \frac{\sum a_j^{-1}}{\max a_j^{-1}} \in [1,n].$$

(These formulas hold if we define $a_j^{-1} = 0$ for $j = k+1, ..., n$.) Clearly the quotient ν_f/ι_f can assume all values in the closed interval $[1,n]$ (we allow $k=1$ and $k=n$). Thus (2.3.8) is sharp.

2.4 Analyticity theorems for sets of plurisubharmonic functions

Let Ω be an open set in $\mathbf{C}^n$ and Φ an arbitrary subset of $PSH(\Omega)$. Let κ be a functional on Φ which is comparable to the integrability index in the sense that (2.3.5) holds for some positive constants s and t. Then we get from (2.3.6) and (2.3.7):

$$\bigcap_{\varphi\in\Phi} E^\kappa_{tc}(\varphi) \subset \bigcap_{\varphi\in\Phi} E^\iota_c(\varphi) \subset \bigcap_{\varphi\in\Phi} E^\kappa_{sc}(\varphi) \subset \bigcap_{\varphi\in\Phi} P(\varphi). \tag{2.4.1}$$

It is convenient to introduce a notation for these sets:

$$E^\kappa_c(\Phi) = \bigcap_{\varphi\in\Phi} E^\kappa_c(\varphi)$$

for any functional κ. If we define the value of the functional on the whole set Φ as

$$\kappa_\Phi(x) = \inf_{\varphi\in\Phi} \kappa_\varphi(x),$$

then $E_c^\kappa(\Phi)$ is just the superlevel set of κ_Φ. We can also define the polar set of Φ as

$$P(\Phi) = \bigcap_{\varphi\in\Phi} P(\varphi).$$

With this notation we can write (2.4.1) as

$$E_{tc}^\kappa(\Phi) \subset E_c^\iota(\Phi) \subset E_{sc}^\kappa(\Phi) \subset P(\Phi). \tag{2.4.2}$$

Theorem 2.4.1 *Let κ be a functional on a subset Φ of $PSH(\Omega)$ which is comparable to the integrability index in the sense that* (2.3.5) *holds for some positive constants s and t. If the superlevel sets $E_c^\kappa(\Phi)$ are independent of c over an interval of sufficiently large logarithmic length, viz. if $E_{sc}^\kappa(\Phi) = E_{tc}^\kappa(\Phi)$, then $E_{sc}^\kappa(\Phi)$ is an analytic variety. A little more generally, if Y is an analytic subset of Ω and $Y \cap E_{sc}^\kappa(\Phi) = Y \cap E_{tc}^\kappa(\Phi)$, then $Y \cap E_{sc}^\kappa(\Phi)$ is analytic.*

Proof Since the result is local, we can assume Ω to be pseudoconvex. The inclusions (2.4.2) then show that $E_{sc}^\kappa(\Phi) = E_c^\iota(\Phi)$, which is the set of common zeros of a family of holomorphic functions in Ω; see (2.3.4). Similarly, $Y \cap E_{sc}^\kappa(\Phi) = Y \cap E_c^\iota(\Phi)$.

Theorem 2.4.2 *Let Ω, κ and Φ be as in Theorem* 2.4.1, *and assume in addition that κ is positively homogeneous, i.e., $\varphi \in \Phi$ and $t > 0$ implies $t\varphi \in \Phi$ and $\kappa_{t\varphi}(a) = t\kappa_\varphi(a)$. Let Y be an analytic subset of Ω, and let X be a subset of Y. Assume that*

$$\inf_{x\in X} \kappa_\varphi(x) > 0 \quad \textit{for all } \varphi \in \Phi; \tag{2.4.3}$$

and

$$\textit{for every } x \in Y \setminus X \textit{ there exists a function } \varphi \in \Phi \textit{ such that } \kappa_\varphi(x) = 0. \tag{2.4.4}$$

Then X is an analytic set.

Proof Define $\psi = \varphi/\varepsilon_\varphi$, where $\varepsilon_\varphi = \inf_{x\in X} \kappa_\varphi(x) > 0$, and let Ψ be the set of all functions ψ obtained in this way. Then $\kappa_\psi(x) \geqslant 1$ for every $\psi \in \Psi$ and every $x \in X$. Using the notation for superlevel sets we can write this as

$$X \subset E_1^\kappa(\Psi) \subset E_c^\kappa(\Psi)$$

for all c with $0 \leqslant c \leqslant 1$.

On the other hand, if $x \in Y \setminus X$, then by (2.4.4) there is a $\varphi \in \Phi$ such that $\kappa_\varphi(x) = 0$. Thus $\psi = \varphi/\varepsilon_\varphi$ is in Ψ and $\kappa_\psi(x) = \kappa_\varphi(x)/\varepsilon_\varphi = 0$. So $x \notin E_c^\kappa(\psi)$, c being any positive number. Thus

$$x \notin E_c^\kappa(\Psi) = \bigcap_{\psi\in\Psi} E_c^\kappa(\psi).$$

Therefore $E_c^\kappa(\Psi)\cap Y\subset X$ for every $c>0$. Combining this with the first part of the proof we see that $E_c^\kappa(\Psi)\cap Y = X$ for all c satisfying $0<c\leqslant 1$. Hence $E_c^\kappa(\Psi)\cap Y$ is constant for these c and Theorem 2.4.1 yields that X is analytic.

A particular case of Theorem 2.4.2 is when we can associate with a given function or current a family of plurisubharmonic functions on which our functional takes values that we can control. The following result is of this character. It holds also for functionals which have only a loose connection to the integrability index or the Lelong number; more precisely functionals which are zero at the same time as the integrability index in a semiuniform way:

Theorem 2.4.3 *Let $\Phi=\{\varphi_\alpha;\ \alpha\in A\}$ be a set of plurisubharmonic functions in an open set Ω, and let κ be a functional on Φ which is weakly comparable to the integrability index in the sense that*

$$\text{for every } \varepsilon>0 \text{ there is a } \delta>0 \text{ such that } \varphi\in\Phi,\ \iota_\varphi(x)<\delta \text{ implies } \kappa_\varphi(x)<\varepsilon,$$

and

$$\varphi\in\Phi,\ \kappa_\varphi(x)=0 \text{ implies } \iota_\varphi(x)=0.$$

Suppose that the values $\kappa_{\varphi_\alpha}(x)$ are given by a formula $\kappa_{\varphi_\alpha}(x) = G(H(x),\alpha)$ for some functions $G\colon[0,+\infty]\times A\to[0,+\infty]$ and $H\colon\Omega\to[0,+\infty]$. We assume that $c\mapsto G(c,\alpha)$ is increasing, and that there exists a number c_0 such that $G(c_0,\alpha)>0$ for all $\alpha\in A$. Finally we suppose that for every $c<c_0$ there is an $\alpha\in A$ such that $G(c,\alpha)=0$. Then the superlevel set $\{x;\ H(x)\geqslant c_0\}$ is analytic.

Proof We shall apply Theorem 2.4.2 to $X=\{x;\ H(x)\geqslant c_0\}$. First we note that

$$\inf_{x\in X}\kappa_{\varphi_\alpha}(x)=\inf_{x\in X}G(H(x),\alpha)\geqslant G(c_0,\alpha)=\varepsilon_\alpha>0$$

for any $\alpha\in A$. Hence $\iota_{\varphi_\alpha}(x)\geqslant\delta_\alpha>0$, which means that (2.4.3) holds for ι. Next, if $x\notin X$, then $c=H(x)<c_0$ and there is an α such that $G(c,\alpha)=0$. We get

$$\kappa_{\varphi_\alpha}(x)=G(H(x),\alpha)=G(c,\alpha)=0;$$

hence also $\iota_{\varphi_\alpha}(x)=0$, so that (2.4.4) holds for ι. Thus Theorem 2.4.2, applied to ι, shows that X is analytic.

This theorem contains the classical theorem of Siu [1974]. For if we let $\kappa_\varphi(x)=\nu_\varphi(x)$, $A=[0,c_0[$, $G(c,\alpha)=(c-\alpha)^+$, $H(x)=\nu_f(x)$, and define φ_α as

$$\varphi_\alpha(x)=\inf_t\,[u_f(x,t)-t\alpha;\ (x,t)\in\Omega_q],\qquad x\in\Omega,\ \alpha\in[0,c_0[,$$

then by Theorem 2.2.1,

$$\nu_{\varphi_\alpha}(x)=(\nu_f(x)-\alpha)^+=G(H(x),\alpha).$$

The function $G(c,\alpha)=(c-\alpha)^+$ satisfies the hypotheses of Theorem 2.4.3, so it follows that the superlevel set $\{x;\ \nu_f(x)\geqslant c_0\}$ is an analytic variety. The singularities of the φ_α are the same as those of f, but attenuated to some degree as shown by the formula. This attenuation is the reason behind their usefulness in proving Siu's theorem.

Chapter 3
Order and type as measures of growth

3.1 Introduction

The notions of order and type of entire functions are classical in complex analysis. They result from a comparison of a given function with standard functions. The purpose of this chapter is to generalize this comparison in such a way that order and type become dual to each other in the sense of convex analysis (section 3.4), and to show that the concept of order so obtained appears as the natural answer to a problem of extrapolation: to extend convex functions from the union of two parallel hyperplanes to as large a set as possible (section 3.7). Then we return to entire functions to consider an analogous extension problem for them (section 3.8).

It is shown that the relative order of one function with respect to another can always be calculated from the growth of its Taylor coefficients (section 3.6). This is true for the type only if the growth is sufficiently regular (see Kiselman [1993]).

In Kiselman [1983] I studied order and type from this point of view, using methods from my paper [1981]. For earlier developments see the references in that paper. See also Kiselman [1984, 1986]. A different approach to the relation between maximum modulus and Taylor coefficients is presented in Freund and Görlich [1985]. Halvarsson [forthc.] has proved an extension theorem for entire functions with estimates both from above and from below. He has also studied the dependence of the order on parameters.

3.2 Order and type in classical complex analysis

Let h be an entire function in $\mathbf{C}^n$, $h \in \mathcal{O}(\mathbf{C}^n)$. Its order and type are defined classically by comparing h with the function $\exp(b|z|^a)$ for various choices of the parameters a and b. More precisely, one considers first estimates

$$|h(z)| \leqslant C_a e^{|z|^a}, \qquad z \in \mathbf{C}^n,$$

and defines the order ρ as the infimum of all numbers a for which such an estimate holds ($0 < a < +\infty$; $0 \leqslant \rho \leqslant +\infty$). In the case where $0 < \rho < +\infty$ one then considers all numbers b such that

$$|h(z)| \leqslant C_b e^{b|z|^\rho}, \qquad z \in \mathbf{C}^n,$$

for some constant C_b. The type (with respect to the order ρ) is then the infimum σ of all such numbers b ($0 < b < +\infty$; $0 \leqslant \sigma \leqslant +\infty$).

For the order we have the formula

$$\rho = \operatorname{order}(h) = \limsup_{r \to +\infty} \sup_{|z|=r} \frac{\log \log |h(z)|}{\log r}. \tag{3.2.1}$$

Now $\sup_{|z|=r} \log |h(z)|$ is a convex function of $\log r$ in view of the Hadamard three-circle theorem, so it is natural to consider the function

$$f(t) = \sup_{|z|=e^t} \log|h(z)|, \qquad t \in \mathbf{R};$$

we shall call it the *growth function* of h. The definition of order then means that we consider all numbers a such that

$$f(t) \leqslant e^{at} + C_a, \qquad t \in \mathbf{R},$$

for some constant C_a, and then define the order as the infimum of all such numbers a. (The role of the constant C_a is to eliminate all influence of values of f at any particular point.) Similarly, the type (for order ρ) is the infimum of all numbers b such that

$$f(t) \leqslant be^{\rho t} + C_b, \qquad t \in \mathbf{R}.$$

Now this leads naturally to the idea of comparing with some other function g instead of the exponential function $g(t) = e^t$. So we might want to consider all numbers a such that

$$f(t) \leqslant g(at) + C_a, \qquad t \in \mathbf{R}, \tag{3.2.2}$$

and then take the infimum of all a.

For reasons which will become clear when we come to the duality between order and type, it is desirable to change this inequality to

$$f(t) \leqslant \frac{1}{a} g(at) + C_a, \qquad t \in \mathbf{R}. \tag{3.2.3}$$

Now in the classical case, when $g(t) = e^t$, the factor $1/a$ does not make any difference whatsoever, for in this case we see that for any $a > 0$ and any $b > a$ there is a constant $C_{a,b}$ such that

$$g(at) \leqslant \frac{1}{b} g(bt) + C_{a,b} \qquad \text{and} \qquad \frac{1}{a} g(at) \leqslant g(bt) + C_{a,b}.$$

This implies that comparisons with $g(at)$ and with $g(at)/a$ give identical infima. But of course there exist functions g such that this is not true (e.g., $g(t) = t$), and then (3.2.2) and (3.2.3) lead to different definitions of the order.

3.3 Relative order and type of convex functions

Definition 3.3.1 Let $f, g\colon E \to [-\infty, +\infty]$ be two functions defined on a real vector space E. We consider inequalities of the form

$$f(x) \leqslant \frac{1}{a} g(ax) + c, \qquad x \in E, \tag{3.3.1}$$

where a is a positive constant and c a real constant. We shall call the infimum of all positive numbers a such that (3.3.1) holds for some constant c the *order of f relative to g*, and denote it by $\rho = \operatorname{order}(f : g)$.

Examples The motivating example is

$$\operatorname{order}(t \mapsto e^{At} : t \mapsto e^t) = A$$

for all positive numbers A. Trivial examples are: $\operatorname{order}(a : b) = 0$ if a and b are finite constants; $\operatorname{order}(f : +\infty) = 0$; $\operatorname{order}(-\infty : g) = 0$; $\operatorname{order}(f : -\infty) = +\infty$ except if f is identically $-\infty$; $\operatorname{order}(+\infty : g) = +\infty$ except if g is identically $+\infty$.

If g is convex, we know that

$$\frac{1}{a}g(ax) \leqslant \frac{1-t}{a}g(0) \dot{+} \frac{t}{a}g(bx) = \Big(\frac{1}{a} - \frac{1}{b}\Big)g(0) \dot{+} \frac{1}{b}g(bx), \qquad x \in E, \tag{3.3.2}$$

if $0 < a < b$ and $ax = (1-t)\cdot 0 + tbx$, i.e., $t = a/b$. Here the sign $\dot{+}$ denotes *upper addition*, which is an extension of the usual addition from $\mathbf{R}^2$ to $[-\infty, +\infty]^2$; it satisfies $(+\infty) \dot{+} (-\infty) = +\infty$. Similarly we define *lower addition* as the extension of $+$ which satisfies $(+\infty) \underset{\cdot}{+} (-\infty) = -\infty$. If $g(0) = +\infty$, the inequality (3.3.2) is without interest, but if $g(0) < +\infty$, it shows that the inequality (3.3.1) for a particular a implies the same inequality with a replaced by b for any $b > a$. The set of all numbers a, $0 < a < +\infty$, such that (3.3.1) holds is therefore an interval, either $[\rho, +\infty[$ or $]\rho, +\infty[$ for some $\rho \in [0, +\infty]$.

So although Definition 3.3.1 has a sense for all f and g, it is often desirable to assume that g is convex with $g(0) < +\infty$: in this case the order determines the set of all a for which (3.3.1) holds, with the exception of one point, the order itself.

Lemma 3.3.2 *Let f_y denote the translate of f by the vector y: $f_y(x) = f(x-y)$. If one of f and g is convex and real valued, then*

$$\operatorname{order}(f_y : g) = \operatorname{order}(f : g_y) = \operatorname{order}(f : g).$$

In particular $\operatorname{order}(f_y : g_y) = \operatorname{order}(f : g)$ so that the order is translation invariant and can be defined on affine spaces as soon as one of the functions is convex and real valued.

Proof If f is convex and real valued, we know that

$$f(x-y) \leqslant \frac{1}{b}f(bx) + \Big(1 - \frac{1}{b}\Big)f(z)$$

for any $b > 1$, if we choose z such that

$$x - y = \frac{1}{b}bx + \Big(1 - \frac{1}{b}\Big)z,$$

i.e., if $z = -y/(1-1/b)$. If $\operatorname{order}(f : g) = \rho$, there are numbers a arbitrarily close to ρ such that

$$f(x) \leqslant \frac{1}{a}g(ax) + c.$$

We then estimate f as follows:

$$f(x-y) \leqslant \frac{1}{b}f(bx) + \Big(1 - \frac{1}{b}\Big)f(z) \leqslant \frac{1}{ab}g(abx) + \frac{1}{b}c + \Big(1 - \frac{1}{b}\Big)f(z).$$

Since $f(z)$ is finite and independent of x, this shows that $\operatorname{order}(f_y : g) \leqslant ab$, and since b is arbitrarily close to 1, we see that $\operatorname{order}(f_y : g) \leqslant \rho$. If we apply this result to f_y, translating by the vector $-y$, we get equality.

Similarly, if g is convex and real valued, we can write

$$f(x-y) \leqslant \frac{1}{a}g(a(x-y)) + c \leqslant \frac{1}{ab}g(abx) + \frac{1}{a}\Big(1 - \frac{1}{b}\Big)g(z) + c,$$

where $z = -ay/(1-1/b)$, thus independent of x. This shows that, in this case also, $\operatorname{order}(f_y : g) \leqslant ab$ with ab arbitrarily close to ρ.

It remains to consider $\operatorname{order}(f : g_y)$. The arguments are the same as for $\operatorname{order}(f_y : g)$; we omit the proof.

It is easy to give examples of functions with values in $]-\infty, +\infty]$ such that the order is not translation invariant:

Example Let f be the indicator function of the ball rB, i.e., let $f(x) = 0$ when $|x| \leqslant r$ and $f(x) = +\infty$ otherwise. Similarly let g be the indicator function of the ball sB. In the case where $0 < s \leqslant r$ we get

$$\frac{s}{r} \leqslant \operatorname{order}(f_y : g_y) = \frac{s+|y|}{r+|y|} \leqslant 1.$$

If $s > r > 0$, we have

$$\operatorname{order}(f_y : g_y) = \begin{cases} \dfrac{s-|y|}{r-|y|} \geqslant \dfrac{s}{r} > 1 & \text{when } |y| < r; \\ +\infty & \text{when } |y| \geqslant r. \end{cases}$$

We now consider a generalization of the notion of type in complex analysis.

Definition 3.3.3 Given two functions $f, g \colon E \to [-\infty, +\infty]$ on a vector space E, we consider inequalities

$$f(x) \leqslant bg(x) + c, \qquad x \in E, \tag{3.3.3}$$

where b is a positive number. We define the *type of f relative to g* as the infimum of all positive numbers b such that (3.3.3) holds for some constant c. We shall denote it by $\sigma = \operatorname{type}(f : g)$.

Example The motivating example is

$$\operatorname{type}(t \mapsto Ae^{\rho t} : t \mapsto e^{\rho t}) = A.$$

The two functions here are the growth functions of the entire functions $\exp(Az^\rho)$ and $\exp(z^\rho)$ if ρ is a natural number, and then A is the classical type with respect to order ρ.

If g is bounded from below, the set of all numbers such that (3.3.3) holds is an interval, for as soon as $b_1 > b$ we have

$$bg(x) + c = b_1 g(x) + c - (b_1 - b)g(x) \leqslant b_1 g(x) + c - (b_1 - b)\inf g = b_1 g(x) + c_1.$$

Therefore, although the definition has a sense for all functions, it is clear that it will often be necessary to assume g bounded from below. In this case the type determines all numbers b for which (3.3.3) holds, except the number σ itself.

Proximate orders are introduced to give functions of finite order normal type ($0 < \sigma < +\infty$); see Lelong & Gruman [1986, Appendix II]. The type with respect to a proximate order is a special case of Definition 3.3.3.

A generalization of the classical order and type has been studied, e.g., by Sato [1963] and Juneja, Kapoor & Bajpai [1976, 1977]. For given integers p and q, they study the (p,q)-order defined as

$$\rho_{pq} = \limsup \frac{\log^{[p]} M(r)}{\log^{[q]} r} = \limsup \frac{\log^{[p-1]} f(t)}{\log^{[q-1]} t},$$

where $M(r) = \exp f(\log r)$. (Sato considered this only for $q = 1$.) Here the brackets indicate iterations of the logarithm function. Now it is easy to see that the (p,q)-order is just $\mathrm{order}(f_q : g_p)$ where $f_q(t) = f(\exp^{[q-1]}(t))$ and $g_p(t) = \exp^{[p-1]}(t)$. Both f_q and g_p are convex. Their generalization of the notion of type is, however, different from that of Definition 3.3.3. The (p,q)-type is

$$T_{pq} = \limsup \frac{\log^{[p-1]} M(r)}{(\log^{[q-1]} r)^{\rho}} = \limsup \frac{\log^{[p-2]} f(t)}{(\log^{[q-2]} t)^{\rho}}.$$

For $p \geqslant 3$ this is not the relative type of one convex function with respect to another, but rather an order: it is the order of $f(\exp^{[q-2]} t^{1/\rho})$ with respect to $\exp^{[p-2]}(t)$. Therefore our results on order generalize those of the authors mentioned, but our type is different, and some of the earlier results on type can be interpreted as orders in the framework of the present chapter.

3.4 Order and type in duality

The notion of order and type as defined in the last section are dual, or conjugate, to each other in the sense of convexity theory. We shall express duality here in terms of *the Fenchel transformation*: for any function $f\colon E \to [-\infty, +\infty]$ we define

$$\tilde{f}(\xi) = \sup_{x \in E}(\xi \cdot x - f(x)), \qquad \xi \in E'. \tag{3.4.1}$$

Here E is a real vector space, and E' is any fixed linear subspace of its algebraic dual E^*. The function $\tilde{f}$ is called the *Fenchel transform* of f; other names are the *Legendre transform* of f, or the *conjugate function*. It is easy to see that $\tilde{f}$ is convex, lower semicontinuous for the weak-star-topology $\sigma(E', E)$ and that it never takes the value $-\infty$ except when it is equal to $-\infty$ identically.

Points where $f(x) = +\infty$ do not influence the supremum in (3.4.1). We shall use this fact in the following way. Let $\mathrm{dom}\, f$ denote the set where $f(x) < +\infty$, the *effective domain* of f. Then for any set M such that $\mathrm{dom}\, f \subset M \subset E$ we have

$$\tilde{f}(\xi) = \sup_{x \in M}(\xi \cdot x - f(x)), \qquad \xi \in E'. \tag{3.4.2}$$

The inequality

$$\xi \cdot x \leqslant f(x) \dotplus \tilde{f}(\xi), \tag{3.4.3}$$

which follows from (3.4.1), is called Fenchel's inequality (here the sign $\dotplus$ denotes upper addition; see section 3.3). Applying the transformation twice we get

$$\tilde{\tilde{f}}(x) = \sup_{\xi \in E'} (\xi \cdot x - \tilde{f}(\xi)) \leqslant f(x), \qquad x \in E.$$

Thus always $\tilde{\tilde{f}} \leqslant f$; the equality $\tilde{\tilde{f}} = f$ holds if and only if f is convex, lower semicontinuous for the weak topology $\sigma(E, E')$, and takes the value $-\infty$ only if it is $-\infty$ identically. More generally, it follows that $\tilde{\tilde{f}}$ is the maximal convex lower semicontinuous minorant of f which never takes the value $-\infty$ except when it is the constant $-\infty$. For these properties of the Fenchel transform see Rockafellar [1970]. Of course $\tilde{\tilde{f}}$ depends on the choice of E'; if $E' = \{0\}$, then $\tilde{\tilde{f}}$ is the constant $\inf f$. If $E = \mathbf{R}^n$ it is natural to take $E' = E^* \cong \mathbf{R}^n$; if E is a topological vector space one usually takes E' as the topological dual of E.

Proposition 3.4.1 *Let $f, g\colon E \to [-\infty, +\infty]$ be two functions on a vector space E. Then*

$$\operatorname{type}(\tilde{g} : \tilde{f}) \leqslant \operatorname{order}(f : g).$$

Proof If $\operatorname{order}(f : g) < A$, then $f(x) \leqslant g(ax)/a + c$ for some number $a < A$, and we deduce that $\tilde{f}(\xi) \geqslant \tilde{g}(\xi)/a - c$, which we write as $\tilde{g}(\xi) \leqslant a\tilde{f}(\xi) + ac$. Therefore $\operatorname{type}(\tilde{g} : \tilde{f}) \leqslant a < A$.

Proposition 3.4.2 *If $f, g\colon E \to [-\infty, +\infty]$ are two functions on a vector space E, then*

$$\operatorname{order}(\tilde{g} : \tilde{f}) \leqslant \operatorname{type}(f : g).$$

Proof If $\operatorname{type}(f : g) < A$ there are numbers $a < A$ and c such that $f(x) \leqslant ag(x) + c$. We take the transformation to obtain $\tilde{f}(\xi) \geqslant a\tilde{g}(\xi/a) - c$, which can be written as $\tilde{g}(\xi) \leqslant \tilde{f}(a\xi)/a + c/a$. Therefore $\operatorname{order}(\tilde{g} : \tilde{f}) \leqslant a < A$.

Theorem 3.4.3 *Let $f, g\colon E \to [-\infty, +\infty]$ be two functions on a vector space E such that $\tilde{\tilde{f}} = f$ and $\tilde{\tilde{g}} = g$. Then*

$$\operatorname{order}(\tilde{g} : \tilde{f}) = \operatorname{type}(f : g) \quad \textit{and} \quad \operatorname{type}(\tilde{g} : \tilde{f}) = \operatorname{order}(f : g).$$

Proof We just combine Propositions 3.4.1 and 3.4.2.

Corollary 3.4.4 *Let $E = \mathbf{R}^n$ and choose $E' = \mathbf{R}^n$. Let $f, g\colon \mathbf{R}^n \to [-\infty, +\infty]$ be two functions satisfying the hypotheses of the theorem. Assume in addition that f is finite in a neighborhood of the origin and grows faster than any linear function, and that g is not the constant $+\infty$. Then*

$$\operatorname{order}(f : g) = \limsup_{\xi \to \infty} \frac{\tilde{g}(\xi)}{\tilde{f}(\xi)}.$$

Proof If $f \leqslant M$ for $|x| < \varepsilon$ we obtain $\tilde{f}(\xi) \geqslant \varepsilon|\xi| - M$. Therefore $0 < \tilde{f} < +\infty$ in a neighborhood of ∞, and $\lim \tilde{f} = +\infty$, so that the type is given by

$$\operatorname{type}(\tilde{g} : \tilde{f}) = \limsup_{\xi \to \infty} \frac{\tilde{g}(\xi)}{\tilde{f}(\xi)}.$$

3.5 The infimal convolution

The infimal convolution is an important operation in convexity theory. It is actually dual to addition, so many problems can be reduced to simple questions using the Fenchel transformation, but it is often preferable to work directly with it. In this section we just recall the definition.

The *infimal convolution* of two functions $f, g \colon E \to [-\infty, +\infty]$ is defined by

$$f \,\square\, g(x) = \inf_y \, (f(y) \dot{+} g(x - y)), \qquad x \in E. \tag{3.5.1}$$

Here the sign $\dot{+}$ denotes upper addition; see section 3.3. The Fenchel transform of an infimal convolution is

$$(f \,\square\, g)^{\tilde{}}(\xi) = \tilde{f}(\xi) \underset{\cdot}{+} \tilde{g}(\xi), \qquad \xi \in E',$$

where $\underset{\cdot}{+}$ is lower addition. (It might seem strange that we get lower addition here, for in general $f \dot{+} g$ is convex when both f and g are convex, but not $f \underset{\cdot}{+} g$. However, in this case $\tilde{f} \underset{\cdot}{+} \tilde{g}$ equals $\tilde{f} \dot{+} \tilde{g}$ except when it is constant, so it is always convex.)

The infimal convolution is sometimes called the epigraphical sum. The explanation is the following formula for the strict epigraph (cf. 1.1.4) of $f \,\square\, g$

$$\operatorname{epi}_s(f \,\square\, g) = \operatorname{epi}_s f + \operatorname{epi}_s g,$$

where the plus sign denotes vector addition in $\mathbf{R}^{n+1}$.

3.6 The order of an entire function

Let $F \in \mathcal{O}(\mathbf{C}^n)$ be an entire function. We shall measure its growth by

$$f(t) = \sup_z [\log|F(z)|;\, z \in \mathbf{C}^n, |z| \leqslant e^t], \qquad t \in \mathbf{R}. \tag{3.6.1}$$

Here $|z|$ can be any norm on $\mathbf{C}^n$, or even an arbitrary function which is complex homogeneous of degree one and positive on the unit sphere. We shall refer to f as the *growth function* of F. In view of Hadamard's three-circle theorem, f is convex and increasing, and we shall write

$$\operatorname{order}(F : G) = \operatorname{order}(f : g)$$

by abuse of language if F, G are two entire functions and f, g are their growth functions.

One may ask which convex increasing functions can appear as growth functions. A necessary condition is that of Hayman [1968]: for a transcendental entire function $F \in \mathcal{O}(\mathbf{C})$, we have

$$\limsup_{t \to +\infty} f''(t) \geqslant H,$$

where H is an absolute constant satisfying $0.18 < H \leqslant 0.25$. Kjellberg [1974] proved that $0.24 < H < 0.25$. (A similar statement holds for polynomials.) Another necessary condition is as follows. Define

$$f_1(t) = \sup_{j \in \mathbf{N}} (jt - \tilde{f}(j)), \qquad t \in \mathbf{R}.$$

(The epigraph of f_1 is the smallest polygon which contains the epigraph of f and whose sides have integer slopes.) Then there is a constant C such that

$$f(t) - C \leqslant f_1(t) \leqslant f(t), \qquad t \in \mathbf{R}.$$

Moreover the best constant C satisfies $\log 2 \leqslant C \leqslant \log 3$ (Kiselman [1984, Proposition 3.5.1]). These two results are not unrelated, for the latter implies that $H \geqslant (8C)^{-1}$. With $C = \log 3$ this gives $H \geqslant (8 \log 3)^{-1} \approx 0.11$, which is much weaker than the Hayman–Kjellberg result. On the other hand, that statement does not say anything about tangents of integer slope.

If two entire functions F and G are given, we consider their expansions in terms of homogeneous polynomials P_j and Q_j:

$$F(z) = \sum_0^\infty P_j(z), \qquad G(z) = \sum_0^\infty Q_j(z),$$

and ask whether we can determine order$(F : G)$ from knowledge of the growth of $|P_j|$ and $|Q_j|$. It turns out that this is so. For the classical order, when $G = \exp$, this is well known. This is not necessarily true for type$(F : G)$; see Kiselman [1993].

So let F be given with an expansion in terms of homogeneous polynomials P_j. Cauchy's inequalites say that

$$|P_j(z)| \leqslant \exp(f(\log |z|)),$$

but the homogeneity of P_j also gives

$$|P_j(z)| = \frac{|z|^j}{e^{jt}} P_j(e^t z/|z|) \leqslant \frac{|z|^j}{e^{jt}} \exp(f(t)) = |z|^j \exp(f(t) - jt)$$

for all real t and all $z \in \mathbf{C}^n$. We take the infimum over all t and get

$$|P_j(z)| \leqslant |z|^j \exp(-\tilde{f}(j)).$$

We define the norm $\|P_j\|$ of the homogeneous polynomial P_j as

$$\|P_j\| = \sup_{|z| \leqslant 1} |P_j(z)|, \qquad j \in \mathbf{N}.$$

(When $n = 1$ we have a Taylor expansion $F(z) = \sum a_j z^j$ and $\|P_j\| = |a_j|$.) We next define a function $p\colon \mathbf{R} \to \,]-\infty, +\infty]$ as

$$p(j) = \begin{cases} -\log \|P_j\| & \text{when } j \in \mathbf{N}; \\ +\infty & \text{when } j \in \mathbf{R} \setminus \mathbf{N}. \end{cases} \tag{3.6.2}$$

We shall call p the *coefficient function* of F. Cauchy's inequalities become just $\|P_j\| \leqslant \exp(-\tilde{f}(j))$, or more concisely

$$p \geqslant \tilde{f} \qquad \text{on } \mathbf{R}. \tag{3.6.3}$$

This implies of course that $\tilde{p} \leqslant \tilde{\tilde{f}} = f$. Note also that

$$\exp \tilde{p}(\log r) = \sup_{j \in \mathbf{N}} \sup_{|z| \leqslant r} |P_j(z)|.$$

We now ask for inequalities in the other direction. To describe this result we need an auxiliary function K which is defined as follows:

$$K(t) = \begin{cases} -\log(1 - e^t), & t < 0; \\ +\infty, & t \geqslant 0. \end{cases} \tag{3.6.4}$$

We have $K(t) \geqslant -\log(-t)$ when $t < 0$ (a good approximation for small $|t|$) and $K(t) \geqslant e^t$ for all t (a good approximation for $t \ll 0$). The Fenchel transform of K is

$$\tilde{K}(\tau) = \begin{cases} \tau \log \tau - (\tau + 1) \log(\tau + 1), & \tau > 0; \\ 0, & \tau = 0; \\ +\infty, & \tau < 0. \end{cases}$$

We note that

$$-1 - \log(\tau + 1) \leqslant \tilde{K}(\tau) \leqslant -\log(\tau + 1), \qquad \tau \geqslant 0. \tag{3.6.5}$$

The inverse of K is given by $K^{-1}(s) = -K(-s)$ for $s > 0$: this means that the graph of K is symmetric around the line $s + t = 0$. This symmetry corresponds to the functional equation $\tilde{K}(1/\tau) = \tilde{K}(\tau)/\tau$, $\tau > 0$, for the transform.

Theorem 3.6.1 *Let F be an entire function in $\mathbf{C}^n$ and define f and p by* (3.6.1) *and* (3.6.2), *respectively. Then*

$$\tilde{p} \leqslant f \leqslant \tilde{p} \,\square\, K \qquad \text{on } \mathbf{R}. \tag{3.6.6}$$

Proof We have just noted that Cauchy's inequalities give $\tilde{p} \leqslant \tilde{\tilde{f}} = f$. To estimate f from above we write

$$|F(z)| \leqslant \sum \|P_j\| \cdot |z|^j \leqslant \sum \exp(-p(j) + jt),$$

where $t = \log |z|$. We shall apply Fenchel's inequality (3.4.3) $jt \leqslant p(j) \dotplus \tilde{p}(t)$ in the form

$$-p(j) + jt \leqslant js + \tilde{p}(t - s).$$

This gives

$$f(t) \leqslant \log \sum \exp(-p(j) + jt) \leqslant \log \sum \exp(js + \tilde{p}(t-s)).$$

We observe that

$$\sum_{j \in \mathbf{N}} e^{js} = \frac{1}{1-e^s} = e^{K(s)}$$

if $s < 0$, which is why we introduced K. Thus $f(t) \leqslant \tilde{p}(t-s) + K(s)$ for all $t \in \mathbf{R}$ and all $s < 0$. Now for $s \geqslant 0$, $K(s) = +\infty$, so then the inequality also holds, and we can let s vary over the whole real axis:

$$f(t) \leqslant \inf_s (\tilde{p}(t-s) \dot{+} K(s)) = (\tilde{p} \,\square\, K)(t), \qquad t \in \mathbf{R}.$$

This proves the theorem.

The inequalities (3.6.6) say that the graph of f is in a strip whose lower boundary is the polygon defined by $\tilde{p}$ and whose upper boundary is given by $\tilde{p} \,\square\, K$. Since $K(-\log 2) = \log 2$, the width of this strip is at most $\sqrt{2} \log 2 \approx 0.98 < 1$.

Applying the Fenchel transformation to all members of (3.6.6) we get:

$$p \geqslant \tilde{\tilde{p}} \geqslant \tilde{f} \geqslant \tilde{\tilde{p}} + \tilde{K}, \tag{3.6.7}$$

where $\tilde{K}$ can be estimated by (3.6.5).

For lacunary series we can state:

Theorem 3.6.2 *Let F be lacunary: $P_j = 0$ for $j \notin J$. Then*

$$\tilde{p} \leqslant f \leqslant \tilde{p} \,\square\, K_J \qquad \text{on } \mathbf{R},$$

where

$$K_J(s) = \log\Big(\sum_{j \in J} e^{js}\Big).$$

Proof Just restrict summation in the proof of Theorem 3.6.1 to $j \in J$.

It could be noted here that for any convex function H which is positive on the negative half-axis and tends to $+\infty$ as $t < 0$, $t \to 0$, there exists an infinite set $J \subset \mathbf{N}$ such that $K_J \leqslant H$.

Theorem 3.6.1 implies that the norms of the homogeneous polynomials P_j can serve just as well as the growth function f to determine the order of F relative to any other function. More precisely we have:

Corollary 3.6.3 *Let F be an entire function on $\mathbf{C}^n$, let f be its growth function defined by (3.6.1), and let p be its coefficient function defined by (3.6.2). Assume that F is not a polynomial. Then*

$$\operatorname{order}(f : \tilde{p}) = \operatorname{order}(\tilde{p} : f) = 1.$$

Proof From (3.6.6) we get immediately

$$\operatorname{order}(\tilde{p} : f) \leqslant 1, \qquad \operatorname{order}(f : \tilde{p} \,\square\, K) \leqslant 1.$$

Now $\tilde{p} \,\square\, K(t) \leqslant \tilde{p}(t+1) + K(-1)$ and Lemma 3.3.2 shows that the translation of $\tilde{p}$ does not influence the order, neither does of course the additive constant $K(-1)$. Therefore

$$\operatorname{order}(f : \tilde{p}) \leqslant \operatorname{order}(f : \tilde{p} \,\square\, K) \leqslant 1.$$

It follows from Corollary 3.4.4 that $\operatorname{order}(f : f) = 1$. By submultiplicativity,

$$1 = \operatorname{order}(f : f) \leqslant \operatorname{order}(f : \tilde{p}) \cdot \operatorname{order}(\tilde{p} : f) \leqslant 1,$$

so that all orders must be one. (When F is a polynomial, $\operatorname{order}(f : f) = 0$ and $\operatorname{order}(f : \tilde{p}) = \operatorname{order}(\tilde{p} : f) = 0$.)

Corollary 3.6.4 *Let F be an entire function in $\mathbf{C}^n$, with expansion*

$$F(z) = \sum P_j(z)$$

in terms of homogeneous polynomials P_j. Let f be its growth function defined by (3.6.1) *and let p be its coefficient function defined by* (3.6.2). *Let $g\colon \mathbf{R} \to [-\infty, +\infty]$ be any function which satisfies $\tilde{\tilde{g}} = g$. Then*

$$\operatorname{order}(f : g) = \operatorname{order}(\tilde{p} : g) = \operatorname{type}(\tilde{g} : \tilde{\tilde{p}}). \tag{3.6.8}$$

When F is not a polynomial and g is bounded from below and not identically $+\infty$, the order is also given by

$$\limsup_{j \to +\infty} \frac{\tilde{g}(j)}{p(j)}. \tag{3.6.9}$$

Proof Using Corollary 3.6.3 we can write

$$\operatorname{order}(f : g) \leqslant \operatorname{order}(f : \tilde{p}) \cdot \operatorname{order}(\tilde{p} : g) = \operatorname{order}(\tilde{p} : g)$$

provided F is not a polynomial. Similarly

$$\operatorname{order}(\tilde{p} : g) \leqslant \operatorname{order}(\tilde{p} : f) \cdot \operatorname{order}(f : g) = \operatorname{order}(f : g).$$

The last equality in (3.6.8) follows from Theorem 3.4.3. If F is a polynomial, one can verify (3.6.8) directly, using $\tilde{p} \leqslant f \leqslant \tilde{p} + \log N$, where N is the number of terms in the expansion (see Theorem 3.6.2). The only possibilities are then $\operatorname{order}(f : g) = 0, +\infty$.

We finally have, if F is not a polynomial and g is bounded from below and not identically $+\infty$,

$$\operatorname{type}(\tilde{g} : \tilde{\tilde{p}}) = \limsup_{\tau \to +\infty} \frac{\tilde{g}(\tau)}{\tilde{\tilde{p}}(\tau)} = \limsup_{j \to +\infty} \frac{\tilde{g}(j)}{p(j)}. \tag{3.6.10}$$

The first equality here is proved like in the proof of Corollary 3.4.4. There is a difference in that $f(t)$ does not go to $+\infty$ when $t \to -\infty$, but if g is bounded from below, the behavior for negative τ in (3.6.10) is unimportant. The last equality in (3.6.10) holds because on the one hand $\tilde{\tilde{p}} \leqslant p$, on the other hand $\tilde{\tilde{p}} = p$ in a sequence of integers tending to plus infinity, and $\tilde{\tilde{p}}$ is affine in between these points.

Formula (3.6.9) generalizes the classical formula for the order

$$\rho = \limsup \frac{j \log j}{-\log |a_j|}$$

of an entire function $\sum a_j z^j$. Indeed, when the comparison function is $g(t) = e^t$, then $\tilde{g}(j) = j \log j - j$.

The (p,q)-order of Juneja, Kapoor & Bajpai [1976, Theorem 1] is determined in terms of the coefficients by the formula

$$\rho_{p,q} = \limsup \frac{\log^{[p-1]} j}{\log^{[q-1]} \left(-(1/j) \log |a_j| \right)};$$

we state it only for $p > q \geqslant 1$ here. Sato [1963] proved this for $q = 1$. In the latter case (3.6.9) is a generalization. For $q \geqslant 2$, however, this is not so, since then $f(\exp^{[q-1]} t)$ is used as the growth function and consequently defines another relation between the coefficients a_j (or $p(j)$) and f.

It could also be noted here that Corollary 3.6.4 generalizes the classical result that the order can be calculated from the dominant term in a series expansion $\sum a_j z^j$. Indeed, with $t = \log |z|$ the maximal term is just

$$\sup_j |a_j z^j| = \exp \sup_j (jt - p(j)) = \exp \tilde{p}(t).$$

When we have two entire functions we can state:

Corollary 3.6.5 *Let F, G be two entire functions in $\mathbf{C}^n$, with expansions*

$$F(z) = \sum P_j(z), \qquad G(z) = \sum Q_j(z),$$

in terms of homogeneous polynomials P_j, Q_j. Let p and q denote their coefficient functions defined by (3.6.2). *Then*

$$\operatorname{order}(F : G) = \operatorname{order}(\tilde{p} : \tilde{q}) = \operatorname{type}(\tilde{\tilde{q}} : \tilde{\tilde{p}}).$$

Proof The proof is analogous to that of Corollary 3.6.4.

We can also define a growth function related to the growth of an entire function on polydisks, and to Taylor expansions in terms of monomials. Let us define

$$f(x) = \sup_{|z_j| \leqslant \exp x_j} \log |F(z)|, \qquad x \in \mathbf{R}^n, \tag{3.6.11}$$

if F is an entire function on $\mathbf{C}^n$. Then f is convex in $\mathbf{R}^n$. The function F has an expansion

$$F(z) = \sum_{k \in \mathbf{N}^n} A_k z^k, \qquad z \in \mathbf{C}^n,$$

where z^k denotes the monomial $z_1^{k_1} \cdots z_n^{k_n}$ of multidegree $k = (k_1, ..., k_n)$ and total degree $k_1 + \cdots + k_n$. Cauchy's inequalities now say that, for $r = (r_1, ..., r_n)$ with $r_j > 0$,

$$|A_k| r^k \leqslant \sup_{|z_j| \leqslant r_j} |F(z)| = e^{f(x)}, \qquad x_j = \log r_j.$$

This gives $|A_k| \leqslant \exp(f(x) - k \cdot x)$ for all $x \in \mathbf{R}^n$, and therefore, after variation of x,

$$|A_k| \leqslant \exp(-\tilde{f}(k)), \qquad k \in \mathbf{N}^n.$$

We introduce in analogy with (3.6.2)

$$a(k) = \begin{cases} -\log |A_k| & \text{when } k \in \mathbf{N}^n; \\ +\infty & \text{when } k \in \mathbf{R}^n \setminus \mathbf{N}^n. \end{cases} \tag{3.6.12}$$

Then $a \geqslant \tilde{f}$ and $\tilde{a} \leqslant \tilde{\tilde{f}} = f$. Next define $K_n(x) = K(x_1) + \cdots + K(x_n)$ for $x \in \mathbf{R}^n$. In complete analogy with Theorem 3.6.1 we have:

Theorem 3.6.6 *Let F be an entire function in $\mathbf{C}^n$ and define the growth function f and the coefficient function a by* (3.6.11) *and* (3.6.12), *respectively. Then*

$$\tilde{a} \leqslant f \leqslant \tilde{a} \,\square\, K_n \qquad \text{on } \mathbf{R}^n.$$

A variant of the growth function can be defined as follows. Let u be a plurisubharmonic function on $\mathbf{C}^n$ which is extremal in the set $a < u(z) < b$: it is the regularized supremum of all plurisubharmonic functions φ in a neighborhood of the closure of $\{z;\, a < u(z) < b\}$ which satisfy $\varphi(z) \leqslant a$ when $u(z) \leqslant a$ and $\varphi(z) \leqslant b$ when $u(z) \leqslant b$. We suppose that $\{z;\, u(z) < b\}$ is bounded, and define for $F \in \mathcal{O}(\mathbf{C}^n)$

$$f_u(t) = \sup_z (\log |F(z)|;\, u(z) < t).$$

Then f_u is easily seen to be convex on $]a, b]$. (The growth function f defined by (3.6.1) is with respect to the extremal plurisubharmonic function $u(z) = \log |z|$ provided $|z|$ is a norm or more generally $\log |z|$ is plurisubharmonic; if not, we can replace it by a suitable plurisubharmonic minorant.)

We can for instance ask whether a holomorphic function on a complex analytic variety X admits an entire extension of the same order: if $F \in \mathcal{O}(X)$, $X \subset \mathbf{C}^n$, does there exist an entire function $G \in \mathcal{O}(\mathbf{C}^n)$ such that order$(G : F) = 1$? Here it might be natural to define the growth functions f_u and g_v of F and G with respect to extremal functions u on X and v on $\mathbf{C}^n$, respectively.

3.7 A geometric characterization of the relative order

In this section we shall give a geometric interpretation of the relative order. Let E be a real vector space. We consider two hyperplanes $E \times \{0\}$ and $E \times \{1\}$ in the Cartesian product $E \times \mathbf{R}$. Now let two functions $f_0, f_1 \colon E \to]-\infty, +\infty]$ be given. We consider them as defined on $E \times \{0\}$ and $E \times \{1\}$ respectively, and want to find a function $F \colon E \times \mathbf{R} \to]-\infty, +\infty]$ extending them, i.e., a function such that

$$F(x, j) = f_j(x), \qquad x \in E, \quad j = 0, 1.$$

If the f_j are convex, a solution is of course the supremum of all convex minorants to the function $f(x,t) = f_t(x)$ if $t = 0$ or $t = 1$, $f(x,t) = +\infty$ otherwise. This solution is the largest possible: it majorizes all others. But it is of no interest outside the slab $\{0 \leqslant t \leqslant 1\}$, since it is always $+\infty$ there.

In general there is no unique solution, for we can always add $t^2 - t$ to any given solution. We can however write down an explicit formula for an extremal solution.

Proposition 3.7.1 *Let E be a real vector space and E' a subspace of its algebraic dual. Let $f_0, f_1 \colon E \to]-\infty, +\infty]$ be two given convex functions which are lower semicontinuous with respect to $\sigma(E, E')$. We assume that they are not identically plus infinity. Then the extrapolation problem*

$$\begin{cases} \text{Find } F \colon E \times \mathbf{R} \to]-\infty, +\infty] \text{ such that} \\ F(x, j) = f_j(x), \qquad x \in E, \quad j = 0, 1, \end{cases} \tag{3.7.1}$$

has a solution

$$\begin{aligned} F(x,t) &= \sup_\xi \left[\xi \cdot x - (1-t)\tilde{f}_0(\xi) - t\tilde{f}_1(\xi);\ \xi \in \operatorname{dom} \tilde{f}_0 \cup \operatorname{dom} \tilde{f}_1\right] \\ &= \sup_\xi \left[\xi \cdot x - ((1-t)\tilde{f}_0(\xi) \dotplus t\tilde{f}_1(\xi));\ \xi \in E'\right], \qquad (x,t) \in E \times \mathbf{R}. \end{aligned} \tag{3.7.2}$$

This solution is extremal in the sense that any convex solution G which is lower semicontinuous in x satisfies $G \leqslant F$ in $\{0 \leqslant t \leqslant 1\}$ and $G \geqslant F$ outside this slab.

Proof First a word about the definition of F. We note that the function $t \mapsto t \cdot (+\infty)$ is convex on the whole real line, if we define $0 \cdot (+\infty) = 0$. We also note that in the first expression defining F at most one of the three terms is infinite, for we have $-\infty < \tilde{f}_j \leqslant +\infty$ everywhere, and at most one of them is allowed to be plus infinity in the set of ξ which we use. Therefore F is well defined, and it is convex as a supremum of functions of (x,t) each of which is an affine function plus possibly one function of the form $(t-1) \cdot (+\infty)$ or $(-t) \cdot (+\infty)$. Moreover, for $t = j$ the function F assumes the values $\tilde{\tilde{f}}_j(x) = f_j(x)$, $j = 0, 1$, in view of (3.4.2). Therefore it is a convex solution to the extension problem. It is of course not lower semicontinuous in all variables, but it is lower semicontinuous in x for fixed t.

Now let G be another convex solution to the problem. Let us consider

$$\tilde{G}_t(\xi) = \sup_{x \in E}(\xi \cdot x - G(x,t)), \qquad t \in \mathbf{R}, \quad \xi \in E'.$$

It is concave in t for fixed ξ, for it is the marginal function of a concave function of (x,t); cf. Theorem 1.3.1. It satisfies moreover $\tilde{G}_j(\xi) = \tilde{f}_j(\xi)$, $j = 0, 1$. If we assume that G is lower semicontinuous in x and $> -\infty$, we also have

$$G(x,t) = \sup_\xi(\xi \cdot x - \tilde{G}_t(\xi)).$$

When $0 < t < 1$ we have

$$\operatorname{dom}((1-t)\tilde{f}_0 \dot{+} t\tilde{f}_1) = \operatorname{dom} \tilde{f}_0 \cap \operatorname{dom} \tilde{f}_1 \subset \operatorname{dom} \tilde{f}_0 \cup \operatorname{dom} \tilde{f}_1.$$

The fact that $\tilde{G}_j = \tilde{f}_j$ for $j = 0, 1$ implies that $\tilde{G}_t \geqslant (1-t)\tilde{f}_0 + t\tilde{f}_1$. This gives $G \leqslant F$.

When $t < 0$ or $t > 1$ the concavity in t gives $\tilde{G}_t \leqslant (1-t)\tilde{f}_0 \dot{+} t\tilde{f}_1$ and then $G \geqslant F$. This establishes the extremal character of the solution F.

We now ask how far outside the slab $\{0 \leqslant t \leqslant 1\}$ we can obtain a real-valued solution to the extrapolation problem. An answer is given by the next theorem.

Theorem 3.7.2 *Let $f_0, f_1 \colon E \to]-\infty, +\infty]$ be two given convex and lower semicontinuous functions. Assume that $f_0(0) < +\infty$. If the extrapolation problem* (3.7.1) *admits a convex solution F which is finite at a point $(0,t)$ with t satisfying $1 < t < +\infty$, then*

$$\operatorname{order}(f_1 : f_0) \leqslant \frac{t}{t-1}.$$

Conversely, if $1 \leqslant \operatorname{order}(f_1, f_0) = \rho < +\infty$, then the extrapolation problem has a lower semicontinuous convex solution F with $F(0,t) < +\infty$ for all t with $0 \leqslant t < \rho/(\rho-1)$. Thus if we denote by b the supremum of all numbers t such that there exists a solution F which is finite at the point $(0,t)$, then

$$\operatorname{order}(f_1 : f_0) = \rho = \frac{b}{b-1} = b'.$$

(*We assume $1 \leqslant \rho < +\infty$ and $1 < b \leqslant +\infty$.*)

Proof If F is convex we have

$$F(x,1) \leqslant \frac{1}{a}F(ax,0) \dot{+} \Big(1 - \frac{1}{a}\Big)F(0,t),$$

where $a > 1$ is chosen so that

$$(x,1) = \frac{1}{a}(ax,0) + \Big(1 - \frac{1}{a}\Big)(0,t) \in E \times \mathbf{R},$$

i.e., $a = t/(t-1)$. Now if $F(0,t) < +\infty$ this inequality shows that

$$f_1(x) \leqslant \frac{1}{a}f_0(ax) + c,$$

in other words that $\operatorname{order}(f_1 : f_0) \leqslant a = t/(t-1)$.

Conversely, if $\operatorname{order}(f_1, f_0) \leqslant \rho$ with $1 \leqslant \rho < +\infty$, then the solution F defined by (3.7.2) has the desired properties. We need only estimate F as follows. For any $a > \rho$ we know that $f_1(x) \leqslant f_0(ax)/a + c$, which gives $\tilde{f}_1 \geqslant a^{-1}\tilde{f}_0 - c$. In particular we see that $\operatorname{dom} \tilde{f}_0 \supset \operatorname{dom} \tilde{f}_1$. For any $t < a/(a-1)$ we can write, letting ξ vary in $\operatorname{dom} \tilde{f}_0$,

$$\begin{aligned} F(x,t) &\leqslant \sup_\xi [\xi \cdot x - (1-t)\tilde{f}_0(\xi) - t(a^{-1}\tilde{f}_0(\xi) - c)] \\ &= \sup_\xi [\xi \cdot x - (1 - t + t/a)\tilde{f}_0(\xi)] + tc \\ &= (1 - t + t/a) \sup_\xi [(1 - t + t/a)^{-1}\xi \cdot x - \tilde{f}_0(\xi)] + tc \\ &= \delta f_0(x/\delta) + tc, \end{aligned}$$

where δ is the positive number $1 - t + t/a$. Now, since we assume that $f_0(0) < +\infty$, this shows that $F(0,t)$ is finite for all $t \in [0, a/(a-1)[$, and since a is any number larger than ρ, the function is finite for all $t \in [0, b[$.

For real-valued functions the geometry is particularly simple:

Corollary 3.7.3 *Let $f_0, f_1 \colon E \to]-\infty, +\infty]$ be two functions as in Proposition 3.7.1 and assume in addition that one of them is real valued. If the extrapolation problem (3.7.1) admits a convex solution F which is finite at some point (x,t) with t satisfying $1 < t < +\infty$, then*

$$\operatorname{order}(f_1 : f_0) \leqslant \frac{t}{t-1}.$$

Conversely, if $1 \leqslant \operatorname{order}(f_1, f_0) = \rho < +\infty$, then the extrapolation problem has a lower semicontinuous convex solution F which is real valued in the slab

$$E \times]0, \rho'[= \{(x,t) \in E \times \mathbf{R};\ 0 < t < \rho'\},$$

where $\rho' = \rho/(\rho - 1)$; $1 < \rho' \leqslant +\infty$.

Therefore the relative order of f_1 with respect to f_0 is determined by, and determines, the maximal slab $E \times]0, b[$ in which our extrapolation problem has a solution.

Proof Suppose f_j is real valued ($j = 0$ or $j = 1$). It is clear that if a solution F is finite at some point (x, s) with $s > 1$, then F is finite in the convex hull of the union of (x, s), some point $(y, 0)$ where f_0 is finite, and the hyperplane $E \times \{j\}$. This convex hull contains the slab $E \times]0, s[$. Thus Theorem 3.7.2 implies Corollary 3.7.3.

It follows again (cf. Lemma 3.3.2) that the notion of relative order is translation invariant for real-valued convex functions (at least when $1 \leqslant \rho + \infty$). Indeed, the slabs are invariant under transformations $(x,t) \mapsto (x - (1-t)y - tz, t)$ for all y and z; these transformations correspond to translations $f_0 \mapsto f_{0,y}$ and $f_1 \mapsto f_{1,z}$.

3.8 An extension theorem for holomorphic functions

In this section we shall first characterize the classical order in terms of an extension property of holomorphic functions. Then we pass to the relative order.

Theorem 3.8.1 *An entire function $F \in \mathcal{O}(\mathbf{C}^n)$ is of order at most ρ $(1 \leqslant \rho < +\infty)$ if and only if there exists a holomorphic function H in the cylinder*

$$\Omega = \{(z, w) \in \mathbf{C}^n \times \mathbf{C};\ |w| < e^{\rho'}\},$$

where $\rho' = \rho/(\rho - 1)$, satisfying

$$|H(z, w)| \leqslant e^{|z|} \quad \textit{for } z \in \mathbf{C}^n,\ |w| \leqslant 1, \tag{3.8.1}$$

and

$$H(z, e) = F(z) \quad \textit{for } z \in \mathbf{C}^n. \tag{3.8.2}$$

Proof Suppose such an H exists. Then putting

$$h(s, t) = \sup\,[\log|H(z, w)|;\ |z| \leqslant e^s,\ |w| \leqslant e^t], \qquad s \in \mathbf{R},\ t < \rho', \tag{3.8.3}$$

we get a convex function of (s, t) which satisfies $h(s, 0) \leqslant e^s$ and $h(s, 1) \geqslant f(s)$. Therefore, applying Corollary 3.7.3 with $f_0(s) = h(s, 0)$ and $f_1(s) = h(s, 1)$, we can write

$$\operatorname{order}(F : \exp) \leqslant \operatorname{order}(f_1 : \exp) \leqslant \operatorname{order}(f_1 : f_0) \cdot \operatorname{order}(f_0 : \exp) \leqslant \rho.$$

In the other direction the results of section 3.7 give only convex, not holomorphic, solutions to the extrapolation problem. But it turns out that there is an explicit solution in terms of power series.

We expand F in a series of homogeneous polynomials:

$$F(z) = \sum P_j(z).$$

Then we just define

$$H(z, w) = \sum P_j(z)(w/e)^{m_j}, \tag{3.8.4}$$

where the integers m_j are chosen large enough to make (3.8.1) true. This means that we take

$$\|P_j\| e^{-m_j} \leqslant \frac{1}{j!}.$$

On the other hand, we do not want to take them unnecessarily large, so we prescribe that

$$\log\|P_j\| + \log j! \leqslant m_j < \log\|P_j\| + \log j! + 1$$

unless $P_j = 0$ in which case the choice of m_j is immaterial, so we may take $m_j = 0$.

Since F is of order ρ, we know that for any $a > \rho$ there is an estimate $f(t) \leqslant e^{at} + C_a$, which implies that

$$\tilde{f}(\tau) \geqslant \frac{\tau}{a}\Big(\log\frac{\tau}{a} - 1\Big) - C_a$$

and

$$-\log\|P_j\| = p(j) \geqslant \tilde{f}(j) \geqslant \frac{j}{a}\Big(\log\frac{j}{a} - 1\Big) - C_a. \tag{3.8.5}$$

This estimate shows that the series defining H converges uniformly on any compact subset of Ω. In fact, the series defining H converges uniformly for $|z| \leqslant R_1 < R$ and $|w| \leqslant r_1 < r$ if $\|P_j\| R^j (r/e)^{m_j} \to 0$. Substituting the expression for m_j we see that this is true if $(\log r - 1)\log j! - p(j)\log r + j\log R \to -\infty$. Now this holds for all positive R if

$$\frac{(\log r - 1)\log j! - p(j)\log r}{j} \to -\infty.$$

Using finally the estimate (3.8.5) for p and the inequality $j! \leqslant j^j$ for the factorial function we see that this follows if

$$(\log r - 1)\log j - \frac{1}{a}\Big(\log\frac{j}{a} - 1\Big)\log r \to -\infty,$$

which in turn is true if $\log r < a/(a-1)$. Here the only condition is that $a > \rho$, so the series defining H converges locally uniformly in the set $\log|w| < \rho/(\rho - 1)$.

We now replace the exponential function in Theorem 3.8.1 to obtain the following result.

Theorem 3.8.2 *Let two transcendental entire functions $F, G \in \mathcal{O}(\mathbf{C}^n)$ be given, and let $1 \leqslant \rho < +\infty$. We define an open set Ω in the space of $n+1$ variables as*

$$\Omega = \{(z, w) \in \mathbf{C}^n \times \mathbf{C};\ |w| < e^{\rho'}\},$$

where $\rho' = \rho/(\rho - 1)$ (thus $1 < \rho' \leqslant +\infty$). For a holomorphic function H in Ω we denote by h_w the growth function of the partial function $z \mapsto H(z, w)$. Let K denote the function defined by (3.6.4). Then the following five conditions are equivalent.

(a) $\mathrm{order}(F : G) \leqslant \rho$.

(b) *There exists a holomorphic function $H \in \mathcal{O}(\Omega)$ satisfying*

$$h_w \leqslant g \,\square\, K \quad \text{when } |w| = 1, \quad \text{and} \quad f \leqslant h_w \,\square\, K \quad \text{when } |w| = e.$$

(b′) *There exists a holomorphic function $H \in \mathcal{O}(\Omega)$ satisfying $H(z, 1) = G(z)$,*

$$g \leqslant h_w \,\square\, K, \quad \text{and} \quad h_w \leqslant g \,\square\, K \quad \text{when } |w| = 1,$$

and

$$f \leqslant h_w \,\square\, K \quad \text{when } |w| = e.$$

(c) *There exists a holomorphic function $H \in \mathcal{O}(\Omega)$ satisfying $H(z, e) = F(z)$ and*

$$h_w \leqslant g \,\square\, K \quad \text{when } |w| = 1.$$

(c′) *There exists a holomorphic function $H \in \mathcal{O}(\Omega)$ satisfying $H(z, e) = F(z)$,*

$$f \leqslant h_w \,\square\, K \quad \text{and} \quad h_w \leqslant f \,\square\, K \quad \text{when } |w| = e,$$

and

$$h_w \leqslant g \,\square\, K \quad \text{when } |w| = 1.$$

In particular, $\operatorname{order}(H(\cdot,w):G) \leqslant 1$ for $|w|=1$ and $\operatorname{order}(F:H(\cdot,w)) \leqslant 1$ for $|w|=e$, if H is the holomorphic function whose existence is guaranteed by (b), (b′) or (c′).

Proof The proof that (b) implies (a) and that (c) implies (a) is just like the easy direction in the proof of Theorem 3.8.1. If H is a holomorphic function satisfying (b) or (c) we let h be the growth function of two real variables defined by (3.8.3); it is related to the h_w by the formula $h(s,t) = \sup_{|w|=e^t} h_w(s)$. By submultiplicativity we then have

$$\operatorname{order}(f:g) \leqslant \operatorname{order}(f:h(\cdot,1)) \cdot \operatorname{order}(h(\cdot,1):h(\cdot,0)) \cdot \operatorname{order}(h(\cdot,0):g) \leqslant \rho.$$

It is also clear that (b′) implies (b) and that (c′) implies (c).

For the proof of (a) implies (b′) we expand G and F in terms of homogeneous polynomials:

$$F(z) = \sum_{j\in\mathbf{N}} P_j(z), \qquad G(z) = \sum_{j\in\mathbf{N}} Q_j(z),$$

and define

$$H(z,w) = \sum_{j\in\mathbf{N}} Q_j(z)w^{n_j} + (w-1)\sum_{j\notin J} Q_j^*(z)w^{n_j},$$

where J is the set of all $j \in \mathbf{N}$ such that $q(j) \leqslant \tilde{\tilde{q}}(j) + \log 3$, q being the coefficient function of G defined by (3.6.2). Moreover n_j are suitable integers and Q_j^* homogeneous polynomials of degree j and norm $\|Q_j^*\| = \frac{1}{3}\exp(-\tilde{\tilde{q}}(j)) \geqslant \|Q_j\|$. Let p_w denote the coefficient function of the entire function $H(\cdot,w)$. Consider first $|w|=1$: when $j \in J$ we have $p_w(j) = q(j)$, and when $j \notin J$ we can estimate as follows:

$$\|Q_j w^{n_j} + (w-1)Q_j^* w^{n_j}\| \leqslant \|Q_j\| + 2\|Q_j^*\| \leqslant 3\|Q_j^*\| = \exp(-\tilde{\tilde{q}}(j)),$$

so that $p_w(j) \geqslant \tilde{\tilde{q}}(j)$ when $j \notin J$. Therefore we have $p_w \geqslant \tilde{\tilde{q}}$ everywhere, and $p_w = q$ in J, which implies $\tilde{\tilde{p}}_w = \tilde{\tilde{q}}$ for $|w|=1$. Thus in view of Theorem 3.6.1,

$$h_w \leqslant \tilde{\tilde{p}}_w \mathbin{\square} K = \tilde{\tilde{q}} \mathbin{\square} K \leqslant g \mathbin{\square} K \quad \text{as well as} \quad g \leqslant \tilde{\tilde{q}} \mathbin{\square} K = \tilde{\tilde{p}}_w \mathbin{\square} K \leqslant h_w \mathbin{\square} K.$$

This far the numbers n_j play no role; we shall now choose them to get the right kind of growth of $H(\cdot,w)$ for $|w|=e$. When $|w|=e$ and $j \in J$ we have $p_w(j) = q(j) - n_j \leqslant \tilde{\tilde{q}}(j) - n_j + \log 3$. The homogeneous part of degree $j \notin J$ in $H(z,w)$ can be estimated as

$$\begin{aligned} \|Q_j w^{n_j} + (w-1)Q_j^* w^{n_j}\| &\geqslant e^{n_j}(\|(w-1)Q_j^*\| - \|Q_j\|) \\ &\geqslant e^{n_j}(e-2)\|Q_j^*\| = \frac{e-2}{3}\exp(n_j - \tilde{\tilde{q}}(j)), \end{aligned}$$

which gives

$$\begin{aligned} p_w(j) &= q(j) - n_j \leqslant \tilde{\tilde{q}}(j) - n_j + \log 3 \leqslant \tilde{\tilde{q}}(j) - n_j + 2, && j \in J, \\ p_w(j) &\leqslant \tilde{\tilde{q}}(j) - n_j - \log\left(\tfrac{e-2}{3}\right) \leqslant \tilde{\tilde{q}}(j) - n_j + 2, && j \notin J. \end{aligned}$$

We shall now choose the integers n_j as follows. If $\tilde{\tilde{q}}(j) = +\infty$ (this can happen for finitely many numbers j only), then also $\tilde{\tilde{p}}(j) = +\infty$ and we choose $n_j = 0$. If $\tilde{\tilde{q}}(j) < +\infty$, we choose n_j as the smallest non-negative integer which is $\geqslant \tilde{\tilde{q}}(j) - \tilde{\tilde{p}}(j) + 2$. Thus in all cases $p_w \leqslant \tilde{\tilde{p}}$ for every w with $|w| = e$, so that $\tilde{p} \leqslant \tilde{p}_w$ and we get

$$f \leqslant \tilde{p} \,\square\, K \leqslant \tilde{p}_w \,\square\, K \leqslant h_w \,\square\, K, \qquad |w| = e.$$

Finally we have to make sure that H is holomorphic in all of Ω. To prove this it is enough to prove that

$$\|Q_j\| R^j r^{n_j} \to 0 \qquad \text{and} \qquad \|Q_j^*\| R^j r^{n_j} \to 0$$

as $j \to \infty$ for all R and all $r < e^{\rho'}$. This in turn follows if we can prove that

$$\frac{n_j \log r - q(j)}{j} \to -\infty \qquad \text{and} \qquad \frac{n_j \log r - q^*(j)}{j} \to -\infty. \tag{3.8.6}$$

We shall use the fact that $\operatorname{type}(\tilde{\tilde{q}} : \tilde{\tilde{p}}) = \operatorname{order}(f : g) \leqslant \rho$, which yields an inequality $\tilde{\tilde{q}} \leqslant a\tilde{\tilde{p}} + C_a$ for every $a > \rho$. If $n_j = 0$, the first expression in (3.8.6) is at most $-\tilde{\tilde{q}}(j)/j$ which certainly tends to $-\infty$. If $n_j > 0$, it can be estimated by (it suffices to consider $r > 1$)

$$\frac{n_j \log r - q(j)}{j} \leqslant \frac{(\tilde{\tilde{q}}(j) - \tilde{\tilde{p}}(j) + 3) \log r - \tilde{\tilde{q}}(j)}{j} \leqslant \frac{\tilde{\tilde{p}}(j)(a \log r - \log r - a) + O(1)}{j},$$

which tends to $-\infty$ if $\log r < a/(a-1)$.

If $n_j = 0$, the second expression in (3.8.6) is $-q^*(j)/j = -(\tilde{\tilde{q}}(j) - \log 3)/j$ which tends to $-\infty$; if $n_j > 0$, it can be estimated by

$$\begin{aligned}\frac{n_j \log r - q^*(j)}{j} &\leqslant \frac{(\tilde{\tilde{q}}(j) - \tilde{\tilde{p}}(j) + 3) \log r - \tilde{\tilde{q}}(j) + \log 3}{j} \\ &\leqslant \frac{\tilde{\tilde{p}}(j)(a \log r - \log r - a) + O(1)}{j},\end{aligned}$$

which tends to $-\infty$ as soon as $\log r < a/(a-1)$; here again a is any number greater than ρ. This proves that the series defining H converges locally uniformly in Ω and finishes the proof of (b′).

The proof that (a) implies (c′) is similar to that of Theorem 3.8.1. As in that proof we define H by (3.8.4):

$$H(z, w) = \sum_{j \in \mathbf{N}} P_j(z)(w/e)^{m_j},$$

where we shall choose integers m_j. Then obviously $H(z, e) = F(z)$. For $|w| = e$ we have $p_w(j) = p(j)$. This gives $\tilde{p}_w = \tilde{p}$ and therefore, for all w with $|w| = e$,

$$h_w \leqslant \tilde{p}_w \,\square\, K = \tilde{p} \,\square\, K \leqslant f \,\square\, K \quad \text{as well as} \quad f \leqslant \tilde{p} \,\square\, K = \tilde{p}_w \,\square\, K \leqslant h_w \,\square\, K.$$

For $|w| = 1$, on the other hand, we obtain

$$\|P_j(w/e)^{m_j}\| = \exp(-m_j - p(j)) \leqslant \exp(-m_j - \tilde{\tilde{p}}(j)).$$

Thus, when $|w| = 1$ we have $p_w(j) = p(j) + m_j \geqslant \tilde{\tilde{p}}(j) + m_j$. We now choose m_j so that $p_w \geqslant \tilde{\tilde{q}}$, which implies $\tilde{p}_w \leqslant \tilde{q}$ and yields the estimate

$$h_w \leqslant \tilde{p}_w \,\Box\, K \leqslant \tilde{q} \,\Box\, K \leqslant g \,\Box\, K.$$

To be explicit, if $\tilde{\tilde{q}}(j) = +\infty$, then $\tilde{\tilde{p}}(j) = +\infty$ and we take $m_j = 0$; if $\tilde{\tilde{q}}(j) < +\infty$, we take m_j as the smallest non-negative integer greater than or equal to $\tilde{\tilde{q}}(j) - \tilde{\tilde{p}}(j)$. This guarantees that $p_w \geqslant \tilde{\tilde{q}}$ and gives the estimate above. On the other hand, m_j is not too large, which will ensure that $\|P_j\| R^j (r/e)^{m_j}$ tends to zero for every R and every $r < e^{\rho'}$ and hence that H is holomorphic in Ω. The calculation is very similar to the one we just carried out in the case of (b′) and is omitted.

References

Abrahamsson, L.

[1988] Microlocal Lelong numbers of plurisubharmonic functions, *J. Reine Angew. Math.* **388**, 116–128.

Demailly, J.-P.

[1987] Nombres de Lelong généralisés, théorèmes d'intégralité et d'analyticité, *Acta Math.* **159**, 153–169.

[1989] *Potential Theory in Several Complex Variables*, lecture notes, École d'été d'analyse complexe, CIMPA, Nice, July 1989.

Freund, M. and Görlich, E.

[1985] On the relation between maximum modulus, maximum term, and Taylor coefficients of an entire function, *J. Approx. Theory* **43**, 194–203.

Halvarsson, S.

[forthc.] Extension of entire functions with controlled growth, to appear in *Math. Scand.*

[forthc.] Growth properties of entire functions depending on a parameter, Uppsala University 1993, manuscript 20pp.

Hayman, W.K.

[1968] Note on Hadamard's convexity theorem, in: *Entire Functions and Related Parts of Analysis*, Proceedings of Symposia in Pure Mathematics **11**, American Mathematical Society, Providence, RI; 210–213.

Hörmander, L.

[1990] *Complex Analysis in Several Variables*, North-Holland, Amsterdam.

[forthc.] *Notions of Convexity*, monograph to appear.

Juneja, O.P., Kapoor, G.P. and Bajpai, S.K.

[1976] On the (p,q)-order and lower (p,q)-order of an entire function, *J. Reine Angew. Math.* **282**, 53–67.

[1977] On the (p, q)-type and lower (p, q)-type of an entire function, *J. Reine Angew. Math.* **290**, 180–190.

Kiselman, C.O.

[1977] Fonctions delta-convexes, delta-sousharmoniques et delta-plurisousharmoniques, *Séminaire Pierre Lelong (Analyse) Année 1975/76* (P. Lelong, ed.), Lecture Notes in Math. **578**, Springer-Verlag, Berlin – Heidelberg – New York; 93–107.

[1978] The partial Legendre transformation for plurisubharmonic functions, *Invent. Math.* **49**, 137–148.

[1979] Densité des fonctions plurisousharmoniques, *Bull. Soc. Math. France* **107**, 295–304.

[1981] The growth of restrictions of plurisubharmonic functions, in: *Mathematical Analysis and Applications,* Part B (L. Nachbin, ed.), Advances in Mathematics Supplementary Studies, vol. 7B, 435–454.

[1983] The use of conjugate convex functions in complex analysis, in: *Complex Analysis* (J. Ławrynowicz and J. Siciak, eds.), Banach Center Publ. **11**, 131–142.

[1984] Croissance des fonctions plurisousharmoniques en dimension infinie, *Ann. Inst. Fourier (Grenoble)* **34**, 155–183.

[1986] *Konvekseco en kompleksa analitiko unu-dimensia*, lecture notes 1986:LN2, Uppsala University, Department of Mathematics.

[1987] Un nombre de Lelong raffiné, in: *Séminaire d'Analyse Complexe et Géométrie 1985–87*, Faculté des Sciences de Tunis & Faculté des Sciences et Techniques de Monastir; 61–70.

[1992] La teoremo de Siu por abstraktaj nombroj de Lelong, *Aktoj de Internacia Scienca Akademio Comenius* **1**, Beijing; 56–65.

[1993] Order and type as measures of growth for convex or entire functions, *Proc. London Math. Soc. (3)* **66**, 152–186.

Kjellberg, B.

[1974] The convexity theorem of Hadamard–Hayman, *Proceedings of the Symposium in Mathematics at the Royal Institute of Technology, June 1973*, The Royal Institute of Technology, Stockholm; 87–114.

Klimek, M.

[1991] *Pluripotential Theory*, Oxford University Press.

Lelong, P.

[1969] *Plurisubharmonic functions and positive differential forms*, Gordon and Breach, London.

[1985] Les objets souples de l'analyse complexe, *Exposition. Math.* **3**, 149–164.

Lelong, P. and Gruman, L.
[1986] *Entire Functions of Several Variables*, Springer-Verlag, Berlin – Heidelberg – New York.

Riihentaus, J.
[1989] On a theorem of Avanissian–Arsove, *Exposition. Math.* **7**, 69–72.

Rockafellar, R.T.
[1970] *Convex Analysis*, Princeton University Press.

Sato, D.
[1963] On the rate of growth of entire functions of fast growth, *Bull. Amer. Math. Soc.* **69**, 411–414.

Siu, Y.-T.
[1974] Analyticity of sets associated the Lelong numbers and the extension of closed positive currents, *Invent. Math.* **27**, 53-156.

Skoda, H.
[1972] Sous-ensemble analytiques d'ordre fini ou infini dans $\mathbf{C}^n$, *Bull. Soc. Math. France* **100**, 353–408.

Wang, X.
[1991] Analyticity theorems for parameter-dependent currents, *Math. Scand.* **69**, 179–198.

Wiegerinck, J.
[1988] Separately subharmonic functions need not be subharmonic, *Proc. Amer. Math. Soc.* **104**, 770–771.

Chebyshev-type quadratures: use of complex analysis and potential theory

Jacob KOREVAAR

Faculty of Mathematics and Computer Science
University of Amsterdam
Plantage Muidergracht 24
NL-1018TV Amsterdam
The Netherlands

Notes by
Arno B.J. KUIJLAARS

Abstract

These lectures are devoted to quadrature formulas with equal coefficients for surfaces in $\mathbf{R}^n$, $n \geq 1$ equipped with normalized area measure. Fundamental results of S.N. Bernstein for the interval $[-1, 1]$ are surveyed and extended. Applications are made to optimal formulas and to quadrature on domains of product type, notably the sphere. It is shown that on the sphere, good N-tuples of nodes for Chebyshev-type quadrature correspond to configurations of N equal point charges $1/N$, for which the electrostatic field is extremely small on compact subsets of the ball. Complex analysis and potential theory are used to support the conjecture that this field can be made as small as $\exp(-c\sqrt{N})$ in the case of $\mathbf{R}^3$. We finally present a logarithmic convexity theorem for supremum norms of harmonic functions. Many of the recent results represent joint work with J.L.H. Meyers. The notes include a dozen open problems.

Chapter 1
Chebyshev-type quadrature and the spherical Faraday cage

General references on Chebyshev-type quadrature: surveys by W. Gautschi [Ga76] and K.-J. Förster [F93].

1.1 Chebyshev-type quadrature

Let E be a compact set in $\mathbf{R}^d$ and let σ be a positive measure on E of total mass $\sigma(E) = 1$. For us, an important example is provided by the unit sphere $S = S(0, 1)$ in $\mathbf{R}^3$ with $\sigma = \lambda/4\pi$, normalized area measure. A *Chebyshev-type quadrature formula* for E and σ (of

P. M. Gauthier (ed.) and G. Sabidussi (techn. ed.), Complex Potential Theory, 325–364.

order N) is a numerical integration formula which gives the same weight to each of the N (not necessarily distinct) nodes $\zeta_1, \ldots, \zeta_N$:

$$\int_E f(x)d\sigma(x) \approx \frac{1}{N}\sum_{j=1}^{N} f(\zeta_j), \qquad \zeta_j \in E. \tag{1.1}$$

In other words, integrals are approximated by arithmetic means of function values. The study of such quadrature formulas was initiated by P.L. Chebyshev [Ch1874].

A "good" N-tuple

$$Z_N = (\zeta_1, \ldots, \zeta_N) \tag{1.2}$$

of Chebyshev nodes would be one for which formula (1.1) is *exact* for all polynomials $f(x) = f(x_1, \ldots, x_d)$ to relatively high degree $p = p_N$. (In this case we speak of formulas "of polynomial degree p".) Alternatively, the quadrature formula should have a very small remainder

$$R(f, Z_N) = \int_E f(x)d\sigma(x) - \frac{1}{N}\sum_{j=1}^{N} f(\zeta_j) \tag{1.3}$$

for such polynomials.

Example 1.1.1 Let E be $C(0,1)$, the unit circle in $\mathbf{R}^2 \simeq \mathbf{C}$. Using the parametrization $z = x + iy = e^{i\theta}$, $0 \le \theta < 2\pi$ we take $d\sigma = d\theta/2\pi$. Letting $\zeta_j = e^{i\theta_j} = \cos\theta_j + i\sin\theta_j$, $1 \le j \le N$, run over the Nth roots of unity, one finds that

$$\begin{aligned}\int_{C(0,1)} z^k d\sigma &= \frac{1}{2\pi}\int_0^{2\pi} e^{ik\theta}d\theta = \frac{1}{2\pi}\int_0^{2\pi}(\cos k\theta + i\sin k\theta)d\theta \\ &= \frac{1}{N}\sum_{j=1}^{N}\zeta_j^k = \frac{1}{N}\sum_{j=1}^{N}(\cos k\theta_j + i\sin k\theta_j) \qquad \text{for } k = 0, 1, \ldots, N-1.\end{aligned}$$

It follows that the corresponding Chebyshev-type quadrature formula

$$\int_{C(0,1)} f(x,y)d\sigma = \frac{1}{2\pi}\int_0^{2\pi} f(\cos\theta, \sin\theta)d\theta \approx \frac{1}{N}\sum_{j=1}^{N} f(\cos\theta_j, \sin\theta_j)$$

is exact for all polynomials $f(x,y)$ of degree $\le p = N-1$. Indeed, for such polynomials, $f(\cos\theta, \sin\theta)$ can be written as a trigonometric polynomial of order $\le N-1$.

One would like to obtain an equally nice result for the unit sphere S in $\mathbf{R}^3$!

The best known quadrature formula for $E = [-1,1]$ is the *classical m-point Gauss formula* for the measure $d\sigma(x) = \frac{1}{2}dx$, cf. [Sz75]:

$$\int_{-1}^{1} f(x)\,\tfrac{1}{2}\,dx \approx \sum_{k=1}^{m} \lambda_k f(\alpha_k). \tag{1.4}$$

Here the nodes $\alpha_1 > \cdots > \alpha_m$ are the zeros of the Legendre polynomial $P_m(x)$. The coefficients or Cotes-Christoffel numbers λ_k are given by

$$\lambda_k = \frac{1}{(1-\alpha_k^2)P_m'(\alpha_k)^2}. \tag{1.5}$$

A very rough estimate would be

$$\lambda_k \approx \tfrac{1}{4}(\alpha_{k-1} - \alpha_{k+1}) \approx (\pi/2m)\sqrt{1-\alpha_k^2}.$$

For $m \geq 3$ formula (1.4) is not a Chebyshev-type formula. However, the Gauss formula for the measure $d\sigma(x) = dx/\pi\sqrt{1-x^2}$ on $[-1,1]$ does provide a quadrature formula with equal coefficients:

$$\int_{-1}^{1} f(x)\frac{dx}{\pi\sqrt{1-x^2}} \approx \frac{1}{m}\sum_{k=1}^{m} f(\xi_k). \tag{1.6}$$

Here the nodes ξ_k are the zeros $\cos((2k-1)\pi/2m)$ of the Chebyshev polynomial $T_m(x) = \cos(m \arccos x)$. The Gauss formulas for $[-1,1]$ are associated with orthogonal polynomials and the m-point formulas are exact to degree $2m-1$. Formula (1.6) occurs in a paper by F.G. Mehler of 1864, cf. [Sz75, Sections 3.4 and 15.3].

A hundred years later, J.L. Ullman [U66] made the surprising discovery that there are other measures on $[-1,1]$ which admit mth order Chebyshev-type formulas of polynomial degree $\geq m$ for every m. References concerning further work in this direction may be found in [F93].

1.2 Minimal number of nodes for exactness to degree p

We are mostly interested in Chebyshev-type quadrature for simple surfaces E (such as the sphere S) and ordinary (normalized) surface measure σ. By the fundamental work of P.D. Seymour and T. Zaslavsky [SeZ84] for arcwise connected sets, there exist Chebyshev-type quadrature formulas for E and σ of every degree p. A central question will be: What is the (order of the) *minimal number* $N = N_E(p)$ of nodes for which one can have Chebyshev-type quadrature (1.1) of polynomial degree p? S.N. Bernstein considered and solved this problem for the interval $[-1,1]$ with $d\sigma = \frac{1}{2}dx$. His results may be formulated as follows; they are basic in our subsequent work. Proofs will be discussed in Sections 2.1 and 3.1.

Theorem A [Be37a, Be38] *Suppose that the nodes $x_1, \ldots, x_N$ on $[-1,1]$ are such that*

$$\int_{-1}^{1} f(x)\,\tfrac{1}{2}\,dx = \frac{1}{N}\sum_{j=1}^{N} f(x_j) \tag{1.7}$$

for all polynomials $f(x)$ of degree $\leq p = 2m-1$. Then $N > \frac{1}{4}m^2$ so that $p < 4\sqrt{N}$.

Theorem B [Be37b] *Let m be a positive integer and let M be any even integer $\geq N_0(m) = 2\left[2\sqrt{2}(m+1)(m+4)+1\right]$ where $[\ ]$ denotes the integral part. Then there exist points*

$t_k \in (-1,1)$, $t_1 > t_2 > \cdots > t_{2m-1}$, $t_{2m-k} = -t_k$ and positive integers $M_k = M_{2m-k}$ with $\sum_{k=1}^{2m-1} M_k = M$ such that

$$\int_{-1}^{1} f(x)\,\tfrac{1}{2}\,dx = \frac{1}{M}\sum_{k=1}^{2m-1} M_k f(t_k) \tag{1.8}$$

for all polynomials $f(x)$ of degree $\leq 2m-1$.

Formula (1.8) is a symmetric Chebyshev-type formula of order M in which the node t_k occurs with multiplicity M_k. The appearance of multiple nodes is inherent in the method of proof, cf. Chapter 3 below. However, it has recently been shown by A.B.J. Kuijlaars [Ku93a] that the multiple nodes in (1.8) can be split into simple nodes without losing the polynomial exactness to degree $2m-1$. Nevertheless the occurrence of multiple nodes seems to be natural here. In fact, using complex analysis we will show that there is massive coalescence of nodes in certain optimal Chebyshev-type formulas (1.7), see Chapter 2.

By Theorems A and B with $p = 2m-1$ and $M = N_0(m)$, the number $N_I(p)$ for the interval $I = [-1,1]$ with $d\sigma = \frac{1}{2}dx$ has order p^2. For the unit circle $C = C(0,1)$ it is not difficult to prove that $N_C(p) = p+1$, cf. Example 1.1.1 and [DGS77, KM93b]. Combination of these results provides information about $N_E(p)$ for sets E that can be represented as Cartesian products of intervals and/or circles by suitable parametrization, see Chapter 3.

Considering the unit sphere S in $\mathbf{R}^3$ with $\sigma = \lambda/4\pi$ as a product $I \times C$, it is found that

$$c_1 p^2 \leq N_S(p) \leq c_2 p^3 \qquad (c_1, c_2 > 0). \tag{1.9}$$

It is an important open problem to determine the precise order of $N_S(p)$ for large p. In subsequent chapters we will present support for the following conjecture.

Conjecture 1.2.1 $N_S(p)$ is of order p^2 as $p \to \infty$.

In combinatorics, J.J. Seidel et al. have written several papers on so-called *spherical p-designs*, see [DGS77] and [Se93]. These are configurations of N *distinct* points $\zeta_j \in S$ for which formula (1.1) with $E = S$ has polynomial degree (at least) p. Section 3.3 exhibits spherical p-designs consisting of $N = \mathcal{O}(p^3)$ points, improving an earlier bound by P. Rabau and B. Bajnok [RB91].

1.3 An electron problem for the sphere and Chebyshev-type quadrature

The classical *Faraday cage phenomenon* of electrostatics may be described as follows. Let E be a bounded hollow conductor in $\mathbf{R}^3$ with smooth outer boundary. Consider a positive charge distribution on E, of total charge 1 (say), in the most stable equilibrium. That is, the potential energy should be minimal. Then there is no (measurable) electrostatic field inside E. The electrostatic potential is constant on E and throughout its interior. Assuming *continuous* charge distribution, classical potential theory (culminating in the work of O. Frostman [Fr35]) provides a very satisfactory explanation.

Question Can one explain the Faraday cage phenomenon on the basis of a model in which the charge 1 on E is made up of N equal *point charges* $1/N$, where N is large?

Although the case of a spherical conductor may not be typical – in the plane, circular conductors are much better than others, cf. [KG71, K74, KK83] – we limit ourselves to the sphere S. The spherical problem is of independent interest and it is the most amenable to calculations. Leaving aside the question of minimal potential energy for the time being, we address the following

"Electron Problem" Let $\mathcal{E}(x) = \mathcal{E}(x, Z_N)$ denote the electrostatic field at x, due to charges $1/N$ at the points ζ_j of an N-tuple Z_N. How small can one make $\sup|\mathcal{E}(x)|$ on the balls $B(0,r)$ with $r < 1$ by appropriate choice of the N-tuple Z_N on S?

A small electrostatic field $\mathcal{E}(x)$ on the balls $B(0,r)$ is equivalent to nearly constant potential

$$U(x) = U(x, Z_N) = \frac{1}{N}\sum_{j=1}^{N}\frac{1}{|\zeta_j - x|} \tag{1.10}$$

on such balls. Indeed, one can go from $\mathcal{E}(x) = -\operatorname{grad} U(x)$ on the ball $B(0,r)$ to $U(y) - U(0) = U(y) - 1$ on the sphere $S(0,r)$ by integration. Conversely, one can go from $U(y) - 1$ on $S(0,r)$ to $\mathcal{E}(x)$ inside that sphere by differentiation of the Poisson integral for $U(x) - 1$ on the ball $B(0,r)$.

Observing that $U(x) - 1$ can be interpreted as a Chebyshev-type quadrature remainder:

$$\int_S \frac{1}{|\zeta - x|}d\sigma(\zeta) - \frac{1}{N}\sum_{j=1}^{N}\frac{1}{|\zeta_j - x|} = 1 - U(x), \qquad (|x| < 1)$$

it is possible to deduce the following

Equivalence Principle Charges $1/N$ at the points ζ_j of Z_N give a nearly constant potential or a small electrostatic field $\mathcal{E}(x, Z_N)$ on a ball $B(0,r)$, $0 < r < 1$ if and only if Z_N forms a good N-tuple of nodes for Chebyshev-type quadrature on S.

For later use we shall prove the following precise form of the principle. Note that we always use normalized area measure $\sigma = \lambda/4\pi$ on S.

Theorem E [KM93c] *Let Z_N be a fixed N-tuple of points on S.*

(i) *If for some $r \in (0,1)$ and positive δ and A,*

$$\sup_{|x|=r} |U(x, Z_N) - 1| \le \delta r^A, \tag{1.11}$$

then for every polynomial $f(x,y,z)$ of degree $\le p \le A$,

$$|R(f, Z_N)| \le (2p+1)\delta r^{A-p} \sup_S |f| \le (2A+1)\delta \sup_S |f|. \tag{1.12}$$

(ii) *If for* $\epsilon \geq 0$ *and all polynomials* f *of degree* $\leq p$

$$|R(f, Z_N)| \leq \epsilon \sup_S |f|, \tag{1.13}$$

then for every $r \in (0,1)$,

$$\sup_{|x|=r} |U(x, Z_N) - 1| \leq (\epsilon + r^p)\frac{r}{1-r}. \tag{1.14}$$

Corollary 1.3.1 *The Chebyshev-type quadrature formula for* S *corresponding to the* N*-tuple* Z_N *is polynomially exact to degree* p *if and only if*

$$U(x, Z_N) = 1 + \mathcal{O}(|x|^{p+1}) \qquad \text{for } 0 \leq |x| \leq r_0 < 1. \tag{1.15}$$

Proof Apply Theorem E with $A = p + 1$ and $\epsilon = 0$, letting $r \to 0$ in (1.12). □

The results in Chapters 3 and 4 imply that for "optimal" N-tuples Z_N with $N \to \infty$, (1.15) holds with

$$c_1 N^{1/3} \leq p \leq c_2 N^{1/2}, \qquad c_1, c_2 > 0.$$

If $N_S(p)$ is indeed of order p^2 (as conjectured in Section 1.2), then p will be of order $N^{1/2}$ for optimal Z_N.

Corollary 1.3.2 *The following statements involving a constant* $\alpha > 0$ *and a family of* N*-tuples* Z_N *on* S *with* $N \to \infty$ *are equivalent:*

(i) *For some (or every)* $r \in (0,1)$ *there is a constant* $c_3 = c_3(r) > 0$ *such that*

$$\sup_{|x|=r} |U(x, Z_N) - 1| = \mathcal{O}\{\exp(-c_3 N^\alpha)\}, \qquad (N \to \infty); \tag{1.16}$$

(ii) *There are positive constants* c_4, c_5 *such that*

$$R(f, Z_N) = \mathcal{O}\{\exp(-c_4 N^\alpha)\} \sup_S |f| \tag{1.17}$$

uniformly for the class of polynomials f *of degree* $\leq c_5 N^\alpha$.

Proof Apply Theorem E with $A \log r = -c_3 N^\alpha$ and $p = \frac{1}{2}A$ in part (i), and with $\epsilon = \exp(-c_4 N^\alpha)$ in part (ii). □

In Chapter 4 we will use "Several complex variables" to attack the Electron Problem. This method supports the conjecture that for optimal Z_N, (1.16) and (1.17) are true with $\alpha = \frac{1}{2}$.

1.4 Spherical harmonics and the proof of Theorem E

For Theorem E and for later applications we recall some facts about spherical harmonics in $\mathbf{R}^3$ that can be found in [SW71] and other books.

A *spherical harmonic* $Y_k(\xi)$ of order k is the restriction to S of a homogeneous harmonic polynomial $h_k(x) = h_k(x_1, x_2, x_3)$ of degree k:

$$Y_k(\xi) = h_k(\xi) = h_k(x)/r^k, \qquad x = r\xi, \quad r > 0, \quad \xi \in S.$$

The spherical harmonics of order k form a rotation invariant linear subspace H_k of $L^2(S)$ of dimension $2k+1$. Let us parametrize S by setting

$$\xi_1 = \sin\theta\cos\phi, \quad \xi_2 = \sin\theta\sin\phi, \quad \xi_3 = \cos\theta, \qquad 0 \le \theta \le \pi, \quad 0 \le \phi < 2\pi,$$

so that $d\sigma = d\lambda/4\pi = \sin\theta\, d\theta d\phi/4\pi$. Then a standard orthonormal basis of H_k is given by the functions

$$Y_{k,s}(\xi) = \left\{(2k+1)\frac{(k-|s|)!}{(k+|s|)!}\right\}^{\frac{1}{2}} (\sin\theta)^{|s|}\left(D^{|s|}P_k\right)(\cos\theta)e^{is\phi}, \quad -k \le s \le k, \tag{1.18}$$

where $P_k(t)$ is the Legendre polynomial of degree k.

Other important spherical harmonics of order k are the so-called zonal harmonics

$$Y_k(\xi) = P_k(\zeta \cdot \xi), \qquad \xi \in S$$

where ζ is a fixed unit vector [and the dot denotes the usual inner product].

The subspaces $H_0, H_1, H_2, \ldots$ are pairwise orthogonal and their direct sum is $L^2(S)$. The orthogonal projection of a function $f(\xi)$ in $L^2(S)$ on H_k has the useful representation

$$Y_k(\xi) = Y_k(f, \xi) = (2k+1)\int_S f(\eta)P_k(\xi\cdot\eta)d\sigma(\eta). \tag{1.19}$$

The orthogonal decomposition $f(\xi) = \sum_{k=0}^{\infty} Y_k(f,\xi)$ is called the *Laplace series* for $f(\xi)$. Observe that one has a Parseval formula

$$\|f\|^2_{L^2(S)} = \sum_{k=0}^{\infty} \|Y_k\|^2_{L^2(S)}.$$

If $f(\xi)$ is the restriction to S of a polynomial $f(x_1, x_2, x_3)$ of degree $\le q$, the Laplace series terminates – it will have no terms of order $> q$:

$$f(\xi) = \sum_{k=0}^{q} Y_k(f, \xi). \tag{1.20}$$

Example 1.4.1 Using the generating function for the Legendre polynomials, we obtain the following Laplace series for the potential $U(r\xi, Z_N)$ of (1.10):

$$\begin{aligned} U(r\xi, Z_N) &= \frac{1}{N}\sum_{j=1}^{N}\frac{1}{|\zeta_j - r\xi|} = \frac{1}{N}\sum_{j=1}^{N}\frac{1}{(1+r^2-2r\xi\cdot\zeta_j)^{1/2}} \\ &= \frac{1}{N}\sum_{j=1}^{N}\sum_{k=0}^{\infty} r^k P_k(\xi\cdot\zeta_j) = 1 + \sum_{k=1}^{\infty} r^k Q_k(\xi), \qquad \xi \in S, \quad 0 \le r < 1. \end{aligned} \tag{1.21}$$

Here $Q_k(\xi)$ may be expressed as a quadrature remainder, cf. (1.3):

$$Q_k(\xi) = \frac{1}{N}\sum_{j=1}^{N} P_k(\xi \cdot \zeta_j) = R(-P_k(\xi \cdot \zeta), Z_N). \tag{1.22}$$

Indeed, for $k \geq 1$, $\int_S P_k(\xi \cdot \zeta) d\sigma(\zeta) = 0$.

Proof of Theorem E, part (i) For $x \in S(0, r)$ we set $x = r\xi$, $\xi \in S$. We now apply Parseval's formula to the orthogonal series for $U(r\xi) - 1$ obtained from (1.21). Using the hypothesis (1.11) (we only need a weaker L^2 form), we find that

$$\begin{aligned}\sum_{k=1}^{\infty} r^{2k}\|Q_k\|^2_{L^2(S)} &= \|U(r\xi, Z_N) - 1\|^2_{L^2(S)} \qquad (1.23)\\ &= \int_S |U(r\xi - 1|^2 d\sigma(\xi) \leq \delta^2 r^{2A}.\end{aligned}$$

Next, let f be any polynomial in three variables of degree $\leq p$. We decompose the restriction $f(\xi)$ of f to S into spherical harmonics as in (1.20):

$$f(\xi) = \sum_{k=0}^{p} Y_k(\xi), \qquad \sum_{k=0}^{p} \|Y_k\|^2_{L^2(S)} = \|f\|^2_{L^2(S)}. \tag{1.24}$$

Here Y_0 will be equal to the average of f over S. Hence by (1.3),

$$R(f, Z_N) = \int_S f(\xi) d\sigma(\xi) - \frac{1}{N}\sum_{j=1}^{N} f(\zeta_j) = -\sum_{k=1}^{p}\frac{1}{N}\sum_{j=1}^{N} Y_k(\zeta_j).$$

Representing $Y_k(\zeta_j)$ by formula (1.19) with $f = Y_k$, we obtain

$$\begin{aligned}-R(f, Z_N) &= \sum_{k=1}^{p}(2k+1)\int_S Y_k(\xi)\frac{1}{N}\sum_{j=1}^{N} P_k(\xi \cdot \zeta_j) d\sigma(\xi)\\ &= \int_S \sum_{k=1}^{p}\left((2k+1)r^{-k}Y_k(\xi) \cdot r^k Q_k(\xi)\right) d\sigma(\xi). \qquad (1.25)\end{aligned}$$

Finally applying Cauchy-Schwarz to the above inner product $(\phi, \psi) = \int_S \sum_{k=1}^{p} \phi_k \overline{\psi_k} d\sigma$, we conclude from (1.23)–(1.25) that

$$\begin{aligned}|R(f, Z_N)|^2 &\leq \int_S \sum_{k=1}^{p}(2k+1)^2 r^{-2k}|Y_k(\xi)|^2 d\sigma(\xi) \int_S \sum_{k=1}^{p} r^{2k}|Q_k(\xi)|^2 d\sigma(\xi)\\ &\leq (2p+1)^2 r^{-2p}\sum_{k=1}^{p}\|Y_k\|^2_{L^2(S)} \cdot \delta^2 r^{2A} \leq (2p+1)^2\delta^2 r^{2A-2p}\|f\|^2_{L^2(S)}\end{aligned}$$

which implies the desired result (1.12).

Proof of Theorem E, part (ii) Setting $x = r\xi$ with $r \in (0,1)$, we derive from formulas (1.21) and (1.22) that

$$U(r\xi, Z_N) - 1 = \sum_{k=1}^{\infty} r^k R(-P_k(\xi \cdot \zeta), Z_N). \tag{1.26}$$

Since $\sup |P_k(t)| = 1$ for $-1 \le t \le 1$, definition (1.3) and the hypothesis (1.13) imply that

$$|R(-P_k(\xi \cdot \zeta)| \le \begin{cases} 1 & \text{for } k \ge 1, \\ \epsilon & \text{for } k \le p. \end{cases}$$

Thus formula (1.26) gives the estimate

$$|U(r\xi, Z_N) - 1| = \left| \sum_{k=1}^{p} + \sum_{k=p+1}^{\infty} \right| \le \sum_{k=1}^{p} r^k \epsilon + \sum_{k=p+1}^{\infty} r^k \le \frac{\epsilon r}{1-r} + \frac{r^{p+1}}{1-r}$$

which proves (1.14). □

Chapter 2
Chebyshev-type quadrature on $[-1,1]$. Massive coalescence of nodes in optimal formulas

As we saw in Chapter 1, Bernstein's Theorems A and B lead to the following conclusion for Chebyshev-type quadrature on $[-1,1]$ with $d\sigma = \frac{1}{2}\,dx$. For polynomial exactness to degree p, the number of nodes N must be at least of order p^2 and this order works. In Bernstein's formula (1.8) only $p \le c\sqrt{N}$ of the nodes are distinct. After proving Theorem A and a supplement, we will use complex analysis to show that in certain natural optimal formulas, also at most $c\sqrt{N}$ of the nodes are distinct.

2.1 Proof of Bernstein's Theorem A

We refer to Section 1.2 for the precise statement of Theorem A. Bernstein discovered an interesting relation between Chebyshev-type quadrature as in (2.1) below and the Gauss formula (1.4). His basic observation was the following, cf. also Gautschi's survey [Ga76].

Theorem 2.1.1 *Suppose that the real nodes $x_1 \ge x_2 \ge \cdots \ge x_N$ are such that the corresponding Chebyshev-type quadrature formula*

$$\int_{-1}^{1} f(x)\,\tfrac{1}{2}\,dx \approx \frac{1}{N}\sum_{j=1}^{N} f(x_j) \tag{2.1}$$

is exact for all polynomials $f(x)$ of degree $\le 2m-1$. Then $x_1 \ge \alpha_1(m)$ and

$$N \ge \frac{1}{\lambda_1(m)}, \tag{2.2}$$

where $\alpha_1(m)$ is the largest zero of the Legendre polynomial $P_m(x)$ and $\lambda_1(m)$ is the corresponding Christoffel number in the Gauss formula (1.4).

Proof Following Bernstein's beautiful method, we define polynomials F_1 and F_2, both of degree $\leq 2m-1$, by

$$F_1(x) = \frac{P_m(x)^2}{(x-\alpha_1)^2}, \qquad F_2(x) = (x-\alpha_1)F_1(x) \tag{2.3}$$

where $\alpha_1 = \alpha_1(m)$. Under the hypothesis of the Theorem and by the Gauss formula

$$\frac{1}{N}\sum_{j=1}^{N} F_1(x_j) = \int_{-1}^{1} F_1(x)\,\tfrac{1}{2}\,dx = \lambda_1 F_1(\alpha_1) > 0, \tag{2.4}$$

$$\frac{1}{N}\sum_{j=1}^{N} F_2(x_j) = \int_{-1}^{1} F_2(x)\,\tfrac{1}{2}\,dx = 0. \tag{2.5}$$

By (2.5) there are two possibilities: either (i) all nodes x_j belong to the zero set $Z(F_2) = \{\alpha_1, \alpha_2, \ldots, \alpha_m\}$, or (ii) they do not. By (2.4) not all x_j belong to the zero set $Z(F_1) = \{\alpha_2, \ldots, \alpha_m\}$. Thus in case (i) at least one x_j must be equal to α_1. In particular then $x_1 = \max x_j = \alpha_1$. In case (ii), relation (2.5) shows that there is at least one positive value $F_2(x_j)$ since the nonzero values add up to zero. Such an x_j must be $> \alpha_1$ since $F_2 > 0$ only for $x > \alpha_1$. Hence in case (ii), $x_1 = \max x_j > \alpha_1$.

Thus in any case $x_1 \geq \alpha_1$. Since $F_1(x)$ is ≥ 0 everywhere and monotone increasing for $x > \alpha_2$, it follows from (2.4) that

$$N\lambda_1 F_1(\alpha_1) = \sum_{j=1}^{N} F_1(x_j) \geq F_1(x_1) \geq F_1(\alpha_1). \qquad \square$$

After establishing Theorem 2.1.1, Bernstein derived estimates for $\lambda_1(m)$. In [Be37a] he found $\lambda_1(m) < 4m^{-2}$ which leads to Theorem A.

Using an appropriate transformation of the Legendre differential equation, the author [K92] obtained the following monotonicity result:

$$m(m+1)\lambda_k(m) \nearrow J_0'(j_k)^{-2} \quad \text{for } (k \leq)\ m \nearrow \infty, \tag{2.6}$$

where j_k is the kth positive zero of the Bessel function $J_0(t)$. Hence in particular

$$m(m+1)\lambda_1(m) < J_0'(j_1)^{-2} \approx 3.71.$$

Corollary 2.1.2 *If formula* (2.1) *is polynomially exact to degree $p = 2m-1$, then*

$$N > J_0'(j_1)^2 m^2 \approx .269m^2, \qquad p < \frac{2}{|J_0'(j_1)|}\sqrt{N} \approx 3.85\sqrt{N}, \tag{2.7}$$

which gives a little more than Theorem A. (The final inequality was obtained earlier by L. Gatteschi and G. Vinardi [GV78] by a different method.)

Remark 2.1.3 W. Gautschi and others have observed that Theorem 2.1.1 has an immediate extension to Chebyshev-type quadrature for ***arbitrary probability measures*** σ on $[-1,1]$. In this case one has to replace the Legendre polynomials by orthogonal polynomials associated with σ and $\lambda_1(m)$ by the first (or last) Christoffel number in the corresponding Gauss formula, cf. the survey [Ga76]. For example, if $d\sigma(x) = (2/\pi)(1-x^2)^{1/2}dx$ one obtains the Chebyshev polynomials $U_m(x)$ of the second kind for which, cf. [Sz75, formula (15.3.4)],

$$\lambda_1(m) = 2(m+1)^{-1}\sin^2\pi/(m+1) < 2\pi^2(m+1)^{-3}. \tag{2.8}$$

Thus Chebyshev-type quadrature of degree p with respect to $(2/\pi)(1-x^2)^{1/2}dx$ requires at least cp^3 nodes. (This order works, see A.B.J. Kuijlaars [Ku93a] and cf. Example 3.2.2.)

2.2 A supplement to Theorem A

For N-tuples $X_N = (x_1, \ldots, x_N)$ of nodes in $[-1,1]$, let

$$R(f, X_N) = \int_{-1}^{1} f(x)\tfrac{1}{2}\,dx - \frac{1}{N}\sum_{j=1}^{N} f(x_j).$$

Theorem A implies that for every N-tuple X_N there is a polynomial p of degree approximately $4\sqrt{N}$ such that $|R(f, X_N)| > 0$. The author and Meyers [KM93a] have refined Bernstein's proof to obtain

Theorem 2.2.1 *For every N and every N-tuple $X_N = (x_1, x_2, \ldots, x_N)$ of real nodes, there is a polynomial f of degree $q < 4\sqrt{N}$ with $\sup_{|x|\le 1}|f| = 1$ for which*

$$|R(f, X_N)| > \frac{1}{100N}. \tag{2.9}$$

In fact, taking $m = \{2\sqrt{N}\}$, the integer closest to $2\sqrt{N}$, they proved that (2.9) is always satisfied for at least one of the polynomials

$$f_1(x) = \frac{F_1(x)}{F_1(1)} = b_1\frac{P_m(x)^2}{P_m'(\alpha_1)^2(x-\alpha_1)^2}, \quad b_1 = b_1(m), \tag{2.10}$$

$$f_2(x) = \frac{F_2(x)}{F_2(1)} = \frac{1-\alpha_1}{x-\alpha_1}P_m(x)^2 = \frac{x-\alpha_1}{1-\alpha_1}f_1(x), \tag{2.11}$$

where F_1 and F_2 are given by (2.3). Observe that $4\sqrt{N} - 3 < 2m - 2 = \deg f_1 < \deg f_2 = 2m - 1 < 4\sqrt{N}$.

The proof is computational and utilizes the following monotonicity results (we refer to [KM93a] for the details):

$$\frac{1-\alpha_1}{1+\alpha_1}m^2 \nearrow \frac{1}{4}j_1^2, \qquad \frac{1-\alpha_1}{1+\alpha_1}m(m+1) \searrow \frac{1}{4}j_1^2, \tag{2.12}$$

$$b_1 = b_1(m) = (1-\alpha_1)^2 P_m'(\alpha_1)^2 = \frac{1-\alpha_1}{1+\alpha_1}\frac{1}{\lambda_1} \searrow \frac{1}{4} j_1^2 J_0'(j_1)^2. \tag{2.13}$$

The first relations follow from Sturm theory applied to solutions of appropriate transformations of the Legendre differential equation. For (2.13) one combines (2.12) and (2.6).

It is not known if the order of the right hand side in (2.9) is sharp.

2.3 Minimum norm Chebyshev-type quadrature

Since Nth order formulas (2.1) can not be polynomially exact to degree $p \geq 4\sqrt{N}$, various authors have proposed to use such formulas with nodes that minimize quadratic expressions involving a number of nonzero remainders

$$\Delta_k = \Delta_k(x_1, \ldots, x_N) = R(x^k, X_N) = \int_{-1}^{1} x^k \tfrac{1}{2}\, dx - \frac{1}{N}\sum_{j=1}^{N} x_j^k, \tag{2.14}$$

cf. the survey [Ga76]. Using linear algebra, W. Gautschi and H. Yanagiwara [GaY74] have shown that for such choices, there will in general be coincident nodes. The author and Meyers [KM93a] have used complex analysis to prove the following result which implies massive coalescence of nodes for large N.

Theorem 2.3.1 *For $0 < r < 1$ and $4\sqrt{N} \leq p = p_N \leq \infty$, let the N-tuple X_N of nodes $x_j \in [-1,1]$ be chosen so as to minimize the function*

$$G_r(x_1, \ldots, x_N) = \sum_{k=1}^{p} \Delta_k^2 r^{2k}. \tag{2.15}$$

Then the number of distinct nodes is $\leq C(r)\sqrt{N}$ with a constant $C(r)$ independent of N and p. Similarly, if the nodes minimize G_r with $r = 1$ and if $0 < \lambda < 1$, then $\leq c(\lambda)\sqrt{N}$ of the approximately λN nodes on $[-\lambda, \lambda]$ are distinct.

The proof depends on Theorem 2.2.1 and some simple lemmas.

Lemma 2.3.2 *Let $\phi(z)$ be a bounded analytic function on the disc $B(0,R)$ in* **C** *which has s zeros $z_1, \ldots, z_s$ on the smaller disc $\bar{B}(0,\rho)$. Then for $|z| \leq \sigma < R$,*

$$|\phi(z)| \leq \sup|\phi| \cdot e^{-bs} \qquad \text{with} \quad b = \log\frac{R^2 + \rho\sigma}{(\rho+\sigma)R} > 0. \tag{2.16}$$

Proof Divide out a Blaschke-type product:

$$\phi(z) = B(z)\psi(z) = \prod_{j=1}^{s} \frac{(z - z_j)R}{R^2 - \bar{z}_j z}\psi(z), \qquad \sup_{B(0,R)} |\psi| = \sup_{B(0,R)} |\phi|.$$

Next estimate each factor of $B(z)$ on the disc $\bar{B}(0,\sigma)$. □

Lemma 2.3.3 (S.N. Bernstein [Be12]) *Let $f(x)$ be a polynomial of degree q such that $|f(x)| \leq 1$ for $x \in [-1,1]$. Then for any $\tau > 1$ and for $z = x + iy$ on or inside the ellipse*

$$E_\tau : \quad \frac{x^2}{\frac{1}{4}(\tau + 1/\tau)^2} + \frac{y^2}{\frac{1}{4}(\tau - 1/\tau)^2} = 1, \tag{2.17}$$

one has $|f(z)| \leq \tau^q$.

Proof Introduce the 1-1 conformal map $z = g(w) = \frac{1}{2}(w + 1/w)$ from the punctured disc $0 < |w| < 1$ onto the exterior of the interval $[-1,1]$ in the z-plane. Under this map, the circle $|w| = 1/\tau$ corresponds to the ellipse E_τ. One now applies the maximum principle to the polynomial $P(w) = w^q f\{g(w)\}$ to conclude that $|P(w)| \leq 1$ for $|w| \leq 1$. Hence $|f\{g(w)\}| \leq \tau^q$ for $|w| = 1/\tau$ so that $|f(z)| \leq \tau^q$ for $z \in E_\tau$, and of course also for z inside E_τ. □

Proof of Theorem 2.3.1 We will prove the second part here (the first part is similar). Accordingly, we take $r = 1$ and choose the N-tuple X_N in $[-1,1]^N$ such that G_1 is minimal. (G_1 has a finite minimum which is attained for nodes inside $(-1,1)$.) Fixing $\lambda \in (0,1)$, we suppose that X_N contains precisely s distinct nodes x_j on $[-\lambda, \lambda]$. For these x_j's, the partial derivatives $\partial G_1/\partial x_j$ exist and must vanish on X_N:

$$\frac{\partial G_1}{\partial x_j} = \sum_{k=1}^{p} 2\Delta_k \frac{\partial \Delta_k}{\partial x_j} = -\frac{2}{N} \sum_{k=1}^{p} \Delta_k k x_j^{k-1} = 0.$$

Thus the auxiliary analytic function

$$\phi(z) = \sum_{k=1}^{p} \Delta_k k z^{k-1} \lambda^k, \qquad |z| < \lambda^{-1/2} \tag{2.18}$$

with $\sup|\phi| \leq A = 2\lambda(1 - \sqrt{\lambda})^{-2}$ has at least s distinct zeros $z_j = x_j/\lambda$ on $\bar{B}(0,1)$. Estimating $|\phi|$ on $\bar{B}(0,1)$ by (2.16), forming $\Phi(z) = \int_0^z \phi$ and using Parseval's theorem, we conclude that

$$\begin{aligned} \left(\sum_{k=1}^{p} \Delta_k^2 \lambda^{2k}\right)^{1/2} &= \left(\frac{1}{2\pi}\int_0^{2\pi} |\Phi(e^{i\theta})|^2 d\theta\right)^{1/2} \leq \sup_{\bar{B}(0,1)} |\Phi| \\ &\leq \sup_{\bar{B}(0,1)} |\phi| \leq A e^{-bs} \quad \text{with } b = \log\{(1+\lambda)/2\sqrt{\lambda}\}. \end{aligned} \tag{2.19}$$

For polynomials $f(x) = \sum_0^q c_k x^k$ with $\sup_{-1 \leq x \leq 1} |f| = 1$, we now estimate the quadrature remainder by Cauchy-Schwarz:

$$|R(f, X_N)| = \left|\sum_{k=1}^{q} c_k \Delta_k\right| \leq \left(\sum_{k=1}^{q} \Delta_k^2 \lambda^{2k}\right)^{1/2} \left(\sum_{k=1}^{q} |c_k|^2 \lambda^{-2k}\right)^{1/2}. \tag{2.20}$$

Again by Parseval, the final factor is bounded by $\sup|f(z)|$ on $\bar{B}(0, 1/\lambda)$. Observe that the disc $\bar{B}(0, 1/\lambda)$ will belong to the closed interior of the ellipse E_τ (2.17) when we choose

$\frac{1}{2}(\tau - 1/\tau) = 1/\lambda$ or $\tau = (1 + \sqrt{1+\lambda^2})/\lambda$. Hence for this τ, Lemma 2.3.3 shows that $\sup |f(z)|$ on $\bar{B}(0, 1/\lambda)$, and thus the final factor in (2.20), is bounded by τ^q.

Using our special N-tuple of nodes X_N, we finally take f equal to a polynomial of degree $q < 4\sqrt{N}$ with a "large" quadrature remainder as in (2.9). Then the inequalities (2.20) and (2.19) imply that $1/100N < |R(f, X_N)| < Ae^{-bs}\tau^{4\sqrt{N}}$, hence

$$bs < 4(\log \tau)\sqrt{N} + \log N + \log(100A). \qquad \square$$

For minimal G_1, the total number of nodes x_j on $[-\lambda, \lambda]$ (counting multiplicity) will be asymptotic to λN as $N \to \infty$. For a proof we refer to [KM93a].

2.4 Open questions related to Theorem 2.3.1

Problem 2.4.1 Investigate the following conjecture. Combination of the multiple nodes for minimal G_r or G_1 will result in a quadrature formula which resembles the m-point Gauss formula for some $m \approx c\sqrt{N}$.

Question 2.4.2 What can one say about coalescence of nodes if G_1 is minimized under the condition $\Delta_1 = \cdots = \Delta_q = 0$ with maximal $q = q_N$? L.A. Anderson and W. Gautschi [AG75] have treated the special case $p = q + 1$.

Question 2.4.3 What sort of coalescence of nodes occurs in minimum norm quadrature formulas of Chebyshev type for other nice probability measures, for example ultraspherical measures? In the latter case there are good estimates for the Christoffel numbers, see for example L. Gatteschi–G. Vinardi [GV78] and [KM93b].

Chapter 3
Chebyshev-type quadrature on product domains including the sphere

General reference for Section 3.1: [Be37b] and for Sections 3.2, 3.3: [KM93b].

3.1 Discussion of Bernstein's Theorem B for $[-1, 1]$

For the convenience of the reader we repeat the statement:

Theorem B *For $m \geq 1$, let M be an even integer*

$$\geq N_0(m) = 2\left[2\sqrt{2}(m+1)(m+4) + 1\right]$$

where [ ] *denotes the integral part. Then there exist points* $t_k \in (-1,1)$, $t_1 > t_2 > \cdots > t_{2m-1}$, $t_{2m-k} = -t_k$ *and positive integers* $M_k = M_{2m-k}$ *such that*

$$\int_{-1}^{1} f(x)\,\tfrac{1}{2}\,dx = \frac{1}{M}\sum_{k=1}^{2m-1} M_k f(t_k) \tag{3.1}$$

for all polynomials $f(x)$ *of degree* $\le 2m-1$.

Bernstein derived this result from a fundamental existence theorem for positive quadrature formulas, i.e., formulas with positive coefficients p_k. His formulas are symmetric: they have the form

$$\int_{-1}^{1} f(x)\,\tfrac{1}{2}\,dx \approx \sum_{k=1}^{2m-1} p_k f(t_k), \qquad t_{2m-k} = -t_k \in (-1,1), \quad p_{2m-k} = p_k \tag{3.2}$$

and they are polynomially exact to degree $2m-1$.

For preassigned distinct symmetric nodes $t_1, \ldots, t_{2m-1}$ there is always a unique formula (3.2) of polynomial degree $2m-1$: substituting $f(x) = 1, x, \ldots, x^{2m-2}$, one obtains a system of linear equations for $p_1, \ldots, p_{2m-1}$ with nonvanishing Vandermonde determinant. The unique solution of this system will be symmetric ($p_{2m-k} = p_k$) and hence formula (3.2) holds also for $f(x) = x^{2m-1}$. However, positivity of the solutions p_k depends critically on the choice of nodes t_k. Bernstein's sufficient condition for a positive formula (3.2) is not necessary but it allows a fair amount of freedom.

To state the precise result we have to introduce a number of canonical (positive) quadrature formulas and related notation. Keeping m fixed, $\alpha_1 > \alpha_2 > \cdots > \alpha_m$ will again be the zeros of the Legendre polynomial $P_m(x)$. We also need the zeros $1 = \beta_0 > \beta_1 > \cdots > \beta_{m-1} > \beta_m = -1$ of $(1-x^2)P_m'(x)$, as well as those of the combinations

$$P_m(x,a) = \begin{cases} P_m(x) - a(1-x)P_m'(x) & \text{for } a \ge 0, \\ P_m(x) - a(1+x)P_m'(x) & \text{for } a \le 0. \end{cases} \tag{3.3}$$

For every $a \in \mathbf{R}$, $P_m(x,a)$ has m simple zeros

$$\xi_1(a) > \cdots > \xi_m(a) \quad \text{in } (-1,1); \tag{3.4}$$

as a decreases from $+\infty$ to $-\infty$, $\xi_k(a)$ decreases from $\beta_{k-1} =$ "$\xi_k(\infty)$" to $\beta_k =$ "$\xi_k(-\infty)$".

The first canonical formulas are the m-point *Gauss formula* (1.4), where we henceforth write $\lambda_k = \lambda(\alpha_k)$, and the *Lobatto formula*

$$\int_{-1}^{1} f(x)\,\tfrac{1}{2}\,dx \approx \sum_{k=0}^{m} \lambda(\beta_k) f(\beta_k).$$

For every $a > 0$ one has a *lower canonical formula*

$$\int_{-1}^{1} f(x)\,\tfrac{1}{2}\,dx \approx \sum_{k=1}^{m} \lambda(\xi_k(a)) f(\xi_k(a)) + \rho_a f(-1),$$

while for every $a < 0$ there is an *upper canonical formula* in which $f(1)$ occurs instead of $f(-1)$; here $\rho_a = \rho_{-a}$. All these quadrature formulas are positive and of degree $2m-1$, but the lower and upper canonical formulas are not symmetric.

The above formulas define $\lambda(x) = \lambda(x, m)$ for every $x \in [-1, 1]$. One can show that $\lambda(x)$ is the maximal weight which a positive quadrature formula of degree $2m - 1$ can have at the point x. More important is the function $\pi(x)$, defined as the maximal weight which a positive quadrature formula of degree $2m - 1$ can have on the interval $[x, 1]$. This $\pi(x) = \pi(x, m)$ may be represented in the following way:

$$\pi(x) = \begin{cases} \sum_{j=1}^{k} \lambda(\xi_j(a)) & \text{if } x = \xi_k(a) \text{ with } 0 \le a \le +\infty, \\ \sum_{j=1}^{k} \lambda(\xi_j(a)) + \rho_a & \text{if } x = \xi_k(a) \text{ with } -\infty \le a \le 0. \end{cases} \tag{3.5}$$

The function $\pi(x)$ is positive, non-increasing and continuous. Furthermore, it is possible to derive the following useful estimate:

$$\pi(x) - \pi(y) \ge \tfrac{1}{2}\left(1 - \frac{1}{m(m+1)}\right)(y - x) \qquad \text{for } 0 < x < y < 1. \tag{3.6}$$

We can now state Bernstein's main result in the following convenient form:

Theorem 3.1.1 [Be37b, Theorem VI] *For simplicity of notation let $m = 2l$ be even, let a_k, $k = 1, \ldots, l$ and b_k, $k = 1, \ldots, l-1$ be two sequences of parameters such that*

$$\infty = a_1 > a_2 > \cdots > a_l \ge 0 \ge b_1 > b_2 > \cdots > b_{l-1} \ge -\infty \tag{3.7}$$

and set

$$\begin{aligned} p_1 &= \pi(\xi_1(a_1)), \\ p_{2k-1} = p_{2m-2k+1} &= \pi(\xi_k(a_k)) - \pi(\xi_{k-1}(b_{k-1})), \quad k = 2, \ldots, l, \\ p_{2k} = p_{2m-2k} &= \pi(\xi_k(b_k)) - \pi(\xi_k(a_k)), \quad k = 1, \ldots, l-1, \\ p_m &= 1 - 2\textstyle\sum_{k=1}^{m-1} p_k. \end{aligned} \tag{3.8}$$

Then $p_1, p_2, \ldots, p_{2m-1}$ are the weights in a symmetric quadrature formula (3.2) which is polynomially exact to degree $2m - 1$. The nodes $t_1, t_2, \ldots, t_{m-1}$ satisfy the inequalities

$$\begin{aligned} \xi_1(a_1) &\ge t_1 \ge \alpha_1, \\ \xi_k(b_k) &\ge t_{2k} \ge \xi_{k+1}(a_k), \quad k = 1, \ldots, l-1, \\ \xi_k(a_k) &\ge t_{2k-1} \ge \xi_k(b_{k-1}), \quad k = 2, \ldots, l. \end{aligned} \tag{3.9}$$

Proof We present the outline of a proof which resembles Bernstein's original demonstration. Only minor changes have been made.

For his proof Bernstein first assumes that there exist parameters a_k^0 and b_k^0 satisfying (3.7) with strict inequalities such that the conclusion of the Theorem holds. It follows from the strict inequalities that the nodes in the resulting quadrature formula are mutually distinct. One may then introduce small perturbations which give the Theorem for all decreasing sequences a_k and b_k sufficiently close to a_k^0 and b_k^0. The perturbations can be continued as long as the nodes in the quadrature formula are distinct, which is the case as long as the sequences a_k and b_k are strictly decreasing. In this way the Theorem is proved for all strictly decreasing sequences a_k and b_k, and a final limit argument proves the Theorem for all sequences satisfying (3.7).

It remains to be shown that the Theorem holds for some strictly decreasing sequences of parameters a_k and b_k.

If we take all a_k's equal to 0 and the b_k's arbitrary decreasing, the relations (3.8) amount to

$$p_1 = \pi(\xi_1(0)) = \pi(\alpha_1) = \lambda(\alpha_1) = \lambda_1,$$

and for $k = 1, \ldots, l-1$,

$$p_{2k} + p_{2k+1} = \pi(\xi_{k+1}(0)) - \pi(\xi_k(0)) = \pi(\alpha_{k+1}) - \pi(\alpha_k) = \lambda(\alpha_{k+1}) = \lambda_{k+1}.$$

Putting $t_1 = \alpha_1$, $t_{2k} = t_{2k+1} = \alpha_{k+1}$, $k = 1, \ldots, l-1$, $t_{2m-k} = -t_k$, we obtain the m-point Gauss formula but with most nodes counted twice. Bernstein proceeds to split the double nodes. He takes $a_1 > 0$ small, $a_2 = \cdots = a_l = 0$, $b_1 = b_2 = \cdots = b_{l-1} = -\infty$, and proves that with this choice of a_k and b_k the Theorem holds and that the resulting symmetric quadrature formula has $2m-1$ distinct nodes. (This is non-trivial.) Having these distinct nodes Bernstein effects a small perturbation of the parameters a_k and b_k such that (3.7) is satisfied with strict inequalities and such that the Theorem continues to hold.

This concludes the proof of Theorem 3.1.1. □

To derive Theorem B, Bernstein went on to show that for M of order m^2 as described, the coefficients p_k may be taken of the form M_k/M with positive integral M_k.

To obtain suitable relations (3.8) with decreasing sequences a_k and b_k, it suffices to make all numbers $\pi(\xi_k(a_k))$, $\pi(\xi_k(b_k))$ in (3.8) equal to integral multiples of $1/M$. This will be possible if there are decreasing sequences $\tilde{a}_k$, $\tilde{b}_k$ such that

$$\begin{aligned} \pi(\xi_k(\tilde{a}_k)) - \pi(\xi_k(\tilde{a}_{k-1})) &\geq 1/M, \qquad k = 1, \ldots, l, \qquad (\tilde{a}_0 = \infty), \\ \pi(\xi_k(\tilde{b}_k)) - \pi(\xi_k(\tilde{b}_{k-1})) &\geq 1/M, \qquad k = 1, \ldots, l-1, \quad (\tilde{b}_0 = 0). \end{aligned}$$

Using (3.6) Bernstein proved the existence of such sequences if M satisfies the inequality given in Theorem B.

Remark 3.1.2 In [Ku93a, Ku93b] Kuijlaars has extended Theorem B to ultraspherical and Jacobi measures.

3.2 Application of Theorem B to square and disc

Let E be a compact set in $\mathbf{R}^d$ furnished with a probability measure σ. *Question* (cf. Section 1.2): What is the smallest number $N = N_E(p)$ for which there is an Nth order Chebyshev-type quadrature formula for (E, σ) which has degree p relative to $\mathbf{R}^d$-polynomials? We are primarily interested in the order of $N_E(p)$ for large p.

Bernstein's Theorems A and B show that $N_I(p)$ is of order p^2 for the interval $I = [-1, 1]$ with $d\sigma = \frac{1}{2}dx$. For the unit circle $C = C(0,1)$ in $\mathbf{R}^2 \simeq \mathbf{C}$ with $d\sigma = ds/2\pi$ one has $N_C(p) = p + 1$. For various pairs (E, σ) which can be represented as *Cartesian products* of intervals and/or circles by suitable parametrization, these results readily imply an upper bound for $N_E(p)$. In a number of cases this upper bound will be of the right order.

Example 3.2.1 For the square $Q = I^2$ with $d\sigma = \frac{1}{4}dxdy$, Theorem B with $m \geq 1$ and $M = N_0(m)$ gives the "product formula"

$$\int_Q f(x,y)\,\tfrac{1}{4}\,dxdy \approx \frac{1}{N_0(m)^2} \sum_{k,l=1}^{2m-1} M_k M_l f(t_k, t_l), \tag{3.10}$$

where t_k and M_k are as in (3.1). This formula will be exact for all monomials $f(x,y) = x^\alpha y^\beta$ with $0 \leq \alpha, \beta \leq 2m-1$, hence for all polynomials $f(x,y)$ of degree $\leq 2m-1$ in x and y separately. It follows that $N_Q(p) = \mathcal{O}(p^4)$.

A result in the other direction may be obtained by the method of Theorem A or more precisely, Theorem 2.1.1. Using the polynomials $F_1(x)F_1(y)$ and $F_1(x)F_2(y) + F_2(x)F_1(y)$, cf. (2.3), one can show that every Chebyshev-type quadrature formula for (Q, σ) of degree p must have order $N \geq cp^4$, $c > 0$ independent of p, see [KM93b].

Example 3.2.2 For the closed unit disc D: $x = r\cos\phi$, $y = r\sin\phi$ $(0 \leq r \leq 1, 0 \leq \phi < 2\pi)$ and every polynomial $f(x,y) \simeq F(r,\phi) \simeq F(\sqrt{\frac{1}{2}(t+1)}, \phi)$ of degree $\leq 4m-1$ in (x,y), the product method leads to the formula

$$\begin{aligned} \int_D f(x,y)\,\tfrac{1}{\pi}\,dxdy &= \frac{1}{\pi}\int_0^1\int_0^{2\pi} F(r,\phi)\,r\,dr\,d\phi \\ &= \frac{1}{2}\int_{-1}^1 \frac{1}{2\pi}\int_0^{2\pi} F(\sqrt{\tfrac{1}{2}(t+1)}, \phi)\,dt\,d\phi \\ &= \frac{1}{4mN_0(m)} \sum_{k=1}^{2m-1} M_k \sum_{l=1}^{4m} F(\sqrt{\tfrac{1}{2}(t_k+1)}, \phi_l) \end{aligned} \tag{3.11}$$

where $N_0(m)$, t_k and M_k are as in Theorem B with $M = N_0(m)$ and $\phi_l = (l-1)2\pi/4m$.

Indeed, every monomial $x^\alpha y^\beta$ with $0 \leq \alpha + \beta \leq 4m-1$ can be written as a linear combination of functions

$$F(r,\phi) = r^\lambda \cos(\lambda - 2\mu)\phi, \qquad F(r,\phi) = r^\lambda \sin(\lambda - 2\mu)\phi$$

with $0 \leq 2\mu \leq \lambda \leq 4m-1$. For such functions $F(r,\phi)$ with $0 < \lambda - 2\mu$ $(< 4m)$, both the integral and the sum in (3.11) are equal to zero. We finally take $\lambda = 2\mu$ $(< 4m)$ so that

$$F(r,\phi) = r^{2\mu} = \{\tfrac{1}{2}(t+1)\}^\mu \qquad \text{with } 0 \leq \mu \leq 2m-1.$$

For these functions formula (3.11) becomes a special case of Theorem B.

It follows from the preceding that $N_D(p) = \mathcal{O}(p^3)$. One can not do better than order p^3: by "projection", an Nth order Chebyshev-type formula of degree $p = 2m-1$ for D

implies one for $[-1,1]$ with $d\sigma = (2/\pi)(1-x^2)^{1/2}dx$. Indeed, if in such a formula for D with nodes (x_j, y_j), $1 \le j \le N$, we take $f(x,y)$ equal to an arbitrary polynomial $g(x)$ of degree $\le 2m-1$, the result is

$$\frac{1}{\pi}\int_D f(x,y)dxdy = \frac{2}{\pi}\int_{-1}^{1} g(x)(1-x^2)^{1/2}dx = \frac{1}{N}\sum_{j=1}^{N} g(x_j).$$

The method of Theorem 2.1.1 applied to the measure $d\sigma = (2/\pi)(1-x^2)^{1/2}dx$ on $[-1,1]$ now implies that

$$N \ge 1/\lambda_1(m) > (m+1)^3/2\pi^2,$$

see Remark 2.1.3.

Remark 3.2.3 The case of the torus T ="$C \times C$" in $\mathbf{R}^3$ is more difficult because the area element of T is not the square of the element of arc length along C. Nevertheless Kuijlaars [Ku93c] has been able to show that $N_T(p)$ is indeed of order p^2, as one would naively predict.

3.3 Application to the sphere, spherical designs

Let S be the unit sphere $S(0,1)$ in $\mathbf{R}^3$ with parametrization

$$x = \sin\theta\cos\phi, \quad y = \sin\theta\sin\phi, \quad z = \cos\theta, \qquad 0 \le \theta \le \pi,\ 0 \le \phi < 2\pi$$

and normalized area element

$$d\sigma = \frac{1}{4\pi}\sin\theta\, d\theta d\phi = \frac{1}{4\pi}|dz|d\phi.$$

On S, every polynomial $f(x,y,z)$ of degree $\le 2m-1$ is equal to a linear combination $F(\cos\theta,\phi) = F(z,\phi)$ of spherical harmonics of order $\le 2m-1$, cf. Section 1.4, formula (1.20).

Theorem 3.3.1 *For $m \ge 1$ and every polynomial $f(x,y,z) \simeq F(z,\phi)$ of degree $\le 2m-1$,*

$$\int_S f(x,y,z)d\sigma = \frac{1}{4\pi}\int_{-1}^{1}\int_0^{2\pi} F(z,\phi)dzd\phi = \frac{1}{2mM}\sum_{k=1}^{2m-1}\sum_{l=1}^{2mM_k} F(t_k,\phi_{kl}) \tag{3.12}$$

where t_k, M_k and $M \ge N_0(m)$ are as in Theorem B and $\phi_{kl} = (l-1)2\pi/2mM_k$.

Proof It is sufficient to verify (3.12) for a set of basis elements $F(z,\phi)$ for the linear space of the spherical harmonics of order $\le 2m-1$:

$$F(z,\phi) = (1-z^2)^{|s|/2}P_n^{(|s|)}(z)e^{is\phi}, \qquad n = 0,1,\ldots,2m-1, \quad -n \le s \le n, \tag{3.13}$$

cf. formula (1.18). Now for a function (3.13) with $0 < |s| \le 2m-1$, both the integral and the sum in (3.12) are equal to zero. In the remaining case $s = 0$ one has $F(z,\phi) = P_n(z)$ and then (3.12) becomes a special case of Theorem B. □

Remark 3.3.2 Observe that for $M = N_0(m)$, (3.12) provides a Chebyshev-type quadrature formula for S of polynomial degree $2m-1$ with $N = 2mN_0(m) \approx 8\sqrt{2}m^3$ nodes. The nodes $\tilde{\zeta}_j, j = 1, \ldots, N$ run through the points on S with spherical coordinates (θ_k, ϕ_{kl}), $0 < \theta_k < \pi$, $0 \leq \phi_{kl} < 2\pi$ given by

$$\cos\theta_k = t_k, \qquad \phi_{kl} = (l-1)2\pi/2mM_k, \qquad 1 \leq k \leq 2m-1, \quad 1 \leq l \leq 2mM_k.$$

Thus the nodes $\tilde{\zeta}_j$ lie on $2m-1$ horizontal circles with z-coordinates $t_1, \ldots, t_{2m-1}$, respectively. The kth circle contains $2mM_k$ evenly distributed nodes.

The $\tilde{\zeta}_j$'s are pairwise distinct; i.e., they form a spherical $(2m-1)$-design, cf. Section 1.2.

Corollary 3.3.3 *For $p \to \infty$, there exist spherical p-designs consisting of $\mathcal{O}(p^3)$ points.*

A lower bound for $N_S(p)$ may be obtained by projection. Indeed, substituting $f(x,y,z) = g(z)$, an Nth order Chebyshev-type quadrature formula of degree $2m-1$ for S with nodes (x_j, y_j, z_j) implies one for the interval $-1 \leq z \leq 1$ with nodes z_j. Thus by Theorem A, $N \geq \frac{1}{4}m^2$ and hence for all $p \geq 1$,

$$c_1p^2 \leq N_S(p) \leq c_2p^3 \qquad (c_1, c_2 > 0). \tag{3.14}$$

We do not know the exact order of $N_S(p)$; cf. however Conjecture 1.2.1 and Problem 3.4.4.

Remark 3.3.4 Our method readily extends to higher dimensions. For the d-dimensional unit sphere S^d in $\mathbf{R}^{d+1}$ the result is

$$c_1p^d \leq N_{S^d}(p) \leq c_2p^{d(d+1)/2}.$$

Here we also expect that the lower bound gives the true order.

3.4 Open questions

Question 3.4.1 For the interval $I = [-1,1]$ with $d\sigma = \frac{1}{2}dx$, $N_I(p)$ is of order p^2. Numerical results indicate that the lower bound $N_0(m) \sim 4\sqrt{2}m^2$ for M in Theorem B can be reduced to about $\frac{1}{2}m^2$, [Ku93b]. Is there a constant c such that $N_I(p) \sim cp^2$ as $p \to \infty$?

Question 3.4.2 Let $E \subset \mathbf{R}^2$ be an ellipse with measure σ defined by normalized arc length. Investigate the conjecture that $N_E(p) = \mathcal{O}(p)$.

Problem 3.4.3 Let σ be a probability measure on $I = [-1,1]$ and let $\lambda_1(m,\sigma)$ be the first Christoffel number in the m-point Gauss formula for σ. We know that a Chebyshev-type quadrature formula for (I,σ) of polynomial degree $p = 2m-1$ must have order $N \geq 1/\lambda_1(m,\sigma)$, cf. Remark 2.1.3. Under what general conditions on σ will there be Chebyshev-type formulas for (I,σ) of degree $2m-1$ which have order N comparable to $1/\lambda_1(m,\sigma)$ as $m \to \infty$?

Kuijlaars [Ku93a, Ku93b] has studied the case of ultraspherical and Jacobi measures. For Jacobi measures

$$c_{\alpha,\beta}(1-x)^\alpha(1+x)^\beta dx, \qquad \alpha \geq \beta > -1,\ \alpha \geq 0,$$

he found that the number $N(2m-1)$ is of the same order as $1/\lambda_1(m)$. It is conjectured that the restriction $\alpha \geq 0$ is not necessary.

Principal problem 3.4.4 What is the true order of $N_S(p)$ as $p \to \infty$? We conjecture that $N_S(p) = \mathcal{O}(p^2)$, and also that there are spherical p-designs consisting of $\mathcal{O}(p^2)$ points. The following observation provides some support for the conjecture. For the square $Q = [-1,1]^2$ with measure $d\mu = \pi^{-2}(1-x^2)^{-1/2}(1-y^2)^{-1/2}dxdy$, there is a Chebyshev-type formula with m^2 nodes which is exact for all polynomials $f(x,y)$ of degree $\leq 2m-1$ (in x and y separately), cf. (1.6). The Chebyshev-type quadrature problem for S is essentially equivalent to the Chebyshev-type problem for the unit disc D with measure $d\nu = (2\pi)^{-1}(1-x^2-y^2)^{-1/2}dxdy$. This measure has about the same boundary behavior as $d\mu$. For further support, see Sections 4.1, 4.2.

Chapter 4
Electron problem for the sphere: use of quadrature results and complex analysis

Let $Z_N = (\zeta_1, \ldots, \zeta_N)$ be an N-tuple of points on the unit sphere $S \subset \mathbf{R}^3$. As in Section 1.3, $U(x, Z_N)$ denotes the potential at x due to point charges $1/N$ at the points of Z_N while $\mathcal{E}(x, Z_N)$ stands for the corresponding electrostatic field:

$$U(x, Z_N) = \frac{1}{N}\sum_{j=1}^{N} \frac{1}{|\zeta_j - x|}, \qquad \mathcal{E}(x, Z_N) = -\operatorname{grad} U(x, Z_N). \tag{4.1}$$

Similarly, $R(f, Z_N)$ denotes the quadrature remainder in the Chebyshev-type quadrature formula for f on S corresponding to Z_N:

$$R(f, Z_N) = \int_S f(\zeta)d\sigma(\zeta) - \frac{1}{N}\sum_{j=1}^{N} f(\zeta_j). \tag{4.2}$$

In the following we will repeatedly use the Equivalence Theorem (Theorem E) for the "Electron Problem" that was discussed in Sections 1.3 and 1.4. After exploring the consequences for that problem of certain quadrature results, we turn things around and start with the Electron Problem. Using "several complex variables", we show that reasonable hypotheses about certain extremal distributions of point charges would lead to a strong result for the Electron Problem. That result would in turn yield a quadrature result which supports the conjecture $N_S(p) = \mathcal{O}(p^2)$ enunciated in Section 1.2.

4.1 A general lower bound for the electrostatic field

Considering arbitrary N-tuples Z_N on S, we will first use part (i) of Theorem E to obtain a good lower bound for $\sup|U(x, Z_N) - 1|$ on spheres about the origin. To this end we fix $r \in (0,1)$ and for suitable A to be chosen later, we define δ by the formula

$$\sup_{|x|=r} |U(x, Z_N) - 1| = \delta r^A. \tag{4.3}$$

For any polynomial f of degree $\leq A$ with nonzero quadrature remainder $R(f, Z_N)$, formula (1.12) now provides a lower bound for δ:

$$\delta \geq \frac{|R(f, Z_N)|}{(2A+1)\sup_S |f|}. \tag{4.4}$$

Appropriate choice of A and f will give

Theorem 4.1.1 [KM93c] *For any N-tuple Z_N of points $\zeta_j \in S$,*

$$\sup_{|x|=r} |U(x, Z_N) - 1| > r^{2\sqrt{N}}/4(\sqrt{N}+1)^3, \qquad 0 < r < 1. \tag{4.5}$$

Proof We will obtain a simple quadrature remainder if f is nonnegative on S and vanishes at the points ζ_j of Z_N: in that case

$$R(f, Z_N) = \int_S f. \tag{4.6}$$

To ensure positivity of the integral we will take $f|_S$ of the form $g\overline{g}$ with $g \not\equiv 0$. For the construction of suitable g it is convenient to introduce the linear space V_q of the spherical harmonics of order $\leq q = [\sqrt{N}]$. Then $\dim V_q = 1 + 3 + \cdots + (2q+1) = (q+1)^2 > N$, hence we can determine $g \in V_q$ with L^2 norm $\|g\|_S = 1$ such that $g(\zeta_j) = 0$, $j = 1, \ldots, N$. Indeed, a homogeneous linear system of N equations in more than N unknowns always has a nonzero solution.

In order to define f we decompose our g into spherical harmonics of different order: $g(\xi) = \sum_{k=0}^{q} Y_k(\xi)$, cf. (1.20). Setting $x = r\xi$ with $\xi \in S$ we now take $f(x)$ equal to the product of the harmonic polynomials corresponding to g and $\overline{g}$:

$$f(x) = \sum_{k=0}^{q} r^k Y_k(\xi) \sum_{l=0}^{q} r^l \overline{Y_l(\xi)}.$$

Thus f is a polynomial of degree $2q \leq 2\sqrt{N}$ and

$$\int_S f = \int_S g\overline{g} = 1. \tag{4.7}$$

We still have to estimate $\sup_S |f|$. To that end we start with g and use the representation (1.19) for Y_k in terms of g:

$$g(\xi) = \sum_{k=0}^{q} Y_k(\xi) = \int_S g(\eta) \sum_{k=0}^{q} (2k+1) P_k(\xi \cdot \eta) d\sigma(\eta).$$

Thus by Cauchy-Schwarz, the Pythagorean formula for the square of the norm of an orthogonal sum and formula (1.18) for the L^2 norm of P_k on S,

$$\begin{aligned} f(\xi) &= |g(\xi)|^2 \le \int_S |g(\eta)|^2 d\sigma(\eta) \int_S \left|\sum_{k=0}^{q}(2k+1)P_k(\xi\cdot\eta)\right|^2 d\sigma(\eta) \\ &= \sum_{k=0}^{q}(2k+1)^2 \int_S |P_k(\xi\cdot\eta)|^2 d\sigma(\eta) = \sum_{k=0}^{q}(2k+1) \\ &= (q+1)^2 < (\sqrt{N}+1)^2, \qquad \forall \xi \in S. \end{aligned} \tag{4.8}$$

Combination of (4.6)–(4.8) gives

$$|R(f, Z_N)| = \int_S f = 1 > \sup_S |f|/(\sqrt{N}+1)^2. \tag{4.9}$$

Having (4.9) we are ready to use (4.3) and (4.4). Taking $A = 2\sqrt{N}$, defining δ by (4.3) and using our present f in (4.4) we conclude from (4.9) that

$$\delta > \frac{1}{(4\sqrt{N}+1)(\sqrt{N}+1)^2}.$$

The desired inequality (4.5) now follows from (4.3). □

The proof implies the following sharper result.

Corollary 4.1.2 *If the N-tuple Z_N contains only M distinct points ζ_j, then*

$$\sup_{|x|=r} |U(x, Z_N) - 1| > r^{2\sqrt{M}}/4(\sqrt{M}+1)^3, \qquad 0 < r < 1.$$

Indeed, the preceding proof goes through with $q = [\sqrt{M}]$ and $A = 2\sqrt{M}$.

More important, we have the following Corollary on the electrostatic field $\mathcal{E}(x, Z_N)$.

Corollary 4.1.3 *Let $Z_N = (\zeta_1, \ldots, \zeta_N)$ be any N-tuple of points on S. Placing point charges $1/N$ at the points of Z_N, the resulting electrostatic field $\mathcal{E}(x, Z_N) = -\operatorname{grad} U(x, Z_N)$ must satisfy the following inequality on every sphere $S(0, r)$ with $0 < r < 1$:*

$$\sup_{|x|=r} |\mathcal{E}(x, Z_N)| > \frac{1}{4(\sqrt{N}+1)^3} r^{2\sqrt{N}-1}. \tag{4.10}$$

Indeed, if for some r the $\le$ sign would hold in (4.10), then by the maximum principle, $|\mathcal{E}(x, Z_N)|$ would be bounded by the right-hand side of (4.10) throughout the ball $B(0, r)$. Integration would then give the inequality

$$|U(r\xi, Z_N) - 1| = \left|\int_0^r \frac{\partial}{\partial s} U(s\xi, Z_N) ds\right| \le \int_0^r |\mathcal{E}(s\xi, Z_N)| ds \le \frac{1}{4(\sqrt{N}+1)^3} r^{2\sqrt{N}}$$

for every $\xi \in S$, contradicting Theorem 4.1.1.

Remark 4.1.4 It is interesting to consider the corresponding result for the unit circle $C = C(0,1)$ in $\mathbf{R}^2 \simeq \mathbf{C}$. Using the logarithmic potential, the electrostatic field due to point charges $1/N$ at the points ζ_j of an N-tuple Z_N on C has the complex representation

$$\mathcal{E}(x, Z_N) = \left\{ -\frac{1}{N} \sum_{j=1}^{N} \frac{1}{\zeta_j - z} \right\}^{-},$$

where the bar stands for complex conjugation. An analog of Theorem E for the circle and a calculation similar to the one in the proof of Theorem 4.1.1 will now show that

$$\sup_{|z|=r} |\mathcal{E}(z, Z_N)| > \frac{r^{N-1}}{2N+2}, \qquad 0 < r < 1 \tag{4.11}$$

for *every* N-tuple Z_N on C. Observe that for large N, the general lower bound in (4.11) is "almost" achieved for the *special* N-tuple Z_N^* consisting of the Nth roots of unity:

$$\sup_{|z|=r} |\mathcal{E}(z, Z_N^*)| = \frac{r^{N-1}}{1-r}, \qquad 0 < r < 1. \tag{4.12}$$

This observation tends to support the conjecture that the general lower bound in (4.10) for the case of the sphere is also of the right order as $N \to \infty$. That is, there should be special N-tuples Z_N on S for which the order $r^{2\sqrt{N}}$ in the right-hand side of (4.10) is "almost" achieved. If this order can indeed be achieved for $N \geq N_0$ and special N-tuples Z_N, it will follow from Theorem E that $N_S(p) = \mathcal{O}(p^2)$, as conjectured in Section 1.2.

4.2 Upper bounds for the case of special N-tuples

Theorem 3.3.1 described special N-tuples Z_N which give relatively good Chebyshev-type quadrature on S. We will begin by combining that result with Theorem E of Section 1.3 to obtain an upper bound for the field corresponding to those special N-tuples, cf. [KM93c].

Theorem 4.2.1 *For $N = 2mN_0(m)$ with $m \geq 1$ and $N_0(m)$ as in Bernstein's Theorem B (Section 1.2), there is a configuration $\tilde{Z}_N$ of N distinct points on S such that for $0 < r < 1$,*

$$\sup_{|x|=r} |U(x, \tilde{Z}_N) - 1| \leq \frac{r^{2m}}{1-r}, \tag{4.13}$$

$$\sup_{|x|=r} |\mathcal{E}(x, \tilde{Z}_N)| < \frac{12emr^{2m-1}}{1-r}, \qquad (m > \frac{r}{1-r}). \tag{4.14}$$

Here $2m \approx .89N^{1/3}$ for large m.

Proof For $N = 2mN_0(m)$ we take $\tilde{Z}_N$ equal to the special N-tuple $(\tilde{\zeta}_1, \ldots, \tilde{\zeta}_N)$ of Remark 3.3.2 which generates a Chebyshev-type quadrature formula for S of polynomial degree $2m - 1$. Now apply part (ii) of Theorem E to $\tilde{Z}_N$, taking $p = 2m - 1$ and $\epsilon = 0$. Inequality

(1.14) then gives inequality (4.13). Inequality (4.14) may be obtained from (4.13) with the aid of the Poisson integral for $U(x)-1$ on a ball of radius slightly larger than r, see [KM93c] for the details. The asymptotic relation for $2m$ follows from the definition of $N_0(m)$ in Theorem B. □

Observe that there is a substantial gap between the upper bounds in Theorem 4.2.1 and the lower bounds in Theorem 4.1.1 and Corollary 4.1.3. As a tool to narrow the gap we will establish the basic Proposition 4.2.3 below. A complex-analytic method of proof for this auxiliary result will be developed in Sections 4.3, 4.4.

It will be convenient to begin with some terminology.

Definition 4.2.2 We will say that the (adjusted) potential $U(x)-1 = U(x, Z_N)-1$ is of *zero type* (or of *stationary type*) (r, δ, M), where $r \in (0,1)$, $\delta > 0$ and $M \geq 10$, if the function $U(r\eta)-1$, $\eta \in S$ vanishes (or is stationary, respectively) at M points $\eta_k \in S$, $1 \leq k \leq M$ which are *well-distributed* and *well-separated* in the following precise sense:

(i) every spherical cap $\Sigma \subset S$ of area $\geq (1/5)$ area S contains $\geq M/10$ points η_k;

(ii) the points η_k admit separation constant $2\delta/\sqrt{M}$:

$$|\eta_j - \eta_k| \geq 2\delta/\sqrt{M}, \qquad j,k = 1,\ldots,M, \quad j \neq k.$$

The numbers 5 and 10 in the above Definition have been chosen for convenience; they could be replaced by other constants.

Basic Proposition 4.2.3 *For given $r \in (0,1)$ and $\delta > 0$, there are positive constants B and c such that for every (adjusted) potential $U(x, Z_N)-1$ of zero type (r, δ, M) with $M \geq 10$,*

$$\sup_{|x|=r} |U(x, Z_N)-1| \leq Be^{-c\sqrt{M}}. \tag{4.15}$$

The constants B and c depend on r and δ but not otherwise on Z_N, nor on M. There is a corresponding result for potentials $U(x, Z_N)-1$ of stationary type (r, δ, M).

For the proof, see Section 4.4. In the remainder of this Section we illustrate how Proposition 4.2.3 might give better results than Theorem 4.2.1.

Conjecture 4.2.4 We conjecture that for $r \in (0,1)$ and all $N \geq N_0(r)$, there exist special N-tuples Z_N for which the potentials $U(x, Z_N)-1$ are of zero type or stationary type $(r, 1, M)$ with $M \geq \gamma N$, where $\gamma = \gamma(r) > 0$ is independent of N.

One would hope to obtain such N-tuples by minimization of suitable functionals. Thus minimization of

$$\begin{aligned} F_r &= F_r(\zeta_1,\ldots,\zeta_N) = \int_S \{U(r\eta, Z_N)-1\}^2 d\sigma(\eta) \\ &= \frac{1}{r}\int_0^r \frac{1}{N^2}\sum_{j,k=1}^{N} \frac{1}{|\zeta_j - t^2\zeta_k|}dt - 1, \qquad \zeta_j \in S \end{aligned} \tag{4.16}$$

(with N large) may well lead to a potential $U(x, Z_N) - 1$ of zero type $(r, 1, [N/6])$. Indeed, the harmonic function $U(x, Z_N) - 1$ has average 0 on $S(0, r)$. For minimal F_r, one would expect that the minimizing points $\zeta_1, \ldots, \zeta_N$ are very evenly distributed. One might further expect that $U(r\eta) - 1$ then vanishes on some closed curve in every spherical triangle formed by adjacent points $\zeta_{j-1}, \zeta_j, \zeta_{j+1}$. Thus one would expect $U(x, Z_N) - 1$ to be of zero type $(r, 1, [N/6])$.

The related function

$$G_r = G_r(\zeta_1, \ldots, \zeta_N) = \frac{1}{N^2} \sum_{j,k=1}^{N} \frac{1}{|\zeta_j - r\zeta_k|}, \qquad \zeta_j \in S \tag{4.17}$$

is perhaps even more promising. Suppose that G_r has been minimized. Then it is again reasonable to expect that the minimizing points $\zeta_1, \ldots, \zeta_N$ are well-distributed and well-separated. Keeping the points ζ_j different from a given point ζ_k fixed for a moment, we observe that the function

$$U_k(r\eta) \stackrel{\text{def}}{=} U(r\eta, Z_N) - \frac{1}{N|\zeta_k - r\eta|} = \frac{1}{N} \sum_{j=1, j\neq k}^{N} \frac{1}{|\zeta_j - r\eta|}, \qquad \eta \in S \tag{4.18}$$

must be minimal for $\eta = \zeta_k$. Hence $U(r\eta, Z_N)$ must be stationary at $\eta = \zeta_k$, and this will be so for each k. Thus one would expect $U(x, Z_N) - 1$ to be of stationary type $(r, 1, N)$.

If Conjecture 4.2.4 is true for some $r \in (0, 1)$, then Proposition 4.2.3 will establish the following.

Basic Conjecture 4.2.5 For $r \in (0, 1)$ there exist positive constants $B = B(r)$ and $c = c(r)$ such that for all $N \geq N_0(r)$,

$$\inf_{Z_N} \sup_{|x|=r} |U(x, Z_N) - 1| \leq Be^{-c\sqrt{N}}. \tag{4.19}$$

Since $|U(x, Z_N) - 1| \leq 2/(1 - R)$ for $|x| = R < 1$, it readily follows from Theorem 5.1.1 in Chapter 5 that the validity of (4.19) for some $r \in (0, 1)$ implies its validity for every $r \in (0, 1)$ with suitable B and $c > 0$.

By Corollary 1.3.2 the preceding conjecture is equivalent to a conjecture on Chebyshev-type quadrature:

Conjecture 4.2.6 There exist positive constants c_1, c_2, c_3 such that for Chebyshev-type quadrature on S based on the special N-tuples Z_N of Conjecture 4.2.5,

$$|R(f, Z_N)| \leq c_1 e^{-c_2\sqrt{N}} \sup_S |f| \tag{4.20}$$

for all polynomials f of degree $\leq c_3\sqrt{N}$.

Thus the plausibility of Conjecture 4.2.4 offers additional support for the conjecture $N_S(p) = O(p^2)$ in Section 1.2.

4.3 Intermezzo: an auxiliary result in complex analysis of several variables

We need a special result on smallness of bounded analytic functions $\phi(z)$ which vanish at many real points. For our application we consider functions in $\mathbf{C}^2$, where we write $z = (z_1, z_2) = x + iy$, but the result easily extends to higher dimensions. Other than in the one-variable Lemma 2.3.2, the given zero points have to be well-separated.

Definition 4.3.1 We say that a function ϕ defined on $D \subset \mathbf{R}^2$ is of zero type (D, δ, s), where $\delta > 0$ and s is a positive integer, if D contains a subset $\{\xi_1, \ldots, \xi_s\}$ of s zero points of ϕ with separation constant $2\delta/\sqrt{s}$:

$$|\xi_j - \xi_k| \geq 2\delta/\sqrt{s}, \qquad j, k = 1, \ldots, s, \quad j \neq k. \tag{4.21}$$

Proposition 4.3.2 *Let D be the closure of a bounded convex domain in $\mathbf{R}^2$ $(y = 0)$, let Ω be a $\mathbf{C}^2$ neighborhood of D, let $E \subset \Omega$ be compact and let δ be a positive constant. Then there is a constant $b > 0$ depending only on D, Ω, E and δ, such that for every bounded analytic function $\phi(z)$ on Ω which is of zero type (D, δ, s) for some s,*

$$\sup_E |\phi| \leq \sup_\Omega |\phi| \cdot e^{-b\sqrt{s}}. \tag{4.22}$$

[*There is a corresponding n-dimensional result with $s^{1/n}$ instead of $\sqrt{n}$ in* (4.21) *and* (4.22).]

The proof will be based on the following known results. We refer to the literature for proofs of these lemmas.

Lemma 4.3.3 (P. Lelong [Le50], H. Rutishauser [Rut50], cf. [Ru80, p.386]) *Let $f(z)$ be an analytic function on the ball $B(0,r)$ in $\mathbf{C}^2$ which vanishes at the origin. Then the zero set $Z(f)$ of f in $B(0,r)$ has area*

$$A_0(r) \geq \pi r^2.$$

Observe that πr^2 is the area of the intersection $B(0,r) \cap \{z_1 = 0\}$.

[*There is a corresponding result for $\mathbf{C}^n$ involving real $(2n-2)$-dimensional volume.*]

Lemma 4.3.4 (B. Berndtsson [B78]) *Let D be a compact convex set in $\mathbf{R}^2$ and let D_λ $(\lambda > 0)$ denote the λ-neighborhood of D in $\mathbf{C}^2$. For an analytic function $\phi(z)$ on a $\mathbf{C}^2$ neighborhood Ω of D, zero set $Z(\phi)$, and for $D_\lambda \subset \Omega$, we let $A_D(\lambda)$ denote the area of $Z(\phi) \cap D_\lambda$. Then $A_D(\lambda)/\lambda$ is a nondecreasing function of λ.*

[*In the n-dimensional case $A_D(\lambda)/\lambda^{n-1}$ is nondecreasing.*]

Lemma 4.3.5 (An extension of Jensen's theorem to $\mathbf{C}^2$, see [Ru80, p.385]) *Let f be analytic on the ball $\bar{B}(0,r)$ in $\mathbf{C}^2$ and let $A_0(t) = A_0(t, f)$ denote the area of $Z(f) \cap B(0,t)$. Then*

$$\int_0^r \frac{A_0(t,f)}{t^3} dt = \int_S \log|f(r\zeta)| d\sigma(\zeta) - \log|f(0)| \tag{4.23}$$

where σ denotes normalized area measure on $S = S(0,1)$ in $\mathbf{C}^2$.
[In $\mathbf{C}^n$, the exponent 3 becomes $2n-1$.]

For the following *three-regions theorem* in $\mathbf{C}^n$ we refer to E. Bishop [Bi63, p.475], cf. also Theorem 5.1.2 below. The continuity of $\gamma(E,\Omega_0)$ in Lemma 4.3.6 follows easily from the standard proof which uses coverings by balls.

Lemma 4.3.6 *Let Ω be a (connected) domain in $\mathbf{C}^n$, Ω_0 a non-empty open subset and E a compact subset of Ω. Then there is a constant $\gamma \in (0,1]$ depending only on E, Ω_0 and Ω such that for all analytic functions $\phi(z)$ on Ω,*

$$\sup_E |\phi| \le (\sup_{\Omega_0} |\phi|)^\gamma (\sup_\Omega |\phi|)^{1-\gamma}. \tag{4.24}$$

For fixed Ω and simple relatively compact open subsets Ω_0 such as balls, the best constant $\gamma = \gamma(E,\Omega_0)$ depends continuously on E and Ω_0.

Proof of Proposition 4.3.2 Let D, Ω, E and δ be as in the Proposition and let $\phi(z)$ be a bounded analytic function on Ω of zero type (D,δ,s). We let $\{\xi_1,\ldots,\xi_s\}$ denote a corresponding set of s zero points of ϕ in D as in Definition 4.3.1. It may be assumed that $\sup_\Omega |\phi| = 1$ and that $\Omega \ne \mathbf{C}^2$ (or there is nothing to prove). We now set $\rho = d(D,\partial\Omega)/5$ and reduce δ if necessary so that $\delta \le \rho$. As in Lemma 4.3.4, D_λ will be the λ-neighborhood of D in $\mathbf{C}^2$, $Z(\phi)$ the zero set of ϕ and $A_D(\lambda)$ the area of $Z(\phi)\cap D_\lambda$. We similarly write $A_z(\lambda)$ or $A_z(\lambda,\phi)$ for the area of $Z(\phi)\cap B(z,\lambda)$.

By the Lelong-Rutishauser Theorem (Lemma 4.3.3), the intersections of $Z(\phi)$ with the balls $B(\xi_k,\delta/\sqrt{s})$ in Ω have area $A_{\xi_k}(\delta/\sqrt{s}) \ge \pi\delta^2/s$, $1 \le k \le s$. By the separation condition (4.21) these balls in $D_{\delta/\sqrt{s}}$ are disjoint, hence

$$A_D(\delta/\sqrt{s}) \ge s\pi\delta^2/s = \pi\delta^2.$$

Next using Berndtsson's Theorem (Lemma 4.3.4), we conclude that

$$A_D(\rho) \ge \frac{A_D(\delta/\sqrt{s})}{\delta/\sqrt{s}}\rho \ge \pi\delta\rho\sqrt{s}. \tag{4.25}$$

The (closure of the) set D_ρ may be covered by a finite (minimal) number $\nu = \nu(D,\rho)$ of balls $B(x,2\rho)$ with center x in D. At least one of these ν balls, say $B(x_0,2\rho)$, must meet $Z(\phi)$ in a set of area

$$A_{x_0}(2\rho) \ge (\pi\delta\rho/\nu)\sqrt{s} \stackrel{\text{def}}{=} a\sqrt{s}.$$

Taking an arbitrary point $z \in B(x_0,\rho)$, we next apply "Jensen's Theorem" (Lemma 4.3.5) to $\phi(w)$ on the ball $\bar{B}(z,4\rho)$ in Ω. Setting $f(w) = \phi(z+w)$ for $|w| \le r = 4\rho$ so that $A_0(t,f) = A_z(t,\phi) = A_z(t)$, relation (4.23) gives

$$\log|\phi(z)| = \log|f(0)| = \int_S \log|\phi(z+4\rho\zeta)|d\sigma(\zeta) - \int_0^{4\rho} A_z(t)t^{-3}dt. \tag{4.26}$$

Observe that $|\phi| \le 1$ while for $(4\rho \ge)\ t \ge 3\rho$, $B(z,t)$ contains $B(x_0, t-\rho)$ so that

$$A_z(t) \ge A_{x_0}(t-\rho) \ge A_{x_0}(2\rho) \ge a\sqrt{s}.$$

Hence (4.26) implies that for every $z \in B(x_0, \rho)$,

$$\log|\phi(z)| \le -\int_{3\rho}^{4\rho} A_z(t)t^{-3}dt \le -\int_{3\rho}^{4\rho} a\sqrt{s}t^{-3}dt \overset{\text{def}}{=} -c\sqrt{s}, \tag{4.27}$$

where $c > 0$ depends only on D, Ω, E and δ.

For the proof of (4.22) we still have to pass from $B(x_0,\rho)$ to E. In this step one may apply the three-regions theorem (Lemma 4.3.6) to the function $\phi(z)$, using the present sets Ω and E while taking $\Omega_0 = B(x_0,\rho)$. By (4.27) and (4.24),

$$\log|\phi(z)| \le -c\sqrt{s}\gamma(E,\Omega_0), \qquad \forall z \in E. \tag{4.28}$$

Here E is fixed but we don't know where in D the center x_0 of $\Omega_0 = B(x_0,\rho)$ is located. However, by Lemma 4.3.6 the function

$$\mu(x) \overset{\text{def}}{=} \gamma(E, B(x,\rho)), \qquad x \in D$$

is positive and continuous. Since D is compact, $\mu(x)$ will have a positive minimum μ. The final conclusion from (4.28) is

$$\log|\phi(z)| \le -\mu c\sqrt{s} \qquad (\text{with } \mu c > 0) \qquad \forall z \in E. \qquad \square$$

4.4 Proof of the Basic Proposition 4.2.3

We will deal in detail with the first part of Proposition 4.2.3. Accordingly, let

$$U(x) - 1 = U(x, Z_N) - 1 = \frac{1}{N}\sum_{j=1}^{N} \frac{1}{|\zeta_j - x|} - 1$$

be a potential of zero type (r,δ,M) with $M \ge 10$. For $U(r\eta) - 1$, $\eta \in S$ we let $\eta_1, \dots, \eta_M$ be M well-distributed, well-separated zero points as in Definition 4.2.2.

Setting $R = \max(r, 4/5) < 1$ we introduce the closed disc D in the (x_1,x_2)-plane given by $x_1^2 + x_2^2 \le R^2$ and we let $\Sigma \subset S$ be the spherical cap which lies above D. Since $R \ge 4/5$, the area $(1-\sqrt{1-R^2})2\pi$ of Σ will be $\ge (1/5)$ area S. Hence by our hypothesis, the cap Σ and each of its rotations about the origin contain $\ge M/10$ points η_k.

Observe that for $\eta = (x_1, x_2, \sqrt{1-x_1^2-x_2^2}) \in \Sigma$ with $(x_1,x_2,0) \in D$,

$$\begin{aligned} U(r\eta) - 1 &= \frac{1}{N}\sum_{j=1}^{N}\left\{1 + r^2 - 2r\left(\zeta_{j1}x_1 + \zeta_{j2}x_2 + \zeta_{j3}\sqrt{1-x_1^2-x_2^2}\right)\right\}^{-\frac{1}{2}} - 1 \\ &= W(x_1,x_2), \end{aligned} \tag{4.29}$$

say. For $z = (z_1, z_2)$ ranging over a $\mathbf{C}^2$ neighborhood Ω of D to be specified below, we now introduce the following analytic function, the "complexified potential"

$$W(z) = \frac{1}{N}\sum_{j=1}^{N}\left\{1 + r^2 - 2r\left(\zeta_{j1}z_1 + \zeta_{j2}z_2 + \zeta_{j3}\sqrt{1 - z_1^2 - z_2^2}\right)\right\}^{-\frac{1}{2}} - 1, \qquad (4.30)$$

where we take the holomorphic branches of the roots that are positive on D. Abusing the notation, we will henceforth write $(x_1, x_2, 0) = (x_1, x_2) = x$ for points of D (until now, x was the point (x_1, x_2, x_3) in $\mathbf{R}^3$).

By the preceding, $U(r\eta) - 1$ vanishes at $s \geq M/10$ points η_k on the cap Σ; it is convenient to rename these points $\eta_1, \ldots, \eta_s$. Then by (4.29)

$$W(x) = 0 \qquad \text{for } x = \xi_k \stackrel{\text{def}}{=} (\eta_{k1}, \eta_{k2}) \in D, \qquad k = 1, \ldots, s. \qquad (4.31)$$

It will follow from the hypothesis (cf. Definition 4.2.2) that our points $\xi_1, \ldots, \xi_s$ are well-separated in the sense of Definition 4.3.1:

$$|\xi_j - \xi_k| \geq 2\delta_1/\sqrt{s}, \qquad j, k = 1, \ldots, s, \quad j \neq k \qquad (4.32)$$

with a constant $\delta_1 > 0$ that depends only on δ and r. In fact, an admissible δ_1 may be obtained from our original δ by two reductions. The projection of Σ on D causes a reduction by a positive factor depending only on R, hence on r, and the change-over from M to s causes an additional reduction by a factor $\sqrt{10}$ $(1/\sqrt{M} \geq 1/\sqrt{10s})$.

In order to obtain a $\mathbf{C}^2$ neighborhood Ω of D on which $W(z)$ is holomorphic and $|W(z)|$ has a finite upper bound independent of N, we may define Ω as the connected open set whose points (z_1, z_2) are constrained by the inequalities

$$|z_1|^2 + |z_2|^2 < R, \qquad (|z_1|^2 + |z_2|^2 + |1 - z_1^2 - z_2^2|)^{1/2} < (1 + r)^2/4r. \qquad (4.33)$$

These inequalities are satisfied by the points $z = x$ of D since $R > R^2$ and $(1 + r)^2/4r > 1$. Moreover, a simple calculation will show that

$$|W(z)| < A = 3/(1 - r), \qquad \forall z \in \Omega. \qquad (4.34)$$

By (4.30)–(4.34) the function $W(z)$ on Ω is of zero type (D, δ_1, s) in the sense of Definition 4.3.1. Hence we may apply Proposition 4.3.2 to $\phi(z) = W(z)$ on Ω. Taking $E = D$ and using the inequality $s \geq M/10$, we conclude that there is a constant $c = b/\sqrt{10} > 0$ depending only on D, Ω and δ_1 such that

$$|W(x)| < Ae^{-c\sqrt{M}}, \qquad \forall x \in D. \qquad (4.35)$$

Thus in view of (4.29),

$$|U(r\eta) - 1| < Ae^{-c\sqrt{M}} \qquad (4.36)$$

for all points $\eta \in \Sigma$. The same inequality will hold on every spherical cap obtained from Σ by rotation about the origin in $\mathbf{R}^3$, hence (4.36) holds for all $\eta \in S$. That is, we have (4.15).

We finally remark that A depends only on r while D, Ω and δ_1, and hence c, depend only on r and δ. This observation completes the proof of the first part of Proposition 4.2.3.

The proof of the second part is similar, but here one would apply Proposition 4.3.2 to $\phi(z) = \partial W/\partial z_1$. Integrating the resulting inequality for $\partial W/\partial x_1$ on D from 0 to $(x_1, 0)$ and rotating about the origin in $\mathbf{R}^2$ one will obtain an analog to (4.35) for $W(x) - W(0)$. Returning to Σ and rotating about the origin in $\mathbf{R}^3$, one may conclude that $|U(r\eta) - 1|$ has its oscillation on S bounded by $Be^{-c\sqrt{M}}$. Since $U(r\eta) - 1$ has average zero on S, (4.15) will follow. □

4.5 Important open questions

Principal problem 4.5.1 Prove (or disprove) the Basic Conjecture 4.2.5!

For a proof it would be sufficient to show that there is a constant $\gamma > 0$ such that for every large N and minimal $G_{\frac{1}{2}}(\zeta_1, \ldots, \zeta_N)$ (4.17), there is a subset of $M \geq \gamma N$ of the N points ζ_j which satisfies a distribution and separation condition as in (i), (ii) of Definition 4.2.2.

Question 4.5.2 (Faraday cage for the case of equal point charges) Consider the special N-tuples Z_N^* on S for which the corresponding system of N point charges $1/N$ has *minimal potential energy*

$$V(Z_N) = \frac{1}{N^2} \sum_{j,k=1, j\neq k}^{N} \frac{1}{|\zeta_j - \zeta_k|}.$$

Let K be a "fat" compact subset of the interior of S, for example a ball. How small does

$$\epsilon_N(K) = \sup_{x \in K} |U(x, Z_N^*) - 1|$$

become as $N \to \infty$?

Several years ago, the author proved that $\epsilon_N(K) = \mathcal{O}(1/\sqrt{N})$, see [K76]. The same order estimate probably holds in the case of general smooth surfaces in $\mathbf{R}^3$ of the topological type of a sphere. Can one do substantially better for the sphere?

Cf. the case of charges $1/N$ at Nth order "Fekete points" on smooth Jordan curves in $\mathbf{R}^2$, where one can do better than $\epsilon_N(K) = \mathcal{O}(1/N)$ only for circles, see [KG71, K74, KK83].

Chapter 5
Propagation of smallness for harmonic functions: logarithmic convexity of supremum norms

References: [KM92] and for background material on (sub)harmonic functions: W.K. Hayman and P.B. Kennedy [HK76].

5.1 Introduction and results

In the course of our research on the Electron Problem for the sphere (Section 1.3), we noticed that bounded harmonic functions which are "exponentially small" on an open part of the unit ball are "exponentially small" on every compact subset of the ball. We found later that related "propagation of smallness" had been observed before, cf. D.H. Armitage, T. Bagby and P.M. Gauthier [ABG85] and the Remarks 5.1.3 below.

For *analytic* functions there have been precise results on propagation of smallness for a long time. The oldest result is *Hadamard's three-circles theorem* for analytic functions $f(z)$ on an annulus $(0 <)\ \rho \le |z| \le R\ (< \infty)$ in $\mathbf{C}$. Denoting $\max_\theta |f(re^{i\theta})|$ by $M(r)$, one has

$$|f(z)| \le M(r) \le M(\rho)^\beta M(R)^{1-\beta}, \qquad \rho < |z| = r < R \tag{5.1}$$

where $\beta \in (0,1)$ is the "Hadamard exponent":

$$\beta = \beta_H(\frac{\rho}{R}, \frac{r}{R}) = \frac{\log r/R}{\log \rho/R}. \tag{5.2}$$

In particular, if $|f(z)| \le 1$ in the annulus and $M(\rho) \le e^{-A}$, then $M(r) \le e^{-\beta A}$. Observe also that $\log M(r)$ is a convex function of $\log r$ on annuli. Hadamard's inequality (5.1), (5.2) has a direct extension to analytic functions on a spherical shell in $\mathbf{C}^n$, cf. [Bi63].

A more important extension is the so-called *two-constants theorem* for analytic functions in $\mathbf{C}$, see R. Nevanlinna [N36, ch.III, §2]. The precise result depends on the concept of *harmonic measure*. To keep things simple, let D be a bounded (connected) domain in $\mathbf{C}$ whose boundary Γ consists of a finite number of Jordan curves. For a nonempty open boundary arc Γ_0, the harmonic measure relative to D is defined as the bounded harmonic function $\omega(z)$ on D with boundary values 1 on Γ_0 and 0 on $\Gamma - \bar{\Gamma}_0$. Here $0 < \omega(z) \le 1$ and for all bounded analytic $f(z)$ on D with boundary values of absolute value $\le M_0$ on Γ_0 and $\le M_1$ on $\Gamma - \bar{\Gamma}_0$,

$$|f(z)| \le M_0^{\omega(z)} M_1^{1-\omega(z)}, \qquad \forall z \in D. \tag{5.3}$$

The proof uses the fact that $v(z) = \log|f(z)|$ is subharmonic. Subharmonic functions are majorized by harmonic functions with the same or larger boundary values. Hence in our case

$$\log|f(z)| \le \omega(z) \log M_0 + \{1 - \omega(z)\} \log M_1$$

which gives (5.3).

There can be no three-circles or three-spheres theorem similar to (5.1) for harmonic functions. Indeed, a harmonic function on the spherical shell $\rho \le |x| \le R$ in $\mathbf{R}^n$ may vanish on the inner sphere $S(0,\rho)$ without vanishing identically. Also, for harmonic u, the use of harmonic measure only gives an arithmetic inequality which is much weaker than (5.1) when $|u|$ is very small on $S(0,\rho)$, cf. Lemma 5.2.3. Thus it may be surprising that there is an analog to (5.1) for harmonic functions on a *ball*:

Theorem 5.1.1 (Three-balls theorem, [KM92]) *Suppose* $0 < \rho < r < R < \infty$ *and* $n \ge 2$. *Then there exists a constant* $\alpha \in (0,1)$ *depending only on* ρ/R, r/R *and* n *such that* *for all complex-valued harmonic functions u on the ball $B(0,R)$ in $\mathbf{R}^n$,*

$$\|u\|_r \le \|u\|_\rho^\alpha \|u\|_R^{1-\alpha} \qquad \text{where } \|u\|_s = \sup_{B(0,s)} |u(x)|. \tag{5.4}$$

A proof will be presented in Section 5.3. We observe here that Theorem 5.1.1 can be extended to a "three-regions theorem" analogous to Lemma 4.3.6 by a standard covering argument, cf. [Bi63].

Theorem 5.1.2 (Three-regions theorem, [KM92]) *Let Ω be a (connected) domain in $\mathbf{R}^n$, $n \ge 2$, $\Omega_0 \subset \Omega$ a nonempty open subset and $E \subset \Omega$ a compact subset (which may be just a point). Then there is a constant $\alpha \in (0,1]$ depending only on E, Ω_0 and Ω such that for all complex-valued harmonic functions u on Ω,*

$$\sup_E |u| \le (\sup_{\Omega_0} |u|)^\alpha (\sup_\Omega |u|)^{1-\alpha}. \tag{5.5}$$

Remarks 5.1.3 Various authors have obtained three-balls theorems with an additional constant $C = C(\rho/R, r/R, n)$:

$$\|u\|_r \le C\|u\|_\rho^\alpha \|u\|_R^{1-\alpha}. \tag{5.6}$$

The first of these was probably E.M. Landis [L63]. A different three-balls theorem for harmonic functions was obtained by V.P. Zahariuta, cf. [Z93] and his earlier papers. Landis actually proved an inequality (5.6) for solutions of second order linear elliptic partial differential equations. Since then one has used methods of functional analysis to obtain inequalities (5.6) in rather general abstract spaces "of power series type", see D. Vogt [V82] and the references given there. Quite recently, R.G.M. Brummelhuis [Br93b] has succeeded in removing the constant C in (5.6) for the partial differential equations case. We finally remark that E.D. Solomentsev [So66] has obtained a fairly complicated three-spheres theorem for harmonic functions in which the right-hand side also involves the normal derivative of u.

5.2 Auxiliary results

The following lemmas list some useful auxiliary results for complex-valued harmonic functions in $\mathbf{R}^n$ ($n \ge 2$).

Lemma 5.2.1 (Poisson integral) *Let u be harmonic on the closed ball $\bar{B}(0,R)$ in $\mathbf{R}^n$. Then*

$$u(x) = \frac{1}{\sigma_n} \int_{S(0,R)} \frac{R^2 - |x|^2}{R|y-x|^n} u(y)\,ds(y), \qquad \forall x \in B(0,r). \tag{5.7}$$

Here ds denotes the ordinary (non-normalized) area element and $\sigma_n = 2\pi^{n/2}/\Gamma(n/2)$ is the surface area of the unit sphere $S = S(0,1)$ in $\mathbf{R}^n$.

Setting $x = r\xi$ where $\xi \in S$, $u(x)$ may also be represented by a Laplace series. Denoting the orthogonal projection of $u(R\xi)$ on H_k by $Y_k(R\xi)$, cf. Section 1.4, one finds that

$$u(r\xi) \sim \sum_{k=0}^{\infty} Y_k(r\xi) = \sum_{k=0}^{\infty} Y_k(\xi) r^k, \qquad 0 \leq r \leq R. \tag{5.8}$$

By this representation one has the following identity for L^2 norms on spheres:

$$\|u\|_{r,2}^2 \stackrel{\text{def}}{=} \frac{1}{\sigma_n r^{n-1}} \int_{S(0,r)} |u(x)|^2 ds(x) = \frac{1}{\sigma_n} \int_S |u(r\xi)|^2 ds(\xi) = \sum_{k=0}^{\infty} \|Y_k\|_{L^2(S)}^2 r^{2k}. \tag{5.9}$$

Hence by Hadamard's three-circles theorem (5.1) applied to $f(z) = \sum \|Y_k\|^2 z^{2k}$ or by direct computation, $\log \|u\|_{r,2}$ is a convex function of $\log r$:

Lemma 5.2.2 *For harmonic functions u on $\bar{B}(0, R)$,*

$$\|u\|_{r,2} \leq \|u\|_{\rho,2}^{\beta} \|u\|_{R,2}^{1-\beta}, \qquad 0 < \rho < r < R \tag{5.10}$$

where β is the Hadamard exponent $\beta_H(\frac{\rho}{R}, \frac{r}{R})$ of (5.2).

The maximum principle for harmonic functions gives a corresponding arithmetic inequality for sup norms $\|u\|_r$ on spheres $S(0, r)$:

Lemma 5.2.3 *For $0 < \rho < r < R$ and harmonic u on the closed spherical shell $\bar{B}(0, R) - B(0, \rho)$,*

$$\|u\|_r \leq \begin{cases} \dfrac{r^{2-n} - R^{2-n}}{\rho^{2-n} - R^{2-n}} \|u\|_\rho + \dfrac{\rho^{2-n} - r^{2-n}}{\rho^{2-n} - R^{2-n}} \|u\|_R & \text{if } n \geq 3, \\[2ex] \dfrac{\log r - \log R}{\log \rho - \log R} \|u\|_\rho + \dfrac{\log \rho - \log r}{\log \rho - \log R} \|u\|_R & \text{if } n = 2. \end{cases} \tag{5.11}$$

We will also need the following special integral:

Lemma 5.2.4 *For $n \geq 2$ and $a > b > 0$,*

$$\frac{\sigma_{n-1}}{\sigma_n} \int_0^\pi \frac{\sin^{n-2}\theta}{(a - b\cos\theta)^n} d\theta = \frac{a}{(a^2 - b^2)^{(n+1)/2}}. \tag{5.12}$$

For the proof one may start with the integral of $(a - b\cos\theta)^{-1}$ and carry out differentiations with respect to a and b.

5.3 Proof of the Three-balls Theorem

We begin with some preliminary reductions. Changing scale in Theorem 5.1.1 we may take $R = 1$, so that $0 < \rho < r < 1$. If u is unbounded on $B = B(0, 1)$ or if $u \equiv 0$ there is nothing to prove, hence we may assume that u is bounded and that $\|u\|_1 = \sup_B |u| = 1$. Setting

$u_\tau(x) = u(\tau x)$ with $\tau \uparrow 1$ one has $u(x) = \lim u_\tau(x)$ on B, uniformly on compact subsets. Thus $\|u_\tau\|_s \to \|u\|_s$ if $s < 1$ while $\|u_\tau\|_1 = \|u\|_\tau \le \|u\|_1$. Hence it is sufficient to prove Theorem 5.1.1 for functions (such as u_τ) which are harmonic on the closed unit ball.

We may finally assume the following normalizations, based on operations which do not increase the norms $\|u\|_s$ $(0 < s \le 1)$ and which leave the norm for $s = r$ unchanged.

(i) For the given r, the norm $\|u\|_r$ is equal to the value of u at the point re_1. (One may rotate about the origin and multiply by a constant of absolute value 1);

(ii) u is real-valued. (With (i) satisfied, u may be replaced by Re u);

(iii) u has axial symmetry about the x_1-axis. (With (i) satisfied, symmetrization of u with respect to the x_1-axis – which replaces u by an average of rotations of u – leads to a function $\tilde{u}$ with $\|\tilde{u}\|_r = u(re_1) = \tilde{u}(re_1)$).

Definition 5.3.1 For fixed $0 < \rho < r < 1$, $n \ge 2$ and $0 < \epsilon \le 1$, $H_\epsilon = H(\epsilon, \rho, r, n)$ will denote the class of those harmonic functions u on the closed unit ball $\bar{B}(0,1)$ in $\mathbf{R}^n$ that satisfy (i) - (iii) and for which

$$\|u\|_1 = 1, \qquad \|u\|_\rho \le \epsilon. \tag{5.13}$$

Setting

$$m(\epsilon) = m(\epsilon, \rho, r, n) \stackrel{\text{def}}{=} \sup_{u \in H_\epsilon} \|u\|_r = \sup_{u \in H_\epsilon} u(re_1), \tag{5.14}$$

it follows from the preceding that Theorem 5.1.1 is equivalent to the following

Proposition 5.3.2 *There is a constant $\alpha \in (0,1)$ (depending on ρ, r, n) such that*

$$m(\epsilon) \le \epsilon^\alpha, \qquad 0 < \epsilon \le 1. \tag{5.15}$$

Proof (a) A *first majorant* $m_1(\epsilon)$ for $m(\epsilon)$ is obtained from Lemma 5.2.3 with $R = 1$: by (5.13),

$$m(\epsilon) = \sup_{u \in H_\epsilon} \|u\|_r \le m_1(\epsilon) \stackrel{\text{def}}{=} c_1\epsilon + 1 - c_1, \tag{5.16}$$

where $c_1 = c_1(\rho, r, n)$ is the constant between 0 and 1 given by

$$c_1 = \frac{r^{2-n} - 1}{\rho^{2-n} - 1} \text{ if } n \ge 3, \qquad c_1 = \frac{\log r}{\log \rho} \text{ if } n = 2.$$

However, the proof of (5.15) requires a better majorant for small ϵ, one that tends to 0 sufficiently rapidly as $\epsilon \downarrow 0$!

(b) To obtain a *second majorant* $m_2(\epsilon)$ for $m(\epsilon)$ we will exploit the L^2 estimate (5.10) in conjunction with the Poisson integral. Let u be any function in H_ϵ. Replacing r in (5.10) by a parameter $t \in (r, 1]$ to be specified later and observing that the L^2 norms on the right-hand side of (5.10) are majorized by the sup norms, Lemma 5.2.2 and (5.13) give

$$\|u\|_{t,2} \le \|u\|_{\rho,2}^{\beta} \|u\|_{1,2}^{1-\beta} \le \|u\|_\rho^\beta \|u\|_1^{1-\beta} \le \epsilon^\beta, \tag{5.17}$$

where $\beta = \beta_H(\rho, t) = \log t/\log \rho$.

We now use the Poisson integral (5.7) for $u(re_1)$ on the ball $\bar{B}(0,t)$. Substituting $r/t = \lambda \in [r, 1)$ and $y = t\eta$ with $\eta \in S = S(0,1)$ so that $ds(y) = t^{n-1}ds(\eta)$, we find

$$\begin{aligned} u(re_1) &= \frac{1}{\sigma_n}\int_{S(0,t)} \frac{t^2 - r^2}{t|y - re_1|^n} u(y)ds(y) \\ &= \frac{1}{\sigma_n}\int_S \frac{1-\lambda^2}{|\eta - \lambda e_1|^n} u(t\eta)ds(\eta) \\ &= (1-\lambda^2)\frac{1}{\sigma_n}\int_S \frac{1}{(1+\lambda^2 - 2\lambda\eta_1)^{n/2}} u(t\eta)ds(\eta). \end{aligned} \tag{5.18}$$

We next apply Cauchy-Schwarz to the final integral to obtain

$$u(re_1) \le (1-\lambda^2)\left\{\frac{1}{\sigma_n}\int_S \frac{ds(\eta)}{(1+\lambda^2-2\lambda\eta_1)^n}\right\}^{\frac{1}{2}} \left\{\frac{1}{\sigma_n}\int_S |u(t\eta)|^2 d\sigma(\eta)\right\}^{\frac{1}{2}}. \tag{5.19}$$

Here the first average A_1 over S may be evaluated by substituting $\eta = (\cos\theta, \sin\theta\cdot\eta')$ where η' runs over the unit sphere S' in $(x_2, \ldots, x_n)$-space, so that $ds(\eta) = (\sin\theta)^{n-2}d\theta\ ds(\eta')$. Integrating over S' and using Lemma 5.2.4 with $a = 1+\lambda^2$, $b = 2\lambda$ we get

$$A_1 = \frac{1}{\sigma_n}\int_0^\pi \frac{(\sin\theta)^{n-2}}{(1+\lambda^2 - 2\lambda\cos\theta)^n} d\theta \cdot \sigma_{n-1} = \frac{1+\lambda^2}{(1-\lambda^2)^{n+1}}. \tag{5.20}$$

The final factor in (5.19) is simply $\|u\|_{t,2}$ for which we have the estimate (5.17).

Combination of (5.17)–(5.20) shows that for every $u \in H_\epsilon$,

$$u(re_1) \le (1-\lambda^2)\left\{\frac{1+\lambda^2}{(1-\lambda^2)^{n+1}}\right\}^{\frac{1}{2}} \epsilon^\beta < \sqrt{2}(1-\lambda^2)^{\frac{1}{2}(1-n)}\epsilon^{(\log\frac{r}{\lambda})/\log\rho}. \tag{5.21}$$

This inequality holds for every $\lambda = r/t \in [r, 1)$; a simple choice will be $\lambda = t = \sqrt{r}$. Thus (5.21) gives a second majorant for $m(\epsilon)$:

$$m(\epsilon) = \sup_{u\in H_\epsilon} u(re_1) \le m_2(\epsilon) \stackrel{\text{def}}{=} c_2\epsilon^{\frac{1}{2}\log r/\log\rho} \tag{5.22}$$

with

$$c_2 = c_2(r, n) = \sqrt{2}(1-r)^{\frac{1}{2}(1-n)}.$$

(c) *Conclusion.* Observe that the majorants $m_1(\epsilon)$ (5.16) and $m_2(\epsilon)$ (5.22) of $m(\epsilon)$ on $(0, 1]$ are both increasing. The graph of $m_1(\epsilon)$ is a straight line and $m_1(0+) = c_1 \in (0,1)$, $m_1(1) = 1$. The graph of $m_2(\epsilon)$ is concave and $m_2(0+) = 0$, $m_2(1) = c_2 > 1$. Thus the graphs intersect at exactly one point $\epsilon_0 \in (0,1)$. It follows that

$$m(\epsilon) \le M(\epsilon) \stackrel{\text{def}}{=} \begin{cases} m_1(\epsilon) = c_1\epsilon + 1 - c_1 & \text{for } \epsilon_0 \le \epsilon \le 1, \\ m_2(\epsilon) = c_2\epsilon^{\frac{1}{2}\log r/\log\rho} & \text{for } 0 < \epsilon \le \epsilon_0. \end{cases} \tag{5.23}$$

Determining $\alpha \in \mathbf{R}$ such that $\epsilon_0^\alpha = M(\epsilon_0)$ one readily sees that $M(\epsilon)$ is majorized by ϵ^α on $(0, 1]$. Indeed, since $\epsilon_0^{\frac{1}{2}\log r/\log\rho} < M(\epsilon_0) = \epsilon_0^\alpha < 1$, α must be between 0 and $\frac{1}{2}\log r/\log\rho$, hence $\epsilon^\alpha > M(\epsilon)$ on $(0, \epsilon_0)$. Also, $\epsilon^\alpha > M(\epsilon)$ on $(\epsilon_0, 1)$ because the graph of ϵ^α is concave. This completes the proof of (5.15). □

5.4 Open problems

Problem 5.4.1 Let H be the class of all harmonic functions on the unit ball $B = B(0,1)$ in $\mathbf{R}^n$ $(n \geq 2)$ such that $\|u\|_1 = \sup_B |u| = 1$. Prove the following conjecture. For arbitrary given $0 < \rho < r < 1$ and β equal to the Hadamard exponent $\log r / \log \rho$, there is no finite constant $K = K(\rho, r, n)$ such that

$$\|u\|_r = \sup_{B(0,r)} |u| \leq K \|u\|_\rho^\beta = K(\sup_{B(0,\rho)} |u|)^\beta$$

for all functions $u \in H$.

Problem 5.4.2 (Notation as in Problem 5.4.1). It follows from the proof of Proposition 5.3.2 (cf. (5.21)) that for every exponent $\alpha \in (0, \beta)$ there is a finite constant $K = K(\alpha, \rho, r, n)$ such that

$$\|u\|_r \leq K \|u\|_\rho^\alpha, \qquad \forall u \in H.$$

Determine the smallest possible constant K for $\alpha = \frac{1}{2}\beta$.

Problem 5.4.3 Determine the optimal (largest possible) constant α in the Three-balls Theorem 5.1.1.

For the subclass of the positive harmonic functions on $B(0, R)$, the paper [KM92] contains an explicit expression for the optimal exponent α.

References

[AG75] Anderson, L.A. and Gautschi, W., Optimal weighted Chebyshev-type quadrature formulas, *Calcolo* **12** (1975), 211–248.

[ABG85] Armitage, D.H., Bagby, T. and Gauthier, P.M., Note on the decay of solutions of elliptic equations, *Bull. London Math. Soc.* **17** (1985), 554–556.

[B78] Berndtsson, B., Zeros of analytic functions of several variables, *Ark. Mat.* **16** (1978), 251–262.

[Be12] Bernstein, S.N., Sur l'ordre de la meilleure approximation des fonctions continues par des polynomes de degré donné, *Mém. Acad. Royale Belgique (2)* **4** (1912), 1–104.

[Be37a] Bernstein, S.N., Sur les formules de quadrature de Cotes et Tchebycheff, *C.R. Acad. Sci. URSS (Dokl. Akad. Nauk SSSR), N.S.* **14** (1937), 323–327.

[Be37b] Bernstein, S.N., On quadrature formulas with positive coefficients, *Izv. Akad. Nauk SSSR, Ser. Mat.* **1**, No. 4 (1937), 479–503 (Russian). See also the announcements in *C.R. Acad. Sci. Paris* **204** (1937), 1294–1296 and 1526–1529.

[Be38] Bernstein, S.N., Sur un système d'équations indéterminées, *J. Math. Pures Appl. (9)* **17** (1938), 179–186.

[Bi63] Bishop, E., Holomorphic completions, analytic continuation, and the interpolation of semi-norms, *Ann. of Math.* **78** (1963), 468–500.

[Br93a] Brummelhuis, R.G.M., Logarithmic convexity of L^2 norms for solutions of linear elliptic equations, *Indag. Math. N.S.* **4** (1993), 423–429.

[Br93b] Brummelhuis, R.G.M., Three-spheres theorems for second order elliptic equations, preprint, Leiden University, 1993.

[Ch1874] Chebyshev, P.L., Sur les quadratures, *J. Math. Pures Appl. (2)* **19** (1874), 19–34; Oeuvres vol. II, Chelsea, New York, 1962, 165–180.

[DGS77] Delsarte, P., Goethals, J.M. and Seidel, J.J., Spherical codes and designs, *Geom. Dedicata* **6** (1977), 363–388.

[F93] Förster, K.-J., Variance in quadrature – a survey, in: *Numerical Integration IV* (H. Brass and G. Hämmerlin, eds.), Birkhäuser, Basel, 1993, 91–100.

[Fr35] Frostman, O., Potentiel d'équilibre et capacité des ensembles avec quelques applications à la théorie des fonctions, dissertation, *Lunds Univ. Mat. Sem.* **3** (1935), 1–118.

[GV78] Gatteschi, L. and Vinardi, G., Sul grado di precisione di formule di quadratura del tipo di Tchebycheff, *Calcolo* **15** (1978), 59–85.

[Ga76] Gautschi, W., Advances in Chebyshev quadrature, in: *Numerical Analysis* (Proc. 6th Dundee Conf., G.A. Watson ed.), Lecture Notes in Math. **506**, Springer, Berlin - Heidelberg - New York, 1976, 100–121.

[GaY74] Gautschi, W. and Yanagiwara, H., On Chebyshev-type quadratures, *Math. Comp.* **28** (1974), 125–134.

[HK76] Hayman, W.K. and Kennedy, P.B., *Subharmonic Functions*, vol. 1, London Math. Soc. Monographs **9**, Academic Press, London, 1976.

[K64] Korevaar, J., Asymptotically neutral distributions of electrons and polynomial approximation, *Ann. of Math.* **80** (1964), 403–410.

[K74] Korevaar, J., Equilibrium distributions of electrons on roundish plane conductors, *Nederl. Akad. Wetensch. Proc. Ser. A* **77** = *Indag. Math.* **36** (1974), 423–456.

[K76] Korevaar, J., Problems of equilibrium points on the sphere and electrostatic fields, Univ. of Amsterdam, Math. Dept. Report 1976-03.

[K92] Korevaar, J., Behavior of Cotes numbers and other constants, with an application to Chebyshev-type quadrature, *Indag. Math. N.S.* **3** (1992), 391–402.

[KG71] Korevaar, J. and Geveci, T., Fields due to electrons on an analytic curve, *SIAM J. Math. Anal.* **2** (1971), 445–453.

[KK83] Korevaar, J. and Kortram, R.A., Equilibrium distributions of electrons on smooth plane conductors, *Nederl. Akad. Wetensch. Proc. Ser. A* **86** = *Indag. Math.* **45** (1983), 203–219.

[KM92] Korevaar, J. and Meyers, J.L.H., Logarithmic convexity for supremum norms of harmonic functions, Univ. of Amsterdam, Math. Dept. Report 1992-13, to appear in *Bull. London Math. Soc.* 1994.

[KM93a] Korevaar, J. and Meyers, J.L.H., Massive coalescence of nodes in optimal Chebyshev-type quadrature on $[-1,1]$, *Indag. Math. N.S.* **4** (1993), 327–338.

[KM93b] Korevaar, J. and Meyers, J.L.H., Chebyshev-type quadrature on multidimensional domains, Univ. of Amsterdam, Math. Dept. Report 1993-01, to appear in *J. Approx. Theory*, 1994.

[KM93c] Korevaar, J. and Meyers, J.L.H., Spherical Faraday cage for the case of equal point charges and Chebyshev-type quadrature on the sphere, *Integral Transforms Special Functions* **1** (1993), 105–117.

[Ku93a] Kuijlaars, A.B.J., The minimal number of nodes in Chebyshev type quadrature formulas, *Indag. Math. N.S.* **4** (1993), 339–362.

[Ku93b] Kuijlaars, A.B.J., Chebyshev type quadrature for Jacobi weight functions, to appear in *J. Comput. Appl. Math.*

[Ku93c] Kuijlaars, A.B.J., Chebyshev type quadrature and partial sums of the exponential series, Univ. of Amsterdam, Math. Dept. Report 1993-07.

[L63] Landis, E.M., A three-spheres theorem, *Dokl. Akad. Nauk SSSR* **148** (1963), 277–279 (Russian). English translation in *Soviet Math. Doklady* **4**, no. 1 (1963), 76–78.

[Le50] Lelong, P., Propriétés métriques des variétés analytiques complexes définies par une équation, *Ann. Sci. École Norm. Sup. (3)* **67** (1950), 393–419.

[N36] Nevanlinna, R., *Eindeutige analytische Funktionen*, Grundlehren Math. Wiss. **46**, Springer-Verlag, Berlin, 1936. English translation of second edition: *Analytic Functions*, Grundlehren Math. Wiss. **162**, Springer-Verlag, Berlin, 1970.

[RB91] Rabau, P. and Bajnok, B., Bounds on the number of nodes in Chebyshev-type quadrature formulas, *J. Approx. Theory* **67** (1991), 199–215.

[Ru80] Rudin, W., *Function Theory in the Unit Ball of* $\mathbf{C}^n$, Grundlehren Math. Wiss. **241**, Springer-Verlag, Berlin - Heidelberg - New York, 1980.

[Rut50] Rutishauser, H., Über Folgen und Scharen von analytischen und meromorphen Funktionen mehrerer Variabeln, sowie von analytischen Abbildungen, *Acta Math.* **83** (1950), 249–325.

[Se93] Seidel, J.J., Isometric embeddings and geometric designs, to appear in *Trends in Discrete Math.*

[SeZ84] Seymour, P.D. and Zaslavsky, T., Averaging sets: a generalization of mean values and spherical designs, *Adv. Math.* **52** (1984), 213–240.

[So66] Solomentsev, E.D., A three-spheres theorem for harmonic functions, *Akad. Nauk Armyan. SSR Dokl.* **42** (1966), 274–278 (Russian).

[SW71] Stein, E.M. and Weiss, G., *Introduction to Fourier Analysis on Euclidean Spaces*, Princeton Univ. Press, Princeton, NJ, 1971.

[Sz75] Szegő, G., *Orthogonal Polynomials*, Amer. Math. Soc. Colloq. Publ. **23**, Providence, RI, fourth edition 1975.

[U66] Ullman, J.L., A class of weight functions that admit Tchebycheff quadrature, *Michigan Math. J.* **13** (1966), 417–423. See also the announcement in *Bull. Amer. Math. Soc.* **72** (1966), 1073–1075.

[V82] Vogt, D., Charakterisierung der Unterräume eines nuklearen stabilen Potenzreihenraumes von endlichem Typ, *Studia Math.* **71** (1982), 251–270.

[Z93] Zahariuta, V.P., Spaces of harmonic functions, in: *Functional Analysis* (K.D. Bierstedt et al., eds.), Lecture Notes in Pure and Appl. Math. **150**, Marcel Dekker, New York, 1993.

General aspects of potential theory with respect to problems of differential equations

Nikolai N. TARKHANOV
Max-Planck-Gesellschaft
Arbeitsgruppe "Analysis"
Universität Potsdam
Postfach 60 15 53
D-14415 Potsdam
Germany

and

Institute of Physics
Siberian Section, Russian Academy of Science
Akademgorodok
660036 Krasnoyarsk
Russia

Abstract

These lectures were intended as an attempt to bring together various topics in partial differential equations related to potential theory. We will restrict our attention to elliptic equations. In section 1 we have compiled some basic facts on solutions regular at infinity. These are nothing but potentials of compactly supported distributions. In section 2 some of the recent results on removable singularities of solutions are discussed. Section 3 is devoted to the study of approximation by solutions in Sobolev spaces. In the last two sections we indicate how these techniques may be used to analyze the Cauchy problem for solutions of elliptic equations. In section 4 we give a brief exposition of the theory of bases with double orthogonality which were elaborated as tools for constructive approximations. In section 5 we proceed with the study of the Cauchy problem by the modified Fischer-Reisz equations method.

Introduction

In the classical theory of DE's the potential method is understood to be a method of investigating boundary value problems for equations of mathematical physics by means of representing solutions in the form of potentials and reducing the problems to integral equations. In the early 70's, a wide class of problems of the theory of DE's connected with

[1]This research was supported by the Alexander-von-Humboldt Foundation.

P. M. Gauthier (ed.) and G. Sabidussi (techn. ed.), Complex Potential Theory, 365–418.

removable singularities and approximation, led to the need for studying not only potentials of measures supported on a smooth hypersurface or in a smoothly bounded domain, but also potentials of arbitrary compactly supported distributions. The kernels of such potentials are fundamental solutions or parametrixes of DO's. So they do not satisfy the non-singularity condition accepted in non-linear potential theory. The purpose of these lectures is to give an introduction, and a survey of parts of the potential theory with respect to problems of DE's. Then the interested reader should be able to find his way through the theory by means of the bibliography. The bibliography does not claim to be complete, but it is not limited to papers mentioned in the text.

The author wishes to thank Prof. V. Zakharyuta for fruitful discussions.

Chapter 1
Solutions regular at infinity

1.1 Fundamental solutions of elliptic equations with real analytic coefficients

Let $P \in \mathrm{do}_p(X)$ be an elliptic DO of order p with real analytic coefficients, defined on an open set $X \subset \mathbb{R}^n$.

We write

$$P(x, D) = \sum_{|\alpha| \leq p} P_\alpha(x) \cdot D^\alpha,$$

where for a multi-index $\alpha = (\alpha_1, \ldots, \alpha_n)$ we set $|\alpha| = \alpha_1 + \cdots + \alpha_n$, and $D^\alpha = D_1^{\alpha_1} \ldots D_n^{\alpha_n}$ with

$$D_j = \frac{1}{\sqrt{-1}} \frac{\partial}{\partial x_j}.$$

Then ellipticity means that

$$\sum_{|\alpha| = p} P_\alpha(x) z^\alpha \neq 0$$

for all $x \in X$ and $z \in \mathbb{R}^n \setminus \{0\}$.

Denote by

$$P'(x, D) = \sum_{|\alpha| \leq p} (-1)^{|\alpha|} D^\alpha (P_\alpha(x) \times .)$$

the transpose operator of P.

Theorem 1.1 *P has a fundamental solution, i.e., there exists a distribution $\Phi(x, y) \in \mathcal{D}'(X \times X)$ such that*

$$P'(y, D)\Phi(x, y) = P(x, D)\Phi(x, y) = \delta(x - y)$$

on $X \times X$, where $\delta(.)$ is the Dirac measure.

Proof See Malgrange [36]. □

The distribution $\Phi(x,y)$ is actually a real analytic function off the diagonal $\Delta = \{(x,x) : x \in X\}$. As to the singularities of Φ near Δ, they are completely described by Seeley's theorem [45] because $\Phi(x,y)$ is the Schwartz kernel of some pseudodifferential operator $\Phi \in \mathrm{pdo}_{-p}(X)$ of order $(-p)$ on X.

Remark 1.2 In particular, $\Phi(.,.) \in L^1_{\mathrm{loc}}(X \times X)$ and for each compact set $K \subset X$ there are constants $c_{\alpha,\beta,\gamma}$ such that

$$\left|D_x^\alpha D_y^\beta (D_x + D_y)^\gamma \Phi(x,y)\right| \leq c_{\alpha,\beta,\gamma} |x-y|^{p-n-|\alpha|-|\beta|}$$

for all $(x,y) \in K \times K$ provided $|\alpha| + |\beta| > p - n$.

Any two fundamental solutions of P differ by a real analytic function on $X \times X$ satisfying $P'(y,D). = P(x,D). = 0$.

1.2 Solutions regular at infinity

Under our assumptions on the DO P, it is analytically hypoelliptic. This means that every weak solution $f \in \mathcal{D}'(\mathcal{O})$ of the equation $Pf = 0$ on an open set $\mathcal{O} \subset X$ is real analytic in $\mathcal{O}$.

Certainly, an open set is the natural domain of a solution of the equation $Pf = 0$. However, some problems require the consideration of solutions on sets $\sigma \subset X$ which are not open. Here we are interested not simply in restrictions of solutions to the given set, but also in the so-called local solutions of the equation $Pf = 0$ on σ, that is, solutions of this equation in a neighborhood of σ.

The space of local solutions of the equation $Pf = 0$ on σ will be denoted by $\mathcal{S}_P(\sigma)$, or simply $\mathcal{S}(\sigma)$ if it is clear which operator is being considered.

Assuming that σ has a relatively compact complement, how to define a solution $f \in \mathcal{S}(\sigma)$ regular at infinity? We are going to do this in such a way that a Liouville type theorem should hold. However, this depends on a compactification of X.

We shall use the so-called one-point compactification, or Alexandrov compactification, of X denoted by $\hat{X}$. That means $\hat{X}$ is the union of X and the symbolic point ∞, and the topology of $\hat{X}$ is given by the following bases of neighborhoods of points. If $x \in X$ is a "usual" point of $\hat{X}$, then we take the usual basis of neighborhoods of x (for example, the family of all balls centered at x). If $x = \infty$, then the basis of neighborhoods of x is defined to be the family $\{\mathcal{O} \cup \infty\}$, where $\mathcal{O}$ is an open subset of X with compact complement.

Now we fix some fundamental solution Φ of the DO P.

Definition 1.3 For a set $\sigma \subset X$ with relatively compact complement, a solution $f \in \mathcal{S}(\sigma)$ is said to be *regular at infinity* if in a neighborhood of ∞ we have $f = \Phi(F)$, where $F \in \mathcal{E}'(X)$.

We emphasize that this definition depends in an essential way on the choice of the fundamental solution Φ.

When a set $\sigma \subset \hat{X}$ contains ∞, $\mathcal{S}_P(\sigma)$ (or $\mathcal{S}(\sigma)$) means the space of solutions of the equations $Pf = 0$ in a neighborhood of σ in $\hat{X}$, regular at infinity.

Example 1.4 $\mathcal{S}(\infty) = \{\Phi(F) : F \in \mathcal{E}'(X)\}$. □

1.3 Green's formula for solutions regular at infinity

Denote by $G_P(.,.)$ a Green operator for the DO P. It is a bidifferential operator of order $(p-1)$ with values in the space of $(n-1)$-forms on X, defined by the equality $dG_P(g,f) = (g(Pf) - (P'g)f)\,dx$ that should hold for all $g, f \in \mathcal{E}(X)$.

Theorem 1.5 *Suppose that $\mathcal{O}$ is an open subset of X with compact complement and piecewise smooth boundary. Then a solution $f \in \mathcal{S}(\hat{\mathcal{O}})$ is regular at infinity if and only if*

$$f(x) = -\int_{\partial\mathcal{O}} G_P(\Phi(x,y), f(y)) \quad (x \in \mathcal{O}). \tag{1.1}$$

Proof Necessity. It suffices to consider the case when $\mathcal{O}$ is connected. Then (1.1) is fulfilled for all $x \in \mathcal{O}$ if and only if it is fulfilled for $x \in \mathcal{O}$ "large" enough. In order to prove (1.1) for $x \in \mathcal{O}$ "large" enough, we may use the equality $f = \Phi(F)$ because the integral on the right hand side of (1.1) is actually independent of the cycle $\partial\mathcal{O}$.

Sufficiency. It is obvious because (1.1) implies that $f = -\Phi([\partial\mathcal{O}]^P f)$ in $\mathcal{O}$ with a distribution $[\partial\mathcal{O}]^P f$ supported on $\partial\mathcal{O}$. □

This theorem was actually proved by Grothendieck [17].

1.4 Separating regular part

What topology is natural in the (vector) space $\mathcal{S}(\sigma)$?

It easily follows from Theorem 1.1 that if $\mathcal{O}$ is an open subset of X, then two topologies in $\mathcal{S}(\mathcal{O})$ coincide, one induced from $\mathcal{D}'(\mathcal{O})$ and another induced from $\mathcal{E}(\mathcal{O})$. Being endowed with one of these topologies $\mathcal{S}(\mathcal{O})$ is a Frechet-Schwartz space. Moreover, if $\mathcal{O}$ is an open subset of X with compact complement, then $\mathcal{S}(\mathcal{O} \cup \infty)$ is a closed subspace of $\mathcal{S}(\mathcal{O})$ being endowed with the induced topology. To prove this, use Theorem 1.5. Now, for an arbitrary set $\sigma \subset \hat{X}$, we endow $\mathcal{S}(\sigma)$ with the topology of inductive limit of the spaces $\mathcal{S}(\mathcal{O})$, where $\mathcal{O}$ is a neighborhood of σ in $\hat{X}$. Then $\mathcal{S}(\sigma)$ is a complete locally convex space.

Remark 1.6 If σ is open in $\hat{X}$, then the topology of $\mathcal{S}(\sigma)$ coincides with the Frechet-Schwartz topology determined above.

Suppose that σ is a relatively compact subset of X. The space $\mathcal{S}(\hat{X}\backslash\sigma)$ seems to depend on the particular fundamental solution Φ that is used in Definition 1.3. However, this is not the case, as follows from Theorem 1.7 below.

Theorem 1.7 *If σ is a relatively compact subset of X, then*

$$\mathcal{S}(X\setminus\sigma) = \mathcal{S}(X) \overset{\text{top.}}{\oplus} \mathcal{S}(\hat{X}\setminus\sigma). \tag{1.2}$$

Proof Choose a sequence $\{\mathcal{O}_\nu\}$ of open subsets of X with piecewise smooth boundaries, such that $\mathcal{O}_\nu \Subset \mathcal{O}_{\nu+1}$ and $\bigcup_{\nu=1}^{\infty} \mathcal{O}_\nu = X$.

For $f \in \mathcal{S}(X\setminus\sigma)$, we set

$$f_e(x) = \lim_{\nu\to\infty} - \int_{\partial\mathcal{O}_\nu} G_P(\Phi(x,y), f(y)), \quad (x \in X).$$

It is easy to see that for each fixed $x \in X$ the sequence

$$\left\{ - \int_{\partial\mathcal{O}_\nu} G_P(\Phi(x,.), f) \right\}$$

stabilizes starting with some ν. Hence it follows that $f_e \in \mathcal{S}(X)$.

Let $f_r = f - f_e$. Then $f_r \in \mathcal{S}(X\backslash\sigma)$. Moreover, if $\mathcal{O}$ is an open subset of X with compact complement and piecewise smooth boundary such that $\bar{\mathcal{O}} \subset X\setminus\sigma$, then $f_r = -\Phi([\partial\mathcal{O}]^P f)$ in $\mathcal{O}$ because of Green's formula. So f_r is regular at infinity.

To complete the proof it suffices to use Theorem 1.5 and the explicit formulas for the functions f_e and f_r. □

This theorem is due to Grothendieck [17].

1.5 Examples

Example 1.8 Let $P = \partial/\partial\bar{z}$ be the Cauchy-Riemann operator in the complex plane $X = \mathbb{C}^1$, where $z = x_1 + \sqrt{-1}\,x_2$ and $\partial/\partial\bar{z} = \frac{1}{2}(\partial/\partial x_1 + \sqrt{-1}\,\partial/\partial x_2)$. Then each distribution

$$\Phi(x,y) = \frac{1}{\pi}\frac{1}{z-\zeta} + \Phi_0(x,\zeta)$$

is a fundamental solution of P, where $\Phi_0(.,.)$ is an entire function in $\mathbb{C}^2$. A function f, holomorphic in the complement of a compact subset of $\mathbb{C}^1$, is said to be *regular at infinity* if

$$f = (\frac{1}{\pi}\frac{1}{\zeta}) * F + \Phi_0(F)$$

outside a ball B for some distribution F supported on $\bar{B}$. In particular, if $\Phi_0 = 0$, then f is regular at infinity if and only if $f(\infty) = 0$. □

Example 1.9 More generally, suppose that P is an open elliptic homogeneous DO of order $p < n$ with constant coefficients in $\mathbf{R}^n$. Such a DO is known to have a fundamental solution of the form

$$\Phi(x,y) = \Phi(\frac{x-y}{|x-y|}) \cdot |x-y|^{p-n},$$

where $\Phi(.)$ is a real analytic function in a neighborhood of the unit sphere in $\mathbf{R}^n$. If we are using this fundamental solution, then in order that a solution f of $Pf = 0$ in the complement of a compact subset of $\mathbf{R}^n$ be regular at infinity it is necessary and sufficient that $f(\infty) = 0$. □

1.6 Dual space of the space of solutions

The following theorem has its roots in the paper of Grothendieck [17].

Theorem 1.10 *For each (open or closed) subset σ of X there is an isomorphism*

$$\mathcal{S}_P(\sigma)' \overset{\text{top.}}{\cong} \mathcal{S}_{P'}(\hat{X} \setminus \sigma). \tag{1.3}$$

Proof Consider the following pairing $[\partial\sigma]^P$ of $\mathcal{S}_{P'}(\hat{X} \setminus \sigma)$ and $\mathcal{S}_P(\sigma)$.

Let $g \in \mathcal{S}_{P'}(\hat{X} \setminus \sigma)$ and $f \in \mathcal{S}_P(\sigma)$. Then there exists an open set $\mathcal{O} \Subset X$ with piecewise smooth boundary, such that $g \in \mathcal{S}_{P'}(\hat{X} \setminus \mathcal{O})$ and $f \in \mathcal{S}_P(\mathcal{O})$. So we are allowed to set

$$[\partial\sigma]^P(g \otimes f) = \int_{\partial\mathcal{O}} G_P(g, f).$$

This bilinear form is separately continuous on $\mathcal{S}_{P'}(\hat{X} \setminus \sigma) \times \mathcal{S}_P(\sigma)$. Moreover, it is nondegenerate in the first argument, as follows from Theorem 1.5 and Runge's theorem of section 3.3.

Thus we have embedded $\mathcal{S}_{P'}(\hat{X} \setminus \sigma)$ into the space of continuous linear functionals on $\mathcal{S}_P(\sigma)$.

To prove that this embedding is an epimorphism, we notice that, for each functional $F \in \mathcal{S}_P(\sigma)'$, its "Fantappie indicatrix" $F(\Phi(.,y)) = \Phi'(F)(y)$ is in $\mathcal{S}_{P'}(\hat{X} \setminus \sigma)$ and satisfies $F(f) = [\partial\sigma]^P(-\Phi'(F) \otimes f)$ for all $f \in \mathcal{S}_P(\sigma)$.

Finally, the topological character of the isomorphism (1.3) follows directly from its explicit form. □

In particular, if σ is a relatively compact subset of X, then (1.3) can be rewritten in a form independent of the compactification of X, namely,

$$\mathcal{S}_P(\sigma)' \overset{\text{top.}}{\cong} \frac{\mathcal{S}_{P'}(X \setminus \sigma)}{\mathcal{S}_{P'}(X)}.$$

1.7 Solutions with compact singularities

The most general interpretation of singularities is to consider functions of $\mathcal{S}(X \setminus \sigma)$ as solutions of the equation $Pf = 0$ with singularities on σ.

A (non-negative) measure m on σ is said to be *massive* if $m(\Sigma) = 0$ for a subset Σ of σ implies that Σ has no interior points on σ.

Example 1.11 Choose a sequence $\{y_\nu\}$ of points of σ dense in σ, and a sequence $\{m_\nu\}$ of positive numbers such that $\sum_{\nu=1}^{\infty} m_\nu < \infty$. For a subset Σ of σ, we set $m(\Sigma) = \sum_{y_\nu \in \Sigma} m_\nu$. Then m is a massive measure on σ. □

Havin [21] defined the class of regular compact sets $K \subset \mathbb{R}^n$ by the condition that the (real) analytic functions on K, whose Taylor expansions converge in balls of radius $\geq r$, may be extended to analytic functions in a neighborhood of K dependent on r only. In particular, each locally connected compact subset of $\mathbb{R}^n$ was proved to be regular.

Suppose we are given some regular compact set K in X, and a massive measure m on K.

Theorem 1.12 *For f to be a solution of $Pf = 0$ with singularities on K, it is necessary and sufficient that there exist a solution $f_e \in \mathcal{S}(X)$ and a sequence $\{c_\alpha\} \subset L^2(m)$ satisfying*

$$\lim_{|\alpha| \to \infty} ||\alpha! c_\alpha||_{L^2(m)}^{1/|\alpha|} = 0,$$

such that

$$f(x) = f_e(x) + \sum_\alpha \int_K D_y^\alpha \Phi(x, y) \cdot c_\alpha(y) dm(y) \quad (x \in X \setminus K). \tag{1.4}$$

Proof Necessity. The idea is to describe the eigen-topology of $\mathcal{S}_{P'}(K)$ by means of radial convergence of germs at points of K, and then to use the duality between $\mathcal{S}_P(\hat{X} \setminus K)$ and $\mathcal{S}_{P'}(K)$ given in the proof of Theorem 1.10.

Sufficiency. It follows directly from the uniform estimates of derivatives of (real) analytic functions. Namely, for each compact set $k \subset (X \times X) \setminus \Delta$, there exist constants c and α such that

$$\sup_{(x,y) \in k} |D_x^\alpha D_y^\beta \Phi(x, y)| \leq c|\alpha + \beta|! \sigma^{|\alpha + \beta|}$$

for all multi-indices α and β of $\mathbb{Z}_+^n$.

See Tarkhanov [52] for details. □

For holomorphic functions of a complex variable, series of the form (1.4) are called *Golubev series*. Theorem 1.12 here was proved by Havin (see his article [21]).

1.8 Separation of singularities into atomic singularities

One of the consequences of Theorem 1.12 is the possibility of separating compact singularities of solutions of the equation $Pf = 0$ into one-point (i.e., atomic) singularities.

Corollary 1.13 *Suppose that K is a regular compact subset of X, and that $\{y_\nu\}$ is a dense sequence of points in K. Then each solution $f \in \mathcal{S}(X \setminus K)$ can be represented in the form $f = f_e + \sum_\nu f_\nu$ where $f_e \in \mathcal{S}(X)$ and $f_\nu \in \mathcal{S}(X \setminus \{y_\nu\})$, and the series converges in the topology of $\mathcal{S}(X \setminus K)$.*

Proof Apply Example 1.11 and Theorem 1.12. □

1.9 Representation of solutions by boundary integrals

Another application of Theorem 1.12 is a representation by a boundary integral of solutions of the equation $Pf = 0$ on an open set $\mathcal{O} \subset X$ with a pretty "nice" boundary, which have no boundary values in the usual sense. This is a generalization of Green's formula (1.1).

We denote by ds the Lebesgue measure on $\partial\mathcal{O}$ induced by the Lebesgue measure on X.

Corollary 1.14 *If $\mathcal{O}$ is an open subset of $\hat{X}$ with piecewise smooth compact boundary (in X!), then for each solution $f \in \mathcal{S}(\mathcal{O})$ there exists a sequence $\{c_\alpha\} \subset L^2(ds)$ satisfying*

$$\lim_{|\alpha|\to\infty} ||\alpha! c_\alpha||^{1/|\alpha|}_{L^2(ds)} = 0,$$

such that

$$f(x) = \sum_\alpha \int_{\partial\mathcal{O}} D_y^\alpha \Phi(x, y) \cdot c_\alpha(y)\, ds(y) \quad (x \in \mathcal{O}).$$

Proof It suffices to use Theorem 1.12 in the case $K = \partial\mathcal{O}$ and $m = ds$, setting $f = 0$ in the complement of $\bar{\mathcal{O}}$. □

1.10 Solutions of finite order of growth

If $K = \{x^0\}$ is an one-point set, then formula (1.4) becomes

$$f(x) = f_e(x) + \sum_\alpha D_y^\alpha \Phi(x, x^0) c_\alpha(x^0) \quad (x \in X \setminus \{x^0\})$$

where $\lim_{|\alpha|\to\infty} |\alpha! c_\alpha(x^0)|^{1/|\alpha|} = 0$.

Under what conditions does a solution $f \in \mathcal{S}(X \setminus \{x^0\})$ admit such a representation with a finite number of summands, i.e.,

$$f = f_e + \sum_{|\alpha|\le\delta} D_y^\alpha \Phi(\,.\,, x^0) c_\alpha(x^0) \text{ in } X \setminus \{x^0\}?$$

For $\delta \ge p - n$, this is the case if and only if $f(x) = o(|x - x^0|^{p-n-\delta-1})$ as $x \to x^0$. That is, for a solution $f \in \mathcal{S}(X \setminus \{x^0\})$ to be representable by the formula with a finite number of summands, it is necessary and sufficient that the singular point x^0 should be a pole for f.

Therefore solutions $f \in \mathcal{S}(X \setminus K)$, whose expansions (1.4) contain only a finite number of summands, are analogous to meromorphic functions in complex analysis.

Theorem 1.15 *A solution $f \in \mathcal{S}(X \setminus K)$ has an expansion* (1.4) *with a finite number of summands if and only if the functional $[\partial K]^P(.\otimes f)$ on $S_{P'}(K)$ is continuous in the topology defined by the family of seminorms $\{\|D_g^\alpha/\alpha!\|_{L^2(m)}\}_{\alpha\in\mathbf{Z}_+^n}$.*

Proof See Tarkhanov [52]. We emphasize that no regularity of K is needed. □

If a solution $f \in \mathcal{S}(X \setminus K)$ has an expansion (1.4) with a finite number of summands, i.e., $f = f_e + \Phi(\sum_{|\alpha|\le\delta}(-1)^{|\alpha|}D^\alpha(c_\alpha m))$ in $X \setminus K$, then the right hand side of this equality gives an extension of f to a distribution $\tilde{f} \in \mathcal{D}'(X)$ with $P\tilde{f} = \sum_{|\alpha|\le\delta}(-1)^{|\alpha|}D^\alpha(c_\alpha m))$ being supported on K. Moreover f has a finite order of growth near K, as follows from Remark 1.2. Namely,

$$f(x) = o(\mathrm{dist}(x, K)^{p-n-\delta-1})$$

uniformly on compact subsets of X provided $p-n-\delta \le 0$. Conversely, if $f \in \mathcal{S}(X\setminus K)$ can be extended to a distribution $\tilde{f} \in \mathcal{D}'(X)$, then $P\tilde{f}$ is supported on K, and $f = f_e + \Phi(P\tilde{f})$ in $X \setminus K$ with $f_e \in \mathcal{S}(X)$. The distribution on $P\tilde{f}$ has a finite order of singularity, and it may be written in the form $P\tilde{f} = D'(ds)$ with some DO D of order δ on X and the Lebesgue measure ds on K, provided K is assumed to be pretty smooth. Hence it follows again that f has a finite order of growth near K. Finally, we suppose that K is sufficiently smooth, and $f \in \mathcal{S}(X \setminus K)$ has a finite order of growth near K. This means that, for some $\gamma \ge 0$, the function $\mathrm{dist}(x, K)^\gamma f$ may be extended to a bounded function $\tilde{f}$ on X. Therefore the function f can also be extended to a distribution of $\mathcal{D}'(X)$ by regularization of

$$\left(\frac{\tilde{f}}{\mathrm{dist}(x, k)^\gamma}\right)$$

in the sense pf Hadamard's finite part. Thus, for pretty smooth compact sets $K \subset X$, the following properties of $f \in \mathcal{S}(X \setminus K)$ are equivalent:

(1) f is representable by (1.4) with a finite number of summands;

(2) f may be extended to a distribution on the whole set X;

(3) f has a finite order of growth near K.

If K is the boundary of a domain $\mathcal{O} \Subset X$, then the last condition is equivalent to the following one:

(4) f belongs to the Sobolev space $W^{s,q}(\mathcal{O})$ $(= W^{-s,q'}(\mathcal{O})')$ with some negative exponent s.

The next theorem relates to the question. Let $\mathcal{O}$ be a relatively compact domain in X with smooth boundary, and $\{B_j\}_{j=0}^{p-1}$ a Dirichlet system of order $p-1$ on $\partial\mathcal{O}$, where $B_j \in \mathrm{do}_{b_j}(U)$ are DO's of order b_j defined in a neighborhood U of $\partial\mathcal{O}$. This means that: (1) $\{B_j\}$ is normal, i.e., $b_i \ne b_j$ for $i \ne j$, and $\sigma(B_j)(y, n(y)) \ne 0$ for all $y \in \partial\mathcal{O}$, where $n(y)$ is the unit vector of the exterior normal to $\partial\mathcal{O}$ at the point y, and (2) $b_j \le p-1$ for all j.

Example 1.16 $B_j = (\partial/\partial n)^j$ $(j = 0, \ldots, p-1)$ is a Dirichlet system of order $(p-1)$ on $\partial\mathcal{O}$. □

Theorem 1.17 *A solution $f \in \mathcal{S}(\mathcal{O})$ is of a finite order of growth near $\partial\mathcal{O}$ if and only if the expressions $B_j f$ $(j = 0, \ldots, p-1)$ have weak limit values $f_j \in \mathcal{D}'(\partial\mathcal{O})$ on $\partial\mathcal{O}$ in the sense that*

$$\lim_{\varepsilon \to +0} \int_{\partial\mathcal{O}} g(y) \cdot B_j f(y - \varepsilon n(y))\, ds(y) = \langle g, f_j \rangle \tag{1.5}$$

for all $g \in \mathcal{D}(\partial\mathcal{O})$.

Proof See Shlapunov and Tarkhanov [47]. This version of the proof used a fundamental result of Rojtberg [43] on boundary values of generalized solutions of elliptic equations. □

For harmonic functions, Theorem 1.17 is due to Straube [51].

1.11 Solutions with closed singularities

For solutions of the equation $Pf = 0$ with singularities on an arbitrary closed subset σ of X, we are able to show only a local version of the expansion (1.4).

Let σ be a smooth closed hypersurface in X. For a point $x^0 \in \sigma$, we choose a relatively compact neighborhood $\mathcal{O}$ in X. Then the part of σ lying in $\mathcal{O}$ has finite Lebesgue measure, i.e., $\int_{\sigma\cap\mathcal{O}} ds < \infty$, where ds is the induced Lebesgue volume form on σ.

Theorem 1.18 *For each solution $f \in \mathcal{S}(\mathcal{O}\setminus\sigma)$, there exists a solution $f_e \in \mathcal{S}(\mathcal{O})$ and a sequence $\{c_\alpha\} \subset L^q(\sigma \cup \mathcal{O})$ $(1 \le q < \infty)$ satisfying*

$$\lim_{|\alpha|\to\infty} ||\alpha! c_\alpha||^{1/|\alpha|}_{L^q(\sigma\cap\mathcal{O})} = 0,$$

such that

$$f(x) = f_e(x) + \sum_\alpha \int_{\sigma\cap\mathcal{O}} D_y^\alpha \Phi(x, y) \cdot c_\alpha(y)\, ds(y) \quad (x \in \mathcal{O}\setminus\sigma). \tag{1.6}$$

Proof The main idea in the proof is the duality between $\mathcal{S}_P(\mathcal{O}\setminus\sigma)$ and $\mathcal{S}_{P'}(\hat{X}\setminus(\mathcal{O}\setminus\sigma))$ given by Theorem 1.10. We use also some facts from the theory of locally convex spaces. See Fischer and Tarkhanov [14] for details. □

The theorem asserts that a solution of $Pf = 0$ in $\mathcal{O}\setminus\sigma$ can be split into two parts, one of which has the only singularity at infinity (of $\hat{\mathcal{O}}$), and another part (the sum of the potentials $\Phi((-1)^{|\alpha|}D^\alpha(c_\alpha ds))$) carries all the finite singularities and has a "weak singularity" at infinity of $\hat{\mathcal{O}}$ (it is bounded on the tangent half-sets $\{x \in \mathcal{O} : \mathrm{dist}(x, \sigma\cap\mathcal{O}) \ge \varepsilon\}$).

For functions holomorphic off the real axis, a representation like (1.6) was proved by Baernstein [5].

1.12 Hyperfunctions

Sato hyperfunctions generalize Schwartz distributions. Loosely speaking, a hyperfunction on $\sigma\cap\mathcal{O}$ is supposed to be like a "boundary jump" $f(y + 0\cdot n(y)) - f(y - 0\cdot n(y))$ of some solution $f \in \mathcal{S}(\mathcal{O}\setminus\sigma)$, where $n(y)$ is the unit vector of the "exterior" normal to $\sigma \cap \mathcal{O}$ at a point $y \in \sigma \cap \mathcal{O}$. Precisely, the vector space of hyperfunctions on $\sigma \cap \mathcal{O}$ is defined to be the quotient of the space of all solutions in $\mathcal{O}\setminus\sigma$ modulo the space of all solutions in $\mathcal{O}$. Usually, the DO P has been chosen to be the Laplace operator in $\mathcal{O}$ (see Schapira [44]).

From Theorem 1.18 we obtain the following result concerning the representation of hyperfunctions by integrals over $\sigma \cap \mathcal{O}$.

Corollary 1.19 *Every hyperfunction on $\sigma \cap \mathcal{O}$ has a representing solution of the form*

$$\sum_{\alpha} \int_{\sigma\cap\mathcal{O}} D_y^{\alpha}\Phi(x,y)\cdot c_{\alpha}(y)\, ds(y) \quad (x \in \mathcal{O}\setminus\sigma)$$

with $\lim_{|\alpha|\to\infty} \|\alpha! c_{\alpha}\|_{L^q(\sigma\cap\mathcal{O})}^{1/|\alpha|} = 0$ $(1 \le q < \infty)$.

Proof This is obvious because of (1.6). □

Chapter 2
Removable singularities of solutions

2.1 Removable singularities

Let $\mathcal{O}$ be a relatively compact set in X, and σ a closed subset of $\mathcal{O}$.

The following concept was elaborated in the papers of Bochner [9], Littman [35], and Harvey and Polking [18].

Definition 2.1 Given a class $\mathfrak{C}$ of distributions in $\mathcal{O}$, the set σ is said to be *removable for $\mathfrak{C}$ with respect to P* if each distribution $f \in \mathfrak{C}$, satisfying $Pf = 0$ in $\mathcal{O}\setminus\sigma$, also satisfies $Pf = 0$ weakly everywhere in $\mathcal{O}$.

Since the operator P is analytically hypoelliptic, each distribution $f \in \mathcal{D}'(\mathcal{O})$ satisfying $Pf = 0$ may be represented by a real analytic function.

Our programme is a more-or-less systematic assault on the question: describe removable sets σ as explicitly as possible for $\mathfrak{C} = W^{s,q}_{\mathrm{loc}}(\mathcal{O})$, where s is a non-negative integer and $1 < q < \infty$.

2.2 Sufficient conditions for Sobolev spaces

We first recall the definition of Hausdorff measure.

Let $\varphi(r)$ $(r > 0)$ be some increasing function satisfying $\varphi(r) \to 0$ as $r \to 0$.

For a set $\sigma \subset X$ and any ε $(0 < \varepsilon \le +\infty)$, we set

$$h_{\varphi}^{(\varepsilon)}(\sigma) = \inf \sum_{\nu} \varphi(r_{\nu}),$$

where the infimum is taken over all (countable) coverings $\{B_\nu\}$ of σ by balls with radius $r_\nu \le \varepsilon$.

Clearly, $h_\varphi^{(\varepsilon_1)}(\sigma) \le h_\varphi^{(\varepsilon_2)}(\sigma)$ if $\varepsilon_2 \le \varepsilon_1$, so

$$h_\varphi(\sigma) = \lim_{\varepsilon\to 0} h_\varphi^{(\varepsilon)}(\sigma)$$

exists. This is the *Hausdorff measure* of σ with respect to φ.

If $\varphi(r) = r^d$ $(d > 0)$, we write $h_d(\sigma)$ instead of $h_\varphi(\sigma)$.

Remark 2.2 The set function $h_\varphi^{(\infty)}(\sigma)$ is sometimes called *Hausdorff content.* One can prove that $h_\varphi^{(\infty)}(\sigma) = 0$ if and only if $h_\varphi(\sigma) = 0$.

For properties of these set functions, we refer to Carleson [11].

Denote by q' the dual index for q, i.e., $\frac{1}{q} + \frac{1}{q'} = 1$.

Theorem 2.3 *Suppose that $s < p$ is an integer, and $1 < q < \infty$. If $h_{n-(p-s)q'}(K) < \infty$ for every compact set $K \subset \sigma$, then σ is removable for $W^{s,q}_{\text{loc}}(\mathcal{O})$ with respect to P.*

Proof Let $f \in W^{s,q}_{\text{loc}}(\mathcal{O})$ satisfy $Pf = 0$ in $\mathcal{O}\setminus\sigma$. Fix an arbitrary test function $g \in \mathcal{D}(\mathcal{O})$, and set $K = \sigma \cap \operatorname{supp} g$.

If a function $\varphi \in \mathcal{D}(\mathcal{O})$ is equal to 1 in a neighborhood of K, then we get

$$|\langle g, Pf\rangle_{\mathcal{O}}| = |\langle \varphi g, Pf\rangle_{\mathcal{O}}| \le c\|\varphi\|_{W^{p-s,q'}(\mathcal{O})} \cdot \|f\|_{W^{s,q}(\operatorname{supp} g)}$$

with some constant c independent of φ.

Therefore

$$|\langle g, Pf\rangle_{\mathcal{O}}| = c \left(\inf_{\substack{\varphi\in\mathcal{D}(\mathcal{O}):\, \varphi\equiv 1 \text{ in a} \\ \text{neighborhood of } K}} \|\varphi\|_{W^{p-s,q'}(\mathcal{O})} \right) \cdot \|f\|_{W^{s,q}(\operatorname{supp} g)}.$$

One can prove that, if $h_{n-(p-s)q'}(K) < \infty$, then the infimum on the right hand side is equal to zero (see Adams and Polking [1]).

So $\langle g, Pf\rangle_{\mathcal{O}} = 0$, as we wanted to show. □

Theorem 2.3 is due to Harvey and Polking [18].

If σ is a subset of a d-dimensional submanifold of $\mathcal{O}$, then the theorem guarantees removability of σ for $W^{s,q}_{\text{loc}}(\mathcal{O})$ with respect to P provided $q \ge \frac{n-d}{n-d-(p-s)}$. This range for q is sharp as is shown by the example $f = \Phi(\chi_\sigma dh_d)$, where χ_σ is the characteristic function of the set σ.

Remark 2.4 For $1 < q < \infty$, the set σ is removable for $W^{p,q}_{\text{loc}}(\mathcal{O})$ with respect to P if and only if $h_n(\sigma) = 0$. A deep result of Nguyen [40] is that this remains valid for $q = \infty$.

2.3 Special capacities

As a rule, it has not been a success to characterize removable singularities in metric terms.

The classical results of Brelot [10] for harmonic functions and Ahlfors [2] for analytic ones have prompted the idea that it could be done in terms of appropriate capacities. This programme was realized by Harvey and Polking [19].

Definition 2.5 Given a normed space L of distributions on $\mathcal{O}$, the *L-capacity* of a compact set $K \subset \mathcal{O}$ with respect to P, denoted by L-cap(K), is the norm of the linear functional $f \to Pf(1)$ on the subspace $L \cap \mathcal{S}(\hat{\mathcal{O}} \setminus K)$ of L.

Notice that $\mathcal{D}'(\mathcal{O}) \cap \mathcal{S}(\hat{\mathcal{O}} \setminus K) = \{\Phi(F) : F \in \mathcal{E}'_K\}$ as follows from section 1.11.

We extend this definition to all sets $\sigma \subset X$ in the following way:

$$L\text{-cap}(\sigma) = \sup L\text{-cap}(K),$$

where the supremum is taken over all compact sets $K \subset \sigma$.

The following properties are obvious or easy to prove.

Proposition 2.6 *If $\sigma_1 \subset \sigma_2$, then $0 \le L\text{-cap}(\sigma_1) \le L\text{-cap}(\sigma_2)$.*

Proposition 2.7

$$L\text{-cap}(\sigma) = \sup \{|F(1)| : F \in \mathcal{E}'_\sigma,\ \Phi(F) \in L,\ \|\Phi(F)\|_L \le 1\}.$$

As to the relation between the capacity L-cap(σ) and the concept of Fuglede [15], Reshetnjak [42], Meyers [38], and Maz'ja and Havin [37], no capacity contains another. They may be compared only if Φ is a kernel of the type considered by Fuglede [15] etc., i.e., positive and lower semicontinuous. If this is the case, then the difference is that these authors considered potentials of (non-negative) measures while Definition 2.5 deals with potentials of compactly supported distributions on σ.

The capacity of Fuglede [15] etc. is a Choquet capacity. On the other hand, it is not even known whether the analytic (i.e., for $P = \partial/\partial\bar{z}$) capacity $L^\infty(\mathbf{C}^1)$-cap(σ) is semiadditive.

We shall be interested in the capacities $W^{s,q}(\mathcal{O})$-cap(σ), which will simply be denoted by cap(σ) when it does not cause a misunderstanding.

2.4 Capacitary potential

As follows from boundedness theorems for pseudodifferential operators in Sobolev spaces,

$$W^{s,q}(\mathcal{O}) \cap \mathcal{S}(\hat{\mathcal{O}} \setminus K) = \Phi(W_K^{s-p,q}),$$

where $W_K^{s-p,q}$ is the set of all distributions $F \in W^{s-p,q}(\mathcal{O})$ supported by a compact set $K \subset \mathcal{O}$.

Theorem 2.8 *For every compact set $K \subset \mathcal{O}$, there is a function $f_K \in W^{s,q}(\mathcal{O}) \cap \mathcal{S}(\hat{\mathcal{O}} \setminus K)$ such that*

$$||f_K||_{W^{s,q}(\mathcal{O})} \leq 1 \quad \text{and} \quad \text{cap}(K) = |Pf_K(1)|.$$

Proof See Tarkhanov [52]. □

A function f_K with the properties listed in the theorem is said to be a $W^{s,q}(\mathcal{O})$*-capacitary potential* of K with respect to P.

The following result states upper semicontinuity of the capacity $\text{cap}(\sigma)$.

Corollary 2.9 *Let K be a compact subset of $\mathcal{O}$, and $\{K_\nu\}$ a decreasing sequence of compact subsets of $\mathcal{O}$ whose intersection is K. Then*

$$\text{cap}(K) = \lim_{\nu \to \infty} \text{cap}(K_\nu).$$

Proof Use Theorem 2.8. See Tarkhanov [52] for details. □

2.5 Capacity mass

For an integer $s < 0$ and $1 < q < \infty$, the space $W^{s,q}(\mathcal{O})$ is defined to be the completion of $L^q(\mathcal{O})$ with respect to the norm

$$||g||_{W^{s,q}(\mathcal{O})} = \sup_{f \in W^{-s,q'}(\mathcal{O})} \frac{|\langle g, f \rangle_{\mathcal{O}}|}{||f||_{W^{-s,q'}(\mathcal{O})}}. \tag{2.1}$$

Then there is a pairing $\langle , \rangle$ between $W^{s,q}(\mathcal{O})$ and $W^{-s,q'}(\mathcal{O})$ determined by

$$\langle g, f \rangle = \lim_{\nu \to \infty} \langle g_\nu, f \rangle_{\mathcal{O}},$$

where $\{g_\nu\} \subset L^q(\mathcal{O})$ approximates g in $W^{s,q}(\mathcal{O})$, and $W^{s,q}(\mathcal{O})$ and $W^{-s,q'}(\mathcal{O})$ are dual spaces with respect to the pairing.

Using Theorem 2.8, it is easy to give a dual description of the capacity $\text{cap}(\sigma)$.

Proposition 2.10 *For a compact set $K \subset \mathcal{O}$,*

$$\text{cap}(K) = \inf_{\substack{g \in \mathcal{D}(\mathcal{O}):\, g \equiv 1 \text{ in a} \\ \text{neighborhood of } K}} ||P'g||_{W^{-s,q'}(\mathcal{O})}.$$

Proof In fact (with all infimums taken over all $g \in \mathcal{D}(\mathcal{O})$ such that $g \equiv 1$ in a neighborhood of K) we have:

$$\begin{aligned} \text{cap}(K) = |Pf_K(1)| &\leq \inf |\langle P'g, f_K \rangle| \\ &\leq \inf ||P'g||_{W^{-s,q'}(\mathcal{O})}, \end{aligned}$$

while

$$\begin{aligned}\inf \|P'g\|_{W^{-s,q'}(\mathcal{O})} &= \inf \sup_{f \in W^{s,q}(\mathcal{O})} \frac{|\langle P'g, f\rangle_{\mathcal{O}}|}{\|f\|_{W^{s,q}(\mathcal{O})}} \\ &\geq \inf \frac{|\langle P'g, f_K\rangle_{\mathcal{O}}|}{\|f_K\|_{W^{s,q}(\mathcal{O})}} \\ &\geq \inf |\langle P'g, f_K\rangle_{\mathcal{O}}| \\ &= \mathrm{cap}(K).\end{aligned}$$

□

Now we have the following result due to Harvey and Polking [19].

Theorem 2.11 *For every compact set $K \subset \mathcal{O}$, there exists a distribution G_K belonging to the closure of the subspace*

$$\{P'g : g \in \mathcal{D}(\mathcal{O}) \text{ and } g \equiv 1 \text{ in a neighborhood of } K\}$$

in $W^{-s,q'}(\mathcal{O})$, such that

$$\mathrm{cap}(K) = \|G_K\|_{W^{-s,q'}(\mathcal{O})}.$$

Proof See Harvey and Polking [19]. □

A distribution G_K with the properties listed in the theorem is said to be a $W^{s,q}(\mathcal{O})$-*capacity mass* of K with respect to P.

Remark 2.12 It is easy to see that $\mathrm{cap}(K) = |\langle G_K, f_K\rangle|$.

2.6 Examples

Example 2.13 Let $P = \partial/\partial\bar{z}$ be the Cauchy-Riemann operator in the complex plane $X = \mathbf{C}^1$, where $z = x_1 + \sqrt{-1}x_2$. We choose the standard fundamental solution $\Phi(x, y) = 1/(\pi(z - \zeta))$ of $\bar{\partial}$. Given a normed space L of distributions on an open set $\mathcal{O} \Subset \mathbf{C}^1$ and a compact subset $K \subset \mathcal{O}$, we have

$$\begin{aligned}L\text{-cap}(K) &= \sup_{\substack{f \in L \cap \mathcal{S}(\hat{\mathbf{C}}^1 \setminus K) \\ \|f\|_L \leq 1}} \left|[\partial K]^{\bar{\partial}'}(f \otimes 1)\right| \\ &= \sup_{\substack{f \in L \cap \mathcal{S}(\hat{\mathbf{C}}^1 \setminus K) \\ \|f\|_L \leq 1}} \lim_{z \to \infty} \left|\frac{f(z)}{\frac{1}{\pi}\frac{1}{z}}\right| \\ &= \sup_{\substack{f \in L \cap \mathcal{S}(\hat{\mathbf{C}}^1 \setminus K) \\ \|f\|_L \leq 1}} \pi|f'(\infty)|.\end{aligned}$$

This is just the well-known L-analytic capacity of K in complex analysis (modulo the non-essential constant π). □

Example 2.14 More generally, suppose P to be an elliptic homogeneous DO of order $p < n$ with constant coefficients in $\mathbf{R}^n$. We take the fundamental solution

$$\Phi(x,y) = \Phi\left(\frac{x-y}{|x-y|}\right)|x-y|^{p-n}$$

of P as in Example 1.9. Given a normed space L of distributions on an open set $\mathcal{O} \Subset \mathbf{R}^n$ and a compact subset K of $\mathcal{O}$, we have

$$\begin{aligned} L\text{-cap}(K) &= \sup_{\substack{f \in L \cap S(\mathring{\mathbf{R}}^n \setminus K) \\ \|f\|_L \leq 1}} \left|[\partial K]^{P'}(f \otimes 1)\right| \\ &= \sup_{\substack{f \in L \cap S(\mathring{\mathbf{R}}^n \setminus K) \\ \|f\|_L \leq 1}} \lim_{x \to \infty} \left|\frac{f(x)}{\Phi(x,0)}\right|. \end{aligned}$$

To get the last equality we applied the Laurent expansion for solutions of $Pf = 0$ (see Tarkhanov [52]). □

2.7 Description of removable singularities in terms of capacity

It is quite easy to describe removable sets σ using the capacities cap(σ).

Theorem 2.15 *A closed set $\sigma \subset \mathcal{O}$ is removable for $W^{s,q}_{\mathrm{loc}}(\mathcal{O})$ if and only if* cap$(\sigma) = 0$.

Proof Necessity is obvious.

Sufficiency. Fix some distribution $f \in W^{s,q}_{\mathrm{loc}}(\mathcal{O})$ satisfying $Pf = 0$ in $\mathcal{O} \setminus \sigma$. If $Pf = 0$ is not fulfilled on the whole set $\mathcal{O}$, then there exists a test function $\varphi \in \mathcal{D}(\mathcal{O})$ such that $\langle \varphi, Pf \rangle_{\mathcal{O}} \neq 0$. Consider the distribution $f' = \Phi(\varphi Pf)$. Since $f' = \varphi f + \Phi([\varphi, P]f)$ and the commutator $[\varphi, P]$ is a DO of order $(p-1)$, we have $f' \in W^{s,q}(\mathcal{O})$. Moreover, f' is a solution of $Pf' = 0$ outside the compact set $K = \sigma \cap \operatorname{supp} \varphi$, regular at infinity. Therefore

$$\text{cap}(\sigma) \geq \text{cap}(K) \geq \frac{|Pf'(1)|}{\|f'\|_{W^{s,q}(\mathcal{O})}} = \frac{|\langle \varphi, Pf \rangle_{\mathcal{O}}|}{\|f'\|_{W^{s,q}(\mathcal{O})}} > 0. \qquad \square$$

Theorem 2.15 is due to Harvey and Polking [19].

2.8 Non-linear capacity associated with Sobolev spaces

For $sq > n$ the elements of $W^{s,q}_{\mathrm{loc}}(\mathcal{O})$ can be represented as continuous functions by Sobolev's theorem. It is a rather natural idea to try to measure the lack of continuity when $sq \leq n$ by means of a set function, the so-called (s,q)-capacity, which is associated to the norm of the space.

Definition 2.16 Let $K \subset \mathcal{O}$ be compact. Then

$$C_{s,q}(K) = \inf_{\substack{\varphi \in \mathcal{D}(\mathcal{O}): \\ \varphi \geq 1 \text{ on } K}} \|\varphi\|^q_{W^{s,q}(\mathcal{O})}.$$

We extend this definition to all subsets of $\mathcal{O}$ in the following way. If $\sigma \subset \mathcal{O}$ is open, then

$$C_{s,q}(\sigma) = \sup C_{s,q}(K),$$

where the supremum is taken over all compact sets $K \subset \sigma$. And if $\sigma \subset \mathcal{O}$ is arbitrary, then

$$C_{s,q}(\sigma) = \inf C_{s,q}(\sigma'),$$

where the infimum is taken over all open sets $\sigma' \subset \mathcal{O}$ containing σ. A capacity extended in this way to all sets $\sigma \subset \mathcal{O}$ is called an *outer capacity.*

A property that holds true for all x except those belonging to a set of zero (s, q)-capacity is said to be true (s, q)-quasi-everywhere.

For $s = 1$ and $q = 2$ the extremal problem in Definition 2.16 immediately leads to a second order linear partial DE and to classical potential theory. For $q \neq 2$, however, the corresponding equation are non-linear and very difficult to handle.

Because of this, the theory of (s, q)-capacities was not developed very far. It was a breakthrough when around 1970 it was realized by several people (Fuglede [15], Reshetnjak [42], Meyers [38], Maz'ja and Havin [37]) that one can get much further by slightly redefining the (s, q)-capacity.

The key of this observation is Calderon's theorem about the representation of $W^{s,q}(\mathbf{R}^n)$ as a space of Bessel potentials.

The Bessel kernel B_s is most easily defined through its Fourier transform, $\hat{B}_s(z) = (1 + |z|^2)^{-s/2}$ (see Stein's book [50] for further information).

Definition 2.17 For a compact set $K \subset \mathbf{R}^n$,

$$C'_{s,q}(K) = \inf_{\substack{f \in L^q(\mathbf{R}^n): \\ B_s * f \geq 1 \text{ on } K}} \|f\|^q_{L^q(\mathbf{R}^n)}.$$

The definition of $C'_{s,q}$ is extended to arbitrary sets $\sigma \in \mathbf{R}^n$ as in Definition 2.16.

It is easy to prove that $C'_{s,q}(\sigma)$ is a Choquet capacity.

In the small, the capacities $C_{s,q}(\sigma)$ and $C'_{s,q}(\sigma)$ behave equally.

Theorem 2.18 *For every compact set $K \subset \mathcal{O}$, there are positive constants c_1 and c_2 such that*

$$c_1 C'_{s,q}(\sigma) \leq C_{s,q}(\sigma) \leq c_2 C'_{s,q}(\sigma)$$

for all $\sigma \subset K$.

Proof This follows from Calderon's theorem which is a consequence of the Calderon-Zygmund theory of singular integrals (see Stein's book [50]). □

These constants are not going to be important for us, so we shall from now on drop the distinguishing notation $C'_{s,q}$, and assume that $C_{s,q}$ is defined by Definition 2.17.

2.9 Comparison with Hausdorff measure

In order to give a more concrete idea of the properties of (s,q)-capacities we give some comparison theorems. These results should be compared to those for classical potentials given by Carleson [11].

Let $B(x,r)$ denote the ball in $\mathbf{R}^n$ with centre at a point x and radius r.

Theorem 2.19 *Let $\varphi(r) = r^{n-sq}$ if $n - sq > 0$, and $\varphi(r) = (\log \frac{2}{r})^{1-q}$ if $n - sq = 0$. Then there are constants c_1 and c_2 such that*

$$c_1\varphi(r) \le C_{s,q}(B(x,r)) \le c_2\varphi(r)$$

for all $0 < r \le 1$.

Proof This is easy to prove. See Meyers [38]. □

The first statement in the following theorem is an immediate consequence of Theorem 2.19.

Theorem 2.20 *Suppose that φ is the function of Theorem 2.19. Then :*

(1) *there is a constant c such that $C_{s,q}(K) \le c\, h_\varphi^{(\infty)}(K)$ for all compact sets $K \subset B(x,1)$;*

(2) *$C_{s,q}(K) = 0$ if $h_\varphi(K) < \infty$.*

Proof See Meyers [38] and Maz'ja and Havin [37]. □

In the converse direction we have the following deeper result.

Theorem 2.21 *Suppose that $\varphi(r)$ is an increasing continuous function with $\varphi(0) = 0$ such that*

$$\int_0^1 \left(\frac{\varphi(r)}{r^{n-sq}}\right)^{q'-1} \frac{dr}{r} < \infty.$$

Then there is a constant c such that

$$h_\varphi^{(\infty)}(K) \le c\, C_{s,q}(K)$$

for all compact sets $K \subset \mathbf{R}^n$.

Proof See Maz'ja and Havin [37]. □

Hence it follows that if M is a smooth d-dimensional manifold in $\mathbf{R}^n$, then $C_{s,q}(M) = 0$ if and only if $n - sq \ge d$.

2.10 Commensurability of special capacities

The main result is that all the capacities $W^{s,q}(\mathcal{O})$-cap(σ) are equivalent in the small independent of the particular choice of a DO P of order p, because each of them is commensurable in the small with the $(1/q')$-th power of $C_{p-s,q'}(\sigma)$. This is actually a consequence of the fact that pseudodifferential operators are bounded in Sobolev spaces with non-extremal indices q.

Theorem 2.22 *For every compact set $K \subset \mathcal{O}$, there exist positive constants c_1 and c_2 such that*

$$c_1(C_{p-s,q'}(\sigma))^{1/q'} \le \text{cap}(\sigma) \le c_2(C_{p-s,q'}(\sigma))^{1/q'} \qquad (2.2)$$

for all $\sigma \subset K$.

Proof It suffices to prove these estimates for compact sets $\sigma \subset K$ only (use Corollary 2.9).

On the one hand, we get for each compact set $\sigma \subset \mathcal{O}$ just as in the proof of Theorem 2.3,

$$\text{cap}(\sigma) = c' \left(\inf_{\substack{\varphi\in\mathcal{D}(\mathcal{O}):\ \varphi\equiv 1 \text{ in a} \\ \text{neighborhood of } \sigma}} ||\varphi||_{W^{p-s,q'}(\mathcal{O})} \right),$$

where the constant c' depends only on P and $\mathcal{O}$.

To obtain the right-hand estimate of (2.2) it remains to apply the result of Adams and Polking [1], according to which we have (with some constant c'' depending on K and $\mathcal{O}$)

$$\inf_{\substack{\varphi\in\mathcal{D}(\mathcal{O}):\ \varphi\equiv 1 \text{ in a} \\ \text{neighborhood of } \sigma}} ||\varphi||^{q'}_{W^{p-s,q'}(\mathcal{O})} \le c'' C_{p-s,q'}(\sigma) \quad \text{for all } \sigma \subset K.$$

On the other hand, using a result of Maz'ja and Havin [37] we get for a compact set $\sigma \subset \mathcal{O}$

$$\begin{aligned} C_{p-s,q'}(\sigma) &= \sup_{\substack{m\ge 0: \\ \text{supp}\, m\subset\sigma}} \left(\frac{m(\sigma)}{||B_{p-s} * m||_{L^q(\mathbb{R}^n)}} \right)^{q'} \\ &= \left(\sup_{\substack{m\ge 0: \\ \text{supp}\, m\subset\sigma}} \frac{||\Phi(m)||_{W^{s,q}(\mathcal{O})}}{||B_{p-s} * m||_{L^q(\mathbb{R}^n)}} \right)^{q'} \left(\sup_{\substack{F\in\mathcal{E}'_\sigma \\ \Phi(F)\in W^{s,q}(\mathcal{O})}} \frac{|F(1)|}{||\Phi(F)||_{W^{s,q}(\mathcal{O})}} \right)^{q'} \end{aligned}$$

To derive the left-hand estimate of (2.2) from here it suffices to use Proposition 2.7 and Calderon's theorem (see Stein [50]), according to which $||\Phi(m)||_{W^{s,q}(\mathcal{O})} \le c||B_{p-s} * m||_{L^q(\mathbb{R}^n)}$ for all $m \in W^{s-p,q}_{\text{comp}}(\mathcal{O})$ with some constant c depending on Φ and $\mathcal{O}$. □

2.11 Semi-additivity of special capacities

Corollary 2.23 *Let K be a compact subset of $\mathcal{O}$. Then there exists a constant c such that*

$$\operatorname{cap}(\sigma_1 \cup \sigma_2) \le c(\operatorname{cap}(\sigma_1) + \operatorname{cap}(\sigma_2))$$

for all $\sigma_1, \sigma_2 \subset K$.

Proof In fact, if $\sigma_1, \sigma_2 \subset K$, then according to Theorem 2.22,

$$\begin{aligned} \operatorname{cap}(\sigma_1 \cup \sigma_2) &\le c_2(C_{p-s,q'}(\sigma_1 \cup \sigma_2))^{1/q'} \\ &\le c_2(C_{p-s,q'}(\sigma_1) + C_{p-s,q'}(\sigma_2))^{1/q'} \\ &\le c_2(C_{p-s,q'}(\sigma_1)^{1/q'} + C_{p-s,q'}(\sigma_2)^{1/q'}) \\ &\le \frac{c_2}{c_1}(\operatorname{cap}(\sigma_1) + \operatorname{cap}(\sigma_2)) \end{aligned}$$

which is what we wanted to prove. □

2.12 Rado' theorem

In conclusion we bring a problem of removable singularities which is still open.

Problem 2.24 Prove that if a function $f \in C^{p-1}_{\mathrm{loc}}(\mathcal{O})$ satisfies $Pf = 0$ outside its zero set $\{x \in \mathcal{O} : f(x) = 0\}$, then $Pf = 0$ everywhere in $\mathcal{O}$.

For holomorphic functions this is a well-known theorem of Rado. For harmonic functions the answer is also "yes", and it was obtained by Kral [30]. Finally, for polyharmonic functions the proof was recently found by Chesnokov (1992, unpublished).

Chapter 3
Approximation in Sobolev spaces

3.1 Quasicontinuous representatives of Sobolev functions

By the definition of $W^{s,q}_{\mathrm{loc}}(\mathcal{O})$, its elements belong to $L^q_{\mathrm{loc}}(\mathcal{O})$, i.e., the elements are equivalence classes of functions defined and equal outside Lebesgue nullsets. However, in applications it is important to be able to give values to these elements on many sets of zero measure, for example on manifolds of dimension $d \le n-1$.

For $sq > n$ there is the Sobolev embedding theorem, which tells us that every element of $W^{s,q}_{\mathrm{loc}}(\mathcal{O})$ contains exactly one continuous representative, so that elements of $W^{s,q}_{\mathrm{loc}}(\mathcal{O})$ can be identified with continuous functions in a natural way.

If $sq \le n$, this is no longer possible, but according to the well-known embedding theorems of Sobolev and others, the elements of $W^{s,q}_{\mathrm{loc}}(\mathcal{O})$ have traces on d-manifolds provided d is large enough, the traces being locally integrable functions with respect to d-dimensional measure.

In the case of $W^{s,q}_{\text{loc}}(\mathcal{O})$, traces can be defined on arbitrary sets of positive (s,q)-capacity in the following way.

Let $f \in W^{s,q}_{\text{comp}}(\mathcal{O})$, and let $\{\omega_\nu\}_{\nu=1,2,\ldots}$ be an approximate identity, i.e., $\omega_\nu(x) = \nu^n\omega(\nu x)$, where $\omega \in \mathcal{D}(B(0,1))$ is a non-negative function normed by the condition $\int \omega(x)\,ds = 1$. Set $f_\nu = \omega_\nu * f$ $(\nu = 1,2,\ldots)$. Then $f_\nu \in \mathcal{D}(\mathcal{O})$ (for ν large enough) and $f_\nu \to f$ in $W^{s,q}(\mathcal{O})$. By Definition 2.16 of capacity, we have

$$C_{s,q}(\{x : |f_{\nu_2}(x) - f_{\nu_1}(x)| \geq \varepsilon\}) \leq c\varepsilon^{-q}\|f_{\nu_2} - f_{\nu_1}\|^q_{W^{s,q}(\mathcal{O})}.$$

One easily proves that a sufficiently sparse subsequence converges to a function $\tilde{f}(x)$ outside a set of zero (s,q)-capacity, and uniformly outside an open set of arbitrarily small (s,q)-capacity. It follows that:

(1) $\tilde{f}(x) = f(x)$ almost everywhere, i.e., the function $\tilde{f}$ is a representative of the element f of $W^{s,q}_{\text{loc}}(\mathcal{O})$;

(2) $\tilde{f}(x)$ is defined (s,q)-quasi-everywhere; and

(3) for every $\varepsilon > 0$ there is an open set σ with $C_{s,q}(\sigma) < \varepsilon$ such that $\tilde{f}|_{\mathcal{O}\setminus\sigma}$ is continuous on $\mathcal{O}\setminus\sigma$.

Definition 3.1 Functions with properties (2) and (3) are called (s,q)-*quasicontinuous.*

What makes this notion interesting is the following uniqueness theorem.

Theorem 3.2 *Let f_1 and f_2 be two (s,q)-quasicontinuous functions such that $f_1(x) = f_2(x)$ almost everywhere. Then $f_1(x) = f_2(x)$ (s,q)-quasi-everywhere.*

Proof See Maz'ja and Havin [37]. □

Corollary 3.3 *Every element in $W^{s,q}_{\text{comp}}(\mathcal{O})$ has an (s,q)-quasicontinuous representative, which is uniquely determined up to sets of zero (s,q)-capacity.*

It is now clear how to define the trace on an arbitrary set $\sigma \Subset \mathcal{O}$ of an element f of $W^{s,q}_{\text{loc}}(\mathcal{O})$.

Definition 3.4 The *trace* $f|_\sigma$ of $f \in W^{s,q}_{\text{loc}}(\mathcal{O})$ is the restriction to σ of any (s,q)-quasicontinuous representative of φf where $\varphi \in \mathcal{D}(\mathcal{O})$ is equal to 1 in a neighborhood of σ.

Thus, $f|_\sigma$ is defined (s,q)-quasi-everywhere on σ. See Hedberg [22] for further information.

3.2 Approximation problem in Sobolev spaces

Let us assume that K is a compact subset of X.

It is natural to try to define the space $W^{s,q}(K)$ as the quotient of $W^{s,q}_{\mathrm{loc}}(X)$ by the subspace of functions whose derivatives up to order s vanish on K. However, such a subspace is not clearly defined in general because it is meaningless to talk about the values at the individual points of K.

We therefore define $W^{s,q}(K)$ as the quotient of $W^{s,q}_{\mathrm{loc}}(X)$ by the closure in $W^{s,q}_{\mathrm{loc}}(X)$ of $\mathcal{D}(X \setminus K)$. According to the spectral synthesis theorem of Hedberg and Wolff [27], this closure can be described in terms of function values assumed (s,q)-quasi-everywhere.

Theorem 3.5 *Let s be a positive integer, and $1 < q < \infty$. Then $f \in W^{s,q}_{\mathrm{loc}}(X)$ belongs to the closure in $W^{s,q}_{\mathrm{loc}}(X)$ of $\mathcal{D}(X \setminus K)$ if and only if $D^\alpha f|_K = 0$ for all $\alpha, |\alpha| \le s-1$.*

Proof Here the necessity is obvious, so the hard part is the sufficiency. This was proved in increasing generality in Hedberg [23], [24], and Hedberg-Wolff [27]. We refer to these papers for the proof. □

With the quotient topology, $W^{s,q}(K)$ is a Banach space. It may be realized, say, as the quotient of $W^{s,q}(\mathcal{O})$ be the closure in $W^{s,q}(\mathcal{O})$ of the subspace of functions equal to zero near K, where $\mathcal{O}$ is a relatively compact neighborhood of K in X.

In the wide sense, one understands by the approximation problem in $W^{s,q}(K)$ by solutions of the equation $Pf = 0$ the following.

Problem 3.6 Describe the closure of the subspace $\mathcal{S}(X)$ in $W^{s,q}(K)$.

Approximation theorems appear to be one of the main technical tools in the theory of DE's. In particular, the best developed method of investigating stability of solutions of boundary value problems for the equation $Pf = 0$ with respect to variation of the boundary manifold consists of applying approximation theorems (see Keldysh [29], Babuška [4], etc.).

3.3 Runge's theorem

Topological difficulties in Problem 3.6 are overcome with the help of a theorem similar to Runge's.

Theorem 3.7 *Let $\mathcal{O}$ be an open subset of X. Then the following conditions are equivalent:*

(1) *the complement of $\mathcal{O}$ has no compact connected components in X;*

(2) *$\mathcal{S}(X)$ is dense in $\mathcal{S}(\mathcal{O})$.*

Proof See Malgrange [36] and Lax [34] for original proofs, and elsewhere. □

Corollary 3.8 *If the complement of K has no relatively compact components in X, then the closures in $W^{s,q}(K)$ of the subspaces $\mathcal{S}(X)$ and $\mathcal{S}(K)$ are identical.*

Thus the principal analytical difficulties in Problem 3.6 can be formulated as

Problem 3.9 Describe the closure of the subspace $\mathcal{S}(K)$ in $W^{s,q}(K)$.

3.4 Approximation by potentials with singularities off a compactum

When considering holomorphic approximations in the plane, one speaks of rational approximations. More generally, for Problem 3.9 we have the following situation.

As already said in section 1.10, solutions of the equation $Pf = 0$ of the form

$$\sum_{|\alpha|\leq\delta} D_y^\alpha \Phi(.\,, y_\alpha)c_\alpha$$

are analogous to rational functions with poles at points $y_\alpha \in X$. In particular, they are in $\mathcal{S}(K)$ provided all the poles y_α ($|\alpha| \leq \delta$) are off K.

From this point of view, the following result may be considered as a theorem of "rational" approximation with a fixed set of singularities.

Theorem 3.10 *Let σ be a finite or countable subset of $X \setminus K$ such that every connected component of $X \setminus K$ contains at least one point of σ. Then the subspace of $\mathcal{S}(K)$ consisting of solutions of the form*

$$\sum_{|\alpha|\leq\delta} D_y^\alpha \Phi(.\,, y_\alpha)c_\alpha,$$

where $y_\alpha \in \sigma$, is dense in $\mathcal{S}(K)$.

Proof It follows from the condition that the annihilator of the subspace of $\mathcal{S}(K)$ mentioned in the theorem is equal to $P'\mathcal{E}'_K$. To finish the proof it remains to apply the Hahn-Banach theorem in the standard way. □

In all probability, this theorem was first found by Tarkhanov [53].

3.5 Description of the annihilator of the subspace of solutions

The main step in the solution of Problem 3.9 is an adequate description of the annihilator of the subspace $\mathcal{S}(K)$ of $W^{s,q}(K)$. Let us denote it by $\mathcal{S}(K)^\perp$. By definition, $\mathcal{S}(K)^\perp$ is the space of continuous linear functionals on $W^{s,q}(K)$ equal to zero on $\mathcal{S}(K)$.

Our definition of $W^{s,q}(K)$ conveniently differs from many others in that the space dual to $W^{s,q}(K)$ can be identified with the subspace of all the continuous linear functionals on $W^{s,q}_{\text{loc}}(X)$ with supports in K, i.e., $W^{s,q}(K)' = W_K^{-s,q'}$. Hence we obtain the following technical result, in which the role of the boundedness theorem for pseudodifferential operators in Sobolev spaces is isolated.

Lemma 3.11 *If K is a compact subset of X and $1 < q < \infty$, then the mapping $P' : W_K^{p-s,q'} \to \mathcal{S}(K)^\perp$ is a topological isomorphism.*

Proof One verifies that the mapping $\Phi' : W_K^{-s,q'} \to W_{\mathrm{loc}}^{p-s,q'}(X)$ maps $\mathcal{S}(K)^\perp$ to $W_K^{p-s,q'}$. □

3.6 Approximation of high order

With the help of Lemma 3.11 and the Hahn-Banach theorem it is now quite easy to describe the closure of $\mathcal{S}(K)$ in $W^{s,q}(K)$ for $s \geq p$.

Theorem 3.12 *Let $s \geq p$ and $1 < q < \infty$. A function $f \in W^{s,q}(K)$ belongs to the closure of $\mathcal{S}(K)$ in $W^{s,q}(K)$ if and only if $D^\alpha(Pf)|_K = 0$ for all α, $|\alpha| \leq s - p$.*

Proof Necessity follows easily from Theorem 3.5.

Sufficiency. Let $g \in \mathcal{S}(K)^\perp$. According to Lemma 3.11, there exists a distribution $v \in W_K^{p-s,q'}$ such that $g = P'v$ on X. If $f \in W_{\mathrm{loc}}^{s,q}(X)$ and $D^\alpha(Pf)|_K = 0$ for all α, $|\alpha| \leq s-p$, then Pf belongs to the closure in $W_{\mathrm{loc}}^{s-p,q}(X)$ of $\mathcal{D}(X \setminus K)$ by Theorem 3.5. Hence we may conclude that $\langle g, f\rangle = \langle v, Pf\rangle = 0$. To finish the proof it suffices to use the Hahn-Banach theorem. □

Thus, for any compact set $K \subset X$, the most transparent necessary condition "$D^\alpha(Pf)|_K = 0$ for $|\alpha| \leq s - p$" is also sufficient in order that a function $f \in W^{s,q}(K)$ $(s \geq p)$ belong to the closure of $\mathcal{S}(K)$ in $W^{s,q}(K)$.

For the iterated Cauchy-Riemann operator $P = (\partial/\partial\bar{z})^p$, Theorem 3.12 was proved by Verdera [55]. In the general case it is due to Tarkhanov [53].

3.7 Reducing approximation of lower order to a problem of spectral synthesis

If $0 \leq s < p$, what are the necessary conditions for a function $f \in W^{s,q}(K)$ to be approximated to an arbitrary degree of accuracy in the norm of this space by elements of $\mathcal{S}(K)$? The most transparent among them appear to be local conditions. Namely, f should satisfy the equation $Pf = 0$ in the sense of distribution in $\mathring{K}$ (the interior of K).

However, the local conditions on $f \in W^{s,q}(K)$, generally speaking, are not sufficient for f to belong to the closure of $\mathcal{S}(K)$ in $W^{s,q}(K)$.

We shall make use of the "principal" index $d = \frac{n}{q'} - (p - s)$.

Example 3.13 Suppose that $0 \leq s < p$ and $d \geq 0$, i.e., $p - n < s$ and $\frac{n}{n-p+s} \leq q < \infty$. Then these exists a nowhere dense compact set $K \subset X$ of positive Lebesgue measure such that the potential $\Phi(\varphi\chi_K)$ does not belong to the closure of $\mathcal{S}(K)$ in $W^{s,q}(K)$ for some function $\varphi \in \mathcal{E}(X)$. The compact set K is just the one constructed by Polking [41] as a modification of the standard "Swiss cheese" or Sierpinski curve in $\mathbf{R}^2$. Note that, if $\varphi \in \mathcal{E}(X)$, the potential $\Phi(\varphi\chi_K)$ lies in $W^{s,q}_{\mathrm{loc}}(X)$ for every $q < \infty$. So the formulation is plausible. See Gauthier and Tarkhanov [16] for details. □

Example 3.14 Let $0 \le s < p$ and $\frac{n}{n-1} \le q < \infty$, and let δ be any non-negative integer with $-d \le \delta \le p - s - 1$. Then there exists a compact set $K \subset X$ with non-empty interior such that the potential $\Phi(\sum_{|\alpha|\le\delta} D^\alpha m_\alpha)$ does not belong to the closure of $\mathcal{S}(K)$ in $W^{s,q}(K)$ for some measures m_α ($|\alpha| \le \delta$) supported on ∂K. The compact set K is a slight modification of the one of Hedberg's example [23]. Note that for any measure m_α ($|\alpha| \le \delta$) supported on ∂K, the potential $\Phi(\sum_{|\alpha|\le\delta} D^\alpha m_\alpha)$ lies in $W^{p-\delta-1,q}_{\mathrm{loc}}(X) \cap \mathcal{S}(\mathring{K})$. Besides that, for our choice of δ, we have $p - \delta - 1 \ge s$. Once again the formulation is plausible. See Gauthier and Tarkhanov [16] for details. □

Hence, there arises the problem of describing the compact sets $K \subset X$ for which the above-mentioned necessary conditions are sufficient.

Problem 3.15 Let $0 \le s < p$. Under what conditions on K is the subspace $\mathcal{S}(K)$ dense in $W^{s,q}(K) \cap \mathcal{S}(\mathring{K})$ in the norm of $W^{s,q}(K)$?

For the Cauchy-Riemann operator in the plane, the decisive progress in studying the compact sets $K \subset \mathbf{C}^1$ for which $\mathcal{S}(K)$ is dense in $L^2(K) \cap \mathcal{S}(\mathring{K})$, was achieved by Havin [20]. He discovered that these compact sets may be described by means of the capacity $C_2(\sigma)$ ($= C_{1,2}(\sigma)$!) of classical potential theory. This revelation was of principal importance because the problem was proved to be of purely "real" character. Papers by Bagby [6], Hedberg [22], and others give further results.

Extensions of the results to general elliptic equations are based on the modern technique of pseudodifferential operators. The main point is that pseudodifferential operators are bounded in Sobolev spaces $W^{s,q}_{\mathrm{loc}}(X)$ with non-extremal exponents $1 < q < \infty$.

It turns out that Problem 3.15 is equivalent to a certain problem about completeness of compactly supported functions in Sobolev spaces, i.e., without any relation to the DO P.

Theorem 3.16 *For $0 \le s < p$ and $1 < q < \infty$, the following conditions are equivalent :*

(1) *$\mathcal{S}(K)$ is dense in $W^{s,q}(K) \cap \mathcal{S}(\mathring{K})$ in the norm of $W^{s,q}(K)$;*

(2) *$\mathcal{D}(\mathring{K})$ is dense in $W^{p-s,q'}_K$ in the norm of this space.*

Proof This easily follows from Lemma 3.11. For details, see Tarkhanov [53]. □

This theorem was proved in increasing generality in Polking [41] ($s = 0$), and Tarkhanov [53] ($0 < s \le p - 1$).

Remark 3.17 As follows from functional analysis, condition (2) of Theorem 3.16 may be reformulated by saying that $\mathcal{D}(\mathring{K})$ is dense in $W^{p-s,q'}_K$ in the $*$-weak topology of $(W^{s-p,q}_{\mathrm{loc}}(X))'$.

3.8 Approximation by potentials with singularities on a compactum

We now present the main theorem of this chapter which affirms that, in fact, in order to successfully solve Problem 3.15, it is sufficient to know how to approximate potentials of derivatives of measures supported on ∂K.

Theorem 3.18 *Let $0 \le s < p$ and $1 < q < \infty$. Then, for each function $f \in W^{s,q}(K) \cap \mathcal{S}(\mathring{K})$ and for each $\varepsilon > 0$, there exists a solution $f_\varepsilon \in \mathcal{S}(K)$ and measures m_α ($|\alpha| \le p-s-1$) on ∂K such that*

$$\|f - (f_\varepsilon + \sum_{|\alpha| \le p-s-1} \Phi(D^\alpha m_\alpha))\|_{W^{s,q}(K)} < \varepsilon.$$

Proof This follows from Lemma 3.11 and Theorem 3.5 by using the Hahn-Banach Theorem in the standard way. For details, see Gauthier and Tarkhanov [16]. □

The preceding theorem has been proved for the Cauchy-Riemann operator in the plane successively by Bers [8] (for $q = 2$), Havin [20] (for $2 < q < \infty$), and Hedberg [23] in the general case. In Hedberg's work [23] a less explicit theorem for solutions of the equation $Pf = 0$ and $s = 0$ was obtained. Namely, the measures m_α ($|\alpha| \le p-1$) were allowed to be supported on the complement of $\mathring{K}$. Then f_ε can be omitted, as follows from Theorem 3.10. In the present form, the theorem is due to Gauthier and Tarkhanov [16].

If the compact set K is assumed to be nowhere dense, one may give a more precise form of the theorem.

Corollary 3.19 *Let K be a compact subset of X without interior, $0 \le s < p$ and $1 < q < \infty$. Then, for each $f \in W^{s,q}(K)$ and each $\varepsilon > 0$, there exists a solution $f_\varepsilon \in \mathcal{S}(K)$ and a function $\varphi \in \mathcal{E}(X)$ such that*

$$\|f - (f_\varepsilon + \Phi(\varphi\chi_K))\|_{W^{s,q}(K)} < \varepsilon.$$

Proof This follows from Lemma 3.11 by using the Hahn-Banach Theorem. For details, see Gauthier and Tarkhanov [16]. □

3.9 Solution in terms of non-linear capacity for nowhere dense compact sets

Compact sets $K \subset X$ for which $\mathcal{D}(\mathring{K})$ is dense in $W_K^{p-s,q'}$ were actively studied in the 70's. A detailed bibliography on these questions can be found in Tarkhanov's book [52].

As a rule, conditions on K were formulated in the language of $(p-s, q')$-capacity. In the language of $(p-s, q')$-capacity one can describe, in particular, those nowhere dense compact sets $K \subset X$ for which $W_K^{p-s,q'} = \{0\}$ (see Hedberg [22] and Polking [41]). Theorem 3.16 allows one to apply these results to Problem 3.15.

Theorem 3.20 *Let K be a compact subset of X with empty interior, and $0 \le s < p$. Then $\mathcal{S}(K)$ is dense in $W^{s,q}(K)$ for $1 \le q < q_0$, where $q_0 = \frac{n}{n-p+s}$ if $s > p - n$, and $q_0 = \infty$ if $s \le p - n$.*

Proof For $d < 0$ it follows from Sobolev's embedding theorem that $W_{\text{loc}}^{p-s,q'}(X) \subset C_{\text{loc}}(X)$. So, if K is nowhere dense, then $W_K^{p-s,q'} = \{0\}$. The theorem now follows from Lemma 3.11 by using the Hahn-Banach Theorem. □

As follows from Example 3.13, the range of q in Theorem 3.20 is sharp.

For $q \ge q_0$, the following is true.

Theorem 3.21 *Assume that K is a compact subset of X with empty interior, and $p - n < s < p$, $\frac{n}{n-p+s} \le q < \infty$. Then the following conditions are equivalent :*

(1) *$\mathcal{S}(K)$ is dense in $W^{s,q}(K)$;*

(2) *for an arbitrary open set $\sigma \subset X$,*

$$C_{p-s,q'}(\sigma \setminus K) = C_{p-s,q'}(\sigma);$$

(3) *for almost all $x \in K$,*

$$\overline{\lim_{r \to 0}} \frac{C_{p-s,q'}(B(x,r) \setminus K)}{r^n} > 0.$$

Proof It suffices to complement Theorem 3.16 with a result of Hedberg [22] and Polking [41]. □

According to Theorem 2.19, the $(p-s, q')$-capacity of the ball $B(x,r)$ for small r behaves like $r^{n-(p-s)q'}$. The jump between the strongest necessary condition (2), analogous to Vitushkin's condition [56], and the weakest sufficient condition (3) in Theorem 3.21 is explained by the instability of the capacity (see Fernström [12]).

3.10 The problem for arbitrary compact sets

If the compact set K is allowed to have interior points, Theorem 3.20 becomes false for q in the same range. In this case, the range of q should be defined not by the existence of the embedding $W_{\text{loc}}^{p-s,q'}(X) \subset C_{\text{loc}}(X)$ but from the existence of the embedding $W_{\text{loc}}^{p-s,q'}(X) \subset C_{\text{loc}}^{p-s-1}(X)$. We have the following result.

Theorem 3.22 *If $0 \le s < p$ and $1 < q < \frac{n}{n-1}$, then $\mathcal{S}(K)$ is dense in $W^{s,q}(K) \cap \mathcal{S}(\mathring{K})$ in the norm of $W^{s,q}(K)$.*

Proof Polking [41] proved that if $1 < q < \frac{n}{n-1}$, then $\mathcal{D}(\mathring{K})$ is dense in $W_K^{p-s,q'}$. To finish the proof, it suffices to apply Theorem 3.16. □

Example 3.14 explains why the range for q in Theorem 3.22 is sharp.

The upper limit for q in Theorem 3.22 corresponds to the number q_0 when $s = p - 1$. For that exponent s it is possible to obtain a criterion of Wiener type [57] for approximation in $W^{s,q}(K)$ by solutions of the equation $Pf = 0$.

Theorem 3.23 *For $p - n < s < p$ and $\frac{n}{n-p+s} \le q < \infty$, conditions (2) and (3) below are equivalent and follow from (1), and moreover, for $s = p - 1$, the converse implication holds as well:*

(1) *$\mathcal{S}(K)$ is dense in $W^{s,q}(K) \cap \mathcal{S}(\mathring{K})$ in the norm of $W^{s,q}(K)$;*

(2) *for any open set $\sigma \subset X$,*

$$C_{p-s,q'}(\sigma \setminus K) = C_{p-s,q'}(\sigma \setminus \mathring{K});$$

(3) *for all points $x \in \partial K$, except perhaps for a set of zero $(p - s, q')$-capacity,*

$$\int_0^1 \left(\frac{C_{p-s,q'}(B(x,r) \setminus K)}{r^{n-(p-s)q'}} \right)^{q-1} \frac{dr}{r} = \infty.$$

Proof As follows from Theorem 3.16 and a result of Polking [41], the implication (1) $\Rightarrow$ (2) holds for all $0 \le s < p$ and $1 < q < \infty$. It is not excluded that also (2) $\Rightarrow$ (1) for the same s and q. At least, for $s = p - 1$, this follows from Theorem 3.16 and a result of Hedberg and Wolff [27]. The implication (2) $\Rightarrow$ (3) was proved in increasing generality in Hedberg [22] ($\frac{n}{n-p+s} \le q < 1 + \frac{n}{n-p+s}$), and Hedberg-Wolff [27] ($1 + \frac{n}{n-p+s} \le q < \infty$). Finally, the converse implication (3) $\Rightarrow$ (2) was proved by Hedberg [22]. □

In the case when $n = 2$ and $P = \partial/\partial\bar{z}$ the equivalence (1) $\Leftrightarrow$ (2) was proved by Bagby [6].

Whether condition (3) is sufficient for other $p-n < s < p$ besides $s = p-1$ is not known. We mention, however, that for $d < 0$ the capacity $C_{p-s,q'}(\sigma)$ behaves pathologically. In this case condition (3) is satisfied for all compact sets K, and so it does not, generally speaking, imply (1) unless $1 < q < \frac{n}{n-1}$.

3.11 Solution in terms of special capacities

In this section we describe the compact sets K in Problem 3.15 in the language of special capacities of section 2.3.

There are two opportunities here.

On the one hand, according to Theorem 3.16 such a description does not depend on the structure of the DO P and is determined only by its order. Consequently, P can be replaced by some DO with a simple structure, provided the order is preserved. It is convenient to take as P a homogeneous elliptic DO of order p with constant coefficients in $\mathbb{R}^n$, say, the principal homogeneous part of the DO P with the coefficients "frozen" at some point $x^0 \in X$.

This way leads to a full solution of Problem 3.15 (see Tarkhanov [53]). However, the solution makes one wish for something better because the included capacities are difficult to work with.

On the other hand, one can directly apply results of sections 2.9 and 3.10, and Theorem 2.23 to obtain an answer to Problem 3.15 in terms of the capacity $\mathrm{cap}(\sigma)$.

The second way seems to be preferable because Theorem 3.16 is impossible for approximations in spaces which are unstable under actions of pseudodifferential operators (of order 0).

So we fix a relatively compact neighborhood $\mathcal{O}$ of K in X, and define the capacity $W^{s,q}(\mathcal{O})$-$\mathrm{cap}(\sigma)$ (or simply $\mathrm{cap}(\sigma)$) as in section 2.3.

Theorem 3.24 *Assume that K is a compact subset of X with empty interior, and $p-n < s < p$, $\frac{n}{n-p+s} \le q < \infty$. Then the following conditions are equivalent :*

(1) *$\mathcal{S}(K)$ is dense in $W^{s,q}(K)$;*

(2) *for an arbitrary open subset σ of $\mathcal{O}$,*

$$\mathrm{cap}(\sigma \setminus K) = \mathrm{cap}(\sigma);$$

(3) *for almost all $x \in K$,*

$$\overline{\lim_{r\to 0}} \frac{\mathrm{cap}(B(x,r)\setminus K)}{r^{d+(p-s)}} > 0.$$

Proof The implication (1) $\Rightarrow$ (2) is standard (see, for example, Tarkhanov [53]). The implication (2) $\Rightarrow$ (3) is obvious because for small r the capacity $\mathrm{cap}(B(x,r))$ behaves like r^d. Finally. it follows from Theorem 3.22 that condition (3) of Theorem 3.24 is equivalent to condition (3) of Theorem 3.21. So (3) $\Rightarrow$ (1), which proves the theorem. □

For the Cauchy-Riemann operator in the plane, Theorem 3.24 is due to Sinanjan [48].

Theorem 3.25 *If $s = p-1$ and $\frac{n}{n-p+s} \le q < \infty$, then the following conditions for a compact set $K \subset X$ are equivalent :*

(1) *the closure of $\mathcal{S}(K)$ in $W^{s,q}(K)$ is equal to $W^{s,q}(K) \cap \mathcal{S}(\mathring{K})$;*

(2) *for any open set $\sigma \subset \mathcal{O}$,*

$$\mathrm{cap}(\sigma \setminus K) = \mathrm{cap}(\sigma \setminus \mathring{K});$$

(3) *for all points $x \in \partial K$, except perhaps for a set of zero capacity,*

$$\int_0^1 \left(\frac{\mathrm{cap}(B(x,r)\setminus K)}{r^d} \right)^q \frac{dr}{r} = \infty.$$

Proof The implication (1) $\Rightarrow$ (2) is again standard (see Tarkhanov [53]). Using Theorem 2.22 one can prove that condition (2) of Theorem 3.25 implies condition (2) of Theorem 3.23 (see Hedberg [22]). Therefore, we get (2) $\Rightarrow$ (3). To complete the proof it suffices to note that conditions (3) of Theorem 3.23 and 3.25 are equivalent because of Theorem 2.22. □

Whether the theorem holds for all $0 \le s < p$, provided $d \ge 0$, is not known. We mention a simpler version of the problem.

Problem 3.26 Prove that, if $d \geq 0$, then condition (1) of Theorem 3.25 follows from the next one:

(3′) for all $x \in \partial K$,

$$\varliminf_{r \to 0} \frac{\operatorname{cap}(B(x,r) \setminus K)}{\operatorname{cap}(B(x,r) \setminus \overset{\circ}{K})} > 0.$$

3.12 Bounded point evaluations

Let D be an elliptic DO of order s on X. We shall in addition assume that for each $x \in X$ there are constants c and δ such that

$$|D_x \Phi(x,y)| \geq c|x-y|^{p-n-s} \quad \text{if } |x-y| < \delta.$$

Remark 3.27 This assumption is valid for a wide class of operators. In fact, the author knows of no counterexample.

As it turns out, the following definition does not depend on the particular choice of the DO D.

Definition 3.28 A point $x \in K$ is called a *bounded point evaluation* (BPE) for $\mathcal{S}(K) \subset W^{s,q}(K)$ if the functional $f \to Df(x)$ is continuous on $\mathcal{S}(K)$ in the topology of $W^{s,q}(K)$.

It seems clear that if a set K has a BPE, then $\mathcal{S}(K)$ is not dense in $W^{s,q}(K)$. This appealingly intuitive statement seems to be difficult to prove, however.

The following theorem shows the correspondence between bounded point evaluations and fundamental solutions.

Theorem 3.29 *Let $0 \leq s < p$ and $1 < q < \infty$. Then x is a* BPE *for $\mathcal{S}(K) \subset W^{s,q}(K)$ if and only if there is a function $g \in W^{p-s,q'}_{\mathrm{loc}}(X)$ such that $g(y) = D_x\Phi(x,y)$ for all $y \in X \setminus K$.*

Proof Necessity. Let x be a BPE for $\mathcal{S}(K) \subset W^{s,q}(K)$. Then there is a function $G \in W_K^{-s,q'}$ such that $Df(x) = \langle G, f\rangle$ for all $f \in \mathcal{S}(K)$. Set $g = \Phi'(G)$. Then, since $\Phi(.\,,y) \in \mathcal{S}(K)$ if $y \in X \setminus K$, we have $g(y) = D_x\Phi(x,y)$ for all $y \in X\setminus K$. On the other hand, $P'g = G$, so the standard regularity results for elliptic operators imply that $g \in W^{p-s,q'}_{\mathrm{loc}}(X)$.

Sufficiency. Now suppose $g \in W^{p-s,q'}_{\mathrm{loc}}(X)$ satisfies $g(y) = D_x\Phi(x,y)$ if $y \in X \setminus K$. Set $G = P'g$. Then $G \in W^{-s,q'}_{\mathrm{loc}}(X)$ and $G(y) = 0$ if $y \notin K$. Let $f \in \mathcal{S}(K)$. Choose an infinitely differentiable function φ such that $\varphi \equiv 1$ in a neighborhood of K, and such that $\varphi \equiv 0$ outside the domain of f. Then

$$\langle G, f\rangle = \langle P'g, \varphi f\rangle = \langle g, P(\varphi f)\rangle.$$

Since $P(\varphi f) = 0$ in a neighborhood of K, and $g(y) = D_x\Phi(x,y)$ for $y \in X \setminus K$, we get

$$\langle G, f\rangle = \langle D_x\Phi(x,.), P(\varphi f)\rangle = (D\Phi P(\varphi f))(x) = Df(x).$$

Hence $Df(x) = \langle G, f\rangle$ for all $f \in \mathcal{S}(K)$. Since $G \in (W^{s,q}(K))'$, the point x is a BPE for $\mathcal{S}(K) \subset W^{s,q}(K)$. □

For $\nu = 0, 1, \ldots$, we denote by $A_\nu(x)$ the annulus

$$\{y : \frac{1}{2^{\nu+1}} < |y - x| \le \frac{1}{2^\nu}\}.$$

Corollary 3.30 *If $p-n < s < p$ and $\frac{n}{n-p+s} \le q < \infty$, then x is a* BPE *for $\mathcal{S}(K) \subset W^{s,q}(K)$ if and only if*

$$\sum_{\nu=0}^{\infty} 2^{(n-p+s)q'\nu} C_{p-s,q'}(A_\nu(x) \setminus K) < \infty.$$

Proof See Fernström and Polking [13] (Theorem 3). □

Note that if $p - n < s < p$ and $1 \le q < \frac{n}{n-p+s}$, then x is a BPE for $\mathcal{S}(K) \subset W^{s,q}(K)$ if and only if x is an interior point of K. If $0 \le s \le p - n$, there are no BPE's. For a proof, see *ibid.*

Fernström and Polking [13] made a complete analysis of the relationship between the existence of bounded point evaluations and the density of $\mathcal{S}(K)$ in $L^q(K)$. The following theorem extends their result to approximation in Sobolev spaces.

Theorem 3.31 *Let K be a complete subset of X without interior, and $q \notin \left[\frac{n}{n-p+s}, \frac{n}{p-s}\right]$. Then $\mathcal{S}(K)$ is dense in $W^{s,q}(K)$ if and only if no point of K is a* BPE *for $\mathcal{S}(K) \subset W^{s,q}(K)$.*

Proof This follows from Theorem 3.21 and Corollary 3.30. □

If $q \in \left[\frac{n}{n-p+s}, \frac{n}{p-s}\right]$, such an answer to Problem 3.15 is impossible.

Example 3.32 Suppose that $p - \frac{n}{2} \le s$ and $\frac{n}{n-p+s} \le q \le \frac{n}{p-s}$. Then there exists a compact set $K \subset X$ without interior such that no point of K is a BPE for $\mathcal{S}(K) \subset W^{s,q}(K)$, while $\mathcal{S}(K)$ is not dense in $W^{s,q}(K)$. For a construction of K, see Theorem 2 of Fernström and Polking [13] with $\alpha = p - s$. □

Chapter 4
Bases with double orthogonality

4.1 The Cauchy problem

Let $\mathcal{O}$ be a relatively compact domain in X with smooth boundary.

We fix a Dirichlet system of order $(p-1)$ on $\partial\mathcal{O}$, say $B_j \in \mathrm{do}_{b_j}(U)$ $(j = 0, \ldots, p-1)$, where U is some neighborhood of $\partial\mathcal{O}$ in X (see section 1.10).

A priori it is not clear whether a solution $f \in \mathcal{S}(\mathcal{O})$ of Lebesgue class $L^1(\mathcal{O})$ has a finite order of growth near $\partial\mathcal{O}$, that is, whether the expressions $B_j f$ $(j = 0, \ldots, p-1)$ have weak limit values on $\partial\mathcal{O}$ (cf. Theorem 1.17).

Let $C_j \in \mathrm{do}_{p-b_j-1}(U)$ $(j = 0, \ldots, p-1)$ be the Dirichlet system of order $(p-1)$ on $\partial\mathcal{O}$ which is adjoint to the system $\{B_j\}$ with respect to Green's formula.

For each $g \in \mathcal{D}(\partial\mathcal{O})$ there exists a function $\tilde{g} \in \mathcal{D}(X)$ such that $C_j\tilde{g} = g$, and $C_i\tilde{g} = 0$ $(i \neq j)$ on $\partial\mathcal{O}$. Then we set

$$\langle g, B_j f\rangle = -\int_{\mathcal{O}} (P'\tilde{g}) f \, dv \qquad (g \in \mathcal{D}(\partial\mathcal{O})). \tag{4.1}$$

Theorem 4.1 *For any solution $f \in L^1(\mathcal{O}) \cap \mathcal{S}(\mathcal{O})$, the weak limit values of the expressions $B_j f$ $(j = 0, \ldots, p-1)$ on $\partial\mathcal{O}$ defined by formula (4.1) exist, and they coincide with the ones defined by (1.5). Moreover, $f \in W^{s,q}(\mathcal{O})$ $(1 < q < \infty)$ if and only if $B_j f \in W^{s-b_j-1/q,q}(\partial\mathcal{O})$ $(j = 0, \ldots, p-1)$.*

Proof See Shlapunov and Tarkhanov [47]. □

We now suppose that $\mathcal{S}$ is a set of positive ($(n-1)$-dimensional) measure on the boundary of $\mathcal{O}$.

The wording of the Cauchy problem for solutions of the equation $Pf = 0$ in $\mathcal{O}$ with data on $\mathcal{S}$ consists of the following.

Problem 4.2 Let $f_j \in W^{s-b_j-1/q,q}(\mathcal{S})$ $(j = 0, \ldots, p-1)$ be known functions on $\mathcal{S}$, where $s \in \mathbf{Z}_+$ and $1 < q < \infty$. It is required to find a solution $f \in W^{s,q}(\mathcal{O}) \cap \mathcal{S}(\mathcal{O})$ such that $B_j f = f_j$ $(j = 0, \ldots, p-1)$ on $\mathcal{S}$.

In order to justify the term "Cauchy problem" for Problem 4.2, we note that, if it is sensible, the values of $B_j f$ $(j = 0, \ldots, p-1)$ on $\mathcal{S}$ determine all the derivatives of f up to order $(p-1)$ on $\mathcal{S}$.

4.2 Green's formula

In order to connect the weak limit values of $B_j f$ $(j = 0, \ldots, p-1)$ on $\partial\mathcal{O}$ with other values (radial, non-tangential, by some norm, etc.), Green's formula and theorems on jumps of the boundary integral in the formula are usually used.

Lemma 4.3 *For any solution $f \in \mathcal{S}(\mathcal{O})$ with a finite order of growth near $\partial\mathcal{O}$, the following formula holds :*

$$-\int_{\partial\mathcal{O}} \sum_{j=0}^{p-1} C_j\Phi(x,.) \cdot B_j f \, ds = \begin{cases} f(x), & x \in \mathcal{O}, \\ 0, & x \in X \setminus \bar{\mathcal{O}}. \end{cases} \tag{4.2}$$

Proof It suffices to represent f by the "classical" Green formula in the domain $\{x \in \mathcal{O} : \mathrm{dist}(x, \partial\mathcal{O}) > 0\}$ where $\varepsilon > 0$ is small enough, and then to make the limit passage for $\varepsilon \to +0$. □

One of the consequences of the lemma is the property of local regularity of solutions of Problem 4.2.

Corollary 4.4 *Let $f \in \mathcal{S}(\mathcal{O})$ have a finite order of growth near $\partial\mathcal{O}$. If $B_j f \in C^{s-b_j}_{\mathrm{loc}}(\mathring{\mathcal{S}})$ $(j = 0, \ldots, p-1)$, then $f \in C^{s}_{\mathrm{loc}}(\mathcal{O} \cup \mathring{\mathcal{S}})$.*

Proof See Shlapunov and Tarkhanov [47]. □

We indicate now a wide class of boundary sets $\mathcal{S}$, for which Problem 4.2 has no more than one solution.

Corollary 4.5 *If, for a solution $f \in \mathcal{S}(\mathcal{O})$ with a finite order of growth near $\partial\mathcal{O}$, the weak boundary values of $B_j f$ $(j = 0, \ldots, p-1)$ vanish on a set $\mathcal{S} \subset \partial\mathcal{O}$ with non-empty interior, then $f \equiv 0$ in $\mathcal{O}$.*

Proof Denote by $G(\oplus B_j f)$ the integral on the left hand side of formula (4.2). Let $x^0 \in \mathcal{S}$, and $B = B(x^0, r)$ be an open ball in X such that $B \cap \partial\mathcal{O} \subset \mathcal{S}$. We set $\mathcal{D} = \mathcal{O} \cup B$. Then $G(\oplus B_j f) \in \mathcal{E}(\mathcal{D})$ satisfies $PG(\oplus B_j f) = 0$ in the domain $\mathcal{D} \subset X$, and it vanishes on the non-empty open subset $B \setminus \bar{\mathcal{O}}$ of this domain. Since P is analytically hypoelliptic, we can conclude that $G(\oplus B_j f) \equiv 0$ in $\mathcal{D}$. Hence $f \equiv 0$ in $\mathcal{O}$, which is what we wanted to prove. □

4.3 Green's integral

Using "initial" data in Problem 4.2, we construct a Green type integral. Exactly, we denote by $\tilde{f}_j \in W^{s-b_j-\frac{1}{q},q}(\partial\mathcal{O})$ some extensions of the functions f_j to the whole boundary.

Example 4.6 If $s = 0$ and the functions f_j are sufficiently "smooth", that is, $f_j \in L^2(\mathcal{S})$ $(j = 0, \ldots, p-1)$, then it is possible to extend them by zero to $\partial\mathcal{O} \setminus \mathcal{S}$. □

In any case the extensions may be so chosen that they will be concentrated in a neighborhood of the closure of $\mathcal{S}$ on $\partial\mathcal{O}$, given in advance.

Set $\tilde{f} = \oplus \tilde{f}_j$, and

$$G(\tilde{f}) = \Phi\left(\sum_{j=0}^{p-1} C'_j([\partial\mathcal{O}]\tilde{f}_j)\right)$$

or

$$G(\tilde{f})(x) = \int_{\partial\mathcal{O}} \sum_{j=0}^{p-1} C_j \Phi(x, .) \cdot \tilde{f}_j \, ds \qquad (x \notin \partial\mathcal{O}). \tag{4.3}$$

Lemma 4.7 *For every relatively compact open set $\mathcal{D} \subset X$, the restriction of the integral $G(\tilde{f})$ to each of the sets $\mathcal{O} \cup \mathcal{D}$ and $\mathcal{D} \setminus \bar{\mathcal{O}}$ is of class $W^{s,q}$, and it satisfies $PG(\tilde{f}) = 0$ outside $\partial\mathcal{O}$.*

Proof This easily follows from boundedness theorems for pseudodifferential operators in Sobolev spaces on manifolds with boundary. □

4.4 Main lemma

Choose some domain $\mathcal{D} \Subset X$ such that $\mathcal{O} \subset \mathcal{D}$ and $\partial\mathcal{O} \cap \mathcal{D} = \mathcal{S}$, i.e., the only part of $\partial\mathcal{O}$ lying in $\mathcal{D}$ is $\mathcal{S}$.

Remark 4.8 This is possible only if $\mathcal{S}$ is open. However, the assumption is not too restrictive in view of Corollary 4.5.

Set $\mathcal{D}^- = \mathcal{O}$ and $\mathcal{D}^+ = \mathcal{D} \setminus \bar{\mathcal{O}}$, and denote by $f^\pm$ the restrictions of a distribution f in $\mathcal{D}$ to the sets $\mathcal{D}^\pm$.

We now fix an arbitrary domain $\sigma \Subset \mathcal{D}^+$. According to Lemma 4.7, we have $G(\tilde{f})^+ \in \mathcal{S}(\bar{\sigma})$.

Lemma 4.9 *In order that Problem 4.2 be solvable, it is necessary and sufficient that the integral $G(\tilde{f})^+$ be extendable from σ to the whole domain $\mathcal{D}$ as a solution of $W^{s,q}(\mathcal{D}) \cap \mathcal{S}(\mathcal{D})$.*

Proof Necessity. If f is a solution of Problem 4.2, then the function F, equal to $G(\tilde{f}) + f$ on $\mathcal{D}^-$ and $G(\tilde{f})$ on $\mathcal{D}^+$, is an extension of $G(\tilde{f})^+$ to $\mathcal{D}$ with the required properties.

Sufficiency. Conversely, if $F \in W^{s,q}(\mathcal{D}) \cap \mathcal{S}(\mathcal{D})$ is equal to $G(\tilde{f})$ on σ, then $f = -G(\tilde{f}) + F$ is a solution of Problem 4.2.

For details, see Shlapunov and Tarkhanov [47]. □

When $n = 2$ and $P = \partial/\partial\bar{z}$ is the Cauchy-Riemann operator, Lemma 4.9 is due to Aizenberg and Kytmanov [3].

4.5 Extension problem

If $q = 2$, Lemma 4.9 reduces Problem 4.2 to a special case of the question whether an element of a Hilbert space belongs to the range of some injective compact operator with dense range.

Suppose that $T : H \to H_0$ is a continuous linear mapping of Hilbert spaces with dense range.

Problem 4.10 Given an element $h_0 \in H_0$, it is required to find an element $h \in H$ such that $Th = h_0$.

As follows from the Open Mapping Theorem, the problem is stable if and only if the operator T is surjective.

Example 4.11 Consider $H = W^{s,2}(\mathcal{D}) \cap \mathcal{S}(\mathcal{D})$ and $H_0 = W^{s,2}(\sigma) \cap \mathcal{S}(\sigma)$. Then H (resp. H_0) is a separable Hilbert space with the metrics induced from $W^{s,2}(\mathcal{D})$ (resp. $W^{s,2}(\sigma)$). The mapping $T : H \to H_0$ is defined to be the restriction from $\mathcal{D}$ to σ, i.e., $Th = h|_\sigma$ for $h \in H$. Then Problems 4.2 and 4.10 are equivalent in view of Lemma 4.9. Under natural conditions on σ, the operator T has dense range. However, T is not surjective, so Problem 4.10 is unstable in this case. □

Thus, Problem 4.10 is of interest for us precisely in the case when T is not surjective. For that case, the problem cannot be investigated in the language of continuous linear functionals on H_0.

4.6 Bases with double orthogonality

We shall obtain a full investigation of Problem 4.10 ($q = 2$) in terms of bases with double orthogonality.

In a paper dated 1927, Bergman [7] developed the remarkable concept of sequences of analytic functions that are pairwise orthogonal according to integration over two domains one of which contains the closure of the other. He used this idea, at least in principle, to get a criterion of analytic continuation.

This beautiful and potentially useful idea did not receive sufficient recognition, probably because its practical application required the preliminary solution of an eigenvalue problem which might be difficult to solve. Apparently, the idea of bases with double orthogonality appeared again in the papers of Slepian et al. [49] in the 60's, independently of Bergman.

Definition 4.12 A sequence $\{b_\nu\} \subset H$ is said to be a *basis with double orthogonality* (BDO) if: (1) $\{b_\nu\}$ is an orthonormal basis in H, and (2) $\{Tb_\nu\}$ is an orthonormal basis in H_0.

For such a basis in H to exist, it is necessary that T should: (1) have dense range; (2) be injective, and (3) be compact. These conditions turn out to be sufficient as well.

Lemma 4.13 *Under the above conditions, a* BDO *exists.*

Proof Consider the operator T^*T in H. This operator is self-adjoint, injective and compact. According to the Spectral Theorem, T^*T has a complete orthonormal system of eigenvectors $\{b_\nu\}_{\nu=1,2,\ldots}$ corresponding to (positive) eigenvalues $\{\lambda_\nu\}$. However, simple calculations show that

$$(Tb_\mu, Tb_\nu)_{H_0} = \lambda_\mu (b_\mu, b_\nu)_H,$$

i.e., the system $\{Tb_\nu\}$ is orthogonal in H_0. Since T has dense range, the system $\{Tb_\nu\}$ is an orthogonal basis in H_0, as we wanted to prove. □

This lemma was proved by Krasichkov [31] (see also Shaprio [46]).

4.7 Investigation of the extension problem

The main property of BDO's consists of the following.

Lemma 4.14 *If* $\{b_\nu\}$ *is a* BDO *in* H, *then for every* $h \in H$,

$$(h, b_\nu)_H = \frac{(Th, Tb_\nu)_{H_0}}{\|Tb_\nu\|^2_{H_0}} \qquad (\nu = 1, 2, \ldots).$$

Proof In fact,

$$(h, b_\nu)_H = (h, \frac{1}{\lambda_\nu} T^* T b_\nu)_H = \frac{1}{\lambda_\nu}(Th, Tb_\nu)_{H_0} = \frac{(Th, Tb_\nu)_{H_0}}{\|Tb_\nu\|^2_{H_0}},$$

as claimed. □

So, in order to know the Fourier coefficients of an element $h \in H$ with respect to the system $\{b_\nu\}$, it suffices to know only the Fourier coefficients of the image $Th \in H_0$ with respect to the system $\{Tb_\nu\}$.

For $h_0 \in H_0$, we denote by $k_\nu(h_0)$ the Fourier coefficients of h_0 with respect to the system $\{Tb_\nu\}$, i.e.,

$$k_\nu(h_0) = \frac{(h_0, Tb_\nu)_{H_0}}{\|Tb_\nu\|^2_{H_0}} \qquad (\nu = 1, 2, \ldots).$$

Lemma 4.15 *If* $\{b_\nu\}$ *is a* BDO *in* H, *then, in order that Problem* 4.10 *be solvable, it is necessary and sufficient that*

$$\sum_{\nu=1}^{\infty} |k_\nu(h_0)|^2 < \infty.$$

Proof This follows from the Fischer-Riesz Theorem and Lemma 4.14. □

Lemma 4.16 *Under the condition of Lemma* 4.15, *the solution of Problem* 4.10 *is given by the formula*

$$h = \sum_{\nu=1}^{\infty} k_\nu(h_0) b_\nu.$$

Proof In fact, the series $\sum_{\nu=1}^{\infty} k_\nu(h_0) b_\nu$ converges in the norm of H to some element $h \in H$, and we get

$$Th = \sum_{\nu=1}^{\infty} k_\nu(h_0) \cdot Tb_\nu = h_0.$$

□

4.8 Existence of bases with double orthogonality

We come back to Problem 4.2, i.e., let $H = W^{s,2}(\mathcal{D}) \cap \mathcal{S}(\mathcal{D})$, $H_0 = W^{s,2}(\sigma) \cap \mathcal{S}(\sigma)$ and the operator $T : H \to H_0$ be given by $Th = h|_\sigma$.

Lemma 4.17 *If $\mathcal{D} \setminus \sigma$ has no compact connected components and $\partial\sigma$ is "regular", then T has dense range.*

Proof Since $\mathcal{D} \setminus \sigma$ has no compact connected components, the subspace $\mathcal{S}(\bar{\mathcal{D}})$ is dense in $\mathcal{S}(\bar{\sigma})$. And, since $\partial\sigma$ is "regular", $\mathcal{S}(\bar{\sigma})$ is dense in $W^{s,2}(\sigma) \cap \mathcal{S}(\sigma)$. Hence it follows that $\mathcal{S}(\bar{\mathcal{D}}) \hookrightarrow H$ is dense in H_0. That is more than we wanted to prove. □

Proving the lemma we can see how one ought to understand the words "regular boundary". If $s \geq p$, the word "regular" means any boundary (see Theorem 3.12). And if $s < p$, then this means that the complement of $\bar{\sigma}$ at each boundary point would be massive enough in the sense of the $W^{s,q}(\mathcal{D})$-capacity (see section 3.11).

Lemma 4.18 *The operator T is injective.*

Proof This is easy to prove because of analytic hypoellipticity of P. □

Lemma 4.19 *The operator T is compact.*

Proof Let σ be a bounded set in H. Since the topology of H is stronger than the one of $\mathcal{S}(\mathcal{D})$, the set σ is bounded in $\mathcal{S}(\mathcal{D})$. Hence it follows that σ is relatively compact because $\mathcal{S}(\mathcal{D})$ is a Frechet-Schwartz space. Finally, since the topology of $\mathcal{S}(\mathcal{D})$ is stronger than the topology of H_0, the set σ is relatively compact in H_0. □

We now can formulate the main result concerning the existence of BDO's.

Theorem 4.20 *If $\mathcal{D} \setminus \sigma$ has no compact connected components and $\partial\sigma$ is "regular", then there is an orthonormal basis $\{b_\nu\}$ in $W^{s,2}(\mathcal{D}) \cap \mathcal{S}(\mathcal{D})$ whose restriction to σ is an orthogonal basis in $W^{s,2}(\sigma) \cap \mathcal{S}(\sigma)$.*

Proof This follows from Lemmas 4.13, and 4.17–4.19. □

4.9 Solvability of the Cauchy problem

We pass to a formulation of solvability conditions for Problem 4.2.

Assuming σ to be a relatively compact subdomain of $\mathcal{D}^+$ whose complement has no compact connected components and whose boundary is "regular", we fix some BDO $\{b_\nu\}$ in $W^{s,2}(\mathcal{D}) \cap \mathcal{S}(\mathcal{D})$.

Let $G(\tilde{f})$ be the Green's integral constructed from the "initial" data of Problem 4.2 via formula (4.3). As already mentioned, the restriction of the function $G(\tilde{f})$ to σ belongs to the space H_0.

Lemma 4.21 *For $\nu = 1, 2, \ldots,$*

$$k_\nu(G(\tilde{f})) = \int_{\partial\mathcal{O}} \sum_{j=0}^{p-1} C_j k_\nu(\Phi(.\,,y)) \cdot \tilde{f}_j \, ds. \tag{4.4}$$

Proof This follows immediately from formula (4.3). □

So, in order to know the coefficients $k_\nu(G(\tilde{f}))$ $(\nu = 1, 2, \ldots)$ it is not necessary to know the basis $\{Tb_\nu\}$ in H_0. It suffices to know only the coefficients of the expansion of the fundamental solution $\Phi(.\,,y)$ $(y \in \partial\mathcal{O})$ into a series with respect to the basis. And these coefficients may often be obtained by indirect considerations.

Theorem 4.22 *In order that Problem 4.2 be solvable, it is necessary and sufficient that*

$$\sum_{\nu=1}^{\infty} \left|k_\nu(G(\tilde{f}))\right|^2 < \infty. \tag{4.5}$$

Proof It suffices to apply Lemmas 4.9 and 4.15. □

This result was first obtained by Shlapunov and Tarkhanov [47].

4.10 Regularization of the Cauchy problem

Bases with double orthogonality give the possibility side by side with solvability conditions for Problem 4.2 (for $q = 2$) to obtain a reasonable formula for approximate solutions (the so-called Carleman formula).

Consider the following kernels $K^{(N)}$, defined for $(x, y) \in \mathcal{D} \times X$ $(x \neq y)$:

$$K^{(N)}(x,y) = \Phi(x,y) - \sum_{\nu=1}^{N} b_\nu(x) \otimes k_\nu(\Phi(.\,,y)) \qquad (N = 1, 2, \ldots).$$

Lemma 4.23 *For any number $N = 1, 2, \ldots,$ the kernel $K^{(N)} \in \mathcal{E}(\mathcal{D} \times (X \setminus \sigma))$ satisfies $P(x)K^{(N)} = 0$ for $x \in \mathcal{D}$, and $P'(y)K^{(N)} = 0$ for $y \in X \setminus \bar{\sigma}$ everywhere except on the diagonal $\{x = y\}$.*

Proof This is immediately seen because of the properties of Φ and $\{b_\nu\}$. □

The sequence of the kernels $\{K^{(N)}\}$, interpolated in some way to all real values $N \geq 1$, provides a special Carleman function for Problem 4.2 (see Lavrentiev [33]).

For a solution $f \in W^{s,2}(\mathcal{O}) \cap \mathcal{S}(\mathcal{O})$, we denote by $\tilde{f}_j \in W^{s-b_j-\frac{1}{2},2}(\partial\mathcal{O})$ $(j = 0, \ldots, p-1)$ some (arbitrary) extensions of the functions $B_j f$ from $\mathcal{S}$ to the whole boundary.

Theorem 4.24 *For every solution $f \in W^{s,q}(\mathcal{O}) \cap \mathcal{S}(\mathcal{O})$, the following formula holds:*

$$f(x) = - \lim_{N \to \infty} \int_{\partial \mathcal{O}} \sum_{j=0}^{p-1} C_j K^{(N)}(x, .) \cdot \tilde{f}_j \, ds \qquad (x \in \mathcal{O}). \tag{4.6}$$

Proof This follows from Lemmas 4.9 and 4.16. □

We emphasize that the integral on the right hand side of formula (4.6) depends on the values of the expressions $B_j f$ $(j = 0, \ldots, p-1)$ on $\mathcal{S}$ only. So this formula is a quantitative expression of Corollary 4.5 (uniqueness). However, it gives a bit more than Corollary 4.5 because there is sufficiently complete information about the kernels $\{K^{(N)}\}$.

Remark 4.25 The limit in (4.6) is reached in the topology of the space $W^{s,2}(\mathcal{O})$.

Formula (4.6) apparently occurs for the first time in the paper of Shlapunov and Tarkhanov [47].

4.11 Example for holomorphic functions

Let $P = \partial/\partial \bar{z}$ be the Cauchy-Riemann operator in the complex plane $X = \mathbf{C}^1$. For a domain $\mathcal{O} \Subset \mathbf{C}^1$, the standard Dirichlet system of order 0 on $\partial\mathcal{O}$ is $\{B_0 = 1\}$ (see Example 1.16). Then Problem 4.2 is just the problem of analytic continuation from a boundary set $\mathcal{S} \subset \partial\mathcal{O}$ to the whole domain within Sobolev class $W^{s,q}(\mathcal{O})$. Let $\mathcal{D}$ be the circle $B(0, R)$ $(R > 0)$ in $\mathbf{C}^1$, and $\mathcal{S}$ a smooth hypersurface in $\mathcal{D}$ which does not contain 0 and divides $\mathcal{D}$ into two domains $\mathcal{D}^{\pm}$. We suppose that zero belongs to $\mathcal{D}^+$, and consider the Cauchy problem for a function $f \in L^2(\mathcal{O}) \cap \mathcal{S}(\mathcal{O})$ $(\mathcal{O} = \mathcal{D}^-)$ with data on $\mathcal{S}$. If $r > 0$ is small enough, the circle $\sigma = B(0, r)$ lies in $\mathcal{D}^+$ as well as its closure. So we have the situation considered in section 4.4. For this case, a BDO may be written in explicit form. Namely, one can easily verify that the system of holomorphic monomials

$$b_\nu = \sqrt{\frac{\nu+1}{\pi}} \frac{1}{R^{\nu+1}} z^\nu \qquad (\nu = 0, 1, \ldots)$$

is an orthonormal basis in $L^2(\mathcal{D}) \cap \mathcal{S}(\mathcal{D})$, and its restriction to σ is an orthogonal basis of $L^2(\sigma) \cap \mathcal{S}(\sigma)$. Obtained this way, Theorem 4.22 is due to Aizenberg and Kytmanov [3].

4.12 Example for harmonic functions

Suppose that $P = \Delta$ is the Laplace operator in $X = \mathbf{R}^n$. For a domain $\mathcal{O} \Subset \mathbf{R}^n$, the standard Dirichlet system of order 1 on $\partial\mathcal{O}$ is $\{B_0 = 1,\ B_1 = \partial/\partial n\}$ (see Example 1.16). Then Problem 4.2 is the classical example of an ill-posed boundary value problem. Let $\mathcal{D}$ be the ball $B(0, R)$ $(R > 0)$ in $\mathbf{R}^n$, and $\mathcal{S}$ a smooth hypersurface in $\mathcal{D} \setminus \{0\}$ dividing $\mathcal{D}$ into two connected components. Denote by $\mathcal{O}$ the component which does not contain zero, and consider the Cauchy problem for a function $f \in L^2(\mathcal{O}) \cap \mathcal{S}(\mathcal{O})$ with data on

$\mathcal{S}$. For $r < \text{dist}(0, \mathcal{S})$, the ball $\sigma = B(0, r)$ lies in $\mathcal{D} \setminus \bar{\mathcal{O}}$ as well as its closure. Again we have the situation considered in section 4.4. For this case, a BDO may be constructed in explicit form. Namely, let $\{h_\nu^{(i)}\}$ be a set of homogeneous harmonic polynomials which form a complete orthonormal system in $L^2(\partial B(0,1))$, where $\nu = 0, 1, \ldots$ is the degree of homogeneity, and $i = 1, \ldots, \mathcal{I}(\nu)$ is the number of the polynomials of degree ν belonging to the basis. Then the system

$$\left\{ \sqrt{\frac{n+2\nu}{R^{n+2\nu}}}\, h_\nu^{(i)} \right\}$$

is proved to be an orthonormal basis in $L^2(\mathcal{D}) \cap \mathcal{S}(\mathcal{D})$ and an orthogonal basis in $L^2(\sigma) \cap \mathcal{S}(\sigma)$, i.e., a BDO.

Remark 4.26 This example is of particular interest for applications to overdetermined systems of the simplest type (see Shlapunov and Tarkhanov [47]).

Chapter 5
The Fischer-Riesz equations method

5.1 Abstract problem in Hilbert spaces

In this section we exhibit another Hilbert space method for boundary value problems for solutions of $Pf = 0$. In ideological respect the approach is akin to the so-called Fischer-Riesz equations method, which entailed considerable progress in applications to elliptic equations through the work of Italian mathematicians of the 50's. A description of this method and a detailed bibliography may be found in the book of Miranda [39].

When considered in Hilbert spaces, many boundary value problems go into the following scheme of functional analysis.

Let $M : H \to \tilde{H}$ be some continuous linear mapping of Hilbert spaces, and assume $\tilde{H}$ to be separable. We suppose that: (1) M is injective, and (2) M has closed range.

Further, for some special reasons, we distinguish a closed subspace $H_0 \subset \tilde{H}$, and consider the orthogonal projection Π onto H_0 in $\tilde{H}$. Denote by T the composition $\Pi \circ M : H \to H_0$.

Problem 5.1 Given an element $h_0 \in H_0$, it is required to find an element $h \in H$ such that $Th = h_0$.

We emphasize that the operator T is not required to have dense image. So, Problem 5.1 goes beyond the range of the problems considered in section 4.5 although it is formulated in the same way as Problem 4.10.

5.2 Special bases

Let $M^* : \tilde{H} \to H$ be the adjoint operator of M in the sense of Hilbert spaces.

Lemma 5.2 *The kernel space of the operator M^* (denoted by $\ker M^*$) is separable in the topology induced from $\tilde{H}$.*

Proof This is a school fact from the theory of metric spaces. □

In applications there are many ways to choose a complete system $\{B_i\}_{i=1,2,...}$ in $\ker M^*$.

Lemma 5.3 *The sequence $\{(1-\Pi)B_i\}$ is complete in $\tilde{H} \ominus H_0$ if and only if the mapping $\Pi M : H \to H_0$ is injective.*

Proof See Karepov and Tarkhanov [28]. □

After removing those elements which are linear combinations of preceding ones, from the system $\{(1-\Pi)B_i\}$ we may apply Gram-Schmidt orthogonalization to the system in $\tilde{H} \ominus H_0$. Then we obtain a sequence $\{b_\nu\}$ in $\ker M^*$ such that the system $\{(1-\Pi)b_\nu\}$ is an orthonormal basis of $\tilde{H} \ominus H_0$, at least, if the mapping $\Pi \circ M$ is injective.

Definition 5.4 (Orthonormal) bases in $\tilde{H} \ominus H_0$ of the form $\{(1-\Pi)b_\nu\}$, where $\{b_\nu\} \subset \ker M^*$, are called *special bases.*

Notice that every element b_ν ($\nu = 1, 2, \ldots$) of the new system has an obvious expression through the elements $\{B_1, \ldots, B_\nu\}$ of the old system in the form of Gram's determinants.

5.3 Abstract form of the Fischer-Riesz equations method

In this section we assume the uniqueness theorem for solutions of Problem 5.1, i.e., the injectivity of T.

Fix some special basis $\{(1-\Pi)b_\nu\}_{\nu=1,2,...}$ in $\tilde{H} \ominus H_0$, where $\{b_\nu\} \subset \ker M^*$.

For $\tilde{h} \in \tilde{H} \ominus H_0$, we denote by $k_\nu(\tilde{h})$ the Fourier coefficients of $\tilde{h}$ with respect to the basis $\{(1-\Pi)b_\nu\}$, i.e.,

$$k_\nu(\tilde{h}) = (\tilde{h}, (1-\Pi)b_\nu)_{\tilde{H}\ominus H_0} \qquad (\nu = 1, 2, \ldots).$$

The main property of special bases is expressed in the following lemma.

Lemma 5.5 *If $h \in H$, then*

$$k_\nu((1-\Pi)Mh) = -(Th, \Pi b_\nu)_{H_0} \qquad (\nu = 1, 2, \ldots).$$

Proof In fact,

$$k_\nu((1-\Pi)Mh) = ((1-\Pi)Mh, (1-\Pi)b_\nu)_{\tilde{H}} = ((1-\Pi)Mh, b_\nu)_{\tilde{H}} = -(Th, \Pi b_\nu)_{H_0}$$

as desired. □

Thus, to find the Fourier coefficients of the projection of Mh onto $\tilde{H} \ominus H_0$ with respect to a special basis in this subspace, it suffices to know only the image Th in H_0.

Lemma 5.6 *In order that Problem* 5.1 *be solvable, it is necessary and sufficient that*

(1)

$$\sum_{\nu=1}^{\infty} |c_\nu|^2 < \infty,$$

where $c_\nu = -(h_0, \Pi b_\nu)_{H_0}$*;*

(2)

$$(h_0, \Pi\tilde{h})_{H_0} = 0$$

for all $\tilde{h} \in \ker M^*$ *such that* $(1 - \Pi)\tilde{h} = 0$.

Proof The necessity follows from Lemma 5.5. For a proof of the sufficiency, see Karepov and Tarkhanov [28]. □

Convergence of the series in (1) guarantees the stability of Problem 5.1. Under this condition, the range of the mapping T is described in terms of continuous linear functionals on the space $\tilde{H}$ (see (2)). This is impossible in the general case.

Corollary 5.7 *If the restriction of* $(1 - \Pi)$ *to* $\ker M^*$ *is injective, then for Problem* 5.1 *to be solvable, it is necessary and sufficient that*

$$\sum_{\nu=1}^{\infty} |c_\nu|^2 < \infty,$$

where $c_\nu = -(h_0, \Pi b_\nu)_{H_0}$.

Proof This follows immediately from Lemma 5.6 because condition (2) is satisfied. □

We are also able to derive a reasonable formula for approximate solutions of Problem 5.1.

Lemma 5.8 *For each continuous linear functional* F *on* H*, there exists an element* $k_F \in \operatorname{im} M$ *such that*

$$F(h) = \lim_{N\to\infty} \left(Th, k_F - \sum_{\nu=1}^{N} (k_F, (1-\Pi)b_\nu)_{\tilde{H}} b_\nu \right)_{\tilde{H}} \qquad (h \in H).$$

Proof See Karepov and Tarkhanov [28]. □

5.4 Generalized Hardy spaces

In order to apply the Fischer-Riesz equations method to boundary value problems for solutions of $Pf = 0$, (generalized) Hardy spaces seem to be a more suitable tool than Sobolev spaces.

Let $\mathcal{O}$ be a relatively compact subdomain of X with smooth boundary, and $\{B_j\}_{j=0}^{p-1}$ a Dirichlet system of order $(p-1)$ on $\partial\mathcal{O}$.

Definition 5.9 For $1 \le q \le \infty$, the space $H^q_{P,B}(\mathcal{O})$ is defined to consist of all solutions $f \in \mathcal{S}(\mathcal{O})$ of finite order of growth near $\partial\mathcal{O}$, such that the weak limit values of the expressions $B_j f$ $(j = 0, \ldots, p-1)$ on $\partial\mathcal{O}$ belong to $L^q(\partial\mathcal{O})$.

When endowed with the norm

$$||f||_{H^q_{P,B}(\mathcal{O})} = \left(\sum_{j=0}^{p-1} \int_{\partial\mathcal{O}} |B_j f|^q \, ds \right)^{1/q},$$

the space $H^q_{P,B}(\mathcal{O})$ is a Banach space, and even a Hilbert space if $q = 2$.

Example 5.10 If $P = \partial/\partial\bar{z}$ is the Cauchy-Riemann operator in the plane (and $B_0 = 1$), then the $H^q_{P,B}(\mathcal{O})$'s are just the classical Hardy spaces of holomorphic functions in the domain $\mathcal{O} \subset \mathbb{C}^1$. □

Notice that the spaces $H^q_{P,B}(\mathcal{O})$ essentially depend on the particular choice of the Dirichlet system $\{B_j\}$ on $\partial\mathcal{O}$ (see Example 1.13 of Tarkhanov [54]).

Theorem 5.11 *For functions $f_j \in L^q(\partial\mathcal{O})$ $(j = 0, \ldots, p-1)$, there exists a solution $f \in H^q_{P,B}(\mathcal{O})$ such that $B_j f|_{\partial\mathcal{O}} = f_j$ $(j = 0, \ldots, p-1)$ if and only if*

$$\int_{\partial\mathcal{O}} \sum_{j=0}^{p-1} C_j \Phi(x, .) \cdot f_j \, ds = 0$$

for all $x \in X \setminus \bar{\mathcal{O}}$.

Proof See Tarkhanov [52] (§29). □

5.5 The Cauchy problem

Suppose that $\mathcal{S}$ is a closed subset of $\partial\mathcal{O}$ of positive measure.

Problem 5.12 Given functions $f_j \in L^2(\mathcal{S})$ $(j = 0, \ldots, p-1)$, it is required to find a solution $f \in H^2_{P,B}(\mathcal{O})$ such that $B_j f|_{\mathcal{S}} = f_j$ $(j = 0, \ldots, p-1)$.

We would like to insert the problem into the abstract scheme of section 5.1.

For this purpose we set

$$H = H^2_{P,B}(\mathcal{O}) \quad \text{and} \quad \tilde{H} = \bigoplus_0^{p-1} L^2(\partial\mathcal{O}),$$

so that H and $\tilde{H}$ are separable Hilbert spaces.

The mapping $M : H \to \tilde{H}$ is defined by

$$M(f) = \bigoplus_{j=0}^{p-1} B_j f|_{\partial \mathcal{O}}.$$

The Hermitian structure in H has been introduced in such a way that M is isometric, and therefore injective.

Lemma 5.13 *The mapping M has closed range.*

Proof This follows from Theorem 5.11. □

For each function $f_0 \in L^2(\mathcal{S})$ we may consider the function $\tilde{f}$ on $\partial\mathcal{O}$, equal f_0 on $\mathcal{S}$ and 0 on $\partial\mathcal{O} \setminus \mathcal{S}$. Clearly $\tilde{f}$ is in $L^2(\partial\mathcal{O})$, so one can consider the space $H_0 = \bigoplus_0^{p-1} L^2(\mathcal{S})$ as a (closed) subspace of $\tilde{H}$. Then the projector $\Pi : \tilde{H} \to H_0$ is simply interpreted as multiplication by the characteristic function of $\mathcal{S}$.

Under these assumptions, Problem 5.1 is just a rewording of Problem 5.12.

Lemma 5.14 *If $g \in \mathcal{S}_{P'}(\bar{\mathcal{O}})$, then*

$$\bigoplus_{j=0}^{p-1} \overline{C_j g}|_{\partial\mathcal{O}} \in \ker M^*.$$

Proof In fact, for $f \in H$, we get by the Green formula

$$\left(Mf, \bigoplus_{j=0}^{p-1} \overline{C_j g} \right)_{\tilde{H}} = \int_{\partial\mathcal{O}} \sum_{j=0}^{p-1} C_j g \cdot B_j f \, ds = 0$$

as desired. □

The subspace of $\tilde{H}$ formed by the elements $\bigoplus_{j=0}^{p-1} \overline{C_j g}|_{\partial\mathcal{O}}$, where $g \in \mathcal{S}_{P'}(\bar{\mathcal{O}})$, is separable. Hence there are many ways to point out a sequence $\{G_i\}$ in $\mathcal{S}_{P'}(\bar{\mathcal{O}})$ so that the system $\left\{\bigoplus_{j=0}^{p-1} \overline{C_j G_i}|_{\partial\mathcal{O}}\right\}$ is complete in this subspace.

Lemma 5.15 *The system $\left\{\bigoplus_{j=0}^{p-1} \overline{C_j G_i}|_{\partial\mathcal{O}}\right\}$ is complete in* $\ker M^*$.

Proof It suffices to apply the Hahn-Banach Theorem and Theorem 5.11. □

The following lemma expresses the most important property of the sequence $\{G_i\}$.

Lemma 5.16 *The restriction of the system $\left\{\bigoplus_{j=0}^{p-1} \overline{C_j G_i}\right\}$ to $\partial\mathcal{O} \setminus \mathcal{S}$ is complete in the space $\bigoplus_0^{p-1} L^2(\partial\mathcal{O} \setminus \mathcal{S})$ if and only if Problem* 5.12 *admits the uniqueness theorem.*

Proof This follows from Lemmas 5.3 and 5.15. □

After removing the elements which are linear combinations of the preceding ones, from the system $\left\{\bigoplus_{j=0}^{p-1}\overline{C_jG_i}|_{\partial\mathcal{O}\setminus\mathcal{S}}\right\}$ we may apply the Gram-Schmidt orthogonalization to the system in $\bigoplus_0^{p-1} L^2(\partial\mathcal{O}\setminus\mathcal{S})$. Then we get a sequence $\{g_\nu\}$ in $\mathcal{S}_{P'}(\bar{\mathcal{O}})$ such that the restriction of the system $\left\{\bigoplus_{j=0}^{p-1}\overline{C_jg_\nu}\right\}$ to $\partial\mathcal{O}\setminus\mathcal{S}$ is an orthonormal basis in $\bigoplus_0^{p-1} L^2(\partial\mathcal{O}\setminus\mathcal{S})$.

Remark 5.17 $\left\{\bigoplus_{j=0}^{p-1}\overline{C_jg_\nu}|_{\partial\mathcal{O}\setminus\mathcal{S}}\right\}$ is just a special basis in $\tilde{H}\ominus H_0$ for this case.

5.6 Example of a special basis

Let $\sigma = \{x_i\}$ be a finite or countable set of points of $X\setminus\bar{\mathcal{O}}$ such that each connected component of $X\setminus\bar{\mathcal{O}}$ contains at least one point of σ. For every multi-index $\alpha\in\mathbf{Z}_+^n$ and $i=1,2,\ldots$, the derivative $D_x^\alpha\Phi(x_i,.)$ belongs to $\mathcal{S}_{P'}(\bar{\mathcal{O}})$. As follows from the Hahn-Banach Theorem and Theorem 5.11, the system

$$\left\{\bigoplus_{j=0}^{p-1}\overline{C_jD_x^\alpha\Phi(x_i,.)}|_{\partial\mathcal{O}}\right\}$$

is complete in the subspace of $\bigoplus_0^{p-1} L^2(\partial\mathcal{O})$ consisting of elements of the form $\left\{\bigoplus_{j=0}^{p-1}\overline{C_jg}|_{\partial\mathcal{O}}\right\}$ where $g\in\mathcal{S}_{P'}(\bar{\mathcal{O}})$. Therefore, the restriction of the system

$$\left\{\bigoplus_{j=0}^{p-1}\overline{C_jD_x^\alpha\Phi(x_i,.)}\right\}$$

to $\partial\mathcal{O}\setminus\mathcal{S}$ is complete in the space $\bigoplus_0^{p-1} L^2(\partial\mathcal{O}\setminus\mathcal{S})$, at least if $\mathcal{S}$ has non-empty interior. However, there may be linearly dependent elements in

$$\left\{\bigoplus_{j=0}^{p-1}\overline{C_jD_x^\alpha\Phi(x_i,.)}|_{\partial\mathcal{O}\setminus\mathcal{S}}\right\}$$

because among the derivatives $D_x^\alpha\Phi(x_i,.)$ there are non-trivial dependences generated by the equation $P(x,D)\Phi(x,y)=\delta(x-y)$. All these dependences must be excluded before applying the Gram-Schmidt orthogonalization in $\bigoplus_0^{p-1} L^2(\partial\mathcal{O}\setminus\mathcal{S})$. So we get a special basis in $\bigoplus_0^{p-1} L^2(\partial\mathcal{O}\setminus\mathcal{S})$ of the form

$$\left\{\bigoplus_{j=0}^{p-1}\overline{C_j\Phi'(F_\nu)}|_{\partial\mathcal{O}\setminus\mathcal{S}}\right\},$$

where $\{F_\nu\}\subset\mathcal{E}'_\sigma$.

5.7 Solvability of the Cauchy problem

In the next two sections we assume the uniqueness theorem for solutions of Problem 5.12. This is the case, in particular, if $\overset{\circ}{\mathcal{S}} \neq \emptyset$.

Fix some special basis $\left\{\bigoplus_{j=0}^{p-1} \overline{C_j g_\nu}|_{\partial\mathcal{O}\setminus\mathcal{S}}\right\}$ in $\bigoplus_0^{p-1} L^2(\partial\mathcal{O}\setminus\mathcal{S})$, where $\{g_\nu\} \subset \mathcal{S}_{P'}(\bar{\mathcal{O}})$.

Theorem 5.18 *Problem* 5.12 *is solvable if and only if the following two conditions hold:*

(1)

$$\sum_{\nu=1}^{\infty} |c_\nu|^2 < \infty, \quad \textit{where} \quad c_\nu = -\int_{\mathcal{S}} \sum_{j=0}^{p-1} C_j g_\nu \cdot f_j \, ds$$

(2)

$$\int_{\mathcal{S}} \sum_{j=0}^{p-1} C_j g \cdot f_j \, ds = 0$$

for all $g \in \mathcal{S}_{P'}(\bar{\mathcal{O}})$ *such that* $C_j g|_{\partial\mathcal{O}\setminus\overset{\circ}{\mathcal{S}}} = 0$ $(j = 0, \ldots, p-1)$.

Proof This follows from Lemma 5.6 and the results of section 5.5. □

The last condition of the theorem is needed only if $\mathcal{S} = \partial\mathcal{O}$. In this case, condition (2) implies condition (1) because all the coefficients c_ν are zero. As follows from Theorem 3.10, Theorem 5.18 ($\mathcal{S} = \partial\mathcal{O}$) is equivalent to Theorem 5.11.

Corollary 5.19 *If* $\mathcal{S} \neq \partial\mathcal{O}$, *then in order that Problem* 5.12 *be solvable it is necessary and sufficient that condition* (1) *of Theorem* 5.18 *be satisfied.*

Proof Since the set $\mathcal{S}$ is assumed to be closed, the condition $\mathcal{S} \neq \partial\mathcal{O}$ means that $\partial\mathcal{O}\setminus\overset{\circ}{\mathcal{S}}$ has at least one interior point (on $\partial\mathcal{O}$!). Therefore, if $g \in \mathcal{S}_{P'}(\bar{\mathcal{O}})$ and $C_j g|_{\partial\mathcal{O}\setminus\overset{\circ}{\mathcal{S}}} = 0$ $(j = 0, \ldots, p-1)$, we may concude by Corollary 4.5 that $g \equiv 0$ in $\mathcal{O}$. So condition (2) of Theorem 5.18 is fulfilled. □

It would be of interest to derive the following result from Theorem 5.18 directly.

Remark 5.20 It follows from the Hahn-Banach Theorem and Lemma 5.16, that if $\mathcal{S} \neq \partial\mathcal{O}$, then Problem 5.12 is dense solvable.

5.8 Regularization of the Cauchy problem

Consider the following kernels $K^{(N)}$, defined for $x \in (X\setminus\partial\mathcal{O}) \cup \overset{\circ}{\mathcal{S}}$ and for y in a neighborhood of the closure of $\mathcal{O}$:

$$K^{(N)}(x,y) = \Phi(x,y) - \sum_{\nu=1}^{N} \left[\int_{\partial\mathcal{O}\setminus\mathcal{S}} \sum_{j=0}^{p-1} C_j \Phi(x,.) \cdot \overline{C_j g_\nu} \, ds\right] \otimes g_\nu(y) \qquad (N = 1, 2, \ldots).$$

Lemma 5.21 *For any number $N = 1, 2, \dots$, the kernel $K^{(N)}$ is infinitely differentiable in $x \in (X \setminus \partial\mathcal{O}) \cup \overset{\circ}{S}$ and in y in a neighborhood of $\bar{\mathcal{O}}$ except on the diagonal $\{x = y\}$, and for such x and y satisfies $P(x)K^{(N)} = 0$ and $P'(y)K^{(N)} = 0$.*

Proof This is obvious. □

The sequence of the kernels $\{K^{(N)}\}$, interpolated in some way to all real values $N \geq 1$, provides a special Carleman function for Problem 5.12 (cf. section 4.10).

Theorem 5.22 *For every solution $f \in H^2_{P,B}(\mathcal{O})$, the following formula holds:*

$$f(x) = -\lim_{N\to\infty} \int_S \sum_{j=0}^{p-1} C_j K^{(N)}(x, .) \cdot B_j f \, ds \qquad (x \in \mathcal{O}).$$

Proof See Karepov and Tarkhanov [28]. □

For holomorphic functions, both Theorem 5.18 and Theorem 5.22 are due to Zin [58].

5.9 Dirichlet problem

Since P is assumed to be elliptic of order p, the number p has to be even unless $n = 1$.

We formulate the Dirichlet problem for solutions of $Pf = 0$ in the following way.

Problem 5.23 Given functions $f_j \in L^2(\partial\mathcal{O})$ $(j = 0, \dots, \frac{p}{2} - 1)$ on $\partial\mathcal{O}$, it is required to find a solution $f \in H^2_{P,B}(\mathcal{O})$ such that $B_j f|_{\partial\mathcal{O}} = f_j$ $(j = 0, \dots, \frac{p}{2} - 1)$.

The simplest examples show that the uniqueness may be absent in Problem 5.23.

Example 5.24 Let $P = (\partial/\partial\bar{z})^2$ be the iterated Cauchy-Riemann operator in the plane, and $\mathcal{O} = \{z \in \mathbb{C}^1 : |z| < 1\}$ the unit circle. For each holomorphic function f_0 in a neighborhood of $\bar{\mathcal{O}}$, the function $f(z) = f_0(z)(1 - |z|^2)$ satisfies $Pf = 0$ and is equal to zero on $\partial\mathcal{O}$. □

We suppose that the formal adjoint problem to Problem 5.23 is solvable in the class of sufficiently smooth functions.

Problem 5.25 Given functions $g_j \in \mathcal{E}(\partial\mathcal{O})$ $(j = \frac{p}{2}, \dots, p-1)$ on $\partial\mathcal{O}$, it is required to find a solution $g \in \mathcal{E}(\bar{\mathcal{O}})$ such that $P'g = 0$ in $\mathcal{O}$ and $C_j g|_{\partial\mathcal{O}} = g_i$ $(j = \frac{p}{2}, \dots, p-1)$.

Under this assumption, the uniqueness theorem holds for Problem 5.23.

Lemma 5.26 *If $f \in S(\mathcal{O})$ is a solution having finite order of growth near $\partial\mathcal{O}$, and $B_j f|_{\partial\mathcal{O}} = 0$ $(j = 0, \dots, \frac{p}{2} - 1)$, then $f \equiv 0$ in $\mathcal{O}$.*

Proof For each point $x \in \mathcal{O}$, we find a function $g(x,.) \in \mathcal{E}(\bar{\mathcal{O}})$ such that $P'g(x,.) = 0$ in $\mathcal{O}$ and $C_j g(x,.)|_{\partial\mathcal{O}} = C_j\Phi(x,.)$ $(j = \frac{p}{2},\dots,p-1)$. Using Lemma 4.3, we get the following (Poisson) formula for every solution $f \in \mathcal{S}(\mathcal{O})$ of finite order of growth near $\partial\mathcal{O}$:

$$f(x) = -\int_{\partial\mathcal{O}} \sum_{j=0}^{\frac{p}{2}-1} C_j\,[\Phi(x,.) - g(x,.)]\cdot B_j f\,ds \qquad (x \in \mathcal{O}). \tag{5.1}$$

Hence the lemma follows. □

Now we would like to include Problem 5.23 into the abstract scheme of section 5.1. For this purpose we set

$$H = H^2_{P,B}(\mathcal{O}) \quad \text{and } \tilde{H} = \bigoplus_{0}^{p-1} L^2(\partial\mathcal{O}),$$

so that H and $\tilde{H}$ are separable Hilbert spaces.

The mapping $M : H \to \tilde{H}$ is defined by

$$M(f) = \bigoplus_{j=0}^{p-1} B_j f|_{\partial\mathcal{O}}.$$

Then M is injective and has closed range (cf. section 5.5).

In $\tilde{H}$ we consider the closed vector subspace H_0 consisting of the elements of the form

$$(f_0 \oplus \cdots \oplus f_{\frac{p}{2}-1}) \oplus (\underbrace{0 \oplus \cdots \oplus 0}_{p/2}),$$

where $f_j \in L^2(\partial\mathcal{O})$ $(j = 0,\dots,\frac{p}{2}-1)$. Then the orthogonal projection Π of $\tilde{H}$ onto H_0 is given by

$$\Pi\left(\bigoplus_{j=0}^{p-1} f_j\right) = \left(\bigoplus_{j=0}^{\frac{p}{2}-1} f_j\right) \oplus \left(\bigoplus_{j=p/2}^{p-1} 0\right).$$

With these conventions, Problem 5.1 is just a reformulation of Problem 5.23.

According to Lemma 5.14, if $g \in \mathcal{S}_{P'}(\bar{\mathcal{O}})$, then

$$\bigoplus_{j=0}^{p-1} \overline{C_j g}|_{\partial\mathcal{O}} \in \ker M^*.$$

Using Lemma 5.15 it is possible to find a sequence $\{G_i\}$ in $\mathcal{S}_{P'}(\bar{\mathcal{O}})$ such that the system $\left\{\bigoplus_{j=0}^{p-1} \overline{C_j G_i}|_{\partial\mathcal{O}}\right\}$ is complete in $\ker M^*$.

Lemma 5.27 *The system* $\left\{\bigoplus_{j=p/2}^{p-1} \overline{C_j G_i}|_{\partial\mathcal{O}}\right\}$ *is complete in the subspace* $\tilde{H} \ominus H_0 = \bigoplus_{p/2}^{p-1} L^2(\partial\mathcal{O})$.

Proof This follows from Lemmas 5.3 and 5.26. □

After removing the elements which are linear combinations of the preceding ones, from the system $\left\{\bigoplus_{j=p/2}^{p-1}\overline{C_jG_i}|_{\partial\mathcal{O}}\right\}$ we may apply the Gram-Schmidt orthogonalization to the system in $\bigoplus_{p/2}^{p-1}L^2(\partial\mathcal{O})$. Then we get a sequence $\{g_\nu\}$ in $\mathcal{S}_{P'}(\bar{\mathcal{O}})$ such that the system $\left\{\bigoplus_{j=p/2}^{p-1}\overline{C_jg_\nu}|_{\partial\mathcal{O}}\right\}$ is an orthonormal basis in $\bigoplus_{p/2}^{p-1}L^2(\partial\mathcal{O})$.

Remark 5.28 $\left\{\bigoplus_{j=p/2}^{p-1}\overline{C_jg_\nu}|_{\partial\mathcal{O}}\right\}$ is just a special basis in $\tilde{H}\ominus H_0$ for this case.

5.10 Example of a special basis

Suppose that $\sigma=\{x_i\}$ is a finite or countable set of points of $X\setminus\bar{\mathcal{O}}$ whose intersection with each connected component of $X\setminus\bar{\mathcal{O}}$ is non-void. According to section 5.6, the system $\left\{\bigoplus_{j=0}^{p-1}\overline{C_jD_x^\alpha\Phi(x,.)}|_{\partial\mathcal{O}}\right\}$ is complete in $\ker M^*$. Hence it follows by Lemma 5.27 that the system $\left\{\bigoplus_{j=p/2}^{p-1}\overline{C_jD_x^\alpha\Phi(x,.)}|_{\partial\mathcal{O}}\right\}$ is complete in the space $\tilde{H}\ominus H_0=\bigoplus_{p/2}^{p-1}L^2(\partial\mathcal{O})$. Removing the elements which are linear combinations of the preceding ones from the system, and applying Gram-Schmidt orthogonalization, we get a special basis in $\bigoplus_{p/2}^{p-1}L^2(\partial\mathcal{O})$ of the form

$$\left\{\bigoplus_{j=p/2}^{p-1}\overline{C_j\Phi'(F_\nu)}|_{\partial\mathcal{O}}\right\},\quad \text{where}\quad \{F_\nu\}\subset\mathcal{E}'_\sigma.$$

For the classical boundary value problems, such bases were first constructed by Kupradze [32] who developed the Fischer-Riesz equations method as a tool for getting effective approximate solutions.

5.11 Solvability of the Dirichlet problem

Choose some special basis $\left\{\bigoplus_{j=p/2}^{p-1}\overline{C_jg_\nu}|_{\partial\mathcal{O}}\right\}$ in $\bigoplus_{p/2}^{p-1}L^2(\partial\mathcal{O})$, where $\{g_\nu\}\subset\mathcal{S}_{P'}(\bar{\mathcal{O}})$.

Theorem 5.29 *In order that Problem* 5.23 *be solvable, it is necessary and sufficient that:*

(1)

$$\sum_{\nu=1}^{\infty}|c_\nu|^2<\infty,\quad \textit{where}\quad c_\nu=-\int_{\partial\mathcal{O}}\sum_{j=0}^{\frac{p}{2}-1}C_jg_\nu\cdot f_j\,ds;$$

(2)

$$\int_{\partial\mathcal{O}}\sum_{j=0}^{\frac{p}{2}-1}C_jg\cdot f_j\,ds=0$$

for all $g\in\mathcal{S}_{P'}(\bar{\mathcal{O}})$ *such that* $C_jg|_{\partial\mathcal{O}}=0$ $(j=\frac{p}{2},\ldots,p-1)$.

Proof This follows from Lemma 5.6 and the results of section 5.9. □

In particular, if Problem 5.25 has a unique solution, then for Problem 5.23 to be solvable it is necessary and sufficient that condition (1) of Theorem 5.29 be satisfied.

Example 5.30 Consider the DO P^*P on X. It is a (formal) selfadjoint elliptic operator of order $2p$. The system of boundary operators $\{B_j, C_j \circ P\}_{j=0}^{p-1}$ is a Dirichlet system of order $(2p-1)$ on $\partial\mathcal{O}$. Problem 5.23 with P^*P instead of P and $\{B_j\}_{j=0}^{p-1}$ instead of $\{B_j\}_{j=0}^{(p/2)-1}$ is formally selfadjoint. It is solvable for all smooth Dirichlet data, and hence the problem is dense solvable and has a unique solution. In this case, condition (2) of Theorem 5.29 is automatically satisfied. □

5.12 Regularization of the Dirichlet problem

Consider the following kernels $K^{(N)}$, defined for $x \in X \setminus \partial\mathcal{O}$ and for y in a neighborhood of the closure of $\mathcal{O}$:

$$K^{(N)}(x,y) = \Phi(x,y) - \sum_{\nu=1}^{N} \left[\int_{\partial\mathcal{O}} \sum_{j=p/2}^{p-1} C_j\Phi(x,.) \cdot \overline{C_j g_\nu}\, ds \right] \otimes g_\nu(y) \qquad (N = 1,2,\ldots).$$

Lemma 5.31 *For every $N = 1,2,\ldots$, the kernel $K^{(N)}$ is infinitely differentiable with respect to $x \in X \setminus \partial\mathcal{O}$ and with respect to y in a neighborhood of $\bar{\mathcal{O}}$ except on the diagonal $\{x = y\}$, and for such x and y satisfies $P(x)K^{(N)} = 0$ and $P'(y)K^{(N)} = 0$.*

Proof This is obvious. □

It follows that the sequence of the kernels $\{K^{(N)}\}$ gives a special approximation of the Poisson kernel of Problem 5.23 (see (5.1)).

Theorem 5.32 *For every solution $f \in H^2_{P,B}(\mathcal{O})$ we have*

$$f(x) = -\lim_{N\to\infty} \int_{\partial\mathcal{O}} \sum_{j=0}^{\frac{p}{2}-1} C_j K^{(N)}(x,.) \cdot B_j f\, ds \qquad (x \in \mathcal{O}).$$

Proof This follows directly from Green's formula (4.2) and the properties of $K^{(N)}$. □

Both Theorem 5.29 and Theorem 5.32 were proved by Tarkhanov [54].

When the existence is known a priori, the Fischer-Riesz equations method may be developed as a tool for getting effective approximate solutions.

Remark 5.33 As Kupradze [32] noted, "calculations for the simplest typical problems, performed in the Computer Centre of the Georgian Academy of Sciences, gave an accuracy quite sufficient for practice. In particular, for the Dirichlet problem, coincidence up to six decimals occurred already when $N = 24$."

References

[1] Adams, D.R. and Polking, J.C., The equivalence of two definitions of capacity, *Proc. Amer. Math. Soc.* **37** (1973), 529–534.

[2] Ahlfors, L., Bounded analytic functions, *Duke Math. J.* **14** (1947), 1–11.

[3] Aizenberg, L.A. and Kytmanov, A.M., On the possibility of the holomorphic extension to a domain of functions given on a connected piece of the boundary, *Mat. Sb.* **182**(4) (1991), 490–507 (Russian).

[4] Babuška, J., Stability of domains with respect to the main problems of PDE's, *Czechoslovak Math. J.* **11** (1961), 76–105, 165–203.

[5] Baernstein, A., A representation theorem for functions holomorphic off the real axis, *Trans. Amer. Math. Soc.* **169** (1972), 159–165.

[6] Bagby, T., Quasi topologies and rational approximation, *J. Funct. Anal.* **10** (1972), 259–268.

[7] Bergman, S., *The Kernel Function and Conformal Mapping*, revised edition, Math. Surveys Monographs **5**, Amer. Math. Soc., Providence, RI, 1970.

[8] Bers, L., An approximation theorem, *J. Analyse Math.* **14** (1965) 1–4.

[9] Bochner, S., Weak solutions of linear partial differential equations, *J. Math. Pures Appl.* **35** (1956), 193–202.

[10] Brelot, M., *Étude des fonctions sousharmoniques au voisinage d'un point*, Actualités scientifiques et industrielles, Hermann, Paris, 134 (1934).

[11] Carleson, L., *Selected Problems on Exceptional Sets*, Van Nostrand, Princeton, NJ, 1972.

[12] Fernström, C., On the instability of capacity, *Ark. Mat.* **15** (1977), 241–252.

[13] Fernström, C. and Polking, J.C., Bounded point evaluation and approximation in L^p by solutions of elliptic partial differential equations, *J. Funct. Anal.* **28** (1978), 1–20.

[14] Fischer, B. and Tarkhanov, N., Local singularities of solutions of elliptic equations, to appear in *Math. Ann.*

[15] Fuglede, B., Applications du théorème minimax à l'étude de diverses capacités, *C.R. Acad. Sci. Paris* **266** (1968), 921–923.

[16] Gauthier, P.M. and Tarkhanov, N., Degenerate cases of approximation by solutions of systems with injective symbols, *Canad. J. Math.* **20** (1993), 1–18.

[17] Grothendieck, A., Sur les espaces de solutions d'une classe générale d'équations aux derivées partielles, *J. Analyse Math.* **2** (1952–53), 243–280.

[18] Harvey, R. and Polking, J.C., Removable singularities of solutions of linear partial differential equations, *Acta Math.* **125** (1970), 39–56.

[19] Harvey, R. and Polking, J.C., A notion of capacity which characterizes removable singularities, *Trans. Amer. Math. Soc.* **169** (1972), 183–195.

[20] Havin, V.P., Approximation in the mean by analytic functions, *Dokl. Akad. Nauk SSSR* **178** (1968), 1025–1028 (Russian); English translation: *Soviet Math. Dokl.* **9** (1968), 245–252.

[21] Havin, V.P., Golubev series and the analyticity on a continuum, in: *Linear and Complex Analysis Problem Book* (V.P. Havin et al., eds.), Lecture Notes in Math. **1043**, Springer-Verlag, Berlin – Heidelberg – New York, 1984; 670–673.

[22] Hedberg, L.I., Non-linear potentials and approximations in the mean, *Math. Z.* **129** (1972), 299–319.

[23] Hedberg, L.I., Two approximation problems in function spaces, *Ark. Mat.* **16** (1978), 51–81.

[24] Hedberg, L.I., Spectral synthesis in Sobolev spaces and uniqueness of solutions of the Dirichlet problem, *Acta Math.* **147** (1981), 235–264.

[25] Hedberg, L.I., Nonlinear potential theory and Sobolev spaces, Linköping University, preprint LITH-MAT-R-86-10, 1986.

[26] Hedberg, L.I., Approximation by harmonic functions, and stability of the Dirichlet problem, to appear in *Exposition. Math.*

[27] Hedberg, L.I. and Wolff, T.H., Thin sets in nonlinear potential theory, *Ann. Inst. Fourier (Grenoble)* **33** (1983), 161–187.

[28] Karepov, O.V. and Tarkhanov, N., The Fischer-Riesz equations method in the Cauchy problem for systems with injective symbols, *Dokl. Akad. Nauk* **326**(5) (1992), 776–780 (Russian).

[29] Keldysh, M.V., On the solvability and the stability of the Dirichlet problem, *Uspekhi Mat. Nauk* **8** (1941), 171–231 (Russian); English translation: *Amer. Math. Soc. Transl.* **51** (1966), 1–73.

[30] Kral, J., Some extension results concerning harmonic functions, *J. London Math. Soc.* **28** (1983), 62–70.

[31] Krasichkov, I.F., Systems of functions with the double orthogonality property, *Mat. Zametki* **4**(5) (1968), 551–556 (Russian).

[32] Kupradze, V.D., On approximate solving the problems of mathematical physics, *Uspekhi Mat. Nauk* **22**(2) (1967), 59–107 (Russian);

[33] Lavrentiev, M.M., On the Cauchy problem for the Laplace equation, *Izv. Akad. Nauk SSSR. Ser. Mat.* **20** (1956), 819–842 (Russian).

[34] Lax, P., A stability theory of abstract differential equations and its applications to the study of local behaviors of solutions of elliptic equations, *Comm. Pure Appl. Math.* **9** (1956), 747–766.

[35] Littman, W., Polar sets and removable singularities of partial differential equations, *Ark. Mat.* **7** (1967), 1–9.

[36] Malgrange, B., Existence et approximation des équations aux derivées partielles et des équations de convolution, *Ann. Inst. Fourier (Grenoble)* **6** (1955–56), 271–355.

[37] Maz'ja, V.G. and Havin, V.P., Nonlinear potential theory, *Uspekhi Mat. Nauk* **26**(6) (1972), 67–138 (Russian);

[38] Meyers, N.G., A theory of capacity for potentials of functions in Lebesgue classes, *Math. Scand.* **26** (1970), 255–292.

[39] Miranda, C., *Equazioni alle derivate parziali di tipo ellittico*, Springer-Verlag, Berlin, 1955.

[40] Nguyen, X.Uy, Removable sets of analytic functions satisfying a Lipschitz condition, *Ark. Mat.* **17** (1979), 19–27.

[41] Polking, J.C., Approximation in L^p by solutions of elliptic partial differential equations, *Amer. J. Math.* **94** (1972), 1231–1244.

[42] Reshetnjak, Ju.G., On the concept of capacity in the theory of functions with generalized derivatives, *Sibirsk. Mat. Zh.* **10**(5) (1969), 1109–1138 (Russian).

[43] Rojtberg, Ja.A., On the boundary values of generalized solutions of elliptic equations, *Mat. Sb.* **86**(2) (1971), 248–267 (Russian).

[44] Schapira, P., *Théorie des hyperfonctions*, Springer-Verlag, Berlin – Heidelberg – New York, 1970.

[45] Seeley, R., Topics in pseudo-differential operators, in: *Pseudo-Differential Operators* (L. Nirenberg, ed.), CIME – Edizioni Cremonese, Roma, 1969; 167–306.

[46] Shapiro, H.S., Stefan Bergman's theory of doubly-orthogonal functions – an operator-theoretic approach, *Proc. Roy. Irish Acad. Sect. A* (1979), 49–56.

[47] Shlapunov, A. and Tarkhanov, N., Bases with double orthogonality in the Cauchy problem for systems with injective symbols, *Dokl. Akad. Nauk* **326**(1) (1992), 45–49 (Russian).

[48] Sinanjan, S.O., The uniqueness property of analytic functions on closed sets without interior, *Sibirsk. Mat. Zh.* **6**(6) (1965), 1365–1381 (Russian).

[49] Slepian, D., et al., Prolate spheroidal wave functions, Fourier analysis and uncertainty (I–IV), *Bell Systems Tech. J.* **40** (1961), 43–63, 65–86; **41** (1962), 1295–1336; **43** (1964), 3009–3057.

[50] Stein, E.M., *Singular Integrals and Differentiability Properties of Functions*, Princeton University Press, Princeton, NJ, 1970.

[51] Straube, E.J., Harmonic and analytic functions admitting a distribution boundary value, *Ann. Scuola Norm. Sup. Pisa Cl. Sci. (4)* **11** (1984), 559–591.

[52] Tarkhanov, N., *Laurent Series for Solutions of Elliptic Systems*, Nauka, Novosibirsk, 1991 (Russian).

[53] Tarkhanov, N., Approximation on compact sets by solutions of systems with surjective symbols, Institute of Physics, Krasnoyarsk, preprint 48 M, 1989 (Russian).

[54] Tarkhanov, N., On approximate solving the Dirichlet problem for the generalized Laplacian, to appear in *Math. Nachr.* (1994).

[55] Verdera, J., Approximation by rational modules in Sobolev and Lipschitz norms, *J. Funct. Anal.* **58** (1984), 267–290.

[56] Vitushkin, A.G., Analytic capacity of sets in problems of approximation theory, *Uspekhi Mat. Nauk* **22** (1967), 141–199 (Russian); English translation: *Russian Math. Surveys* **22** (1967), 139–200.

[57] Wiener, N., The Dirichlet problem, *J. Math. Phys. Massachusetts Inst. Technology* **3** (1924), 127–147.

[58] Zin, G., Esistenza e reppresentazione di funzioni analitiche, le quali, su una curva di Jordan, si riducono a una funzione assegnata, *Ann. Mat. Pura Appl.* **34** (1953), 365–405.

Removability, capacity and approximation

Joan VERDERA
Departament de Matemàtiques
Universitat Autònoma de Barcelona
E-08193 Bellaterra (Barcelona)
Spain

Abstract

In this paper we are primarily interested in problems of qualitative approximation by holomorphic functions of one complex variable belonging to some fixed class, that is defined by restricting the growth of the functions (L^p, $1 < p \leq \infty$) or by requiring certain smoothness (Lip s or C^m). Part of the approximation problem consists in understanding the removable sets for the class under consideration and its associated capacity.

In Chapter 1 we deal with bounded analytic functions. We are thus led to the Painlevé problem and analytic capacity. We discuss the solution of the Denjoy conjecture via L^2-estimates for the Cauchy integral on Lipschitz graphs. We then show that the same ideas can be applied to describe removable sets for Lipschitz analytic functions, the role of the Cauchy integral being played by the Beurling transform.

Chapter 2 is devoted to Vitushkin's Theorem on uniform approximation by rational functions; the simplest available proof is described in detail.

In Chapter 3, problems of approximation by analytic functions in Lipschitz and C^m classes are considered. Vitushkin's scheme and the mapping properties of the Beurling transform are combined to obtain satisfactory answers to the main questions.

In Chapter 4 we discuss the relationship between L^p-approximation by analytic functions and spectral synthesis for Sobolev spaces.

Chapter 5 is a survey of recent results about approximation by solutions of elliptic equations in classical Banach spaces.

Contents

P. M. Gauthier (ed.) and G. Sabidussi (techn. ed.), Complex Potential Theory, 419–473.

Introduction

This paper grew out of a series of lectures given by the author at the 1993 Montreal Summer School on Complex Potential Theory.

Our starting point is the notion of set of removable singularities for functions satisfying a given partial differential equation, and subject to some previously specified growth or smoothness conditions. For example, one can consider removable sets for L^p analytic functions, $1 \le p \le \infty$, or for harmonic functions in the Lipschitz classes. Associated to a fixed removability problem there is a set function, called capacity, which enjoys the property of vanishing exactly on removable sets. In fact one should think of capacity as a way of quantifying the notion of a non-removable set. It turns out that capacity is the key tool to understand qualitative approximation problems, in the class under consideration, by solutions of the given partial differential equation. This explains the title.

Since we do not want to present the theory in full generality, we start in Chapter 1 by considering bounded analytic functions, in accordance with the historical development. We are thus led to Painlevé's problem and analytic capacity. After introducing some basic results on analytic capacity, we give an outline of the proof of one of the most relevant achievements in the subject, namely, the solution of the Denjoy conjecture on the relationship between analytic capacity and arc-length on rectifiable curves. The connection with L^p and weak L^1 estimates for the Cauchy integral on Lipschitz curves is explored. Next we describe two nice applications of the techniques introduced to prove the Denjoy conjecture. First we characterize removable sets for Lipschitz analytic functions as those having zero area. Then we consider another problem of a complex potential theoretic flavor: the existence, in the principal value sense, of the Cauchy transform of a finite Borel measure.

Our intention has not been to present an exhaustive survey on analytic capacity. Although we mention the more relevant recent contributions, we skip important parts of the subject. For example, duality is not dealt with at all. For this and other classical topics the reader is referred to Garnett's book [Ga], which is still the best introduction to analytic capacity.

Chapter 2 deals with uniform approximation by rational functions. Vitushkin's Theorem and some of its partial results, including Mergelyan's Theorem, are described in detail. The core of the Chapter is the proof of Vitushkin's Theorem, which is discussed thoroughly, in the shortest way known. For instance, following Davie [Da1], we avoid mentioning analytic center and diameter. This simplifies considerably the classical expositions of Gamelin [G] and Zalcman [Z]. The reader is also referred to [Ko] and [OF5] for two interesting surveys on rational approximation.

In Chapter 3 we deal with smooth approximation by rational functions. The underlying theme here is the use, as a technical device, of the mapping properties of the Beurling transform (the Calderón-Zygmund operator in the plane with kernel z^{-2}).

We start by presenting a rather compact proof of O'Farrell's Theorem on rational approximation in $\mathrm{Lip}\, s$ norm, $0 < s < 1$. The invariance of $\mathrm{Lip}\, s$ under the Beurling transform explains why the Vitushkin's approximation scheme of the previous chapter works even better in the $\mathrm{Lip}\, s$ case than in the uniform case.

We then show how one can apply the weak L^1 inequality for the Beurling transfom to solve the C^1-approximation problem by rational functions.

Higher order smooth approximation is also discussed.

In Chapter 4, L^p approximation by rational functions is considered. As shown by Lindberg [Li] one can also solve the L^p problem using Vitushkin's constructive technique. We prefer, however, to present the original approach suggested by Havin [H] and fully exploited by Bagby [B1] and Hedberg [He2]. The idea is to use duality to reduce matters to the spectral synthesis problem for the Sobolev space W_1^q, q being the exponent conjugate to p. In the context at hand, spectral synthesis turns out to be a rather elementary result because W_1^q is stable under truncation. A conceptually clear, concise proof of the main result follows.

Finally, in Chapter 5 we present a survey of recent work in qualitative approximation by solutions of fairly general elliptic equations.

Our notational conventions will be standard. For example, we will denote by C a constant which may be different in different occurrences and which is independent of the relevant variables under consideration.

The author would like to express his gratitude to P. Gauthier for offering him the opportunity of lecturing in Montreal. Thanks are also due to D. Adams, T. Gamelin, L. Hedberg and P. Mattila for revising the first draft of the manuscript.

Chapter 1
Analytic capacity

1.1 The Painlevé problem

Last century Riemann proved a result on the removability of isolated singularities which has become part of the background of the whole mathematical community. It reads as follows. Let f be a function which is analytic on some domain Ω of the complex plane except (perhaps) at some point $a \in \Omega$. Suppose that $f(z)$ stays bounded as z approaches a. Then f has a removable singularity at the point a, which means that f can be defined at the point a in such a way that the resulting function is analytic on the whole of Ω.

Here is the proof of this simple fact. Expand f in a Laurent series on some small punctured disc centered at a, which we assume to be the origin:

$$f(z) = \cdots + c_{-n}z^{-n} + \cdots + c_{-1}z^{-1} + c_0 + c_1 z + \cdots .$$

Then, for ε small enough, $c_{-n} = \frac{1}{2\pi i}\int_{|z|=\varepsilon} f(z)z^{n-1}\,dz$, and so $|c_{-n}| \leq \|f\|_\infty \varepsilon^n$. Therefore $c_{-n} = 0$ for $n = 1, 2, \ldots$ and f turns out to be analytic at the origin, as desired.

This raises the question of understanding the nature of those sets that share with points the property of being removable singularities for bounded analytic functions. To get a rigorous formulation of the problem, we need a definition.

Definition A compact subset K of the complex plane is said to be *removable for bounded analytic functions* if given an open set Ω containing K and a bounded analytic function f on $\Omega \backslash K$, there exists $\tilde{f}$ analytic on Ω such that $f = \tilde{f}$ on $\Omega \backslash K$.

In checking that a given set is removable it is enough to take $\Omega = \mathbb{C}$ in the definition above, as a simple application of the Cauchy integral formula shows. Then, K is removable if and only if any bounded analytic function on $\mathbb{C}\backslash K$ can be extended to a (bounded) entire function. In other words, K is removable if and only if bounded analytic functions on $\mathbb{C}\backslash K$ are constant.

Painlevé Problem Describe, in geometric or metric terms if possible, removable sets for bounded analytic functions.

We must say right now that Painlevé's problem has not been solved yet and that is not clear at this time what geometric or metric conditions (if any) could characterize removability.

The following examples are intended to illustrate the notion of removable set.

Example 1 Finite sets are removable because of Riemann's result. Indeed, an application of Baire's category Theorem readily gives that compact countable sets are removable.

Example 2 A disc is non-removable. In fact, assuming the disc centered at the origin, the function $f(z) = \frac{1}{z}$ is bounded and analytic outside the disc and clearly is not constant.

Example 3 Compact sets of positive area are non-removable. The following proof of this fact illustrates one of the most simple and powerful methods to show that a set K is non-removable: take an adequate non-zero measure μ supported on K and try to prove that $f = \frac{1}{z} * \mu$ is bounded outside K. Since $\bar{\partial} f = \pi\mu$ in the distributions sense, because $\frac{1}{\pi z}$ is a fundamental solution of $\bar{\partial} = \frac{1}{2}\left(\frac{\partial}{\partial x} + i\frac{\partial}{\partial y}\right)$, f is analytic outside K and, on the other hand, there is no analytic continuation of f to the whole plane (otherwise f would be constant by Liouville's Theorem and thus μ would vanish identically). When K has positive area one can take $\mu = \chi_K(\zeta)\, dx\, dy$, where $\zeta = x + iy$ and χ_K is the characteristic function of K. Setting $A = K\backslash\Delta(z,1)$ and $B = \Delta(z,1)$, $\Delta(z,1)$ being the disc with center z and radius 1, we obtain

$$|f(z)| \leq \int_A \frac{dx\, dy}{|z-\zeta|} + \int_B \frac{dx\, dy}{|z-\zeta|} \leq \operatorname{area}(K) + 2\pi.$$

Example 4 An interval I in the real axis is non-removable. A quick way to see that is to consider the conformal mapping of the complement of I in the extended complex plane onto the unit disc. That function is non-constant, bounded and analytic on $\mathbb{C}\backslash I$. Alternatively, one can apply the general method outlined in the preceding example: choose as μ the measure $\varphi(x)\, dx$, where $\varphi \in C_0^\infty(I)$, $\varphi \not\equiv 0$.

Example 5 A compact subset of the real axis of zero (one dimensional) Lebesgue measure is removable. In fact a stronger statement is true, namely, that for each $f \in H^\infty(\mathbb{C}\backslash K)$, $K \subset \mathbf{R}$, $\bar{\partial} f$ is an absolutely continuous measure with respect to Lebesgue measure on K. To show this, take an interval I containing K and let R_ε be the rectangle with sides parallel to the coordinate axis, whose horizontal sides are $I \pm i\varepsilon$. Let $r > 0$ be so big that $\Delta(0,r) \supset I$. Then for $z \in \Delta(0,r)\backslash I$,

$$f(z) = \frac{1}{2\pi i}\int_{|\zeta|=r} \frac{f(\zeta)}{\zeta - z}\, d\zeta - \frac{1}{2\pi i}\int_{\partial R_\varepsilon} \frac{f(\zeta)}{\zeta - z}\, d\zeta.$$

Letting $r \to \infty$ and then $\varepsilon \to 0$ we get

$$f(z) = f(\infty) - \frac{1}{2\pi i} \int_{-\infty}^{\infty} (f^+(x) - f^-(x)) \frac{dx}{x - z},$$

where $f^{\pm}(x) = \lim_{\varepsilon \to 0} f(x \pm i\varepsilon)$ exist dx-a.e. by Fatou's Theorem. Thus

$$\bar{\partial} f = \frac{f^+(x) - f^-(x)}{2i} \, dx.$$

The simplest idea that comes to mind after consideration of the above examples is that small sets should be removable and big sets not. This turns out to be true, and the only difficulty to get precise sharp statements is to find an adequate way of measuring the size of sets. This can be achieved using the notion of Hausdorff measure (and content) and of Hausdorff dimension, which we introduce in the next section.

1.2 Hausdorff content

Let $h(t)$ be a non-decreasing continuous function of the non-negative real variable t such that $h(0) = 0$. For each $\delta > 0$ and each $E \subset \mathbf{R}^n$ set

$$M_\delta^h(E) = \inf_{\delta_j \leq \delta} \sum_j h(\delta_j),$$

where the infimum is taken over all coverings of E by cubes of side length $\delta_j \leq \delta$, with sides parallel to the coordinate axes. When $\delta = \infty$, that is, when there is no restriction on the size of the cubes covering E, we get a quantity $M^h(E) \equiv M_\infty^h(E)$, which is usually called *Hausdorff content* or *Hausdorff capacity* of E relative to h. On the other hand,

$$\Lambda^h(E) = \sup_{\delta > 0} M_\delta^h(E) = \lim_{\delta \to 0} M_\delta^h(E)$$

turns out to be a Borel measure (not necessarily locally finite), which is known as the Hausdorff measure associated to the measure function h. The set function M^h is subadditive but is not a measure. Its main advantage over Λ^h is that it is finite on compact sets and therefore becomes often a quantity suitable to get sharp estimates.

When $h(t) = t^\alpha$, $M^h = M^\alpha$ and $\Lambda^h = \Lambda^\alpha$ are respectively called *α-dimensional Hausdorff content and measure.*

A simple exercise shows that M^h and Λ^h vanish on the same sets and that

$$\begin{aligned} \sup\{\alpha : M^\alpha(E) > 0\} &= \inf\{\alpha : M^\alpha(E) = 0\} \\ &= \sup\{\alpha : \Lambda^\alpha(E) = \infty\} \\ &= \inf\{\alpha : \Lambda^\alpha(E) = 0\}. \end{aligned}$$

The Hausdorff dimension $d_H(E)$ of a set E is the common value of the expressions in the above identities.

The next five examples illustrate how Hausdorff content captures measure theoretic and geometric features of sets.

1. $M^\alpha(Q) = l(Q)^\alpha$ for each cube Q with side length $l(Q)$.
2. If I and J are intervals in the real line whose lengths satisfy $l(I) \geq l(J)$, then, for the rectangle $R = I \times J$, we have $M^\alpha(R) = l(I)l(J)^{\alpha-1}$, provided $\alpha \geq 1$.
3. Let I be a line segment (not reduced to a point) in $\mathbf{R}^n$. Then $d_H(I) = 1$.
4. The Hausdorff dimension of the familiar ternary Cantor set is $\log 2/\log 3$.
5. Let K be a continuum (compact connected set not reduced to a point) in $\mathbf{R}^n$. Projecting K into an appropriate straight line one obtains

$$M^1(K) \geq C\mathrm{diam}(K),$$

where the constant C is independent of K.

The main reason why Hausdorff content becomes a very useful tool in analysis is that, as Frostman discovered, it can be defined in terms of finite Borel measures satisfying a growth condition.

Let μ be a positive Borel measure supported on some compact set K and satisfying

$$\mu(B(x,r)) \leq h(r), \text{ for all } x \in \mathbf{R}^n \text{ and } r > 0, \tag{1}$$

$B(x,r)$ being the ball with center x and radius r. Let $K \subset \bigcup_j Q_j$ be a covering of K by cubes of sidelength δ_j. Each Q_j can be covered by N balls with radius δ_j, where N depends only on the dimension. Thus $\mu(Q_j) \leq Nh(\delta_j)$ and therefore $\sup \mu(K) \leq NM^h(K)$, the supremum being taken on those measures supported on K and satisfying (1). The Frostman Lemma asserts that the reverse inequality, with a different constant, is also true (see [Ga]).

Frostman Lemma *For any compact set K there exists a positive measure μ supported on K, satisfying* (1) *and $\mu(K) \geq CM^h(K)$, where the constant $C = C(n)$ depends only on n.*

It is now not difficult to prove the first significant result in the subject, namely that dimension 1 is critical for removability.

1.2.1 Theorem *Let $K \subset \mathbf{C}$ be compact.*

(i) *If $d_H(K) < 1$ then K is removable.*

(ii) *If $d_H(K) > 1$ then K is not removable.*

Proof Instead of (i) we prove the stronger result, due to Painlevé, that zero length (one dimensional Hausdorff measure) implies removability. Assume that K has zero length and let f be a bounded analytic function on $\mathbf{C}\setminus K$. Given $\varepsilon > 0$, one can cover K by a finite family $\{Q_j\}$ of non overlapping squares such that $\sum l(Q_j) < \varepsilon$. If z is not in K and ε is chosen small enough depending on z, then

$$|f(z) - f(\infty)| = \left|\frac{1}{2\pi i}\int_{\partial(\cup Q_j)} \frac{f(\zeta)}{\zeta - z}\, d\zeta\right| \leq C\|f\|_\infty d(z,K)^{-1}\varepsilon.$$

Thus f is constant.

For (ii) suppose $M^{1+\varepsilon}(K) > 0$ for some $\varepsilon > 0$. By the Frostman Lemma there is a measure μ supported on K such that $\mu(K) \geq CM^{1+\varepsilon}(K) > 0$ and $\mu(\Delta(z,r)) \leq r^{1+\varepsilon}$, $z \in \mathbb{C}$, $r > 0$. Set $f(z) = \int \frac{d\mu(\zeta)}{z-\zeta}$, $z \notin K$. Clearly f is analytic on $\mathbb{C}\setminus K$ and non-constant (because $\mu \not\equiv 0$). We are left with the task of showing that f is bounded, which is not hard. In fact

$$\begin{aligned} |f(z)| &\leq \int_{|\zeta-z|>1} \frac{d\mu(\zeta)}{|\zeta-z|} + \sum_{j=0}^{\infty} \int_{2^{-j-1}<|\zeta-z|\leq 2^{-j}} \frac{d\mu(\zeta)}{|\zeta-z|} \\ &\leq \mu(K) + \sum_{j=0}^{\infty} 2^{j+1}\mu(\Delta(z,2^{-j})) \\ &\leq \mu(K) + \sum_{j=0}^{\infty} 2^{j+1}2^{-j(1+\varepsilon)} \leq \mu(K) + 2(1-2^{-\varepsilon})^{-1}. \end{aligned}$$

□

In dimension 1 one can find both removable and non-removable sets. For example, we already remarked that an interval is a non removable set of dimension 1. On the other hand one can construct linear Cantor sets of dimension 1 and zero length, that are removable according to Painlevé's result mentioned in the proof of the above theorem or to Example 5 in the preceding section.

1.3 Analytic capacity

Analytic capacity is a set function, introduced by Ahlfors in [A], that vanishes exactly on removable sets. Its definition, on a compact set K, is given by

$$\gamma(K) = \sup |f'(\infty)|,$$

where the supremum is taken over those bounded analytic functions on $\mathbb{C}\setminus K$ satisfying the normalization condition $|f(z)| \leq 1$, $z \notin K$. We recall that $f'(\infty) = \lim_{z\to\infty} z(f(z) - f(\infty))$ is the coefficient a_1 in the expansion

$$f(z) = a_0 + \frac{a_1}{z} + \cdots, \text{ as } z \longrightarrow \infty.$$

For an arbitrary set $E \subset \mathbb{C}$ we set

$$\gamma(E) = \sup\{\gamma(K) : K \text{ compact}, K \subset E\}.$$

It is clear from the definition that $\gamma(K) = 0$ whenever K is removable. Conversely, if K is non-removable and if f is a non-constant bounded analytic function on $\mathbb{C}\setminus K$ then, for some positive integer n, we have

$$f(z) = a_0 + \frac{a_n}{z^n} + \cdots, \text{ as } z \longrightarrow \infty,$$

and $a_n \neq 0$. Thus the bounded analytic function $g(z) = z^{n-1}(f(z) - a_0)$ has a nonzero derivative at ∞ and so $\gamma(K) > 0$.

Therefore, one should view $\gamma(K)$ as a numerical indicator measuring the size of the unit ball of $H^\infty(K^c)$, the space of bounded analytic functions on the complement of K, and hence measuring the non removability of K. Thus analytic capacity is the appropriate tool to formulate quantitative relationships involving the notion of removability. For example, a continuum K is non-removable, as shows consideration of the conformal mapping f of the complement of K in the Riemann sphere onto the unit disc. The 1/4 Theorem of Koebe shows that $|f'(\infty)| \geq \frac{1}{4}\mathrm{diam}(K)$. Thus $\mathrm{diam}(K) \leq 4\gamma(K)$ if K is a continuum, which is a quantified version of the statement that K is non-removable. As another example, we mention a quantified version of Theorem 1.2.1, which essentially follows from the proof we presented before.

1.3.1 Theorem *For any compact $K \subset \mathbb{C}$ and $\varepsilon > 0$*

$$C_\varepsilon M^{1+\varepsilon}(K)^{1/(1+\varepsilon)} \leq \gamma(K) \leq \frac{2}{\pi} M^1(K), \tag{2}$$

where C_ε is a constant depending only on ε.

Notice that for an interval I one has $M^{1+\varepsilon}(I) = 0$ for all $\varepsilon > 0$, but $\gamma(I) > 0$, which shows that the first inequality in (2) is far from being reversible. It took some effort to prove that also the second inequality in (2) cannot be reversed, that is, to construct a set with zero analytic capacity and positive length. This was achieved by Vitushkin, and shortly afterwards Garnett and Ivanov independently showed that the planar Cantor set one gets by taking the "corner quarters" enjoys that property (see [Ga]). The conclusion is that, in spite of the fact that γ is a one-dimensional object, one-dimensional Hausdorff content is not sharp enough to describe analytic capacity.

Vitushkin suggested in [Vi] another candidate to characterize removable sets. Let P_θ be the orthogonal projection onto the straight line through the origin forming an angle θ with the real axis. Set

$$CR(K) = \frac{1}{\pi} \int_0^\pi \mathrm{length}(P_\theta(K))\, d\theta.$$

According to [M], this quantity was introduced by Crofton in 1868 in connection with the solution of the Buffon needle problem (see [M] for the reference to Crofton's article). Vitushkin asked if $CR(K) = 0$ is equivalent to $\gamma(K) = 0$. The question was answered in the negative by Mattila [Ma] in an astonishing way. He proved that the condition $CR(K) = 0$ is not conformally invariant and so, since $\gamma(K) = 0$ clearly is, we deduce that the conditions under consideration are not equivalent. Nevertheless, it has not been possible to decide which implication is false from Mattila's method. Recently Jones and Murai [J-M] constructed a set K (of infinite length) for which $\gamma(K) > 0$ and $CR(K) = 0$. It is worth remarking that Vitushkin's conjecture is still open for sets of finite length. The interested reader should consult the forthcoming book by David and Semmes [D-S], where connections with singular integrals are explored.

1.4 The Denjoy conjecture

We have mentioned that length and capacity are not comparable. However, Pommerenke [Po] showed that $\gamma(K) = \frac{1}{4}|K|$ for subsets of the real line, where $|K|$ is the length of K. It is much easier to prove the following weaker statement.

Proposition *For* $K \subset \mathbf{R}$,

$$\frac{1}{4}|K| \leq \gamma(K) \leq \frac{2}{\pi}|K|. \tag{3}$$

Proof The second inequality in (3) is contained in (2) because $M^1(K) = |K|$ for $K \subset \mathbf{R}$. To prove the first observe that the real part of $i/\pi z$ is the Poisson kernel $y/\pi(x^2+y^2)$. Then the real part of $h(z) = \frac{1}{\pi i}\int_K \frac{dt}{t-z}$ is bounded in absolute value by 1, and so the function

$$(e^{\pi i h(z)/2} - 1)(e^{\pi i h(z)/2} + 1)^{-1} = \frac{|K|}{4}\frac{1}{z} + \cdots, \text{ as } z \longrightarrow \infty,$$

maps the complement of K into the unit disc, which gives the desired inequality. □

Denjoy tried to prove that analytic capacity and length vanish simultaneously on subsets of a rectifiable curve. He believed he had found a proof [De], but unfortunately a gap in the argument left the question unsolved. It has been known since then as the Denjoy Conjecture.

Havin and Havinson already knew in the fifties that the Denjoy Conjecture would follow from the L^2 boundedness of the Cauchy Integral on Lipschitz graphs (with small Lipschitz constant). In the seventies harmonic analysts were led to the problem of estimating the Cauchy integral on a Lipschitz graph by their interest in the role played by singular integrals in the study of partial differential equations with minimal smoothness conditions. In 1977, Calderón [C] obtained L^2-estimates for Lipschitz graphs with small constant, thus completing the proof of the Denjoy conjecture (see [M]). Later L^2-estimates were shown to hold on any Lipschitz graph [C-Mc-M] and finally David [D] characterized those rectifiable curves for which the Cauchy integral is bounded on L^2. These are the curves Γ satisfying

$$\text{length}(\Delta \cap \Gamma) \leq C \text{ radius}(\Delta), \text{ for all discs } \Delta,$$

and also the curves for which

$$\text{length}(K) \leq C\gamma(K), \text{ for all } K \subset \Gamma.$$

We proceed now to describe the path towards the solution of the Denjoy Conjecture. The first step consists in reducing the problem to Lipschitz graphs. By a Lipschitz graph we understand a curve which is a rotation of the graph

$$\Gamma = \{x + iA(x) : x \in \mathbf{R}\} \tag{4}$$

of a Lipschitz function A. The number $\|A'\|_\infty$ is called the Lipschitz constant of the Lipschitz graph.

1.4.1 Lemma *Let K be a compact subset of positive length of a rectifiable curve. Then, given $\varepsilon > 0$ there exists a Lipschitz graph Γ, with Lipschitz constant less than ε, such that $K \cap \Gamma$ has positive length.*

Proof Let $z(s)$, $s \in [0, L]$, be the arc-length parametrization of a piece, which already contains K, of the given rectifiable curve. In other words, $z([0, L]) \supset K$, $z(s)$ is a Lipschitz function on $[0, L]$ and $|z'(s)| = 1$ a.e. in $[0, L]$. Set, for those s at which $z(s)$ is differentiable and for $n = 1, 2, \ldots$,

$$Q_n(s) = \sup\left\{\left|\frac{z(t) - z(s)}{t - s} - z'(s)\right| : 0 < |t - s| \leq n^{-1}\right\}.$$

Then $Q_n(s) \to 0$ as $n \to \infty$ a.e. on $[0, L]$. By Egorov's Theorem, given $\eta > 0$ there exists a closed $F \subset [0, L]$ such that $|[0, L] \backslash F| < \eta$ and $Q_n(s) \to 0$ uniformly on F as $n \to \infty$. Thus the restriction of z' to F is continuous on F and

$$z(t) - z(s) - (t - s)z'(s) = o(t - s)$$

where the small "o" is uniform in t, $s \in F$. Applying Whitney's extension theorem [St, Chapter VI], we get some $\tilde{z}$ continuously differentiable on $\mathbf{R}$ such that $\tilde{z} = z$ on F. Set $K^* = z^{-1}(K)$ and choose $\eta = |K^*|/2$. Then $|K^* \cap F| \geq |K^*| - \eta = \eta > 0$. Let s_0 be a point of density of $K^* \cap F$. Thus $\tilde{z}'(s_0) = z'(s_0)$ and so we can assume, performing a rotation if necessary, that $\tilde{z}'(s_0) = 1$. It is then clear that for δ small enough the C^1 arc $\tilde{z}(s_0 - \delta, s_0 + \delta)$ coincides with the graph Γ of some C^1 function A with $\|A'\|_\infty$ as small as desired. Moreover $K \cap \Gamma$ contains $K \cap z(F \cap (s_0 - \delta, s_0 + \delta))$ whose length $|K^* \cap F \cap (s_0 - \delta, s_0 + \delta)|$ is positive owing to the choice of s_0. □

Remark The Lipschitz graph constructed in the proof of the above lemma is in fact C^1, i.e., A' is continuous. However, the continuity of A' does not help at all in what follows and it turns out that the mere boundedness of A' is the right hypothesis to work with.

Let Γ be a Lipschitz graph as in (4). The Cauchy integral on Γ is the operator

$$Cf(z) = \lim_{\varepsilon \to 0} \int_{|\zeta - z| > \varepsilon} \frac{f(\zeta)}{\zeta - z}\, ds \equiv \text{P.V.} \int_\Gamma \frac{f(\zeta)}{\zeta - z}\, ds, \quad z \in \Gamma, \tag{5}$$

where P.V. stands for principal value and ds for the arc-length measure on Γ. An elementary argument shows that $Cf(z)$ exists for almost all $z \in \Gamma$ when $f \in C_0^\infty(\mathbb{C})$. However the existence a.e. of the principal values for f in some $L^p(\Gamma)(= L^p(ds))$ is a deep result and in fact a consequence of L^2 estimates. For the maximal Cauchy operator of an $L^p(\Gamma)$ function f, $1 \leq p < \infty$, defined by

$$C^*f(z) = \sup_{\varepsilon > 0} \left| \int_{|\zeta - z| > \varepsilon} \frac{f(\zeta)}{\zeta - z}\, ds \right|, \quad z \in \Gamma,$$

such existence problems disappear completely because the truncated integrals $\int_{|\zeta - z| > \varepsilon} \frac{f(\zeta)}{\zeta - z}\, ds$ clearly make sense. The L^2 estimates for the Cauchy Integral we have been referring to are the inequalities

$$\int_\Gamma |C^*f(z)|^2\, ds \leq C \int_\Gamma |f(z)|^2\, ds, \quad f \in L^2(\Gamma). \tag{6}$$

Once (6) is proven, it is a standard fact that the principal value integral in (5) exists a.e. for $f \in L^2(\Gamma)$ and that (6) holds with C^*f replaced by Cf.

The discussion of the proof of (6) would take us too far away and so we refer the interested reader to the original papers mentioned above and to [C-J-S] where two elementary proofs of (6) are presented. Our next task will be to describe the connections between (6) and analytic capacity.

Given a compact $K \subset \Gamma$ of positive length, we want to construct a non-constant bounded analytic function outside K. Assume now that we can find $f \in L^\infty(K)$, $f \not\equiv 0$, such that $Cf \in L^\infty(\Gamma)$. It then follows without pain [Ch, p.109] that $|Cf(z)| \le \text{Const}$ for $z \notin K$, so that Cf is the nontrivial function in $H^\infty(K^c)$ we are looking for. The difficulty is that the operator C (as the Hilbert transform on the line) does not map $L^\infty(\Gamma)$ into $L^\infty(\Gamma)$. Even the most obvious candidate for f, namely χ_K, fails, because $C(\chi_K) \notin L^\infty(\Gamma)$ unless K has zero length. By duality, C does not map $L^1(\Gamma)$ into $L^1(\Gamma)$. But in the L^1 context a substitute key result is available, which turns out to be almost exactly what we need. We are refering to the weak L^1 inequality

$$|\{z \in \Gamma : C^*\mu(z) > t\}| \le \frac{C}{t}\|\mu\|, \tag{7}$$

where the bars stand for arc-length measure, μ is a complex finite Borel measure on Γ, and

$$\begin{aligned} C^*\mu(z) &= \sup_{\varepsilon>0} |C_\varepsilon\mu(z)|, \\ C_\varepsilon\mu(z) &= \int_{|\zeta - z|>\varepsilon} \frac{d\mu(\zeta)}{\zeta - z}. \end{aligned} \tag{8}$$

That (7) is a consequence of (6) is part of the standard Calderón-Zygmund theory on homogeneous spaces. We plan to show that (6) yields

$$|K| \le C \sup \int f\,ds, \tag{9}$$

where the supremum is taken over those $f \in L^\infty(K)$, $0 \le f \le 1$ such that $|C(f\,ds)| \le 1$ a.e. on Γ. Here $C\mu(z) = \lim_{\varepsilon\to 0} C_\varepsilon\mu(z)$. Clearly (9) implies

$$|K| \le C\gamma(K), \quad K \subset \Gamma,$$

which solves the Denjoy conjecture on Lipschitz graphs.

The next lemma, due to Davie and Oksendal [Da-O], tells us how to dualize a weak L^1 inequality for a linear operator. Modulo technicalities, it proves (9). Since the lemma has other surprisingly striking applications, some of which will be presented in the next two sections, we will discuss its proof in detail. We follow closely [Ch, p.107].

For a locally compact Hausdorff space X, $M(X)$ stands for the set of finite complex Radon measures and $C_0(X)$ for the set of continuous functions on X vanishing at ∞.

1.4.2 Lemma *Let X and Y be locally compact Hausdorff spaces and T a linear operator from $M(X)$ into $C_0(Y)$ such that its transpose T^* sends $M(Y)$ into $C_0(X)$. Let m be a positive Radon measure on X. Then the following are equivalent.*

(i) $m\{x \in X : |T^*\mu(x)| > t\} \le At^{-1}\|\mu\|$, $\mu \in M(Y)$.

(ii) *For each compact* $K \subset X$,

$$m(K) \le B \sup\{\int f\,dm : 0 \le f \le 1, \operatorname{spt} f \subset K, |T(f\,dm)| \le 1\}.$$

Proof (ii) $\Rightarrow$ (i). This is the easy part. It is enough to prove that

$$m\{x \in X : \operatorname{Re} T^*\mu(x) > 1\} \le A\|\mu\|, \quad \mu \in M(Y). \tag{10}$$

Let K be a compact subset of $\{x \in X : \operatorname{Re} T^*\mu(x) > 1\}$ and let f be a function supported on K, $0 \le f \le 1$, $|T(f\,dm)| \le 1$ and such that $m(K) \le 2B \int f\,dm$. Clearly

$$\int f\,dm \le \int f \operatorname{Re} T^*\mu\,dm = \operatorname{Re}\int f\, T^*\mu\,dm = \operatorname{Re}\int T(f\,dm)\,d\mu \le \|\mu\|,$$

and thus (10) follows.

(ii) $\Rightarrow$ (i). Set $B = 8A$ and assume that (ii) is not true for some K. Set

$$B_1 = \{T(f\,dm) : 0 \le f \le 1, \operatorname{spt} f \subset K, 8A \int f\,dm \ge m(K)\}$$

and

$$B_2 = \{g \in C_0(Y) : \|g\|_\infty < 1\}.$$

Then B_1 and B_2 are convex disjoint sets and B_2 is open. By the separation theorem there exists $\mu \in M(Y)$, $\|\mu\| = 1$, such that

$$\operatorname{Re}\int g\,d\mu < \operatorname{Re}\int T(f\,dm)\,d\mu.$$

Maximizing the left hand side on B_2 we get

$$1 \le \operatorname{Re}\int T^*\mu(x) f(x)\,dm(x). \tag{11}$$

Set $F = \{x \in K : |T^*\mu(x)| \le 2A/m(K)\}$. Applying (i) to $t = 2A/m(K)$, we obtain $m(F) \ge \frac{1}{2}m(K)$. Define $f = (4A)^{-1}\chi_F$. Then $8A\int f\,dm \ge m(K)$ and (10) yields a contradiction. □

Let us now prove (9). Unfortunately Lemma 1.4.2 cannot be applied to the truncated Cauchy integral $C_\varepsilon\mu$, because $C_\varepsilon\mu(z)$ is not a continuous function of z. This minor difficulty can be overcome by regularizing the kernel $1/z$. Set $K_\varepsilon = \frac{1}{z} * \frac{\chi_{\Delta(0,\varepsilon)}}{\pi\varepsilon^2}$. Thus $K_\varepsilon(z) = 1/z$ for $|z| \ge \varepsilon$ and $\bar{z}/\varepsilon^2$ for $|z| \le \varepsilon$, as can be easily seen by computing $\bar\partial$ of both sides. Define

$$K_\varepsilon\mu(z) = \int K_\varepsilon(\zeta - z)\,d\mu(\zeta).$$

Then

$$|K_\varepsilon\mu(z) - C_\varepsilon\mu(z)| \le \sup_{\varepsilon>0}\frac{|\mu|(\Delta(z,\varepsilon))}{\varepsilon} \equiv M\mu(z). \tag{12}$$

Since the maximal Hardy-Littlewood type operator $M\mu$ clearly satisfies

$$|\{z \in \Gamma : M\mu(z) > t\}| \leq \frac{C}{t}\|\mu\|, \quad \mu \in M(\Gamma),$$

we can apply Lemma 1.4.2 to $K_\varepsilon\mu$. Thus for some constant C and each ε we can find a function f_ε supported on K that satisfies $0 \leq f_\varepsilon \leq 1$, $|K| \leq C \int f_\varepsilon \, ds$ and $|K_\varepsilon(f_\varepsilon \, ds)| \leq 1$ on Γ. In view of (12) we can replace K_ε by C_ε in the last condition. Let f be a weak-star cluster point of $\{f_\varepsilon\}$ in $L^\infty(K)$. Then $0 \leq f \leq 1$ and $|K| \leq C \int f \, ds$. It remains to show that $|C(f \, ds)| \leq 1$ a.e. on Γ.

For $g \in L^2(\Gamma)$ we have

$$\int C_\varepsilon(f_\varepsilon \, ds) g \, ds = -\int C_\varepsilon(g \, ds) f_\varepsilon \, ds.$$

The right hand side tends to $-\int C(g \, ds) f \, ds$ as ε tends to zero because $C_\varepsilon(g \, ds)$ tends to $C(g \, ds)$ in $L^2(\Gamma)$ and hence in $L^1(K)$. Therefore $C_\varepsilon(f_\varepsilon \, ds) \to C(f \, ds)$ weakly in $L^2(\Gamma)$, and so convex combinations of the $C_\varepsilon(f_\varepsilon \, ds)$ tend to $C(f \, ds)$ in the norm of $L^2(\Gamma)$. Passing to a subsequence we can assume that the convergence is pointwise a.e.. Consequently $|C(f \, ds)| \leq 1$ a.e. on Γ.

Let's finally mention that Murai [Mu1] has recently proved a quantitative version of the Denjoy conjecture. His result is as follows. Let $\Gamma = \{x + iA(x) : \alpha \leq x \leq \beta\}$ be a rectifiable graph. Then, for each compact subset K of Γ,

$$\gamma(K) \geq C|p(K)|^{3/2}|\Gamma|^{-1/2},$$

where $p(K)$ is the projection of K into the real axis and C is an absolute constant. It can be shown that the exponent $3/2$ is best possible [Mu2].

1.5 Removable sets of Lipschitz analytic functions

A compact $K \subset \mathbb{C}$ is said to be *α-removable*, $0 < \alpha \leq 1$, if given an open set $\Omega \supset K$ and a function f analytic on $\Omega \setminus K$ and in $\operatorname{Lip}(\alpha, \Omega)$, i.e., satisfying

$$|f(z) - f(w)| \leq C|z - w|^\alpha, \quad z, w \in \Omega,$$

then f is analytic on the whole of Ω.

As in the case of bounded analytic functions it is enough to test the above condition for $\Omega = \mathbb{C}$. A neat result of Carleson and Dolzenko states that for $0 < \alpha < 1$, K is α-removable if and only if $M^{1+\alpha}(K) = 0$. The reader can consult [Ga] for a proof of this simple fact.

The case $\alpha = 1$ offered much stronger resistance. It is easy to show that the condition area$(K) = 0$ is sufficient for 1-removability. For, if $f \in \operatorname{Lip}(1, \mathbb{C})$ then the first order distributional derivatives of f are in $L^\infty(\mathbb{C})$. If moreover f is analytic outside K, then $\bar{\partial} f = 0$ a.e. on $\mathbb{C}$, and $\bar{\partial} f = 0$ as a distribution. Weyl's Lemma tells us that f is entire.

The converse is much subtler. We must prove that if area$(K) > 0$, then there exists $f \in \operatorname{Lip}(1, \mathbb{C})$, analytic on $\mathbb{C} \setminus K$ and not entire. This was proved by Uy in 1979 [U] by dualizing the weak L^1 inequality for the Beurling transform. To the best of our knowledge, this was the first time a weak L^1 inequality for a linear operator was dualized. Here is the proof of Uy's Theorem.

Take $h \in L^\infty(K)$, $h \not\equiv 0$ and set ($\zeta = x + iy$)

$$f(z) = \int \frac{h(\zeta)}{z-\zeta}\, dx\, dy.$$

Then f is analytic on $\mathbb{C}\setminus K$, and f is not entire. On the other hand, the first partial derivatives of f are given by $\bar{\partial} f = \pi h$ and

$$\partial f = -\ \text{P.V.} \int \frac{h(\zeta)}{(z-\zeta)^2}\, dx\, dy \equiv -Bh,$$

where the last identity defines the Beurling transform Bh of h. The Beurling transform is a Calderón-Zygmund singular integral of convolution type. As such, B does not map $L^\infty(\mathbb{C})$ into $L^\infty(\mathbb{C})$ and thus we cannot guarantee that $\partial f \in L^\infty(\mathbb{C})$ or, what is the same, that $f \in \text{Lip}\,(1, \mathbb{C})$. The idea is to show that a function h as above can be carefully chosen so that $Bh \in L^\infty(\mathbb{C})$. It becomes now quite clear that we only have to mimic the argument that was used in the preceding section to obtain the Denjoy conjecture from L^2 estimates for the Cauchy Integral.

Set, for $\mu \in M(\mathbb{C})$ and $\varepsilon > 0$,

$$B_\varepsilon \mu(z) = \int_{|\zeta - z| > \varepsilon} \frac{d\mu(\zeta)}{(z-\zeta)^2}$$

and

$$B^*\mu(z) = \sup_{\varepsilon > 0} |B_\varepsilon \mu(z)|.$$

The classical Calderón-Zygmund theory tells us that

$$\text{area}\{z \in \mathbb{C} : B^*\mu(z) > t\} \le \frac{C}{t}\|\mu\|, \quad \mu \in M(\mathbb{C}).$$

Applying Lemma 1.4.2 to a smoothed version of $B_\varepsilon\mu$ and then using a limiting process (as in the argument given after the proof of Lemma 1.4.2) we obtain a function h supported on K satisfying $0 \le h \le 1$, $\text{area}(K) \le C \int h\, dx\, dy$ and $|B(h)| \le 1$ a.e. on $\mathbb{C}$.

We will give in section 3.2 an interesting application of Nguyen's Theorem to C^1 approximation by rational functions. For another striking application, see [B-G].

1.6 Cauchy potentials of measures

We would now like to describe an interesting recent development of a potential theoretic nature, which shows again the relevance of L^2 estimates for the Cauchy integral on Lipschitz curves.

One facet of classical potential theory is the study of the continuity properties of logarithmic or newtonian potentials of finite measures. For example, the notion of regular point for the Dirichlet problem and the notion of thin set are related to the lack of continuity of these potentials. The continuity problem for the Cauchy potential $C\mu = \frac{1}{z} * \mu$ of a finite Borel measure μ leads to ask how "strictly" the locally integrable function $C\mu$ can be defined. For instance, $C\mu(z)$ is clearly well defined at points z for which

$$\int \frac{d|\mu|(\zeta)}{|\zeta - z|} < \infty, \tag{13}$$

and this happens almost everywhere with respect to newtonian capacity (the classical capacity associated to the kernel $1/|z|$). Principal values provide a subtler way of defining $C\mu(z)$. Call $E = E_\mu$ the set of points z where

$$\text{P.V.} \int \frac{d\mu(\zeta)}{\zeta - z} = \lim_{\varepsilon \to 0} \int_{|\zeta - z| > \varepsilon} \frac{d\mu(\zeta)}{\zeta - z} \tag{14}$$

does not exist or is infinite. We define $C\mu(z)$ as the limit in (14) for $z \notin E$. The question is now how to estimate the size of the exceptional set E. Since E is contained in the set of points where (13) fails, E has zero newtonian capacity. The following result of Mattila and Melnikov [M-M] improves substantially the preceding statement.

1.6.1 Theorem *For any rectifiable curve Γ, $E \cap \Gamma$ has zero length, that is, the principal value integral* (14) *exists almost everywhere with respect to arc length on Γ.*

Proof We present a sketch of the proof given in [Ve1]. We already know that we can assume the given curve Γ to be a Lipschitz graph. The result will now follow from standard methods from real analysis if we can prove that

$$|\{z \in \Gamma : C^*\mu(z) > t\}| \le Ct^{-1}\|\mu\|, \quad \mu \in M(\mathbb{C}), \tag{15}$$

where the absolute value stands for arc length and $C^*\mu$ was defined in (8).

The inequality (15) is in turn equivalent to

$$|\{z \in \Gamma : |C_\varepsilon\mu(z)| > t\}| \le Ct^{-1}\|\mu\|, \quad \mu \in M(\mathbb{C}), \tag{16}$$

where C is independent of ε. This equivalence is a well known consequence of Cotlar's inequality, a familiar result in the theory of singular integrals.

As we mentioned in section 1.4, the L^2-boundedness of the Cauchy integral plus standard Calderón-Zygmund theory implies that for some C independent of ε

$$|\{z \in \Gamma : |C_\varepsilon\mu(z)| > t\}| \le Ct^{-1}\|\mu\|, \quad \mu \in M(\Gamma). \tag{17}$$

Instead of working directly with (16) and (17), we will compare their dual statements. Is not hard to show from Lemma 1.4.2 that the dual to (17) is

$$|K| \le C \sup_{b \in B} \left| \int b(s)\, ds \right|, \quad K \subset \Gamma, \tag{18}$$

where $B = \{b \in L^\infty(K) : 0 \le b \le 1, |C_\varepsilon(b(s)\,ds)| \le 1$ a.e. on $\Gamma\}$, and the dual to (16) is again (18) with B replaced by

$$B' = \{b \in L^\infty(K) : 0 \le b \le 1, |C_\varepsilon(b(s)\,ds)(z)| \le 1 \text{ for } z \in \mathbf{C}\backslash\Gamma\}.$$

A simple calculation [Ch, p.109] now shows that $b \in B$ implies $|C_\varepsilon(b(s)\,ds)(z)| \le \text{Const}$, $z \in \mathbf{C}\backslash\Gamma$. Since all constants were independent of ε (18) with B replaced by B' follows from (18). The proof is complete. □

1.7 Some open problems

We give a list of four unsolved problems involving analytic capacity. The second is old and very well known. The third is a qualitative version of the last, which is taken from [Vel].

Problem 1 Let $T : \mathbb{R}^2 \to \mathbb{R}^2$ be the linear transformation $T(x,y) = (x, 2y)$. Show that $\gamma(T(K)) = 0$ if and only if $\gamma(K) = 0$.

This problem was raised by O'Farrell. It is the simplest instance of the following more general question. Let Φ be a bilipschitz mapping of the plane into itself, that is,

$$C^{-1}|z-w| \le |\Phi(z) - \Phi(w)| \le C|z-w|.$$

Is it true that for some constant A and all compact sets K,

$$A^{-1}\gamma(K) \le \gamma(\Phi(K)) \le A\gamma(K)?$$

A positive answer would obviously mean that analytic capacity has a metrical nature.

Problem 2 Does there exist a constant C such that for each compact K one can find a positive measure μ supported on K satisfying

$$\left|\int \frac{d\mu(\zeta)}{\zeta - z}\right| \le 1, \quad z \notin K \text{ and } \gamma(K) \le C\|\mu\|?.$$

Problem 3 Let μ be a finite Borel measure. Does $\lim_{\varepsilon\to 0} \int_{|\zeta - z|>\varepsilon} \frac{d\mu(\zeta)}{\zeta - z}$ exist except for a set of zero analytic capacity? A positive answer would improve Theorem 1.6.1 because $\gamma(E) = 0$ implies $\text{length}(E \cap \Gamma) = 0$ for all rectifiable curves Γ.

Problem 4 Does a constant C exist such that

$$\gamma\{z \in \mathbf{C} : C^*\mu(z) > t\} \le Ct^{-1}\|\mu\|$$

for any finite Borel measure μ and any $t > 0$?

Chapter 2
Uniform approximation by rational functions

2.1 The main results

Let X be a compact subset of $\mathbb{C}$ and f an analytic function on a neighbourhood of X. A weak form of Runge's Theorem tells us that f can be uniformly approximated on X by rational functions with poles off X. We remark that the above version of Runge's Theorem is very easily proved: take a system Γ of closed piecewise smooth curves surrounding X, express $f(z)$, for z in X, by means of the Cauchy integral formula on Γ and then approximate the contour integral by Riemann sums. We wish to face the much subtler problem of describing the most general function with the property of being uniformly approximable on X by rational functions with poles off X. Let $R(X)$ be the algebra formed by such functions. Notice that there are two obvious necessary conditions for f to belong to $R(X)$: f must be continuous on X and analytic on the interior $\mathring{X}$ of X. Is not clear at first glance that these two necessary conditions are not always sufficient, but a famous example due to Mergelyan and Roth shows that this is indeed the case. The compact set constructed by Mergelyan and Roth has no interior points, and is called the Swiss Cheese for reasons which will become apparent after its definition. The Swiss cheese is $X = \bar{\Delta}_0 \setminus \bigcup_{n=1}^{\infty} \Delta_n$, where Δ_0 is the open unit disc and the Δ_n are open discs of radius r_n, satisfying

(i) $\bar{\Delta}_n \subset \Delta_0$ and $\bar{\Delta}_n \cap \bar{\Delta}_m = \emptyset$ for $n \neq m$,

(ii) $\bigcup \Delta_n$ is dense in Δ_0,

(iii) $\sum_{n=1}^{\infty} r_n < \infty$.

A sequence of such discs Δ_n can be readily defined inductively from an enumeration of the rational points in the unit disc. The measure

$$\lambda = dz_0 - \sum_{n\geq 1} dz_n,$$

where dz_n, $n \geq 0$, is dz on $\partial\Delta_n$, vanishes on $R(X)$ but is not identically zero. Thus some continuous function on X not belonging to $R(X)$ must exist. Such a function can be explicitly constructed if we know that X has positive area, which is indeed the case by the Hartogs-Rosenthal Theorem (to be stated and proved later) or by a direct argument [Br, p.163]. Setting $f = \frac{1}{z} * \chi_X$ we get $\int f(z)\,d\lambda = -2\pi i$ area $(X) \neq 0$, which shows that f is as desired.

Let $A(X)$ be the algebra of those continuous functions on X which are analytic on $\mathring{X}$. Then $R(X) \subset A(X)$ for all X, and the inclusion can be strict. A "collective" variant of the problem we are envisaging is that of characterizing those X for which $R(X) = A(X)$.

In the sixties Vitushkin introduced a powerful new method to attack the above approximation problems leading to fairly complete solutions, thus culminating work previously done by Lavrentiev, Mergelyan, Walsh and others. We state now the "individual" form of Vitushkin's Theorem.

2.1.1 Theorem *Let $f \in C(\mathbb{C})$ and let $X \subset \mathbb{C}$ be compact. Then the following statements* *are equivalent.*

(i) $f \in R(X)$.

(ii) *For some function* $\varepsilon(\delta)$ *tending to zero with* δ,

$$\left|\int f(z)\bar{\partial}\varphi(z)\,dx\,dy\right| \le \varepsilon(\delta)\delta\|\nabla\varphi\|_\infty\gamma(\Delta\setminus X),$$

for all open discs Δ *of radius* δ *and all* $\varphi \in C_0^\infty(\Delta)$.

Notice that if we take $\Delta \subset X$ then (ii) gives $\int f(z)\bar{\partial}\varphi(z)\,dx\,dy = 0$, $\varphi \in C_0^\infty(\Delta)$, which means that f is analytic on $\mathring{X}$. Therefore, (ii) should be regarded as a weak analyticity condition quantified by γ.

To state the solution to the collective problem we need to introduce a slight variant of analytic capacity. The *continuous analytic capacity* of a compact $K \subset \mathbb{C}$ is $\alpha(K) = \sup|f'(\infty)|$, where the supremum is taken over all functions f which are continuous on the complex plane, analytic on $\mathbb{C}\setminus K$ and satisfy $|f(z)| \le 1$, $z \in \mathbb{C}$. If E is not compact, $\alpha(E)$ is defined as the supremum of $\alpha(K)$ over all compact subsets K of E. Clearly $\alpha \le \gamma$. If I is an interval, Morera's Theorem yields $\alpha(I) = 0$ and thus α and γ are different set functions. They coincide, however, on open sets. As we did with γ, it is easy to see that $\alpha(K) = 0$ if and only if K is removable for continuous analytic functions.

2.1.2 Theorem *For a compact* $X \subset \mathbb{C}$ *the following are equivalent.*

(i) $R(X) = A(X)$.

(ii) $\alpha(\Delta\setminus\mathring{X}) = \alpha(\Delta\setminus X)$, *for all discs* Δ.

(iii) *For some constant* C,

$$\alpha(\Delta\setminus\mathring{X}) \le C\alpha(\Delta\setminus X), \text{ for all discs } \Delta.$$

Notice that $\Delta\setminus\mathring{X} = (\Delta\setminus X) \cup (\Delta \cap \partial X)$ and so $\alpha(\Delta\setminus\mathring{X}) \ge \alpha(\Delta\setminus X)$ because α is an increasing set function. Thus conditions (i) and (ii) should be interpreted as local negligibility conditions on ∂X quantified by α.

We will now deduce some consequences from 2.1.2 to test its power. First, we explicitly state 2.1.2 for sets without interior.

2.1.3 Corollary *For a compact* $X \subset \mathbb{C}$ *the following are equivalent.*

(i) $R(X) = C(X)$.

(ii) $\gamma(\Delta\setminus X) = \delta$, *for each open disc* Δ *of radius* δ.

(iii) *For some constant* C,

$$\gamma(\Delta\setminus X) \ge C\delta, \text{ for each open disc } \Delta \text{ of radius } \delta.$$

Being more careful with the estimates in Example 3 of section 1.1 one gets $\gamma(E) \geq \left(\frac{\text{area}(E)}{\pi}\right)^{1/2}$ (see [G, p.200]). When X has zero area we then have $\gamma(\Delta\backslash X) \geq \left(\frac{\text{area}(\Delta\backslash X)}{\pi}\right)^{1/2} = \delta$, for any disc Δ of radius δ, which proves the next result.

2.1.4 Theorem (Hartogs and Rosenthal) *If X has zero area, then $R(X) = C(X)$.*

Let $P(X)$ be the uniform closure on X of the set of analytic polynomials. A well known elementary result in complex analysis tells us that $P(X) = R(X)$ if and only if X^c is connected.

2.1.5 Theorem (Lavrentiev) *We have $P(X) = C(X)$ if and only if X^c is connected and X has no interior.*

2.1.6 Theorem (Mergelyan) *We have $P(X) = A(X)$ if and only if X^c is connected.*

Proof of Theorems 2.1.5 and 2.1.6 Only the sufficiency of the conditions needs an explanation. Take a disc $\Delta = \Delta(z, \delta)$, $z \in \partial X$, and a point w in $X^c \cap \Delta(z, \delta/2)$. We can join w and ∞ by a curve not intersecting X, because X^c is connected. Then $\Delta\backslash X$ contains a continuum of diameter at least $\delta/2$. Therefore the capacity of $\Delta\backslash X$ is at least $\delta/8$, and so $\alpha(\Delta\backslash X) = \gamma(\Delta\backslash X) \geq \delta/8 \geq \alpha(\Delta\backslash \mathring{X})/8$. Hence $\alpha(\Delta\backslash \mathring{X}) \leq 8\alpha(\Delta\backslash X)$ provided Δ is an open disc centered at ∂X. It can be shown with almost no additional effort that this condition already implies $R(X) = A(X)$. □

A little modification of the argument just presented gives the following.

2.1.7 Corollary *Assume that X^c has finitely many connected components or, more generally, that the diameters of the connected components of X^c are bounded from below away from zero. Then $R(X) = A(X)$.*

The simplest example where the hypothesis of 2.1.7 fails is $X = \bar{\Delta}\backslash\bigcup_n \Delta_n$ where Δ is the unit disc and the Δ_n are open discs with mutually disjoint closures contained in $\bar{\Delta}$ and not containing 0, and with centers tending to 0. However, it can be easily proven that $R(X) = A(X)$ is also true in this case. This fact is a simple instance of a much more general result due to Davie and Oksendal [Da-O] which we describe next.

The *inner boundary* $\partial_i X$ of X is the set of boundary points which do not belong to the boundary of a connected component of X^c. In the previous example $\partial_i X = \{0\}$.

2.1.8 Theorem (Davie-Oksendal) *If $d_H(\partial_i X) < 1$, then $R(X) = A(X)$.*

Let us remark that given a compact K without interior, there exists a compact X such that $\partial_i X = K$. Indeed, we can take $X = \bar{\Delta}\backslash\bigcup_n \Delta_n$ where Δ is a disc containing K and the Δ_n are open discs with mutually disjoint closures not intersecting K and accumulating at each point of K. Consequently, the inner boundary of a compact set can be extremely complicated.

Recall that $\alpha(E) = 0$ follows from $d_H(E) < 1$. Assume now that $\alpha(\partial_i X) = 0$ and let us try to extend 2.1.8 by proving that $R(X) = A(X)$. For each disc Δ, $\Delta \setminus \mathring{X} = (\Delta \setminus X) \cup (\Delta \cap \partial_e X) \cup (\Delta \cap \partial_i X)$, where $\partial_e X = \partial X \setminus \partial_i X$ is the exterior boundary of X. The proof of the Mergelyan Theorem we will present in the next section can be adapted to show that $\alpha((\Delta \setminus X) \cup (\Delta \cap \partial_e X)) = \alpha(\Delta \setminus X)$. Therefore, if we knew that α were subadditive as a set function, we would get $\alpha(\Delta \setminus \mathring{X}) \leq \alpha(\Delta \setminus X)$ and thus $R(X) = A(X)$. Unfortunately the subadditivity problem for α (or for γ) is still unsolved, and in fact is the most outstanding open problem in the area. We state separately, for emphasis, the two open problems we just mentioned (see [Vi]).

The semiadditivity problem Show that for some constant C and all pairs of disjoint compact sets K_1 and K_2 one has

$$\alpha(K_1 \cup K_2) \leq C(\alpha(K_1) + \alpha(K_2)).$$

The inner boundary conjecture If $\alpha(\partial_i X) = 0$, then $R(X) = A(X)$.

A special instance of the inner boundary conjecture arises when one requires $\partial_i X$ to be part of a rectifiable arc. In this case one is led to consider a very interesting inequality which is only known to hold under rather severe restrictions.

Let Γ be a closed rectifiable Jordan curve bounding a simply connected domain Ω. Let $K \subset \bar{\Omega}$ be compact and assume that f is continuous on $\bar{\Omega}$ and analytic on $\Omega \setminus K$. We would like to characterize those curves Γ for which the inequality

$$\left| \int_\Gamma f(z)\, dz \right| \leq C \|f\|_\infty \alpha(K), \tag{1}$$

holds for some $C = C(\Gamma)$ independent of f and K. If $\partial_i X$ is a subset of a curve satisfying (1) then $R(X) = A(X)$ follows rather directly and this explains our interest in (1).

Melnikov proved (1) when Γ is a circle and, more generally, when Γ is real analytic [Me]. Later Vitushkin [Vi] extended (1) to $C^{1+\epsilon}$ curves, using conformal mapping (see also [Da1] where a Dini type condition on the tangent vector is shown to be sufficient), but his arguments cannot be pushed to cover the C^1 case, which is still open. As a consequence, we do not know whether the inner boundary conjecture is true when $\partial_i X$ is a C^1 arc.

It is shown in [Da1] that a necessary condition for (1) is

$$\text{length}(K) \leq C\gamma(K), \quad K \subset \Gamma,$$

which is equivalent to the Ahlfors-David condition

$$\text{length}(\Delta \cap \Gamma) \leq C \text{ radius } \Delta, \text{ for all discs } \Delta,$$

as mentioned in section 1.4. In fact Vitushkin suggested in [Vi, p.159] that the curves satisfying (1) might be exactly those satisfying the Ahlfors-David condition, which we now know to be those for which the Cauchy integral satisfies L^2 estimates.

2.2 Proof of Vitushkin's theorems

The most interesting step in the proof of Theorems 2.1.1 and 2.1.2 is (ii) ⇒ (i) in 2.1.1, and so we will concentrate on the approximation of the function f in the statement of 2.1.1.

There are two basic ingredients in the main argument. The first is a localization procedure which consists in decomposing f as $\sum_j f_j$, where f_j is a replica of f associated to some disc of radius δ, a small positive number given in advance. The number of terms in the above sum becomes arbitrarily large as δ decreases. The difficulty lies now in finding a suitable way of modifying f_j to get a function g_j, analytic on some neighbourhood of X, and such that $\|f - \sum_j g_j\|_\infty$ tends to zero with δ. The second ingredient of the proof is such a modification method which, as the reader will see, is extremely clever and effective.

To discuss the localization method we first introduce the Vitushkin localization operator. Given a disc Δ of radius δ, $\varphi \in C_0^1(\Delta)$ and a distribution f, set

$$V_\varphi f = \frac{1}{\pi z} * \varphi\bar{\partial} f,$$

which makes sense, because it is the convolution of the compactly supported distribution $\varphi\bar{\partial} f$ with the locally integrable function $1/\pi z$. Since $1/\pi z$ is the fundamental solution of $\bar{\partial}$, $\bar{\partial}(V_\varphi f) = \varphi\bar{\partial} f$ and then $V_\varphi f$ is analytic wherever f is, and off Δ. Using $\bar{\partial}(\varphi f) = \varphi\bar{\partial} f + f\bar{\partial}\varphi$ we obtain

$$V_\varphi f = \varphi f - \frac{1}{\pi z} * f\bar{\partial}\varphi, \tag{2}$$

from which it follows immediately that $V_\varphi f$ is continuous when f is. Since V_φ vanishes on constants, replacing f by $f - f(z)$ in (2) and evaluating at z we get

$$V_\varphi f(z) = \frac{1}{\pi} \int \frac{f(\zeta) - f(z)}{\zeta - z} \bar{\partial}\varphi(\zeta)\, dx\, dy. \tag{3}$$

If f is uniformly continuous and $z \in \Delta$, then (3) readily gives

$$|V_\varphi f(z)| \le C w(\delta)\delta \|\nabla\varphi\|_\infty, \tag{4}$$

where $w(\delta) = \sup\{|f(z) - f(w)| : |z - w| \le \delta\}$ is the modulus of continuity of f. Since $V_\varphi f$ is analytic outside Δ, including ∞, (4) holds for each $z \in \mathbb{C}$.

To perform the localization we are looking for, we need to construct certain partitions of unity subordinated to special coverings of the plane by open discs. We will use the following simple fact.

2.2.1 Lemma *Given any $\delta > 0$ there exist a countable family of open discs (Δ_j) of radius δ and a family of functions $\varphi_j \in C_0^\infty(\Delta_j)$ such that*

(i) $\mathbb{C} = \bigcup_j \Delta_j$.

(ii) *The family (Δ_j) is almost disjoint, that is, for some constant C each $z \in \mathbb{C}$ belongs to at most C discs Δ_j (in fact we can take $C = 21$).*

(iii) $\sum_j \varphi_j = 1$, $0 \le \varphi_j$ *and* $\|\nabla\varphi_j\|_\infty \le C\delta^{-1}$, *where* C *is some absolute constant.*

Proof Take a grid (Q_j) of squares of side length $\delta/2$, with sides parallel to the coordinate axes and covering $\mathbb{C}$. Clearly $\sum_j \chi_{Q_j} = 1$, where χ_{Q_j} is the characteristic function of Q_j. Now we only need to regularize the non-smooth partition of unity (χ_{Q_j}). To this end take $0 \le \chi \in C_0^\infty(\mathbb{C})$ with support $\chi \subset \Delta\left(0, \frac{1}{2}\right)$ and $\int \chi = 1$. Set $\chi_\delta(z) = \delta^{-2}\chi(\delta^{-1}z)$ and define $\varphi_j = \chi_\delta * \chi_{Q_j}$. If Δ_j is the disc of radius δ concentric with Q_j (i), (ii) and (iii) above hold, as is easily checked. □

Now let f satisfy condition (ii) of Theorem 2.1.1. Replacing f by φf, where $\varphi \in C_0^\infty(\mathbb{C})$ takes the value 1 on a neighbourhood of X we can assume that f has compact support. Fix $\delta > 0$ and consider discs Δ_j and functions φ_j given by 2.2.1. Then $\bar\partial f = \sum_j \varphi_j \bar\partial f$ and only finitely many terms in the above sum are non-zero. Set $f_j = V_{\varphi_j} f$. The identity $f = \sum_j f_j$ follows from the definition of f_j and the fact that $f = \frac{1}{\pi z} * \bar\partial f$, which in turn is a consequence of Liouville's Theorem. On the other hand, due to the properties of the localization operator, f_j is continuous on $\mathbb{C}$ and analytic outside a compact subset of $\Delta_j \backslash \mathring{X}$, and $\|f_j\|_\infty \le Cw(\delta)$.

A *singularity* of a function f is a point z such that f is not analytic in any neighbourhood of z. In other words, z is a singularity of f if z belongs to the support of $\bar\partial f$. Then the effect of the decomposition $f = \sum_j f_j$ is the *distribution of the singularities* of f among the small discs Δ_j.

We turn now our attention to the modification of f_j. Since f_j is analytic outside Δ_j, f_j can be expanded in a Laurent series

$$f_j(z) = \frac{a_1}{z - z_j} + \frac{a_2}{(z - z_j)^2} + \cdots, \quad |z - z_j| > \delta, \tag{5}$$

z_j being the center of Δ_j. Assume that we can find a function g_j analytic outside a compact subset of $\Delta_j \backslash X$, with $\|g_j\|_\infty \le \varepsilon(\delta)$, where $\varepsilon(\delta) \to 0$ as $\delta \to 0$, such that

$$g_j(z) = \frac{a_1}{z - z_j} + \frac{a_2}{(z - z_j)^2} + O(|z|^{-3}), \text{ as } z \to \infty.$$

Namely, we are assuming that it is possible to push the singularities of f_j out of X but still inside Δ_j, keeping the new function small and with the same Laurent expansion up to order 2. Then the next lemma, applied to $h_j = f_j - g_j$, tells us that $g \equiv \sum_j g_j$ satisfies

$$\|f - g\|_\infty \le C \max_j(\|f_j\|_\infty + \|g_j\|_\infty) \le C(w(\delta) + \varepsilon(\delta)).$$

Hence g will be the desired approximant.

2.2.2 The Triple Zero Lemma *Let* h_j *be a function which is bounded on* $\mathbb{C}$, *analytic outside* Δ_j *and assume that* $h_j = O(|z|^{-3})$ *as* $z \to \infty$. *Then*

$$\|\sum_j h_j\|_\infty \le C \max_j \|h_j\|_\infty,$$

for some absolute constant C.

Proof Consider the expansion of h_j outside Δ_j,

$$h_j(z) = \sum_{n\geq 3} \frac{a_n}{(z-z_j)^n}, \quad |z-z_j| > \delta.$$

Then $a_n = \frac{1}{2\pi i}\int_{|z-z_j|=\delta} h_j(z)(z-z_j)^{n-1}\,dz$, and so $|a_n| \leq \|h_j\|_\infty \delta^n$. Consequently

$$|h_j(z)| \leq 2\|h_j\|_\infty \frac{\delta^3}{|z-z_j|^3}, \quad |z-z_j| > 2\delta.$$

Given $z \in \mathbf{C}$, there is some numerical constant N such that for at most N indices j one has $z \in 2\Delta_j (= \Delta(z_j, 2\delta))$. Hence

$$\sum_j |h_j(z)| \leq \left(N + 2\sum_{z\notin 2\Delta_j} \frac{\delta^3}{|z-z_j|^3}\right) \max_j \|h_j\|_\infty .$$

The almost disjointness of the family (Δ_j) now gives

$$\sum_{z\notin 2\Delta_j} \frac{\delta^3}{|z-z_j|^3} \leq C \sum_{z\notin 2\Delta_j} \delta \int_{\Delta_j} \frac{dx\,dy}{|\zeta - z|^3} \leq C\delta \int_{|\zeta - z|>\delta} \frac{dx\,dy}{|\zeta - z|^3} = C,$$

and the lemma follows. □

Remark If the hypothesis in 2.2.2 is weakened to $h_j(z) = O(|z|^{-2})$ as $z \to \infty$, then a simple modification of the above argument yields

$$\|\sum_j h_j\|_\infty \leq C \log\frac{d}{\delta} \max_j \|h_j\|_\infty, \tag{6}$$

where d is the diameter of $\bigcup_j \Delta_j$.

With the material introduced so far we can prove with little additional effort two significant results, which are particular instances of our main theorems. Namely, we can prove (iii) ⇒ (i) in 2.1.3, and Mergelyan's Theorem. The proofs will provide examples in which the construction of the perturbed functions g_j is specially transparent.

Assume first that (iii) of 2.1.3 holds. Using the expression (2) for f_j, with φ replaced by φ_j we get a formula for the coefficient a_n in the expansion (5),

$$a_n = -\frac{1}{\pi}\int (\zeta - z_j)^{n-1} f(\zeta)\bar\partial\varphi_j(\zeta)\,dx\,dy. \tag{7}$$

In particular

$$a_1 = -\frac{1}{\pi}\int f(\zeta)\bar\partial\varphi_j(\zeta)\,dx\,dy = -\frac{1}{\pi}\int (f(\zeta) - f(z_j))\bar\partial\varphi_j(\zeta)\,dx\,dy,$$

and thus

$$|a_1| \leq Cw(\delta)\delta \leq Cw(\delta)\gamma(\Delta_j \backslash X),$$

by (iii) of 2.1.3. Choose a function h analytic outside a compact subset of $\Delta_j\backslash X$ such that $\|h\|_\infty \le 1$ and $2h'(\infty) \ge \gamma(\Delta_j\backslash X)$. Set $g_j = a_1 h/h'(\infty)$, so that $f_j - g_j = O(|z|^{-2})$ and $\|g_j\|_\infty \le Cw(\delta)$. By (6) we get

$$\|f - \sum_j g_j\|_\infty \le C\left(\log\frac{C}{\delta}\right) w(\delta)$$

which tends to zero with δ provided $w(\delta) \le C\delta$. This is true if $f \in C_0^1(\mathbb{C})$, which we can take for granted without loss of generality, in view of the Stone-Weierstrass Theorem and the fact that X has no interior when (iii) of 2.1.3 holds.

Suppose now that the complement of X is connected. If $\Delta_j \subset X$ then $f_j \equiv 0$. So let us assume that Δ_j contains a point z in X^c. Joining z to ∞ by a curve in X^c we see that $2\Delta_j\backslash X$ contains a continuum J of diameter at least δ and consequently of analytic capacity at least $\delta/4$. We can then find a function h, analytic outside J, with $\|h\|_\infty \le 1$ and such that $|h'(\infty)| \ge \delta/4$. We want now to show the existence of complex numbers λ_1 and λ_2 such that

$$\lambda_1 h + \lambda_2 h^2 = \frac{a_1}{z - z_j} + \frac{a_2}{(z - z_j)^2} + O(|z|^{-3}), \text{ as } z \to \infty. \tag{8}$$

If

$$h(z) = \frac{\alpha_1}{z - z_j} + \frac{\alpha_2}{(z - z_j)^2} + O(|z|^{-3}), \text{ as } z \to \infty,$$

then the solution to (8) is

$$\lambda_1 = a_1/\alpha_1, \quad \lambda_2 = (a_2 - a_1\alpha_2/\alpha_1)/\alpha_1^2,$$

from which we get the estimate

$$|\lambda_j| \le Cw(\delta), \quad j = 1, 2.$$

Set $g_j = \lambda_1 h + \lambda_2 h^2$. Then g_j is analytic outside a compact subset of $2\Delta_j\backslash X$, $\|g_j\|_\infty \le Cw(\delta)$ and $f_j - g_j = O(|z|^{-3})$ as $z \to \infty$. Applying 2.2.2 to the almost disjoint family $(2\Delta_j)$ we get

$$\|f - \sum_j g_j\|_\infty \le Cw(\delta),$$

which completes the proof of Mergelyan's Theorem.

We turn now our attention to describing the modification procedure of f_j in full generality, thus completing the proof of (ii) $\Rightarrow$ (i) in 2.1.1.

We need two lemmas.

2.2.3 Lemma *Let Δ be a disc of center a and radius δ and let h be a continuous function on $\mathbb{C}$, analytic outside a compact subset K of Δ. If $h(\infty) = 0$ then*

$$|h(z)| \le C\frac{\alpha(K)}{|z - a|}, \quad |z - a| > 2\delta. \tag{9}$$

Proof Assume $a = 0$ and write

$$h(z) = \sum_{n=1}^{\infty} \frac{a_n}{z^n}, \quad |z| > \delta.$$

Then $|a_n| \le \|h\|_\infty \delta^n$ and so the function

$$g(z) = z^{n-1}\left(h(z) - \sum_{j=1}^{n-1} \frac{a_j}{z^j}\right),$$

which is continuous on $\mathbb{C}$ and analytic outside K, satisfies $\|g\|_\infty \le \|h\|_\infty n\delta^{n-1}$.

Hence

$$|a_n| = |g'(\infty)| \le \alpha(K)\|h\|_\infty n\delta^{n-1}, \tag{10}$$

from which (9) follows easily. □

2.2.4 Lemma *Let $E \subset \mathbb{C}$, $\delta = \alpha(E)$ and (Δ_j) a finite, almost disjoint family of discs of radius δ covering E. Set $E_j = E \cap \Delta_j$. If f_j is continuous on $\mathbb{C}$, analytic outside a compact subset of E_j and $f_j(z) = O(|z|^{-1})$ as $z \to \infty$, then*

$$\|\sum_j f_j\|_\infty \le C \max_j \|f_j\|_\infty. \tag{11}$$

In particular

$$\sum_j \alpha(E_j) \le C\alpha(E). \tag{12}$$

Proof Let z_j be the center of Δ_j. By the preceding lemma,

$$\sum_j |f_j(z)| \le C \max \|f_j\|_\infty \left(1 + \sum_{z \notin 2\Delta_j} \frac{\alpha(E_j)}{|z - z_j|}\right).$$

An application of Hölder's inequality with exponents 3 and 3/2 gives

$$\sum_{z\notin 2\Delta_j} \frac{\alpha(E_j)}{|z-z_j|} \le \left(\sum_j \left(\frac{\alpha(E_j)}{\delta}\right)^{3/2}\right)^{2/3} \left(\sum_{z\notin 2\Delta_j} \frac{\delta^3}{|z-z_j|^3}\right)^{1/3} \le C\left(\sum_j \frac{\alpha(E_j)}{\alpha(E)}\right)^{2/3}.$$

Thus

$$\left\|\sum f_j\right\|_\infty \le C \max \|f_j\|_\infty \left(1 + \sum_j \frac{\alpha(E_j)}{\alpha(E)}\right)^{2/3}. \tag{13}$$

Assume now that g_j is such that $g_j'(\infty) = \alpha(E_j) - \varepsilon$ and $\|g_j\|_\infty \le 1$. Then from the definition of $\alpha(E)$, and (13) with f_j replaced by g_j,

$$\sum_j \alpha(E_j) - \varepsilon = \left(\sum_j g_j\right)'(\infty) \le C\left(1 + \sum_j \frac{\alpha(E_j)}{\alpha(E)}\right)^{2/3} \alpha(E).$$

Letting $\varepsilon \to 0$ we obtain

$$\sum_j \frac{\alpha(E_j)}{\alpha(E)} \le C + C\left(\sum_j \frac{\alpha(E_j)}{\alpha(E)}\right)^{2/3}$$

which implies (12). Clearly (11) follows from (12) and (13). □

Proof of (ii) $\Rightarrow$ (i) **in Theorem 2.1.1** Fix $\delta > 0$ and consider a δ-Vitushkin scheme $(\Delta_j, \varphi_j, f_j)$ for the approximation of f. That is, (Δ_j) is an almost disjoint covering of $\mathbf{C}$ by open discs of radius δ, (φ_j) a partition of unity subordinated to (Δ_j) as in Lemma 2.2.1, and $f_j = V_{\varphi_j}(f)$. For the second coefficient a_2 in the expansion (5) of f_j we have the estimate $|a_2| \le C\delta^2$, which can be improved to $|a_2| \le C\alpha(\Delta_j \setminus \mathring{X})\delta$. This is not good enough for our purposes (except when $\gamma(2\Delta_j \setminus X) \ge C\delta$) and so we will perform a second localization of f to overcome that difficulty.

Set $\gamma = \gamma(\Delta_j \setminus X)$ and let (D_k, ψ_k, f_{jk}) be a γ-Vitushkin scheme for f_j. Thus $f_{jk} = V_{\psi_k}(f_j)$. In fact $f_{jk} = V_{\varphi_j \psi_k}(f)$, because both sides have the same $\bar\partial$ derivative and vanish at ∞. Hence $\|f_{jk}\|_\infty \le C\omega(\gamma)$. Let $a_1(f_{jk})$ be the first coefficient in the expansion at ∞ of f_{jk}. Then

$$|a_1(f_{jk})| = \left|\frac{1}{\pi}\int f(z)\bar\partial(\varphi_j\psi_k)(z)\,dx\,dy\right| \le C\varepsilon(\gamma)\gamma(D_k \setminus X),$$

by our hypothesis (ii). Let h be analytic outside a compact subset of $D_k \setminus X$ such that $\|h\|_\infty \le 1$ and $2h'(\infty) \ge \gamma(D_k \setminus X)$. Set $G_{jk} = a_1(f_{jk})h/h'(\infty)$ and $G_j = \sum_k G_{jk}$. Then G_{jk} is analytic outside a compact subset of $D_k \setminus X$, $\|G_{jk}\|_\infty \le C\varepsilon(\gamma)$ and $f_{jk} - G_{jk} = O(|z|^{-2})$ as $z \to \infty$. Clearly $f_j - G_j = O(|z|^{-2})$, and $\|G_j\|_\infty \le C\varepsilon(\delta)$ according to Lemma 2.2.4. Since f_j is analytic outside Δ_j, $f_{jk} \equiv 0$ whenever D_k does not intersect Δ_j, and so we can consider only indices k with $D_k \cap \Delta_j \ne \emptyset$. Thus G_j is analytic outside a compact subset of $3\Delta_j \setminus X$. Let z_j be the center of Δ_j and z_{jk} the center of D_k.

Consider the expansions

$$f_j(z) - G_j(z) = \frac{a_2}{(z - z_j)^2} + O(|z|^{-3})$$

and

$$f_{jk}(z) - G_{jk}(z) = \frac{a_2^k}{(z - z_{jk})^2} + O(|z|^{-3}).$$

We have $a_2 = \sum_k a_2^k$ and

$$a_2^k = -\frac{1}{\pi}\int f(z)\frac{\partial}{\partial \bar z}(\varphi_j(z)\psi_k(z)(z - z_{jk}))\,dx\,dy - a_2(G_{jk}),$$

where $a_2(G_{jk})$ is the second coefficient in the expansion of G_{jk} in negative powers of $z - z_{jk}$. Applying (ii) and (10) for $n = 2$ we get $|a_2^k| \le C\varepsilon(\gamma)\gamma\gamma(D_k \setminus X)$, and therefore

$$|a_2| \le C\varepsilon(\gamma)\gamma\sum_k \gamma(D_k \setminus X) \le C\varepsilon(\delta)\gamma(3\Delta_j \setminus X)^2,$$

by 2.2.4. Let now h be analytic outside a compact subset of $3\Delta_j\backslash X$ and such that $\|h\|_\infty \leq 1$ and $2h'(\infty) \geq \gamma(3\Delta_j\backslash X)$. The function

$$H_j = a_2(h/h'(\infty))^2 = \frac{a_2}{(z-z_j)^2} + O(|z|^{-3}) \text{ as } z \longrightarrow \infty,$$

satisfies $\|H_j\|_\infty \leq C\varepsilon(\delta)$. If we set $g_j = G_j + H_j$, then g_j is analytic outside a compact subset of $3\Delta_j\backslash X$, $\|g_j\|_\infty \leq C\varepsilon(\delta)$ and $f_j - g_j = O(|z|^{-3})$ as $z \to \infty$. Thus

$$\|f - \sum_j g_j\|_\infty \leq C(\omega(\delta) + \varepsilon(\delta)),$$

and the proof is complete. □

Proof of (iii) ⇒ (i) in Theorem 2.1.2 Let $f \in A(X)$. We can assume that f is the restriction to X of a compactly supported function in $C(\mathbb{C})$. If Δ is an open disc of radius δ and $\varphi \in C_0^\infty(\Delta)$ then

$$\left|\int f(z)\bar\partial\varphi(z)\,dx\,dy\right| = \pi|(V_\varphi f)'(\infty)| \leq \pi\alpha(\Delta\backslash\mathring{X})\|V_\varphi f\|_\infty \leq Cw(\delta)\delta\|\nabla\varphi\|_\infty\alpha(\Delta\backslash X),$$

where in the last inequality we applied (4) and (iii). Hence f is in $R(X)$ by 2.1.1. □

Proof of (i) ⇒ (ii) in Theorem 2.1.1 Let $\|g\|_E$ stand for the sup norm of g on the set E. Take functions f_n, analytic on some neighbourhood (depending on n) of X, such that $\|f - f_n\|_X \to 0$ as $n \to \infty$. We will now show that the f_n can be chosen so that $\|f - f_n\|_{\mathbb{C}} \to 0$ as $n \to \infty$. There is a neighbourhood U_n of X, on which f_n is analytic and such that $\|f - f_n\|_{U_n} \leq 2\|f - f_n\|_X$.

Let d_n be a continuous extension of $f - f_n$ from U_n to $\mathbb{C}$ with $\|d_n\|_{\mathbb{C}} = \|f - f_n\|_{U_n}$. Redefine f_n outside U_n so that $f_n = f - d_n$. Thus each new f_n is analytic on U_n and $\|f - f_n\|_{\mathbb{C}} = \|f - f_n\|_{U_n} \to 0$ as $n \to \infty$.

Given an open disc of radius δ and $\varphi \in C_0^\infty(\Delta)$ we have

$$\left|\int f_n\bar\partial\varphi\,dx\,dy\right| = \pi|V_\varphi(f_n)| \leq \pi\alpha(\Delta\backslash X)\|V_\varphi f_n\|_\infty$$

and so, letting $n \to \infty$ and using (4),

$$\left|\int f\bar\partial\varphi\,dx\,dy\right| \leq \pi\alpha(\Delta\backslash X)\|V_\varphi f\|_\infty \leq Cw(\delta)\delta\|\nabla\varphi\|_\infty\gamma(\Delta\backslash X),$$

which is (ii). □

Proof of (i) ⇒ (ii) in Theorem 2.2.1 Given an open disc Δ and $\varepsilon > 0$, let f be continuous on $\mathbb{C}$, analytic outside a compact subset of $\Delta\backslash\mathring{X}$, satisfying $\|f\|_\infty \leq 1$ and $f'(\infty) > \alpha(\Delta\backslash\mathring{X}) - \varepsilon$. Choose f_n analytic on some neighbourhood of X such that $f_n \to f$ uniformly on X. Using the modification argument described above we can assume that f_n is continuous on $\mathbb{C}$, analytic on a neighbourhood of X and $f_n \to f$ uniformly on $\mathbb{C}$. Hence $\|f_n\|_\infty \to \|f\|_\infty$ and $f_n'(\infty) \to f'(\infty)$ as $n \to \infty$. Thus $\alpha(\Delta\backslash\mathring{X}) - \varepsilon \leq \alpha(\Delta\backslash X)$ for each $\varepsilon > 0$, which gives (ii). □

Chapter 3
Smooth approximation by rational functions, and the Beurling transform

3.0 Introduction

To gain some perspective on Vitushkin's Theorem, in this chapter we consider approximation problems by rational functions in norms which are stronger than the uniform norm, and in fact require some smoothness of the functions under consideration. We focus our attention on two representative cases, namely, Lip s, $0 < s < 1$, and C^1, and in the last section we briefly describe the results one gets for higher orders of smoothness.

As in the uniform case, the possibility of Lip s approximation, $0 < s < 1$, is controlled by an appropriate capacity, called s-analytic capacity. The novelty is that s-analytic capacity is comparable to $(1+s)$-dimensional Hausdorff content and then satisfies some (weak) density theorems, which allow specially satisfactory characterizing conditions for approximation.

In the C^1 context the situation changes because the degree of smoothness of the approximation coincides with the order of the operator. One gets necessary conditions for the vanishing of $\bar{\partial}$ of the approximable functions, which turn out to be also sufficient.

As far as technique is concerned, we exploit throughout the mapping properties of the Beurling transform to match the second coefficients in the expansion at ∞ of the localized functions. This idea, originally introduced by Lindberg [Li] in the L^p context, has turned out to be really fruitful when one is dealing with Beurling invariant spaces such as Lip s, $0 < s < 1$. Of course C^1 is not Beurling invariant, but Uy's Theorem described in section 1.5 allows us to control the bad behaviour of the Beurling transform on L^∞.

The results of section 3.1 are basically due to O'Farrell [OF1], who proved them without making use of the Beurling transform, and the results of section 3.2 are due to the author [Ve2].

3.1 Lip s approximation, $0 < s < 1$

We start with some definitions. A function f defined on some set $E \subset \mathbb{C}$ is said to be in Lip (s, E), $0 < s < 1$, if

$$\|f\|_{s,E} \equiv \sup \frac{|f(z) - f(w)|}{|z - w|^s} < \infty,$$

the sup being taken over $z, w \in E$, $z \neq w$. When $E = \mathbb{C}$ we write $\| \quad \|_s$ for $\| \quad \|_{s,\mathbb{C}}$.

The quantity

$$\|f\|'_{s,E} = |f(z_0)| + \|f\|_{s,E},$$

where z_0 is a fixed point in E, is a Banach space norm on Lip (s, E). When X is compact, the closure of $C^\infty(\mathbb{C})_{|X}$ in Lip (s, X) is precisely the subspace of Lip (s, X) consisting of those functions f satisfying $\omega(\delta)\delta^{-s} \to 0$ as $\delta \to 0$, $\omega(\delta)$ being the modulus of continuity of f. Denote this space by lip (s, X).

Let now $R_s(X)$ be the closure in Lip (s, X) of rational functions with poles off X. Since Runge's Theorem works in the norm of Lip s, as can be easily ascertained, any function analytic on some neighbourhood of X belongs to $R_s(X)$.

Clearly,

$$R_s(X) \subset A_s(X) \equiv \{f \in \operatorname{lip}(s, X) : f \text{ is analytic in } \mathring{X}\},$$

but, as in the uniform case, it is not difficult to produce examples where equality does not hold.

The following two results give complete answers to the individual and collective analytic approximation problems in the class Lip s.

3.1.1 Theorem *Let $f \in \operatorname{lip}(s, \mathbb{C})$ and $X \subset \mathbb{C}$ compact. Then the following are equivalent.*

(i) $f \in R_s(X)$.

(ii) *For some function $\varepsilon(\delta) \to 0$ as $\delta \to 0$,*

$$\left|\int f(z)\bar{\partial}\varphi(z)\, dx\, dy\right| \le \varepsilon(\delta)\delta\|\nabla\varphi\|_\infty M^{1+s}(\Delta\setminus X),$$

for all open discs Δ of radius δ and all $\varphi \in C_0^\infty(\Delta)$.

To state the collective theorem we need to introduce a variant of M^{1+s}. The *lower β-dimensional Hausdorff content* of $E \subset \mathbb{C}$ is defined by

$$M_*^\beta(E) = \sup M^h(E),$$

where the sup is taken over those h such that $h(t) \le t^\beta$ and $h(t)t^{-\beta} \to 0$ as $t \to 0$.

3.1.2 Theorem *Let $X \subset \mathbb{C}$ be compact and $0 < s < 1$. Then the following are equivalent.*

(i) $R_s(X) = A_s(X)$.

(ii) $M_*^{1+s}(\Delta\setminus\mathring{X}) \le CM^{1+s}(\Delta\setminus X)$, *for all open discs Δ.*

(iii) $\displaystyle\limsup_{r\to 0} \frac{M^{1+s}(\Delta(z,r)\setminus X)}{r^{1+s}} > 0$, *for M_*^{1+s}-almost all $z \in \partial X$.*

The above two statements are much more satisfactory than the corresponding results for uniform approximation. This stems from the fact that Hausdorff content is a metric quantity of a very simple nature. For example, M^β and M_*^β are subadditive and so $M_*^{1+s}(\partial X) = 0$ is sufficient to give $R_s(X) = A_s(X)$.

Let us define s-analytic capacity. Given a compact subset K of $\mathbb{C}$ set $\gamma_s(K) = \sup|f'(\infty)|$ where the sup is taken over the functions f which are analytic in K^c and $\|f\|_s \le 1$. If moreover we require that the functions f belong to lip $(s, \mathbb{C})$ then we get a different quantity $\alpha_s(K)$, which should be viewed as the analogue of $\alpha(K)$. For an arbitrary set E, $\gamma_s(E)$ (respectively $\alpha_s(E)$) is defined as the supremum of $\gamma_s(K)$ (respectively $\alpha_s(K)$) for all compact subsets K of E.

Using the Frostman Lemma and elementary estimates one obtains [OF1]:

3.1.3 Lemma *For some absolute constant C and all sets E,*

$$C^{-1}M^{1+s}(E) \le \gamma_s(E) \le CM^{1+s}(E)$$

and

$$C^{-1}M_*^{1+s}(E) \le \alpha_s(E) \le CM_*^{1+s}(E).$$

This proof of 3.1.3 and some additional simple considerations give

3.1.4 Lemma *For each disc Δ of radius δ and each $\varphi \in C_0^\infty(\Delta)$ one has*

$$\|V_\varphi f\|_s \le C\delta \|\nabla\varphi\|_\infty \|f\|_{s,\Delta}. \tag{1}$$

Recall that the Beurling transform is the singular integral operator

$$B(f)(z) = \lim_{\varepsilon\to 0} \frac{1}{\pi} \int_{|\zeta - z|>\varepsilon} f(\zeta)(z-\zeta)^{-2}\, dx\, dy, \tag{2}$$

defined for appropriate functions f (a different normalization was used in Chapter 1). It turns out [St, p.50] that

$$\|Bf\|_s \le C_s \|f\|_s, \tag{3}$$

where C_s depends only on s. It is also important to recall that

$$\frac{\partial}{\partial z} Cf = -Bf, \tag{4}$$

where

$$Cf(z) = \frac{1}{\pi} \int f(\zeta)(z-\zeta)^{-1}\, dx\, dy$$

is the Cauchy transform of f.

Our main technical device for the proof of 3.1.1 and 3.1.2 is the following double zero lemma for the class Lip s, first proved in [B-V].

3.1.5 Lemma *Let Δ_j be a sequence of open discs of radius δ_j such that for some $\lambda > 1$, the family $(\lambda\Delta_j)$ is almost disjoint. Let $f_j \in$ Lip $(s, \mathbb{C})$ be analytic outside Δ_j such that $f_j(z) = O(|z|^{-2})$ as $z \to \infty$. Then*

$$\|\sum_j f_j\|_s \le C \max_j \|f_j\|_s.$$

Proof It is a simple exercise to prove that the conclusion of the lemma is true when spt $f_j \subset \Delta_j$ for all j. The idea is to perform a reduction to this case.

Assume that for each j we are able to construct functions F_j and G_j such that

$$f_j = B(F_j) + G_j, \quad \|F_j\|_s,\ \|G_j\|_s \le C\|f_j\|_s$$

and spt F_j, spt $G_j \subset \lambda\Delta_j$. Then

$$\left\|\sum f_j\right\|_s \le \|B(\sum_j F_j)\|_s + \left\|\sum G_j\right\|_s \le C\left\|\sum F_j\right\|_s + \left\|\sum G_j\right\|_s \le C \max\|f_j\|_s,$$

where we applied (3) in the second inequality, and the opening remark in the last inequality.

To show the existence of F_j and G_j, fix j, set $\Delta = \Delta_j$, $\delta = \delta_j$ and assume that Δ is centered at the origin. Consider the expansion $f(z) = \sum_{n=2}^{\infty} a_n z^{-n}$, $|z| > \delta$. We have

$$|a_n| = \left|\frac{1}{2\pi i}\int_{|z|=\delta}(f(z)-f(0))z^{n-1}\,dz\right| \le \|f\|_s\delta^{n+s}.$$

Set $\lambda_1 = 1+\frac{1}{3}(\lambda-1)$ and $\lambda_2 = 1+\frac{2}{3}(\lambda-1)$. Take $\varphi \in C^\infty(\mathbb{C})$ such that $\varphi(z) = 1$, $|z| > \lambda_2\delta$, $\varphi(z) = 0$, $|z| < \lambda_1\delta$ and $|\partial^\beta\varphi| \le C\delta^{-|\beta|}$, $0 \le |\beta| \le 2$. Clearly

$$\frac{\varphi(z)}{z^{n-1}} = C\left(\frac{\bar\partial\varphi(z)}{z^{n-1}}\right)$$

for $n \ge 2$, and so

$$\partial\left(\frac{\varphi(z)}{z^{n-1}}\right) = -B\left(\frac{\bar\partial\varphi(z)}{z^{n-1}}\right)$$

by (4). Therefore

$$\frac{1}{z^n} = B\left(\frac{\bar\partial\varphi(z)}{(n-1)z^{n-1}}\right), \quad |z| > \lambda_2\delta, \quad n \ge 2.$$

Define

$$F(z) = \sum_{n\ge2} a_n\frac{\bar\partial\varphi(z)}{(n-1)z^{n-1}}.$$

To show the absolute convergence of the above series in Lip $(s, \mathbb{C})$ we need a uniform estimate of the gradient of the n-th term. We easily get

$$\left|\nabla\left(\frac{\bar\partial\varphi(z)}{z^{n-1}}\right)\right| \le C\delta^{-2}(\lambda_1\delta)^{-(n-1)} = C\lambda_1^{-(n-1)}\delta^{-(n+1)},$$

and so

$$\left\|\frac{\bar\partial\varphi(z)}{z^{n-1}}\right\|_s \le C\lambda_1^{-(n-1)}\delta^{-(n+s)}.$$

Since $\lambda_1 > 1$ we obtain

$$\|F\|_s \le \sum_{n=2}^{\infty} C\|f\|_s\delta^{n+s}\frac{\lambda_1^{-(n-1)}}{n-1}\delta^{-(n+s)} = C\|f\|_s.$$

Set $G = f - B(F)$, so that $\|G\|_s \le C\|f\|_s$ by (3). Since for $|z| > \lambda_2\delta$,

$$f(z) = \sum_{n=2}^{\infty} a_n z^{-n} = \sum_{n=2}^{\infty} a_n B\left(\frac{\bar\partial\varphi(z)}{z^{n-1}}\right) = B(F),$$

we see that $\operatorname{spt} G \subset \lambda\Delta$. □

Proof of (ii) ⇒ (i) in Theorem 3.1.1 Let f be a function in Lip$(s, \mathbb{C})$ which is analytic on $\mathring{X}$. Without loss of generality f can be assumed to be compactly supported. Let $(\Delta_j, \varphi_j, f_j)$ be a δ-Vitushkin scheme for f. Thus $f_j = V_{\varphi_j}(f)$ and, by (1), $\|f_j\|_s \le C\eta(\delta)$, with $\eta(\delta) \to 0$ as $\delta \to 0$. Let z_j be the center of Δ_j and consider the expansion

$$f_j(z) = \frac{a_1}{z - z_j} + \frac{a_2}{(z - z_j)^2} + \cdots, \quad |z - z_j| > \delta.$$

Clearly (ii) gives

$$|a_1| = \left|\frac{1}{\pi}\int f(z)\bar{\partial}\varphi_j(z)\,dx\,dy\right| \le C\varepsilon(\delta)M^{1+s}(\Delta_j\backslash X)$$

and hence, in view of 3.1.3,

$$|a_1| \le C\varepsilon(\delta)\gamma_s(\Delta_j\backslash X).$$

Let now h be analytic outside a compact subset of $\Delta_j\backslash X$, satisfying $\|h\|_s \le 1$ and $2h'(\infty) \ge \gamma_s(\Delta_j\backslash X)$. Set $g_j = a_1(h/h'(\infty))$.

Then $f_j - g_j = O(|z|^{-2})$, as $z \to \infty$, $\|g_j\|_s \le C\varepsilon(\delta)$ and g_j is analytic outside a compact subset of $\Delta_j\backslash X$. Applying 3.1.5 we obtain

$$\|f - \sum_j g_j\|_s \le C(\eta(\delta) + \varepsilon(\delta)),$$

which completes the proof. □

The implications (i) ⇒ (ii) in 3.1.1 and (i) ⇒ (ii) in 3.1.2 can be proved following the pattern described in Chapter 2. The fact that (iii) follows from (ii) in 3.1.2 is just the density theorem for M_*^{1+s} (see [M-O, p.728]). The first proof that (iii) ⇒ (ii) was given in [M-O], in which a slightly different problem is considered. In [Ma-O] a simple, purely measure theoretic proof of (iii) ⇒ (ii) is presented.

3.2 C^1 approximation

Let f be a continuously differentiable function on $\mathbb{C}$ and let $X \subset \mathbb{C}$ be compact. We ask under what conditions there exists a sequence of rational functions r_n without poles on X with the property that

$$r_n \longrightarrow f \text{ and } \nabla r_n \longrightarrow \nabla f, \text{ uniformly on } X.$$

A necessary condition is obviously that $\bar{\partial} f(z) = 0$, $z \in X$. The content of the main result of this section is that the above condition is also sufficient [Ve2].

3.2.1 Theorem *Let $f \in C^1(\mathbb{C})$ and let $X \subset \mathbb{C}$ be compact. Then the following are equivalent.*

(i) *Given* $\varepsilon > 0$ *there exists a rational function* r *without poles on* X *such that for all* $z \in X$

$$|f(z) - r(z)| < \varepsilon \text{ and } |\nabla f(z) - \nabla r(z)| < \varepsilon.$$

(ii) $\bar{\partial} f(z) = 0$, $z \in X$.

Proof We only need to prove that (ii) $\Rightarrow$ (i), and to this end we can assume, without loss of generality, that f is compactly supported. Given $\delta > 0$, let $(\Delta_j, \varphi_j, f_j)$ be a δ-Vitushkin scheme for the approximation of f. If $\Delta_j \cap X \neq \emptyset$, hypothesis (ii) yields

$$\|\bar{\partial} f_j\|_\infty = \|\varphi_j \bar{\partial} f\|_\infty \le C\omega(\nabla f, \delta),$$

where $\omega(\nabla f, \delta)$ is the modulus of continuity of ∇f at level δ. Now we would like to get a similar estimate for ∂f_j. Since f_j is given by (3) of the previous chapter with φ replaced by φ_j, differentiation under the integral sign gives ($\zeta = x + iy$)

$$\partial f_j(z) = \frac{1}{\pi} \int \frac{f(\zeta) - f(z) - \partial f(z)(\zeta - z)}{(\zeta - z)^2} \bar{\partial} \varphi_j(\zeta)\, dx\, dy. \tag{5}$$

Applying Taylor's formula we get

$$f(\zeta) - f(z) - \partial f(z)(\zeta - z) = \bar{\partial} f(z)(\zeta - z) + O(\omega(\nabla f, \delta)(\zeta - z)), \quad \zeta, z \in \Delta_j,$$

which, inserted in (5), yields

$$|\partial f_j(z)| \le C\omega(\nabla f, \delta), \quad z \in \Delta_j.$$

Hence, by the maximum principle,

$$\|\partial f_j\|_\infty \le C\omega(\nabla f, \delta).$$

We have thus shown that f_j is small in the C^1 norm.

As $z \to \infty$, $f_j(z) = \frac{a_1}{z} + \ldots$, where

$$a_1 = \frac{1}{\pi} \int \varphi_j(\zeta) \bar{\partial} f(\zeta)\, dx\, dy.$$

Thus, using (ii) again,

$$|a_1| \le C\omega(\nabla f, \delta) \text{area}(\Delta_j \setminus X),$$

provided $\Delta_j \cap X \neq \emptyset$. Take a compact $K \subset \Delta_j \setminus X$ such that $\text{area}(\Delta_j \setminus X) \le 2\,\text{area}(K)$. According to Uy's Theorem (see section 1.5) there exists a function g in $\text{Lip}(1, \mathbb{C})$ such that $\|\nabla g\|_\infty \le 1$, g is analytic outside K and $\text{area}(K) \le C|g'(\infty)|$. Set $g_j = a_1(g/g'(\infty))$. Then g_j is analytic outside K, $\|\nabla g_j\|_\infty \le C\omega(\nabla f, \delta)$, $f_j - g_j = O(|z|^{-2})$ as $z \to \infty$ and $\partial f_j - \partial g_j = O(|z|^{-3})$ as $z \to \infty$. Set

$$g = \sum{}' g_j + \sum{}'' f_j,$$

where $\sum'$ denotes summation over those indices j such that $\Delta_j \cap X \neq \emptyset$ and $\sum''$ summation over the remaining indices. Clearly g is analytic on a neighbourhood of X. Since $\bar\partial f_j$ and $\bar\partial g_j$ are supported in Δ_j,

$$\|\bar\partial f - \bar\partial g\|_\infty = \left\|\sum{}' \bar\partial f_j - \bar\partial g_j\right\|_\infty \leq C\omega(\nabla f, \delta).$$

On the other hand, by the triple zero lemma 2.2.2 we have

$$\|\partial f - \partial g\|_\infty = \left\|\sum{}' \partial f_j - \partial g_j\right\|_\infty \leq C\omega(\nabla f, \delta).$$

Noticing that

$$|f_j(z) - g_j(z)| \leq C\omega(\nabla f, \delta)\delta, \quad z \in \mathbb{C},$$

and using the double zero lemma (of Chapter 2) we obtain

$$\|f - g\|_\infty = \left\|\sum{}' f_j - g_j\right\|_\infty \leq C\omega(\nabla f, \delta)\delta \log \frac{1}{\delta},$$

provided $\delta \leq 1/2$.

Finally, to get an approximating rational function from g it is enough to apply (the C^1 version of) Runge's Theorem. □

It is a well established fact that the kind of convergence involved in condition (i) of the above theorem, that is, uniform convergence of the function and the first order derivatives on X, is not always the notion of convergence associated to the norm of some Banach space. On the other hand, there are at least two Banach space versions of C^1 on X for which it is interesting to look at the rational approximation problem. The first is the "jet" version

$$C^1_{\text{jet}}(X) = C^1(\mathbb{C})/J(X),$$

endowed with the quotient norm, where

$$J(X) = \{f \in C^1(\mathbb{C}) : \partial^\alpha f(z) = 0, \quad z \in X, \quad 0 \leq |\alpha| \leq 1\}.$$

Of course, $C^1(\mathbb{C})$ is equipped with its natural Banach space structure given by the norm $\|f\| = \sum_{|\alpha|\leq 1} \|\partial^\alpha f\|_\infty$.

The Whitney Extension Theorem identifies $C^1_{\text{jet}}(X)$ with a space of jets. A C^1-jet is a triple (f_0, f_1, f_2) of continuous functions on X such that

$$f_0(w) - f_0(z) - f_1(z)(w - z) - f_2(z)(\bar w - \bar z) = o(|w - z|),$$

the small o being uniform in $z, w \in X$.

Clearly, each $f \in C^1(\mathbb{C})$ gives rise to a jet, namely $(f_{|X}, \partial f_{|X}, \bar\partial f_{|X})$. Notice that different jets can have the same f_0. For example, if $X = [0, 1]$ then the jets associated to the functions 0 and y are $(0,0,0)$ and $\left(0, \frac{1}{2i}, -\frac{1}{2i}\right)$ respectively. The least constant C such that

$$|f_j(z)| \leq C, \quad z \in X, \quad 0 \leq j \leq 2$$

and

$$|f_0(w) - f_0(z) - f_1(z)(w - z) - f_2(z)(\bar{w} - \bar{z}) \leq C|w - z|, \; w, z \in X,$$

is a Banach space norm on the set of C^1-jets.

The Whitney Extension Theorem asserts, in our context, that given a C^1-jet (f_0, f_1, f_2) there exists a function f in $C^1(\mathbb{C})$ such that $f_0 = f$, $f_1 = \partial f$ and $f_2 = \bar{\partial} f$ on X. This shows that the linear mapping from $C^1_{\text{jet}}(X)$ into the space of C^1-jets given by $\tilde{f} \to (f_{|X}, \partial f_{|X}, \bar{\partial} f_{|X})$ is a Banach space isomorphism.

Let $R^1_{\text{jet}}(X)$ be the closure in $C^1_{\text{jet}}(X)$ of the set of $\widetilde{\varphi r}$ where r is a rational function, $\varphi \in C_0^\infty(\mathbb{C})$ takes the value 1 on a neighbourhood of X, and no pole of r belongs to the support of φ. The proof of 3.2.1 gives the following result, which solves the problem of C^1_{jet} approximation by rational functions.

3.2.2 Theorem *For each compact $X \subset \mathbb{C}$ one has*

$$R^1_{\text{jet}}(X) = \{\tilde{f} \in C^1_{\text{jet}}(X) : \bar{\partial} f(z) = 0, \quad z \in X\}.$$

Therefore, $R^1_{\text{jet}}(X) = C^1_{\text{jet}}(X)$ if and only if X is the closure of its interior.

There is a second Banach space version of C^1 on X, namely, the "function" version

$$C^1(X) = C^1(\mathbb{C})/I(X),$$

endowed with the quotient norm, where

$$I(X) = \{f \in C^1(\mathbb{C}) : f(z) = 0, \quad z \in X\}.$$

It is obvious that $C^1(X)$ can be identified with the set of restrictions to X of functions in $C^1(\mathbb{C})$. Define $R^1(X)$ as the closure in $C^1(X)$ of the set of restrictions to X of rational functions with poles off X.

Notice that it does not make sense to ask whether $\bar{\partial} f(z) = 0$, $z \in X$, for $f \in R^1(X)$. The reason is that, for some X, there is no way of defining pointwise some first order derivative of a function $f \in C^1(X)$. In other words, it can happen that $f = 0$ on X but $\nabla f(z) \neq 0$, $z \in X$. The typical example is given by the function $f(x + iy) = y$ and the set $X = [0, 1]$.

The key idea to understand $R^1(X)$ was introduced by O'Farrell in [OF2], where he showed that $R^1(X) = C^1(X)$ if and only if X is a subset of a finite union of simple C^1 curves (i.e., Jordan C^1 arcs or Jordan closed C^1 curves). O'Farrell introduced, as a tool in proving the above result, a special notion of tangent vector to a set at a point. Given $a \in X$ and a complex number v with $|v| = 1$, we say that $v \in \text{Tan}\,(X, a)$ if there exist two sequences (z_n) and (w_n) tending to a, with $z_n \neq w_n$ for all n, and such that $(z_n - w_n)/|z_n - w_n| \to v$ as $n \to \infty$. For instance, if $X = [0, 1] \cup \{(x, y) : 0 \leq x \leq 1 \text{ and } y = x^2\}$, then 1 and i belong to $\text{Tan}\,(X, 0)$. Define X^* to be the set of points $z \in X$ such that the span of $\text{Tan}\,(X, z)$ is $\mathbb{C}$. The relevance of the set X^* is due to the fact that if $z \in X^*$ then $\bar{\partial} f(z)$ is determined by the values of f on X, and defines a continuous linear functional on $C^1(X)$. Another important property of X^* is that area $(X \backslash X^*) = 0$. This follows from the fact that almost all points of a measurable subset of the plane are points of density in the directions of the coordinate axes [S, p.208].

3.2.3 Theorem *For all compact $X \subset \mathbb{C}$*

$$R^1(X) = \{f \in C^1(X) : \bar{\partial} f(z) = 0, \quad z \in X^*\}.$$

Proof That the left hand side is included in the right is a consequence of the preceding discussion. To prove the reverse inclusion take $f \in C_0^1(\mathbb{C})$ and assume that $\bar{\partial} f(z) = 0$, $z \in X^*$. Given $\delta > 0$ let $(\Delta_j, \varphi_j, f_j)$ be a δ-Vitushkin scheme for f. We write $j \in I$ if $2\Delta_j \cap X^* \neq \emptyset$. Otherwise we write that $j \in II$.

The proof of the main result of [OF2] shows that if $j \in II$ then $\bar{\Delta}_j \cap X$ is contained in a finite union of simple C^1 curves. Then $f_j \in R^1(\bar{\Delta}_j \cap X)$, again from [OF2]. Since f_j is analytic outside Δ_j, for each $z \in X$ there is a neighbourhood U of z such that $f_j \in R^1(\bar{U} \cap X)$. But this implies that $f_j \in R^1(X)$ because $R^1(X)$ is local [OF3].

If $j \in I$ then, since $\text{area}(X \backslash X^*) = 0$, we can apply the argument used in the proof of 3.2.1 to construct a $g_j \in C^1(\mathbb{C})$, analytic outside a compact subset of $\Delta_j \backslash X$, such that $f_j - g_j = O(|z|^{-2})$ and

$$\|\nabla g_j\|_\infty \leq C\|\nabla f_j\|_\infty \leq Cw(\nabla f, \delta).$$

Therefore

$$\|f - \sum_{j\in I} g_j - \sum_{j \in II} f_j\|_{C^1(\mathbb{C})} \to 0 \text{ as } \delta \to 0.$$

□

3.3 C^s approximation, $s > 1$

Given a positive real number s, choose $m \in \mathbb{Z}$ and σ, $0 \leq \sigma < 1$, such that $s = m + \sigma$. We say that $f \in C^s(\mathbb{C})$ if f has continuous bounded derivatives up to order m and, if $\sigma \neq 0$, $\partial^\alpha f \in \text{lip}\,(\sigma, \mathbb{C})$ for $|\alpha| = m$.

3.3.1 Theorem *Let $X \subset \mathbb{C}$ be compact and let s be a real number of the form $s = m + \sigma$, where m is a positive integer and $0 \leq \sigma < 1$. Then, given $f \in C^s(\mathbb{C})$, the following are equivalent.*

(i) *There exists a sequence (r_n) of rational functions with poles off X such that*

$$\partial^\alpha r_n \to \partial^\alpha f \text{ uniformly on } X, \quad |\alpha| \leq m,$$

and, if $\sigma \neq 0$,

$$\partial^\alpha r_n \to \partial^\alpha f \text{ in Lip}\,(\sigma, X), \quad |\alpha| = m.$$

(ii) *$\bar{\partial} f$ vanishes up to order $m-1$ on X, i.e.,*

$$\partial^\alpha(\bar{\partial} f)(z) = 0, \quad z \in X, \quad |\alpha| \leq m - 1.$$

For non-integer s larger than 1 the result was proven in [OF4] and for integer s in [Ve2].

The jet version of 3.3.1, that is, the higher order analog of 3.2.2 is still true. However, the analog of 3.2.3 is not known for $s \geq 2$. The main difficulty seems to be finding the right analog for X^*.

We present the proof of 3.3.1 only for $s = 1+\sigma$, $0 < \sigma < 1$. The argument can be adapted to any non-integer $s > 1$. For integer $s \geq 2$ the reader is referred to [Ve2].

Proof of (ii) $\Rightarrow$ (i) **(for** $s = 1+\sigma$, $0<\sigma<1$**)** Let $f \in C^s(\mathbb{C})$ satisfy $\bar\partial f(z) = 0$, $z \in X$. For $n = 1, 2, \ldots$ set

$$U_n = \{z \in \mathbb{C} : \operatorname{dist}(z, X) < n^{-1}\}.$$

Take $\varphi_n \in C_0^\infty(U_n)$, $\varphi_n = 1$ on U_{n+1}, $0 \leq \varphi_n \leq 1$ and $|\nabla \varphi_n| \leq cn$. For $z \in X$ we have $f(z) = f_n(z) + R_n(z)$, where

$$\pi f_n(z) = \int \frac{f(\zeta)\bar\partial\varphi_n(\zeta)}{z-\zeta}\, dx\, dy$$

and

$$\pi R_n(z) = \int \frac{\varphi_n(\zeta)\bar\partial f(\zeta)}{z-\zeta}\, dx\, dy.$$

Since f_n is analytic on U_{n+1} it is enough to show that $\nabla R_n \to 0$ in $\operatorname{Lip}(\sigma, X)$.

Clearly

$$\bar\partial R_n = \varphi_n \bar\partial f$$

and

$$\partial R_n = -B(\bar\partial R_n),$$

B being the Beurling transform. Thus

$$\|\nabla R_n\|_\sigma \leq C\|\varphi_n\bar\partial f\|_\sigma,$$

because B preserves $\operatorname{Lip}(\sigma, \mathbb{C})$. It is now very simple to show that

$$\|\varphi_n\bar\partial f\|_\sigma \to 0 \text{ as } n \to \infty$$

using the hypothesis (ii). □

Chapter 4
L^p approximation by rational functions and spectral synthesis

4.0 Introduction

Let $X \subset \mathbb{C}$ be compact and let $R^p(X)$, $1 < p < \infty$, be the closure in $L^p(X)$ $(= L^p(X, dx\, dy))$ of the set of rational functions with poles off X. Clearly, $R^p(X)$ is always contained in $A^p(X)$, which is defined as the set of functions in $L^p(X)$ that are analytic on $\mathring{X}$. We wish to describe those X for which $R^p(X) = A^p(X)$.

Following Lindberg [Li], one can combine Vitushkin's constructive technique with mapping properties of the Beurling transform, to build a proof of the main result which parallels that described in Chapter 3 in dealing with Lipschitz classes. We prefer, however, to present a new general approach to the problem which has been successfully applied to approximation by solutions of quite general elliptic equations in different norms. The idea, which goes back to Havin [H] and has been exploited by many other people (e.g. Bagby [B1], Hedberg [He2], Polking [Pl] and Gauthier and Tarkhanov [G-T]), consists in using duality to reduce matters to an approximation problem of a real analytic nature called spectral synthesis; see also Mateu and Verdera [M-V] and Mateu and Orobitg [M-O] where one goes the other way. The terminology stems from its resemblance with the classical spectral synthesis problems in group algebras. It turns out that the kind of spectral synthesis problem we are led to, can be solved rather easily. This produces a compact clever proof of our main theorem.

In section 4.1 we deal with approximation in L^p, $1 < p < 2$. The restriction on the range of p makes the problem much simpler from the technical point of view. For instance, no capacity is involved in the range under consideration, and thus the corresponding spectral synthesis problem admits a particularly clear formulation and treatment.

In section 4.2 we deal with the range $2 \le p < \infty$. Here a capacity associated to the index p must be considered, and the main result involves capacitary conditions like those in Vitushkin's or O'Farrell's Theorems in Chapters 2 and 3. The proof we present, taken from [He2], is basically that of section 4.1, complicated by the technical difficulties introduced by the presence of capacity.

4.1 The range $1 < p < 2$

Here is the main result of this section.

4.1.1 Theorem *For $1 < p < 2$ and any compact X, we have $R^p(X) = A^p(X)$.*

Proof Let $q = p(p-1)^{-1}$ be the exponent dual to p. Let $h \in L^q(X)$ satisfy $\int fh\,dx\,dy = 0$, $f \in R^p(X)$. We must show that h annihilates $A^p(X)$. Set $\hat{h}(z) = \frac{1}{\pi}\int \frac{h(\zeta)}{z-\zeta}\,dx\,dy$, so that $\hat{h}(z) = 0$, $z \notin X$, because of the orthogonality assumption on h. Notice that $\hat{h}$ is continuous, as convolution of the compactly supported L^q function h with the locally L^p integrable function $1/\pi z$. Therefore $\hat{h}(z) = 0$, $z \notin (\mathring{X})^c$.

On the other hand, we have $\bar{\partial}\hat{h} = h$ and $\partial\hat{h} = -B(h)$, B being the Beurling transform; see (2) of Chapter 3. Since B preserves $L^q(\mathbb{C})$ [St], we conclude that $\hat{h}$ belongs to the Sobolev space W_1^q, which is defined as the set of $L^q(\mathbb{C})$ functions whose first order partial derivatives, in the sense of distributions, are again $L^q(\mathbb{C})$ functions. Later on we will prove the spectral synthesis theorem for W_1^q, $q > 2$ (cf. 4.1.2 below), which tells us that, since $\hat{h}$ vanishes on the closed set $(\mathring{X})^c$, there are $\varphi_j \in C_0^\infty(\mathring{X})$ such that $\varphi_j \to \hat{h}$ in W_1^q. Hence $\bar{\partial}\varphi_j \to h$ in L^q and so, if $f \in A^p(X)$,

$$\int fh\,dx\,dy = \lim_{j\to\infty}\int f\bar{\partial}\varphi_j\,dx\,dy.$$

Now we only have to notice that the integrals in the right hand side of the last identity vanish, because f is analytic on $\mathring{X}$ and φ_j is supported there. □

By the Sobolev Imbedding Theorem [St, p.124], functions in W_1^q, $q > 2$, are continuous on $\mathbb{C}$, and consequently the vanishing condition in the statement below is meaningful.

4.1.2 Theorem (Spectral synthesis in W_1^q, $q > 2$) *Let $F \subset \mathbb{C}$ be closed and let $f \in W_1^q$, $q > 2$, satisfy $f(z) = 0$, $z \in F$. Then there exist $\varphi_j \in C_0^\infty(F^c)$ such that $\varphi_j \to f$ in W_1^q.*

We need a lemma.

4.1.3 Lemma *Let $f \in W_1^q$, $1 \leq q \leq \infty$, and set $f^+ = \max(f, 0)$. Then $f^+ \in W_1^q$ and $\|\nabla f^+\|_q \leq \|\nabla f\|_q$.*

Proof The argument is simple and depends on the following two easily proven facts.

First, if F is an absolutely continuous function on the real line then F^+ is also absolutely continuous (because $|a^+ - b^+| \leq |a - b|$, $a, b \in \mathbb{R}$) and $(F^+)'(t) = F'(t)\chi_{\{F>0\}}(t)$ a.e. on the line.

The second fact is that a function f in $L^1_{\text{loc}}(\mathbb{R}^n)$ has partial derivatives (in the sense of distributions) which are in $L^1_{\text{loc}}(\mathbb{R}^n)$ if and only if f coincides a.e. with some function which is absolutely continuous on almost all lines parallel to the coordinate axis and whose partial derivatives, computed in the ordinary sense, are in $L^1_{\text{loc}}(\mathbb{R}^n)$.

Combining the above two statements one gets $\nabla f^+(x) = \nabla f(x)\chi_{\{f>0\}}(x)$ a.e., from which the conclusions of the lemma follow. □

Proof of Theorem 4.1.2 Replacing f by f^+ and f^- we can assume that $f \geq 0$. Set $f_n = \left(f - \frac{1}{n}\right)^+$. Then f_n vanishes on $\{x : f(x) < 1/n\}$ which is an open set containing F. Clearly $f_n \to f$ in L^q. Since $\nabla(f - f_n) = \nabla f \chi_{\{f \leq 1/n\}}$,

$$\|\nabla(f - f_n)\|_q^q \to \int_{\{f=0\}} |\nabla f|^q \text{ as } n \to \infty.$$

Now, almost all points of a measurable subset of $\mathbb{R}^n$ are points of density in the directions of the coordinate axes, and so $\nabla f(x) = 0$ a.e. on the set $\{f = 0\}$. Thus $f_n \to f$ in W_1^q. Regularization and truncation show that in fact one can take $f_n \in C_0^\infty(E^c)$. □

4.2 The range $2 \leq p < \infty$

The key fact explaining why the range under consideration is more involved than the range $1 < p < 2$ is that $1/z \in L^p_{\text{loc}}$ if and only if $1 \leq p < 2$. As a consequence, for $2 \leq p < \infty$ there is a non-trivial capacity associated to the operator $\bar{\partial}$ and L^p, which is defined for $2 < p < \infty$ by

$$\gamma_p(E) = \sup |f'(\infty)|,$$

the supremum being taken over those functions f which are analytic outside a compact subset of E and satisfy $\|f\|_p^p = \int_{\mathbb{C}} |f(z)|^p \, dx\, dy \leq 1$. For $p = 2$ the above definition is not useful because $1/z$ is not in L^2 at ∞, and this would force γ_2 to vanish identically. One has to replace the normalization condition by $\|f\|_{2,D}^2 \equiv \int_D |f(z)|^2 \, dx\, dy \leq 1$, where D is a fixed disc containing E. If we adapt to a p, $1 < p < 2$, the definition just given for $p = 2$ we easily get, using that $1/z \in L^p(D)$,

$$0 < c < \gamma_p(E) \leq (2\pi \mathrm{dist}\,(E, B^c))^{-1} \mathrm{area}(D)^{p^{-1}(p-1)},$$

for all subsets E of D, and for some constant c depending only on D. In other words, γ_p is trivial for $1 < p < 2$.

There is another capacity relevant for our purposes, which turns out to be a dual version of γ_p and is very closely related to the Sobolev space W_1^q, where $q = p(p-1)^{-1}$ is the exponent dual to p. It is defined, for $2 < q < \infty$, by

$$C_q(E) = \inf \|f\|_q^q \tag{1}$$

where the infimum is taken over those non-negative L^q functions f such that $|z|^{-1} * f \geq 1$ on E. For $q = 2$, $\|f\|_2$ must be replaced in (1) by $\|f\|_{2,D}$, as before.

In [He1] and [He2] one can find a proof of the following lemma.

4.2.1 Lemma *For all $1 < q < 2$ (respectively $q = 2$) there exists a positive constant C such that*

$$C^{-1} C_q(E)^{1/q} \leq \gamma_p(E) \leq C C_q(E)^{1/q},$$

for all sets E (respectively, for all $E \subset D$).

As we said before, by the Sobolev Imbedding Theorem, functions in W_1^q are continuous if $q > 2$, but this is not the case for $1 < q \leq 2$, as simple examples show. The capacity C_q provides the right tool to describe the sharp continuity properties of functions in W_1^q, $1 < q \leq 2$, which is the task we plan to undertake next. It is more convenient to work with the space of potentials $|z|^{-1} * L^q = \{|z|^{-1} * g : g \in L^q\}$, which is larger than W_1^q as the following lemma shows.

4.2.2 Lemma $W_1^q \subset |z|^{-1} * L^q$, $1 < q < \infty$.

Proof Let $f \in W_1^q$, and assume that we already have $f = |z|^{-1} * g$ for some g in L^q. Taking $\bar\partial$ of both sides we have

$$\bar\partial f = \bar\partial(|z|^{-1}) * g = -\frac{1}{2}\ \mathrm{P.V.}\ (z/|z|^3) * g = R(g),$$

where R is the Calderón-Zygmund operator with kernel $-\frac{1}{2}(z/|z|^3)$. It turns out that R is invertible and the inverse is again a Calderón-Zygmund operator, say T, with kernel of the form $c(\bar z/|z|^3)$ for some appropriate numerical constant c. This can be checked easily by examining the Fourier transforms of the kernels. Therefore $g = T(\bar\partial f)$. If we define g as $T(\bar\partial f)$ when $f \in W_1^q$ is given, then a simple Fourier transform computation shows that $f = |z|^{-1} * g$. Since T preserves L^q, $1 < q < \infty$, g is in L^q and so the lemma is proved. □

Recall that the Hardy-Littlewood maximal operator of a function $f \in L^1_{\text{loc}}(\mathbb{C})$ is defined by

$$M(f)(z) = \sup_{\delta>0} \frac{1}{\pi\delta^2} \int_{\Delta(z,\delta)} |f(z)|\, dx\, dy.$$

A simple computation gives

$$M(|z|^{-1})(z) \le C|z|^{-1}, \quad z \in \mathbb{C}. \tag{2}$$

Write $P(f)$ for the newtonian potential $|z|^{-1} * f$ of a function f in L^q. From (2) we obtain

$$M(Pf) \le M(P(|f|)) \le CP(|f|),$$

and so

$$C_q\{z : M(Pf)(z) > t\} \le Ct^{-q}\|f\|_q^q,$$

which together with the density of smooth functions in L^q yields

$$Pf(z) = \lim_{\delta\to 0} \frac{1}{\pi\delta^2} \int_{\Delta(z,\delta)} P(f)(z)\, dx\, dy,\ C_q\text{-a.e.},$$

where by C_q-a.e. we understand "except for a set of zero C_q-capacity". The conclusion is that the Lebesgue differentiation theorem can be improved in the class W_1^q, in the sense that the exceptional set is, generically speaking, smaller than in the L^q class (because $C_q(E) = 0$ implies area$(E) = 0$).

Write $P_nf(z)$ for the mean of Pf on the disc of center z and radius $1/n$. Then $P_nf \to Pf$ C_q-a.e. Imitating the proof of Egorov's Theorem we can show that, given $\varepsilon > 0$, there exists an open set G with $C_q(G) < \varepsilon$ such that $P_nf \to Pf$ uniformly on the closed set G^c. Since P_nf is continuous for each n, we have proved the following statement [My], [H-M].

4.2.3 Theorem *Given $f \in L^q$ and $\varepsilon > 0$ there exists an open set G with $C_q(G) < \varepsilon$ such that the restriction of Pf to G^c is continuous on G^c.*

A function satisfying the conclusion of 4.2.3 is called C_q-quasicontinuous. In particular, each function in W_1^q is (a.e. equal to) a C_q-quasicontinuous function.

The notion of C_q-quasiopen set is naturally associated to that of C_q-quasicontinuity. A set U is said to be C_q-quasiopen if, given $\varepsilon > 0$, there exists an open set G with $C_q(G) < \varepsilon$ such that $U \cap G^c$ is open in G^c. For example, if f is C_q-quasicontinuous then $\{f \ne 0\}$ is C_q-quasiopen.

We are now ready to prove the spectral synthesis theorem in W_1^q, $1 < q \le 2$.

4.2.4 Theorem *Let $F \subset \mathbb{C}$ be closed and let $f \in W_1^q$ be C_q-quasicontinuous. Then the following are equivalent.*

(i) *$f(z) = 0$, C_q-a.e. on F.*

(ii) *There exists $\varphi_n \in C_0^\infty(F^c)$ such that $\varphi_n \to f$ in W_1^q.*

Remark Given f in W_1^q, we know that f can be modified on a set of zero area so that the new function $\tilde{f}$ is C_q-quasicontinuous. Moreover, it can be shown that this C_q-quasicontinuous representative is unique in the sense that if f and g are C_q-quasicontinuous and $f = g$ area-a.e., then $f = g$ C_q-a.e. [H-M]. Therefore the C_q-quasicontinuity assumption in the statement of 4.2.4 is not actually a restriction.

Proof of Theorem 4.2.4 (ii) $\Rightarrow$ (i). Since

$$C_q\{z : |\varphi_n(z) - f(z)| > \varepsilon\} \leq C\varepsilon^{-q}\|\nabla\varphi_n - \nabla f\|_q^q$$

we see that for each $\varepsilon > 0$

$$C_q\{z : |\varphi_n(z) - f(z)| > \varepsilon\} \to 0 \text{ as } n \to \infty.$$

Thus, standard reasoning tells us that, passing to a subsequence if necessary, we have $\varphi_n \to f$ C_q-a.e., which gives (i).

(i) $\Rightarrow$ (ii). Using 4.1.3 we can assume that $f \geq 0$. Replacing f by $\min(f, n)$, $n = 1, 2, \ldots$, and applying 4.1.3 again we can further suppose that f is bounded. Set $U_n = \{z : f(z) < 1/n\}$ and $f_n = \left(f - \frac{1}{n}\right)^+$. Then f_n vanishes on U_n, and $f_n \to f$ in W_1^q (because of the proof of 4.1.2), but now U_n is a C_q-quasiopen set, not necessarily open. Let G_n be an open set such that $C_q(G_n) < 1/n$ and $U_n \cap G_n^c$ is open in G_n^c. Then $U_n \cap G_n^c = V_n \cap G_n^c$, for some open set V_n. Set $\Omega_n = G_n \cup U_n$. Since $\Omega_n = G_n \cup V_n$, Ω_n is open, and clearly $\Omega_n \supset F$. Let $g_n \in L^q$, $g_n \geq 0$, such that $P_n \equiv |z|^{-1} * g_n \geq 1$ on G_n and $\|g_n\|_q^q < 1/n$. Set $\omega_n = \min(P_n, 1)$. Then $\omega_n = 1$ on G_n and $\nabla\omega_n = \nabla P_n \chi_{\{P_n<1\}}$ by 4.1.3. Clearly we have $\bar{\partial}P_n = -\frac{1}{2}(z/|z|^3) * g_n$ and thus $\|\bar{\partial}P_n\|_q \leq C\|g_n\|_q$ by the Calderón-Zygmund theory. Hence $\nabla\omega_n \to 0$ in L^q. From this, it follows readily that $\omega_n \to 0$ in measure. For example, if $2 < q < \infty$ the L^p inequalities for Riesz potentials [St, p.119] tell us that $\omega_n \to 0$ in L^r, $\frac{1}{r} = \frac{1}{q} - \frac{1}{2}$. Set $\varphi_n = f_n(1 - \omega_n)$, so that $\varphi_n = 0$ on Ω_n. We want to prove that $\varphi_n \to f$ in W_1^q. Clearly

$$\nabla\varphi_n = \nabla f_n - \omega_n\nabla f_n - f_n\nabla\omega_n,$$

and we know that $\nabla f_n \to \nabla f$ in L^q. For each $\delta > 0$,

$$\int |\omega_n\nabla f_n|^q \leq \delta^q \int |\nabla f|^q + \int_{\{\omega_n > \delta\}} |\nabla f|^q,$$

from which we get $\omega_n\nabla f_n \to 0$ in L^q.

On the other hand,

$$\|f_n\nabla\omega_n\|_q \leq \|f\|_\infty\|\nabla\omega_n\|_q \to 0.$$

Therefore $\nabla\varphi_n \to \nabla f$ in L^q. Similarly, we can show that $\varphi_n \to f$ in L^q. □

The main result of this section is the following.

4.2.5 Theorem *Let $X \subset \mathbb{C}$ be compact, $2 \leq p < \infty$ and $q = p(p-1)^{-1}$. Then the following are equivalent.*

(i) $R^p(X) = A^p(X)$.

(ii) *For all open sets* Ω, $\gamma_p(\Omega\backslash\mathring{X}) = \gamma_p(\Omega\backslash X)$.

(iii) *For some constant* C,

$$C_q(\Omega\backslash\mathring{X}) \le CC_q(\Omega\backslash X), \text{ for all open sets } \Omega.$$

Remarks (1) In (ii) and (iii) one can replace open sets by open discs [Li].

(2) The capacities γ_p and C_q do not satisfy density theorems. However there is a Wiener type condition which plays the role of a density condition. In [H-W] it is proved that

$$\int_0 \left(\frac{C_q(\Delta(z,\delta)\backslash X)}{C_q(\Delta(z,\delta))}\right)^{p-1} \frac{d\delta}{\delta} = \infty, \ C_q\text{-a.e. on } \partial X, \tag{3}$$

is equivalent to (i). Thus (3) can be thought of as an analogue of condition (iii) in 3.1.2. On the other hand, it was established in [Da2] that nothing similar can be found in the uniform approximation setting.

Proof of Theorem 4.2.5 (i) $\Rightarrow$ (ii). This can be proved as in Chapter 2.

(ii) $\Rightarrow$ (iii). Just apply Lemma 4.2.1.

(iii) $\Rightarrow$ (i). Let $g \in L^q(X)$ be an annihilator of $R^p(X)$. Then $\hat{g} = \frac{1}{\pi z} * g$ vanishes outside X. Since $\hat{g} \in W_1^q$, $\hat{g}$ has a C_q-quasicontinuous representative which we denote again by $\hat{g}$. We claim that $\hat{g} = 0$ C_q-a.e. on $(\mathring{X})^c$. To show the claim, set $U = \{z : \hat{g}(z) \neq 0\}$. Then U is C_q-quasiopen and so, given $\varepsilon > 0$, we can find an open set G such that $C_q(G) < \varepsilon$ and $U \cap G^c$ is open in G^c. Let V be an open set such that $V \cap G^c = U \cap G^c$. Since $\Omega \equiv U \cup G = V \cup G$, Ω is open. Then, by (iii) and the subadditivity of C_q,

$$C_q(U\backslash\mathring{X}) \le C_q(\Omega\backslash\mathring{X}) \le CC_q(\Omega\backslash X) \le CC_q(U\backslash X) + C\varepsilon,$$

and so, letting $\varepsilon \to 0$, $C_q(U\backslash\mathring{X}) \le CC_q(U\backslash X)$. Clearly $U\backslash X = \emptyset$ and hence $C_q(U\backslash\mathring{X}) = 0$, which shows the claim.

We can now apply the Spectral Synthesis Theorem to the closed set $(\mathring{X})^c$ and the function $\hat{g}$. We get $\varphi_n \in C_0^\infty(\mathring{X})$ such that $\bar{\partial}\varphi_n \to g$ in L^q. Thus if $f \in A^p(X)$

$$\int fg\,dx\,dy = \lim_n \int f\bar{\partial}\varphi_n = 0. \qquad \square$$

Chapter 5
A survey of recent results in qualitative approximation by solutions of elliptic equations

5.0 Introduction

Some of the approximation theorems of Chapters 2, 3 and 4 have been extended to a fairly wide class of partial differential operators including the powers of $\bar{\partial}$ in the plane and the powers of Δ in $\mathbb{R}^d$. As for the analytic case considered in the preceding chapters, there are two main approaches: the first is based on Vitushkin's constructive scheme and the second on spectral synthesis. Although neither of them can be applied to all cases, a recent announcement of Netrusov [N] concerning spectral synthesis on Triebel-Lizorkin spaces seems to indicate that L^p and Lipschitz spaces can be dealt with simultaneously by means of the second approach.

We describe now precisely the context in which we will be working. We denote by L a constant coefficient, homogeneous, elliptic operator in $\mathbb{R}^d$, $d \geq 2$.

We are interested in obtaining approximation theorems by functions f that satisfy the equation $L(f) = 0$ on some neighbourhood of a given compact set X. Thus, if L is $\bar{\partial}$ we are dealing with analytic functions (in one variable) and if L is Δ with harmonic functions.

The approximation will take place in the norm of a Banach space V belonging to the following list.

1. $C_0^m(\mathbb{R}^d)$, $0 \leq m \in \mathbb{Z}$, the closure in $C^m(\mathbb{R}^d)$ of $C_0^\infty(\mathbb{R}^d)$. Recall that $f \in C^m(\mathbb{R}^d)$ if and only if f has bounded continuous derivatives up to order m. The norm in $C^m(\mathbb{R}^d)$ is $\sum_{|\alpha|\leq m} \|\partial^\alpha f\|_\infty$.

2. $\lambda^s(\mathbb{R}^d)$, $0 < s \notin \mathbb{Z}$, the closure of $C_0^\infty(\mathbb{R}^d)$ in $\Lambda^s(\mathbb{R}^d)$. If $0 < s < 1$, $\Lambda^s(\mathbb{R}^d)$ is just $\mathrm{Lip}\,(s, \mathbb{R}^d)$, the space of Lipschitz functions of order s. If m is the integer part of s and $\sigma = s - m$ then $f \in \Lambda^s(\mathbb{R}^d)$ if and only if $f \in C^m(\mathbb{R}^d)$ and $\partial^\alpha f \in \Lambda^\sigma(\mathbb{R}^d)$, $|\alpha| = m$. The norm in $\Lambda^s(\mathbb{R}^d)$ is

$$\sum_{|\alpha|\leq m} \|\partial^\alpha f\|_\infty + \sum_{|\alpha|=m} \|\partial^\alpha f\|_\sigma.$$

If we write $\lambda^m(\mathbb{R}^d) = C_0^m(\mathbb{R}^d)$ for $0 \leq m \in \mathbb{Z}$, we get a scale of spaces covering all positive degrees of smoothness.

3. $L^p(\mathbb{R}^d)$, $1 < p < \infty$.

4. $\mathrm{CMO}(\mathbb{R}^d)$, the closure of $C_0^\infty(\mathbb{R}^d)$ in $\mathrm{BMO}(\mathbb{R}^d)$, the familiar space of functions of bounded mean oscillation (CMO stands for continuous mean oscillation). Recall that the "norm" in $\mathrm{BMO}(\mathbb{R}^d)$ is

$$\sup_Q \frac{1}{|Q|} \int_Q |f - f_Q|,$$

where $f_Q = \frac{1}{|Q|} \int_Q f$ is the mean of f on the cube Q.

Given a compact $X \subset \mathbb{R}^d$, we have a natural version $V(X)$ of V on X, defined in most cases by restriction. Being a quotient, $V(X)$ inherits a Banach space structure from V. For the spaces $\lambda^s(\mathbb{R}^d)$, $s \geq 1$, we have actually two possible definitions for λ^s on X. One is the "function" version

$$\lambda^s_{\text{func}}(X) = \lambda^s(\mathbb{R}^d)/I(X) \simeq \lambda^s(\mathbb{R}^d)_{|X},$$

endowed with the quotient norm, where

$$I(X) = \{f \in \lambda^s(\mathbb{R}^d) : f = 0 \text{ on } X\}.$$

The second is the "jet" version

$$\lambda^s_{\text{jet}}(X) = \lambda^s(\mathbf{R}^d)/J(X),$$

also endowed with the quotient norm, where

$$J(X) = \{f \in \lambda^s(\mathbf{R}^d) : \partial^\alpha f = 0 \text{ on } X,\ |\alpha| \le [s]\}.$$

Whitney extension theorem [St, Chapter VI] provides a description of the "jets" (f_α) such that there exists a function $f \in C^s(\mathbf{R}^d)$ with $f_\alpha = \partial^\alpha f$ on X, $|\alpha| \le [s]$.

Define $H(X, L, V)$ as the closure in $V(X)$ of those equivalence classes with a representative f satisfying $L(f) = 0$ on some neighbourhood of X. The space $h(X, L, V)$ is defined as the set of $\tilde{f} \in V(X)$ such that $L(f) = 0$ on $\mathring{X}$. Hence $H(X, L, V) \subset h(X, L, V)$ for all X. The V-approximation problem for the operator L consists in describing those X for which we have equality in the above inclusion.

5.1 Uniform approximation

The uniform approximation problem for L seems to be a hard problem for any L (at least for $d \ge 2$). We already encountered in Chapter 2 a sample of the difficulties one must overcome in dealing with sup norm approximation. For $L = \Delta$ the uniform approximation problem was considered and solved independently by Deny [Dy] and Keldysh [K] about 20 years before the work of Vitushkin. Deny and Keldysh used a classical potential theoretic approach which has no counterpart in the analytic case. This explains why Vitushkin had to invent a completely new technique to solve the problem in the analytic setting. It is a remarkable fact that nobody has been able to give a proof of the Deny-Keldysh Theorem using the constructive scheme of Vitushkin. The interested reader is referred to the recent delightful exposition of Hedberg [He4] on the work of Deny and Keldysh.

Let C denote the Wiener capacity of classical potential theory in $\mathbf{R}^d$. Write $H(X)$ and $h(X)$ instead of $H(X, \Delta, C^0_0(\mathbf{R}^d))$ and $h(X, \Delta, C^0_0(\mathbf{R}^d))$ respectively.

5.1.1 Theorem (Deny and Keldysh) *For a compact $X \subset \mathbf{R}^d$ the following are equivalent.*

(i) $H(X) = h(X)$.

(ii) *For some constant $C > 0$,*

$$C(B\backslash\mathring{X}) \le C\,C(B\backslash X), \text{ for all open balls } B.$$

(iii) *For C-almost all $x \in \partial X$,*

$$\int_0^1 \frac{C(B(x,\delta) \cap X^c)}{C(B(x,\delta))}\frac{d\delta}{\delta} = \infty.$$

Some remarks on the above statement are in order. The equivalence of (i) and (iii) is the main result of Deny and Keldysh. That (ii) is equivalent to the other conditions was observed in [He2] and in [L]. This shows the strong formal analogy between the Deny-Keldysh and Vitushkin Theorems. Condition (iii) can be rephrased by saying that X^c is thick at C-almost all points of ∂X. Statements (iii) in 3.1.2 of and (3) of Chapter 4 are the appropriate counterparts of this thickness condition.

No complete results are known for other operators besides $\bar\partial$ or Δ. The case $L = \bar\partial^2$ in the plane is particularly intriguing. Since the fundamental solution $\frac{1}{\pi}\frac{\bar z}{z}$ of $\bar\partial^2$ is bounded, the natural capacity associated to $\bar\partial^2$ and L^∞ is trivial, in the sense that the capacity of a non-empty set stays in between two absolute positive constants. This and other facts support the following conjecture of the author [Ve5]. Write $H(X, \bar\partial^2)$ and $h(X, \bar\partial^2)$ for $H(X, \bar\partial^2, C_0^0(\mathbf{C}))$ and $h(X, \bar\partial^2, C_0^0(\mathbf{C}))$ respectively.

Conjecture *For any compact $X \subset \mathbf{C}$, $H(X, \bar\partial^2) = h(X, \bar\partial^2)$.*

Here is a list of cases in which the conjecture has been verified.

1. X has no interior points [T-W].
2. $\gamma(\Delta(z, r)\backslash X) \geq Cr$, for all $z \in \partial X$ and sufficiently small r [Ca].
3. The inner boundary of X is countable [W].
4. The inner boundary of X is a subset of a straight line [Me].

The author recently proved in [Ve5] that for arbitrary X any Dini-continuous function in $h(X, \bar\partial^2)$ belongs to $H(X, \bar\partial^2)$.

5.2 Smooth approximation

Let r be the order of the operator L. For a degree of smoothness $s \geq r$ one has the following solution to the λ^s_{jet} approximation problem for L.

5.2.1 Theorem *Let $X \subset \mathbf{R}^d$ be compact and $r \leq s$. Then*

(i) $H(X, L, \lambda^s_{\text{jet}}) = \{\tilde f \in \lambda^s_{\text{jet}}(X) : \partial^\alpha (Lf)(x) = 0,\ x \in X,\ |\alpha| \leq [s] - r\}$.

(ii) $H(X, L, \lambda^s_{\text{jet}}) = \lambda^s_{\text{jet}}(X)$ *if and only if X is the closure of its interior.*

Clearly (ii) follows readily from (i). In (i) the inclusion of the left hand side in the right is obvious and so the interesting part is the reverse inclusion. For s non-integer the result was proven by O'Farrell [OF4]. A quick proof using the invariance of $\Lambda^\sigma(\mathbf{R}^d)$, $0 < \sigma < 1$, under Calderón-Zygmund operators is now available (see section 3.3 for a particular case). For $r < s \in \mathbf{Z}$, 3.2.1 was proven in [Ve3] by a simple application of the Vitushkin scheme. For $s = r$ the proof is subtler because one must use a version of Uy's Theorem for finitely many Calderón-Zygmund operators.

For the function version λ^s_{func} of λ^s the corresponding approximation problem is open even for the Laplacean and $s = 2$. One likely needs to find the appropriate substitute of X^* in 3.3.2.

For $s < r$ the λ^s_{jet} and λ^s_{func} problems are essentially the same and so we use λ^s to denote any of them. We let $H_s(X)$ and $h_s(X)$ stand for $H(X, L, \lambda^s)$ and $h(X, L, \lambda^s)$ respectively. Here is another complete result.

5.2.2 Theorem *Let $X \subset \mathbb{R}^d$ be compact and let $r-2 < s < r$, $s \neq r-1$. Then the following are equivalent.*

(i) $H_s(X) = h_s(X)$.

(ii) *There exists a constant $C > 0$ such that*

$$M_*^{d-r+s}(B\backslash \mathring{X}) \leq C M^{d-r+s}(B\backslash X), \text{ for all open balls } B.$$

(iii) *For M_*^{d-r+s}-almost all $x \in \partial X$,*

$$\limsup_{r\to 0} \frac{M^{d-r+s}(B(x,r)\cap X^c)}{r^{d-r+s}} > 0.$$

Remark The statement is true also for $s = r-1$ provided we change the definition of λ^m for integer m using the Zygmund class (see [M-O]).

For $r-1 < s < r$, 5.2.2 is an extension of O'Farrell's Theorem discussed in Chapter 3 and the proof is the same modulo technical difficulties [Ve3]. For $r-2 < s < r-1$ the proof is much more elaborated [M-O]. There is a first constructive step in which an appropriate covering lemma is used in applying a refinement of Vitushkin's scheme. In the second part a duality argument completes the proof via a differentiability result for Riesz potentials of distributions in the Hardy classes $H^p(\mathbb{R}^n)$, $0 < p \leq 1$.

We come now to the discussion of the case $s = r-1$. We state the result only for $L = \Delta$, although it seems rather clear that the method of proof should work for a general L. The natural capacity associated to C^1 and Δ is

$$\chi(E) = \sup |\langle \Delta f, 1\rangle|$$

where the sup is taken over all functions which are harmonic outside a compact subset of E and satisfy $|\nabla f(x)| \leq 1$, $x \in \mathbb{R}^d$. If we add the requirement of $\nabla f(x)$ being continuous on $\mathbb{R}^d$ then we get a smaller quantity which we denote by $\chi_*(E)$. In the plane $\chi \leq \gamma$ and $\chi_* \leq \alpha$ and it is not known whether the reverse inequalities (with a multiplicative constant) hold. Even worse, it is not known whether $\chi(K) \geq \text{Const}\,\text{diam}(K)$ for all plane continua K.

The C^1 approximation problem for Δ was solved recently by Paramonov [Pa] using still another refinement of the Vitushkin scheme.

5.2.3 Theorem *Let $X \subset \mathbb{R}^d$ be compact. Then the following are equivalent.*

(i) $H_1(X) = h_1(X)$.

(ii) *For some constant C, $\chi_*(B\backslash \mathring{X}) \leq C\chi(B\backslash X)$, for all open balls B.*

The case $s \leq r-2$ has not been yet well understood. Very likely one should be able to adapt a well known example of Hedberg [He3] in the L^p context to generate some counterexamples in the spirit of [G-T].

5.3 BMO approximation

It is well know that often BMO is a good substitute for L^∞. On the other hand BMO can be thought of as the limit as $s \to 0$ of the Lipschitz classes Λ^s. Thus we expect that in the context of qualitative approximation the BMO-theorems should be obtained by replacing s by 0 in the Λ^s-theorems, $0 < s < 1$. This is true for $L = \bar{\partial}$ or for $L = \Delta$ in dimension 2 but nothing else is known.

Write $H_*(X, L)$ and $h_*(X, L)$ for $H(X, L, CMO)$ and $h(X, L, CMO)$ respectively.

5.3.1 Theorem [Ve4] *Let $X \subset \mathbb{C}$ be compact. Then the following are equivalent.*

(i) $H_*(X, \bar{\partial}) = h_*(X, \bar{\partial})$.

(ii) *For some constant C, $M_*^1(\Delta \backslash \mathring{X}) \le CM^1(\Delta \backslash X)$, for all open discs Δ.*

(iii) *For M_*^1-almost all $z \in \partial X$,*

$$\limsup_{r \to 1} \frac{M^1(\Delta(z, r) \backslash X)}{r} > 0.$$

Clearly 5.3.1 is the BMO counterpart of 3.1.2.

5.3.2 Theorem [M-V] *Let $X \subset \mathbb{C}$ be compact. Then*

$$H_*(X, \Delta) = h_*(X, \Delta) \quad \textit{and} \quad H_*(X, \bar{\partial}^2) = h_*(X, \bar{\partial}^2).$$

The proof of this striking result uses a rather technical covering lemma and a differentiability result for Riesz potentials of $H^1(\mathbb{R}^2)$ functions.

The higher dimensional version of 5.3.2 for Δ is that $H_*(X, \Delta) = h_*(X, \Delta)$ occurs if and only if $M_*^{d-2}(B \setminus \mathring{X}) \le \text{Const}\, M^{d-2}(B \setminus X)$ for each open ball B. This natural conjecture is still open, confirming the fact that the case in which the smoothness degree of the approximation coincides with the order of the operator minus 2 is critical.

5.4 L^p approximation

Before discussing the known results in the L^p context it is convenient to introduce a capacity associated to the Sobolev space W_k^q, $q = p(p-1)^{-1}$. Recall that W_k^q consists of those L^q functions whose derivatives up to order k are again in L^q. Given $E \subset \mathbb{R}^d$ and a positive integer k one defines

$$C_{q,k}(E) = \inf \|\nabla^k \varphi\|_q^q,$$

where the infimum is taken over those $\varphi \in C_0^\infty$ such that $\varphi \ge 1$ on E, and $\nabla^k \varphi$ stands for the vector valued function $(\partial^\alpha \varphi)_{|\alpha| = k}$.

Write $H^p(X)$ and $h^p(X)$ for $H(X, L, L^p)$ and $h^p(X, L, L^p)$ respectively. We then have [Pl].

5.4.1 Theorem *For $1 < p < d(d-1)^{-1}$ and any X one has $H^p(X) = h^p(X)$.*

In general is not difficult to prove that

$$C_{q,r}(B\backslash \mathring{X}) = C_{q,r}(B\backslash X) \text{ for each open ball } B, \tag{1}$$

is a necessary condition for $H^p(X) = h^p(X)$. On the other hand, Hedberg [H-W] has shown, using spectral synthesis, that the set of conditions

$$C_{q,k}(B\backslash \mathring{X}) \leq CC_{q,k}(B\backslash X), \text{ for each open ball } B,\ 1 \leq k \leq r,$$

is sufficient for $H^p(X) = h^p(X)$. An example was constructed [He3] for the range $d(d-1)^{-1} \leq p < d(d-r)^{-1}$ showing that (1) is not enough for $H^p(X) = h^p(X)$. The same statement for the range $d(d-r)^{-1} \leq p < \infty$, $d \geq 3$, has been recently proved by Mateu [Mt] using a modification of Hedberg's idea.

Necessary and sufficient conditions in terms of infinite families of appropriate capacitary conditions have been given by Bagby in [B2]. See also [G-T], where this idea is applied to another context. It seems, however, that Bagby's conditions are not easily applied to concrete situations.

5.5 Summary

One can summarize the results reviewed in the preceding sections in a rather compact way, provided some exceptions are allowed. Let's associate to each of the Banach spaces V considered in section 0 a "degree of smoothness" $\sigma = \sigma(V)$. We set $\sigma(\lambda^s) = s$, $0 < s$, $\sigma(L^\infty) = \sigma(\text{BMO}) = 0$ and $\sigma(L^p) = -d/p$. The negative smoothness attributed to L^p comes from consideration of the Sobolev imbedding theorem. Then $\sigma(V)$ generates four ranges in which different things happen.

1. $r \leq \sigma(V)$: the conditions for individual approximability are expressed in terms of the vanishing of $L(f)$ on the set X. No capacitary conditions are involved.

2. $r-2 \leq \sigma(V) < r$: there is a natural capacity characterizing sets with the approximation property. Unsolved cases are still $\bar{\partial}^2 - L^\infty$ and Δ-BMO in dimension > 2.

3. $-(d-1) \leq \sigma(V) < r-2$: the characterization of sets with the approximation property involves a family of capacitary conditions and only one capacitary condition is not enough.

4. $\sigma(V) < -(d-1)$: all compact sets enjoy the approximation property.

Finally, we would like to say that it would be very interesting to improve the existing techniques to have a better insight into uniform approximation. Also, may be the time has come to have a general approximation theorem, at least for $\bar{\partial}$ in the plane, involving an abstract Banach space V in the spirit of [Pa-Ve].

References

[A] Ahlfors, L., Bounded analytic functions, *Duke Math. J.* **14** (1947), 1–11.

[B1] Bagby, T., Quasi topologies and rational approximation, *J. Funct. Anal.* **10** (1972), 259–268.

[B2] Bagby, T., Approximation in the mean by solutions of elliptic equations, *Trans. Amer. Math. Soc.* **281** (1984), 761–784.

[B-G] Bagby, T. and Gauthier, P., An arc of finite 2-measure that is not rationally convex, *Proc. Amer. Math. Soc.* **114** (1992), 1033–1034.

[B-V] Boivin, A. and Verdera, J., Approximation par fonctions holomorphes dans les espaces L^p, Lip α et BMO, *Indiana Univ. Math. J.* **40** (1991), 393–418.

[Br] Browder, A., *Rational Approximation and Function Algebras*, Benjamin, New York, 1969.

[C] Calderón, A.P., Cauchy integrals on Lipschitz curves and related operators, *Proc. Nat. Acad. Sci. USA* **74** (1977), 1324–1327.

[Ca] Carmona, J.J., Mergelyan's approximation theorem for rational modules, *J. Approx. Theory* **44** (1985), 113–125.

[Ch] Christ, M., *Lectures on Singular Integral Operators*, CBMS Regional Conf. Ser. in Math. **77**, Amer. Math. Soc., Providence, RI, 1990.

[C-J-S] Coifman, R.R., Jones, P.W. and Semmes, S., Two elementary proofs of the L^2 boundedness of the Cauchy integral on Lipschitz curves, *J. Amer. Math. Soc.* **2** (1989), 553–564.

[C-Mc-M] Coifman, R.R., McIntosh, A. and Meyer, Y., L'integral de Cauchy définit un opérateur borné sur L^2 pour les courbes lipschitziennes, *Ann. of Math.* **115** (1982), 361–387.

[D] David, G., Opérateurs integraux singuliers sur certaines courbes du plan complexe, *Ann. Sci. École Norm. Sup.* **17** (1984), 157–189.

[D-S] David, G. and Semmes, S., *Analysis of and on Uniformly Rectifiable Sets*, book to appear.

[Da1] Davie, A.M., Analytic capacity and approximation problems, *Trans. Amer. Math. Soc.* **171** (1972), 409–444.

[Da2] Davie, A.M., An example on rational approximation, *Bull. London Math. Soc.* **2** (1970), 83–86.

[Da-O] Davie, A.M. and Øksendal, B., Analytic capacity and differentiability properties of finely harmonic functions, *Acta. Math.* **149** (1982), 127–152.

[De] Denjoy, A., Sur les fonctions analytiques uniformes à singularités discontinues, *C.R. Acad. Sci. Paris* **149** (1909), 258–260.

[Dy] Deny, J., Systèmes totaux de fonctions harmoniques, *Ann. Inst. Fourier (Grenoble)* **1** (1949), 103–113.

[G] Gamelin, T.W., *Uniform Algebras*, Prentice-Hall, Englewood Cliffs, NJ, 1969.

[Ga] Garnett, J.B., *Analytic Capacity and Measure*, Lecture Notes in Math. **297**, Springer-Verlag, Berlin and New York, 1972.

[G-T] Gauthier, P. and Tarkhanov, N., Degenerate cases of uniform approximation by solutions of systems with surjective symbols, *Canad. J. Math.* **45** (1993), 740–757.

[H] Havin, V.P., Approximations in the mean by analytic functions, *Soviet Math. Dokl.* **9** (1968), 245–248.

[H-M] Havin, V.P. and Maz'ja, V.G., Nonlinear potential theory, *Russian Math. Surveys* **27** (1972), 71–148.

[He1] Hedberg, L.I., Approximation in the mean by analytic functions, *Trans. Amer. Math. Soc.* **153** (1972), 157–171.

[He2] Hedberg, L.I., Non-linear potentials and approximation in the mean by analytic functions, *Math. Z.* **129** (1972), 299–319.

[He3] Hedberg, L.I., Two approximation problems in function spaces, *Ark. Mat.* **16** (1978), 51–81.

[He4] Hedberg, L.I., Approximation by harmonic functions, and stability of the Dirichlet problem, *Exposition. Math.* **11** (1993), 193–259.

[H-W] Hedberg, L.I. and Wolff, T.H., Thin sets in non-linear potential theory, *Ann. Inst. Fourier (Grenoble)* **33** (1983), 161–187.

[J-M] Jones, P.W. and Murai, T., Positive analytic capacity but zero Buffon needle probability, *Pacific J. Math.* **133** (1988), 99–114.

[K] Keldysh, M.V., On the solvability and stability of the Dirichlet problem, *Uspekhi Mat. Nauk* **8** (1941), 171–231 (Russian); English translation: *Amer. Math. Soc. Transl.* **51** (1966), 1–73.

[Ko] Korevaar, J., Polynomial and rational approximation in the complex domain, in: *Aspects of Contemporary Complex Analysis* (D.A. Brannan and J.G. Clunie, eds.), Academic Press, London, 1980; 251–292.

[L] Labrèche, M., De l'approximation harmonique uniforme, Thèse de doctorat, Université de Montréal, 1982.

[Li] Lindberg, P., A constructive method for L^p approximation by analytic functions, *Ark. Mat.* **20** (1982), 61–68.

[M] Marshall, D.E., Removable sets for bounded analytic functions, in: *Linear and Complex Analysis Problem Book* (V.P. Havin et al., eds.), Lecture Notes in Math. **1043**, Springer-Verlag, Berlin and New York, 1984; 485–490.

[Mt] Mateu, J., An example on L^p approximation by harmonic functions, in preparation.

[M-O] Mateu, J. and Orobitg, J., Lipschitz approximation by harmonic functions and some applications to spectral synthesis, *Indiana Univ. Math. J.* **39** (1990), 703–736.

[M-V] Mateu, J. and Verdera, J., BMO harmonic approximation in the plane and spectral synthesis for Hardy-Sobolev spaces, *Rev. Mat. Iberoamericana* 4 (1988), 291–318.

[Ma] Mattila, P., Smooth maps, null-sets for integralgeometric measure and analytic capacity, *Ann. of Math.* **123** (1986), 303–309.

[M-M] Mattila, P. and Melnikov, M., Existence and weak type inequalities for Cauchy integrals of general measures on rectifiable curves and sets, to appear in *Proc. Amer. Math. Soc.*

[Ma-O] Mattila, P. and Orobitg, J., On some properties of Hausdorff content related to instability, to appear in *Ann. Acac. Sci. Fenn. Ser. AI Math.* **19** (1994).

[Me] Melnikov, M., A bound for the Cauchy integral along an analytic curve, *Mat. Sb.* **71(113)** (1966), 503–515.

[Mel] Melnikov, P., personal communication.

[My] Meyers, N.G., A theory of capacities for functions in Lebesgue classes, *Math. Scand.* **26** (1970), 255–292.

[Mu1] Murai, T., Comparison between analytic capacity and the Buffon needle probability, *Trans. Amer. Math. Soc.* **304** (1987), 501–514.

[Mu2] Murai, T., The power 3/2 appearing in the estimate of analytic capacity, *Pacific J. Math.* **143** (1990), 313–340.

[N] Netrusov, Y.V., Spectral synthesis in spaces of smooth functions, *Soviet Math. Dokl.* **46** (1993), 135–138.

[OF1] O'Farrell, A.G., Hausdorff content and rational approximation in fractional Lipschitz norms, *Trans. Amer. Math. Soc.* **228** (1977), 187–206.

[OF2] O'Farrell, A.G., Lip 1 rational approximation, *J. London Math. Soc.* **11** (1975), 159–164.

[OF3] O'Farrell, A.G., Localness of certain Banach modules, *Indiana Univ. Math. J.* **24** (1975), 1135–1141.

[OF4] O'Farrell, A.G., Rational approximation in Lipschitz norms II, *Proc. Roy. Irish Acad. Sect. A* **79** (1979), 103–114.

[OF5] O'Farrell, A.G., Qualitative rational approximation on plane compacta, in: *Banach Spaces, Harmonic Analysis and Probability Theory* (R.C. Blei and S.J. Sidney, eds.), Lecture Notes in Math. **995**, Springer-Verlag, Berlin and New York, 1983; 103–122.

[Pa] Paramonov, P.V., On harmonic approximation in the C^1-norm, *Math. USSR Sb.* **71** (1992), 183–207.

[Pa-Ve] Paramonov, P.V. and Verdera, J., Approximation by solutions of elliptic equations on closed subsets of Euclidean space, to appear in *Math. Scand.*

[Po] Pommerenke, C., Über die analytische Kapazität, *Archiv Math. (Basel)* **11** (1960), 270–277.

[Pl] Polking, J., Approximation in L^p by solutions of elliptic partial differential equations, *Amer. J. Math.* **94** (1972), 1231–1244.

[S] Saks, S., *Theory of the Integral*, Dover, New York, 1964.

[St] Stein, E.M., *Singular Integrals and Differentiability Properties of Functions*, Princeton Univ. Press, Princeton, 1970.

[T-W] Trent, T. and Wang, J.L., Uniform approximation by rational modules on nowhere dense sets, *Proc. Amer. Math. Soc.* **81** (1981), 62–64.

[U] Uy, N.X., Removable sets of analytic functions satisfying a Lipschitz condition, *Ark. Mat.* **17** (1979), 19–27.

[Ve1] Verdera, J., A weak type inequality for Cauchy transforms of finite measures, *Publ. Mat.* **36** (1992), 1029–1034.

[Ve2] Verdera, J., On C^m rational approximation, *Proc. Amer. Math. Soc.* **97** (1986), 621–625.

[Ve3] Verdera, J., C^m approximation by solutions of elliptic equations, and Calderón-Zygmund operators, *Duke Math. J.* **55** (1987), 157–187.

[Ve4] Verdera, J., BMO rational approximation and one dimensional Hausdorff content, *Trans. Amer. Math. Soc.* **297** (1986), 283–304.

[Ve5] Verdera, J., On the uniform approximation problem for the square of the Cauchy-Riemann operator, *Pacific J. Math.* **159** (1993), 379–396.

[Vi] Vitushkin, A.G., Analytic capacity of sets in problems of approximation theory, *Russian Math. Surveys* **22** (1967), 139–200.

[W] Wang, J.L., A localization operator for rational modules, *Rocky Mountain J. Math.* **19** (1989), 999–1002.

[Z] Zalcman, L., *Analytic Capacity and Rational Approximation*, Lecture Notes in Math. **50**, Springer-Verlag, Berlin and New York, 1968.

Semigroups of holomorphic isometries

Edoardo VESENTINI
Scuola Normale Superiore
Piazza dei Cavalieri
I-56100 Pisa
Italy

Abstract

The aim of the lectures was that of giving a reasonably self-contained exposition of the current status of the theory of semigroups of holomorphic isometries and of groups of holomorphic automorphisms acting on hyperbolic domains, assuming as a prerequisite only a rather rudimentary knowledge of Fréchet holomorphy in complex Banach spaces. This character – quite far from any purpose of providing an exhaustive exposition of the geometry of hyperbolic domains – has been preserved in the present report, which is an expanded version of the preparatory notes of the lectures. Systematic expositions of the general theory of invariant metrics, of homogeneous domains and of homogeneous Banach manifolds may be found in some of the treatises quoted in the list of references at the end of this report. After reviewing, in the first chapter, the basic theory of invariant metrics on the unit disc of the complex field and on domains in complex Banach spaces, the second chapter deals with holomorphic isometries, holomorphic automorphisms and holomorphic families thereof. These topics are investigated, in the third and fourth chapter, in the case of unit balls of complex Hilbert spaces and, more generally, of Cartan factors of type one. This study sets the stage for the theory of strongly continuous semigroups of linear isometries acting on spaces endowed with an indefinite metric and, more specifically, on Kreĭn and Pontryagin spaces. This theory is exposed in the second part of the third chapter of the present report.

Contents

P. M. Gauthier (ed.) and G. Sabidussi (techn. ed.), Complex Potential Theory, 475–548.

Chapter 1
Invariant distances and invariant differential metrics on domains in complex Banach spaces

In this chapter some aspects of the theory of invariant distances will be reviewed. Main references for these topics are, e.g. [KOB 70], [FR-VE 80], [DIN 89], [HAR 79].

1.1 The Poincaré metric on the unit disc of C

Let $\Delta = \{\zeta \in \mathbb{C} : |\zeta| < 1\}$ be the open unit disc of $\mathbb{C}$, and let $\mathrm{Hol}(\Delta, \Delta)$ be the semigroup of all holomorphic maps of Δ into Δ.

Lemma 1.1 (Schwarz lemma) *If $f(0) = 0$, then*

$$|f(\zeta)| \leq |\zeta| \quad \textit{for all } \zeta \in \Delta \tag{1.1.1}$$

and

$$|f'(0)| \leq 1.$$

Moreover, if either $|f(\zeta_0)| = |\zeta_0|$ *for some* $\zeta_0 \in \Delta \setminus \{0\}$ *or* $|f'(0)| = 1$*, then there exists* $\theta \in \mathbf{R}$ *such that* $f(\zeta) = e^{i\theta}\zeta$ *for all* $\zeta \in \Delta$.

Proof Let

$$f(\zeta) = a_1\zeta + a_2\zeta^2 + \dots \qquad (a_\nu \in \mathbf{C},\ \nu = 1, 2, \dots)$$

be the power series expansion of f in Δ. The function $h : \zeta \mapsto \frac{f(\zeta)}{\zeta} = a_1 + a_2\zeta + \dots$ is holomorphic on Δ, and, for all $0 < r < 1$ and $|\zeta| = r$, then

$$|h(\zeta)| = \frac{|f(\zeta)|}{r} \leq \frac{1}{r}.$$

By the maximum principle, this inequality holds also when $|\zeta| \leq r$. Letting r tend to 1 yields (1.1.1). If $|f(\zeta_0)| = |\zeta_0|$ for some $\zeta_0 \in \Delta \setminus \{0\}$, then $|h(\zeta_0)| = 1$. Thus, by the maximum principle $h(\zeta) = e^{i\theta}\zeta$ for some $\theta \in \mathbf{R}$ and all $\zeta \in \Delta$.

Since $f'(0) = a_1 = h(0)$, the statement concerning $f'(\zeta)$ is a further consequence of the maximum principle. □

Let $\operatorname{Aut}\Delta$ be the group of all holomorphic automorphisms of Δ, and, for $\zeta_0 \in \Delta$, let $(\operatorname{Aut}\Delta)_{\zeta_0}$ be the stability group of ζ_0 in $\operatorname{Aut}\Delta$:

$$(\operatorname{Aut}\Delta)_{\zeta_0} = \{g \in \operatorname{Aut}\Delta : g(\zeta_0) = \zeta_0\}.$$

Corollary 1.1 *The group* $(\operatorname{Aut}\Delta)_0$ *consists of the restrictions to* Δ *of all rotations of* $\mathbf{C}$ *around* 0.

For $\tau \in \Delta$, let $g_\tau \in \operatorname{Hol}(\Delta, \Delta)$ be defined by $g_\tau(\zeta) = \frac{\zeta+\tau}{1+\bar{\tau}\zeta}$. Since

$$1 - |g_\tau(\zeta)|^2 = \frac{|\tau|^2(1 - |\zeta|^2)}{|1 + \bar{\tau}\zeta|^2} > 0,$$

then $g_\tau \in \operatorname{Hol}(\Delta, \Delta)$. Furthermore

$$\zeta = \frac{g_\tau(\zeta) - \tau}{1 - \bar{\tau}g_\tau(\zeta)}$$

and therefore $g_\tau \circ g_{-\tau} = g_{-\tau} \circ g_\tau$ is the identity map in Δ. Thus $g_\tau \in \operatorname{Aut}\Delta$, and $g_\tau^{-1} = g_{-\tau}$.

For $f \in \operatorname{Hol}(\Delta, \Delta)$ and any $\zeta_0 \in \Delta$, $g_{-f(\zeta_0)} \circ f \circ g_{\zeta_0} \in \operatorname{Hol}(\Delta, \Delta)$ and $g_{-f(\zeta_0)} \circ f \circ g_{-\zeta_0}(0) = 0$. Hence Lemma 1.1 implies that

$$|g_{-f(\zeta_0)} \circ f(g_{\zeta_0}(\zeta))| \leq |\zeta| \quad \text{for all } \zeta \in \Delta,$$

and

$$|(g_{-f(\zeta_0)} \circ f \circ g_{\zeta_0})'(0)| \leq 1;$$

if either this second inequality is an equality, or if the first one becomes an equality for some $\zeta \in \Delta \setminus \{0\}$, then $f \in \operatorname{Aut}\Delta$.

As a consequence, the following lemma holds:

Lemma 1.2 (Schwarz-Pick lemma) *For any $f \in \mathrm{Hol}(\Delta, \Delta)$ and for any choice of ζ_0, ζ_1, ζ_2 in Δ, then*

$$\left|\frac{f(\zeta_1) - f(\zeta_2)}{1 - f(\zeta_1)\overline{f(\zeta_2)}}\right| \leq \left|\frac{\zeta_1 - \zeta_2}{1 - \zeta_1\bar{\zeta}_2}\right|$$
$$\left|\frac{f'(\zeta_0)}{1 - |f(\zeta_0)|^2}\right| \leq \frac{1}{1 - |\zeta_0|^2}$$

If $f \in \mathrm{Aut}\,\Delta$, both inequalities become equalities. Conversely, if equality holds for some $\zeta_0 \in \Delta$ or for a pair of distinct points ζ_1 and ζ_2 in Δ, then $f \in \mathrm{Aut}\,\Delta$.

Since
$$g_{\tau_1} \circ g_{\tau_2}(\zeta) = g_{g_{\tau_2}(\tau_1)}(\zeta),$$
the family $\{g_\tau : \tau \in \Delta\}$ is a group, which acts transitively on Δ (because $g_\tau(0) = \tau$).

For any $g \in \mathrm{Aut}\,\Delta$, $g_{-g(0)} \circ g \in (\mathrm{Aut}\,\Delta)_0$. As a consequence of Corollary 1.1, we have

Proposition 1.1 *The group $\mathrm{Aut}\,\Delta$ consists of all Moebius transformations*

$$\zeta \mapsto e^{i\theta}\frac{\zeta + \tau}{1 + \bar{\tau}\zeta} \tag{1.1.2}$$

with $\theta \in \mathbb{R}, \tau \in \Delta$.

Let J be the 2×2 matrix
$$J = \begin{pmatrix} 1 & 0 \\ 0 & -1 \end{pmatrix}.$$

The group $SU(1,1)$ of all 2×2 complex matrices $A = \begin{pmatrix} a & b \\ c & d \end{pmatrix}$ with determinant equal to one, such that ${}^t\bar{A}JA = J$, i.e.

$$\overline{\begin{pmatrix} a & c \\ b & d \end{pmatrix}}\begin{pmatrix} 1 & 0 \\ 0 & -1 \end{pmatrix}\begin{pmatrix} a & b \\ c & d \end{pmatrix} = \begin{pmatrix} 1 & 0 \\ 0 & -1 \end{pmatrix},$$

is given by

$$SU(1,1) = \{A = \begin{pmatrix} a & b \\ \bar{b} & \bar{a} \end{pmatrix} : a, b \in \mathbb{C}, |a|^2 - |b|^2 = 1\}.$$

For any $A \in SU(1,1)$, the meromorphic function on $\mathbb{C}$

$$\zeta \mapsto \frac{a\zeta + b}{\bar{b}\zeta + \bar{a}}$$

is holomorphic in a neighborhood of the closure of Δ. Let $\phi(A)$ be its restriction to Δ. Since

$$\phi(A)(\zeta) = \frac{a}{\bar{a}}\frac{\zeta + \frac{b}{a}}{1 + \frac{\bar{b}}{\bar{a}}\zeta}$$

and $|a/\bar{a}| = 1$, $|b/a| < 1$, $\phi(A)$ is a Moebius transformation. Conversely, the Moebius transformation (1.1.2) is the image $\phi(A)$ of the element $A \in SU(1,1)$ defined by

$$a = (1 - |\tau|^2)^{-\frac{1}{2}} e^{i\frac{\theta}{2}}, \qquad b = \tau a.$$

Since $\phi(A_1 A_2) = \phi(A_1) \circ \phi(A_2)$ for all A_1, A_2 in $SU(1,1)$, the following theorem has been proved.

Theorem 1.1 *The map ϕ is a homomorphism of $SU(1,1)$ onto* $\operatorname{Aut}\Delta$.

It is immediately checked that $\operatorname{Ker}\phi$ consists of $\pm$ the identity matrix.

The *Poincaré metric* on Δ is the Riemannian metric

$$ds^2 = \frac{|d\zeta|^2}{(1 - |\zeta|^2)^2}.$$

Its Gaussian curvature is equal to -4.

For $\tau \in \mathbb{C}$ and $\zeta \in \Delta$, let

$$\langle \tau \rangle_\zeta := \frac{|\tau|}{1 - |\zeta|^2}.$$

Lemma 1.2 implies

Theorem 1.2 *For any $f \in \operatorname{Hol}(\Delta, \Delta)$ and all $\zeta \in \Delta$*

$$\langle f'(\zeta) \rangle_{f(\zeta)} \leq \langle 1 \rangle_\zeta.$$

If $f \in \operatorname{Aut}\Delta$, equality holds for all $\zeta \in \Delta$. If equality holds for some $\zeta \in \Delta$, then $f \in \operatorname{Aut}\Delta$.

Let $\omega : \Delta \times \Delta \to \mathbb{R}_+$ be the Poincaré distance on Δ, i.e. the distance defined by the Poincaré metric:

$$\omega(\zeta_1, \zeta_2) = \inf \int_l ds$$

where $\zeta_1, \zeta_2 \in \Delta$, and inf is taken over all piecewise C^1 curves $l([0,1]) \to \Delta$ such that $l(0) = \zeta_1$, $l(1) = \zeta_2$.

The invariance of the Poincaré metric under the action of $\operatorname{Aut}\Delta$ implies the invariance of the Poincaré distance.

Then, for $\zeta \in \Delta$,

$$\omega(0, \zeta) = \omega(0, |\zeta|) = \int_0^{|\zeta|} \frac{dt}{1 - t^2} = \operatorname{tgh}^{-1}(|\zeta|) = \frac{1}{2} \log \frac{1 + |\zeta|}{1 - |\zeta|}.$$

and therefore, for $\zeta_1, \zeta_2 \in \Delta$,

$$\begin{aligned} \omega(\zeta_1, \zeta_2) &= \omega\left(0, \frac{\zeta_2 - \zeta_1}{1 - \bar{\zeta}_1 \zeta_2}\right) = \operatorname{tgh}^{-1}\left(\frac{|\zeta_2 - \zeta_1|}{|1 - \bar{\zeta}_1 \zeta_2|}\right) \\ &= \frac{1}{2} \log \frac{1 + \left|\frac{\zeta_2 - \zeta_1}{1 - \bar{\zeta}_1 \zeta_2}\right|}{1 - \left|\frac{\zeta_2 - \zeta_1}{1 - \bar{\zeta}_1 \zeta_2}\right|}. \end{aligned}$$

Since the function $t \mapsto \log \frac{1+t}{1-t}$ is strictly increasing, Lemma 1.2 yields

Theorem 1.3 *For any $f \in \mathrm{Hol}(\Delta, \Delta)$ and for any choice of ζ_1, ζ_2 in Δ,*

$$\omega(f(\zeta_1), f(\zeta_2)) \leq \omega(\zeta_1, \zeta_2). \tag{1.1.3}$$

Hence any $f \in \mathrm{Aut}\,\Delta$ is an isometry for ω. Conversely, if $f \in \mathrm{Hol}(\Delta, \Delta)$ is such that (1.1.3) *becomes an equality at two distinct points ζ_1, ζ_2 in Δ, then $f \in \mathrm{Aut}\,\Delta$.*

The following theorem has been established in [VES 82a].

Theorem 1.4 *Let U be a domain in $\mathbf{C}$ and let f, h be two holomorphic maps of U into Δ. The function*

$$z \mapsto \log \omega(f(\zeta), h(\zeta)) \tag{1.1.4}$$

is subharmonic on U.

Proof Let $\zeta_0 \in U$. Since $\mathrm{Aut}\,\Delta$ acts transitively and isometrically, transforming subharmonic functions into subharmonic functions, there is no restriction in assuming $h(\zeta_0) = 0$. It will be shown first that, if $f(\zeta_0) \neq h(\zeta_0)$, then

$$\left(\frac{\partial^2}{\partial\zeta \partial\bar{\zeta}} \log \omega(f(\zeta), h(\zeta)) \right)_{\zeta = \zeta_0} \geq 0. \tag{1.1.5}$$

It will be assumed, with no restriction, that $\zeta_0 = 0$. Since $f(0) \neq h(0)$, the function (1.1.4) is a real analytic function of ζ and $\bar{\zeta}$ in a neighborhood V_0 of 0. As such, it has a power series expansion in ζ and $\bar{\zeta}$ absolutely and uniformly convergent in a neighborhood of 0.

Proving that (1.1.5) holds amounts to showing that the coefficient of $|\zeta|^2$ in the power series expansion is non-negative.

Let

$$f(\zeta) = a_0 + a_1\zeta + a_2\zeta^2 + \cdots, \quad h(\zeta) = b_1\zeta + b_2\zeta^2 + \cdots$$

be the power series expansions of f and h in a neighborhood of 0. Here and in the following the dots stand for terms of higher order. Since $f(0) \in \Delta \setminus \{0\}$, then $0 < |a_0| < 1$.

The function $\lambda : U_0 \to \mathbf{C}$ defined by

$$\lambda(\zeta) = \frac{f(\zeta) - h(\zeta)}{1 - \overline{h(\zeta)} f(\zeta)},$$

which is an analytic function of ζ and $\bar{\zeta}$, has the following power series expansion in ζ and $\bar{\zeta}$, uniformly and absolutely convergent in a neighborhood of 0:

$$\begin{aligned} \lambda(\zeta) &= a_0 + (a_1 - b_1)\zeta + a_0{}^2\bar{b_1}\bar{\zeta} + (a_2 - b_2)\zeta^2 + a_0(2a_1\bar{b_1} - |b_1|^2)|\zeta|^2 \\ &\quad + a_0{}^2(\bar{b_2} - a_0\bar{b_1}{}^2)\bar{\zeta}^2 + \cdots . \end{aligned}$$

Thus

$$|\lambda(\zeta)|^2 = |a_0|^2(1 + 2Re(\alpha_1\zeta) + (|\alpha_1|^2 + 2Re(a_1\bar{b_1}))|\zeta|^2 + 2Re(\alpha_2\zeta^2) + \cdots),$$

where α_1 and α_2 are complex constants,

$$\alpha_1 = \frac{1}{a_0}(a_1 - (1 - |a_0|^2)b_1), \tag{1.1.6}$$

and the convergence is absolute and uniform on a neighborhood of 0. Hence the function $\zeta \mapsto |\lambda(\zeta)|$ is given by the following power series, absolutely and uniformly convergent on a neighborhood of 0:

$$\begin{aligned}|\lambda(\zeta)| &= |a_0|(1 + Re(\alpha_1\zeta) + (\frac{1}{4}|\alpha_1|^2 + Re(a_1\bar{b_1}))|\zeta|^2 \\ &\quad + Re((\alpha_2 - \frac{1}{4}{\alpha_1}^2)\zeta^2) + \cdots).\end{aligned} \tag{1.1.7}$$

Since

$$\omega(f(\zeta), h(\zeta)) = \sum_{n=0}^{+\infty} \frac{1}{2n+1}|\lambda(\zeta)|^{2n+1},$$

then $\omega(f(\zeta), h(\zeta))$ is expressed by the following power series, absolutely and uniformly convergent on a neighborhood of 0:

$$\omega(f(\zeta), h(\zeta)) = \beta_0 + 2Re(\beta_1\zeta) + \beta_{11}|\zeta|^2 + Re(\beta_2\zeta^2) + \cdots,$$

where $\beta_0, \beta_1, \beta_{11}, \beta_2$ are complex coefficients, and

$$\left.\begin{aligned}\beta_0 &= \omega(0, a_0), \\ \beta_1 &= \tfrac{|a_0|\alpha_1}{2(1-|a_0|^2)}, \\ \beta_{11} &= \tfrac{|a_0|}{4(1-|a_0|^2)^2}((1 + |a_0|^2)|\alpha_1|^2 + 4(1 - |a_0|^2)Re(a_1\bar{b_1})).\end{aligned}\right\} \tag{1.1.8}$$

The function (1.1.4) is expressed by the following power series, absolutely and uniformly convergent in a neighborhood of 0,

$$\begin{aligned}\log\omega(f(\zeta), h(\zeta)) &= \log\beta_0 + 2Re(\frac{\beta_1}{\beta_0}\zeta) + \frac{1}{{\beta_0}^2}(\beta_{11}\beta_0 - |\beta_1|^2)|\zeta|^2 \\ &\quad + Re((\frac{\beta_2}{\beta_0} - (\frac{\beta_1}{\beta_0})^2)\zeta^2) + \cdots)\end{aligned}$$

The inequality (1.1.5) is equivalent to

$$\beta_0\beta_{11} - |\beta_1|^2 \geq 0. \tag{1.1.9}$$

By (1.1.6) and (1.1.8),

$$\begin{aligned}\beta_0\beta_{11} - |\beta_1|^2 &= \frac{|a_0|}{4(1 - |a_0|^2)^2}\{((1 + |a_0|^2)\beta_0 - |a_0|)|\alpha_1|^2 + 4(1 - |a_0|^2)\beta_0 Re(a_1\bar{b_1})\} \\ &= \frac{1}{4|a_0|(1 - |a_0|^2)^2}\{((1 + |a_0|^2)\beta_0 - |a_0|)(|a_1|^2 + (1 - |a_0|^2)^2|b_1|^2) \\ &\quad + 2(|a_0| - (1 - |a_0|^2)\beta_0)(1 - |a_0|^2)Re(a_1\bar{b_1})\}.\end{aligned} \tag{1.1.10}$$

Because $0 < |a_0| < 1$, then, by (1.1.8),

$$\begin{aligned}|a_0| - (1 - |a_0|^2)\beta_0 &= |a_0| - (1 - |a_0|^2)(|a_0| + \frac{|a_0|^3}{3} + \frac{|a_0|^5}{5} + \cdots) \\ &= (1 - \frac{1}{3})|a_0|^3 + (\frac{1}{3} - \frac{1}{5})|a_0|^5 + (\frac{1}{5} - \frac{1}{7})|a_0|^7 + \cdots > 0,\end{aligned}$$

and therefore the coefficient of $Re(a_1\bar{b_1})$ in the last term of (1.1.10) is positive. Since

$$2(1 - |a_0|^2)Re(a_1\bar{b_1}) = Re(a_1\overline{(1 - |a_0|^2)b_1)} \geq -|a_1|^2 - (1 - |a_0|^2)^2|b_1|^2,$$

(1.1.10) implies that $\beta_0\beta_{11} - |\beta_1|^2$ can be bounded from below in the following way:

$$\begin{aligned}\beta_0\beta_{11} - |\beta_1|^2 &\geq \frac{1}{4|a_0|(1 - |a_0|^2)^2}((1 + |a_0|^2)\beta_0 - |a_0| - |a_0| + \beta_0(1 - |a_0|^2) \\ &\times (|a_1|^2 + (1 - |a_0|^2)^2|b_1|^2) \\ &= \frac{1}{2|a_0(1 - |a_0|^2)^2}(\beta_0 - |a_0|)(|a_1|^2 + (1 - |a_0|^2)^2|b_1|^2).\end{aligned}$$

Because $0 < a_0 < 1$, then $\beta_0 - |a_0| > 0$ and therefore (1.1.9) holds. Hence the function (1.1.4) is a real analytic function on the domain $U \setminus Z$, where Z is the discrete set:

$$Z = \{\zeta \in U : f(\zeta) = h(\zeta)\}.$$

If $\zeta_0 \in Z$, then, for a sufficiently small $R > 0$,

$$\frac{1}{2\pi}\int_0^{2\pi} \log\omega(f(\zeta_0 + re^{i\theta}), h(\zeta_0 + re^{i\theta}))d\theta > -\infty$$

whenever $0 < r < R$. These inequalities, coupled with the continuity of the function $\zeta \mapsto \omega(f(\zeta), g(\zeta))$, complete the proof of Theorem 1.4. □

Corollary 1.2 *The function $\zeta \mapsto \omega(f(\zeta), h(\zeta))$ is a real analytic subharmonic function on U.*

It turns out that the function $\zeta \mapsto \omega(0, \zeta)$ is submultiplicative [VES 83].

Lemma 1.3 *For ζ_1, ζ_2 in Δ*

$$\omega(0, \zeta_1\zeta_2) \leq \omega(0, \zeta_1)\omega(\zeta_2),$$

equality holding only when $\zeta_1\zeta_2 = 0$.

As a consequence

Corollary 1.3 *For all $\zeta \in \Delta$, the sequence $\{\omega(0, \zeta^n)^{\frac{1}{n}} : n = 1, 2, \ldots\}$ converges decreasingly to $|\zeta|$.*

1.2 The Schwarz lemma in complex Banach spaces

The main tool in the proof of the classical Schwarz lemma is the maximum principle for complex valued holomorphic functions. This principle admits a weak and a strong extension to vector valued holomorphic functions.

First some notations, which will be used systematically throughout this report.

If D and D' are domains in two complex Banach spaces $\mathcal{E}$ and $\mathcal{E}'$, the set of all (Fréchet) holomorphic maps of D into D' will be denoted by $\mathrm{Hol}(D, D')$; $\mathrm{Aut}\, D$ will denote the group of all holomorphic automorphisms of D.

Let U be a domain in $\mathbb{C}$ and let $f \in \mathrm{Hol}(U, \mathcal{E})$.

The function $z \mapsto \|f(z)\|$ (where $\| \; \|$ is the norm in $\mathcal{E}$) is subharmonic. Hence it cannot have a local maximum on U without $\|f(z)\|$ being constant on U. This is the so called *weak maximum principle.* Its weakness is underlined by the fact that there are non-constant holomorphic functions $f \in \mathrm{Hol}(U, \mathcal{E})$ for which $z \mapsto \|f(z)\|$ is constant. (cf., e.g. [VES 70]).

Let $K \subset \mathcal{E}$. A point $x \in K$ is called a *complex extreme point* of K if $y = 0$ is the only vector in $\mathcal{E}$ such that $x + \zeta y \in K$ for all $\zeta \in \Delta$. Obviously a (real) extreme point is a complex extreme point. However, a complex extreme point may fail to be a real extreme point. For example, every point of norm 1 of the space $\mathcal{E} = L^1(\mathbb{R}, dx)$ is a complex extreme point of the closure $\bar{B}$ of the open unit ball B of $\mathcal{E}$. On the other hand $\bar{B}$ has no real extreme point (cf. [TH-WH 67], or also [VES 70]). The following theorem is due to E. Thorp and R. Whitley [TH-WH 67]. Let D be an open convex neighborhood of 0 in $\mathcal{E}$, and let $f \in \mathrm{Hol}(\Delta, \mathcal{E})$ be such that $f(\Delta) \subset \bar{D}$ (the closure of D in $\mathcal{E}$).

Theorem 1.5 *If $f(\Delta) \cap \partial D \neq \emptyset$, then $f(\Delta) \subset \partial D$. If $f(\Delta) \cap \partial D$ contains a complex extreme point of $\bar{D}$, then f is constant.*

Remarks The above theorem holds in the more general case in which $\mathcal{E}$ is any locally convex complex vector space. The proof, given in [VES 81], follows the original idea by Thorp and Whitley. A simplified proof, due to L.A. Harris, yields Theorem 1.5 in the case in which the convex neighborhood D of 0 in the complex Banach space $\mathcal{E}$ is also balanced (cf. [FR-VE 80] also for bibliographical references).

Let B' be the open unit ball of a complex Banach space $\mathcal{E}'$.

Corollary 1.4 *Let $f \in \mathrm{Hol}(D, \mathcal{E}')$ be such that $f(D) \subset \bar{B}'$. If every vector of norm one in $\mathcal{E}'$ is a complex extreme point of $\bar{B}'$, then either $f(D) \subset B'$ or f is constant.*

The hypothesis on B' is satisfied, for example, when $\mathcal{E}'$ is a complex Hilbert space.

Let B the open unit ball of $\mathcal{E}$.

Theorem 1.6 (Schwarz lemma) *Let $f \in \mathrm{Hol}(B, \mathcal{E}')$ be such that $f(0) = 0$ and $f(B) \subset \bar{B}'$. Then*

$$\|f(x)\| \leq \|x\| \quad \text{for all } x \in B. \tag{1.2.1}$$

If $||f(x_0|| = ||x_0||$ *at some point* $x_0 \in B \setminus \{0\}$, *then*

$$||f(\zeta x_0)|| = |\zeta|\,||x_0|| \quad \textit{whenever } |\zeta| < 1/||x_0||. \tag{1.2.2}$$

If moreover the set $\{e^{i\theta}||f(x_0||^{-1}f(x_0) : \theta \in \mathbf{R}\}$ *contains a complex extreme point of* $\bar{B}'$, *then*

$$f(\zeta x_0) = \zeta f(x_0) \quad \textit{whenever } |\zeta| < 1/||x_o||.$$

Proof For $x \in B \setminus \{0\}$, the function $\varphi : \zeta \mapsto \frac{1}{\zeta}f(\zeta x)$ maps holomorphically the open disc $\{\zeta \in \mathbf{C} : |\zeta| < \frac{1}{||x||}\}$ into $\mathcal{E}'$. For $0 < r < \frac{1}{||x||}$ and $|\zeta| = r$

$$||\frac{1}{\zeta}f(\zeta x)|| \leq \frac{1}{r},$$

and, by the maximum principle, for all $|\zeta| \leq r$. Letting $r \uparrow \frac{1}{||x||}$, that implies

$$||f(\zeta x)|| \leq |\zeta|||x||,$$

and, for $|\zeta| = 1$, yields (1.2.1). If $||f(x_0)|| = ||x_0||$ at some $x_0 \in B \setminus \{0\}$ the weak maximum principle applied to the function φ implies (1.2.2) whenever $|\zeta| \leq r$, $0 < r < \frac{1}{||x_0||}$, and therefore, letting $r \uparrow \frac{1}{||x_0||}$, implies (1.2.2) in its full generality.

The final part of the statement follows easily from the strong maximum principle. □

Let $\varphi \in \mathrm{Hol}\,(\Delta, B)$ be such that $\varphi(0) = 0$ and $||\varphi'(0)|| = 1$. By the Hahn-Banach theorem, there is a continuous linear form λ on $\mathcal{E}$ such that $||\lambda|| = 1$ and $\lambda(\varphi'(0)) = ||\varphi'(0)|| = 1$. Since $\lambda \circ \varphi \in \mathrm{Hol}\,(\Delta, \Delta)$, by the Schwarz lemma,

$$\lambda \circ \varphi(\zeta) = \zeta$$

for all $\zeta \in \Delta$. Thus, by Theorem 1.6,

$$|\zeta| = |\lambda(\varphi(\zeta))| \leq ||\varphi(\zeta)|| \leq |\zeta|,$$

and therefore

$$||\varphi(\zeta)|| = |\zeta| \tag{1.2.3}$$

for all $\zeta \in \Delta$. That proves

Proposition 1.2 *If* $\varphi \in \mathrm{Hol}\,(\Delta, B)$ *is such that* $\varphi(0) = 0$, *and* $||\varphi'(0)|| = 1$, *then* (1.2.3) *holds for all* $\zeta \in \Delta$. *If moreover* $\varphi'(0)$ *is a complex extreme point of* $\bar{B}$, *then*

$$\varphi(\zeta) = \zeta\varphi'(0)$$

for all $\zeta \in \Delta$.

The final part of this statement follows from the strong maximum principle applied to the function $\zeta \mapsto \frac{1}{\zeta}\varphi(\zeta)$.

Lemma 1.4 (H. Cartan's uniqueness theorem) *Let D be a bounded domain in $\mathcal{E}$, and let $f : D \to D$ be a holomorphic map fixing a point $x_0 \in D$. If $df(x_0) = Id$, then $f(x) = x$ for all $x \in D$.*

Proof Assuming $x_0 = 0$, f is represented in a neighborhood U of 0 by a convergent power series

$$f(x) = x + P_{q_0}(x) + P_{q_0+1}(x) + \cdots$$

where P_q is a continuous homogeneous polynomial $P_q : \mathcal{E} \to \mathcal{E}$ of degree $q \geq q_0 \geq 2$. If $f^n = f \circ \cdots \circ f$ is the n-th iterate of f, then

$$f^n(x) = x + nP_{q_0}(x) + o_{q_0+1}(||x||)$$

in a neighborhood of 0. If $r_0 > 0$ is the distance of 0 from $\mathcal{E} \setminus D$, then, for any $r \in (0, r_0)$, and any $x \in \mathcal{E} \setminus \{0\}$,

$$nP_{q_0}(x) = n(\frac{||x||}{r})^{q_0} P_{q_0}(\frac{r}{||x||}x) = (\frac{||x||}{r})^{q_0} \frac{1}{2\pi} \int_0^{2\pi} e^{-iq_0\theta} f(\frac{r}{||x||}e^{i\theta}x)d\theta,$$

whence

$$||P_{q_0}(x)|| \leq \frac{||x||^{q_0}}{r^{q_0}} \sup\{||f^n(y)|| : y \in D\}.$$

Since D is bounded, that implies, letting $n \to +\infty$, that $P_{q_0}(x) = 0$ for all $x \in \mathcal{E}$, i.e. $P_{q_0} = 0$. □

Let B and B' be the open unit balls of two complex Banach spaces $\mathcal{E}$ and $\mathcal{E}'$, and let $f \in \mathrm{Hol}(B, B')$ be such that $df(0) \in \mathcal{L}(\mathcal{E}, \mathcal{E}')$ is a surjective isometry. Replacing f by $df(0)^{-1} \circ f$, there is no restriction in assuming $f \in \mathrm{Hol}(B, \mathcal{E}')$ and $df(0) = Id$.

It will be shown now that $f(0) = 0$, so that, by H. Cartan's uniqueness theorem, $f = Id$.

If $x_0 := f(0) \neq 0$, then by the Hahn-Banach theorem there is a continuous linear form $\lambda : \mathcal{E} \to \mathbb{C}$ with norm $||\lambda|| = 1$, such that $\lambda(x_0) = ||x_0||$. Let $h \in \mathrm{Hol}(\Delta, \mathbb{C})$ be defined by

$$h(\zeta) = \lambda(f(\frac{\zeta}{||x_0||}x_0)).$$

Since

$$|h(\zeta)| \leq ||\lambda|| \cdot ||f(\frac{\zeta}{||x_0||}x_0)|| \leq 1,$$

then $h(\Delta) \subset \bar{\Delta}$. Being $h'(0) = \lambda(\frac{1}{||x_0||}x_0) = 1$, then h is non-constant, and therefore $h(\Delta) \subset \Delta$. Since $h(0) = \lambda(f(0)) = \lambda(x_0) = ||x_0||$, the Schwarz-Pick lemma yields

$$\frac{|h'(0)|}{1 - |h(0)|^2} \leq 1,$$

i.e.

$$1 \leq 1 - ||x_0||^2,$$

contradicting the assumption $x_0 \neq 0$.

In conclusion, the following theorem holds, due to L.A. Harris:

Theorem 1.7 *If $f \in \mathrm{Hol}\,(B, B')$ is such that $df(0) \in \mathcal{L}(\mathcal{E}, \mathcal{E}')$ is a surjective isometry, then f is the restriction to B of $df(0)$.*

1.3 The Kobayashi and Carathéodory pseudodistances

The Kobayashi pseudodistance is defined on the domain $D \subset \mathcal{E}$ as follows. By definition, an "analytic chain" in D joining two points x, y in D consists of a finite number ν of functions $\varphi_j \in \mathrm{Hol}\,(\Delta, D)$ and of $\nu + 1$ points $\zeta_0, \zeta_1, \ldots, \zeta_\nu$ in Δ such that

$$\varphi_1(\zeta_0) = x, \quad \varphi_j(\zeta_j) = \varphi_{j+1}(\zeta_j) \ \ (j = 1, \ldots, \nu - 1), \quad \varphi_\nu(\zeta_\nu) = y.$$

The open set D being connected, analytic chains joining x and y in D do exist. The Kobayashi pseudo-distance $k_D(x, y)$ is defined by

$$k_D(x, y) = \inf \sum_{j=1}^{\nu} \omega(\zeta_{j-1}, \zeta_j),$$

where the infimum is taken over all analytic chains joining x and y in D.

Note that k_D is indeed a pseudodistance, and that

$$k_{D'}(f(x), f(y)) \leq k_D(x, y) \tag{1.3.1}$$

for all $f \in \mathrm{Hol}\,(D, D')$, $x, y \in D$. A simple application of the Schwarz-Pick lemma yields

$$k_\Delta(\zeta_1, \zeta_2) = \omega(\zeta_1, \zeta_2) \qquad (\zeta_1, \zeta_2 \in \Delta). \tag{1.3.2}$$

Hence, by (1.3.1),

$$\omega(\psi(x), \psi(y) \leq k_D(x, y) \tag{1.3.3}$$

for all $x, y \in D$ and $\psi \in \mathrm{Hol}\,(D, \Delta)$. Thus, letting

$$c_D(x, y) := \sup\{\omega(\psi(x), \psi(y)) : \psi \in \mathrm{Hol}\,(D, \Delta)\}, \tag{1.3.4}$$

then

$$c_D(x, y) \leq k_D(x, y) \qquad (x, y \in D). \tag{1.3.5}$$

$c_D(x, y)$ is, by definition, the *Carathéodory pseudo-distance* of x and y in D (it is in fact a pseudo-distance). c_D is contracted by all holomorphic maps, i.e.

$$c_{D'}(f(x), f(y)) \leq c_D(x, y) \tag{1.3.6}$$

for all $f \in \mathrm{Hol}\,(D, D')$, $x, y \in D$.

Hence both c_D and k_D are invariant under the action of $\mathrm{Aut}\, D$, i.e. the elements of $\mathrm{Aut}\, D$ are isometries for c_D and k_D. Since $\omega(\zeta_1, \zeta_2) \leq c_\Delta(\zeta_1, \zeta_2)$, then (1.3.5) and (1.3.2) imply that

$$c_\Delta(\zeta_1, \zeta_2) = k_\Delta(\zeta_1, \zeta_2) = \omega(\zeta_1, \zeta_2)$$

for all ζ_1, ζ_2 in Δ.

Example Let $|| \ ||$ be the norm in $\mathcal{E}$, and, for $x_0 \in \mathcal{E}, r > 0$, let

$$B(x_0, r) = \{x \in \mathcal{E} : ||x - x_0|| < r\}$$

By the Hahn-Banach theorem, for any $x \in B(x_0, r)$ there is a continuous linear form λ on $\mathcal{E}$ such that $||\lambda|| = 1$ and $\lambda(x - x_0) = ||x - x_0||$. The holomorphic function $\psi : y \mapsto \frac{1}{r}\lambda(y - x_0)$ maps $B(x_0, r)$ into Δ. For $x \in B(x_0, r)$, $x \neq x_0$, let $\varphi \in \mathrm{Hol}\,(\Delta, B(x_0, r))$ be defined by

$$\varphi(\zeta) = x_0 + r\frac{\zeta}{||x - x_0||}(x - x_0).$$

Since $\psi(x_0) = 0$, $\psi(x) = \frac{||x - x_0||}{r}$ and $\varphi(0) = x_0$, $\varphi(\frac{||x - x_0||}{r}) = x$, then

$$\omega(0, \frac{||x - x_0||}{r}) \leq c_{B(x_0,r)}(x_0, x) \leq k_{B(x_0,r)}(x_0, x) \leq \omega(0, \frac{||x - x_0||}{r}),$$

whence

$$c_{B(x_0,r)}(x_0, x) = k_{B(x_0,r)}(x_0, x) = \omega(0, \frac{||x - x_0||}{r}),$$

for all $x \in B(x_0, r)$. Letting $r \to +\infty$, then

$$c_{\mathcal{E}} = k_{\mathcal{E}} \equiv 0.$$

Let D be a domain and, for any $x_0 \in D$, let $r > 0$ and $s > 0$ be such that

$$B(x_0, r + s) \subset D.$$

Then, for any $x \in \overline{B(x_0, r)}$,

$$\begin{aligned} c_D(x_0, x) &\leq k_D(x_0, x) \leq k_{B(x_0,r+s)}(x_0, x) = \omega(o, \frac{||x - x_0||}{r + s}) \\ &= \frac{||x - x_0||}{r + s} + \frac{1}{3}(\frac{||x - x_0||}{r + s})^3 + \frac{1}{5}(\frac{||x - x_0||}{r + s})^5 + \cdots \\ &\leq \frac{||x - x_0||}{r + s}(1 + (\frac{||x - x_0||}{r + s})^2 + (\frac{||x - x_0||}{r + s})^4 + \cdots) \\ &= \frac{||x - x_0||}{r + s}\frac{1}{1 - (\frac{||x - x_0||}{r+s})^2} \leq \frac{||x - x_0||}{r + s}\frac{1}{1 - (\frac{r}{r+s})^2} \\ &\leq \frac{||x - x_0||}{r + s}\frac{1}{1 - \frac{r}{r+s}} = \frac{||x - x_0||}{s}. \end{aligned} \tag{1.3.7}$$

Thus the functions $x \mapsto c_D(x, x_0)$, $x \mapsto k_D(x, x_0)$ are continuous on D. Since, for x_0, y_0, x, y in D,

$$\begin{aligned} |k_D(x, y) - k_D(x_0, y_0)| &\leq k_D(x_0, x) + k_D(y_0, y), \\ |c_D(x, y) - c_D(x_0, y_0)| &\leq c_D(x_0, x) + c_D(y_0, y), \end{aligned}$$

the functions $c_D : D \times D \to \mathbf{R}_+$, $k_D : D \times D \to \mathbf{R}_+$ are continuous.

If D is bounded, there is $R > 0$ such that

$$D \subset B(x_0, R).$$

For any $x \in \overline{B(x_0, r)}$

$$\frac{\|x - x_0\|}{R} \le \omega(0, \frac{\|x - x_0\|}{R}) = c_{B(x_0,R)} \le c_D(x_0, x) \le k_D(x_0, x).$$

These inequalities, together with (1.3.7) prove the

Theorem 1.8 *If the domain D is bounded, the distances c_D and k_D are equivalent to the norm distance on any closed ball C completely interior to D (i.e. such that* $\inf\{\|x - y\| : x \in C, y \notin D\} > 0$.)

By Riemann's extension theorem, the Carathéodory distance of the domain $\Delta \setminus \{0\}$ is the restriction to $\Delta \setminus \{0\}$ of the Poincaré = Carathéodory = Kobayashi distance of Δ. On the other hand, Kobayashi has shown that the Kobayashi distance of $\Delta \setminus \{0\}$ is complete. Thus $\Delta \setminus \{0\}$ offers an example of a bounded domain on which the Carathéodory and the Kobayashi distances are not equivalent.

The fact that $\Delta \setminus \{0\}$ is not convex plays a crucial role. Indeed the following theorem holds.

Theorem 1.9 *If the domain D is convex, then $c_D = k_D$.*

This theorem was first proved by L. Lempert [LEM 82] and H. Royden, P. Wong in the finite dimensional case, and by S. Dineen, R. Timoney, J.P. Vigué [D-T-V 85] for all domains in any locally convex, Hausdorff complex vector space.

Theorem 1.10 *The function*

$$(x, y) \mapsto \log c_D(x, y)$$

is plurisubharmonic on $D \times D$.

Proof Let $x_1, x_2 \in D$, $u_1, u_2 \in \mathcal{E}$. The set

$$V = \{\zeta \in \mathbf{C} : x_1 + \zeta u_1 \in D,\ x_2 + \zeta u_2 \in D\}$$

is open in $\mathbf{C}$. Let U be a connected component of V.

For any $\psi \in \mathrm{Hol}\,(D, \Delta)$ the function

$$\zeta \mapsto \log \omega(\psi(x_1 + \zeta u_1), \psi(x_2 + \zeta u_2))$$

is subharmonic on U by Theorem 1.4. Since c_D is a continuous function on $D \times D$, the conclusion follows from (1.3.4). □

Corollary 1.5 *The function*

$$(x, y) \mapsto c_D(x, y)$$

is plurisubharmonic on $D \times D$.

Let U be a domain in $\mathbf{C}$ and let $f \in \operatorname{Hol}(U \times D, D)$. By (1.3.6)

$$c_D(f(\tau, x), f(\tau, y)) \leq c_D(x, y)$$

for all $\tau \in U, x, y \in D$.

Lemma 1.5 *If for every pair of points x, y in D there is some $\zeta \in U$ such that*

$$c_D(f(\zeta, x), f(\zeta, y)) = c_D(x, y), \tag{1.3.8}$$

then $f(\zeta, \bullet)$ is an isometry for c_D.

Proof By Corollary 1.5, the maximum principle for subharmonic functions implies that $c_D(f(\tau, x), f(\tau, y)$ does not depend on τ, and thus (1.3.8) holds for all $\zeta \in U$. Since that happens for all x, y in D, the lemma is proved. □

Corollary 1.6 *If $f(\zeta_0, \bullet)$ is an isometry for c_D for some $\zeta_0 \in U$, then $f(\zeta, \bullet)$ is an isometry for c_D for all $\zeta \in U$.*

For $x \in D$, let

$$T(x) = \{\zeta \in U : f(\zeta, x) = x\}.$$

By Theorem 1.10 and a classical result of H. Cartan on the singular sets of subharmonic functions, either $T(x)$ is a G_δ with outer capacity zero or $c_D(f(\zeta, x), x) = 0$ for all $\zeta \in U$. That proves

Lemma 1.6 *If c_D is a distance and if $T(x)$ has positive outer capacity, then $f(\zeta, x) = x$ for all $\zeta \in U$.*

1.4 The Carathéodory and Kobayashi differential pseudometrics

The *Kobayashi differential pseudometric* $\kappa_D : D \times \mathcal{E} \to \mathbf{R}_+$ of the domain $D \subset \mathcal{E}$ is given for $x \in D$, $v \in \mathcal{E}$, by

$$\begin{aligned} \kappa_D(x; v) &= \inf\{\langle\tau\rangle_\zeta : \zeta \in \Delta, \tau \in \mathbf{C}, \varphi \in \operatorname{Hol}(\Delta, D), \varphi(\zeta) = x, \tau\varphi'(\zeta) = v\} \\ &= \inf\{|\tau| : \varphi \in \operatorname{Hol}(\Delta, D), \varphi(0) = x, \tau\varphi'(0) = v\}. \end{aligned}$$

Note that functions φ satisfying the above conditions do exist. For $\varphi \in \operatorname{Hol}(\Delta, D)$, such that $\varphi(\zeta) = x, \tau\varphi'(\zeta) = v$, and $\psi \in \operatorname{Hol}(D, \Delta)$, then

$$d\psi(x)v = d\psi(\varphi(\zeta))d\varphi(\zeta)\tau = d(\psi \circ \varphi)(\zeta)\tau.$$

Hence by the Schwarz-Pick lemma

$$\langle d\psi(x)v\rangle_{\psi(x)} \leq \langle \tau\rangle_\zeta.$$

Since this inequality holds for all $\varphi \in \mathrm{Hol}(\Delta, D)$ such that $\varphi(\zeta) = x$, $\tau\varphi'(\zeta) = v$, then

$$\langle d\psi(x)v\rangle_{\psi(x)} \leq \kappa_D(x, v).$$

Thus, setting

$$\gamma_D(x, v) = \sup\{\langle d\psi(x)v\rangle_{\psi(x)} : \psi \in \mathrm{Hol}(D, \Delta)\}, \tag{1.4.1}$$

then

$$\gamma_D(x, v) \leq k_D(x, v) \quad \text{for all } x \in D, v \in \mathcal{E}. \tag{1.4.1'}$$

The function $\gamma_D : D \times \mathcal{E} \to \mathbf{R}_+$ is called the *Carathéodory differential pseudometric* of D.

The definitions imply readily that, if $f \in \mathrm{Hol}(D, D')$, then

$$\begin{aligned}\kappa_{D'}(f(x), df(x)v) &\leq \kappa_D(x, v),\\ \gamma_{D'}(f(x), df(x)v) &\leq \gamma_D(x, v)\end{aligned}$$

for all $x \in D$, $v \in \mathcal{E}$. In particular, both κ_D and γ_D are invariant under the action of $\mathrm{Aut}\, D$.

Taking $\psi : \zeta \mapsto \zeta$, then

$$\langle \tau\rangle_\zeta \leq \gamma_\Delta(\zeta, \tau) \leq \kappa_\Delta(\zeta, \tau) \leq \langle \tau\rangle_\zeta,$$

whence

$$\gamma_\Delta = \kappa_\Delta = \langle\ \rangle. \tag{1.4.1''}$$

Let B be the open unit ball of $\mathcal{E}$; let $v \in \mathcal{E} \setminus \{0\}$ and let λ be a continuous linear form on $\mathcal{E}$ such that $||\lambda|| = 1, \lambda(v) = ||v||$.

Then, $\lambda \in \mathrm{Hol}(B, \Delta)$. Let $\varphi \in \mathrm{Hol}(\Delta, B)$ be defined by

$$\varphi(\zeta) = \frac{\zeta}{||v||}v.$$

Then

$$\begin{aligned}||v|| = \lambda(v) &= \gamma_\Delta(0, \lambda(v)) = \gamma_\Delta(\lambda(0), d\lambda(0)v)\\ &\leq \gamma_B(0, v) \leq \kappa_B(0, v) = \kappa_B(\varphi(0), d\varphi(0)||v||)\\ &\leq \kappa_\Delta(0, ||v||) = ||v||,\end{aligned}$$

i.e.

$$\gamma_B(0, v) = \kappa_B(0, v) = ||v|| \tag{1.4.2}$$

and therefore, for all $x_0 \in \mathcal{E}$, $r > 0$,

$$\gamma_{B(x_0,r)}(x_0, v) = \kappa_{B(x_0,r)}(x_0, v) = \frac{||v||}{r}. \tag{1.4.3}$$

Letting $r \to +\infty$, this implies that

$$\gamma_{\mathcal{E}} = \kappa_{\mathcal{E}} \equiv 0. \tag{1.4.4}$$

The definition implies that γ_D is a seminorm, i.e.

$$\gamma_D(x, a_1v_1 + a_2v_2) \leq |a_1|\gamma_D(x, v_1) + |a_2|\gamma_D(x, v_2)$$

for all $x \in D$, $a_1, a_2, \in \mathbb{C}, v_1, v_2 \in \mathcal{E}$; (1.4.3) implies then that $\gamma_D(x, \bullet)$ is a continuous seminorm on $\mathcal{E}$.

Lemma 1.7 *If D is bounded, then, for all $x \in D$, $\gamma_D(x, \bullet)$ is an equivalent norm to $\| \ \|$.*

Proof Let $r > 0$, $R > 0$ be such that

$$B(x, r) \subset D \subset B(x, R).$$

Then, for any $v \in \mathcal{E}$,

$$\gamma_{B(x,R)}(x, v) \leq \gamma_D(x, v)) \leq \gamma_{B(x,r)}(x, v),$$

i.e.

$$\frac{\|v\|}{R} \leq \gamma_D(x, v) \leq \frac{\|v\|}{r}.$$

□

Proposition 1.3 *The function $\gamma_D : D \times \mathcal{E} \to \mathbb{R}_+$ is locally Lipschitz.*

Proof a) Let $x_0 \in D, r > 0$ be such that $B(x_0, r) \subset D$, and let $v \in \mathcal{E}$. Let $x_1, x_2 \in B(x_0, \frac{r}{4})$ and suppose that $\gamma_D(x_2, v) \geq \gamma_D(x_1, v)$. Then

$$\begin{aligned} \gamma_D(x_2, v) - \gamma_D(x_1, v) &= \sup\{|d\psi(x_2)v| : \psi \in \mathrm{Hol}\,(D, \Delta), \psi(x_2) = 0\} \qquad (1.4.5) \\ &\quad - \sup\left\{\frac{|d\psi(x_1)v|}{1 - |\psi(x_1)|^2} : \psi \in \mathrm{Hol}\,(D, \Delta), \psi(x_2) = 0\right\} \\ &\leq \sup\{|d\psi(x_2)v| - \frac{|d\psi(x_1)v|}{1 - |\psi(x_1)|^2} : \psi \in \mathrm{Hol}\,(D, \Delta), \psi(x_2) = 0\} \\ &\leq \sup\{|d\psi(x_2)v| - |d\psi(x_1)v| : \psi \in \mathrm{Hol}\,(D, \Delta), \psi(x_2) = 0\} \\ &\leq \sup\{|d\psi(x_2)v - d\psi(x_1)v| : \psi \in \mathrm{Hol}\,(D, \Delta), \psi(x_2) = 0\}. \end{aligned}$$

Now, for $y \in B(x_0, \frac{r}{2})$ and $v \in \mathcal{E} \setminus \{0\}$, $y + \frac{r}{2}\frac{1}{\|v\|}v \in B(x_0, r)$. Therefore, it follows from

$$d\psi(y)v = \frac{1}{2\pi}\int_0^{2\pi} e^{-i\theta}\psi(y + e^{i\theta}v)d\theta,$$

that

$$|d\psi(y)v| \leq \frac{2\|v\|}{r} \tag{1.4.6}$$

for all $v \in \mathcal{E}$. Applying this inequality to the function $y \mapsto d\psi(y)v$ one has

$$|dd\psi(y)(v)(x_2 - x_1)| \leq \frac{2\|x_2 - x_1\|}{r}\frac{2\|v\|}{r} = \frac{4\|x_2 - x_1\|\|v\|}{r^2}.$$

Since, by the mean value theorem,

$$|d\psi(x_2)v - d\psi(x_1)v| \leq \sup\{|dd\psi(y)(v)(x_2 - x_1)| : y = x_1 + t(x_2 - x_1), 0 \leq t \leq 1\},$$

then

$$|d\psi(x_2)v - d\psi(x_1)v| \leq \frac{4}{r^2}\|x_2 - x_1\|\|v\|,$$

and (1.4.1) yields

$$|\gamma_D(x_2, v) - \gamma_D(x_1, v)| \leq \frac{4}{r^2}\|x_2 - x_1\|\|v\|, \tag{1.4.7}$$

for all $x_1, x_2 \in B(x_0, \frac{r}{4})$ and $v \in \mathcal{E}$.

b) For $v_1, v_2 \in \mathcal{E}$, (1.4.6) yields

$$\begin{aligned}|\gamma_D(x_2, v_2) - \gamma_D(x_2, v_1)| &\leq \gamma_D(x_2, v_2 - v_1)\\ &= \sup\{|d\psi(x_2)(v_2 - v_1)| : \psi \in \mathrm{Hol}\,(D, \Delta), \psi(x_2) = 0\} \leq \frac{2}{r}\|v_2 - v_1\|.\end{aligned}$$

c) For $x_1, x_2 \in B(x_0, \frac{r}{4}), v_1, v_2 \in \mathcal{E}$, a) and b) imply

$$\begin{aligned}|\gamma_D(x_2, v_2) - \gamma_D(x_1, v_1)| &\leq |\gamma_D(x_2, v_2) - \gamma_D(x_2, v_1)| + |\gamma_D(x_2, v_1) - \gamma_D(x_1, v_1)|\\ &\leq \frac{2}{r}\|v_1 - v_2\| + \frac{4}{r^2}\|x_2 - x_1\|\|v_1\|.\end{aligned}$$

□

Given two points x, y in D, there is an admissible curve l joining x and y in D, i.e. there is a piecewise C^1 function $l : [0, 1] \to D$ such that $l(0) = x$, $l(1) = y$. Let $L_\gamma(l)$ be the "lenght" of l with aspect to the line element $\gamma(l(t)), \dot{l}(t))\, dt$, i.e.

$$L_\gamma(l) = \int_0^1 \gamma_D(l(t), \dot{l}(t))dt.$$

Given $x, y \in D$, let

$$\tilde{c}_D(x, y) = \inf\{L_\gamma(l) : l \text{ admissible curve joining } x \text{ and } y \text{ in } D\}.$$

Clearly $\tilde{c}_D : D \times D \to \mathbf{R}_+$ is a pseudo-distance. It will be called the *integrated form* of γ_D. Furthermore

$$\tilde{c}_\Delta = \omega = c_\Delta.$$

For any $\psi \in \mathrm{Hol}\,(D, \Delta)$, and any admissible curve l joining x and y in D,

$$\omega(\psi(x), \psi(y)) \leq \int_0^1 \langle \overset{\bullet}{\overbrace{\psi(l(t))}} \rangle_{\psi(l(t))} dt \leq \int_0^1 \gamma_D(l(t), \dot{l}(t))dt,$$

and therefore

$$\omega(\psi(x), \psi(y)) \le \tilde{c}_D(x, y)$$

for all $\psi \in \mathrm{Hol}\,(D, \Delta)$. Thus

$$c_D(x, y) \le \tilde{c}_D(x, y) \tag{1.4.8}$$

for all $x, y \in D$.

It turns out that $\tilde{c}_D$ does not necessarily coincide with c_D.

Example Let D be the domain in $\mathbb{C}^2$ defined, for a given a with $0 < a < \frac{1}{2}$, by

$$D = \Delta \times \Delta \setminus \{(z_1, z_2) : |z_1| \le a, |z_2| \le 1 - a\}.$$

In this case, $\tilde{c}_D \ne c_D$. [FR-VE 80, pp. 89–90 and 137].

As an immediate consequence of the definition, γ_D is the derivative of $\tilde{c}_D$, in the sense that

$$\lim_{\tau \to 0} \frac{1}{|\tau|} \tilde{c}_D(x, x + \tau v) = \gamma_D(x, v).$$

Furthermore, the limit exists locally uniformly in $x \in D, v \in \mathcal{E}$. It will be shown now that γ_D is also the derivative of c_D, i.e. that

$$\lim_{\tau \to 0} \frac{1}{|\tau|} c_D(x, x + \tau v) = \gamma_D(x, v). \tag{1.4.9}$$

for all $x \in D, v \in \mathcal{E}$.

For $x \in D$, let $r > 0$ be such that $B(x, r) \subset D$. For $0 < \|v\| < \frac{r}{2}$, for $\psi \in \mathrm{Hol}\,(D, \Delta)$ with $\psi(x) = 0$, and for $\tau \in \Delta$, set

$$f(\tau) = \psi(x + \frac{\tau r}{\|v\|} v).$$

Since $\|x + \frac{\tau r}{\|v\|} v) - x\| = |\tau| r < r$, then $f \in \mathrm{Hol}\,(\Delta, \Delta)$. If

$$f(\tau) = a_1 \tau + a_2 \tau^2 + \cdots$$

is the power series expansion of f, then, by the Cauchy inequalities,

$$|a_\nu| \le 1 \qquad (\nu = 1, 2, \ldots).$$

Hence

$$|f(\tau) - f'(0)\tau| = |\tau|^2 |a_2 + a_3 \tau + \cdots| \le |\tau|^2 (1 + |\tau| + |\tau|^2 + \cdots) = \frac{|\tau|^2}{1 - |\tau|}.$$

In particular, for $\tau = \frac{\|v\|}{r}$,

$$|f(\frac{\|v\|}{r}) - f'(0)\frac{\|v\|}{r}| \le \frac{\|v\|^2}{r^2} \Big/ (1 - \frac{\|v\|}{r})$$

i.e.

$$|\psi(x+v) - d\psi(x)v| \leq \frac{\|v\|^2}{r^2} \Big/ (1 - \frac{\|v\|}{r}),$$

or also

$$-\frac{\|v\|^2}{r^2} \Big/ (1 - \frac{\|v\|}{r}) \leq |d\psi(x)v| - |\psi(x+v)| \leq \frac{\|v\|^2}{r^2} \Big/ (1 - \frac{\|v\|}{r}).$$

Since

$$\frac{\|v\|}{r} < \frac{1}{2},$$

then

$$\frac{\|v\|^2}{r^2} \Big/ (1 - \frac{\|v\|}{r}) > 2\frac{\|v\|^2}{r^2},$$

and therefore

$$|d\psi(x)v| - 2\frac{\|v\|^2}{r^2} \leq |\psi(x+v)| \leq \omega(0, \psi(x+v)) = \omega(\psi(x), \psi(x+v)).$$

Hence

$$\gamma_D(x,v) - 2\frac{\|v\|^2}{r^2} \leq \omega(\psi(x), \psi(x+v)) \leq c_D(x, x+v). \tag{1.4.10}$$

For $\|v\| < \frac{r}{4}$, let $l(t) = x + tv$. Then $l([0,1]) \subset B(x, \frac{r}{4})$, and

$$\begin{aligned} \tilde{c}_D(x, x+v) &\leq \int_0^1 \gamma_D(l(t), \dot{\ell}(t))dt \\ &= \gamma_D(x,v) + \int_0^1 (\gamma_D(x+tv, v) - \gamma_D(x,v))dt. \end{aligned}$$

Thus, by (1.4.7),

$$\tilde{c}_D(x, x+v) \leq \gamma_D(x,v) + \frac{4}{r^2}\|v\|^2 \int_0^1 tdt = \gamma_D(x,v) + \frac{2}{r^2}\|v\|^2,$$

whence, by (1.4.8)

$$c_D(x, x+v) \leq \gamma_D(x,v) + \frac{2}{r^2}\|v\|^2.$$

Hence, by (1.4.10)

$$\gamma_D(x,v) - \frac{2}{r^2}\|v\|^2 r^2 \leq c_D(x, x+v) \leq \gamma_D(x,v) + \frac{2}{r^2}\|v\|^2. \tag{1.4.11}$$

For $u \in D$, let $r > 0$ be such that $B(u, 2r) \subset D$. Given $\varepsilon > 0$, let $\alpha = \min\{\frac{r}{2}, \varepsilon\frac{r^2}{2}\}$. Then, if $x \in B(u, \alpha)$, and $\|y - x\| < r$,

$$\|y - u\| \leq \|y - x\| + \|x - u\| < r + \frac{r}{2} < 2r,$$

i.e.

$$B(x,r) \subset B(u,2r) \subset D.$$

Moreover, if $\|v\| < \alpha$, then $\|v\| < \varepsilon\frac{r^2}{2}$, i.e. $\varepsilon\|v\| > \frac{2}{r^2}\|v\|^2$, and (1.4.11) yields

$$\gamma_D(x,v) - \varepsilon\|v\| \leq c_D(x, x+v) \leq \gamma_D(x,v) + \varepsilon\|v\|$$

for all $x \in B(u,\alpha)$ and all $v \in B(0,\alpha)$. That proves that (1.4.9) holds.

The behaviour of the Kobayashi differential metric diverges considerably from that of Carathéodory's metric.

While

$$\kappa_D(x, av) = |a|\kappa_D(x,v)$$

for all $a \in \mathbb{C}$, $\kappa_D(x,\bullet)$ is not necessarily subadditive, i.e. is not a seminorm. W. Kaup [KAU 82] has shown that $\kappa_D(x,\bullet)$ is a seminorm if D is convex.

Furthermore κ_D does not share the regularity properties of γ_D. All one can prove so far is the following result, due to H.L. Royden [FR-VE 80]:

The infinitesimal Kobayashi metric $\kappa_D : D \times \mathcal{E} \to \mathbb{R}_+$ is upper semicontinuous. The Kobayashi pseudodistance k_D is the integrated form of κ_D.

By (1.4.1′)

$$\tilde{c}_D(x,y) \leq k_D(x,y), \qquad (x, y \in D).$$

Let D_1 and D_2 be two domains in two Banach spaces $\mathcal{E}_1$ and $\mathcal{E}_2$, and let

$$x_j \in D_j, v_j \in \mathcal{E}_j \qquad (j = 1,2).$$

By the contraction property of holomorphic maps,

$$\kappa_{D_1\times D_2}((x_1,x_2),(v_1,v_2)) \geq \kappa_{D_j}(x_j,v_j) \qquad (j=1,2).$$

If

$$\kappa_{D_1\times D_2}((x_1,x_2),(v_1,v_2)) > \max\{\kappa_{D_1}(x_1,v_1), \kappa_{D_2}(x_2,v_2)\},$$

there exist $\varphi_j \in \mathrm{Hol}(\Delta, D_j), \tau_j \in \mathbb{C}$, such that

$$\varphi_j(0) = x_j, \quad \tau_j\varphi_j'(0) = v_j,$$

$$\kappa_{D_1\times D_2}((x_1,x_2),(v_1,v_2)) > \tau_j \geq \kappa_{D_j}(x_j,v_j) \tag{1.4.12}$$

for $j=1,2$. Let $\varphi(\zeta_1,\zeta_2) = (\varphi_1(\zeta_1), \varphi_2(\zeta_2))(\zeta_j \in \Delta)$. Then $\varphi \in \mathrm{Hol}(\Delta\times\Delta, D_1\times D_2)$, and

$$\varphi(0) = (x_1,x_2), \quad d\varphi(0)\,(\tau_1,\tau_2) = (v_1,v_2).$$

Thus

$$\begin{aligned}\kappa_{D_1\times D_2}((x_1,x_2),(v_1,v_2)) &\leq \kappa_{\Delta\times\Delta}((0,0), {}^t(\tau_1,\tau_2)) \\ &= \max\{|\tau_1|,|\tau_2|\}\end{aligned}$$

(because $\Delta \times \Delta$ is the open unit disc of $\mathbb{C}^2$ for the norm $\|(\zeta_1, \zeta_2)\| = \max\{|\zeta_1|, |\zeta_2|\}$), contradicting (1.4.12).

That proves that

$$\kappa_{D_1 \times D_2}((x_1, x_2), (v_1, v_2)) = \max\{\kappa_{D_1}(x_1, v_1), \kappa_{D_2}(x_2, v_2)\} \tag{1.4.13}$$

for all $x_j \in D_j, v_j \in \mathcal{E}_j$ $(j = 1, 2)$.

1.5 Hyperbolic domains

By Theorem 1.8 the relative topology of D in $\mathcal{E}$ is finer than the topology defined by the pseudodistances k_D and c_D; if the domain D is bounded, all these topologies coincide.

If k_D is a distance and if the topology defined on D by k_D coincides with the relative topology, D is called a *hyperbolic domain.* According to a result of T.J. Barth (cf., e.g., [FR-VE 80]), if $\dim_{\mathbb{C}} \mathcal{E} < \infty$ and if k_D is a distance, then D is hyperbolic. Examples (cf., e.g., [FR-VE 80]) show that Barth's result does not extend to infinite dimensional domains.

The question arises whether Barth's result remains true when k_D is replaced by the Carathéodory distance c_D. The answer is negative as was shown in [J-P-V 91] by exhibiting the example of a domain of holomorphy D in $\mathbb{C}^n$, for any $n \geq 3$, on which c_D is a distance defining a topology that is not equivalent to the relative topology.

Proposition 1.4 *If D is a hyperbolic domain, then, for any $x_0 \in D$, there is a constant $c > 0$ such that*

$$\kappa_D(x_0, v) \geq c\|v\|$$

for all $v \in \mathcal{E}$.

Proof Let $r > 0$ and $s > 0$ be such that

$$B(x_0, r) \subset D, \qquad B_k(x_0, s) \subset B(x_0, r),$$

where $B_k(x_0, s)$ is the open ball with center x_0 and radius s for the Kobayashi distance in D.

If there is no constant $c > 0$ satisfying the theorem, there is a sequence $\{v_\nu\}$ in $\mathcal{E}$, such that $\|v_\nu\| = 1$ for all ν, and

$$\lim_{\nu \to +\infty} \kappa_D(x_0, v_\nu) = 0.$$

Assume $\kappa_D(x_0, v_\nu) < 1$ for all ν, and select a sequence $\{\epsilon_\nu\}$ for which

$$\kappa_D(x_0, v_\nu) < \epsilon_\nu < 1, \qquad \lim_{\nu \to +\infty} \epsilon_\nu = 0.$$

Let $\varphi_\nu \in \mathrm{Hol}(\Delta, D)$ and $\tau_\nu \in \mathbb{C}$ be such that

$$\varphi_\nu(0) = x_0, \quad \tau_\nu \varphi_\nu'(0) = U_\nu, \quad |\tau_\nu| < \epsilon_\nu.$$

Thus

$$\|\varphi'_\nu(0)\| = \|v_\nu\|/|\tau_\nu| = 1/|\tau_\nu| > 1/\epsilon_\nu. \tag{1.5.1}$$

Let $\delta_\nu = \sqrt{\epsilon_\nu}$. Since for $\nu \gg 0$

$$\varphi'_\nu(0) = \frac{1}{2\pi\delta_\nu}\int_0^{2\pi} e^{-i\theta}\varphi_\nu(\delta_\nu e^{i\theta})d\theta,$$

then, by the maximum principle,

$$\|\varphi'_\nu(0)\| \leq \frac{1}{\delta_\nu}\sup\{\|\varphi_\nu(\zeta)\| : |\zeta| \leq \delta_\nu\}.$$

Hence (1.5.1) implies

$$\sup\{\|\varphi_\nu(\zeta)\| : |\zeta| \leq \delta_\nu\} > \frac{\delta_\nu}{\epsilon_\nu} = \frac{1}{\sqrt{\epsilon_\nu}}.$$

Since $\lim_{\nu\to+\infty} 1/\sqrt{\epsilon_\nu} = +\infty$, there is ν_0 such that, whenever $\nu > \nu_0$,

$$\|\varphi_\nu(\zeta_\nu)\| > r$$

for some ζ_ν with $|\zeta_\nu| \leq \delta_\nu$. Thus

$$k_D(x_0, \varphi_\nu(\zeta_\nu)) > s$$

for all $\nu > \nu_0$, contradicting the fact that

$$k_D(x_0, \varphi_\nu(\zeta_\nu)) \leq \omega(0, \zeta_\nu) =\to 0$$

as $\nu \to +\infty$. □

If D is hyperbolic, $B_k(x_0, r)$ is a bounded open domain in $\mathcal{E}$. Hence Lemma 1.4 extends to hyperbolic domains:

Theorem 1.11 *Let D be a hyperbolic domain in $\mathcal{E}$, and let $f : D \to D$ be a holomorphic map fixing a point $x_0 \in D$. If $df(x_0) = Id$, then $f(x) = x$ for all $x \in D$.*

Theorem 1.12 (H. Cartan's linearity theorem) *If the hyperbolic domain D is circular around a point $x_0 \in D$, (i.e. such that $x_0 + e^{i\theta}(x - x_0) \in D$ for all $x \in D$ and all $\theta \in \mathbf{R}$) and if $g \in \operatorname{Aut} D$ fixes x_0, then g is the restriction to D of a bounded linear map.*

Proof Choose $x_0 = 0$, and let

$$g(x) = P_1(x) + P_2(x) + \cdots$$

be the power series expansion of g in a neighbourhood of 0, where $P_q : \mathcal{E} \to \mathcal{E}$ is a continuous homogeneous polynomial of degree $q = 1, 2, \ldots$. Denoting by f_θ the holomorphic automorphism of D defined by

$$f_\theta(x) = e^{i\theta}x$$

for any fixed $\theta \in \mathbb{R}$, let $h \in \operatorname{Aut} D$ be defined by

$$h(x) = g^{-1} \circ f_{-\theta} \circ g \circ f_{\theta}.$$

Then $h(0) = 0$ and

$$dh(0) = Id.$$

By Theorem 1.11 , h is the identity, i.e.

$$g \circ f_{\theta} = f_{\theta} \circ g.$$

Thus, there is a circular neighbourhood U of 0 such that

$$e^{i\theta}(P_1(x) + P_2(x) + \cdots) = e^{i\theta} P_1(x) + e^{2i\theta} P_2(x) + \cdots$$

for all $\theta \in \mathbb{R}$ and all $x \in U$. Hence $P_q = 0$ for $q \geq 2$. □

1.6 Injective hyperbolicity

Let $S(\Delta, D)$ be the set of all injective holomorphic maps of Δ into D. In [HAH 81] K.T. Hahn has introduced a differential metric on D, setting, for $x \in D, v \in \mathcal{E}$,

$$\eta_D(x, v) = \inf\{|\tau| : \tau \in \mathbb{C}, \varphi \in S(\Delta, D), \varphi(0) = x, \tau\varphi'(0) = v\}.$$

In a similar way to the approach followed by S. Kobayashi, Hahn has defined a pseudodistance h_D on D, in the following way.

Given x and y in D, an injective analytic chain joining x and y in D consists of : a positive integer n; $n+1$ points $\zeta_0, \zeta_1, \ldots, \zeta_n \in \Delta$; n functions $\varphi_j \in S(\Delta, D)$ $(j = 1, \ldots, n)$ such that $\varphi_1(\zeta_0) = x$, $\varphi_j(\zeta_j) = \varphi_{j+1}(\zeta_j)$ for $j = 1, \ldots, n-1$, $\varphi_n(\zeta_n) = y$. Injective analytic chains joining x and y in D are easily seen to exist for every choice of $x, y \in D$. Then, $h_D(x, y)$ is, by definition,

$$h_D(x, y) = \inf \sum_{j=1}^{n} \omega(\zeta_{j-1}, \zeta_j),$$

where the infimum is taken over all injective analytic chains joining x and y in D.

If D' is a domain in a complex Banach space $\mathcal{E}'$ and if $f : D \to D'$ is an injective holomorphic map, then

$$\begin{aligned} \eta_{D'}(f(x), df(x)v) &\leq \eta_D(x, v), \\ h_{D'}(f(x), f(y) &\leq h_D(x, y) \end{aligned}$$

for every choice of x, y in D and $v \in \mathcal{E}$. In particular, the holomorphic automorphisms of D are isometries for h_D. Furthermore,

$$k_D(x, y) \leq h_D(x, y).$$

Example Let $D = B(x_0, r)$ $(x_0 \in \mathcal{E}, r > 0)$. Then, for $x \in B(x_0, r)$,

$$h_{B(x_0,r)}(x_0, x) = k_{B(x_0,r)}(x_0, x) = \omega(0, \frac{\|x - x_0\|}{r}).$$

Using this result and arguing as in section 1.3, one shows that $h_D : D \times D \to \mathbb{R}_+$ is continuous, and that, if D is bounded, then h_D is a distance defining the relative topology of D in $\mathcal{E}$.

If turns out that η_D degenerates less often than κ_D. In fact, let D be any domain in $\mathbb{C}$, $D \neq \mathbb{C}$, and, for $z \in D$, let $\delta(z)$ be the distance from z to $\mathbb{C} \setminus D$. K.T. Hahn has shown, as a consequence of the rotation theorem for univalent functions, that

$$\eta_D(z, 1) \geq \frac{1}{4\delta(z)}.$$

In particular,

$$\eta_{\mathbb{C}^*}(z, 1) \geq \frac{1}{4|z|}.$$

Indeed, according to Hahn, equality holds, i.e.

$$\eta_{\mathbb{C}^*}(z, 1) = \frac{1}{4|z|}.$$

Since, on the other hand, the holomorphic function $z \mapsto e^z$ maps $\mathbb{C}$ onto $\mathbb{C}^*$, and since $\kappa_{\mathbb{C}} \equiv 0$, then, by the contractibility property of the Kobayashi metric,

$$\kappa_{\mathbb{C}^*} \equiv 0.$$

The domain $D \subset \mathcal{E}$ will be said to be *S-hyperbolic or injective-hyperbolic* if h_D is a distance defining the relative topology on $D \subset \mathcal{E}$. In particular, every hyperbolic domain is injective hyperbolic; $\mathbb{C}^*$ is injective hyperbolic but not hyperbolic. However, the following theorem seems to indicate that $\mathbb{C}^*$ is, in some sense, an isolated case, and that injective hyperbolicity disjoint from Kobayashi's hyperbolicity does not appear as often as one would expect.

Theorem 1.13 *If D is any domain in $\mathcal{E}$, then $D \times \mathbb{C}^*$ is not injective-hyperbolic.*

This theorem was proved in [VES 87c] when $\dim_{\mathbb{C}}\mathcal{E} \geq 2$, and in [GIG 89], [VIG 89], in the general case.

The concept of injective hyperbolicity yields a characterization of the complex field.

Let $\mathcal{A}$ be a complex unital Banach algebra, and let $\mathcal{A}^{-1}$ be the open set of all invertible elements of $\mathcal{A}$.

Theorem 1.14 [VES 87c] *The complex field is the only complex unital Banach algebra $\mathcal{A}$ such that one of the connected components of $\mathcal{A}^{-1}$ is injective-hyperbolic.*

Mutatis mutandis, the pseudodistance h_D can be defined and the notion of injective hyperbolicity can be introduced for a domain D in any locally convex, Hausdorff, complex vector space $\mathcal{E}$. In this more general context, injective hyperbolicity yields a characterisation of normed spaces:

Theorem 1.15 [VES 87c] *Let D be a domain in a locally convex, Hausdorff, complex vector space $\mathcal{E}$. If D is injective-hyperbolic, $\mathcal{E}$ is equivalent to a normed space.*

This theorem may be viewed as an extension of the classical result whereby any locally convex, locally bounded, Hausdorff vector space is normable.

Chapter 2
Holomorphic automorphisms and holomorphic isometries

In this chapter, D will be, as before, a domain in a complex Banach space $\mathcal{E}$, and $\text{Iso}\, D$ will indicate the semigroup of all holomorphic maps $D \to D$ which are isometries for κ_D. When D is convex, $\text{Iso}\, D$ will coincide with the semigroup of all holomorphic isometries for γ_D.

By the invariance of κ_D under the action of $\text{Aut}\, D$, this group is a subgroup (actually the maximum subgroup) of $\text{Iso}\, D$. In the infinite dimensional case, holomorphic isometries may exist, which are not surjective, and a fortiori are not automorphisms of D.

If D is a finite-dimensional hyperbolic domain or, more in general, a finite-dimensional hyperbolic manifold, then $\text{Aut}\, D$ is a real Lie transformation group.

Is there a complex Lie group structure on $\text{Aut}\, D$?

The answer is negative, as was shown by Kobayashi [KOB 70; p. 70]. However this question generates a more general problem: are there non-trivial holomorphic families of holomorphic automorphisms of D? In other words, does there exist a domain U in $\mathbb{C}$ and a non-trivial holomorphic map $f : U \times D \to D$ such that $f(\zeta, \bullet) \in \text{Aut}\, D$ for all $\zeta \in U$. The answer – a negative answer – to that question was given in [AN-VE 64] for any bounded domain $D \subset \mathbb{C}^n$, and in [FR-VE 80] for any hyperbolic domain $D \subset \mathcal{E}$, under the weaker hypothesis that $f(\zeta, \bullet) \in \text{Aut}\, D$ for some $\zeta \in U$.

Similar questions arise naturally for $\text{Iso}\, D$. These questions – which motivated some of the investigations illustrated in this report – will be discussed in the present section. If turns out that the answer is not always negative, as the example of a non-trivial family of holomorphic isometries of the unit ball of a C^* algebra will show. However the answer becomes negative if the family contains a semigroup, at least under an additional assumption on the set of fixed points of the isometries. This assumption will be removed in chapter 4, in the case of a class of Cartan domains.

2.1 Holomorphic automorphisms and holomorphic isometries

Theorem 2.1 [FR-VE 80] *Let D be a hyperbolic domain in $\mathcal{E}$, and let $g \in \text{Hol}\,(\Delta \times D, D)$ be such that $g(\zeta_0, \bullet) \in \text{Aut}\, D$ for some $\zeta_0 \in \Delta$. Then g is independent of ζ.*

Proof Let h and h_0 be the holomorphic maps $\Delta \times D \to \Delta \times D$ defined by

$$h(\zeta, x) = (\zeta, g(\zeta, x)), h_0(\zeta, x) = (\zeta, g(\zeta_0, x)).$$

Then $h_0 \in \mathrm{Aut}\,(\Delta \times D)$ and for any $x_0 \in D, h \circ h_0{}^{-1} \in \mathrm{Hol}\,(\Delta \times D, \Delta \times D)$ is expressed, in a neighbourhood of (ζ_0, x_0), by

$$\zeta \mapsto \zeta, \quad x \mapsto x + (\zeta - \zeta_0)(u + \sum_{q=1}^{+\infty} P_q(\zeta - \zeta_0, x - x_0)), \tag{2.1.1}$$

where $u \in \mathcal{E}$ and P_q is a continuous, homogeneous polynomial $\mathbb{C} \times \mathcal{E} \to \mathcal{E}$ of degree q.

By (1.4.13), with $D_1 = \Delta,\ D_2 = D$, and by (1.4.1″), the inequality

$$\kappa_{\Delta \times D}((\zeta_0, x_0), (d(h \circ h_0{}^{-1})(\zeta_0, x_0)(\tau, v)) \leq \kappa_{\Delta \times D}((\zeta_0, x_0), (\tau, v)),$$

holding for all $\tau \in \mathbb{C}, v \in \mathcal{E}$, becomes

$$\max\{\frac{|\tau|}{1 - |\zeta|^2}, \kappa_D(x_0; v + \tau u)\} \leq \max\{\frac{|\tau|}{1 - |\zeta|^2}, \kappa_D(x_0, v)\}.$$

Choosing $v = \tau^2 u$, then, for all $\tau \neq 0$,

$$\max\{\frac{1}{1 - |\zeta_0|^2}, |\tau + 1|\kappa_D(x_0, u\} \leq \max\{\frac{1}{1 - |\zeta_0|^2}, |\tau|\kappa_D(x_0, u)\}.$$

For τ real and $\gg 0$, that implies

$$\kappa_D(x_0, u) = 0,$$

and therefore, by Proposition 1.4, $u = 0$. Hence

$$d(h \circ h_0{}^{-1}(\zeta_0, x_0)) = Id,$$

and Theorem 1.11 implies that $h \circ h_0{}^{-1} = Id$, i.e., $h = h_0$. Thus $g(\zeta, x) = g(\zeta_0, x)$ for all $x \in D$. □

Remark As a consequence of the fundamental theorem of algebra and of the Casorati-Weierstrass theorem on the behaviour of a holomorphic function of one variable near an essential singularity, the group $\mathrm{Aut}\,\mathbb{C}^*$ is generated by $\mathbb{C}^*$, acting linearly on itself, and by the map $z \mapsto \frac{1}{z}$. As a consequence, $\mathrm{Aut}\,\mathbb{C}^*$ is a complex Lie group. As was seen in section 1.6, $\mathbb{C}^*$ is injective-hyperbolic. Hence Theorem 2.1 does not necessarily hold when hyperbolicity is replaced by injective hyperbolicity.

Theorem 2.1, which was established in [FR-VE 80], extends an older result concerning the case in which D is a bounded domain in $\mathbb{C}^n$ [AN-VE 64].

For $g \in \mathrm{Hol}\,(\Delta \times D, D)$, $d_1 g(\zeta, x)$ and $d_2 g(\zeta, x)$ will indicate the partial Fréchet differentials with respect to the first and the second variable, evaluated at the point (ζ, x).

Lemma 2.1 *Let D be a hyperbolic domain in $\mathcal{E}$. If $g \in \mathrm{Hol}(\Delta \times D, D)$ is such that, for every $\zeta \in \Delta$, $g(\zeta, \bullet) \in \mathrm{Hol}(D, D)$ is an isometry for the Kobayashi differential metric κ_D, and if, for some $\zeta_0 \in \Delta$ and for every $x \in D$, there is a vector $v \in \mathcal{E} \setminus \{0\}$ such that $d_2 g(\zeta_0, x)v$ is a complex extreme point of the set*

$$\{w \in \mathcal{E} : \kappa_D(g(\zeta_0, x), w) \leq 1\}, \tag{2.1.2}$$

then $d_1 g(\zeta_0, x) = 0$ for all $x \in D$.

Proof For any $x_0 \in D$, letting

$$y_0 = g(\zeta_0, x_0), \qquad d_2 g(\zeta_0, x_0) = A \in \mathcal{L}(\mathcal{E}),$$

then

$$\kappa_D(y_0, Av) = \kappa_D(x_0, v)$$

for all $v \in \mathcal{E}$.

Let

$$g(\zeta, x) = y_0 + (\zeta - \zeta_0)u + A(x - x_0) + o(|\zeta - \zeta_0|, \|x - x_0\|),$$

be the power series expansion of g in a neighbourhood of (ζ_0, x_0) in $\Delta \times D$ with $u = d_1(\zeta_0, x_0) \in \mathcal{E}$.

Let $h \in \mathrm{Hol}(\Delta \times D, \Delta \times D)$ be defined by $h(\zeta, x) = (\zeta, g(\zeta, x))$. Since $dh(\zeta_0, x_0)$ is given by the matrix:

$$dh(\zeta_0, x_0) = \begin{pmatrix} 1 & 0 \\ u & A \end{pmatrix},$$

then, for all $\tau \in \mathbb{C}, v \in \mathcal{E}$

$$\kappa_{\Delta \times D}((\zeta_0, y_0)(\tau, \tau u + Av)) \leq \kappa_{\Delta \times D}((\zeta_0, x_0), (\tau, v)),$$

i.e., by (1.4.12) and (1.4.1″)

$$\max\{\frac{|\tau|}{1 - |\zeta_0|^2}, \kappa_D(y_0, \tau u + Av)\} \leq \max\{\frac{|\tau|}{1 - |\zeta_0|^2}, \kappa_D(x_0, v)\}$$

Hence, for all $\tau \in \mathbb{C}$ for which

$$\begin{aligned} \frac{|\tau|}{1 - |\zeta_0|^2} &\leq \kappa_D(x_0, v), \\ \kappa_D(y_0, \tau u + Av) &\leq \kappa_D(x_0, v) = \kappa_D(y_0, Av). \end{aligned} \tag{2.1.3}$$

Since D is hyperbolic, by Proposition 1.4, $\kappa_D(x_0, v) > 0$, and therefore (2.1.3) is satisfied by all ζ contained in an open disc with positive radius and center 0. Thus, if Av is a complex extreme point of the set (2.1.2), with $x = x_0$, then $u = 0$. □

If B is the open unit ball of $\mathcal{E}$, then, by Theorem 1.9, c_D and k_D, γ_D and κ_D coincide. Hence, by Corollary 1.6, if $g \in \mathrm{Hol}(\Delta \times B, B)$ is such that $g(\zeta_0, \bullet) \in \mathrm{Iso}\, B$ for some $\zeta_0 \in \Delta$, then $g(\zeta, \bullet) \in \mathrm{Iso}\, B$ for all $\zeta \in \Delta$.

Since $\kappa_B(0, v) = \|v\|$ for all $v \in \mathcal{E}$, Lemma 2.1 yields

Proposition 2.1 *Let S be the set of all complex extreme points of the closure $\bar{B}$ of B, and let $g \in \mathrm{Hol}(\Delta \times B, B)$ be such that $g(\zeta_0, \bullet) \in \mathrm{Iso}\, B$ for some $\zeta_0 \in \Delta$. If, for every $\zeta \in \Delta$ and every $x \in B$, there exists $h \in \mathrm{Aut}\, B$ such that*

$$h(g(\zeta, x)) = 0,$$

and

$$S \cap dh(g(\zeta, x))d_2 g(\zeta, x)\mathcal{E} \neq \emptyset,$$

then g is independent of $\zeta \in \Delta$.

Since the differential at 0 of any holomorphic isometry of B fixing 0, is a linear isometry of $\mathcal{E}$, Proposition 2.1 yields

Theorem 2.2 *Let B be homogeneous and let*

$$S \cap U\mathcal{E} \neq \emptyset \tag{2.1.4}$$

for every linear isometry U of $\mathcal{E}$. If $g \in \mathrm{Hol}(\Delta \times B, B)$ is such that $g(\zeta_0, \bullet) \in \mathrm{Iso}\, B$ for some $\zeta_0 \in \Delta$, then g is independent of ζ.

2.2 Holomorphic isometries in Hilbert spaces

The hypotheses of Theorem 2.2 are all fulfilled when B is the open unit ball of a complex Hilbert space $\mathcal{H}$. Hence

Theorem 2.3 *Every holomorphic map $g : \Delta \times B \to B$ for which $g(\zeta_0, \bullet) \in \mathrm{Iso}\, B$ for some $\zeta_0 \in \Delta$ is independent of ζ.*

A direct proof of this theorem, based on the explicit expression of κ_B in the case of Hilbert spaces, can be found in [VES 87, pp. 293-294].

It turns out that H. Cartan's linearity theorem extends to $\mathrm{Iso}\, B$ when B – as will be the case throughout this section – will be the open unit ball of a complex Hilbert space $\mathcal{H}$.

Let $f \in \mathrm{Hol}(B, B)$ with $f(0) = 0$, and let $M \subset B$ be defined by

$$M = \{x \in B : \|f(x)\| = \|x\|\}. \tag{2.2.1}$$

Let

$$f(x) = \sum_{\nu=1}^{+\infty} Q_\nu(x)$$

be the power series expansion of f in B, where $Q_\nu : \mathcal{H} \to \mathcal{H}$ is a continuous homogeneous polynomial of degree $\nu = 1, 2, \ldots$, and $Q_1 = df(0)$.

It will be shown now that M is described also by

$$M = \{x \in B : \|Q_1(x)\| = \|x\|\}. \tag{2.2.2}$$

Since every boundary point of B is a real extreme point of $\bar{B}$, Theorem 1.6 implies that

$$f(\zeta x) = \zeta f(x)$$

for all $x \in M$ and all $\zeta \in \mathbf{C}$ such that $|\zeta|\|x\| < 1$. Hence $Q_\nu(x) = 0$ for all $x \in M$ and $\nu = 2, 3, \ldots$; so

$$f(\zeta x) = \zeta Q_1(x),$$

and therefore

$$\|Q_1(x)\| = \|x\|$$

for all $x \in M$. Hence M is contained in the set defined by the right hand side of (2.2.2). It will be shown now that, conversely, (2.2.2) implies (2.2.1). For all $x \in B$

$$Q_1(x) = \frac{1}{2\pi}\int_0^{2\pi} e^{-i\theta} f(e^{i\theta}x)d\theta,$$

whence

$$\|Q_1(x)\| \leq \sup\{\|f(y)\| : y \in B\},$$

i.e.

$$\|Q_1\| \leq 1.$$

The function $g = f - Q_1 \in \mathrm{Hol}\,(B, \mathcal{H})$ has the power series expansion

$$g(x) = \sum_{\nu=2}^{+\infty} Q_\nu(x) \quad (x \in B).$$

Since

$$\|f(\zeta x)\|^2 = \|Q_1(\zeta x)\|^2 + \|g(\zeta x)\|^2 + 2Re(g(\zeta x)|Q_1(\zeta x)),$$

the Schwarz lemma (Theorem 1.6), whereby

$$\|f(\zeta x)\| \leq |\zeta|\|x\|,$$

implies that, if

$$\|Q_1(x)\| = \|x\|,$$

for some $x \in B \setminus \{0\}$, and therefore $\||Q_1(\zeta x)\| = \|\zeta x\|$, then

$$\|g(\zeta x)\|^2 + 2Re(g(\zeta x)|Q_1(\zeta x)) \leq 0. \tag{2.2.3}$$

Thus

$$Re(g(\zeta x)|Q_1(\zeta x)) \leq 0$$

whenever $|\zeta| < \frac{1}{\|x\|}$. Since

$$(g(\zeta x)|Q_1(\zeta x)) = |\zeta|^2(\zeta(Q_2(x)|Q_1(x)) + \zeta^2(Q_3(x)|Q_1(x)) + \cdots),$$

then $(Q_2(x)|Q_1(x)) = (Q_3(x)|Q_1(x)) = \cdots = 0$. Hence $(g(\zeta x)|Q_1(\zeta x)) = 0$, and therefore, by (2.2.3), $g(\zeta x) = 0$ whenever $|\zeta| < 1/\|x\|$, and in particular $g(x) = 0$. Thus, (2.2.2) implies that $\|f(x) = \|x\|$, and this proves that (2.2.2) implies (2.2.1).

Hence, if $Q_1 = df(0)$ is a linear isometry, then $M = B$, i.e.

$$\|f(x)\| = \|x\| \quad \text{for all } x \in B$$

Since

$$\|f(x)\|^2 = \|Q_1(x)\|^2 + 2Re(g(x)|Q_1(x)) + \|g(x)\|^2,$$

then

$$2Re(g(x)|Q_1(x)) + \|g(x)\|^2 = 0, \quad \text{for all } x \in B. \tag{2.2.4}$$

If $Re(g(x)|Q_1(x)) = 0$ for all $x \in B$, then $g(B) = \{0\}$. Now assume that $Re(g(x)|Q_1(x)) \neq 0$ for some $x \in B$, and let $p \geq 1$ be the least integer such that $Re(Q_{p+1}(x)|Q_1(x))$ does not vanish identically on $\mathcal{H}$, while, if $p \geq 2$, $Re(Q_p(x)|Q_1(x)) \equiv 0$ on $\mathcal{H}$ for $j = 2, \ldots, p$. The sets

$$\{x \in \mathcal{H} : Re(Q_{p+1}(x)|Q_1(x)) = 0\}, \quad \{x \in \mathcal{H} : Im(Q_{p+1}(x)|Q_1(x)) = 0\}$$

are closed subset of $\mathcal{H}$, with no interior point (as a consequence of the identity principle for (scalar valued) polynomials in two real variables). Hence the set

$$N = \{x \in \mathcal{H} : Re(Q_{p+1}(x)|Q_1(x)) \neq 0, Im(Q_{p+1}(x)|Q_1(x)) \neq 0\}$$

is an open dense subset of $\mathcal{H}$.

Let $x \in N$. Since

$$\begin{aligned} Re(g(\zeta x)|Q_1(\zeta x)) &= Re(\zeta^{p+1}\bar{\zeta}(Q_{p+1}(x) + \zeta Q_{p+2}(x) + \cdots)|Q_1(x))) \\ &= |\zeta|^2 Re(\zeta^p(Q_{p+1}(x) + \zeta Q_{p+2}(x) + \cdots)|Q_1(x))), \end{aligned}$$

there exists $\epsilon > 0$ such that, whenever $|\zeta| < \epsilon$, then

$$\begin{aligned} &|\zeta||(Q_{p+2}(x) + \zeta Q_{p+3}(x) + \cdots |Q_1(x))| \\ &\qquad < \frac{1}{2}\min\{|Re(Q_{p+1}(x)|Q_1(x))|, |Im(Q_{p+1}(x)|Q_1(x))|\}. \end{aligned}$$

Therefore, if $|\zeta| < \epsilon$, $Re(Q_{p+1}(x)|Q_1(x))$ and $Im(Q_{p+1}(x)|Q_1(x))$ have the same sign as $Re(Q_{p+1}(x) + \zeta(Q_{p+2}(x) + \cdots)|Q_1(x))$ and of $Im(Q_{p+1}(x) + \zeta Q_{p+2}(x) + \cdots)|Q_1(x))$, respectively. But, letting $\zeta = |\zeta|e^{i\theta} (\theta \in \mathbf{R})$, then

$$\begin{aligned} Re(\zeta^p(Q_{p+1}(x)|Q_1(x))) &= |\zeta|^p[Re(Q_{p+1}(x)|Q_1(x))\cos p\theta \\ &\quad - Im(Q_{p+1}(x)|Q_1(x))\sin p\theta], \end{aligned}$$

and a suitable choice of θ yields

$$Re(\zeta^p(Q_{p+1}(x)|Q_1(x)) > 0,$$

contradicting (2.2.4) with x replaced by ζx. Hence, if $f(0) = 0$ and $df(0)$ is a linear isometry of $\mathcal{H}$, then

$$f(x) = df(0)x \tag{2.2.5}$$

for all $x \in B$. Thus, if $f(0) = 0$ then $k_B(0, f(x)) = k_B(0, x)$ for all $x \in B$ if, and only if, $\kappa_B(0, df(0)v) = \kappa_B(0, v)$ for all $v \in \mathcal{H}$. Since $\operatorname{Aut} B$ acts transitively on B (cf., e.g., [FR-VE 80; Proposition VI. 1.5, pp. 148–149]) the hypothesis $f(0) = 0$ can be removed, proving thereby the following theorem, which extends the Schwarz-Pick lemma to the unit ball of any complex Hilbert space.

Theorem 2.4 *If* $f \in \operatorname{Hol}(B, B)$ *is such that*

$$\kappa_B(f(x), df(x)v) = \kappa_B(x, v)$$

for some $x \in B$ *and all* $v \in \mathcal{H}$, *then* $f \in \operatorname{Iso} B$.

Remarks (1) (2.2.5) proves that H. Cartan's linearization theorem holds for all holomorphic isometries of the unit ball of a complex Hilbert space.

(2) For any $f \in \operatorname{Hol}(B, B)$ the set M defined by (2.2.1) is the intersection of B with a closed linear subspace $\mathcal{F}$ of $\mathcal{H}$, and the restriction of f to M is the restriction to M of a partial isometry with initial space $\mathcal{F}$ and final space $df(0)\mathcal{F}$. For this result and for further details, see [VES 87a].

In a forthcoming paper the conclusions of Theorem 2.3 will be extended to the Cartan domains of type one considered in chapter 4 and to the Cartan domains of type four. The ideas of the proof will now be illustrated in the case in which B is the open unit ball of the complex Hilbert $\mathcal{H}$, thus establishing, by a different argument, Theorem 2.3.

Since Δ is homogeneous, there is no restriction in assuming $\zeta_0 = 0$. Hence (2.2.5) entails that

$$f(0, x) = d_2 f(0, 0)x \quad \text{for all } x \in B. \tag{2.2.6}$$

In view of Theorem 2.1, it will be assumed that

$$f(0, \bullet) \in \operatorname{Iso} B \setminus \operatorname{Aut} B,$$

i.e., that the linear isometry $d_2 f(0, 0)$ is not surjective. Hence the range of $d_2 f(0, 0)$ is a closed linear subspace $\mathcal{K}$ of $\mathcal{H}$, and $d_2 f(0, 0)$ defines a bijective isometry of $\mathcal{H}$ onto $\mathcal{K}$. Let $F \in \mathcal{L}(\mathcal{K}, \mathcal{H})$ be the inverse of this isometry, and let $P \in \mathcal{L}(\mathcal{H}, \mathcal{K})$ be the linear map defined by the orthogonal projection of $\mathcal{H}$ onto $\mathcal{K}$.

Then $P(B) = \mathcal{K} \cap B = d_2 f(0, 0)B$, and $F \circ P \circ d_2 f(0, 0)x = x$ for all $x \in \mathcal{H}$. By H. Cartan's uniqueness theorem (Lemma 1.4), the function $F \circ P \circ f \in \operatorname{Hol}(\Delta \times B, B)$ is such that

$$F \circ P \circ f(0, \bullet) = Id. \tag{2.2.7}$$

Thus, by Theorem 2.1, $F \circ P \circ f(\zeta, \bullet)$ is independent of $\zeta \in \Delta$, i.e., by (2.2.6)

$$F \circ P \circ f(\zeta, x) = x$$

for all $\zeta \in \Delta$ and all $x \in B$. Let

$$f(\zeta, x) = \sum_{\nu=0}^{+\infty} Q_\nu(\zeta, x) \tag{2.2.8}$$

be the power-series expansion of $f(\zeta, \bullet)$ in B, where $Q_\nu(\zeta, \bullet) : \mathcal{H} \to \mathcal{H}$ is a continuous homogeneous polynomial of degree $\nu = 0, 1, 2, \dots$, given by

$$Q_\nu(\zeta, x) = \frac{1}{2\pi} \int_0^{2\pi} e^{-i\nu\theta} f(\zeta, e^{i\theta} x) d\theta.$$

By (2.2.7),

$$F \circ P \circ Q_\nu(\zeta, x) = 0 \quad \text{for } \nu = 0, 2, 3, \dots \tag{2.2.9}$$

and

$$F \circ P \circ Q_1(\zeta, x) = x,$$

for all $x \in \mathcal{H}$. Since, by the Cauchy inequalities,

$$\|Q_1(\zeta, \bullet)\| \leq 1,$$

the chain of inequalities

$$\|x\| = \|F \circ P \circ Q_1(\zeta, x)\| \leq \|F\| \|P\| \|Q_1(\zeta, x)\| \leq \|Q_1(\zeta, x)\| = \|x\|$$

implies that

$$\|Q_1(\zeta, x)\| = \|x\|$$

for all $x \in \mathcal{H}$. Since every boundary point of B is a (real, hence) complex extreme point of $\bar{B}$, the strong maximum principle implies that $Q_1(\zeta, x)$ is independent of $\zeta \in \Delta$. Thus, by (2.2.6),

$$Q_1(\zeta, x) = d_2 f(0, 0) x \quad \text{for all } \zeta \in \Delta \text{ and all } x \in B,$$

while, by (2.2.9), $Q_\nu(\zeta, x) = (I - P) Q_\nu(\zeta, x)$ for $\nu = 0, 2, 3, \dots$.

Since the orthogonal projectors P and $I - P$ are mutually orthogonal and $d_2 f(0, 0)$ is an isometry, the condition

$$\|f(\zeta, x)\| \leq 1$$

reads

$$\|x\|^2 + \|Q_0(\zeta) + \sum_{\nu=2}^{+\infty} Q_\nu(\zeta, x)\|^2 \leq 1.$$

Letting $\|x\|$ tend to 1, then

$$Q_0(\zeta) + \sum_{\nu=2}^{+\infty} Q_\nu(\zeta, x) = 0$$

for all $x \in B$. Hence $Q_0(\zeta) = 0$ and $Q_\nu(\zeta, x) = 0$ for $\nu = 2, 3, \dots$, i.e.

$$f(\zeta, x) = d_2 f(0, 0) x$$

for all $x \in B$ and all $\zeta \in \Delta$.

2.3 Holomorphic families of holomorphic isometries in a C^*-algebra

The hypothesis (2.1.4) in Theorem 2.2 is crucial, as the following considerations will show.

Let $\mathcal{K}_0$ be a complex Hilbert space, and let $\mathcal{K}$ be the Hilbert space, direct sum $\mathcal{K} = \ell^2_+(\mathcal{K}_0) = \mathcal{K}_0 \oplus \mathcal{K}_0 \oplus \cdots$. If $\mathcal{E}$ is the C^*-algebra $\mathcal{E} = \mathcal{L}(\mathcal{K})$ of all bounded linear operators on $\mathcal{K}$, then every $X \in \mathcal{E}$ has a matrix representation

$$X = \begin{pmatrix} X_{00} & X_{01} & \cdots \\ X_{10} & X_{11} & \cdots \\ \vdots & \vdots & \end{pmatrix}$$

where $X_{\alpha\beta} \in \mathcal{L}(\mathcal{K}_0)$. If $F \in \mathcal{E}$ is the right shift operator

$$(x_0, x_1, \ldots) \mapsto (0, x_0, x_1, \ldots),$$

then its adjoint F^* is the left shift

$$(x_0, x_1, x_2, \ldots) \mapsto (x_1, x_2, \ldots),$$

and FXF^* is expressed by the matrix

$$FXF^* = \begin{pmatrix} 0 & 0 \\ 0 & X \end{pmatrix} = \begin{pmatrix} 0 & 0 & 0 & \cdots \\ 0 & X_{00} & X_{01} & \cdots \\ 0 & X_{10} & X_{11} & \cdots \\ \vdots & \vdots & \vdots & \end{pmatrix}.$$

The orthogonal projector P of $\mathcal{K}$ onto $\mathcal{K}_0$ is given by

$$P = \begin{pmatrix} I_0 & 0 \\ 0 & 0 \end{pmatrix} = \begin{pmatrix} I_0 & 0 & \cdots \\ 0 & 0 & \cdots \\ \vdots & \vdots & \end{pmatrix},$$

where I_0 is the identity operator on $\mathcal{K}_0$, and therefore

$$PXP = \begin{pmatrix} X_{00} & 0 \\ 0 & 0 \end{pmatrix} = \begin{pmatrix} X_{00} & 0 & \cdots \\ 0 & 0 & \cdots \\ \vdots & \vdots & \end{pmatrix}.$$

Hence, the holomorphic function $f \in \mathrm{Hol}\,(\mathbf{C} \times \mathcal{E}, \mathcal{E})$ defined by

$$f(\zeta, X) = \zeta (PXP)^2 + FXF^* \tag{2.3.1}$$

has also the matrix representation

$$f(\zeta, X) = \begin{pmatrix} \zeta {X_{00}}^2 & 0 \\ 0 & X \end{pmatrix} = \begin{pmatrix} \zeta X_{00}^2 & 0 & 0 & \cdots \\ 0 & X_{00} & X_{01} & \cdots \\ 0 & X_{10} & X_{11} & \cdots \\ \vdots & \vdots & \vdots & \end{pmatrix}.$$

Since $f(\zeta, 0) = 0$, and

$$\|f(\zeta, X)\| = \max(|\zeta| \|(PXP)^2\|, \|X\|)$$

for all $\zeta \in \mathbb{C} X \in \mathcal{E}$, then $f(\Delta \times B) \subset B$, where B is the open unit ball of $\mathcal{E}$.

Theorem 2.5 *Whenever* $|\zeta| < \sqrt{2}-1$, $f(\zeta, \bullet)$ *is an isometry for the Kobayashi differential metric* κ_B.

The proof of this theorem will be split into a series of lemmas whose proofs can be found in [VES 87a].

Any $Y \in \mathcal{E}$ is represented by a matrix

$$Y = \begin{pmatrix} Y_{00} & Y_{01} & \cdots \\ Y_{10} & Y_{11} & \cdots \\ \vdots & \vdots & \end{pmatrix}$$

with $Y_{\alpha\beta} \in \mathcal{L}(\mathcal{K}_0)$.

Lemma 2.2 *If* $|\zeta| < \sqrt{2} - 1$ *and* $\|X\| < 1$, *then*

$$|\zeta| \|(I_0 - |\zeta|^2 X_{00}{}^2 X_{00}{}^{*2})^{-\frac{1}{2}} (X_{00} Y_{00} + Y_{00} X_{00})(I_0 - |\zeta|^2 X_{00}{}^{*2} X_{00}{}^2)^{-\frac{1}{2}}\| \leq \|Y_{00}\|$$

for all $Y \in \mathcal{E}$.

Lemma 2.3 *If* $\|X\| < 1$, *then*

$$\|(I - XX^*)^{-\frac{1}{2}} Y (I - X^*X)^{-\frac{1}{2}}\| \geq \|Y (I - X^*X)^{-\frac{1}{2}}\|$$

for all $Y \in \mathcal{E}$.

By [VES 87a],

$$\|Y_{00}\| \leq \|Y (I - X^*X)^{-\frac{1}{2}}\|$$

for all $Y \in \mathcal{E}$, whenever $\|X\| < 1$. Therefore Lemmas 2.2 and 2.3 yield

Lemma 2.4 *If* $|\zeta| < \sqrt{2} - 1$ *and* $\|X\| < 1$, *then*

$$|\zeta| \|(I_0 - |\zeta|^2 X_{00}{}^2 X_{00}{}^{*2})^{-\frac{1}{2}} (X_{00} Y_{00} + Y_{00} X_{00})(I_0 - |\zeta|^2 X_{00}{}^{*2} X_{00}{}^2)^{-\frac{1}{2}}\|$$

$$\leq \|(I - XX^*)^{-\frac{1}{2}} Y (I - X^*X)^{-\frac{1}{2}}\|$$

for all $Y \in \mathcal{E}$.

The Kobayashi differential metric κ_B of the open unit ball B of the C^*-algebra $\mathcal{E}$ was computed in [HAR 74] using the fact that Aut B – which acts transitively on B – consists essentially of operator-valued Moebius transformations. For all $X \in B$, $Y \in \mathcal{E}$, κ_B is given by

$$\kappa_B(X,Y) = \|(I - XX^*)^{-\frac{1}{2}} Y (I - X^*X)^{-\frac{1}{2}}\| \tag{2.3.2}$$

A direct computation yields then [VES 87a]

$$\kappa_B(f(\zeta, X); d_2 f(\zeta, X)) = \tag{2.3.3}$$

$$= \max\{|\zeta| \|I_0 - |\zeta|^2 X_{00}{}^2 X_{00}{}^{*2})^{-\frac{1}{2}} (X_{00}Y_{00} + Y_{00}X_{00})(I_0 - |\zeta|^2 X_{00}{}^{*2} X_{00}{}^2)^{-\frac{1}{2}}\|,$$
$$\|(I - XX^*)^{-\frac{1}{2}} Y (I - X^*X)^{-\frac{1}{2}}\|\}.$$

Hence Lemma 2.4 and (2.3.3) complete the proof of Theorem 2.5.

Remarks (1) Theorem 2.5 shows that H. Cartan's linearity theorem does not hold necessarily for holomorphic isometries.

(2) Let $\mathcal{H}$ be any complex Hilbert space, and let F be any pure isometry of $\mathcal{H}$, i.e. a linear isometry $F : \mathcal{H} \to \mathcal{H}$ such that

$$\bigcap_{n=0}^{\infty} F^n \mathcal{H} = \{0\},$$

As is well known (cf. e.g. [FIL 70]), setting $\mathcal{K}_0 = (F\mathcal{H})^\perp$, F is then unitarily equivalent to a unilateral shift operator on the Hilbert space $\ell^2_+(\mathcal{K}_0)$. Hence, denoting by P the orthogonal projector on $\mathcal{H}$ with range $\mathcal{K}_0$ and by B the open unit ball of $\mathcal{L}(\mathcal{H})$. Theorem 2.5 can be re-stated as follows:

Theorem 2.6 *For any pure isometry F of the Hilbert space $\mathcal{H}$, the function $f \in$ Hol $(\mathbb{C} \times B, B)$ defined by* (2.3.1) *is such that $f(\zeta, \bullet) \in$ Iso B whenever $|\zeta| < \sqrt{2} - 1$.*

2.4 Holomorphic semigroups of holomorphic isometries

Throughout the present section, D will be a bounded domain in the complex Banach space $\mathcal{E}$. Let $f : \mathbb{R}_+ \times D \to D$ be such that: $f(t, \bullet)$ is a holomorphic isometry for the Carathéodory distance c_D of D; $f(0, x) = x$,

$$f(t_1 + t_2, x) = f(t_1, f(t_2, x)) \tag{2.4.1}$$

for all $t_1 > 0, t_2 > 0$ and all $x \in D$.

The reason why the Carathéodory distance will be investigated here, instead of the Kobayashi distance, as was done before, is that the logarithmic subharmonicity of c_D (Theorem 1.10) will play a crucial role in the forthcoming considerations. However, if D is convex, as will be the case in all concrete examples that will be investigated in the sequel, then, by Theorem 1.9, $c_D = k_D$.

The following question will now be investigated: does there exist a domain in $\mathbb{C}$, $U \supset \mathbb{R}_+^*$ and a holomorphic map $g \in \mathrm{Hol}\,(U \times D, D)$ whose restriction to $\mathbb{R}_+^* \times D$ is f?

By Theorem 2.3 the answer is negative when D is the open unit ball of a complex Hilbert space. This conclusion will be extended to a larger class of domains in chapter 3.

It turns out that the answer to that question, for any bounded domain D, is negative, at least under the additional hypothesis, that there is some $x_0 \in D$ such that $f(t, x_0) = x_0$ for all values or for "many" values of $t > 0$. More precisely, the following theorems hold for any bounded domain D in $\mathcal{E}$, for any domain U in $\mathbb{C}$, such that $\mathbb{R}_+^* \subset U$, and for a holomorphic map $g : U \times D \to D$.

Put $g(0, x) = x$ and $d_2 g(0, x) = I$ for all $x \in D$.

Theorem 2.7 *If there exists $x_0 \in D$ such that $g(t, x_0) = x_0$ for all $t > 0$ and if the map $T : \mathbb{R}_+ \to \mathcal{L}(\mathcal{E})$ defined by $T(t) = d_2 g(t, x_0)$ is a strongly continuous semigroup of linear isometries for the norm $\gamma_D(x_0, \bullet)$, then $g(\zeta, x) = x$ for all $x \in D, \zeta \in U$.*

Theorem 2.8 *If the restriction of g to $\mathbb{R}_+ \times D$ defines a semigroup of holomorphic isometries for c_D, for which*

$$\lim_{t \downarrow 0} g(t, x) = x$$

for all $x \in D$, and if there exists $x_0 \in D$ such that $g(t, x_0) = x_0$ when t varies on a subset of $\mathbb{R}_+^*$ having positive outer capacity, then $g(\zeta, x) = x$ for all $\zeta \in U, x \in D$.

The proofs of these two theorems, given in [VES 88], go as follows. For any linear operator X on $\mathcal{E}$ with domain $\mathcal{D}(X)$, $\sigma(X), p\sigma(X), c\sigma(X), r\sigma(X), r(X)$ will stand respectively, for the spectrum, the point-spectrum, the continuous spectrum, the residual spectrum, and the resolvent set of X.

Proposition 2.2 *Let $S \in \mathcal{L}(\mathcal{E})$ be a linear isometry. Then either S is surjective and $\sigma(S) \subset \partial\Delta$, or $\sigma(S) = \bar{\Delta}$. In this latter case $\Delta \subset r\sigma(S)$.*

Proof a) The first part of the statement follows from the fact that, if $\zeta \in \partial\sigma(S)$, then $|\zeta| = 1$, as will be shown now. The operator $S - \zeta I$ is either zero (in which case $\zeta \in \partial\Delta$) or a generalized left divisor of zero, i.e. there exists a sequence $\{A_\nu\}$ of operators $A_\nu \in \mathcal{L}(\mathcal{E})$ such that $\|A_\nu\| = 1$, and

$$\lim_{\nu \to \infty} (S - \zeta I) A_\nu = 0.$$

Hence

$$\lim_{\nu \to \infty} \|S A_\nu\| = |\zeta|.$$

On the other hand,

$$\begin{aligned} \|S A_\nu\| &= \sup\{\|S A_\nu x\| : x \in \mathcal{E}, \|x\| = 1\} \\ &= \sup\{\|A_\nu x\| : x \in \mathcal{E}, \|x\| = 1\} = \|A_\nu\| = 1. \end{aligned}$$

b) If the isometry is not surjective, then by a), $\sigma(S) = \bar{\Delta}$. Clearly $\Delta \cap p\sigma(S) = \emptyset$.

Suppose that $\zeta \in \Delta$ belongs to the continuous spectrum of S. Then for any $y \in \mathcal{E}$ there is a sequence $\{x_\nu\}$ in $\mathcal{E}$ such that

$$\lim_{\nu\to\infty}(S - \zeta I)x_\nu = y.$$

Since

$$\|(S - \zeta I)(x_\nu - x_\mu)\| \geq |\|S(x_\nu - x_\mu)\| - |\zeta|\|x_\nu - x_\mu\|| = (1 - |\zeta|)\|x_\nu - x_\mu\|,$$

the fact that $\{(S - \zeta I)x_\nu\}$ is a Cauchy sequence implies that $\{x_\nu\}$ is a Cauchy sequence and therefore converges to some $x \in \mathcal{E}$. But then $(S - \zeta I)x = \lim_{\nu\to\infty}(S - \zeta I)x_\nu = y$, and so $(S - \zeta I)\mathcal{E} = \mathcal{E}$, contradicting the fact that $\zeta \in c\sigma(S)$. □

Now, let $T : \mathbf{R}_+ \to \mathcal{L}(\mathcal{E})$ be a strongly continuous semigroup of linear isometries of $\mathcal{E}$, and let X be the infinitesimal generator of T. Since $T(t)$ is a contraction, then $\sigma(X)$ is contained in the closure of the left half-plane

$$\Pi_\ell = \{\zeta \in \mathbf{C} : Re\zeta < 0\}.$$

If $T(t)$ is not surjective for some $t > 0$, then, by Proposition 2.2, $\sigma(T(t)) = \bar{\Delta}$ and every $\zeta = \rho e^{i\theta}$ with $0 < \rho < 1, \theta \in \mathbf{R}$, belongs to $r\sigma(T(t))$. Hence [PAZ 83; Theorem 2.5, p. 47], there exist $n \in \mathbf{Z}, \tau = \tau' + i\tau'' \in r\sigma(X)(\tau', \tau'' \in \mathbf{R})$ such that

$$\tau' = \frac{1}{t}\log\rho, \quad \tau'' = \frac{\theta + 2n\pi}{t}.$$

This implies that all vertical lines contained in Π_ℓ have a non-empty intersection with $r\sigma(X)$. On the other hand, if $T(s)$ is surjective for some $s > 0$, then $\sigma(T(s)) \subset \partial\Delta$. The inclusion

$$e^{s\sigma(X)} \subset \sigma(T(s))$$

implies that $\sigma(X) \subset i\mathbf{R}$. That proves

Lemma 2.5 *If $T(t)$ is not surjective for some $t > 0$, then $T(t)$ is not surjective for any $t > 0$.*

Let $T(t)$ be non-surjective for all $t > 0$. Completing a previous argument, it will be shown now that the absolute values of the ordinates of the intersections of $r\sigma(X)$ with the vertical lines in Π_ℓ do not stay bounded. More exactly, it will be shown that, given $\tau' < 0$ and $K > 0$, there is some $\tau = \tau' + i\tau'' \in r\sigma(X)$ with $\tau'' \in \mathbf{R}$ and $|\tau''| > K$. Indeed, if that is not the case, there is $M > 0$ such that, if $\tau' + i\tau'' \in r\sigma(X)$, then $|\tau''| \leq M$. For any $t > 0$, the image of the segment $\{\tau' + i\xi : -M \leq \xi \leq M\}$, by the map $\zeta \mapsto e^{t\zeta}$ is the arc $\{e^{\tau' t}e^{i\xi t} : -M \leq \xi \leq M\}$, whose lenght tends to zero as $t \to 0$. But this contradicts the fact that, since $\Delta \subset r\sigma(T(t))$ for all $t > 0$, the image of $r\sigma(X) \cap (\tau' + i\mathbf{R})$ by the map $\zeta \mapsto e^{t\zeta}$ must cover the entire circle $\{\zeta \in \mathbf{C} : |\zeta| = e^{\tau' t}\}$.

Let T be eventually differentiable, i.e. such that there is some $t_0 \geq 0$ for which the function $t \mapsto T(t)x$ is of class C^1 on $(t_0, +\infty)$ for all $x \in \mathcal{E}$. According to A. Pazy [PAZ 83], there exist $a \in \mathbf{R}$, $b \in \mathbf{R}_+^*$ such that

$$\{\zeta \in \mathbf{C} : Re\zeta \geq a - b \log |Im\zeta|\} \subset r(X).$$

In conclusion the following lemma holds:

Lemma 2.6 *If T is a strongly continuous, eventually differentiable semigroup of linear isometries, then $T(t)$ is surjective for all $t \geq 0$.*

Now, let $h : \mathbf{R}_+ \times D \to D$ be such that:

(i) $t \mapsto h(t, \bullet)$ is a homomorphism of $(\mathbf{R}_+, +)$ into the semigroup $\mathrm{Hol}\,(D, D)$, and

$$\lim_{t \downarrow 0} h(t, x) = x$$

for all $x \in D$;

(ii) there exists $x_0 \in D$ such that $h(t, x_0) = x_0$ for all $t > 0$.

This latter condition implies that $T : t \mapsto d_2h(t, x_0)$ is a semigroup of bounded linear operators in $\mathcal{E}$. It will be shown now that T is strongly continuous. Since D is bounded, the Carathéodory distance c_D defines in $D \subset \mathcal{E}$ the relative topology. For $x_0 \in D$ and $r > 0$, let

$$B_c(x_0, r) = \{x \in D : c_D(x_0, x) < r\}.$$

There exist $r_0 > 0, r_1 > r_0$ such that

$$B(x_0, r_0) \subset B_c(x_0, r) \subset B(x_0, r_1).$$

Since, for $\|v\| < r_0$,

$$d_2h(t, x_0)v = \frac{1}{2\pi} \int_0^{2\pi} e^{-i\theta} h(t, x_0 + e^{i\theta} v) d\theta,$$

then

$$\begin{aligned} \|d_2h(t, x_0)v - v\| &= \frac{1}{2\pi} \left\| \int_0^{2\pi} e^{-i\theta} (h(t, x_0 + e^{i\theta} v) - (x_0 + e^{i\theta} v)) d\theta \right\| \\ &\leq \frac{1}{2\pi} \int_0^{2\pi} \|h(t, x_0 + e^{i\theta} v) - (x_0 + e^{i\theta} v)\| d\theta, \end{aligned}$$

whence

$$\lim_{t \downarrow 0} \|d_2h(t, x_0)v - v\| = 0,$$

proving thereby that T is strongly continuous. From (1.4.9) then follows

Lemma 2.7 *If conditions* (i) *and* (ii) *hold and if $h(t, \bullet)$ is a holomorphic isometry for c_D for all $t \geq 0$, then $T : t \mapsto d_2h(t, x_0)$ is a strongly continuous semigroup of linear isometries for the norm $\gamma_D(x_0, \bullet)$.*

Proof of Theorem 2.7 Let C be the open unit ball for the norm $\gamma_D(x_0, \bullet)$. For every $\zeta \in U$, $d_2g(\zeta, x_0)$ defines a holomorphic map of C into C. Since $\mathbb{R}_+^* \to \mathcal{L}(\mathcal{E}), T : t \mapsto d_2g(t, x_0)$, is the restriction to $\mathbb{R}_+^*$ of the holomorphic map $\zeta \mapsto d_2g(\zeta, x_0)$ of U into $\mathcal{L}(\mathcal{E})$, then the semigroup T is differentiable. Hence, by Lemma 2.6, $T(t) = d_2g(t, x_0)$ is a surjective linear isometry for $\gamma_D(x_0, \bullet)$ and thus defines a holomorphic automorphism of C. By Theorem 2.1, $d_2g(\zeta, x_0)$ is independent of $\zeta \in U$, and therefore $d_2g(\zeta, x_0) = Id$ for all $\zeta \in U$. By H. Cartan's uniqueness theorem, $g(\zeta, x) = x$ for all $x \in D$ and all $\zeta \in U$. □

Proof of Theorem 2.8 By Lemma 1.6, $g(\zeta, x_0) = x_0$ for all $\zeta \in U$, and in particular for all $\zeta \in \mathbb{R}_+$. Then (i) and (ii) hold, and, all the requirements of Theorem 2.7 are fulfilled. □

Chapter 3
Hilbert spaces endowed with an indefinite metric

Let D be a bounded domain in a complex Banach space $\mathcal{E}$. If no further hypotheses on the geometry of D are introduced, our knowledge of the structure of $\mathrm{Aut}\, D$ is poor. Even when $\dim_{\mathbb{C}} \mathcal{E} < \infty$, it does not go beyond the result of H. Cartan whereby $\mathrm{Aut}\, D$ carries the structure of a Lie transformations group of D, which is compatible with the topology of uniform convergence on compact sets of D. When the dimension is finite, the situation changes if D is a bounded symmetric domain of $\mathbb{C}^n$: in which case, thanks to É. Cartan's classification and to the work of C.L. Siegel, H. Klingen and U. Hirzebruch, our knowledge of $\mathrm{Aut}\, D$ is essentially complete.

The fact that any (finite dimensional) bounded symmetric domain belonging to one of the four main classes of É. Cartan's classification can be realized as the open unit ball of a JC^*-triple [HAR 74] [1] offers the possibility of extending É. Cartan's description (although not in an exhaustive way) to the case where $\dim_{\mathbb{C}} \mathcal{E} = \infty$.

Here is the definition of a JC^*-triple. Let $\mathcal{K}$ and $\mathcal{K}'$ be complex Hilbert spaces. A *J^*-algebra* (or a *JC^*-triple*) is a closed linear subspace $\mathcal{E}$ of the complex Banach space $\mathcal{L}(\mathcal{K}, \mathcal{K}')$ such that, if $A \in \mathcal{E}$, then $AA^*A \in \mathcal{E}$.

Examples *Cartan factors.* The Banach space $\mathcal{E} = \mathcal{L}(\mathcal{K}, \mathcal{K}')$ is a J^*-algebra, which – extending the terminology used in the finite dimensional case – is called a *Cartan factor of type one.* If, in particular, $\mathcal{K} \simeq \mathbb{C}$, $\mathcal{E} = \mathcal{L}(\mathbb{C}, \mathcal{K}')$ can be identified with the Hilbert space $\mathcal{K}'$, which thus turns out to be a Cartan factor of type one. If $\mathcal{K} = \mathcal{K}'$ then $\mathcal{L}(\mathcal{K}) = \mathcal{L}(\mathcal{K}, \mathcal{K})$ is a Cartan factor of type one. By the Gelfand-Naimark theorem [SAK 71], every C^*-algebra is a J^*-algebra.

To describe other examples, suppose that a conjugation (i.e. a continuous, antilinear involutory map $x \mapsto \bar{x}$ with norm ≤ 1) is fixed in the complex Hilbert space $\mathcal{K}$, and, for $A \in \mathcal{L}(\mathcal{K})$, define the transposed ${}^tA \in \mathcal{L}(\mathcal{K})$ of A by $\overline{{}^tAx} = A^*\bar{x}$.

[1] This realization has been shown to be impossible by O. Loos and K. MacCrimmon [LO-MA 77] (and later on by M. Meschiari [MES 85]) for the two exceptional bounded symmetric domains of $\mathbb{C}^{16}$ and $\mathbb{C}^{27}$.

The spaces

$$\begin{aligned} \mathcal{E} &= \{A \in \mathcal{L}(\mathcal{K}) : {}^tA = A\}, \\ \mathcal{E} &= \{A \in \mathcal{L}(\mathcal{K}) : {}^tA = -A\}, \end{aligned}$$

are J^*-algebras which are called *Cartan factors of type two* and *three* respectively.

A *Cartan factor of type four* is a self-adjoint (i.e. $*$-invariant) closed subspace $\mathcal{E}$ of $\mathcal{L}(\mathcal{K})$ such that, if $A \in \mathcal{E}$, then A^2 is a scalar multiple of the identity operator on $\mathcal{K}$.

The open unit ball of a Cartan factor is called a *Cartan domain.*

The open unit ball D of a J^*-algebra $\mathcal{E}$ is a bounded homogeneous domain. In fact, for any operator $E \in D$, the operator-valued Moebius transformation expressed by

$$A \mapsto (I_{\mathcal{K}'} - E\,E^*)^{-\frac{1}{2}}(A + E)(I_{\mathcal{K}} + E^*\,A)^{-1}(I_{\mathcal{K}} - E^*\,E)^{\frac{1}{2}}$$

(where $I_{\mathcal{K}}$ and $I_{\mathcal{K}'}$ are the identity operators in $\mathcal{K}$ and in $\mathcal{K}'$) defines a holomorphic automorphism of B. The set of all Moebius transformations is a subgroup of $\operatorname{Aut} B$, which acts transitively on B. Hence the description is complete once the stability group $(\operatorname{Aut} B)_0$ of the origin 0 in $\operatorname{Aut} B$ is known.

Since B is a homogeneous ball, the Carathéodory and Kobayashi metrics of B coincide. Let $\operatorname{Iso} B$ be the semigroup of all holomorphic isometries for any one of the two metrics. Since $\operatorname{Aut} B \subset \operatorname{Iso} B$, then also the description of $\operatorname{Iso} B$ is complete, once the stability semigroup $(\operatorname{Iso} B)_0$ of 0 in $\operatorname{Iso} B$ is known. In the infinite dimensional case, the existence of non-surjective linear isometries implies that $\operatorname{Aut} B$ may be a proper subgroup of $\operatorname{Iso} B$.

Contrary to what happens in the finite dimensional case as a consequence of Vitali's theorem on normal families, if $\dim_{\mathbb{C}} \mathcal{E} = \infty$ different types of convergence on B yield radically different topologies on $\operatorname{Iso} B$ and on $\operatorname{Aut} B$.

The topology induced on $\operatorname{Aut} B$ by the locally uniform convergence on B has been investigated by J.-P. Vigué in his thesis [VIG 76].

In the following chapter some preliminary investigations that have been carried out for the unit ball of a Cartan factor of type one will be briefly reviewed.[2] In the present chapter, some facts concerning Kreĭn spaces and, more in general, Hilbert spaces endowed with an indefinite metric, and semigroups of isometries acting on them will be described. These results are important tools in the study of Cartan domains of type one.

3.1 Vector spaces endowed with an indefinite metric

Let $\mathcal{H}$ be a complex vector space and let $x, y \mapsto [x, y]$ be a sesquilinear hermitian form on $\mathcal{H}$. Let $\wp^+, \wp^{++}, \wp^-, \wp^{--}, \wp^0$ be the subsets of $\mathcal{H}$ defined by

$$\begin{aligned} \wp^+ &= \{x \in \mathcal{H} : [x, x] \geq 0\}, \quad \wp^{++} = \{x \in \mathcal{H} : [x, x] > 0\}, \\ \wp^- &= \{x \in \mathcal{H} : [x, x] \leq 0\}, \quad \wp^{--} = \{x \in \mathcal{H} : [x, x] < 0\}, \\ \wp^0 &= \{x \in \mathcal{H} : [x, x] = 0\}. \end{aligned}$$

[2]For a parallel investigation on Cartan domains of type four, cf: [HAR 74], [HAR 79], [HAR 81], [HE-IS 92], [VES 89], [VES 92], [VES 93].

A vector $x \in \mathcal{H}$ is called positive, negative, neutral if, respectively, $[x,x] > 0$, $[x,x] < 0$, $[x,x] = 0$. A linear subspace $\mathcal{K} \subset \mathcal{H}$ is called: positive, strictly positive, negative, strictly negative, neutral if, respectively, $\mathcal{K} \subset \wp^+$, $\mathcal{K} \subset \wp^{++} \cup \{0\}$, $\mathcal{K} \subset \wp^-$, $\mathcal{K} \subset \wp^{--} \cup \{0\}$, $\mathcal{K} \subset \wp^0$.

If $\wp^{--} = \emptyset$, or if $\wp^{++} = \emptyset$, the form [,] is positive or negative semi-definite (positive or negative definite if $\wp^- = \{0\}$ or if $\wp^+ = \{0\}$) and the Schwarz inequality holds:

$$|[x,y]|^2 \leq [x,x][y,y] \qquad (x, y \in \mathcal{H}). \tag{3.1.1}$$

If $\wp^{++} \neq \emptyset$ and $\wp^{--} \neq \emptyset$, the form [,] is said to be indefinite. It will be also said that [,] defines, or is, an indefinite metric.

Two vectors $x, y \in \mathcal{H}$ for which $[x,y] = 0$ are said to be [,] -orthogonal: in symbols $x[\perp]y$. For a non-empty set $K \subset \mathcal{H}$, the set

$$K^{[\perp]} := \{x \in \mathcal{H} : [x,y] = 0 \ \forall y \in K\}$$

is a linear subspace of $\mathcal{H}$. For $z \in \mathcal{H}$, $\{\mathbb{C}z\}^{[\perp]}$ will be denoted also by $z^{[\perp]}$. A vector $x \neq 0$ such that $x^{[\perp]} = \mathcal{H}$ is called an *isotropic vector*. The linear subspace $\mathcal{I}^0 = \mathcal{H}^{[\perp]}$ of all isotropic vectors is called the *isotropic subspace* for [,]. Clearly $\mathcal{I}^0 \subset \wp^0$. If the form [,] is semi-definite, (3.1.1) implies that the opposite inclusion holds, and therefore $\mathcal{I}^0 = \wp^0$.

If $\mathcal{I}^0 \neq \{0\}$, the form [,] (or the metric defined by [,]) is said to be *degenerate*.

Let $\mathcal{K}$ be a linear subspace of $\mathcal{H}$. If $\mathcal{K} \subset \wp^+$, the function $x \mapsto [x,x]^{\frac{1}{2}}$ defines a semi-norm on $\mathcal{K}$, which becomes a norm if $\mathcal{K} \subset \wp^{++} \cup \{0\}$. Similar conclusions hold for the function $x \mapsto -[x,x]$ when $\mathcal{K} \subset \wp^-$ or $\mathcal{K} \subset \wp^{--} \cup \{0\}$. Since on a finite dimensional vector space all norms are equivalent, the following lemma holds:

Lemma 3.1 *Let the linear subspace $\mathcal{K}$ be finite dimensional. If $\mathcal{K}$ is either strictly positive or strictly negative, then, for any norm $\| \ \|$ on $\mathcal{K}$, there exist positive constants h and k such that, respectively,*

$$h[x,x] \leq \|x\|^2 \leq k[x,x] \tag{3.1.2}$$

or

$$-h[x,x] \leq \|x\|^2 \leq -k[x,x] \tag{3.1.3}$$

for all $x \in \mathcal{K}$.

A linear subspace $\mathcal{K} \subset \mathcal{H}$ for which there exist a norm $\| \ \|$ and two positive constants h an k such that (3.1.2) or (3.1.3) holds, will be said to be, respectively, *uniformly positive* or *uniformly negative*. Lemma 3.1 can be re-stated saying that any strictly positive or strictly negative finite-dimensional subspace $\mathcal{K} \subset \mathcal{H}$ is, respectively, uniformly positive or uniformly negative.

Now let $\mathcal{H}$ be a locally convex, Hausdorff, complex vector space and suppose that $[\ ,\ ] : \mathcal{H} \times \mathcal{H} \to \mathbb{C}$ is continuous (and – to avoid trivialities – that it does not vanish identically). Then $\wp^0, \wp^-, \wp^+, \mathcal{I}^0$ are closed, and $\wp^{++}, \wp^{--}$ are open. If $x \in \wp^+$ has a neighbourhood $U \subset \wp^-$ (i.e. $U \cap \wp^{++} = \emptyset$), there is a convex, symmetric neighborhood V of 0 such that

$$[x + tv, x + tv] \leq 0$$

for all $t \in [-1, 1]$, $v \in V$. Then $x \in \wp^0$ and

$$t(2Re[x, v] + t[v, v]) \leq 0$$

whenever $-1 \leq t \leq 1$, whence $2Re[x, v] + t[v, v] = 0$ for all $t \in [-1, 1]$, and, in conclusion, $Re[x, v] = 0$ and $[v, v] = 0$ for all $v \in V$, hence for all $v \in \mathcal{H}$.

Thus $[\ ,\ ]$ vanishes identically on $\mathcal{H} \times \mathcal{H}$. That proves that $\wp^+ = \overline{\wp^{++}}$ and similarly $\wp^- = \overline{\wp^{--}}$. As a consequence $\dot{\wp}^0 = \emptyset$.

3.2 Indefinite forms in Hilbert spaces

Let $\mathcal{H}$ be a complex Hilbert space with inner product $(\ |\ )$ and associated norm $\|\ \|$, and let $[\ ,\ ]$ be continuous. By the representation theorem, there exists a unique self-adjoint operator $L \in \mathcal{L}(\mathcal{H})$ such that

$$[x, y] = (Lx|y) \quad \text{for all } x, y \in \mathcal{H}. \tag{3.2.1}$$

If $x \in \mathcal{I}^0 \setminus \{0\}$, then $(Lx|y) = 0$ for all $y \in \mathcal{H}$, i.e. $Lx = 0$. Thus $[\ ,\ ]$ is degenerate if, and only if, $0 \in p\sigma(L)$. If $[\ ,\ ]$ is non-degenerate, then $[\ ,\ ]$ is said to be singular or regular if, respectively, $0 \in \sigma(L)$ (hence $0 \in c\sigma(L)$) or $0 \in r(L)$.

Since $(\operatorname{Ker} L)^\perp$ is invariant under the action of L, if P stands for the orthogonal projector onto $(\operatorname{Ker} L)^\perp$, then

$$(Lx|y) = (L(Px + (I - P)x)|Py + (I - P)y) = (LPx|Py) = (PLPx|y)$$

for all x, y in $\mathcal{H}$, i.e. $L = PLP$. Thus, the restriction of $[\ ,\ ]$ to $(\operatorname{Ker} L)^\perp$ defines on this closed subspace a non-degenerate sesquilinear, hermitian, continuous form.

Example *Kreĭn space.* Let $\mathcal{H}^+$ and $\mathcal{H}^-$ be two closed, orthogonal subspaces of $\mathcal{H}$, such that $\mathcal{H}^- = \mathcal{H}^{+\perp}$, i.e.

$$\mathcal{H} = \mathcal{H}^+ \oplus \mathcal{H}^-. \tag{3.2.2}$$

If P^+, P^- project orthogonally $\mathcal{H}$ onto $\mathcal{H}^+$ and onto $\mathcal{H}^-$, let $J \in \mathcal{L}(\mathcal{H})$ be the continuous self-adjoint operator defined by

$$J := P^+ - P^-, \tag{3.2.3}$$

or also by the matrix

$$J = \begin{pmatrix} I^+ & 0 \\ 0 & -I^- \end{pmatrix} \tag{3.2.4}$$

(where I^+ and I^- are the identity operators in $\mathcal{H}^+$ and $\mathcal{H}^-$) with respect to the orthogonal decomposition (3.2.2). The operator J defines a continuous hermitian sesquilinear form $[\ ,\ ]$ on $\mathcal{H}$ by

$$[x, y] = (Jx|y), \tag{3.2.5}$$

or also by

$$[x, y] = (x^+|x^+) - (x^-|x^-), \tag{3.2.6}$$

where $x^\pm = P^\pm x, y^\pm = P^\pm y$.

The form (3.2.5) is called a *Kreĭn form*, or a *Kreĭn metric*, and $\mathcal{H}$ a *Kreĭn space.* If $\dim_{\mathbb{C}}\mathcal{H}^+$ or $\dim_{\mathbb{C}}\mathcal{H}^-$ is finite, $\mathcal{H}$ is called a *Pontryagin space*.

Note that $\mathcal{H}^+$ and $\mathcal{H}^-$ are also [,]-orthogonal.

If both projectors P^+ an P^- are different from 0 and I (the identity operator on $\mathcal{H}$), then $\sigma(J) = p\sigma(J) = \{-1, 1\}$, and the Kreĭn space $\mathcal{H}$ is non-degenerate and regular.

If the linear subspace $\mathcal{K}$ is positive, and if $x \in \mathcal{K}$ is such that $P^+x = 0$, then, by (3.2.6), $[x, x] = -\|P^-x\|^2$ and therefore $P^-x = 0$, hence $x = 0$. That proves

Lemma 3.2 *The restriction of P^+ to any positive or of P^- to any negative linear subspace of the Kreĭn space $\mathcal{H}$ is injective. Hence any positive or negative linear subspace of $\mathcal{H}$ has dimension less than or equal to the dimension of $\mathcal{H}^+$ or $\mathcal{H}^-$, respectively.*

The following proposition will now be established.

Proposition 3.1 *If $\mathcal{H}$ is a Pontryagin space, the dimension of any complex vector space contained in $\wp^0$ does not exceed* $\min\{\dim_{\mathbb{C}}\mathcal{H}^+, \dim_{\mathbb{C}}\mathcal{H}^-\}$.

Let $n = \dim_{\mathbb{C}}\mathcal{H}^- < \infty$ and suppose that there exist $m > n$ vectors $x_1, \ldots, x_m$ in $\mathcal{H}$, all different from zero and such that

$$(Jx_\alpha|x_\beta) = 0 \qquad \text{for } \alpha, \beta = 1, \ldots, n.$$

Setting $x_\alpha^\pm = P^\pm x_\alpha$, this condition is equivalent to

$$(x_\alpha^+|x_\beta^+) = (x_\alpha^-|x_\beta^-). \tag{3.2.7}$$

Since $m > n$, there is in $\{x_1^-, \ldots, x_m^-\}$ a maximal set of $p \leq n$ elements, say $\{x_1^-, \ldots, x_p^-\}$, which are linearly independent over $\mathbb{C}$. For $\beta = p+1, \ldots, m$, there exist $a_\beta^\mu \in \mathbb{C}$ $(\mu = 1, \ldots, p)$ such that

$$x_\beta^- = \sum_{\mu=1}^p a_\beta^\mu x_\mu^-.$$

Since, by (3.2.7),

$$\begin{aligned}
(x_\beta^+|\sum_{\mu=1}^p a_\beta^\mu x_\mu^+) &= \sum_{\mu=1}^p \overline{a_\beta^\mu}(x_\beta^+|x_\mu^+) = \sum_{\mu=1}^p \overline{a_\beta^\mu}(x_\beta^-|x_\mu^-) \\
&= (x_\beta^-|x_\beta^-) \leq \|x_\beta^-\| \, \|\sum_{\mu=1}^p a_\beta^\mu x_\mu^-\| = \|x_\beta^+\| \, \|\sum_{\mu=1}^p a_\beta^\mu x_\mu^+\|,
\end{aligned}$$

there is a $t > 0$ such that

$$x_\beta^+ = t\sum_{\mu=1}^p a_\beta^\mu x_\mu^+.$$

It turns out that $t = 1$, because

$$t\|\sum_{\mu=1}^{p} a_\beta^\mu x_\mu^+\| = \|x_\beta^+\| = \|x_\beta^-\| = \|\sum_{\mu=1}^{p} a_\beta^\mu x_\mu^-\| = \|\sum_{\mu=1}^{p} a_\beta^\mu x_\mu{}^+\|.$$

Hence

$$x_\beta = \sum_{\mu=1}^{p} a_\beta^\mu x_\mu,$$

proving thereby the following lemma, which implies Proposition 3.1.

Lemma 3.3 *If* $\dim_{\mathbb{C}}\mathcal{H}^- < \infty$, *the number of linearly independent vectors in the set* $\{x_1, \ldots, x_m\}$ *is less than or equal to* $\dim_{\mathbb{C}}\mathcal{H}^-$.

The continuous form [,] represented by (3.2.1), is strictly indefinite if, and only if, the bounded self-adjoint operator L is such that

$$\sigma(L) \cap \mathbb{R}_-^* \neq \emptyset, \quad \sigma(L) \cap \mathbb{R}_+^* \neq \emptyset.$$

The spectral measure associated to L,

$$L = \int t dE(t)$$

is a regular spectral measure on the Borel sets of $\mathbb{R}$. Setting, as usual, $E(t) = E((-\infty, t])$ for any $t \in \mathbb{R}$, then $E(t)$ is an orthogonal projector such that

$$a \leq b \Rightarrow E(a) \leq E(b)$$

and

$$\lim_{t \downarrow a} E(t) = E(a), \quad \lim_{t \to -\infty} E(t) = 0, \quad \lim_{t \to +\infty} E(t) = I,$$

for the strong operator topology. Setting

$$E(a-0) = \lim_{t \uparrow a} E(t)$$

then $E(a-0) \leq E(a)$, and $E(a-0) \neq E(a)$ if, and only if, a is an eigenvalue of L, in which case $E(a) - E(a-0)$ is the image $E(\{a\})$ of the singleton $\{a\}$; furthermore $\operatorname{Ran} E(\{a\}) = \operatorname{Ker}(L - aI)$. Thus $\operatorname{Ran} E(\{a\})$ is the eigenspace of L corresponding to a, and is different from zero if, and only if, $a \in p\sigma(L)$.

The operators $P^- = E((-\infty, 0])$, $P^0 = E(0) - E(0-0) = E(\{0\})$, $P^+ = E(0, +\infty))$ are mutually orthogonal, orthogonal projectors, such that

$$P^+ \oplus P^0 \oplus P^- = I.$$

The continuous form (3.2.1) is non-degenerate, if, and only if, $P^0 = 0$, i.e. $0 \in r(L) \cup c\sigma(L)$. In this case the mutually orthogonal, closed subspaces $\mathcal{H}^+ := P^+\mathcal{H}$, $\mathcal{H}^- := P^-\mathcal{H}$, span $\mathcal{H}$, i.e. (3.2.2) holds.

From now on, it will be assumed that $0 \in r(L)$. Since P^+ and P^- commute with L, the bounded self-adjoint operators $L^+ \in \mathcal{L}(\mathcal{H}^+)$, $L^- \in \mathcal{L}(\mathcal{H}^-)$ defined by

$$L^+ = P^+LP^+, \quad -L^- = P^-LP^-$$

are both strictly positive. Being $P^+P^- = P^-P^+ = 0$, then L is expressed by the matrix

$$L = \begin{pmatrix} L^+ & 0 \\ 0 & -L^- \end{pmatrix} \tag{3.2.8}$$

with respect to the orthogonal decomposition (3.2.2).

Suppose that [,] is indefinite. Since $0 \in r(L)$, then $\sigma(L) \cap \mathbf{R}^*_-$ and $\sigma(L) \cap \mathbf{R}^*_+$ are non-empty compact subset of $\mathbf{R}_-{}^*$ of $\mathbf{R}^*_+$, and their union is $\sigma(L)$. The projector P^+ is expressed by the norm-convergent integral

$$P^+ = \frac{1}{2\pi i}\int_\ell (\zeta I - L)^{-1} d\zeta$$

where ℓ is any simple, closed rectifiable curve contained in the right half-plane $\Pi_r = \{\zeta \in \mathbf{C} : \ Re\ \zeta > 0\}$, enclosing $\sigma(L) \cap \mathbf{R}^*_+$ in its interior, and oriented counterclockwise; P^- is given by $P^- = I - P^+$ and can be described in a similar way in terms of $\sigma(L) \cap \mathbf{R}^*_-$.

Suppose now that the Hilbert space $\mathcal{H}^-$ is separable and let $\{e_1, e_2, \ldots\}$ be an orthonormal basis of $\mathcal{H}^-$. Let $\mathcal{D}$ be a dense linear subspace of $\mathcal{H}$.

Given any sequence $\{r_1, r_2, \ldots\}$ of positive numbers r_ν, there exists a sequence $\{w_1, \ldots\}$ of points $w_\nu \in \mathcal{D}$ such that

$$\|w_\nu - e_\nu\| < r_\nu \qquad \text{for } \nu = 1, 2, \ldots. \tag{3.2.9}$$

Under which conditions on $\{w_\nu\}$ does the formal sum

$$\sum (x|e_\nu) w_\nu$$

define a linear operator $T_w \in \mathcal{L}(\mathcal{H})$?

The projector $P^- : \mathcal{H} \to \mathcal{H}^-$ is expressed by

$$P^- x = \sum (x|e_\nu) e_\nu,$$

and, by the Schwarz inequality,

$$\|\sum (x|e_\nu)(w_\nu - e_\nu)\| \le \sum |(x|e_\nu)| \|w_\nu - e_\nu\| \le \|x\| \sum \|w_\nu - e_\nu\| \le \|x\| \sum r_\nu,$$

for all $x \in \mathcal{H}$. As a consequence, if $r := \sum r_\nu < \infty$, then $T_w \in \mathcal{L}(\mathcal{H})$ and

$$\|(T_w - P^-)x\| \le r\|x\|, \tag{3.2.10}$$

so that

$$\|(T_w\| \le \|T_w - P^-\| + \|P^-\| \le r + 1.$$

Note that $T_w = T_w \circ P^-$.

It will be shown now that the points $w_\nu \in \mathcal{D}$ can be so chosen that $\overline{T_w(\mathcal{H}^-)}$ is uniformly negative.

Setting $w_\nu^+ = P^+ w_\nu$, $w_\nu^- = P^- w_\nu$, (3.2.9) implies

$$\|w_\nu^+\| < r_\nu, \quad \|w_\nu^- - e_\nu\| < r_\nu \qquad (\nu = 1, 2, \ldots). \tag{3.2.11}$$

The second set of inequalities, together with (3.2.10) yields

$$\|P^- \circ T_w x^- - x^-\| \leq r\|x^-\|$$

for all $x^- \in \mathcal{H}^-$. Hence, if $0 < r < 1$, $P^- \circ T_{w|\mathcal{H}^-} \in \mathcal{L}(\mathcal{H}^-)$ has an inverse in $\mathcal{L}(\mathcal{H}^-)$, for which

$$\begin{aligned} \|(P^- \circ T_{w|\mathcal{H}^-})^{-1}\| &= \|(I^- - (I^- - P^- \circ T_{w|\mathcal{H}^-}))^{-1}\| \\ &\leq 1 + \|I^- - P^- \circ T_{w|\mathcal{H}^-}\| + \|I^- - P^- \circ T_{w|\mathcal{H}^-}\|^2 + \cdots \\ &\leq 1 + r + r^2 + \cdots = \frac{1}{1-r}, \end{aligned}$$

so that

$$\|P^- \circ T_w(x^-)\| \geq (1-r)\|x^-\| \qquad \text{for all } x^- \in \mathcal{H}^-. \tag{3.2.12}$$

Since L^- is strictly positive, there are two real numbers a and b, $0 < a \leq b$, such that

$$a\|x^-\|^2 \leq (L^- x^- | x^-) \leq b\|x^-\|^2 \qquad \text{for all } x^- \in \mathcal{H}^-.$$

Hence, if $0 < r < 1$, (3.2.12) yields

$$(1-r)^2 a\|x^-\|^2 \leq (L^-(P^- \circ T_w(x^-)) | P^- \circ T_w(x^-))$$

i.e.

$$(1-r)^2 a\|x^-\|^2 \leq \sum_{\mu,\nu=1}^{+\infty} (L^- w_\nu^- | w_\mu^-)(x|e_\nu)\overline{(x|e_\mu)} \quad \text{for all } x^- \in \mathcal{H}^-. \tag{3.2.13}$$

As for $\sum_{\mu,\nu=1}^{+\infty} (L^+ w_\nu^+ | w_\nu^+)(x^-|e_\nu)\overline{(x^-|e_\mu)}$, by the first set of inequalities in (3.2.11),

$$\begin{aligned} |\sum_{\mu,\nu=1}^{+\infty} (L^+ w_\nu^+ | w_\mu^+)(x^-|e_\nu)\overline{(x^-|e_\mu)}| &= |(\sum_{\nu=1}^{+\infty}(x^-|e_\nu)L^+ w_\nu^+ | \sum_{\mu=1}^{+\infty}(x^-|e_\mu)w_\mu^+)| \\ &\leq \|\sum_{\mu,\nu=1}^{+\infty}(x^-|e_\nu)L^+ w_\nu^+\| \, \|\sum_{\mu=1}^{+\infty}(x^-|e_\mu)w_\nu^+\| \\ &\leq \|L^+\|(\sum_{\nu=1}^{+\infty}|(x^-|e_\nu)| \, \|w_\nu^+\|)^2 \\ &\leq \|L^+\| \, \|x^-\|^2(\sum_{\nu=1}^{+\infty}\|w_\nu\|)^2 \\ &\leq \|L^+\| \, \|x^-\|^2 r^2. \end{aligned} \tag{3.2.14}$$

Because

$$
\begin{aligned}
(LT_wx|T_wx) &= \sum_{\nu,\mu=1}^{+\infty} (Lw_\nu|w_\mu)(x^-|e_\nu)\overline{(x^-|e_\mu)} \\
&= \sum_{\nu,\mu=1}^{+\infty} ((L^+w_\nu^+|w_\mu^+) - (L^-w_\nu^-|w_\mu^-))(x^-|e_\nu)\overline{(x^-|e_\mu)},
\end{aligned}
$$

then, by (3.2.13) and (3.2.14),

$$
\begin{aligned}
(LT_wx|T_wx) &\leq \left|\sum_{\nu,\mu=1}^{+\infty} (L^+w_\nu^+|w_\mu^+)(x^-|e_\nu)\overline{(x^-|e_\mu)}\right| \\
&\quad - \sum_{\nu,\mu=1}^{+\infty} (L^-w_\nu^-|w_\mu^-)(x^-|e_\nu)\overline{(x^-|e_\mu)} \\
&\leq (\|L^+\|r^2 - (1-r)^2a)\|x^-\|^2
\end{aligned}
$$

for all $x \in \mathcal{H}$, whenever $0 < r < 1$.

Choosing $r_0 > 0$ so small that $0 < r_0 < 1$ and that, if $r \in (0, r_0)$,

$$\|L^+\|r^2 - (1-r)^2a \leq -\frac{a}{2},$$

then

$$(LT_wx|T_wx) \leq -\frac{a}{2}\|x^-\|^2$$

for all $x \in \mathcal{H}$. This proves

Lemma 3.4 *If $\bar{\mathcal{D}} = \mathcal{H}$, the points $w_1, w_2, \ldots$ can be chosen in $\mathcal{D}$ in such a way that $T_w(\mathcal{H}^-)$ is uniformly negative.*

Since $T_we_\nu = w_\nu$, that implies

Corollary 3.1 *If $\mathcal{D}$ is dense in $\mathcal{H}$ and if $\mathcal{H}^-$ is separable, then the points $w_1, w_2, \ldots$ can be chosen in $\mathcal{D}$ in such a way that their closed linear span is uniformly negative.*

In particular, the closed linear span is contained in $\wp^{--} \cup \{0\}$. Of course, it is not necessarily contained in $\mathcal{D}$.

The above results hold in particular for Kreĭn spaces and extend to the separable case a result established in [AZ-IO 89; p. 65] for a Pontryagin space with $\dim_{\mathbb{C}}\mathcal{H}^- \leq \dim_{\mathbb{C}}\mathcal{H}^+$. In this case the linear span is closed.

3.3 Linear isometries for indefinite metrics

As in section 3.2, $\mathcal{H}$ will be a complex Hilbert space endowed with a non-degenerate, regular form, [,] associated by (3.2.1) to a continuously invertible, self-adjoint operator $L \in \mathcal{L}(\mathcal{H})$, represented by the matrix (3.2.8) with respect to the orthogonal decomposition (3.2.2). Let $\Lambda(L)$ be the semigroup

$$\Lambda(L) = \{A \in \mathcal{L}(\mathcal{H}) : A^*LA = L\}.$$

Since for

$$A = \begin{pmatrix} A_{11} & A_{12} \\ A_{21} & A_{22} \end{pmatrix} \tag{3.3.0}$$

with $A_{11} \in \mathcal{L}(\mathcal{H}^+)$, $A_{12} \in \mathcal{L}(\mathcal{H}^-, \mathcal{H}^+)$, $A_{21} \in \mathcal{L}(\mathcal{H}^+, \mathcal{H}^-)$, $A_{22} \in \mathcal{L}(\mathcal{H}^-)$,

$$A^*LA = \begin{pmatrix} A_{11}{}^*L^+A_{11} - A_{21}{}^*L^-A_{21} & A_{11}{}^*L^+A_{12} - A_{21}{}^*L^-A_{22} \\ A_{12}{}^*L^+A_{11} - A_{22}{}^*L^-A_{21} & A_{12}{}^*L^+A_{12} - A_{22}{}^*L^-A_{22} \end{pmatrix}$$

then $A \in \Lambda(L)$ if, and only if,

$$A_{11}{}^*L^+A_{11} - A_{21}{}^*L^-A_{21} = L^+, \tag{3.3.1}$$

$$A_{11}{}^*L^+A_{12} - A_{21}{}^*L^-A_{22} = 0, \tag{3.3.2}$$

$$A_{22}{}^*L^-A_{22} - A_{12}{}^*L^-A_{12} = L^-. \tag{3.3.3}$$

Let $\Gamma(L)$ be the (maximum) subgroup of $\Lambda(L)$, consisting of all elements $A \in \Lambda(L)$ which are invertible in $\mathcal{L}(\mathcal{H})$. Any $A \in \Lambda(L)$ has a left inverse, given by $L^{-1}A^*L \in \mathcal{L}(\mathcal{H})$. This is a right inverse if, and only if,

$$AL^{-1}A^*L = I$$

Since

$$\begin{aligned} L^{-1}A^*L &= \begin{pmatrix} L^{+^{-1}} & 0 \\ 0 & -L^{-^{-1}} \end{pmatrix} \begin{pmatrix} A_{11}{}^* & A_{21}{}^* \\ A_{12}{}^* & A_{22}{}^* \end{pmatrix} \begin{pmatrix} L^+ & 0 \\ 0 & -L^- \end{pmatrix} \\ &= \begin{pmatrix} L^{+^{-1}}A_{11}{}^*L^+ & -L^{+^{-1}}A_{21}{}^*L^- \\ L^{-^{-1}}A_{12}{}^*L^+ & L^{-^{-1}}A_{22}{}^*L^- \end{pmatrix}, \end{aligned}$$

we obtain

$$AL^{-1}A^*L =$$
$$\begin{pmatrix} A_{11}L^{+^{-1}}A_{11}{}^*L^+ - A_{12}L^{-^{-1}}A_{12}{}^*L^+ & -A_{11}L^{+^{-1}}A_{21}{}^*L^- + A_{12}L^{-^{-1}}A_{22}{}^*L^- \\ A_{21}L^{+^{-1}}A_{11}{}^*L^+ - A_{22}L^{-^{-1}}A_{12}{}^*L^+ & -A_{21}L^{+^{-1}}A_{21}{}^*L^- + A_{22}L^{-^{-1}}A_{22}{}^*L^- \end{pmatrix}.$$

Hence $A \in \Gamma(L)$ if, and only if, besides (3.3.1), (3.3.2) and (3.3.3), the following conditions hold:

$$A_{11}L^{+^{-1}}A_{11}{}^* - A_{12}L^{-^{-1}}A_{12}{}^* = L^{+^{-1}}, \tag{3.3.4}$$

$$A_{11}L^{+^{-1}}A_{21}{}^* - A_{12}L^{-^{-1}}A_{22}{}^* = 0, \tag{3.3.5}$$

$$A_{22}L^{-^{-1}}A_{22}{}^* - A_{21}L^{+^{-1}}A_{21}{}^* = L^{-^{-1}}. \tag{3.3.6}$$

Let $L_1 \in \mathcal{L}(\mathcal{H}^+)$, $L_2 \in \mathcal{L}(\mathcal{H}^-)$ be the positive square roots of L^+ and L^- respectively. Then, both L_1 and L_2 are continuously invertible, self-adjoint operators. Setting

$$\tilde{A}_{\alpha\beta} = L_\alpha A_{\alpha\beta} {L_\beta}^{-1},$$

we have $\tilde{A}_{11} \in \mathcal{L}(\mathcal{H}^+)$, $\tilde{A}_{12} \in \mathcal{L}(\mathcal{H}^-, \mathcal{H}^+)$, $\tilde{A}_{21} \in \mathcal{L}(\mathcal{H}^+, \mathcal{H}^-)$, $\tilde{A}_{22} \in \mathcal{L}(\mathcal{H}^-)$, and (3.3.1), ..., (3.3.6) become respectively

$$\tilde{A}_{11}^*\tilde{A}_{11} - \tilde{A}_{21}^*\tilde{A}_{21} = I^+, \tag{3.3.1'}$$

$$\tilde{A}_{11}^*\tilde{A}_{12} - \tilde{A}_{21}^*\tilde{A}_{22} = 0, \tag{3.3.2'}$$

$$\tilde{A}_{22}^*\tilde{A}_{22} - \tilde{A}_{12}^*\tilde{A}_{12} = I^-, \tag{3.3.3'}$$

$$\tilde{A}_{11}\tilde{A}_{11}^* - \tilde{A}_{12}\tilde{A}_{12}^* = I^+, \tag{3.3.4'}$$

$$\tilde{A}_{11}\tilde{A}_{21}^* - \tilde{A}_{12}\tilde{A}_{22}^* = 0, \tag{3.3.5'}$$

$$\tilde{A}_{22}\tilde{A}_{22}^* - \tilde{A}_{21}\tilde{A}_{21}^* = I^-. \tag{3.3.6'}$$

The first three conditions characterize the elements of the semigroup $\Lambda(J)$, and all six the group $\Gamma(J)$.

Let $M \in \mathcal{L}(\mathcal{H})$ be the linear bijective map defined by the matrix

$$\begin{pmatrix} {L_1}^{-1} & 0 \\ 0 & {L_2}^{-1} \end{pmatrix}$$

i.e.

$$Mx = {L_1}^{-1}P^+x + {L_2}^{-1}P^-x. \tag{3.3.7}$$

Since

$$(LMx|My) = (MLMx|y) = (Jx|y) \tag{3.3.8}$$

for all $x, y \in \mathcal{H}$, the following lemma holds:

Lemma 3.5 *The map $M \in \mathcal{L}(\mathcal{H})$ transforms the Hilbert space $\mathcal{H}$, endowed with the form $[\ ,\ ]$ given by* (3.2.1) *into the Kreĭn space $\mathcal{H} = \mathcal{H}^+ \oplus \mathcal{H}^-$ defined by* (3.2.5), *and defines an isomorphism $A \mapsto MAM^{-1}$ of the semigroup $\Lambda(L)$ onto the semigroup $\Lambda(J)$, such that* (3.3.8) *holds for all x, y in $\mathcal{H}$.*

The image of $\Gamma(L)$ is the group $\Gamma(J)$.

The following lemma was established in [VES 90a].

Lemma 3.6 *Let $\mathcal{H}$ be a Pontryagin space with $\dim_{\mathbb{C}}\mathcal{H}^- \leq \dim_{\mathbb{C}}\mathcal{H}^+$ and let $\mathcal{A}$ be a maximal linear subspace contained in $\wp^{--} \cup \{0\}$. Then there exists an $S \in \Gamma(J)$ such that $S\mathcal{H}^- = \mathcal{A}$.*

By Lemma 3.5, Lemma 3.4 extends to the case where $\mathcal{H}$ is any Hilbert space endowed with a non-degenerate, regular form defined by (3.2.1) and such that $P^-\mathcal{H}$ is separable.

Corollary 3.1 yields:

Proposition 3.2 *If $\mathcal{H}$ is a Kreĭn space such that $\mathcal{H}^-$ is separable, and if $\mathcal{D}$ is a dense linear subspace of $\mathcal{H}$, there is an element $S \in \Gamma(L)$ such that $S(\mathcal{H}^-)$ is a uniformly negative subspace of $\mathcal{D}$.*

3.4 Strongly continuous semigroups of linear isometries for an indefinite metric

Let $T : \mathbf{R}_+ \to \mathcal{L}(\mathcal{H})$ be a strongly continuous semigroup; its infinitesimal generator is a closed operator X with dense domain $\mathcal{D}(X)$.

If $T(t) \in \Lambda(L)$, then $M \circ T(t) \circ M^{-1} \in \Lambda(J)$; the semigroup $t \mapsto M \circ T(t) \circ M^{-1}$ is strongly continuous and its infinitesimal generator is $M \circ X \circ M^{-1}$ with domain $\mathcal{D}(M \circ X \circ M^{-1}) = M(\mathcal{D}(X))$. Conversely, if $H : \mathbf{R}_+ \to \Lambda(J)$ is a strongly continuous semigroup, and if $Y : \mathcal{D}(Y) \subset \mathcal{H} \to \mathcal{H}$ is its infinitesimal generator, then $t \mapsto M^{-1} \circ H(t) \circ M$ is a strongly continuous semigroup with values in $\Lambda(L)$, whose infinitesimal generator is $M^{-1} \circ Y \circ M : M^{-1}(\mathcal{D}(Y)) \subset \mathcal{H} \to \mathcal{H}$. The same conclusions hold if T and H are strongly continuous groups $\mathbf{R} \to \Gamma(L)$ and $\mathbf{R} \to \Gamma(J)$ respectively.

Replacing T by H, it will be assumed henceforth that T is a strongly continuous semigroup (or a strongly continuous group) with values in $\Lambda(J)$ (or in $\Gamma(J)$).

Theorem 3.1 *Let $T : \mathbf{R}_+ \to \mathcal{L}(\mathcal{H})$ be a strongly continuous semigroup and let X be its infinitesimal generator, with domain $\mathcal{D}(X)$. Then $T(\mathbf{R}_+) \subset \Lambda(J)$ if, and only if, iJX is a symmetric operator, i.e.*

$$J\mathcal{D}(X) \subset \mathcal{D}(X^*) \tag{3.4.1}$$

and

$$X^*J + JX = 0 \qquad \text{on } \mathcal{D}(X). \tag{3.4.2}$$

Proof If $T(\mathbf{R}_+) \subset \Lambda(J)$, i.e., if

$$T(t)^*JT(t) = J$$

for all $t \geq 0$, then, for x, y in $\mathcal{H}$,

$$(T(t)^*JT(t)x|y) = (JT(t)x|T(t)y) =$$
$$(J(T(t)-I)x|(T(t)-I)y) + (J(T(t)-I)x|y) + (Jx|(T(t)-I)y) + (Jx|y),$$

whence

$$(J(T(t)-I)x|(T(t)-I)y) + (J(T(t)-I)x|y) + (Jx|(T(t)-I)y) = 0.$$

Thus, for $x, y \in \mathcal{D}(X)$, since

$$\lim_{t\downarrow 0} \frac{1}{t}(T(t)-I)x = Xx, \quad \lim_{t\downarrow 0} \frac{1}{t}(T(t)-I)y = Xy,$$

then

$$(JXx|y) + (Jx|Xy) = 0.$$

Hence the linear form $y \mapsto (Xy|Jx)$ is continuous on $\mathcal{D}(X)$. Therefore $Jx \in \mathcal{D}(X^*)$ for all $x \in \mathcal{D}(X)$, and

$$((JX + X^*J)x|y) = 0$$

for all $y \in \mathcal{D}(X)$. This is equivalent to $(JX + X^*J)x = 0$ for all $x \in \mathcal{D}(X)$.

Conversely, if (3.4.1) and (3.4.2) hold, since $T(t)\mathcal{D}(X) \subset \mathcal{D}(X)$ for all $t \geq 0$ and because

$$\frac{d}{dt}T(t)x = T(t)Xx = XT(t)x$$

for all $x \in \mathcal{D}(X)$ and all $t \geq 0$ (the derivative is a right derivative when $t = 0$), then for $x, y \in \mathcal{D}(X)$ and $t \geq 0$,

$$\begin{aligned}\frac{d}{dt}(T(t)^*JT(t)x|y) &= \frac{d}{dt}(JT(t)x|T(t)y)\\ &= (JXT(t)x|T(t)y) + (JT(t)x|XT(t)y)\\ &= ((JX + X^*J)T(t)x|T(t)y)\\ &= 0.\end{aligned}$$

Hence

$$\frac{d}{dt}(T(t)^*JT(t)x) = 0$$

for all $x \in \mathcal{D}(X)$ and all $t \geq 0$, i.e. $T(t)^*JT(t)x$ is independent of $t \geq 0$. Thus

$$T(t)^*JT(t)x = T(0)^*JT(0)x = Jx$$

for all $t \geq 0$ and for any $x \in \mathcal{D}(X)$, hence for any $x \in \mathcal{H}$. □

There are constants $a \in \mathbf{R}, b > 0$ such that

$$||T(t)|| \leq be^{at}$$

for $t \geq 0$ and

$$||(\zeta I - X)^{-n}|| \leq b(\zeta - a)^{-n} \tag{3.4.3}$$

for all real $\zeta > a$, $n = 1, 2, \ldots$. Moreover,

$$\{\zeta \in \mathbf{C} : Re\zeta > a\} \subset r(X).$$

Theorem 3.2 *If $T(\mathbf{R}_+) \subset \Lambda(J)$, then T is the restriction to $\mathbf{R}_+$ of a strongly continuous group $R : \mathbf{R} \to \mathcal{L}(\mathcal{H})$ if and only if iJX is self-adjoint. If iJX is self-adjoint, then $R(\mathbf{R}) \subset \Gamma(J)$.*

Proof Let iJX be self-adjoint. Since

$$\zeta I - X^* = \zeta I + JXJ = J(\zeta I + X)J, \tag{3.4.4}$$

then $\sigma(X^*)$ is the image of $\sigma(X)$ by the map $\zeta \mapsto -\zeta$.

On the other hand $\zeta \in r(X)$ if, and only if, $\bar{\zeta} \in r(X^*)$, and furthermore,

$$(\zeta I - X^*)^{-1} = (\bar{\zeta} I - X)^{-1*}.$$

Thus, by (3.4.4),

$$(\bar{\zeta}I - X)^{-1*} = (\zeta I + JXJ)^{-1} = J \circ (\zeta I + X)^{-1} \circ J,$$

and therefore, by (3.4.3),

$$\begin{aligned} \|(\zeta I + X)^{-n}\| &= \|J(\zeta I - X)^{-n\,*}J\| = \|(\zeta I - X)^{-n\,*}\| \\ &= \|(\zeta I - X)^{-n}\| \le b(\zeta - a)^{-n}, \end{aligned}$$

for all real $\zeta > a$ and for all $n = 1, 2, \ldots$.

Hence $-X$ is the infinitesimal generator of a strongly continuous semigroup $T_1 : \mathbf{R}_+ \to \mathcal{L}(\mathcal{H})$ such that $T_1{}^*(t)JT_1(t) = J$ for all $t \ge 0$.

In conclusion, X generates a strongly continuous group $R : \mathbf{R} \to \mathcal{L}(\mathcal{H})$ such that

$$R(t)^* JR(t) = J$$

for all $t \ge 0$, i.e. $R(\mathbf{R}) \subset \Gamma(J)$.

If X is the infinitesimal generator of a strongly continuous group, then there exists some $a \ge 0$ such that

$$\{\zeta \in \mathbf{C} : |Re\zeta| > a\} \subset r(X).$$

If iJX is symmetric but not self-adjoint, then $J\mathcal{D}(X) \subset \mathcal{D}(X^*)$ and X^* is a proper extension of the closed operator $-JXJ$. As a consequence

$$\begin{aligned} \{\zeta \in \mathbf{C} : Re\zeta < -a\} &\subset r(-JXJ) \subset p\sigma(X^*) \\ &\subset \{\zeta \in \mathbf{C} : \bar{\zeta} \in \sigma(X)\} \\ &\subset \{\zeta \in \mathbf{C} : |Re\,\zeta| \le a\}. \end{aligned}$$

This contradiction shows that iJX is self-adjoint and completes the proof of the theorem. □

Let $\mathcal{H}$ be a Pontryagin space with $\dim_{\mathbf{C}}\mathcal{H}^- \le \dim_{\mathbf{C}}\mathcal{H}^+$. Let $T(\mathbf{R}_+) \subset \Lambda(J)$. Since $\mathcal{D}(X)$ is dense in $\mathcal{H}$, replacing $T(t)$ by $S \circ T(t) \circ S^{-1}$, for a suitable choice of $S \in \Gamma(J)$, it can and will be assumed – with no restriction – that $\mathcal{H}^- \subset \mathcal{D}(X)$. Then: $\mathcal{D}(X) \cap \mathcal{H}^+ = P^+\mathcal{D}(X)$ is dense in $\mathcal{H}^+$;

$$\mathcal{D}(X) = (\mathcal{D}(X) \cap \mathcal{H}^+) \oplus \mathcal{H}^-;$$

X is represented by the matrix

$$\begin{pmatrix} X_{11} & X_{12} \\ X_{21} & X_{22} \end{pmatrix} \tag{3.4.5}$$

whose entries are linear operators

$$\begin{aligned} X_{11} &= P^+XP^+{}_{|\mathcal{D}(X)\cap\mathcal{H}^+} : \mathcal{D}(X) \cap \mathcal{H}^+ \to \mathcal{H}^+, & X_{12} &= P^+XP^-{}_{|\mathcal{H}^-} : \mathcal{H}^- \to \mathcal{H}^+, \\ X_{21} &= P^-XP^+{}_{|\mathcal{D}(X)\cap\mathcal{H}^+} : \mathcal{D}(X) \cap \mathcal{H}^+ \to \mathcal{H}^-, & X_{22} &= P^-XP^-{}_{|\mathcal{H}^-} : \mathcal{H}^- \to \mathcal{H}^-. \end{aligned}$$

If a sequence $\{x_\nu\} \subset \mathcal{D}(X)$ converges to some x, the sequence $\{P^- x_\nu\}$ converges to $P^- x$. Since X is closed, if $\{XP^- x_\nu\}$ converges to some y, then ($P^- x \in \mathcal{D}(X)$ and) $XP^- x = y$. Hence, if X is closed, also XP^- is closed. By the closed graph theorem $XP^- \in \mathcal{L}(\mathcal{H})$, and therefore

$$X_{12} \in \mathcal{L}(\mathcal{H}^-, \mathcal{H}^+), \quad X_{22} \in \mathcal{L}(\mathcal{H}^-).$$

The fact that iJX is symmetric reads, for $x = (x^+, x^-)$, $y = (y^+, y^-)$ with $x^+, y^+ \in \mathcal{D}(X_{11}) = \mathcal{D}(X) \cap \mathcal{H}^+$, $x^-, y^- \in \mathcal{H}^-$,

$$(X_{11}x^+|y^+) + (X_{12}x^-|y^+) + (X_{21}x^+|y^-) + (X_{22}x^-|y^-) =$$
$$-(x^+|X_{11}y^+) + (x^+|X_{12}y^-) + (x^-|X_{21}y^+) - (x^-|X_{22}y^-),$$

and is equivalent to the following conditions,

$$(X_{11}x^+|y^+) = -(x^+|X_{11}y^+), \quad (X_{22}x^-|y^-) = -(x^-|X_{22}y^-), \tag{3.4.6}$$

$$(X_{12}x^-|y^+) = (x^-|X_{21}y^+), \tag{3.4.7}$$

holding for all $x^+, y^+ \in \mathcal{D}(X_{11})$, $x^-, y^- \in \mathcal{H}^-$.

By (3.4.1), iX_{11} is a symmetric operator, and iX_{22} is a continuous self-adjoint operator. By (3.4.7),

$$X_{21} = X_{12}^*|_{\mathcal{D}(X_{11})}. \tag{3.4.8}$$

Since X is closed then X_{11} is closed. The operator

$$X' = \begin{pmatrix} X_{11} & 0 \\ 0 & X_{22} \end{pmatrix} \tag{3.4.9}$$

is a bounded perturbation of X and therefore generates a strongly continuous semigroup $\mathbf{R}_+ \to \mathcal{L}(\mathcal{H})$. Hence, either $\sigma(X_{11}) \subset i\mathbf{R}$ and iX_{11} is self-adjoint, or $\sigma(X_{11}) = \overline{\Pi_\ell} := \{\zeta \in \mathbf{C} : Re\,\zeta \leq 0\}$.

Conversely, let X be given by (3.4.5), where: $iX_{22} \in \mathcal{L}(\mathcal{H}^-)$ is a self-adjoint operator; $X_{12} \in \mathcal{L}(\mathcal{H}^-, \mathcal{H}^+)$; X_{21} is given by (3.4.8), and iX_{11} is a closed symmetric operator such that $\{\zeta \in \mathbf{C} : Re\,\zeta > 0\} \subset r(X_{11})$. Then the operator X' defined by (3.4.9) on the domain $\mathcal{D}(X_{11}) \oplus \mathcal{H}^-$ generates a strongly continuous semigroup $\mathbf{R}_+ \to \mathcal{L}(\mathcal{H})$ of linear isometries. Hence the operator X defined by (3.4.5), as a bounded perturbation of X', generates a strongly continuous semigroup $T : \mathbf{R}_+ \to \mathcal{L}(\mathcal{H})$.

By the equivalence of (3.4.1), (3.4.2) with iJX being symmetric, and by Theorem 3.1, $T(\mathbf{R}_+) \subset \Lambda(J)$. If moreover iX_{11} is self-adjoint, then, by Theorem 3.2, X generates a strongly continuous group $T : \mathbf{R} \to \Gamma(J)$.

Hence the following theorem holds:

Theorem 3.3 *Let $\mathcal{H}$ be a Pontryagin space with* $\dim_{\mathbf{C}}\mathcal{H}^- \leq \dim_{\mathbf{C}}\mathcal{H}^+$. *If T is a strongly continuous semigroup $T : \mathbf{R}_+ \to \Lambda(J)$, there is some $S \in \Gamma(J)$ such that the infinitesimal generator X of T is expressed by the matrix*

$$X = S \begin{pmatrix} X_{11} & X_{12} \\ X_{21} & X_{22} \end{pmatrix} S^{-1}, \tag{3.4.10}$$

where: $iX_{22} \in \mathcal{L}(\mathcal{H}^-)$ is self-adjoint, iX_{11} is a closed symmetric operator with dense domain $\mathcal{D}(X_{11}) \subset \mathcal{H}^+$, such that

$$r(X_{11}) \supset \{\zeta \in \mathbf{C} : Re\zeta > 0\}, \quad X_{12} \in \mathcal{L}(\mathcal{H}^-, \mathcal{H}^+),$$

and X_{21} is given by (3.4.8).

Conversely, if the operators $X_{\alpha,\beta}$ appearing in the matrix (3.4.5) satisfy the above conditions, then, for any $S \in \Gamma(J)$, the operator X defined by (3.4.10), with domain $S(\mathcal{D}(X_{11}) \oplus \mathcal{H}^-)$, generates a strongly continuous semigroup $T : \mathbf{R}_+ \to \Lambda(J)$.

Furthermore, T is the restriction to $\mathbf{R}_+$ of a strongly continuous group $\mathbf{R} \to \Gamma(J)$ if, and only if, iX_{11} is self-adjoint.

Using Lemma 3.5 as a the beginning of this number, Theorem 3.3 generates an equivalent statement for a Hilbert space endowed with the form given by (3.2.1).

Let X be given by the matrix (3.4.5) whose entries satisfy the conditions stated in Theorem 3.3. For $\zeta \in r(X)$, the operator $(\zeta I - X)^{-1} \in \mathcal{L}(\mathcal{H})$ is represented by a matrix

$$Z = Z(\zeta) = (\zeta I - X)^{-1} = \begin{pmatrix} Z_{11} & Z_{12} \\ Z_{21} & Z_{22} \end{pmatrix} \tag{3.4.11}$$

where $Z_{11} \in \mathcal{L}(\mathcal{H}^+)$, $Z_{12} \in \mathcal{L}(\mathcal{H}^-, \mathcal{H}^+)$, $Z_{21} \in \mathcal{L}(\mathcal{H}^+, \mathcal{H}^-)$, $Z_{22} \in \mathcal{L}(\mathcal{H}^-)$, and Ran $(Z_{11}) \subset \mathcal{D}(X_{11})$, Ran $(Z_{12}) \subset \mathcal{D}(X_{11})$.

The fact that $\zeta \in r(X)$ is then equivalent to the following conditions

$$(\zeta I^+ - X_{11}) \circ Z_{11} - X_{12} \circ Z_{21} = I^+ \quad \text{on} \quad \mathcal{H}^+, \tag{3.4.12}$$

$$(\zeta I^+ - X_{11}) \circ Z_{12} - X_{12} \circ Z_{22} = 0 \quad \text{on} \quad \mathcal{H}^-, \tag{3.4.13}$$

$$\begin{aligned}
-X_{21} \circ Z_{11} + (\zeta I^- - X_{22}) \circ Z_{21} &= 0 \quad \text{on} \quad \mathcal{H}^+, \\
-X_{21} \circ Z_{12} + (\zeta I^- - X_{22}) \circ Z_{22} &= I^- \quad \text{on} \quad \mathcal{H}^-, \qquad (3.4.14) \\
Z_{11} \circ (\zeta I^+ - X_{11}) - Z_{12} \circ X_{21} &= I^+ \quad \text{on} \quad \mathcal{D}(X_{11}), \\
-Z_{11} \circ X_{12} + Z_{12} \circ (\zeta I^- - X_{22}) &= 0 \quad \text{on} \quad \mathcal{H}^-, \\
Z_{21} \circ (\zeta I^+ - X_{11}) - Z_{22} \circ X_{21} &= 0 \quad \text{on} \quad \mathcal{D}(X_{11}), \\
-Z_{21} \circ X_{12} + Z_{22} \circ (\zeta I^- - X_{22}) &= I^- \quad \text{on} \quad \mathcal{H}^-,
\end{aligned}$$

For $\zeta \in r(X_{11})$, let $\phi(\zeta) \in \mathcal{L}(\mathcal{H}^-)$ be defined by

$$\phi(\zeta) = \zeta I^- - X_{22} - X_{21} \circ (\zeta I^+ - X_{11})^{-1} \circ X_{12}, \tag{3.4.15}$$

and let G be the set of points $\zeta \in r(X_{11})$ for which $\phi(\zeta)$ is invertible and $\phi(\zeta)^{-1} \in \mathcal{L}(\mathcal{H}^-)$.

A direct computation shows that, if $G \neq \emptyset$, then for any $\zeta \in G$ the continuous linear operators

$$Z_{22} = \phi(\zeta)^{-1} \in \mathcal{L}(\mathcal{H}^-), \tag{3.4.16}$$

$$Z_{12} = (\zeta I^+ - X_{11})^{-1} \circ X_{12} \circ \phi(\zeta)^{-1} \in \mathcal{L}(\mathcal{H}^-, \mathcal{H}^+), \tag{3.4.17}$$
$$Z_{21} = \phi(\zeta)^{-1} \circ X_{21} \circ (\zeta I^+ - X_{11})^{-1} \in \mathcal{L}(\mathcal{H}^+, \mathcal{H}^-), \tag{3.4.18}$$
$$Z_{11} = (\zeta I^+ - X_{11})^{-1} + (\zeta I^+ - X_{11})^{-1} \circ X_{12} \circ Z_{21} \in \mathcal{L}(\mathcal{H}^+), \tag{3.4.19}$$

satisfy all eight conditions written above. This proves

Lemma 3.7 *The set G is contained in $r(X)$. If $G \neq \emptyset$, for any $\zeta \in G$, $(\zeta I - X)^{-1}$ is expressed by the matrix* (3.4.11) *whose entries are given by* (3.4.16), (3.4.17), (3.4.18), (3.4.19).

Note that, by (3.4.16),
$$r(X) \cap r(X_{11}) \subset G.$$

Let
$$C = r(X_{11}) \setminus G. \tag{3.4.20}$$

If $\zeta \in \sigma(X) \setminus \sigma(X_{11})$, then $\zeta \in r(X_{11}) \cap C \subset C$. Conversely, if $\zeta \in C$, then $\zeta \in r(X_{11})$ and $\zeta \notin G$, whence $\zeta \in \sigma(X) \setminus \sigma(X_{11})$.

In conclusion,
$$C = \sigma(X) \setminus \sigma(X_{11}). \tag{3.4.21}$$

Let $\zeta_0 \in G$. Since $\zeta \mapsto \phi(\zeta)$ is a continuous map $r(X_{11}) \to \mathcal{L}(\mathcal{H}^-)$, by the upper semicontinuity of the spectrum there is a neighborhood U of ζ_0 in $r(X_{11})$ such that $0 \notin \sigma(\phi(\zeta))$ for all $\zeta \in U$, i.e. $U \subset G$. Hence G is open, and $\zeta \mapsto \phi(\zeta)^{-1}$ is a holomorphic map of G into $\mathcal{L}(\mathcal{H}^-)$. Since G is open in $r(X_{11})$, then C is closed in $r(X_{11})$. Its boundary in $\mathbb{C}$, ∂C, is contained in $\sigma(X_{11})$:
$$\partial C \subset \sigma(X_{11}).$$

If iX_{11} is self-adjoint, then $r(X_{11}) \supset \{\zeta \in \mathbb{C} : Im\zeta \neq 0\}$. Furthermore, for any $\zeta \in r(X_{11})$

$$\begin{aligned} \phi(\zeta)^* &= \bar{\zeta} I^- + X_{22} - X_{21} \circ (\bar{\zeta} I^+ + X_{11})^{-1} \circ X_{12} \\ &= -(-\bar{\zeta} I^- - X_{22} - X_{21} \circ (-\bar{\zeta} I^+ - X_{11})^{-1} \circ X_{12}) \\ &= -\phi(-\bar{\zeta}). \end{aligned}$$

That proves

Lemma 3.8 *If iX_{11} is self-adjoint, the set C is symmetric with respect to the imaginary axis.*

3.5 Strongly continuous semigroups of linear isometries in a Pontryagin space. Spectral properties

If $\mathcal{H} = \mathcal{H}^+ \oplus \mathcal{H}^-$ is a Pontryagin space with $\dim_{\mathbb{C}} \mathcal{H}^- \leq \dim_{\mathbb{C}} \mathcal{H}^+$. The set C defined by (3.4.20) is expressed by
$$C = \{\zeta \in r(X_{11}) : \det \phi(\zeta) = 0\}.$$

Choosing an orthonormal basis $\{e_1, e_2, \ldots\}$ in $\mathcal{H}^-$, for $\zeta \in r(X_{11})$, $\phi(\zeta)$ is represented by a square matrix $(\phi(\zeta)_{\alpha\beta})$ whose elements are

$$\phi(\zeta)_{\alpha\beta} = (\phi(\zeta)e_\beta|e_\alpha) = \zeta\delta_{\alpha\beta} - (X_{22})_{\alpha\beta} - ((\zeta I^+ - X_{11})^{-1}X_\beta|X_\alpha),$$

where

$$X_\alpha := X_{12}e_\alpha \in \mathcal{L}(\mathcal{H}^+).$$

The point $\zeta \in r(X_{11})$ has an open neighborhood U in $r(X_{11})$ such that for all $\tau \in U$,

$$\begin{aligned}((\tau I^+ - X_{11})^{-1}X_\beta|X_\alpha) &= (((\tau - \zeta)I^+ + (\zeta I^+ - X_{11}))^{-1}X_\beta|X_\alpha) \\ &= ((I^+ + (\tau - \zeta)(\zeta I^+ - X_{11})^{-1})^{-1}(\zeta I^+ - X_{11})^{-1}X_\beta|X_\alpha) \\ &= ((\zeta I^+ - X_{11})^{-1}X_\beta|X_\alpha) \\ &\quad + (\zeta - \tau)\sum_{\nu=0}^{+\infty}(\zeta - \tau)^\nu((\zeta I^+ - X_{11})^{-(\nu+2)}X_\beta|X_\alpha),\end{aligned}$$

the convergence being uniform when τ varies on compact subsets of U. Hence, there are $a_\nu \in \mathbf{C}$, not all vanishing when $\nu = 1, 2, \ldots$, such that

$$\det\phi(\tau) = \det\phi(\zeta) + \sum_{\nu=1}^{+\infty}(\zeta - \tau)^\nu a_\nu,$$

the convergence being uniform on compact subsets of U. That proves

Lemma 3.9 *If $\mathcal{H} = \mathcal{H}^+ \oplus \mathcal{H}^-$ is a Pontryagin space, the set C is a discrete subset of $r(X_{11})$. Each point of C is an isolated polar singularity of the holomorphic map $\zeta \mapsto \phi(\zeta)^{-1}$ of G into $\mathcal{L}(\mathcal{H}^-)$.*

If the closed operator iX_{11} is (symmetric but) not self-adjoint, then $C \subset \Pi_r = \{\zeta \in \mathbf{C} : Re\,\zeta > 0\}$, because $\sigma(X_{11}) = \bar{\Pi}_\ell = \{\zeta \in \mathbf{C} : Re\,\zeta \le 0\}$.

Let $Re\,\zeta < 0$. Then ζ and $\bar{\zeta}$ belong to $r(-X_{11})$. Since $-X_{11} \subset X_{11}{}^*$ and since $X_{11}{}^*$ is a proper extension of the closed operator $-X_{11}$, then $\bar{\zeta} \in p\sigma(X_{11}{}^*)$, i.e. there is $x^+ \in \mathcal{D}(X_{11}{}^*) \setminus \{0\}$ such that

$$(\bar{\zeta}I^+ - X_{11}{}^*)x^+ = 0.$$

If $\zeta \in r(X)$, by (3.4.13)

$$Z_{22}{}^* \circ X_{21}x^+ = 0, \tag{3.5.1}$$

while (3.4.12) yields

$$Z_{21}{}^* \circ X_{21}x^+ = -x^+,$$

implying that $X_{21}x^+ \neq 0$. Hence, by (3.5.1), $Z_{22}{}^*$ is not injective. Since $\dim_{\mathbf{C}}\mathcal{H}^- < \infty$, that is equivalent to $\det Z_{22}{}^* = 0$, and therefore also to $\det Z_{22} = 0$. Thus, there is a vector $y^- \in \mathcal{H}^- \setminus \{0\}$ such that $Z_{22}y^- = 0$. Hence (3.4.13) implies that

$$(\zeta I^+ - X_{11}) \circ Z_{12}y^- = 0.$$

Because $Im\,\zeta < 0$, ζ is an interior point of the spectrum of the skew-symmetric operator X_{11}, and therefore belongs to the residual spectrum. Hence $Z_{12}y^- = 0$. But then (3.4.14) implies

$$y^- = -X_{21} \circ Z_{12}y^- + (\zeta I^- - Z_{22}) \circ Z_{22}y^- = 0.$$

This contradiction shows that, if iX_{11} is symmetric but not self-adjoint, then the closure of the left half-plane Π_ℓ is contained in $\sigma(X)$, so that, by (3.4.21),

$$\sigma(X) = \bar{\Pi}_\ell \cup C.^3 \tag{3.5.2}$$

Let $\tilde{C}$ be the image of C by the reflection $\zeta \mapsto -\bar{\zeta}$ around the imaginary axis, and let E be (either $\emptyset$ or) $\Pi_\ell \setminus \tilde{C}$.

Lemma 3.10 *If iX_{11} is symmetric but not self-adjoint, then $E \subset r\sigma(X)$.*

Proof By (3.5.2), $E \subset \sigma(X)$. If $\zeta \in \Pi_\ell$ is an eigenvalue of $X : Xx = \zeta x$ for some $x \in \mathcal{D}(X) \setminus \{0\}$, then $Jx \in \mathcal{D}(X^*)$ and

$$X^*Jx = -JXx = -\zeta Jx.$$

Since X is closed, $r(X^*)$ is the image of $r(X)$ by the reflection around the real axis. Thus, being $-\zeta \in p\sigma(X^*)$, then $-\bar{\zeta} \in \sigma(X) \cap \Pi_r = C$, and therefore $\zeta \notin E$, i.e.

$$E \cap p\sigma(X) = \emptyset. \tag{3.5.3}$$

If $\zeta \in E$, then $-\bar{\zeta} \in r(X) = r(JXJ)$. Since X^* is a proper extension of the closed operator $-JXJ$, then $\bar{\zeta} \in p\sigma(X^*)$. Thus $\zeta \in p\sigma(X) \cup r\sigma(X)$, and (3.5.3) yields the conclusion. □

Theorem 3.4 *If $\dim_{\mathbb{C}}\mathcal{H}^- < \infty$, $C \cap \Pi_r$ is a finite set of eigenvalues of X. The sum of the dimensions of the corresponding eigenspaces does not exceed $\dim_{\mathbb{C}}\mathcal{H}^-$.*

If iX_{11} is self-adjoint (i.e. if X generates a strongly continuous group $\mathbb{R} \to \Gamma(J)$), then $\sigma(X)$ is the union of C and of a closed subset of $i\mathbb{R}$.

If iX_{11} is symmetric but not self-adjoint, then $C \subset \Pi_r$ and (3.5.2) holds.

The second part of the theorem summarizes some of the facts that have been established before. As a consequence of the first part, the cardinality of $C \cap \Pi_r$ does not exceed $\dim_{\mathbb{C}}\mathcal{H}^-$. This fact was established in [VES 87b] for $\mathcal{H}^- \simeq \mathbb{C}$ and in [VES 90a] in general. The statement concerning the eigenspaces follows also from a general result on J-dissipative operators which was proved in [AZ-IO 89] (cf. also [I-K-L 82]).

[3] The above considerations simplify a proof given in [VES 90a; pp. 24–25]

3.6 Holomorphic families and holomorphic semigroups of linear isometries for an indefinite metric

The questions discussed in sections 2.3 and 2.4 raise similar problems concerning holomorphic functions and holomorphic semigroups with values in $\Lambda(L)$ – or, equivalently by Lemma 3.5 – with values in $\Lambda(J)$.

Let U be a domain in $\mathbf{C}$ containing $\mathbf{R}_+^*$ and let $f \in \mathrm{Hol}\,(U, \mathcal{L}(\mathcal{H}^+ \oplus \mathcal{H}^-))$ be such that

$$f(z)^* J f(z) = J \qquad \text{for all } z \in U. \tag{3.6.1}$$

If

$$f(z) = f_0 + (z-w) f_1 + (z-w)^2 f_2 + \cdots$$

is the power series expansion of f in a neighbourhood V of a point $w \in U$, with $f_\nu \in \mathcal{L}(\mathcal{H}^+ \oplus \mathcal{H}^-)$ for $\nu = 0, 1, 2, \ldots$, then $f_0{}^* J f_0 = J$ and $f_0{}^* J f_\nu = 0$ for all $\nu = 1, 2, \ldots$, whence $f_0{}^* J f(z) = J$, i.e.

$$f(w)^* J f(z) = J \tag{3.6.2}$$

for all $z \in V$, and in conclusion for all z, w in U. If

$$f(t+s) = f(t) f(s)$$

for all $t, s \in \mathbf{R}_+^*$, then (3.6.2) yields

$$J = f(w)^* J f(t+s) = f(w)^* J f(t) f(s) = J f(s)$$

and thus $f(s) = I$ for all $s > 0$. Hence $f(z) = I$ for all $z \in U$, because f is holomorphic on U. That proves the following theorem.

Theorem 3.5 *Let T be a non-trivial semigroup $\mathbf{R}_+ \to \mathcal{L}(\mathcal{H}^+ \oplus \mathcal{H}^-)$. If*

$$T(t)^* J T(t) = J$$

for all $t \geq 0$, the function T cannot be extended to a holomorphic function $\hat{T}$ on a connected open neighborhood U of $\mathbf{R}_+^*$ in $\mathbf{C}$, with values in $\mathcal{L}(\mathcal{H}^+ \oplus \mathcal{H}^-)$ and such that

$$\hat{T}(z)^* J \hat{T}(z) = J \qquad \text{for all } z \in U.$$

Does condition (3.6.1) alone suffice to exclude the existence of non-costant holomorphic functions on U? The answer to this question is negative, as the following example will show [VES 90a].

In the Hilbert space $\ell^2(\mathbf{Z})$ with the natural basis $\{e_\nu : \nu = 0, \pm 1, \pm 2, \ldots\}$ let A_{11}, A_{22}, B_{11}, B_{22} be the linear operators defined by

$$A_{11}(e_\nu) = e_{4\nu}, \quad A_{22}(e_\nu) = e_{4\nu+1}, \quad B_{11}(e_\nu) = e_{4\nu+3}, \quad B_{22}(e_\nu) = e_{8\nu+2},$$

and let $A, B \in \mathcal{L}(\ell^2(\mathbf{Z}) \oplus \ell^2(\mathbf{Z}))$ be defined by

$$A = \begin{pmatrix} A_{11} & 0 \\ 0 & A_{22} \end{pmatrix}, \quad B = \begin{pmatrix} B_{11} & B_{22} \\ B_{11} & B_{22} \end{pmatrix}.$$

The equations

$$A_{22}{}^* A_{11} = 0, \quad B_{11}{}^* A_{11} = 0, \quad B_{22}{}^* A_{11} = 0, \quad B_{11}{}^* A_{22} = 0, \quad B_{22}{}^* A_{22} = 0$$

and the fact that $A_{11}, A_{22}, B_{11}, B_{22}$ are linear isometries imply that

$$B^* J B = 0, \quad A^* J A = J, \quad B^* J A = A^* J B = 0,$$

and therefore the function $f \in \mathrm{Hol}\,(\mathbf{C}, \mathcal{L}(\ell^2(\mathbf{Z}) \oplus \ell^2(\mathbf{Z})))$, $f(z) = A + zB$, satisfies (3.6.1) for $U = \mathbf{C}$.

The answer turns out to be negative when $\mathcal{H}$ is a Pontryagin space. Let U be a domain in $\mathbf{C}$ and suppose that $f \in \mathrm{Hol}\,(U, \mathcal{L}(\mathcal{H}))$ satisfies (3.6.1), where $\mathcal{H} = \mathcal{H}^+ \oplus \mathcal{H}^-$ is a Pontryagin space with $\dim_{\mathbf{C}} \mathcal{H}^- < \infty$.

For $z_0 \in U$ there is an open neighborhood V of z_0 in U such that

$$f(z) = f(z_0) + (z - z_0) g(z)$$

for all $z \in V$, where $g \in \mathrm{Hol}\,(V, \mathcal{L}(\mathcal{H}))$. Denoting by

$$f(z_0) = \begin{pmatrix} f(z_0)_{11} & f(z_0)_{12} \\ f(z_0)_{21} & f(z_0)_{22} \end{pmatrix}, \quad g(z) = \begin{pmatrix} g(z)_{11} & g(z)_{12} \\ g(z)_{21} & g(z)_{22} \end{pmatrix},$$

the matrix representations of $f(z_0)$ and $g(z)$, where $f(z_0)_{11}, g(z)_{11} \in \mathcal{L}(\mathcal{H}^+)$, $f(z_0)_{12}, g(z)_{12} \in \mathcal{L}(\mathcal{H}^-, \mathcal{H}^+)$, $f(z_0)_{21}, g(z)_{21} \in \mathcal{L}(\mathcal{H}^+, \mathcal{H}^-)$, $f(z_0)_{22}, g(z)_{22} \in \mathcal{L}(\mathcal{H}^-)$, then the identities

$$f(z_0)^* J g(z) = 0, \quad g(z)^* J g(z) = 0,$$

which follow from (3.6.1), are equivalent, respectively, to

$$f(z_0)_{11}{}^* g(z)_{11} - f(z_0)_{21}{}^* g(z)_{21} = 0,$$

$$f(z_0)_{22}{}^* g(z)_{22} - f(z_0)_{12}{}^* g(z)_{12} = 0, \tag{3.6.3}$$

$$f(z_0)_{12}{}^* g(z)_{11} - f(z_0)_{22}{}^* g(z)_{21} = 0, \tag{3.6.4}$$

and to

$$g(z)_{11}{}^* g(z)_{11} - g(z)_{21}{}^* g(z)_{21} = 0, \tag{3.6.5}$$

$$g(z)_{22}{}^* g(z)_{22} - g(z)_{12}{}^* g(z)_{12} = 0, \tag{3.6.6}$$

$$g(z)_{12}{}^* g(z)_{11} - g(z)_{22}{}^* g(z)_{21} = 0.$$

Since $\dim_{\mathbf{C}} \mathcal{H}^- < \infty$, Lemma 4.1, which will be established in chapter 4, implies that $f(z_0)_{22}$ is invertible in $\mathcal{L}(\mathcal{H}^-)$. Hence (3.6.4) gives

$$g(z)_{21} = f(z_0)_{22}{}^{*-1} f(z_0)_{12}{}^* g(z)_{11}$$

and (3.6.5) becomes then

$$g(z)_{11}{}^*(I^+ - f(z_0)_{12}f(z_0)_{22}{}^{-1}f(z_0)_{22}{}^{*-1}f(z_0)_{12}{}^*)g(z)_{11} = 0. \qquad (3.6.7)$$

For any $A \in \mathcal{L}(\mathcal{H})$, let $\rho(A)$ denote the spectral radius of A. Then

$$\begin{aligned} \|f(z_0)_{12}f(z_0)_{22}{}^{-1}f(z_0)_{22}{}^{*-1}f(z_0)_{12}{}^*\| &= \rho(f(z_0)_{12}f(z_0)_{22}{}^{-1}f(z_0)_{22}{}^{*-1}f(z_0)_{12}{}^*) \\ &= \rho(f(z_0)_{22}{}^{-1}f(z_0)_{22}{}^{*-1}f(z_0)_{12}{}^*f(z_0)_{12}). \end{aligned}$$

Since $f(z_0) \in \Lambda(J)$, then, by (3.3.3′),

$$f(z_0)_{12}{}^*f(z_0)_{12} = f(z_0)_{22}{}^*f(z_0)_{22} - I^-,$$

and therefore,

$$\begin{aligned} \|(f(z_0)_{12}f(z_0)_{22}{}^{-1}f(z_0)_{22}{}^{*-1}f(z_0)_{12}{}^*\| &= \rho(I^- - f(z_0)_{22}{}^{-1}f(z_0)_{22}{}^{*-1}) \\ &= \rho(I^- - (f(z_0)_{22}{}^*f(z_0)_{22})^{-1}) \\ &= \rho(I^- - (I^- + f(z_0)_{12}{}^*f(z_0)_{12})^{-1}) \\ &= \max\{1 - t : t \in \sigma((I^- + f(z_0)_{12}{}^*f(z_0)_{12})^{-1})\} \\ &= \max\{1 - \frac{1}{t} : t \in \sigma(I^- + f(z_0)_{12}{}^*f(z_0)_{12})\} \\ &= 1 - (\max\{t : t \in \sigma(I^- + f(z_0)_{12}{}^*f(z_0)_{12})\})^{-1} \\ &= 1 - (1 + \rho(f(z_0)_{12}{}^*f(z_0)_{12}))^{-1} \\ &= 1 - (1 + \|f(z_0)_{12}{}^*f(z_0)_{12}\|)^{-1} \\ &= 1 - (1 + \|f(z_0)_{12}\|^2)^{-1} < 1. \end{aligned}$$

Thus

$$((I^+ - f(z_0)_{12}f(z_0)_{22}{}^{-1}f(z_0)_{22}{}^{*-1}f(z_0)_{12}{}^*)x^+|x^+) > 0 \qquad (3.6.8)$$

for all $x \in \mathcal{H}^+ \setminus \{0\}$, and (3.6.7) implies that $g(z)_{11} = 0$ and therefore, by (3.6.5), also $g(z)_{21} = 0$ for all $z \in U$. Since, by (3.6.3)

$$g(z)_{22} = f(z_0)_{22}{}^{*-1}f(z_0)_{12}{}^*g(z)_{12},$$

then (3.6.6) becomes

$$g(z)_{12}{}^*(I^+ - f(z_0)_{12}f(z_0)_{22}{}^{-1}f(z_0)_{22}{}^{*-1}f(z_0)_{12}{}^*)g(z)_{12} = 0.$$

Thus, by (3.6.8), $g(z)_{12} = 0$, and therefore, by (3.6.3) and the invertibility of $f(z_0)_{22}$, also $g(z)_{22} = 0$ for all $z \in U$.

In conclusion $g = 0$ and the following theorem holds [VES 90a].

Theorem 3.6 *If $\mathcal{H}$ is a Pontryagin space there are no non-costant holomorphic maps f of U into $\mathcal{L}(\mathcal{H})$ satisfying* (3.6.1).

Chapter 4
Cartan domains of type one

Going back to the beginning of chapter 3, consider the Cartan factor $\mathcal{L}(\mathcal{H}^-, \mathcal{H}^+)$, and let B be the open unit ball of $\mathcal{L}(\mathcal{H}^-, \mathcal{H}^+)$.

4.1 The unit ball of $\mathcal{L}(\mathcal{H}^-, \mathcal{H}^+)$

Let B be the open unit ball of $\mathcal{L}(\mathcal{H}^-, \mathcal{H}^+)$, and let $A \in \Lambda(J)$ be given by (3.3.0). For any $Z \in B$ and any $x^- \in \mathcal{H}^-$, (3.3.1′), (3.3.2′), (3.3.3′) yield

$$\begin{aligned}
\|(A_{11}Z + A_{12})x^-\|^2 \; - \; \|A_{21}Z + A_{22})x^-\|^2 \; &= \qquad (4.1.1)\\
&= ((Z^*A_{11}{}^* + A_{12}{}^*)(A_{11}Z + A_{12})x^-|x^-)\\
&\quad - ((Z^*A_{21}{}^* + A_{22}{}^*)(A_{21}Z + A_{22})x^-|x^-)\\
&= ((Z^*(A_{11}{}^*A_{11} - A_{21}{}^*A_{21})Z + (A_{12}{}^*A_{12} - A_{22}{}^*A_{22})\\
&\quad + Z^*(A_{11}{}^*A_{12} - A_{21}{}^*A_{22}) + (A_{12}{}^*A_{11} - A_{22}{}^*A_{21})Z)x^-|x^-)\\
&= ((Z^*Z - I^-)x^-|x^-) = \|Zx^-\|^2 - \|x^-\|^2\\
&\leq (\|Z\|^2 - 1)\|x^-\|^2.
\end{aligned}$$

As a consequence, for every $Z \in B$ the linear map

$$A_{21}Z + A_{22} \in \mathcal{L}(\mathcal{H}^-)$$

is injective. Thus, if $\dim_{\mathbb{C}}\mathcal{H}^- < \infty$, the maps is bijective.

If $A \in \Gamma(J)$, (3.3.3′) and (3.3.6′) show that A_{22} has a continuous right inverse and a continuous left inverse, and therefore is invertible, with $A_{22}{}^{-1} \in \mathcal{L}(\mathcal{H}^-)$. Since, by (3.3.6′)

$$\begin{aligned}
\|A_{22}{}^{-1}A_{21}\|^2 &= \|A_{22}{}^{-1}A_{21}A_{21}{}^*A_{22}{}^{*-1}\|\\
&= \|A_{22}{}^{-1}(A_{22}A_{22}{}^* - I^-)A_{22}{}^{*-1}\|\\
&= \|I^- - A_{22}{}^{-1}A_{22}{}^{*-1}\|\\
&= \|I^- - (A_{22}{}^*A_{22})^{-1}\| = \rho(I^- - (A_{22}{}^*A_{22})^{-1})\\
&= \sup\{t : t \in \sigma(I^- - (A_{22}{}^*A_{22})^{-1})\}\\
&= \sup\{1 - t : t \in \sigma((A_{22}{}^*A_{22})^{-1})\}\\
&= \sup\{1 - \frac{1}{t} : t \in \sigma(A_{22}{}^*A_{22})\}\\
&= 1 - (\max\{t : t \in \sigma(A_{22}{}^*A_{22})\})^{-1}\\
&= 1 - (\rho(A_{22}{}^*A_{22}))^{-1}\\
&= 1 - \|A_{22}{}^*A_{22}\|^{-1}\\
&= 1 - \|A_{22}\|^{-2} < 1,
\end{aligned}$$

then, if $Z \in B$,

$$\|A_{22}{}^{-1}A_{21}Z\| \leq \|Z\| < 1.$$

Therefore $A_{21}Z + A_{22}$ is invertible, and $(A_{21}Z + A_{22})^{-1} \in \mathcal{L}(\mathcal{H}^-)$ for all $Z \in B$.

This argument completes the proof of the following lemma ([FRA 81], [VES 90a]).

Lemma 4.1 *If $A \in \Gamma(J)$ or if $A \in \Lambda(J)$ and, in this latter hypothesis, $\dim_{\mathbb{C}}\mathcal{H}^- < \infty$, then $A_{21}Z + A_{22}$ is invertible in $\mathcal{L}(\mathcal{H}^-)$ for all $Z \in B$.*

Corollary 4.1 *Let $\dim_{\mathbb{C}}\mathcal{H}^- < \infty$. If $Z_1 \in \mathcal{L}(\mathcal{H}^-, \mathcal{H}^+)$ and $Z_2 \in \mathcal{L}(\mathcal{H}^-)$ are such that Z_2 is invertible, and $\|Z_1Z_2{}^{-1}\| < 1$, then $A_{21}Z_1 + A_{22}Z_2$ is invertible.*

This corollary follows from the fact that $A_{21}Z_1 + A_{22}Z_2$ is invertible if, and only if, $\det(A_{21}Z_1 + A_{22}Z_2) \neq 0$, and that $\det(A_{21}Z_1 + A_{22}Z_2) = \det(A_{21}Z_1Z_2{}^{-1} + A_{22}) \det Z_2$.

Proposition 4.1 *If the hypotheses of Lemma 4.1 are fulfilled, then, for any $Z \in B$, $(A_{11}Z + A_{12})(A_{21}Z + A_{22})^{-1} \in B$.*

Proof For all $x^- \in \mathcal{H}^-$, setting $y^- = (A_{21}Z + A_{22})^{-1}x^-$, and proceeding as in (4.1.1) one has

$$\|(A_{11}Z + A_{12})(A_{21}Z + A_{22})^{-1}x^-\|^2 - \|x^-\|^2 =$$
$$\|(A_{11}Z + A_{12})y^-\|^2 - \|(A_{21}Z + A_{22})y^-\|^2 \leq (\|Z\|^2 - 1)\|y^-\|^2. \tag{4.1.2}$$

Since

$$\|x^-\| = \|(A_{21}Z + A_{22})y^-\| \leq \|A_{21}Z + A_{22}\|\|y^-\|,$$

and $\|Z\| < 1$, the latter term in (4.1.1) is less than or equal to

$$(\|Z\|^2 - 1)\|A_{21}Z + A_{22}\|^{-2}\|x^-\|^2,$$

and in conclusion

$$\|(A_{11}Z + A_{12})(A_{21}Z + A_{22})^{-1}x^-\|^2 \leq ((1 - (1 - \|Z\|^2)(\|A_{21}Z + A_{22}\|^{-2}))\|x^-\|^2.$$

for all $x^- \in \mathcal{H}^-$. □

As a consequence of Proposition 4.1, setting, for $Z \in B$,

$$\widehat{A}(Z) = (A_{11}Z + A_{12})(A_{21}Z + A_{22})^{-1},$$

one has $\widehat{A} \in \mathrm{Hol}\,(B, B)$.

A direct computation shows that if A' and A'' are in $\Lambda(J)$ and $\dim_{\mathbb{C}}\mathcal{H}^- < \infty$, then $\widehat{A'A''} = \widehat{A'} \circ \widehat{A''}$. Since the image of I is the identity map on B, then $A \mapsto \widehat{A}$ is a homomorphism of $\Gamma(J)$ into $\mathrm{Aut}\, B$, which – when $\dim_{\mathbb{C}}\mathcal{H}^- < \infty$ – extends to a homomorphism of $\Lambda(J)$ into $\mathrm{Iso}\, B$.

The domain B, as the open unit ball of a J^*-algebra, is homogeneous. This fact will now be established directly, showing that $\Gamma(J)$, via the homomorphism $A \mapsto \widehat{A}$, acts transitively on B.

For $Z_0 \in B$, let

$$A_{11} = (I^+ - Z_0 Z_0{}^*)^{-\frac{1}{2}}, \qquad A_{12} = -(I^+ - Z_0 Z_0{}^*)^{-\frac{1}{2}} Z_0,$$
$$A_{21} = -(I^- - Z_0{}^* Z_0)^{-\frac{1}{2}} Z_0{}^*, \quad A_{22} = (I^- - Z_0{}^* Z_0)^{-\frac{1}{2}}.$$

Since

$$(I^+ - Z_0 Z_0{}^*)^{-\frac{1}{2}} Z_0 = Z_0 (I^- - Z_0{}^* Z_0)^{-\frac{1}{2}},$$
$$(I^- - Z_0{}^* Z_0)^{-\frac{1}{2}} Z_0{}^* = Z_0{}^* (I^+ - Z_0 Z_0{}^*)^{-\frac{1}{2}},$$

as the power series expansion of $(I^+ - Z_0 Z_0{}^*)^{-frac12}$ and $(I^- - Z_0{}^* Z_0)^{-\frac{1}{2}}$ show, a direct computation proves that

$$A = \begin{pmatrix} A_{11} & A_{12} \\ A_{21} & A_{22} \end{pmatrix}$$

is an element of $\Gamma(J)$. Since

$$A_{11} Z_0 + A_{12} = (I^+ - Z_0 Z_0{}^*)^{-\frac{1}{2}} Z_0 - (I^+ - Z_0 Z_0{}^*)^{-\frac{1}{2}} Z_0 = 0,$$

then $\widehat{A}(Z_0) = 0$. This proves that $\widehat{\Gamma}(J)$ acts transitively on B.

Suppose that $\dim_{\mathbb{C}} \mathcal{H}^- < \infty$ and let $A \in \Lambda(J)$. Since $\widehat{A}(0) = A_{12} A_{22}{}^{-1}$, then $\widehat{A}(0) = 0$, if, and only if, $A_{12} = 0$. By (3.3.2′), $A_{21}{}^* A_{22} = 0$, and therefore $A_{21} = 0$, so that (3.3.1′) and (3.3.3′) read $A_{11}{}^* A_{11} = I^+$, and $A_{22}{}^* A_{22} = I^-$. But then, for all $W \in \mathcal{L}(\mathcal{H}^-, \mathcal{H}^+)$,

$$\begin{aligned} \|d\widehat{A}(0)W\|^2 &= \|A_{11} W A_{22}{}^{-1}\|^2 \\ &= \|A_{22}{}^{*-1} W^* A_{11}{}^* A_{11} W A_{22}{}^{-1}\| \\ &= \|A_{22}{}^{*-1} W^* W A_{22}{}^{-1}\| \\ &= \rho(A_{22}{}^{*-1} W^* W A_{22}{}^{-1}) \\ &= \rho(A_{22}{}^{-1} A_{22}{}^{*-1} W^* W) \\ &= \rho((A_{22}{}^* A_{22})^{-1} W^* W) = \rho(W^* W) \\ &= \|W^* W\| = \|W\|^2. \end{aligned} \tag{4.1.3}$$

In other terms, under the hypotheses of Lemma 4.1, if $\widehat{A}(0) = 0$, then $d\widehat{A}(0)$ is a linear isometry of $\mathcal{L}(\mathcal{H}^-, \mathcal{H}^+)$. By (3.3.4′), which now reads $A_{11} A_{11}{}^* = I^+$, this isometry is surjective if, and only if, $A \in \Gamma(J)$.

By (4.1.3) and by the homogeneity of B (or also by Theorem 1.9) the Kobayashi and Carathéodory differential metrics of B coincide. Denoting by $|W|_Z$ the lenght of the vector $W \in \mathcal{L}(\mathcal{H}^-, \mathcal{H}^+)$, with respect to any one of those metrics, at the point $Z \in B$, then (1.4.2) reads

$$|W|_0 = \|W\|.$$

Hence for any $A \in \Lambda(J)$, $d\widehat{A}(0)$ is an isometry for $|\ |_0$. Since the Kobayashi-Carathéodory metric is invariant under $\operatorname{Aut} B$, and since $\Gamma(J)$ acts transitively on B via the map $A \mapsto \widehat{A}$, then

$$|d\widehat{A}(Z)W|_{\widehat{A}(Z)} = |W|_Z$$

for all $A \in \Lambda(J)$, $Z \in B$, $W \in \mathcal{L}(\mathcal{H}^-, \mathcal{H}^+)$. Hence, denoting again by $\operatorname{Iso} B$ the semigroup of all holomorphic isometries of B into itself, the following theorem summarizes some of the results that have been obtained so far.

Theorem 4.1 *The map $A \mapsto \widehat{A}$ defines a homomorphisms $\Lambda(J) \to \operatorname{Iso} B$. The image of $\Gamma(J)$ is contained in $\operatorname{Aut} B$ and acts transitively on B.*

The description of $\operatorname{Aut} B$ and $\operatorname{Iso} B$ will be complete once the stability group $(\operatorname{Aut} B)_0 = \{f \in \operatorname{Aut} B : f(0) = 0\}$ and the stability semigroup $(\operatorname{Iso} B)_0 = \{f \in \operatorname{Iso} B : f(0) = 0\}$ of 0 in $\operatorname{Aut} B$ and in $\operatorname{Iso} B$ are known.

In fact, since $\Gamma(J)$ acts transitively on B, for any g in $\operatorname{Aut} B$ or in $\operatorname{Iso} B$ there is some $A \in \Gamma(J)$ such that $g(0) = \widehat{A}(0)$. Hence $f := g \circ \widehat{A}^{-1}$ is contained in $(\operatorname{Aut} B)_0$ or in $(\operatorname{Iso} B)_0$ respectively.

The group $(\operatorname{Aut} B)_0$ – which, by H. Cartan's linearity theorem is contained in $\mathcal{L}(\mathcal{H}^-, \mathcal{H}^+)$ – was characterized by T. Franzoni [FRA 81] who proved the following theorem.

Theorem 4.2 *Let $f \in (\operatorname{Aut} B)_0$. Then*

(a) *If $\dim_{\mathbb{C}} \mathcal{H}^- \neq \dim_{\mathbb{C}} \mathcal{H}^+$, there exist unitary operators U on $\mathcal{H}^+$ and V on $\mathcal{H}^-$ such that*

$$f(Z) = UZV \tag{4.1.4}$$

for all $Z \in B$.

(b) *If $\mathcal{H}^- = \mathcal{H}^+$, for any conjugation $x^- \mapsto \overline{x^-}$ of $\mathcal{H}^-$, there are unitary operators U and V on $\mathcal{H}^-$ such that either* (4.1.4) *or*

$$f(Z) = U\,{}^tZV$$

hold for all $Z \in B$, where ${}^tZ \in \mathcal{L}(\mathcal{H}^-)$ is defined by ${}^tZx^- = \overline{Z^*\overline{x^-}}$ $(x^- \in \mathcal{H}^-)$.

The example exhibited in section 2.3 of a non-linear element $f \in (\operatorname{Iso} B)_0$ indicates that the structure of $\operatorname{Iso} B$ is more complicated. So far, the only case in which a characterization of $\operatorname{Iso} B$ is available is that in which at least one of two spaces $\mathcal{H}^-$ and $\mathcal{H}^+$ has finite dimension, as the following theorem [VES 91] shows:

Theorem 4.3 *If $\dim_{\mathbb{C}} \mathcal{H}^- < \infty$, for any $f \in (\operatorname{Iso} B)_0$ there exist a unitary operator V on $\mathcal{H}^-$ and a linear isometry U in $\mathcal{H}^+$ such that* (4.1.4) *holds for all $Z \in B$.*

4.2 Semigroups on Cartan domains of type one

With the same notations as in section 3.4, let $T : \mathbb{R}_+ \to \Lambda(J)$ be a strongly continuous semigroup. There are constants $\alpha \in \mathbb{R}$ and $\beta > 0$ such that

$$\|T(t)\| \leq \beta e^{\alpha t} \qquad (t \in \mathbb{R}_+),$$

i.e. the semigroup $\tilde{T} : t \mapsto e^{-\alpha t}T(t)$ is bounded.

Consider the space $\mathcal{L}(\mathcal{L}(\mathcal{H}^-, \mathcal{H}))$ of all bounded linear operators on the complex Banach space $\mathcal{L}(\mathcal{H}^-, \mathcal{H})$. The strong topology on $\mathcal{L}(\mathcal{H}^-, \mathcal{H})$ is the locally convex topology defined by the family of seminorms

$$W \mapsto \|W(x^-)\| \qquad (W \in \mathcal{L}(\mathcal{H}^-, \mathcal{H})) \tag{4.2.1}$$

when x^- varies in $\mathcal{H}^-$.

Let

$$\check{T} : \mathbf{R}_+ \to \mathcal{L}(\mathcal{L}(\mathcal{H}^-, \mathcal{H}))$$

be the semigroup defined by

$$\check{T}(t)W = T(t) \circ W \qquad (t \geq 0). \tag{4.2.2}$$

Since

$$\|(\check{T}(t)W)x^-\| = \|T(t) \circ Wx^-\| \leq \beta e^{\alpha t}\|Wx^-\|,$$

the semigroup $t \mapsto \check{T}(t)$ acting on $\mathcal{L}(\mathcal{H}^-, \mathcal{H})$ is continuous on the locally convex space $\mathcal{L}(\mathcal{H}^-, \mathcal{H})$.[4] Its infinitesimal generator is the linear operator $\check{X}$ on $\mathcal{L}(\mathcal{H}^-, \mathcal{H})$ defined by

$$\check{X}(W) = \lim_{h \downarrow 0} \frac{1}{h}(\check{T}(h) - I)W \tag{4.2.3}$$

for the locally convex topology on $\mathcal{L}(\mathcal{H}^-, \mathcal{H})$, i.e.

$$\check{X}(W)(x^-) = \lim_{h \downarrow 0} \frac{1}{h}(\check{T}(h) - I)W(x^-) \tag{4.2.4}$$

for all $x^- \in \mathcal{H}^-$. The domain $\mathcal{D}(\check{X})$ is the linear subspace consisting of those operators $W \in \mathcal{L}(\mathcal{H}^-, \mathcal{H})$ for which the limit (4.2.4) exists. The space $\mathcal{D}(\check{X})$ is dense in $\mathcal{L}(\mathcal{H}^-, \mathcal{H})$ [YOS 80; pp.237–238]. By (4.2.4),

$$\mathcal{D}(\check{X}) = \{W \in \mathcal{L}(\mathcal{H}^-, \mathcal{H}) : Wx^- \in \mathcal{D}(X) \text{ for all } x^- \in \mathcal{H}^-\}$$

and

$$\check{X}(W) = X \circ W$$

for all $W \in \mathcal{D}(\check{X})$. If $\dim_{\mathbf{C}}\mathcal{H}^- < \infty$, the locally convex topology defined on $\mathcal{L}(\mathcal{H}^-, \mathcal{H})$ by the family of semi-norms (4.2.1) is equivalent to the uniform Banach space topology on $\mathcal{L}(\mathcal{H}^-, \mathcal{H})$, and the semigroup $\check{T}$ is then a strongly continuous semigroup of bounded linear operators acting on the complex Banach space $\mathcal{L}(\mathcal{H}^-, \mathcal{H})$.

Does this conclusion hold also when $\dim_{\mathbf{C}}\mathcal{H}^- = \infty$?

More generally, let $T : \mathbf{R}_+ \to \mathcal{L}(\mathcal{H})$ be a strongly continuous semigroup acting on a complex Banach space $\mathcal{H}$. If $\mathcal{H}^-$ is a complex Banach space, let $\check{T} : \mathbf{R}+ \to \mathcal{L}(\mathcal{L}(\mathcal{H}^-, \mathcal{H}))$

[4] The semigroup $t \mapsto e^{-\alpha t}\check{T}(t)$ is equicontinuous with respect to t. This class of semigroups was first considered by L. Schwartz in [SCH 58], and was investigated also by K. Yosida [YOS 80].

be the semigroup defined by (4.2.2). Under which conditions is $\check{T}$ a strongly continuous semigroup acting on the complex Banach space $\mathcal{L}(\mathcal{H}^-, \mathcal{H})$?

A partial answer to this question, in the particular case $\mathcal{H}^- = \mathcal{H}$, is provided by the following results [VES 90b].

Let $\mathcal{A}$ be a complex unital Banach algebra. Every $a \in \mathcal{A}$ defines a bounded linear map $\lambda_a : x \mapsto ax$. Let $\mathcal{M}(\mathcal{A}) \subset \mathcal{L}(\mathcal{A})$ be the image of $\mathcal{A}$ by the map $a \mapsto \lambda_a$.

Theorem 4.4 *Let $T : \mathbf{R}_+ \to \mathcal{L}(\mathcal{A})$ be a strongly continuous semigroup. If there is $t_1 > 0$ such that, for every $t \in (0, t_1), T(t) \in \mathcal{M}(\mathcal{A})$, then the semigroup T is uniformly continuous, and there exists $a \in \mathcal{A}$ such that*

$$T(t)x = (\exp ta)x \tag{4.2.5}$$

for every $t \geq 0$ and every $x \in \mathcal{A}$.

Proof Let 1 be the identity element of $\mathcal{A}$, and let $\mathcal{A}^{-1}$ be the set of all invertible elements of $\mathcal{A}$. Since $\mathcal{A}^{-1}$ is open and non-empty, and $\mathcal{D}(X)$ is dense, then $\mathcal{D}(X) \cap \mathcal{A}^{-1} \neq \emptyset$. For any $y \in \mathcal{A}^{-1} \cap \mathcal{D}(X)$ the limit

$$\lim_{t \downarrow 0} \frac{1}{t}(T(t) - I)y$$

exists. Since, for $t \in (0, t_1)$,

$$\frac{1}{t}(T(t) - I)1 = \frac{1}{t}(T(t) - I)(yy^{-1}) = (\frac{1}{t}(T(t) - I)y)y^{-1},$$

and since the product in $\mathcal{A}$ is continuous, also the limit

$$a := \lim_{t \downarrow 0} \frac{1}{t}(T(t) - I)1$$

exists. Hence, for any $x \in \mathcal{A}$ the limit

$$\lim_{t \downarrow 0} \frac{1}{t}(T(t) - I)x = \lim_{t \downarrow 0} \frac{1}{t}(T(t) - I)(1x) = (\lim_{t \downarrow 0} \frac{1}{t}(T(t) - I)1)x$$

exists. Therefore $\mathcal{A} = \mathcal{D}(X)$, and – the generator X of T being a closed operator – then $X \in \mathcal{L}(\mathcal{A})$. Thus T is uniformly continuous, and (4.2.5) holds. □

Corollary 4.2 *If the semigroup $\check{T} : \mathbf{R}_+ \to \mathcal{L}(\mathcal{L}(\mathcal{H}))$ is strongly continuous on the Banach space $\mathcal{L}(\mathcal{H})$, the semigroup $T : \mathbf{R}_+ \to \mathcal{L}(\mathcal{H})$ is uniformly continuous (and therefore $\check{T}$ is uniformly continuous).*

4.3 A Riccati equation

From now on $\mathcal{H}^-$ will be assumed to be finite dimensional.

By Theorem 3.3 the closed symmetric operator X_{11} appearing in (3.4.5) is such that $r(X_{11}) \supset \Pi_r = \{\zeta \in \mathbf{C} : Re\ \zeta > 0\}$, and thus generates a strongly continuous semigroup $R : \mathbf{R}_+ \to \mathcal{L}(\mathcal{H}^+)$ of linear isometries for $\mathcal{H}^+$. Setting

$$\check{R}(t) : Z_1 \to R(t) \circ Z_1$$

for all $Z_1 \in \mathcal{L}(\mathcal{H}^-, \mathcal{H}^+)$, then $\check{R} : \mathbf{R}_+ \to \mathcal{L}(\mathcal{L}(\mathcal{H}^-, \mathcal{H}^+))$ is a strongly continuous semigroup which is generated by the operator $\check{X}_{11} : Z_1 \mapsto X_{11} \circ Z_1$ with domain

$$\mathcal{D}(\check{X}_{11}) = \{Z_1 \in \mathcal{L}(\mathcal{H}^-, \mathcal{H}^+) : Z_1 x^- \in \mathcal{D}(X_{11}) \text{ for all } x^- \in \mathcal{H}^-\}.$$

The spaces $\mathcal{D}(\check{X})$ and $\mathcal{D}(\check{X}_{11})$ are dense in $\mathcal{L}(\mathcal{H}^-, \mathcal{H})$ and in $\mathcal{L}(\mathcal{H}^-, \mathcal{H}^+)$, and are complete for the norms

$$W \mapsto \|W\| + \|X \circ W\|,$$
$$Z_1 \mapsto \|Z_1\| + \|X_{11} \circ Z_1\|. \tag{4.3.1}$$

Note that $W \in \mathcal{D}(\check{X})$ if, and only if, $P^+ \circ W \in \mathcal{D}(\check{X}_{11})$. If $W^0 \in \mathcal{D}(\check{X}_{11})$, the function defined for $t \geq 0$ by

$$W(t) := T(t) \circ W^0,$$

is the only solution in $C^1(\mathbf{R}_+, \mathcal{L}(\mathcal{H}^-, \mathcal{H}))$ of the Cauchy problem

$$\dot{W}(t) = X \circ W(t) \quad (t > 0) \qquad W(0) = W^0. \tag{4.3.2}$$

Let $Z_1(t) = P^+ \circ W(t)$, $Z_2(t) = P^- \circ W(t)$. Then the Cauchy problem (4.3.2) can be written

$$\dot{Z}_1(t) = X_{11} \circ Z_1(t) + X_{12} \circ Z_2(t), \tag{4.3.3}$$
$$\dot{Z}_2(t) = X_{21} \circ Z_1(t) + X_{22} \circ Z_2(t), \tag{4.3.4}$$
$$Z_1(0) = Z_1{}^0 := P^+ \circ W^0, \quad Z_2(0) = Z_2{}^0 := P^- \circ W^0. \tag{4.3.3'}$$

Since $X_{22} \in \mathcal{L}(\mathcal{H}^-)$ and $\dot{Z}_2(t), X_{22} \circ Z_2(t)$ are continuous functions of $t \in \mathbf{R}_+$, the linear equation (4.3.4) can be integrated, yielding

$$Z_2(t) = \exp(tX_{22}) \circ (Z_2(0) + \int_0^t (\exp(-sX_{22})) \circ X_{21} \circ Z_1(s) ds).$$

Hence the Cauchy problem (4.3.2) is equivalent to the integro-differential equation

$$\begin{aligned} \dot{Z}_1(t) &= X_{11} \circ Z_1(t) + X_{12} \circ (\exp tX_{22}) \circ (Z_2(0) \\ &\quad + \int_0^t (\exp(-sX_{22})) \circ X_{21} \circ Z_1(s) ds) \end{aligned}$$

with the initial condition (4.3.3′).

If $T(t)$ is represented by the matrix

$$T(t) = \begin{pmatrix} T_{11}(t) & T_{12}(t) \\ T_{21}(t) & T_{22}(t) \end{pmatrix}$$

with respect to the orthogonal decomposition (3.2.2), where $T_{11}(t) \in \mathcal{L}(\mathcal{H}^+)$, $T_{12}(t) \in \mathcal{L}(\mathcal{H}^-, \mathcal{H}^+)$, $T_{21}(t) \in \mathcal{L}(\mathcal{H}^+, \mathcal{H}^-)$, $T_{22}(t) \in \mathcal{L}(\mathcal{H}^-)$, then

$$Z_1(t) = T_{11}(t) \circ Z_1{}^0 + T_{12}(t) \circ Z_2{}^0, \tag{4.3.5}$$

$$Z_2(t) = T_{21}(t) \circ Z_1{}^0 + T_{22}(t) \circ Z_2{}^0, \tag{4.3.6}$$

The equation (4.3.3) coupled with the fact that $t \mapsto Z_1(t)$ is contained in $C^1(\mathbf{R}_+, \mathcal{L}(\mathcal{H}^-, \mathcal{H}^+))$, shows that, for any $\epsilon > 0$ and any $t^0 \geq 0$, there is a $\delta_1 > 0$ such that

$$\|X_{11} \circ (Z_1(t) - Z_1(t^0)) + X_{12} \circ (Z_2(t) - Z_2(t_0))\| < \epsilon \tag{4.3.7}$$

whenever $|t - t_0| < \delta_1$, $t \geq 0$. Since $t \mapsto Z_2(t)$ is continuous and $X_{12} \in \mathcal{L}(\mathcal{H}^-, \mathcal{H}^+)$, (4.3.7) implies that there is a $\delta_2 > 0$ such that, if $|t - t_0| < \delta_2$ and $t \geq 0$, then

$$\|X_{11} \circ (Z_1(t) - Z_1(t_0))\| < 2\epsilon, \tag{4.3.8}$$

proving thereby that the function $t \mapsto Z_1(t)$ is continuous for the graph-norm (4.3.1).

Suppose now that $Z_2{}^0$ is invertible in $\mathcal{L}(\mathcal{H}^-)$ and let $Z_1{}^0 \in \mathcal{D}(\check{X}_{11})$ be such that $Z^0 := Z_1{}^0 \circ (Z_2{}^0)^{-1} \in B$. Then by Corollary 4.1 and Proposition 4.1, $Z_2(t)$ is invertible and $Z(t) := Z_1(t) \circ (Z_2(t))^{-1} \in B \cap \mathcal{D}(\check{X}_{11})$ for all $t \geq 0$. Moreover the function $Z : t \mapsto Z(t)$ belongs to $C^1(\mathbf{R}_+, \mathcal{L}(\mathcal{H}^-, \mathcal{H}^+))$ and for $t \geq 0$ satisfies the Riccati equation

$$\dot{Z}(t) = X_{11} \circ Z(t) - Z(t) \circ X_{22} - Z(t) \circ X_{21} \circ Z(t) + X_{12} \tag{4.3.9}$$

with the initial condition

$$Z(0) = Z^0. \tag{4.3.10}$$

Since $t \mapsto Z_2(t)^{-1}$ is continuous and $t \mapsto Z_1(t)$ is continuous for the graph-norm (4.3.1), for $t_0 \geq 0$ the inequality

$$\begin{aligned} \|X_{11} \circ (Z(t) - Z(t_0))\| &\leq \|X_{11} \circ (Z_1(t) - Z_1(t_0))\| \|Z_2(t)^{-1}\| \\ &\quad + \|X_{11} \circ Z_1(t_0)\| \|Z_2(t)^{-1} - Z_2(t_0)^{-1}\| \end{aligned}$$

implies that, for every $\epsilon > 0$ there is $\delta_3 > 0$ such that, if $|t - t_0| < \delta_3$ and $t \geq 0$, then

$$\|X_{11} \circ (Z(t) - Z(t_0))\| < 2\epsilon.$$

Summing up, the following lemma holds.

Lemma 4.2 *For* $Z^0 = Z_1{}^0 \circ (Z_2{}^0)^{-1} \in B \cap \mathcal{D}(\check{X}_{11})$, *the function* $Z : t \mapsto Z(t) = Z_1(t) \circ Z_2(t)^{-1}$ *expressed by* (4.3.5) *and* (4.3.6) *for* $t \geq 0$, *is a solution of the differential equation* (4.3.9) *with the initial condition* (4.3.10), *which is continuous for the graph norm* (4.3.1).

It turns out that the solution satisfying the above conditions is unique, as the following theorem states.

Theorem 4.5 *Let* $\dim_{\mathbb{C}} \mathcal{H}^- < \infty$. *For any* $\gamma > 0$ *and any choice of* $Z^0 = Z_1{}^0 \circ (Z_2{}^0)^{-1} \in B \cap \mathcal{D}(\check{X}_{11})$, *the function* $t \mapsto Z(t) = Z_1(t)(Z_2(t))^{-1}$ *is the unique solution, on the interval* $[0,\gamma]$, *of the differential equation* (4.3.9) *with initial condition* (4.3.10) *and with* $Z([0,\gamma]) \subset B \cap \mathcal{D}(\check{X}_{11})$, *which belongs to* $C^1([0,\gamma], \mathcal{L}(\mathcal{H}^-, \mathcal{H}^+))$ *and is continuous for the graph norm* (4.3.1).

Proof [VES 87b] Let $H : [0,\gamma] \to \mathcal{D}(\check{X}_{11})$ be a solution of (4.3.9) satisfying all the requirements of the theorem. Setting $K(t) = H(t) \circ Z_2(t)$, then (4.3.9), (4.3.3), (4.3.4) show that the function

$$Y : [0,\gamma] \to \mathcal{D}(\check{X}_{11})$$

defined by

$$Y(t) = K(t) - Z_1(t)$$

satisfies the evolution equation

$$Y(t) = \Omega(t)Y(t), \tag{4.3.11}$$

with the initial condition

$$Y(0) = 0, \tag{4.3.12}$$

where the linear operator

$$\Omega(t) := \check{X}_{11} - H(t) \circ X_{21},$$

with domain $\mathcal{D}(\Omega(t)) = \mathcal{D}(\check{X}_{11})$, is a perturbation of $\check{X}_{11}$ by the bounded operator $H(t) \circ X_{21} \in \mathcal{L}(\mathcal{H}^-, \mathcal{H}^+)$, whose norm is

$$\|X_{21}\| \max\{\|H(t)\| : 0 \le t \le \gamma\} = \|X_{12}\| \max\{\|H(t)\| : 0 \le t \le \gamma\}.$$

Furthermore $y \in C^1([0,\gamma], \mathcal{L}(\mathcal{H}^-, \mathcal{H}^+))$ is continuous for the graph norm (4.3.1).

Since $\check{X}_{11}$ generates a strongly continuous semigroup of linear bounded operators on $\mathcal{L}(\mathcal{H}^+)$ and therefore defines a stable family of generators, then [PAZ 83; Theorem 2.3, p.132], $\{\Omega(t) : 0 \le t \le \gamma\}$ is a stable family of generators of continuous semigroups, with stability constants $h > 0$ and $k = \|X_{12}\| \max\{\|H(t)\| : 0 \le t \le \gamma\}$. Because $H \in C^1([0,\gamma], \mathcal{L}(\mathcal{H}^-, \mathcal{H}^+))$, for any $y^0 \in \mathcal{D}(\check{X}_{11})$, the function $t \mapsto \Omega(t)y^0$ from $[0,\gamma]$ to $\mathcal{H}^+$ is continuously differentiable. Thus [PAZ 83; Theorem 4.8 and 4.3, pp. 145, 141], there exists a unique evolution system $\{\Xi(t,s) : 0 \le s \le t \le \gamma)$ such that:

$$\begin{aligned} \Xi(t,s) &\le he^{k(t-s)} \quad \text{whenever } 0 \le s \le t \le \gamma; \\ \frac{\partial^+}{\partial t}\Xi(t,s)Y^0{}_{|t=s} &= \Omega(s)Y^0 \quad \text{for } Y^0 \in \mathcal{D}(\check{X}_{11}),\ 0 \le s \le \gamma; \\ \frac{\partial}{\partial s}\Xi(t,s)Y^0 &= -\Xi(t,s)\Omega(s)Y^0 \quad \text{for } Y^0 \in \mathcal{D}(\check{X}_{11}),\ 0 \le s \le t \le \gamma \end{aligned}$$

($\frac{\partial^+}{\partial t}$ and $\frac{\partial}{\partial s}$ are in the strong sense);

$$\Xi(t,s)\mathcal{D}(\check{X}_{11}) \subset \mathcal{D}(\check{X}_{11}) \quad \text{for } 0 \le s \le t \le \gamma;$$

for $Y^0 \in \mathcal{D}(\check{X}_{11})$,

$$t \mapsto \Xi(t,s)Y^0$$

is continuous on $\mathcal{D}(\check{X}_{11})$ for $0 \leq s \leq t \leq \gamma$ for the graph norm (4.3.1) (which is equivalent to the norm appearing in Theorems 4.8 and 4.3 of [PAZ 83]); if $Y^0 \in \mathcal{D}(\check{X}_{11})$, the function $Y : t \mapsto \Xi(t,s)Y^0$ is the unique solution of (4.3.11) on $[s,\gamma]$ with the initial condition $Y(s) = Y^0$, which is continuous on $\mathcal{D}(\check{X}_{11})$ for the graph norm (4.3.1).

Hence $Y = 0$ is the unique solution of (4.3.11), with initial condition (4.3.12), which is contained in $C^1([0,\gamma],\mathcal{L}(\mathcal{H}^-,\mathcal{H}^+))$, is continuous for the graph-norm (4.3.1), and whose values belong to $\mathcal{D}(\check{X}_{11})$. In conclusion $K = Z_1$ on $[0,\gamma]$. □

References

[AN-VE 64] Andreotti, A., Vesentini, E., On deformations of discontinuous groups, *Acta Math.* **114** (1964), 249–298.

[AZ-IO 89] Azizov, T.Ya., Iokhvidov, I.S., *Linear Operators in Spaces With an Indefinite Metric*, Wiley, New York, 1989.

[CAE 35] Cartan, E., Sur les domaines bornés homogènes de l'espace de n variables complexes, *Abh. Math. Sem. Univ. Hamburg* **11** (1935), 116–162.

[CAH 35] Cartan, H., *Sur les groupes de transformations analytiques*, Herman et Cie., Paris, 1935.

[CHA 85] Soo Bong Chae, *Holomorphy and Calculus in Normed Spaces*, Marcel Dekker, New York and Basel, 1985.

[D-T-V 85] Dineen, S., Timoney, R.M., Vigué, J.-P., Pseudodistances invariantes sur les domaines d'un espace localement convexe, *Ann. Scuola Norm. Sup. Pisa Cl. Sci. (4)* **12** (1985), 515–529.

[DIN 89] Dineen, S., *The Schwarz Lemma*, Oxford Math. Monographs, Clarendon Press, Oxford, 1989.

[FIL 70] Fillmore, P A., *Notes on Operator Theory*, Van Nostrand, New York, 1970.

[FR-VE 80] Franzoni, T., Vesentini, E., *Holomorphic Maps and Invariant Distances*, North-Holland, Amsterdam, 1980.

[FRA 81] Franzoni, T., The group of holomorphic automorphisms in certain J^*-algebras, *Ann. Mat. Pura Appl. (4)* **127** (1981), 51–66.

[GIG 89] Gigante, G., A remark on injective hyperbolicity, unpublished.

[HAH 81] Hahn., K.T., Some remarks on a new pseudo-differential metric, *Ann. Polon. Math.* **39** (1981), 71–81.

[HAL 67] Halmos, P.R., *A Hilbert Space Problem Book*, Van Nostrand, New York, 1967.

[HAR 74] Harris, L.A., Bounded symmetric homogeneous domains in infinite dimensional spaces, Lecture Notes in Math. **634**, Springer-Verlag, Berlin, 1974; 13-40.

[HAR 79a] Harris, L.A., Analytic invariants and the Schwarz-Pick inequality, *Israel J. Math.* **34** (1979), 177–197.

[HAR 79b] Harris, L.A., Schwarz-Pick systems of pseudometrics for domains in normed linear spaces, in: *Advances in Holomorphy* (J.A. Barroso, ed.), North-Holland, Amsterdam, 1979; 345–405.

[HAR 81] Harris, L.A., A generalization of C^*-algebras, *Proc. London Math. Soc. (3)* **42** (1981), 331–361.

[HE-IS 92] Herves, F.J., Isidro, J.M., Isometries and automorphisms of the spaces of spinors, *Rev. Mat. Univ. Complut. Madrid* **5** (1992), 193–200.

[HI-PH 57] Hille, E., Phillips, R.S., *Functional Analysis and Semigroups*, Amer. Math. Soc. Colloq. Publ. **31**, Providence, RI, 1957.

[HIL 62] Hille, E., *Analytic Function Theory*, Vol. II, Ginn and Co., Boston, 1962.

[I-K-L 82] Iohvidov, I.S., Kreĭn, M.G., Langer, H., *Introduction to the Spectral Theory of Operators in Spaces With an Indefinite Metric*, Akademie-Verlag, Berlin, 1982.

[IS-ST 75] Isidro, J.M., Stachó, L.L., *Holomorphic Automorphism Groups in Banach Spaces: an Elementary Introduction*, North-Holland, Amsterdam, 1985.

[J-P-V 91] Jarnicki, M., Pflug, P., Vigué, J.-P., The Carathéodory distance does not define the topology – the case of domains, *C. R. Acad. Sci. Paris, Sér. I Math.* **312** (1991), 77–79.

[JA-PF 93] Jarnicki, M., Pflug, P., *Invariant Distances and Metrics in Complex Analysis*, de Gruyter, Berlin - New York, 1993.

[KAU 82] Kaup, W., Bounded symmetric domains in complex Hilbert spaces, *Sympos. Math.* **26** (1982), 11–21.

[KOB 70] Kobayashi, S., *Hyperbolic Manifolds and Holomorphic Mappings*, Marcel Dekker, New York, 1970.

[KOB 76] Kobayashi, S., Intrinsic distances, measures and geometric function theory, *Bull. Amer. Math. Soc.* **82** (1976), 357–416.

[LEM 82] Lempert, L., Holomorphic retracts and intrinsic metrics in convex domains, *Anal. Math.* **8** (1982), 257–261.

[LO-MA 77] Loos, O., MacCrimmon, K., Speciality of Jordan triple systems, *Comm. Algebra* **5** (1977), 1057–1082.

[MES 85] Meschiari, M., A classification of real and complex finite dimensional J^*-algebras, in: *Geometry Seminar "Luigi Bianchi" II* (E. Vesentini, ed.), Lecture Notes in Math. **1164**, Springer-Verlag, Berlin - Heidelberg - New York, 1985; 1–84.

[PAZ 83] Pazy, A., *Semigroups of Linear Operators and Applications to Partial Differential Equations*, Springer-Verlag, Berlin - Heidelberg - New York, 1983.

[REI 63] Reiffen, H.-J., Die differentialgeometrischen Eigenschaften der invarianten Distanzfunktion von Carathéodory, *Schriftenreihe Math. Inst. Univ. Münster*, 1963.

[REI 65] Reiffen, H.-J., Die Caratheodorysche Distanz und ihre zugehörige Differentialmetrik, *Math. Ann.* **161** (1965), 315–324.

[RO-WO] Royden, H., Wong, P., Carathéodory and Kobayashi metric on convex domains, Preprint.

[SAK 71] Sakai, S., *C^*-Algebras and W^*-Algebras*, Springer-Verlag, Berlin - Heidelberg - New York, 1971.

[SCH 58] Schwartz, L., *Lectures on Mixed Problems in Partial Differential Equations and Representation of Semigroups*, Tata Institute of Fundamental Research, Bombay, 1958.

[TH-WH 67] Thorp, E., Whitley, R., The strong maximum modulus theorem for analytic functions into Banach spaces, *Proc. Amer. Math. Soc.* **18** (1967), 640–646.

[UPM 85a] Upmeier, H., Jordan Algebras in Analysis, Operator Theory, and Quantum Mechanics, CBMS Regional Conf. Ser. in Math. **67**, Amer. Math. Soc., Providence, RI, 1985.

[UPM 85b] Upmeier, H., *Symmetric Banach Manifolds and Jordan C^*-Algebras*, North-Holland, Amsterdam, 1985.

[VEN 89] Venturini, G., On holomorphic isometries for the Kobayashi and Carathéodory distances on complex manifolds, *Atti Accad. Naz. Lincei Rend. (8)* **83** (1989), 139–145.

[VES 70] Vesentini, E., Maximum theorems for vector-valued holomorphic functions, University of Maryland, Technical Report TR 69-132, 1969, also in *Rend. Sem. Mat. Fis. Milano* **40** (1970), 1–34.

[VES 81] Vesentini, E., Complex geodesics, *Compositio Math.* **44** (1981), 375–394.

[VES 82a] Vesentini, E., Complex geodesics and holomorphic maps, *Sympos. Math.* **26** (1982), 211–230.

[VES 82b] Vesentini, E., Invariant distances and invariant differential metrics in locally convex spaces, *Banach Center Publ.* **8** (1982), 439–512.

[VES 83] Vesentini, E., Carathéodory distances and Banach algebras, *Adv. Math.* **47** (1983), 50–73.

[VES 86] Vesentini, E., Hyperbolic domains in Banach spaces and Banach algebras, in: *Aspects of Mathematics and Its Applications* (J.A. Barroso, ed.), North-Holland, Amsterdam, 1986; 859–871.

[VES 87a] Vesentini, E., Holomorphic families of holomorphic isometries, in: *Complex Analysis III. Proceedings 1985-86* (C.A. Berenstein, ed.), Lecture Notes in Math. **1277**, Springer-Verlag, Berlin - Heidelberg - New York, 1987; 290–302.

[VES 87b] Vesentini, E., Semigroups of holomorphic isometries, *Adv. Math.* **65** (1987), 272–306.

[VES 87c] Vesentini, E., Injective hyperbolicity, *Ricerche Mat.* **36** (1987), 99–109.

[VES 88] Vesentini, E., Holomorphic semigroups of holomorphic isometries, *Atti Accad. Naz. Lincei Rend. (8)* **82** (1988), 203–217.

[VES 89] Vesentini, E., Semigroups on Cartan domains of type four, *Note Mat.* **9** - Suppl. (1989), 123–144.

[VES 90a] Vesentini, E., Semigroups in Kreĭn spaces, *Mem. Mat. Accad. Lincei (9)* **1** (1990), 3–29.

[VES 90b] Vesentini, E., Semigruppi fortemente continui in algebre di Banach ed in sistemi di spin, *Rend. Sem. Mat. Fis. Milano* **60** (1990), 157–165.

[VES 91] Vesentini, E., Holomorphic isometries of Cartan domains of type one, *Atti Accad. Naz. Lincei Rend. (9)* **2** (1991), 65–72.

[VES 92] Vesentini, E., Holomorphic isometries of Cartan domains of type four, *Atti Accad. Naz. Lincei Rend. (9)* **3** (1992), 287–294.

[VES 93] Vesentini, E., Holomorphic isometries of spin factors, to appear.

[VIG 76] Vigué, J.-P., Le groupe des automorphismes analytiques d'un domaine borné d'un espace de Banach complexe. Application aux domaines bornés symetriqués, *Ann. Sci. École Norm. Sup. (4)* **9** (1976), 203–282.

[VIG 83] Vigué, J.-P., La distance de Carathéodory n'est pas intérieure, *Resultate Math.* **6** (1983), 100–104.

[VIG 89] Vigué, J.-P., The Carathéodory distance does not define the topology, *Proc. Amer. Math. Soc.* **91** (1984), 223–224.

[YOS 80] Yosida, K., *Functional Analysis*, (6th edition), Springer-Verlag, Berlin - Heidelberg - New York, 1980.

Index